COLLABORATIVE RECOMMENDATIONS

Algorithms, Practical Challenges
and Applications

COLLABORATIVE RECOMMENDATIONS

Algorithms, Practical Challenges and Applications

Shlomo Berkovsky
CSIRO, Australia

Iván Cantador
Universidad Autónoma de Madrid, Spain

Domonkos Tikk
Gravity Research & Development Zrt, Hungary

World Scientific

NEW JERSEY • LONDON • SINGAPORE • BEIJING • SHANGHAI • HONG KONG • TAIPEI • CHENNAI • TOKYO

Published by

World Scientific Publishing Co. Pte. Ltd.
5 Toh Tuck Link, Singapore 596224
USA office: 27 Warren Street, Suite 401-402, Hackensack, NJ 07601
UK office: 57 Shelton Street, Covent Garden, London WC2H 9HE

British Library Cataloguing-in-Publication Data
A catalogue record for this book is available from the British Library.

COLLABORATIVE RECOMMENDATIONS
Algorithms, Practical Challenges and Applications

ISBN 978-981-3275-34-8

For any available supplementary material, please visit
https://www.worldscientific.com/worldscibooks/10.1142/11131#t=suppl

Preface

Shlomo Berkovsky[1], Iván Cantador[2] and Domonkos Tikk[3]
[1]*CSIRO, Australia,* [2]*Universidad Autónoma de Madrid, Spain,*
[3]*Gravity Research & Development, Hungary*
shlomo.berkovsky@gmail.com, ivan.cantador@uam.es,
domonkos.tikk@yusp.com

Recommender systems are very popular nowadays, as both an academic research field and services provided by numerous companies for e-commerce, multimedia and Web content. Collaborative-based methods have been the focus of recommender systems research for more than two decades, evolving over time from neighbor-based methods to matrix factorization and deep learning. During this time, thousands of papers have been published in various conferences and journals, and the area has matured significantly.

Nevertheless, there is no single place that present an encompassing and at the same time practical information on collaborative recommender systems. Therefore, we envisage the "Collaborative Recommendations: Algorithms, Practical Challenges, and Applications" book as a dedicated reference on collaborative recommendations, which thoroughly discusses recommendation algorithms, challenges related to their application on the Web, existing recommender systems leveraging collaborative methods, and future veins of work in this flourishing field, including contributions from the academia and industry researchers alike.

The unique feature of the book is the special attention paid to technical details of collaborative recommenders. The book chapters include details of algorithm implementations, elaborate on practical issues faced when deploying these algorithms in large-scale systems, describe various optimizations and decisions made, list parameters of the algorithms, and so on. These details are usually not covered by academic papers, which widen the gap between the academic works and their uptake by the industry. We feel that this book makes an important step towards bridging this gap and providing a valuable point of reference for recommender systems practitioners.

Hence, we believe that the broad coverage and the attention paid to the practical aspects of collaborative recommendations are the main advantages of this book over other existing publications. It is likely to be attractive not only for researchers and students, but also for a broad range of practitioners and information technology professionals, interested in incorporating recommendation technologies into their systems and services. We thus hope the readers greatly benefit from learning about the algorithms and practical aspects of collaborative recommender systems.

The book is partitioned into three parts, each including a number of chapters written by experts in the respective areas of research and practice. The content of the book is as follows.

- ***Part I: Algorithms*** overviews major algorithmic approaches exploited for collaborative recommendation purposes, namely collaborative filtering, matrix factorization, and deep learning. Next, it presents principal directions aiming at enhancing collaborative recommenders through hybridization with content-based methods and exploitation of contextual information, as well as ways to provide collaborative recommendations targeting groups of users.
- ***Part II: Practical challenges*** surveys common issues to be addressed when deploying a collaborative recommender system. First, it analyzes existing types of user preferences and related problems, such as user preference elicitation, sparsity, and cold start. Second, it delves into topics related to the evaluation of recommender systems, both in the offline and online paradigms, and to recommendation biases and beyond-accuracy objectives. Finally, several chapters are dedicated to practical challenges, such as scalability, distribution, robustness, and privacy in collaborative recommenders.
- ***Part III: Applications*** elaborates on technical details, such as data processing, algorithm selection and setting, and evaluation, taking into consideration a number of real-life use cases; namely, recommendations of TV shows, movies, music, contacts, job offers, academic documents, online courses, food, and clothes. All these chapters are characterized by their focus on practical issues, including the deployment of the algorithms outlined in Part I and the attention paid to the challenges discussed in Part II.

Part I is composed of six chapters, which are summarized next. In Chapter 1, "Collaborative filtering for matrix completion and session-based recommendation tasks," Dietmar Jannach and Markus Zanker provide an overview of the basics of collaborative filtering systems. They equally discuss the classical matrix completion problem formulation, which is used for longer-term relevance predictions given a user-item rating matrix, and the session-based recommendation scenarios, where the goal is to predict relevant items given a user's observed short-term behavior. The chapter particularly focuses on the traditional, yet still often used, neighborhood based methods; describing the user- and item-based nearest neighbor algorithms. The chapter addresses the entire life-cycle of algorithm development, and discusses libraries, datasets, implementation aspects, as well as evaluation issues.

In Chapter 2, "Matrix factorization for collaborative recommendations," Evgeny Frolov and Ivan Oseledets thoroughly discuss matrix factorization (MF), one of the most successful and widely used collaborative filtering techniques, applicable both for the explicit and implicit user feedback. The authors first introduce the problem formulation of matrix factorization as a dimensionality reduction technique. Then, they present singular-value decomposition (SVD) based techniques, including PureSVD, emphasizing practical aspects, like handling online updates. Some advanced MF methods are also discussed, including learning to rank and hybrid factorization models. Finally, the chapter addresses practical aspects of implementation, such as parallel implementation of algorithms, and hyper-parameter tuning, which both play a crucial role in the performance of a real-world application of MF algorithms.

In Chapter 3, "Cutting-edge collaborative recommendation algorithms: Deep learning," Balázs Hidasi provides a detailed overview of recent advances in the application of deep learning to generate personalized recommendations. The chapter starts with a short introduction to deep learning, a complex neural network technology that already revolutionized complex domains of machine learning, such as computer vision, natural language processing, and speech recognition, and recently turned towards recommender systems as well. After a short historical overview of deep learning applications for recommender systems, four main lines of current research are discussed in detail: learning item embeddings, deep collaborative filtering, direct use of content, and session-based recommendations. The chapter also gives a short guide for practitioners who would like to delve into the topic, by discussing the most popular frameworks with their pros and cons, and providing hints on best practice for hyper-parameter optimization, and scalability questions.

In Chapter 4, "Hybrid collaborative recommendations: Practical considerations and tools to develop a recommender," Michal Kompan, Peter Gašpar, and Maria Bielikova overview hybrid collaborative recommender systems, which aim to enhance collaborative filtering by exploiting additional information about ratings, such as content-based item features. The chapter focuses on how these approaches are beneficial with respect to the standard collaborative filtering techniques, and how they can be utilized to solve main limitations of collaborative methods. Next, the chapter provides examples of frameworks and libraries that implement existing recommendation approaches, and a comprehensive list of available datasets (including state-of-the-art results reported on these datasets). The evaluation process of recommender systems is discussed, followed by examples of evaluation frameworks. Finally, pros and cons of several hybrid approaches are summarized with suggestions for practical implementations.

In Chapter 5, "Context-aware recommendations," Yong Zheng and Bamshad Mobasher provide a broad introduction to Context-Aware Recommender Systems (CARS). The chapter especially focuses on algorithmic approaches for integrating context into the recommendation framework, including approaches based on context selection and context-aware collaborative filtering. The chapter starts with clarifying the definition of context — an often challenged topic in the field of recommender systems — and gives a categorization of context types. The three main approaches, contextual pre- and post-filtering, and contextual modeling, are then introduced. Also, particular evaluation strategies for CARS, available datasets, and open-source libraries are presented. Finally, the authors discuss a number of lessons learned from practical applications.

In Chapter 6, "Group recommendations," Ludovico Boratto and Alexander Felfernig overview the area of group recommender systems, which are designed to provide suggestions in scenarios, where instead of a single user, a group of users is engaged. Group recommendation can be naturally adopted in application scenarios that involve groups, e.g., suggesting a restaurant to a group of people willing to dine together. This area of recommender systems exhibits several additional aspects, compared to those focused on single users, like preference acquisition, group modeling, group building, rating prediction for groups, helping group members to achieve consensus, and explanation of recommendations. The chapter illustrates existing families of approaches to produce group recommendations, showing the different high-level architectural representations of a system. It then presents the strategies adopted in the literature to define groups. Next, the authors discuss the group modeling task, aiming at creating a unique representation of the preferences of a group, and illustrate how collaborative rating prediction algorithms are employed in group recommender systems. Finally, a case-study is presented that compares the different families of approaches by employing a user-based collaborative approach.

Then, **Part II** contains the following seven chapters. In Chapter 7, "User preference sources: Explicit vs. implicit feedback," Paolo Cremonesi, Franca Garzotto, and Maurizio Ferrari Dacrema present an exhaustive analysis of the two main types of user preferences a recommender system may deal with; namely, explicit and implicit feedback. Regarding explicit preferences, the authors distinguish between numerical (rating-based) and categorical preferences, and discuss a number of issues to consider in the design and deployment of a collaborative recommender, such as the used rating scale (defined in terms of the ratings value granularity and labels), and the differences in the users' profile length, among others. With respect implicit preferences, the authors focus on the available mechanisms to acquire and process non-explicit user feedback coming from the users' behavior in the system, such as product purchases and views, music listening counts, and unary and binary likes. A number of future research directions are finally given.

In Chapter 8, "User preference elicitation, rating sparsity, and cold start," Mehdi Elahi, Matthias Braunhofer, Tural Gurbanov, and Francesco Ricci address a particular task related to user preferences in recommender systems: the elicitation of the users' preferences. In this context, they discuss the well-known problems of sparsity, i.e., the very low density of the user-item rating matrix, and cold start, i.e., the lack of insufficient volume of ratings for users or items that are new or recent in the system. The authors first survey the research literature on these topics, and then present main algorithmic solutions, such as active learning, cross-domain recommendations, exploiting implicit user feedback and item content-based information, and hybridization techniques. Next, as practical considerations, existing software tools, popular datasets, and reported performance results are provided. The chapter ends with a number of guidelines and future work directions.

In Chapter 9, "Offline and online evaluation of recommendations," Alejandro Bellogín and Alan Said overview some of the most popular methodologies and metrics used to evaluate recommender systems. More specifically, the authors present algorithmic solutions to evaluate particular types of collaborative filtering methods, including social, context-aware, and group-oriented recommendation approaches. They also provide a summary of available resources to conduct and compare recommender system evaluations, namely APIs and software libraries, datasets, and past competitions and challenges. For some of the cited software frameworks and datasets, the authors report the achieved experimental results with several metrics. Finally, they discuss practical considerations and outline future directions around the evaluation of recommender systems.

In Chapter 10, "Recommendations biases and beyond-accuracy objectives in collaborative filtering," Pasquale Lops, Fedelucio Narducci, Cataldo Musto, Marco De Gemmis, Cataldo Musto, Marco Polignano, and Giovanni Semeraro pay attention to particular aspects for evaluating the perceived quality and usefulness of recommendations. They first discuss the effect of item popularity on the accuracy of personalized recommendations. Then, they survey important dimensions beyond accuracy; namely, diversity, novelty, and serendipity. Algorithmic solutions to deal with the popularity bias, and increasing diversity, novelty, and serendipity in recommendation lists are discussed. For these aspects, specialized evaluation resources are also described. Finally, the chapter ends with related challenges and potential future directions.

In Chapter 11, "Scalability and distribution of collaborative recommenders," Evangelia Christakopoulou, Shaden Smith, Mohit Sharma, Alex Richards, David Anastasiu, and George Karypis elaborate on the design of recommender systems that can efficiently handle the huge, ever growing amount of data by the means of modern parallel architectures. In particular, they first present several methods to scale k-nearest neighbor collaborative filtering via parallel computing, by speeding the identification of user neighborhood and the computation of item similarities. Next, they present methods to scale up matrix and tensor factorization collaborative filtering models, by parallelizing the Alternating Least Squares and Stochastic Gradient Descent algorithms. In both cases, the authors also report experimental results on the well known MovieLens-10M dataset, showing the runtimes and speedup achieved in comparison to serial implementations.

In Chapter 12, "Robustness and attacks on recommenders," Neil J. Hurley analyzes to what extent recommendation algorithms are affected to the presence of corrupt data, which is injected to deliberately distort the recommendation outputs for particular users and items. His analysis focuses on the creation of fake identities and of user privacy preservation.

The author reviews research carried out since the early 2000s on robustness attacks and practical defences, and introduces a new library for robustness analysis incorporated into the RankSys recommender systems framework, where major attacks proposed in the literature are implemented, and can be tested against a range of recommendation algorithms.

In Chapter 13, "Privacy in collaborative recommenders," Qiang Tang focuses on the privacy issues associated with recommender systems. More specifically, the author addresses the relationships between privacy, robustness, and transparency. After defining and categorizing privacy concerns in recommender systems, the author provides an analytical comparison of existing privacy-preserving solutions, identifying their strengths and weaknesses. Then, he discusses research directions aimed at designing pragmatic privacy-preserving recommender systems, which can also support other important properties, such as robustness and transparency.

Finally, **Part III** contains eight chapters, some of them written by researchers working for leading companies in recommender systems technologies and services. In Chapter 14, "TV and movie recommendations: The Comcast case," Shahin Sefati, Jan Neumann, and Hassan Sayyadi present the recommender system behind the Comcast X1 platform, which serves personalized recommendations to nearly 30 million users. Since in any given moment there are hundreds to thousands of entertainment choices available on different devices (TV screen, computer, smart phone, tablet), the system makes automatic personalized recommendations that help consumers deal with the often overwhelming volume of choices they face. The chapter reviews a number of developed algorithms and building blocks that power the X1 recommender system. Next, a number of product features and how the challenges associated with each feature are addressed. Further, several offline and online metrics deployed for evaluation are discussed, and, finally, some hints on the potential future developments for X1 recommender system are given.

In Chapter 15, "Music recommendations," Dietmar Jannach, Iman Kamehkhosh, and Geoffray Bonnin survey the area of music recommender systems. The chapter first presents main tasks and specific challenges in music recommendations. It then discusses major algorithmic approaches, such as implicit matrix factorization for item search and discovery, and hybrid methods for adaptive playlist generation, some of which are based on sequence-aware recommendation techniques. Next, practical implementation challenges and evaluation issues of music recommender systems are analyzed, considering both music-oriented and purpose-oriented quality criteria, describing performance assessment in real-world and academic environment settings, and surveying the evaluation protocols, metrics and datasets followed in the literature. The chapter concludes with several lessons learned and open issues.

In Chapter 16, "Contact recommendations in social networks," Javier Sanz-Cruzado and Pablo Castells focus on the confluence of recommender systems and online social networks, to survey academic literature and industry examples on recommending people to connect with in social networks. After discussing the specifics of such a recommendation task, the chapter overviews the existing algorithmic approaches for contact recommendation, proposing a rich taxonomy, which includes neighborhood-based, matrix factorization, path-based, and random walk methods, among others. Related resources, such as datasets, APIs, and software libraries are also described. Next, the authors overview practical aspects for building a contact recommender system, such as the directionality of network edges, the scalability of the algorithms, and the evaluation of generated recommendations. Finally, the authors present and analyze results achieved in exhaustive experiments, and conclude with future research directions.

In Chapter 17, "Job recommendations: The XING case," Katja Niemann, Daniel Kohlsdorf, and Fabian Abel present, using the XING case, why people are interested in receiving job recommendations and what challenges real world job recommendation systems should solve to provide relevant recommendations. People might be actively looking for a new job or be waiting for a new job to find them, they might want to stay informed about their own value on the job market or about the development of the domain they are working in. Additionally, recruiters and companies aim for the right people to see their job postings to find suitable candidates for open positions. This chapter first introduces the professional social network XING, and discusses the challenges it has to face when generating job recommendations. Thereafter, the functionality of XING job recommender system is presented, and it is also discussed how new features are evaluated by XING using offline and online evaluation techniques. Finally, an insight into XING's recommender system architecture is given, which has to serve thousands of requests per minute.

In Chapter 18, "Academic recommendations: The Mendeley case," Maya Hristakeva, Daniel Kershaw, Benjamin Pettit, Saúl Vargas, and Kris Jack present the recommender system developed for Mendeley, Elsevier's editorial tool for managing and sharing research papers, where millions of users compiled personal libraries of relevant articles. The authors first survey previous work on academic recommender systems, discussing the sources of information commonly used, such as collaborative ratings, content-based data (e.g., article texts, metadata, and social tags), and citation networks. The Mendeley recommender consists of a user-based collaborative filtering method that takes advantage of the implicit feedback in the users' libraries, which is complemented by content- and discipline-based recommendations for new users. As explained in the chapter, the authors observed that adding recommendations from the citation network did not result in significant gains to either recommendation quality or coverage, and expect to incorporate additional domain-specific feature to supplement collaborative filtering. When describing the system, the authors provide a number of practical considerations they took into account, such as recommendation post-processing, time decays, impression discounting, and fresh content provision. Then, they discuss important challenges they faced, such as the recommender's scalability, the scientific quality of the recommended items, and the users' privacy. They also describe the utilized resources, and offline and online evaluations followed. To conclude, lessons learned and future work directions are given.

In Chapter 19, "MoocRec.com: Massive Open Online Courses recommender system," Panagiotis Symeonidis and Dimitrios Malakoudis present the recommender system of MoocRec.com, which aims to provide personalized suggestions of online courses. The recommender is based on a matrix factorization model that exploits heterogeneous data about user skills, descriptions of job positions, and course topics. The system extracts such data from external resources, namely EdX and Coursera MOOC websites, and LinkedIn professional social network. In the chapter, the authors describe the system architecture, some of its components, such as the Web crawler and the data proccesser, and its underlying recommendation algorithms.

In Chapter 20, "Food recommendations", Christoph Trattner and David Elsweiler provide an extensive survey of food recommender systems, which aim to suggest food that a user might want to eat and food that may help them to nourish more healthily. They also describe certain specializations, such as food recommender systems for groups of users. The authors examine which collaborative recommendation algorithms have been used in the food domain, the ways the systems are typically evaluated, and the resources available to those interested in building or studying recommender systems in practice. Finally, they discuss a number of related challenges, and offer some lessons learned from past research and potential future work directions.

In Chapter 21, "Clothing recommendations: The Zalando case," Antonino Freno presents the recommender system developed by Zalando, one of the most popular online fashion retails in Europe. Being an excellent representative example of a large-scale on-line retail platform, the system poses a number of challenges that go far beyond selecting the most accurate recommendation algorithm. In particular, the author discusses problems that usually arise from operational constraints, such as the adaptation to novel use cases, the cost and complexity of system maintenance, the capability of reusing pre-existing signals, and integration of heterogeneous data sources. Specifically, he explains how moving from a collaborative filtering approach to a learning-to-rank model helped to effectively tackle the above challenges, while improving the quality of the recommendations. He also provides a description of the system architecture and algorithmic approaches, and conducted experiments, by the means of both offline and online evaluations.

We thank the authors of all the chapters for contributing the content of this book and helping each other improve chapters through peer-reviewing. We believe the readers will find this book valuable, informative, and thought-provoking, and hope it will lead to further advancements in the area of collaborative recommender systems.

Shlomo Berkovsky
Iván Cantador
Domonkos Tikk

Contents

Preface v

1. Collaborative Filtering: Matrix Completion and Session-Based Recommendation Tasks 1
   Dietmar Jannach and Markus Zanker

2. Matrix Factorization for Collaborative Recommendations 35
   Evgeny Frolov and Ivan Oseledets

3. Cutting-Edge Collaborative Recommendation Algorithms: Deep Learning 79
   Balázs Hidasi

4. Hybrid Collaborative Recommendations: Practical Considerations and Tools to Develop a Recommender 127
   Michal Kompan, Peter Gašpar and Maria Bielikova

5. Context-Aware Recommendations 173
   Yong Zheng and Bamshad Mobasher

6. Group Recommendations 203
   Ludovico Boratto and Alexander Felfernig

7. User Preference Sources: Explicit vs. Implicit Feedback 233

Paolo Cremonesi, Franca Garzotto and Maurizio Ferrari Dacrema

8. User Preference Elicitation, Rating Sparsity and Cold Start 253

Mehdi Elahi, Matthias Braunhofer, Tural Gurbanov and Francesco Ricci

9. Offline and Online Evaluation of Recommendations 295

Alejandro Bellogín and Alan Said

10. Recommendations Biases and Beyond-Accuracy Objectives in Collaborative Filtering 329

Pasquale Lops, Fedelucio Narducci, Cataldo Musto, Marco de Gemmis, Marco Polignano and Giovanni Semeraro

11. Scalability and Distribution of Collaborative Recommenders 369

Evangelia Christakopoulou, Shaden Smith, Mohit Sharma, Alex Richards, David Anastasiu and George Karypis

12. Robustness and Attacks on Recommenders 405

Neil J. Hurley

13. Privacy in Collaborative Recommenders 431

Qiang Tang

14. TV and Movie Recommendations: The Comcast Case 465

Shahin Sefati, Jan Neumann and Hassan Sayyadi

15. Music Recommendations 481

Dietmar Jannach, Iman Kamehkhosh and Geoffray Bonnin

16. Contact Recommendations in Social Networks 519
Javier Sanz-Cruzado and Pablo Castells

17. Job Recommendations: The XING Case 571
Katja Niemann, Daniel Kohlsdorf and Fabian Abel

18. Academic Recommendations: The Mendeley Case 599
Maya Hristakeva, Daniel Kershaw, Benjamin Pettit, Saúl Vargas and Kris Jack

19. MoocRec.com: Massive Open Online Courses Recommender System 627
Panagiotis Symeonidis and Dimitrios Malakoudis

20. Food Recommendations 653
Christoph Trattner and David Elsweiler

21. Clothing Recommendations: The Zalando Case 687
Antonino Freno

Index 713

Chapter 1

Collaborative Filtering: Matrix Completion and Session-Based Recommendation Tasks

Dietmar Jannach[a] and Markus Zanker[b]

[a]*AAU Klagenfurt, Austria,* [b]*Free University of Bozen-Bolzano, Italy*
Email: dietmar.jannach@aau.at, markus.zanker@unibz.it

This chapter provides a self-contained overview on the basics of collaborative filtering recommender systems. It covers two main classes of recommendation scenarios. In the classical matrix completion problem formulation, the task of an algorithm is to make longer-term relevance predictions given a user-item rating matrix. In session-based recommendation scenarios, the goal is to predict relevant items given a user's observed short-term behavior. From an algorithmic perspective, the chapter particularly focuses on neighborhood-based methods, which were proposed in the early days of collaborative filtering and which are still relevant today. The chapter addresses the entire life-cycle of algorithm development and also discusses libraries, datasets, and implementation aspects. Furthermore, it covers evaluation issues and reflects on today's research methodology in the field. Overall, the chapter shall serve as a starting point for readers, providing pointers to more detailed discussion of the various aspects regarding the design and evaluation of collaborative filtering recommender systems.

1.1. Introduction

Collaborative filtering (CF) is the predominant technical approach in the field of recommender systems in the academic literature (Jannach *et al.*, 2012). It is also utilized by large companies in productive use since clearly more than 15 years, with the system of Amazon.com being one of the most prominent early examples of wide-scale deployment (Linden *et al.*, 2003). The basic idea of this class of algorithms is to exploit the "wisdom of the crowds" and to use patterns within the collective behavior of a larger user community to determine suitable recommendations for an individual user or in the context of a given reference item.

In contrast to other possible approaches to determine recommendations, collaborative filtering techniques do not rely on content features and metadata of the items when determining the relevance of a recommendable item for a user like content-based or knowledge-based recommendation strategies (Jannach *et al.*, 2011). In its pure form[1], collaborative-filtering relies solely on a collection of explicit or implicit *preference signals* of users towards items. Given the past preference signals of an individual user, the goal is then to determine the (current) relevance of each recommendable item by combining these individual signals with the preference patterns of the community. The corresponding output of CF algorithms usually is either a set of relevance predictions for each item or a ranked list of items.

1.1.1. *Historical Background*

Probably the first work that used the term collaborative filtering in today's meaning in the context of recommender systems was that of Goldberg *et al.* (1992). In their *Tapestry* system, users of an experimental corporate email system could define personal *filters* (using a special query language) for incoming messages that referred to different features of emails, e.g., sender or content. "Collaborative" filters were a special class of filters, which could refer to so-called *annotations* by other users and one could therefore express that only messages should be retained that were voted positively by other users.

Soon afterwards different proposals were made of how to *automate* the task of filtering (news) items based on personal preferences and the opinions of other people. Among these proposals was the *GroupLens* system from Resnick *et al.* (1994), which proposed the comparably simple heuristic that users who shared similar preferences in the past can be exploited to predict a relevance score (rating) for incoming *netnews* messages (i.e., a user's *nearest neighbors*). While during the past two decades hundreds of different sophisticated algorithms were proposed for the rating prediction task, the problem formalization and the basic heuristic underlying this early work has influenced academic research in the field up to today.

1.1.2. *Collaborative Filtering as a Matrix Completion Task*

In Resnick *et al.* (1994), the recommendation problem is considered one of *matrix completion* (or "matrix filling" as termed in the original work).

[1]A variety of additional types of data can be combined with CF approaches, e.g., data that describes the users's context or features of items. See also Chapter 4 of this book.

The input is a matrix where rows and columns represent users and items, respectively, and the cells of the matrix are the known preference statements (ratings) for user-item pairs.

Table 1.1 shows an example matrix where five users (*u1* to *u5*) rated five items (*i1* to *i5*). The recommendable items for user *u1* are items *i4* and *i5* since we assume that the *u1* already knows the other items. The question is now if we should recommend *i4* and *i5* at all, and if so, in which order the items should be recommended.

Table 1.1. User-item Rating Matrix.

	i1	**i2**	**i3**	**i4**	**i5**
u1	3	4	3	?	?
u2		4	3		5
u3		1	3	1	
u4	4		3	2	3
u5	3		3	2	

While the ultimate computational task in many applications is to filter and rank the items, many matrix completion approaches solve this indirectly by estimating a relevance score (or, in this case, predicting a rating) for each unknown entry in the cell. While the ranking is determined by these scores, items that do not surpass a minimum threshold need to be filtered.

Over the years, a huge variety of algorithmic approaches has been proposed to accurately predict the missing matrix values, and many of the more advanced ones will be discussed in Chapter 2 of this book. In this chapter, we will mainly focus on early and comparably simple algorithms, including the one proposed by Resnick *et al.* (1994), which is based on a nearest-neighbor scheme.

Abstracting the recommendation problem to a matrix completion task has different advantages from an academic perspective, as discussed in Jannach and Adomavicius (2016). The computational task is very well defined and the generic nature of the problem formulation allows researchers to design algorithms that are not specific to a certain application domain. Furthermore, different mathematical concepts for data analysis or noise reduction, including principal component analysis or singular value decomposition, can be directly applied on the given data. Finally, over the years, a number of public datasets have become available and agreed-upon evaluation procedures were established in the community (see Section 1.4 in this chapter). These developments, together with the Netflix prize competition[2], geared research efforts on the basis of a matrix completion problem formulation during the last decade. Therefore, most of the chapters in this book will focus on these algorithmic approaches addressing this problem formulation. Nonetheless, as pointed out, e.g., in Jannach *et al.* (2016b) and Jannach and Adomavicius (2016), there are a number of aspects of practical problems for which matrix completion is not the best problem abstraction. Consequently, we will discuss an alternative problem formulation later in Section 1.1.4.

1.1.3. *Basic Algorithms for Matrix Completion*

In this section, we will review two basic collaborative filtering algorithms. Both of them are called "memory-based" (in contrast to "model-based" ones) because their recommendations are not based on learning an abstract representation of the data in a pre-processing step. Instead, they load the existing preference signals of the community into memory and implement neighborhood-based strategies to determine suitable recommendations.

1.1.3.1. *User-based Nearest-Neighbor Algorithms*

This algorithm scheme, which is often referred to simply as "user-based CF", implements the general idea that users "*who agreed in the past will probably agree again*" (Resnick *et al.*, 1994). To make a relevance prediction under that scheme, i.e., predicting the relevance of item $i5$ (called "target item") for user $u1$ (called the "active user") in Table 1.1, two main steps have to be done.

(1) Identify a set of N users who exhibit similar rating patterns as $u1$ (which are often called "neighbors" or "peers") and for whom a relevance signal for item $i5$ is known.
(2) Given this set of N similar users and their ratings for item $i5$, combine their ratings to predict the relevance of $i5$ for $u1$.

Consider the following example. When looking at the rating matrix in Table 1.1, we can see that users $u2$ and $u4$ have rated item $i5$, and the question arises if their ratings for $i5$ would be good predictors for the unknown rating of user $u1$. The basic assumption is if the ratings of user

[2] `http://www.netflixprize.com`

$u1$ were highly similar to those of users $u2$ and $u4$ then these neighbor's ratings for $i5$ should be useful predictors. User $u2$ rated items $i3$ and $i4$ exactly like $u1$ and user $u4$ rated them similarly, but did not use identical values.

Now, when implementing the general idea of user-based collaborative filtering in practice, two basic design choices have to be made.

a) How do we assess the similarity of two users?
b) How do we combine the ratings of the similar users?

To answer the first question, Resnick *et al.* (1994) proposed to use Pearson's correlation coefficient as a measure to assess if users have similar tastes. Given two users a and b, their similarity is computed as follows, see also Jannach *et al.* (2011),

$$sim(a,b) = \frac{\Sigma_{p \in P}(r_{a,p} - \overline{r_a})(r_{b,p} - \overline{r_b})}{\sqrt{\Sigma_{p \in P}(r_{a,p} - \overline{r_a})^2}\sqrt{(r_{b,p} - \overline{r_b})^2}} \tag{1.1}$$

where P is the set of items that were rated both by a and b, $r_{a,p}$ refers to the known rating of user a for item p, and the symbol $\overline{r_a}$ corresponds to the average rating of user a. The resulting values range between -1 and $+1$, where a value $+1$ indicates that two users have identical tastes and -1 expresses that users have opposite tastes. Values close to 0 are indicators of no or only an insignificant correlation between tastes of two users. One characteristic of the correlation coefficient worth noting is that the measure accounts for different interpretations of the rating scale by two users. Instead of comparing absolute rating values, it compares how a user's rating for an item deviated from the user's average rating value.

With respect to the second question of how to combine the relevance predictions given a set of similar users N, Resnick *et al.* (1994) proposed to use the following function $pred(a,p)$, to predict the rating of user a for item p, see also Jannach *et al.* (2011).

$$pred(a,p) = \overline{r_a} + \frac{\Sigma_{b \in N} sim(a,b) * (r_{b,p} - \overline{r_b})}{\Sigma_{b \in N}|sim(a,b)|} \tag{1.2}$$

The idea of this prediction function is to start with the average rating of user a and then consider for each neighbor if item p has been rated above or below the individual average rating. These rating signals of the nearest neighbors are consequently weighted with the similarity between users. Thus the ratings of more similar neighbors have a stronger influence on the derived rating predictions.

Different variations of the scheme proposed by Resnick *et al.* (1994) are possible along the following dimensions.

- Similarity function: Several standard similarity measures have been explored in the literature including cosine similarity, Spearman's rho, Jaccard index or Dice coefficient (Herlocker *et al.*, 2004), where the latter measures are only applicable in case the ratings are binary or unary. Furthermore, considering only the user's deviation from their average rating value for ordinal and continuous rating scales accounts for a consistent under- or over-biasing. In addition, the similarity function could also take the number of co-rated items into account and e.g., apply significance weighting, where the similarity of users with only few co-ratings is reduced (Breese *et al.*, 1998; Herlocker *et al.*, 1999).
- Neighborhood formation: The most common strategies are to select a fixed number of nearest neighbors or to include only those neighbors above a specific similarity threshold. Alternatively, an adaptive neighborhood formation strategy can be implemented that, for instance, could dynamically adjust the similarity threshold to keep the number of nearest neighbors within a predefined range.
- Prediction function: Its task is to aggregate the ratings of neighboring users depending on their similarities and additional characteristics in order to predict a rating value. Besides the standard weighted average like in Equation 1.2, also specific modifications of the aggregation function were proposed, like adjusting the importance weight of controversial items with a high rating variance or amplifying the importance of very similar users (i.e., those close to the highest similarity) (Breese *et al.*, 1998).

Which algorithm configuration works best in practice has to be determined empirically based on the specifics of the given application domain.

1.1.3.2. *Item-based Nearest-Neighbor Algorithms*

The idea of "Item-based CF" is closely related to the user-based variant already discussed with the difference that we compute similarities between items based on co-ratings by individual users. When we seek a prediction for user $u1$ and an item $i5$, we do not scan the data for users that are similar to $u1$ but for items that are similar to $i5$ and then aggregate the ratings $u1$ gave to these similar items.

For instance, turning to our example in Table 1.1, we can compute the similarity between those items that were rated by user $u1$ ($i1$, $i2$, and $i3$) with item $i5$ and then again compute the prediction for $i5$ as a similarity-weighted sum of the ratings for $i1$, $i2$, and $i3$. While user-based and item-based CF are technically similar, in practice item-based CF has some advantages over the user-based method, as discussed, e.g., by Linden *et al.* (2003). Consider that there are on average more ratings (or, more generally, preference signals) per item than there are per user. The similarity computations are therefore often based on a larger number of common ratings and the similarity *models* are therefore more stable. Note that in the user-based case a few more ratings can lead to a largely different set of neighbors. Given that item similarities have a tendency to be more stable, Linden *et al.* (2003) proposed to pre-compute the item similarities in an offline process to speed up the predictions at runtime. Their pre-processing method is computationally expensive in theory, but scales well as there are typically many more customers than catalogue items. Furthermore, for customers who only purchase a few best-selling items, a sampling strategy can be applied to reduce the computational effort.

1.1.4. *Collaborative Filtering as Session-Based Recommendation*

So far, our discussion was limited to scenarios where the recommendation problem is considered a matrix completion task and item relevance predictions are independent of the current usage context, users' intents, or recent observable behavior. However, in many practical applications users can have different intents each time they visit, for instance, an e-commerce site. Such short-term or ephemeral preferences are not explicitly covered in the basic problem formulation and thus specific techniques have been developed to realize context-awareness of RS.[3] In addition, visitors of commercial websites may also not be logged in, which means that in such situations no long-term preferences are known. However, individualized recommendations can be based on click-stream data of the current user session. While several works exploit other types of preference signals than explicit ratings, one of the main assumptions of standard collaborative filtering approaches is still that there is only a single type of preference signal relating users and items. Thus, advanced techniques capable of coping with multiple

[3]Context-aware recommenders use additional sources of information that describe the user's current situation and will be discussed in more detail in Chapter 5 of this book.

different signal types or with multi-dimensional ratings are discussed in later chapters. For instance, Jannach *et al.* (2018) provide a deeper discussion of recommendation algorithms based on implicit feedback.

Next, we will sketch an alternative problem formulation for session-based recommendation, where we are interested in recommendations that suit the context of a specific user session. An in-depth discussion of sequence-aware recommender systems and a more formal characterization of the problem can be found in Quadrana *et al.* (2018).

1.1.4.1. *Inputs and Outputs*

The central input to a session-based recommender is an ordered or time-stamped list of past user actions. User actions can have different types (such as "item-view" or "purchase") and are typically associated with one of the items. The users themselves can be known, i.e. recognized returning users, or anonymous. The (user, item, action) tuples can furthermore be enhanced with additional attributes like user demographics or metadata features. The central part of the inputs can be seen as a collection of enriched clickstream data, which can be easily collected on web platforms. A main difference to the matrix completion formulation is that we can potentially observe multiple interactions of a user with the same item over time.

The second part of the inputs is the context of an "active" session, which consists of an ordered list of the user's actions in the session for which a recommendation is sought for. In the literature, sometimes a differentiation between "session-based" and "session-aware" recommendations is made (Quadrana *et al.*, 2018). In "session-based" scenarios, no longer-term user history is known, e.g., because we have to deal with anonymous users. In "session-aware" scenarios, in contrast, also past sessions of the active user might be known.

The output of a session-based recommender is an ordered lists of predicted next user actions, where these actions are usually related to items, e.g., a list of items for which we predict a view event or a purchase event. In the case of "next-basket predictions", the predictions can also refer to an entire set of items that are bought together. In the general form, the output is however similar to that of a traditional "item-ranking" recommendation setup.

1.1.4.2. *Goals and Computational Tasks*

In contrast to the matrix-completion task, where the goal is usually to predict a context-independent relevance score based on past preference signals, in session-based scenarios one has to also consider the short-term intents of the user. The goal is therefore to determine a ranked list of items that are relevant given the user's actions in the current session. Since the relevance of the items can largely depend on the users current contextual situation, the computational task can therefore involve balancing the users' long term preferences and their short-term situation and goals. Figure 1.1 illustrates the main idea of collaborative filtering as session-based recommendation.

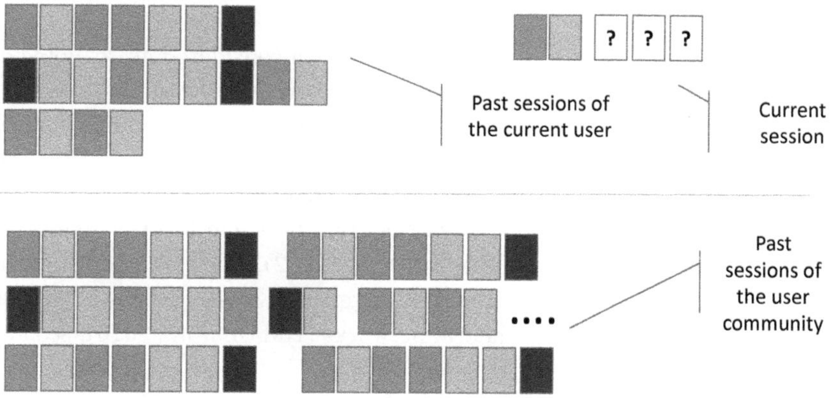

Fig. 1.1. General principle of collaborative filtering as session-based recommendation. We are given a current user session containing actions of different types with the goal of finding suitable recommendations for the session. Both, past sessions of the individual user and the user community can be considered when determining the recommendations. Different colors in the figure indicate different types of actions.

Besides the user's intent, in session-based recommendation scenarios, the *intended purpose* of the recommendations also influences the contextual relevance of individual items. If, for instance, the goal of a recommender is to show the user alternatives for a given product of interest, items that are *similar* to those that the user has recently inspected are probably good candidates. If, in contrast, the recommender should help the user find *accessories*, a very different set of shop items becomes relevant. Finally, also the *repeated suggestion* of items that are already known to the user can be a task of for a session-based recommender, which can lead to additional computations in order to select such items and to determine the best point in time to make such recommendations.

Overall, there is no unique concept of a "good" recommendation in session-based recommendation scenarios, which also makes the evaluation of such systems more complex in academic environments as there are domain-dependent factors. With the matrix-completion problem formulation, on the other hand, several important aspects of real-world recommendation problems are not addressed at all, like short-term trends, the repeated consumption of items, or the consideration of multiple interactions of a user with an item over time.

1.1.5. *A Nearest-Neighbor Algorithm for Session-Based Recommendation*

While user-based CF and item-based CF can be considered as "canonical" approaches to the matrix completion problem that are popular baselines for algorithm comparisons, no standard baseline algorithm (and evaluation protocol) yet exists for session-based recommendation. Historically, typical technical approaches that consider the sequences of events in the recommendation process include works that rely on some form of sequential pattern mining techniques (Mobasher *et al.*, 2002), Markov models (Shani *et al.*, 2005), explicit user feedback (Zanker and Jessenitschnig, 2009b,a) and, more recently, recurrent neural networks (Hidasi *et al.*, 2016), see also Chapter 3 of this book.

An alternative to attempting to mine sequence information from the logs, which is often a computationally costly process, is to focus only on co-occurrence patterns within the different user sessions. One of the most visible examples of this approach in practice are Amazon's "Customers who bought ... also bought" recommendations. Technically, these recommendations can be implemented by considering pair-wise item co-occurrences in past transactions to predict additional items for the current shopping session.

This pairwise approach can be extended to a *session-based nearest neighbor algorithm*. The idea is to take the elements of the current user session and look for past sessions — including those by other users — who are similar to this session. We then examine which other elements appeared frequently within this set of "neighbor sessions". These elements then finally represent our recommendation candidates. In Zanker and Jessenitschnig (2009b) this similarity computation is even designed as a stepwise feature-combination approach, where the feedback categories representing different types of user actions such as "view events", "menu navigation" or explicit feedback are prioritized and low-priority feedback categories are only exploited if not sufficient recommendation candidates can be identified otherwise.

More formally, let s be the current user session, where a session is defined as an ordered list of recorded user actions. Each list entry can be a complex object describing, e.g., the respective item ID and the action type, or simply a set of item IDs if only actions of one type (e.g., "view events") are considered. Given the set S of past sessions of all users and a function $sim(s1, s2)$ that returns a similarity score for two sessions $s1$ and $s2$, compute the set N that contains those k sessions from S, which have the highest similarity score according to the function sim.

The goal is now to compute a relevance score for each recommendable item i for a given session s and a set of neighbor sessions N. We can compute this score as follows, see also Bonnin and Jannach (2014).

$$score_{\mathrm{KNN}}(i, s) = \Sigma_{n \in N} sim(s, n) \times 1_n(i) \tag{1.3}$$

where $1_n(i) = 1$ if session n contains item i and 0 otherwise. The final list of recommendations is then determined as usual by sorting the recommendable items in decreasing order of the relevance score.

Although this method does not take the order of the items within a session into account, it turns out that it leads to competitive results, e.g., in the domains of next-track music recommendation, next-item recommendation in e-commerce, and next-task prediction in workflow modeling, see, e.g., Lerche *et al.* (2016); Jannach *et al.* (2016a, 2017); Jannach and Ludewig (2017b); Ludewig and Jannach (2018).

Similar to the already mentioned user-based and item-based CF methods, different similarity functions can be applied for session-based recommendation methods. Since we typically have to cope with binary or actually unary data, i.e., an item appears or does not appear within a session, we can apply, for instance, set-based similarity measures like the Jaccard index or binary cosine similarity. Recently, different *sequence-aware similarity functions* were explored in Ludewig and Jannach (2018). The obtained results indicate that using such similarity functions in many cases further increase the prediction accuracy of neighborhood-based models. As a result, these comparably simple models often even outperform some of today's sophisticated deep learning based algorithms.

Finally, the number of neighbors is another parameter to be considered. It depends on the application domain, where, for instance, for next-track music recommendation reasonable values range from less than 10 neighbors up to hundreds of neighbors (Bonnin and Jannach, 2014; Jannach and Ludewig, 2017b).

1.2. Recommendation Paradigms

In this section, we will first discuss the general pros and cons of the collaborative filtering recommendation paradigm — also in comparison to other recommendation mechanisms — and will then focus on the specific advantages and limitations of neighborhood-based methods.

1.2.1. *Pros and Cons of Collaborative Filtering*

1.2.1.1. *Existing Recommender Systems Paradigms*

Recommender systems are typically categorized into collaborative filtering methods, as discussed in this book, content-based methods, knowledge-based approaches, and hybrids that are different combinations of the aforementioned paradigms (Jannach *et al.*, 2011).

The idea of content-based methods is to estimate the preference of an individual user towards the specific characteristics of items. In the movie domain, for instance, a corresponding content-based user profile could comprise to which extent a user likes action movies or an actor/actress. Thus, the profile itself is learned by analyzing the features of items that the user liked or disliked in the past.

Knowledge-based approaches also consider item features, but are usually based on explicit, formalized, and domain-dependent information about which items are a good match for a given set of user preferences. These user preferences are typically elicited in an interactive process and are specific for a given recommendation problem instance. The logic of how to match user preferences and items can, for example, be expressed in the forms of logical rules, constraints, or utility functions.

1.2.1.2. *Comparison of CF Methods with Other Paradigms*

All of these methods have their advantages and limitations, which is why various proposals to build hybrid systems were made over the years that aim at overcoming the shortcomings of individual methods.

One main advantage of collaborative filtering methods is that no knowledge about the recommendable items is required. In many application domains, including e-commerce, product catalogs can comprise tens of thousands of items, often leading to significant challenges in terms of product data management for companies. This is particularly the case when the online platform where the recommendation system is deployed is a big market place.

Pure CF-based approaches and the corresponding algorithms can furthermore be applied to a variety of domains and product categories. A large number of different algorithms was proposed over the years and a multitude of off-the-shelf implementations in different programming languages are publicly available. When compared to most knowledge-based systems, CF-based methods (and content-based ones usually as well) can learn over time when additional preference signals become available. In many knowledge-based systems, in contrast, the underlying rules have to be updated when new products with new features should be recommended.

On the downside, CF-based recommenders require the existence of a large community of users in order to be able to identify patterns in their preference relations and behavior. This requirement for a large and ideally returning user community can be particularly challenging for smaller e-commerce sites. The other main problem of CF-based systems is its limited capability of dealing with first-time users and novel items. For new or cold-start users, no preference signals are initially available, making it impossible to find neighbors in the user-based CF approach described above. Similar problems exist for new items that have been recently added to the item database. A huge number of academic research papers made proposals to deal with these aspects, e.g., by using initially an alternative recommendation paradigm or to recommend mostly popular items to new users, see, e.g., Bobadilla *et al.* (2012); Huang *et al.* (2004); Said *et al.* (2012a).

1.2.2. *Pros and Cons of Nearest-Neighbor Methods*

Nearest-neighbor methods in general have the advantage that they are easy to understand, implement, debug, and maintain. More sophisticated algorithms — as will be discussed in later chapters — can be challenging to implement for a typical software engineer. Furthermore, debugging unexpected outputs becomes challenging and significant engineering effort is required to deploy such algorithms in a production environment. For instance, the winning algorithms of the Netflix prize never made it into production, partly due to the involved engineering efforts (Amatriain and Basilico, 2012). In contrast to model-based approaches, which, e.g., rely on learning a prediction function from historical data in an offline process, nearest-neighbor methods are able to immediately consider new preference signals once they become available.

At least from an academic standpoint, the outputs of nearest-neighbor methods are considered to be "easy to explain", as done, e.g., in Herlocker *et al.* (2000) and a number of ways of explaining recommendations based on user-based CF methods were proposed in the literature. To which extent these academic approaches are suited in practice, is, however, still a largely open question.

Finally, nearest-neighbor methods are considered to lead to "reasonably good" " recommendations in many domains. Many more sophisticated algorithms exist which outperform nearest-neighbor methods in offline experiments in terms of predictive accuracy measure. However, it is not clear if these — often only slightly better — results can be actually noticed by users or translate into business value for the provider.

The main challenge of nearest-neighbor methods often lies in their computational complexity. When using a naive implementation, the required computation times very soon exceed the narrow time frames of online recommendation scenarios. Obviously, one cannot scan thousands or even millions of possible neighbors in real-time, whenever a new recommendation should be served to the user. Therefore, offline pre-processing, data-sampling, or parallelization techniques have to be applied to ensure the scalability of neighborhood-based methods, see, e.g., Jannach and Ludewig (2017a).

1.3. Practical Implementation Considerations

Recommendation systems, and collaborative filtering in particular, are a key application of data mining and machine learning at a large scale in industry. The research field of recommender systems is traditionally closely linked to industry needs and application scenarios. Industry challenges like the famous one million dollar Netflix Prize (Bell and Koren, 2007) led to additional research momentum and the ACM conference series on recommender systems, as a result, usually attracts a high share of industry participants. Nevertheless, specific aspects of recommender system implementations traditionally only receive limited attention in academia as will be discussed next.

Scalability Amazon is one of the earliest adopters of large-scale recommendation techniques to personalize their customers' shopping experience.

Thus, in an early paper on item-based collaborative filtering employed at Amazon (Linden *et al.*, 2003), the focus was put on the scalability of this neighborhood method and short response times. The importance of scalability is, for instance, also stressed in a blog post from Netflix (Amatriain and Basilico, 2012), where the strongest algorithms from the *first progress prize* could be put to practice only with significant engineering effort since they had to operate on 5 billion instead of 100 million ratings. Questions of scalability will also be discussed in Chapter 11 of this book.

Freshness and Continuous Updates In addition to the ability to scale to large amounts of data, also the freshness of the data and models is an important aspect for real-world applications. In particular, for cold-start users being able to continuously update the user model and the recommendations based on the most recent user actions within milliseconds is an important system requirement. The concept of a three-tier pipeline architecture consisting of offline, nearline and online tiers has been proposed for this purpose (Amatriain and Basilico, 2012). The offline tier performs the traditional batch processing to rebuild models at predefined intervals, the nearline tier is responsible for incremental model updates that should be very close to real-time. The online tier finally performs the actual filtering of pre-computed recommendations based on the most recent user feedback and general trends.

User Interface Francisco Martin, at that time CEO of the recommendation service provider Strands, stated in his keynote at ACM RecSys 2009 that the user interface can be much more important than the recommendation algorithm itself (Martin, 2009). Also Amatriain and Basilico (2015) mention that the user interaction design is often disregarded in the literature, even though it can have a major impact in practical systems. A recent survey of existing research on user interaction aspects of recommender systems can be found in Jugovac and Jannach (2017); aspects of user-centric evaluation approaches for recommender systems are also discussed in depth in Knijnenburg *et al.* (2012); Knijnenburg and Willemsen (2015); Pu *et al.* (2011)

Data Integration When it comes to interacting with users, almost everything that is on display at a commercial site can in principle be subject to personalization (Amatriain and Basilico, 2012). This, as a result, means that multiple categories of data sources can be considered as a potential input to algorithms, which in turn leads to a need for engineering the most appropriate features. In addition, it becomes more and more visible that explicit item ratings, as used in the context of the Netflix prize, are not the most helpful source to predict the users' next actions, e.g., because there can be a gap between what items users rate highly and which items they actually consume. Therefore, many different types of implicit feedback signals about their behavior, transactions and social connections can be exploited, as discussed in Jannach *et al.* (2018). A particular challenge in that context is that there are not only various potential types of signals but that there can be huge amounts of data to be processed. In Basilico (2013), the author for example mentions that Netflix processes over 100 billion events per day.

Business metrics Finally, when it comes to assessing the quality or value of the recommendation functionality, obviously the well-known RMSE has to be seen as just one proxy measure for the recommendation quality, but in reality more encompassing *business metrics* have to be used in practice. For instance, click-through rates, choice time, user engagement, or low churn rates are measured as success indicators in practical applications (Gomez-Uribe and Hunt, 2015).

Frameworks and Libraries From an engineering perspective, substantial progress was made in the last years and developers can today build their solutions based on several industry-strength frameworks and infrastructures that were not around a decade ago. Examples of such frameworks are the Apache PredictionIO ML server[4] or the Tensorflow[5] open source library for numeric computations using data flow graphs that has been successfully used for building deep learning and factorization-based algorithms. From a more research oriented perspective, a variety of mature open-source recommendation libraries on different language platforms are available today as well, like LensKit (Ekstrand *et al.*, 2011) for Java[6], MyMediaLite (Gantner *et al.*, 2011) for the .NET platform[7], different ML libraries such as scikit-learn (Pedregosa *et al.*, 2011) for Phyton or the *rrecsys* package (Çoba *et al.*, 2017a,b) for R[8]. Public datasets and libraries will be discussed in more depth in Chapter 9 of this book.

[4]https://predictionio.incubator.apache.org
[5]https://www.tensorflow.org
[6]http://lenskit.org
[7]http://www.mymedialite.net
[8]https://cran.r-project.org/web/packages/rrecsys/index.html

1.4. Practical Evaluation Considerations

In the following section, we will discuss common ways of evaluating collaborative filtering recommender systems. An in-depth discussion of practical challenges of evaluating recommenders can be found in Chapter 9 of this book.

1.4.1. *General Considerations*

Algorithmic approaches to build recommender systems, including collaborative filtering based ones, can in general be evaluated in different ways. The most informative way of assessing the effects of different recommendation strategies on users and the corresponding business value is to run A/B tests using websites or applications that host recommender systems. In this context, A/B testing corresponds to a randomized experiment, where the users of the system are split into two or more groups and each group is served by recommendations generated in different ways, i.e., usually the groups see different recommendations or different forms of how the recommendations are presented. At the end of the experiment period, one can then compare the effects of the different strategies, e.g., in terms of the number of clicks on the recommended items or on sales. For instance, Zanker (2012) compares two explanation strategies on a spa tourism platform or chapter 8 in Jannach *et al.* (2011) compare recommendation.

While A/B tests are the most informative measurement instrument, there are also pitfalls when designing these experiments and interpreting their results. For instance, one has to make sure in advance to have a large enough sample to ensure that any observed differences of method A compared to method B are statistically significant. Also, there can be unexpected or unconsidered periodical effects that might lead to distorted results and wrong conclusions.

At the same time, one usually cannot run a virtually infinite number of A/B tests in practice in order to evaluate hundreds of different algorithm configurations. Therefore, companies might resort to cheaper offline tests to determine those candidate configurations that most probably will lead to good results in the production system (Gomez-Uribe and Hunt, 2015). We discuss such offline evaluation procedures in the next section.

1.4.2. *Offline Evaluation*

Since academic researchers usually cannot run experiments on real platforms, they commonly use methodologies to compare algorithms based on historical data. Academic research in collaborative filtering recommender systems is therefore largely dominated by such offline experimental designs.

The methodology to evaluate collaborative filtering system is mainly adopted from related research fields, in particular from the fields of information retrieval and machine learning (Herlocker *et al.*, 2004).

1.4.2.1. *Protocols and Measures for Matrix Completion Problems*

The most common procedure to evaluate recommenders is based on rating datasets. Such rating datasets contain at most one explicit or implicit preference signal for a set of users and recommendable items and can thus be considered as a typically very sparse rating matrix. The general approach is then to split the data into a training and test part, and to predict the preferences in the (held-out) test set based on the patterns that were identified in the training data. Usually, this process is repeated several times in a cross-validation process.

Rating prediction accuracy In the matrix completion task, the goal is to predict the held-out ratings and the quality of the recommendations can, for instance, be measured in terms of the average deviation of the predictions from the true (hidden) values. This measure is referred to as the Mean Absolute Error (MAE). Particularly in the last decade, researchers more often report the Root Mean Squared Error (RMSE), which penalizes larger deviations stronger than smaller ones. Details of how to calculate the measures mentioned in this section are provided, e.g., in Herlocker *et al.* (2004). Many modern collaborative filtering algorithms that will be discussed in later chapters of this book in fact try to minimize the squared error based on an offline training phase.

Looking at results that are obtained for datasets that contain movie ratings on a five-point scale, we can see that modern algorithms are able to achieve MAE values slightly below 0.7 and RMSE values that are around 0.85. The absolute values reported in the literature can however not always be directly compared even when the same dataset is used, because researchers often apply data-filtering procedures, e.g., to filter out inactive users, or use different cross-validation configurations, which leads to larger or smaller training sets, see also Said and Bellogín (2014).

Classification and ranking accuracy While rating prediction can be considered a classical machine learning (function learning) task, it is more "natural" to consider recommendation as a classification or ranking task. Accordingly, the evaluation procedures and measures from this field can be applied.

The probably most common approach according to the literature (Jannach *et al.*, 2012) is to measure and report *precision* and *recall.* Instead of rating prediction, the task of the recommender is to compute a size-restricted ranked list of recommended items for a given user. Usually, the length of the recommendation lists and correspondingly the measurement is limited to a top-10 or top-20 list of items.

The quality of such a recommendation list is then numerically quantified by considering the amount and position of "relevant" items in the top-ranked lists of all users of the test set. The precision measure counts how many items in the top-n list are relevant. With recall, we measure how many of the relevant items actually made it into the top-n list. Since there is often a trade-off between precision and recall — longer list sizes will lead to higher recall and lower precision — researchers often report the F-measure, which is the harmonic mean of precision and recall.

The values that are reported for precision and recall, even for the same dataset, vary strongly from research paper to research paper. The main reason is that the absolute values depend on how one treats those items in the ranked lists for which no "ground truth" is known, i.e., the items for which no rating information existed in the test set. If these items with unknown ground truth are removed from the recommendation lists before determining precision and recall, the resulting values are usually very high and above 70 to 90 percent. Otherwise, if the items without unknown ground truth are kept in the list, precision and recall are usually below 10%. The low numbers for the latter case are not surprising, given that we often have thousands of recommendable items but only know the ground truth for a small subset of these items. Additional factors that influence the absolute values of the measures are the chosen list size and the number of cross-validation splits. In research works that are based on explicit (1-to-5 star) rating datasets, the outcomes are furthermore influenced by the choice of the threshold that is used to discriminate relevant from irrelevant items. Some authors only consider items to be relevant that obtained 5-stars, whereas others consider all items as relevant that have a rating that is higher than the user's average rating.

Precision and recall do not account for the position of the relevant items in the result set. Obviously, however, having a good matching proposition at position 1 is favorable over having it at a later position. Researchers therefore often use measures like the Mean Reciprocal Rank (MRR) that also consider the position of the relevant elements. More details about this and additional rank measures like NDCG or AUC etc. can be found in Herlocker *et al.* (2004).

Beyond-accuracy measures In recent years, researchers investigated a number of quality measures beyond prediction or ranking accuracy. Specifically, the question of recommendation *diversity* was analysed in depth in many works. Being able to produce a set of diverse recommendations is important in many application domains, since the recommendation of many items that are very similar (e.g., movies of a certain series) can be of limited value for the user. The diversity of a recommendation list can be quantified in different ways, e.g., based on the pairwise similarity of the items in terms of their features.

Other possibly desirable features of a recommendation algorithm can be to not only recommending popular items or, more generally, not focusing too much on a small set of items that is recommended to everyone. Furthermore, depending on the application domain, it can be desirable that the algorithm points the user to items outside of his or her usual taste. This is often considered as quality factors like *novelty* and *serendipity* (Castells *et al.*, 2015).

In most cases, considering additional quality factors comes at the price of reduced accuracy values. To be able to judge the quality of recommendations in many domains therefore requires an understanding of the specifics of the domain and a corresponding algorithmic approach to balance the often competing quality goals.

1.4.2.2. *Protocols and Measures for Session-Based Recommendation*

When evaluating session-based algorithms offline, the general principles apply as when evaluating based on rating datasets. The existing datasets, which in this case contain user action logs instead of item ratings, have to be split into training and test sets and the main computational task is to predict the hidden actions.

From the computational perspective this is however slightly different. First, as described above, rating prediction is not meaningful in this setting and our goal is usually to predict the next user action(s), e.g., the next item-view event. Second, in session-based recommendation we are provided with additional information about the user's most recent actions, which have to be considered in the computations.

In Jannach *et al.* (2015b), the authors proposed a general evaluation scheme for session-based recommendation. The overall idea is visualized in Figure 1.2. In a first step, the sessions of each user are split into a training and a test part, e.g., by considering the most recent 20% of the sessions of a user as test sessions. The training data represents the inputs for algorithms to learn long-term preference models. The sessions in the test set are then evaluated individually according to the given prediction task, which could be, e.g., to predict the purchase event in the session. To avoid random effects, one can repeatedly apply the protocol using, e.g., a random subsampling method.

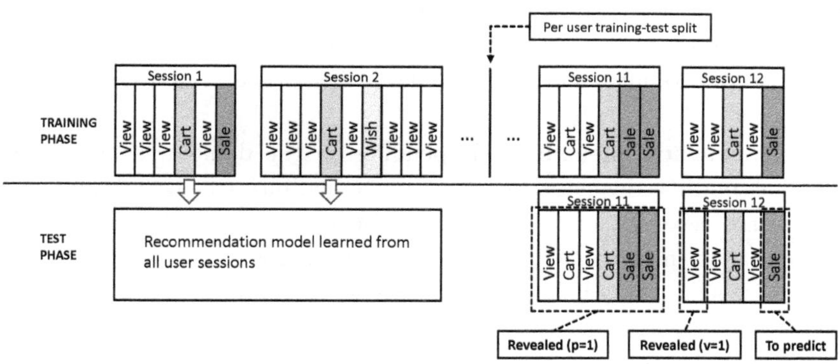

Fig. 1.2. Session-based evaluation protocol as proposed in (Jannach *et al.*, 2015b).

The proposed protocol has two special parameters that can be set for an evaluation. First, there is a parameter v that describes how many items of the current session should be revealed to the algorithm when making a prediction. By varying this parameter, one can measure how quickly different algorithms are able to adapt their recommendations to the user's current goal. With the second parameter p, we can determine how many sessions that *precede* the current user session should be revealed to the session-based algorithm. By changing this parameter, we can test to which extent algorithms are able to leverage the information in these sessions, e.g., by reminding users of items that they have inspected on previous days.

While the protocol was designed for e-commerce scenarios, it is not limited to this domain. Furthermore, it can also be used in case of anonymous sessions. In this case, the training and test datasets obviously cannot be created per user and revealing previous sessions (by changing parameter p) is not possible as well.

Accuracy measures In general, the same classification and ranking measures described above (e.g., precision, recall, and MRR) can be applied. Depending on the chosen evaluation setup, sometimes only one item (the immediate next user action) is the relevant one and precision and recall are proportional in this case.

Since we have user actions of different types, it can furthermore be meaningful to define new measures that assess an algorithm's prediction capabilities in different dimensions. In the ACM RecSys 2015 challenge[9], for example, the task was to first predict whether or not a user will make a purchase in the current session, and if so, to predict which item will be purchased. The evaluation measure used in the challenge correspondingly considered both aspects. An in-depth discussion of evaluation aspects for session-based recommenders can be found in Quadrana *et al.* (2018).

Beyond accuracy measures Depending on the domain, again other quality factors like diversity, novelty, and serendipity can be relevant for session-based recommendation scenarios. In some domains it can also be important that the recommendations represent a good "continuation" for the given session. This holds in particular when the problem is to find a suitable next track to play in a music recommendation scenario. In this application, it is typically desirable that the next played tracks are similar to the last played ones, e.g., in terms of the tempo or mood. Also in the music recommendation domain, it can furthermore be important that the set of recommended tracks is coherent in itself and the transitions between the individual tracks are smooth, see also Chapter 15 of this book. A deeper discussion of beyond-accuracy measures can be found in Chapter 10.

1.4.2.3. *Discussion*

In this section, we reviewed basic and common approaches to evaluate recommendation algorithms in offline experiments. In the academic literature, a rich variety of alternative evaluation measures and protocol variants, e.g.,

[9] http://2015.recsyschallenge.com

to simulate cold-start situations, are used. However, even though the protocols and the measures are in principle well-defined, there are many subtle differences that can have an impact on the absolute values that are measured. Comparing the results across different research papers is therefore difficult, e.g., because different versions of precision and recall are used, datasets are pre-processed, a number of different cross-validation runs are made, or because different ways of splitting the data were applied.

The main problem when applying offline evaluation protocols however often lies on the choice of the quality measure(s). Academic research is mostly focused on prediction accuracy, but in many applications it might not be clear that higher accuracy translates into practical (business) success of the application. It is therefore important to find adequate measures that are chosen in line with the purpose of the recommender system (Jannach and Adomavicius, 2016) and which represent good proxies of the true success measures (e.g., increased sales or customer retention), which often cannot be directly measured in offline experiments.

1.4.3. *The Role of User Studies*

The aforementioned offline evaluation methodology is well-accepted in the research community. Measuring the error in terms of RMSE or MAE when assessing the rating prediction capability resembles the evaluation practice in the field of machine learning, while the methodology for assessing the classification and rank accuracy is borrowed from the field of information retrieval. Nevertheless, more and more works are challenging the view that algorithmic contributions may be solely evaluated based on offline experiments (Konstan and Riedl, 2012; Jannach *et al.*, 2016b) for several reasons. First of all, recommendation techniques aim at helping users by avoiding situations of information overload and by supporting them in decision-making tasks. Moreover, in many situations recommendation functionality should simply facilitate an enjoyable interaction of a user with an information system. Thus, recommendation systems are about enhancing the user experience and the ability of a system to accurately assess users' preferences and needs. Based on these assumptions, making correct predictions or providing many good recommendations is therefore an important component in order to make users happy and to explain their satisfaction as has been proposed by Knijnenburg *et al.* (2012). However, being able to select and present items that are relevant for users is not the only component that influences user satisfaction. Experimental user studies therefore play an important role in determining how different aspects of a recommender contribute to user satisfaction and system usage (Knijnenburg and Willemsen, 2015).

In general, following the above-mentioned works, offline evaluations can only help to assess a small part of the whole picture and therefore need to be complemented with user studies. This is in particular important since a number of recent works indicate that the quality perception or business value of a recommender system does not necessarily correlate with the accuracy measures commonly used in offline experiments (Garcin *et al.*, 2014; Jannach and Hegelich, 2009; Kirshenbaum *et al.*, 2012; Cremonesi *et al.*, 2012).

The recent study of Rossetti *et al.* (2016), for instance, shows that an algorithm ranking based on offline experiments can contradict the outcome of ranking algorithms based on user feedback in an experimental user study, where Precision@k was measured both offline and online. For these reasons, industrial leaders often employ a development pipeline that involves three steps: traditional offline experiments to identify failing approaches from the very beginning; user studies to identify and better understand promising candidates, and A/B testing in the field as the third step in the assessment cycle.

Beyond this aforementioned pragmatic approach of employing user studies as an intermediate step before field tests, further developments of the practice how user studies are conducted can be envisioned. Specifically, since recommendation systems support decision making tasks, additional mechanisms for measurement and feedback collection can be used in the future. For example, the emerging field of NeuroIS[10] employs sensors and methodologies from neuroscience in order to more accurately understand the cognitive processes of users when interacting with technology and information systems. For instance, Rook *et al.* (2018) observed that users with specific personality traits (such as an anxiety-related behavioral inhibition) tend to be more engaged when confronted with proactive recommendations and that the accuracy of provided recommendations moderates this effect, i.e., inaccurate recommendations make them ruminate about them.

In many cases, user studies also have their limitations and the obtained insights have to be interpreted with care. The common potential threats to the validity of the obtained findings include the following. The participant population, for example, which are often students, might not be

[10]See `http://www.neurois.org/` for more information.

representative for the general user population or too small; often, also the study participants do not face a true decision situation and, e.g., do not actually make a real purchase. Questions about their behavioral intentions (e.g., "willingness to buy") might not reflect the findings that one would obtain in a real shopping environment. Finally, some studies suggest that there are familiarity biases in the user perception and that study participants consider items as good recommendations when they already know them (Jannach *et al.*, 2015a; Kamehkhosh and Jannach, 2017).

1.4.4. *Datasets*

Public datasets are a key ingredient when performing academic research in the field recommender systems. Such datasets, like the one from the MovieLens (ML) movie rating and recommendation platform and from companies like Netflix (Amatriain and Basilico, 2015), have led to significant algorithmic improvements in the fields over the years. In combination with public libraries that implement various algorithms, they also led to a certain level of reproducibility of the obtained research results (Ekstrand *et al.*, 2011). However, the availability of datasets also heavily biases the focus of research contributions of the community (Jannach *et al.*, 2012). The MovieLens rating datasets have been available for almost two decades (Harper and Konstan, 2016) and have significantly contributed to the fact that collaborative filtering has become the dominant recommendation paradigm in academia and that movies are the most popular application domain investigated by researchers. Table 1.2 shows the characteristics of a few popular datasets from the movie domain.

Many other application domains for recommender systems beyond movies have been investigated over the years as well. Often, these research works are based on datasets that were published by companies in the context of research challenges and competitions. A number of comparably popular datasets are summarized in Table 1.3, including data from the domains of jokes, dating, music, or general e-commerce. Overall, the increasing availability of data for different domains, different product categories and different sources will in the future allow to further intensify research on cross-domain recommendations (Cantador *et al.*, 2015), on the combination of a variety of user feedback sources such as different categories of implicit transaction traces (Zanker *et al.*, 2007; Zanker and Jessenitschnig, 2009b) and on the enrichment of user preferences from social relationships.

Table 1.2. Characteristics of popular movie datasets.

Dataset	Users	Items	Ratings	Scale
ML 20M[a]	138,493	26,744	20,000,263	[0.5, 5]
ML Latest Small[a]	671	9,066	100,004	[0.5, 5]
MovieLens Latest[a]	259,137	39,443	24,404,096	[0.5, 5]
ML 100K[a]	943	1,682	100,000	[0.5, 5]
ML 1M[a]	6,040	3,706	1,000,209	[0.5, 5]
ML 10M[a]	69,878	10,677	10,000,054	[0.5, 5]
ML Tag Genome[a]	1,100	9,734	12,000,000	binary
MovieTweetings 2,014[b]	24,924	15,142	212,857	[1, 10]
FilmTrust[c]	1,508	2,071	35,497	[0.5, 4]
CiaoDvd[c]	7,375	99,746	278,483	[1, 5]
Douban[d]	129,490	58,541	16,830,839	[1, 5]
Netflix[e]	480,189	17,770	100,480,507	[1, 5]

Remarks and links for download

[a] age, gender, occupation, zip, timestamp, free-text tags
http://grouplens.org/datasets/movielens

[b] https://github.com/sidooms/MovieTweetings/tree/master/recsyschallenge2014

[c] trust relationships among users
https://www.librec.net/datasets.html

[d] friendship relations among users
https://www.cse.cuhk.edu.hk/irwin.king.new/pub/data/douban

[e] dataset is officially retired, but still used in publications

1.5. Summary and Further Reading

Summary and Outlook This chapter provided an introduction on the topic of collaborative filtering with a focus on neighborhood models, since these comparably simple models even today often serve as the basis for evaluating the more sophisticated methods that will be discussed in the subsequent chapters.

Due to its introductory nature and the focus on the main underlying problem of finding items that are relevant for user, many additional academic and practical aspects of collaborative filtering have not been addressed to a large extent. Additional topics and questions to consider when designing a CF-based recommendation system include the following and a number of them will be discussed in later chapters of the book.

- Historically, academic research has focused on the rating prediction problem, which however seems to be of limited value in practice, where the main goal is to determine item rankings based on implicit feedback traces as discussed in Chapter 7 of this book. Session-

Table 1.3. Characteristics of public datasets from different domains.

Dataset	Domain	Users	Items	Ratings	Scale
Jester1[a]	jokes	73,421	100	4,136,356	[-10, 10]
Jester2[a]	jokes	59,132	140	1,761,439	[-10, 10]
Jester3[a]	jokes	50,692	140	1,728,847	[-10, 10]
Book-Crossings (BX)[b]	books	278,858	271,379	1,149,780	[1, 10]
ISL (Views-All)[c]	cigars	1,260	142	18,434	unary
www.libimseti.cz[d]	dating	135,359	168,791	17,359,346	[1, 10]
Retailrocket[e]	e-commerce	1,407,580			unary
Scholar recs[f]	academia	50	100,531		
YOW[g]	news	24	5,921	10,010	multidim.
Epinions (665K)[h]	multiple	40,163	139,738	664,824	[1, 5]
Amazon reviews[i]	multiple	6,643,669	2,441,053	34,686,770	[0.5, 5]
last.fm[j]	music	358,868	292,375	17,535,655	unary
OSDC [k]	music		1,000,000		

Remarks and links for download

[a] http://eigentaste.berkeley.edu/dataset
[b] unary, demographic
http://www2.informatik.uni-freiburg.de/~cziegler/BX
[c] search & purchases
http://isl.ifit.uni-klu.ac.at
[d] gender http://www.occamslab.com/petricek/data
[e] user sessions
https://www.kaggle.com/retailrocket/ecommerce-dataset
[f] researcher interests
http://www.comp.nus.edu.sg/~sugiyama/SchPaperRecData.html
[g] novelty, readability, etc.
https://users.soe.ucsc.edu/~yiz/papers/data/YOWStudy
[h] trust relations
http://www.trustlet.org/epinions.html
[i] textual reviews
https://snap.stanford.edu/data/web-Amazon.html
[j] play counts, demographic
http://www.dtic.upf.edu/~ocelma/MusicRecommendationDataset
[k] tags, artists, track info
https://labrosa.ee.columbia.edu/millionsong/pages/getting-dataset

based recommendations, as briefly discussed in this chapter, are an important area for future research in a field which is still dominated by research based on datasets where only one user-item interaction was recorded.

- A general issue when performing offline experiments with historical data in that context is the question how accurately collaborative filtering can actually predict user tastes, given the inherent noise in the data. The discussion that algorithms cannot be accurate beyond a specific point brought up the concept of a "magic barrier" that is seen as a natural lower bound for all efforts to optimize an algorithms' accuracy (Said *et al.*, 2012b).

- In many application domains, the effectiveness of a recommendation system can depend on whether the system is able to *explain* the reasons for its recommendations (Tintarev and Masthoff, 2011; Friedrich and Zanker, 2011; Nunes and Jannach, 2017), in particular when the goal is to design fair, accountable, and transparent systems.
- When it comes to online choice situations and the decisions users make (Chen *et al.*, 2013) a myriad of different factors like position biases, decoy and framing effects (Teppan and Zanker, 2015) or the characteristics of the rating summary statistics (Çoba *et al.*, 2018) have been shown to measurably influence the choices users make, which are mostly not (yet) considered in actual algorithms.
- From a research perspective, the reproducibility of the obtained research results is still limited in many cases despite the existence of public datasets and recommendation libraries (Ekstrand *et al.*, 2011; Said and Bellogín, 2014; Beel *et al.*, 2016; Çoba and Zanker, 2017). These topics will be discussed in Chapter 9.
- Finally, from a societal perspective, collaborative filtering mechanisms could *fracture the global village into tribes* — a point that was already raised in the original work of Resnick *et al.* (1994) — and thus exhibit a certain tendency of reaffirming users in their beliefs and creating a filter bubble around them (Pariser, 2011). Such societal implications were not discussed so far to a large extent in the research community and represent another area where research in the field has to go beyond computer science.

Further Reading There are several highly cited works on collaborative filtering that are marking milestones of the topic like the early work on user-based automated CF (Resnick *et al.*, 1994) in an application domain (i.e., netnews) that was initially mainly addressed with content-based techniques. The proposition of an item-based neighborhood (Sarwar *et al.*, 2001) and its adoption by Amazon (Linden *et al.*, 2003) was another step into a direction towards the wider adoption of CF techniques. This obviously necessitated methodological questions about the evaluation of collaborative filtering systems as they had been addressed early in this seminal paper of Herlocker *et al.* (2004). Another important milestone for the development of the topic was the Netflix challenge and the development of many variants of scalable matrix factorization techniques, where the reader is, for instance, referred to Koren and Bell (2015) and later chapters.

Another early seminal article of Herlocker *et al.* (2000) addresses issues of algorithmic transparency and accountability by providing explanations, i.e., additional information about the recommendations and how they were derived. Later, due to the maturing of the field, a first introductory textbook appeared Jannach *et al.* (2011) that provides a comprehensive reference on collaborative filtering and its relation to other recommendation paradigms like content-based and knowledge-based techniques. Since then, even more encompassing literature like the handbook on RS with chapters on advancements in collaborative filtering algorithms (Koren and Bell, 2015) and their evaluation (Gunawardana and Shani, 2015) have been published. Finally, a recent article challenges the traditional problem formulation of collaborative filtering as a matrix completion task and advocates to also consider the user interaction and the optimization of conversational moves over time into the problem formulation (Jannach *et al.*, 2016b).

References

Amatriain, X. and Basilico, J. (2012). Netflix recommendations: beyond the 5 stars (part 1), *Netflix Tech Blog* **6**, `https://medium.com/netflix-techblog/netflix-recommendations-beyond-the-5-stars-part-1-55838468f429`.

Amatriain, X. and Basilico, J. (2015). Recommender systems in industry: A netflix case study, in *Recommender Systems Handbook*, 2nd edn. (Springer), pp. 385–419.

Basilico, J. (2013). Recommendation at Netflix Scale, Talk at the first Workshop on Large Scale Recommendation Systems.

Beel, J., Breitinger, C., Langer, S., Lommatzsch, A., and Gipp, B. (2016). Towards reproducibility in recommender-systems research, *User Modeling and User-Adapted Interaction* **26**, 1, pp. 69–101.

Bell, R. M. and Koren, Y. (2007). Lessons from the Netflix prize challenge, *ACM SIGKDD Explorations Newsletter* **9**, 2, pp. 75–79.

Bobadilla, J., Ortega, F., Hernando, A., and Bernal, J. (2012). A collaborative filtering approach to mitigate the new user cold start problem, *Knowledge-Based Systems* **26**, pp. 225–238.

Bonnin, G. and Jannach, D. (2014). Automated generation of music playlists: Survey and experiments, *Computing Surveys* **47**, 2, pp. 26:1–26:35.

Breese, J. S., Heckerman, D., and Kadie, C. (1998). Empirical analysis of predictive algorithms for collaborative filtering, in *Proceedings of the 14th Conference on Uncertainty in Artificial Intelligence*, pp. 43–52.

Cantador, I., Fernández-Tobías, I., Berkovsky, S., and Cremonesi, P. (2015). Cross-domain recommender systems, in *Recommender Systems Handbook*, 2nd edn. (Springer), pp. 919–959.

Castells, P., Hurley, N. J., and Vargas, S. (2015). Novelty and diversity in recommender systems, in *Recommender Systems Handbook*, 2nd edn. (Springer), pp. 881–918.

Chen, L., de Gemmis, M., Felfernig, A., Lops, P., Ricci, F., and Semeraro, G. (2013). Human decision making and recommender systems, *ACM Transactions on Interactive Intelligent Systems (TiiS)* **3**, 3, p. 17.

Çoba, L., Symeonidis, P., and Zanker, M. (2017a). Reproducing and prototyping recommender systems in R, in *Proceedings of the 8th Italian Information Retrieval Workshop*, pp. 84–91.

Çoba, L., Symeonidis, P., and Zanker, M. (2017b). Visual analysis of recommendation performance, in *Proceedings of the 11th ACM Conference on Recommender Systems*, pp. 362–363.

Çoba, L. and Zanker, M. (2017). Replication and reproduction in recommender systems research - evidence from a case-study with the rrecsys library, in *Proceedings 30th International Conference on Industrial Engineering and Other Applications of Applied Intelligent Systems (IEA/AIE)*, pp. 305–314.

Çoba, L., Zanker, M., Rook, L., and Symeonidis, P. (2018). Exploring users' perception of collaborative explanation styles, in *Proceedings 20th IEEE International Conference on Business Informatics (CBI)*.

Cremonesi, P., Garzotto, F., and Turrin, R. (2012). Investigating the persuasion potential of recommender systems from a quality perspective: An empirical study, *Transactions on Interactive Intelligent Systems* **2**, 2, pp. 11:1–11:41.

Ekstrand, M. D., Ludwig, M., Konstan, J. A., and Riedl, J. T. (2011). Rethinking the recommender research ecosystem: reproducibility, openness, and LensKit, in *Proceedings of the 5th ACM Conference on Recommender Systems*, pp. 133–140.

Friedrich, G. and Zanker, M. (2011). A taxonomy for generating explanations in recommender systems, *AI Magazine* **32**, 3, pp. 90–98.

Gantner, Z., Rendle, S., Freudenthaler, C., and Schmidt-Thieme, L. (2011). MyMediaLite: A free recommender system library, in *Proceedings of the 5th ACM Conference on Recommender Systems*, pp. 305–308.

Garcin, F., Faltings, B., Donatsch, O., Alazzawi, A., Bruttin, C., and Huber, A. (2014). Offline and online evaluation of news recommender systems at swissinfo.ch, in *Proceedings of the 8th ACM Conference on Recommender Systems*, pp. 169–176.

Goldberg, D., Nichols, D., Oki, B. M., and Terry, D. (1992). Using collaborative filtering to weave an information tapestry, *Communications of the ACM* **35**, 12, pp. 61–70.

Gomez-Uribe, C. A. and Hunt, N. (2015). The Netflix recommender system: Algorithms, business value, and innovation, *Transactions on Management Information Systems* **6**, 4, pp. 13:1–13:19.

Gunawardana, A. and Shani, G. (2015). Evaluating recommender systems, in *Recommender Systems Handbook*, 2nd edn. (Springer), pp. 265–308.

Harper, F. M. and Konstan, J. A. (2016). The MovieLens datasets: History and Context, *ACM Transactions on Interactive Intelligent Systems* **5**, 4, p. 19.

Herlocker, J. L., Konstan, J. A., Borchers, A., and Riedl, J. (1999). An algorithmic framework for performing collaborative filtering, in *Proceedings of the 22nd Annual International ACM SIGIR Conference on Research and Development in Information Retrieval*, pp. 230–237.

Herlocker, J. L., Konstan, J. A., and Riedl, J. (2000). Explaining collaborative filtering recommendations, in *Proceedings of the 2000 ACM Conference on Computer Supported Cooperative Work*, pp. 241–250.

Herlocker, J. L., Konstan, J. A., Terveen, L. G., and Riedl, J. T. (2004). Evaluating collaborative filtering recommender systems, *Transactions on Information Systems* **22**, 1, pp. 5–53.

Hidasi, B., Karatzoglou, A., Baltrunas, L., and Tikk, D. (2016). Session-based recommendations with recurrent neural networks, in *Proc. ICLR '16*.

Huang, Z., Chen, H., and Zeng, D. (2004). Applying associative retrieval techniques to alleviate the sparsity problem in collaborative filtering, *ACM Transactions on Information Systems* **22**, 1, pp. 116–142.

Jannach, D. and Adomavicius, G. (2016). Recommendations with a purpose, in *Proceedings of the 10th ACM Conference on Recommender Systems*, pp. 7–10.

Jannach, D. and Hegelich, K. (2009). A case study on the effectiveness of recommendations in the mobile internet, in *Proceedings of the 3rd ACM Conference on Recommender Systems*, pp. 205–208.

Jannach, D., Jugovac, M., and Lerche, L. (2016a). Supporting the design of machine learning workflows with a recommendation system, *ACM Transactions on Interactive Intelligent Systems* **6**, 1.

Jannach, D., Lerche, L., and Jugovac, M. (2015a). Item familiarity as a possible confounding factor in user-centric recommender systems evaluation, *i-com Journal of Interactive Media* **14**, 1, pp. 29–39.

Jannach, D., Lerche, L., Kamehkhosh, I., and Jugovac, M. (2015b). What recommenders recommend: an analysis of recommendation biases and possible countermeasures, *User Modeling and User-Adapted Interaction* **25**, 5, pp. 427–491.

Jannach, D., Lerche, L., and Zanker, M. (2018). Recommending based on implicit feedback, in *Social Information Access - Systems and Technologies* (Springer), pp. 510–569.

Jannach, D. and Ludewig, M. (2017a). Determining characteristics of successful recommendations from log data - a case study, in *Proceedings of the ACM Symposium on Applied Computing (SAC)*, pp. 1643–1648.

Jannach, D. and Ludewig, M. (2017b). When recurrent neural networks meet the neighborhood for session-based recommendation, in *Proceedings of the 11th ACM Conference on Recommender Systems*, pp. 306–310.

Jannach, D., Ludewig, M., and Lerche, L. (2017). Session-based item recommendation in e-commerce: On short-term intents, reminders, trends, and discounts, *User-Modeling and User-Adapted Interaction* **27**, 3–5, pp. 351–392.

Jannach, D., Resnick, P., Tuzhilin, A., and Zanker, M. (2016b). Recommender systems — beyond matrix completion, *Communications of the ACM* **59**, 11, pp. 94–102.

Jannach, D., Zanker, M., Felfernig, A., and Friedrich, G. (2011). *Recommender Systems – An Introduction* (Cambridge University Press).

Jannach, D., Zanker, M., Ge, M., and Gröning, M. (2012). Recommender systems in computer science and information systems - a landscape of research, in *Proceedings of the 13th International Conference on E-Commerce and Web Technologies*, pp. 76–87.

Jugovac, M. and Jannach, D. (2017). Interacting with recommenders - overview and research directions, *ACM Transactions on Intelligent Interactive Systems* **7**, p. 10.

Kamehkhosh, I. and Jannach, D. (2017). User perception of next-track music recommendations, in *Proceedings of the 25th Conference on User Modeling Adaptation and Personalization*, pp. 113–121.

Kirshenbaum, E., Forman, G., and Dugan, M. (2012). A live comparison of methods for personalized article recommendation at Forbes.com, in *Proceedings of the European Conference on Machine Learning and Principles and Practice of Knowledge Discovery in Databases*, pp. 51–66.

Knijnenburg, B. P. and Willemsen, M. C. (2015). Evaluating recommender systems with user experiments, in *Recommender Systems Handbook* (Springer), pp. 309–352.

Knijnenburg, B. P., Willemsen, M. C., Gantner, Z., Soncu, H., and Newell, C. (2012). Explaining the user experience of recommender systems, *User Modeling and User-Adapted Interaction* **22**, 4-5, pp. 441–504.

Konstan, J. A. and Riedl, J. (2012). Recommender systems: from algorithms to user experience, *User Modeling and User-Adapted Interaction* **22**, 1, pp. 101–123.

Koren, Y. and Bell, R. (2015). Advances in collaborative filtering, in *Recommender Systems Handbook*, 2nd edn. (Springer), pp. 77–118.

Lerche, L., Jannach, D., and Ludewig, M. (2016). On the value of reminders within e-commerce recommendations, in *Proceedings of the 24th Conference on User Modeling Adaptation and Personalization*, pp. 27–25.

Linden, G., Smith, B., and York, J. (2003). Amazon.com recommendations: Item-to-item collaborative filtering, *IEEE Internet Computing* **7**, 1, pp. 76–80.

Ludewig, M. and Jannach, D. (2018). Evaluation of session-based recommenation algorithms, `arXiv:1803.09587 [cs.IR]`, `https://arxiv.org/abs/1803.09587`.

Martin, F. J. (2009). RecSys '09 Industrial Keynote: Top 10 Lessons Learned Developing, Deploying and Operating Real-world Recommender Systems, in *Proceedings of the 3rd ACM Conference on Recommender Systems*, pp. 1–2.

Mobasher, B., Dai, H., Luo, T., and Nakagawa, M. (2002). Using sequential and non-sequential patterns in predictive web usage mining tasks, in *Proceedings of the 2002 IEEE International Conference on Data Mining*, pp. 669–672.

Nunes, I. and Jannach, D. (2017). A systematic review and taxonomy of explanations in decision support and recommender systems, *User-Modeling and User-Adapted Interaction* **27**, 3–5, pp. 393–444.

Pariser, E. (2011). *The filter bubble: What the Internet is hiding from you* (Penguin UK).

Pedregosa, F., Varoquaux, G., Gramfort, A., Michel, V., Thirion, B., Grisel, O., Blondel, M., Prettenhofer, P., Weiss, R., Dubourg, V., Vanderplas, J., Passos, A., Cournapeau, D., Brucher, M., Perrot, M., and Duchesnay, E. (2011). Scikit-learn: Machine learning in Python, *Journal of Machine Learning Research* **12**, pp. 2825–2830.

Pu, P., Chen, L., and Hu, R. (2011). A user-centric evaluation framework for recommender systems, in *Proceedings of the 5th ACM Conference on Recommender Systems*, pp. 157–164.

Quadrana, M., Cremonesi, P., and Jannach, D. (2018). Sequence-aware recommender systems, *ACM Computing Surveys* `https://arxiv.org/abs/1802.08452`.

Resnick, P., Iacovou, N., Suchak, M., Bergstrom, P., and Riedl, J. (1994). Grouplens: An open architecture for collaborative filtering of netnews, in *Proceedings of the 1994 ACM Conference on Computer Supported Cooperative Work*, pp. 175–186.

Rook, L., Sabic, A., and Zanker, M. (2018). Reinforcement sensitivity and engagement in proactive recommendations: Experimental evidence, in *Information Systems and Neuroscience - Gmunden Retreat on NeuroIS 2017*, pp. 9–15.

Rossetti, M., Stella, F., and Zanker, M. (2016). Contrasting offline and online results when evaluating recommendation algorithms, in *Proceedings of the 10th ACM Conference on Recommender Systems*, pp. 31–34.

Said, A. and Bellogín, A. (2014). Comparative recommender system evaluation: Benchmarking recommendation frameworks, in *Proceedings of the 8th ACM Conference on Recommender Systems*, pp. 129–136.

Said, A., Jain, B. J., and Albayrak, S. (2012a). Analyzing weighting schemes in collaborative filtering: Cold start, post cold start and power users, in *Proceedings of the 27th Annual ACM Symposium on Applied Computing*, pp. 2035–2040.

Said, A., Jain, B. J., Narr, S., Plumbaum, T., Albayrak, S., and Scheel, C. (2012b). Estimating the magic barrier of recommender systems: a user study, in *Proceedings of the 35th ACM SIGIR Conference on Research and Development in Information Retrieval*, pp. 1061–1062.

Sarwar, B., Karypis, G., Konstan, J., and Riedl, J. (2001). Item-based collaborative filtering recommendation algorithms, in *Proceedings of the 10th International Conference on World Wide Web*, pp. 285–295.

Shani, G., Heckerman, D., and Brafman, R. I. (2005). An MDP-Based Recommender System, *The Journal of Machine Learning Research* **6**, pp. 1265–1295.

Teppan, E. C. and Zanker, M. (2015). Decision biases in recommender systems, *Journal of Internet Commerce* **14**, 2, pp. 255–275.

Tintarev, N. and Masthoff, J. (2011). Designing and evaluating explanations for recommender systems, *Recommender Systems Handbook*, pp. 479–510.

Zanker, M. (2012). The influence of knowledgeable explanations on users' perception of a recommender system, in *Proceedings of the 6th ACM Conference on Recommender Systems* (ACM), pp. 269–272.

Zanker, M. and Jessenitschnig, M. (2009a). Case-studies on exploiting explicit customer requirements in recommender systems, *User Modeling and User-Adapted Interaction* **19**, 1-2, pp. 133–166.

Zanker, M. and Jessenitschnig, M. (2009b). Collaborative feature-combination recommender exploiting explicit and implicit user feedback, in *Proceedings of the 2009 IEEE Conference on Commerce and Enterprise Computing*, pp. 49–56.

Zanker, M., Jessenitschnig, M., Jannach, D., and Gordea, S. (2007). Comparing recommendation strategies in a commercial context, *IEEE Intelligent Systems* **22**, 3.

Chapter 2

Matrix Factorization for Collaborative Recommendations

Evgeny Frolov[a] and Ivan Oseledets[a,b]

[a]*Skolkovo Institute of Science and Technology, Nobel St. 3, Skolkovo Innovation Center, 143025 Moscow, Russia*

[a,b]*Institute of Numerical Mathematics of the Russian Academy of Sciences, Gubkina St. 8, 119333 Moscow, Russia*

evgeny.frolov@outlook.com

2.1. Introduction

Matrix factorization (MF) is one of the most successful and widely used collaborative filtering (CF) techniques. One of the key advantages of MF models is the ability to reduce an initial problem's complexity and provide a compact representation of interaction data generated from an observed collective human behavior. With this approach users and items are embedded as vectors in a lower dimensional space of *latent features*. This procedure is known as a *dimensionality reduction* task. As the result, both user tastes and relevant item characteristics can be described by a relatively small set of parameters.

With this representation the relations between users and items follow general linear algebra rules and vector arithmetic. Utility of a particular item to a particular user can be simply estimated via a scalar product of their vectors in the obtained lower dimensional latent feature space. *This is the key concept that connects various MF models presented in this chapter.* From the geometric point of view, the angle between user and item vectors is smaller for relevant items and is larger for irrelevant ones (see [Koren *et al.* (2009), Figure 2] for an illustration). This can be conveniently expressed in terms of the cosine similarity measure and also used for building more efficient neighborhood-based models discussed in Chapter 1 (see also [Adomavicius and Tuzhilin (2005)]).

Unlike the neighborhood-based techniques, MF is less susceptible to the so called limited coverage problem [Desrosiers and Karypis (2011)]. For example, the lack of common preferences information for a pair of users may lead to unreliable correlations in prediction mechanism for the neighborhood models. In contrast, MF models build a more meaningful conceptual description of user interests in terms of latent features, which to a certain extent allows to alleviate that problem. A better expressiveness of the MF models also makes them less sensitive to the data sparsity problem, typically observed in many real applications.

As has been already stated, the concept of *utility*, is one of the key ingredients of MF models. Its purpose is to adequately represent an undetermined decision making process driven by hidden motives behind a particular user choice. The decision making is indirectly observed via partially available interactions data, expressed in the form of a feedback provided by users to some (not all) items. The goal of any MF approach is, given that data, to estimate the corresponding *utility function* f_u, which will not only agree with the observed part of user preferences but will also help to make predictions on the unobserved part. Schematically, this can be denoted as follows:

$$f_u : User \times Item \rightarrow Relevance\, Score, \tag{2.1}$$

where $User$ is a domain of all users, $Item$ is a domain of all items and $Relevance\, Score$ is a measure of utility.

In the simplest case the *Relevance Score* can be directly related to the user feedback. Consider a movie recommendation system, where users express how satisfied they are with a certain movie by providing an *explicit* rating value on some likert scale. The problem of finding f_u can then be transformed into a well studied *matrix completion* problem, already mentioned in Chapter 1, which has become especially popular in the recommender systems community after the famous Netflix Prize competition[1]. Even though the rating values are subjective in their nature [Amatriain *et al.* (2009a)], it is often neglected as the task of recovering the unknown entries of the rating matrix enables a number of very practical and quite efficient methods of solving the problem of recommendations. A considerable part of Sec. 2.4 is devoted to such methods not only because of their

[1] https://www.netflixprize.com

popularity, but also as it helps to provide the necessary background for understanding of more elaborate models introduced in the later sections and later chapters of the book.

Standard matrix completion, however, may not be the best choice in the *implicit* case, where the feedback is not intentionally provided by users and is collected via an indirect observation of their actions, such as clicks on product web-pages, amount of product purchases, time spent reading a product description, etc. Note, that the lack of feedback from a user for a certain item does not immediately imply a negative preference, which is true for both explicit and implicit cases. However, in the implicit case the fact that a user has interacted with an item may not necessarily correspond to a positive preference. Taking that into account requires a more thoughtful problem formulation, which may lead to an alternative definition of the *Relevance Score*, abstracted away from the observable feedback (see Sec. 2.4.3). Note that one of the corner cases of implicit feedback when it simply denotes the fact of interaction is often referred as One-Class Collaborative Filtering (OCCF) [Pan *et al.* (2008); Verstrepen *et al.* (2017)].

In addition to that, in many practical applications it is often more important to return an *ordered list of correctly ranked recommendations*, rather than simply predict rating values. This is known as a *top-n recommendation problem*, where n is the number of recommended items. At first glance, this may seem like a trivial task: once the rating predictions are available, one can simply select the items with the highest predicted score. However, being able to accurately recover rating values does not necessarily guarantee the best performance in terms of generating a ranked list of the most relevant recommendations [Konstan and Riedl (2012)] (also see Chapter 9). This opens the doors for the so called *learning to rank* models with a substantially different objective (see Sec. 2.5.1), more coherent with the task of top-n recommendations. Such models are typically not even suitable for the completion task.

As the matrix completion models can be tuned and evaluated in terms of the ranking problem as well, we will distinguish between the two major types of recommendation tasks — the *rating prediction* and the *top-n recommendation* — and provide a view on factorization models through the lense of this distinction where necessary.

2.2. Problem formulation

As has been already noted, the dimensionality reduction approach in recommender systems allows to describe any user preferences and any item characteristics in terms of a small set of model parameters. Along with a compact representation, it also helps to uncover non-trivial patterns within the data and use them to generate meaningful recommendations. Generally speaking, this can be achieved with the help of various methods, such as neural networks, markov decision processes, latent dirichlet allocation and some others. However, in this chapter we focus specifically on the matrix factorization approach.

Let us start from the matrix completion case as it provides a good illustration of some major concepts and serves as a ground for further improvements. Consider an imaginary scenario in which all known users of some recommendation system have provided their preferences for all available items. This can be conveniently represented in the form of a *complete* matrix of interactions $A \in \mathbb{R}^{M \times N}$. The rows of the matrix correspond to users and its columns correspond to items. Its elements would correspond to some form of a feedback provided by users and this would represent a snapshot of a real "noisy" data. The "noise" may have different nature. It can be caused by variations in individuals' behavior and their tastes or by occasional changes in a context of an interaction, or it can be the result of some other uncontrollable and mostly unpredictable factors. All of it leads to a certain level of unavoidable randomness making the problem of recommendations very complex.

Nevertheless, at a large scale the collective behavior may reveal some regularities and exhibit some common patterns that could be potentially described with a relatively small set of parameters. Therefore, while the dimensionality reduction may lead to a loss of some information, it can still help to uncover and generalize at least some of the hidden commonalities in users' behavior. With this assumption the observed data can be modelled as:

$$A = R + E,$$

where the matrix E denotes the "noise" and R is an approximate *utility matrix* which accommodates the behavioral patterns and have a certain inner structure. *The task of building a recommendation model then translates into the task of recovering R.*

The solution to this problem in the case of a matrix factorization approach can be generally represented in the form of a matrix product:

$$R = PQ^T, \tag{2.2}$$

where matrices $P \in \mathbb{R}^{M\times r}$ and $Q \in \mathbb{R}^{N\times r}$ represent users and items respectively. Each row $\boldsymbol{p}_i^T$ of the matrix P reflects a preference vector of a user i, described in terms of r latent features. In other words it gives a representation of user i in the latent feature space. Similarly, each row $\boldsymbol{q}_j^T$ of the matrix Q describes an association of item j with those latent features, i.e. it gives a representation of an item in the latent feature space.

Vectors $\boldsymbol{p}_i$ and $\boldsymbol{q}_j$ are also called an *embedding of users and items onto the latent feature space.* The utility function f_u of an item j to a user i is, therefore, represented by a scalar product $\boldsymbol{p}_i^T\boldsymbol{q}_j$. The number of latent features r is called the *rank of an approximation.* This number is typically much smaller than the number of items or users. Such a representation of a matrix as a product of two other matrices of smaller sizes is also called a *low rank approximation* and the resulting matrix R is said to have a *low rank structure.*

The final form of the matrices P and Q depends on the formulation of a corresponding optimization problem described in terms of a specific loss function $\mathcal{L}$:

$$\min_{\Theta} \mathcal{L}\left(A, R(\Theta)\right), \tag{2.3}$$

where $\Theta := \{P, Q\}$ is a set of model parameters and $\mathcal{L}$ penalizes deviation of the model from the observations. Worth noting here, that the term *deviation should be treated in a broad sense.* As we discussed in the introduction, a particular form of the function $\mathcal{L}$ may go far beyond standard matrix completion formulation (see Sec. 2.5.1).

Also recall, that in the majority of real systems the observed interactions are typically very scarce and the vast amount of data is missing, which makes the matrix A overly *incomplete.* Therefore, *the optimization problem described by Eq. (2.3) remains ambiguous unless we explicitly define how to deal with the missing values of A* or at least define in what sense a *complete* matrix R approximates an incomplete matrix A. Due to this reason we prefer to avoid the commonly used and intuitive notation $A \approx R$.

We also note, that in some cases an additional processing of the data may help to create a better representation of the observed user behavior and potentially help to improve the quality of recommendations. As an example, in a music recommendation service the logarithmic scaling of a listening frequency (i.e. the number of times a user has listened to a track) may help to generate more accurate recommendations comparing to a naive use of raw counts data as a measure of utility. There is a number of transformation techniques such as data centering and normalization, value binarization, cutting by a threshold, tf-idf transformation and many others which may help to build more accurate recommendation models.

Even in the systems with a fixed explicit feedback, such as a 5-star rating scale, used in many movie recommendation services, a transformation of that scale may improve recommendations. It has an intrinsic connection to a subjective nature of a perceived utility of goods. For example, some users may assign a rating value of 3 to a movie they believe is "OK", i.e. nothing special but still "watchable", whilst for other users this can be a way to indicate that the movie is completely uninteresting, a total waste of time. In addition to that, some empirical studies show that even for a single user the perceived "distance" between different ratings may vary and the uniform rating scale from 1 to 5 used as a measure of a user enjoyment may not be that accurate [Amatriain *et al.* (2009b)]. All of this, along with the fact that the unobserved data is *missing not at random* (MNAR) [Steck (2010); Schnabel *et al.* (2016)], may potentially introduce unintended biases in both recommendation models and evaluation measures.

Both described aspects — the way missing data is handled and the choice of a data preprocessing technique — create additional degrees of freedom for a model construction. Sometimes it may directly affect an optimization procedure and lead to very different factorization algorithms. In other cases it may lead to several variations of the same method. In order to explicitly signify the role of these degrees of freedom we will formulate the optimization problem Eq. (2.3) not in terms of an approximation of the matrix A, but rather as an approximation of some function of A:

$$\min_{\Theta} \mathcal{L}\left(T(A), R(\Theta)\right), \tag{2.4}$$

where $T(A)$ denotes a problem-dependent transformation of the data which may include missing values imputation and/or various data preprocessing steps. In the next sections we will cover some of the most famous factorization models resulting from a combination of different data transformation techniques, various loss functions and optimization algorithms.

2.3. SVD-based models

One of the first factorization algorithms used in the field of recommender systems is the singular value decomposition (SVD) [Golub and Van Loan

(2012)]. It has a straightforward relation to the latent semantic indexing/analysis (LSI/LSA) [Furnas *et al.* (1988); Deerwester *et al.* (1990)] and principal component analysis (PCA). In fact, the SVD-based approach has been adopted directly from the field of document retrieval [Ekstrand *et al.* (2011)], which is not surprising. Indeed, the task of revealing words' semantics based on their occurrence in text documents is in some sense similar to the task of finding alike items based on users' consumption patterns.

The first SVD-based implementations in the recommender systems field had an *enabling* role in a sense that it was used as an intermediate dimensionality reduction step and in order to generate a final list of recommendations its output was fed into a different algorithm based on, for example, a neural network [Billsus and Pazzani (1998)] or a nearest neighbors approach [Sarwar *et al.* (2000)]. The authors of the latter work also used SVD in a *standalone* regime (with a certain preliminary data normalization) for the *rating prediction* task with a little to no improvement over the competing CF algorithm. However, an even simpler SVD-based model, named *PureSVD* [Cremonesi *et al.* (2010)], has been later demonstrated to outperform some state-of-the-art algorithms in terms of the *top-n recommendation* task.

We find it necessary to also introduce here some formal definitions and common results from linear algebra, which will help in further explanations. Any *complete* matrix $A \in \mathbb{R}^{M \times N}$ can be represented in the form:

$$A = U \Sigma V^T,$$

where $U \in \mathbb{R}^{M \times M}$ and $V \in \mathbb{R}^{N \times N}$ are orthogonal matrices, their columns are called the left and the right singular vectors respectively; $\Sigma \in \mathbb{R}^{M \times N}$ is a diagonal matrix with non-negative elements $\sigma_1 \geq \ldots \geq \sigma_K$ on its main diagonal called singular values; $K = \min(M, N)$ is a *rank* of SVD. According to the Eckart-Young theorem [Eckart and Young (1936)], the truncated SVD of rank $r < K$ with $\sigma_{r+1}, \ldots, \sigma_K$ set to 0 *gives the best rank-r approximation of the matrix A.*

2.3.1. *PureSVD*

Unfortunately, the result of the Eckart-Young theorem cannot be directly applied in recommender systems settings as *SVD is undefined for incomplete matrices.* As a workaround the PureSVD model uses a simple imputation technique: to *replace the missing entries of A with zeroes.* Hence, an incomplete matrix A is transformed into a *sparse* matrix A_0 with zero values inplace of the unknown elements, i.e. $T(A) = A_0$. The corresponding loss function can then be expressed as

$$\mathcal{L}(T(A), R) = \|A_0 - R\|_F^2, \tag{2.5}$$

where $\|\cdot\|_F$ denotes the Frobenius norm. As the loss function is now well defined, we can apply the Eckart-Young theorem to find a *globally optimal* solution to the resulting optimization task defined by Eq. (2.4):

$$R = U_r \Sigma_r V_r^T, \tag{2.6}$$

where factor matrices $U_r \in \mathbb{R}^{M \times r}$ and $V_r \in \mathbb{R}^{N \times r}$ have orthonormal columns and represent users and items in the reduced latent space with $r \ll \min(M, N)$ distinct latent features. Square diagonal matrix $\Sigma_r \in \mathbb{R}^{r \times r}$ has r largest singular values on its main diagonal. Equation (2.6) can be equivalently rewritten in the form of Eq. (2.2) simply by allowing $P = U_r \Sigma_r^\beta$ and $Q = V_r \Sigma_r^{1-\beta}$, where β is some real number in the interval [0, 1], typically assigned to 1/2 or 1.

As was noticed by the authors of the PureSVD model, the orthonormality of columns of the factor matrices allows to rewrite Eq. (2.6) in a more convenient form. Assuming that $A_0 = U \Sigma V^T$ is the full SVD of the completed matrix, we have $A_0 V_r V_r^T = U \Sigma V^T V_r V_r^T = U_r \Sigma_r V_r^T$. The last equality is due to the fact that $V^T V_r = [I_r \, 0]^T$, where I_r is the identity matrix of size r and 0 denotes a matrix of all zeros with a conforming size. From here it reads:

$$R = A_0 V_r V_r^T. \tag{2.7}$$

This induces a natural geometrical interpretation: once the right singular vectors are determined, *every row of the prediction matrix R can be computed as an orthogonal projection of the corresponding user preferences onto the latent feature space.* Note, that this eliminates the need for the matrix of user factors U_r, as it can be restored by the means of $U_r \Sigma_r = A_0 V_r$. This can be used to reduce both computational overhead and storage requirements of the model. From now on for brevity we will omit the subscript r in the equations, always assuming a low-rank approximation.

Taking into account that in many practical cases a typical sparsity of the matrix is higher than 99%, setting zeros inplace of the missing data introduces a strong bias of the model prediction towards zero values. This makes the model very bad from the matrix completion perspective and totally *unsuitable for the rating prediction task.* Nevertheless, despite such a bias, the PureSVD approach is known to serve as a *strong baseline in the top-n recommendation problem* outperforming even more elaborate state-of-the-art methods [Cremonesi *et al.* (2010); Lee *et al.* (2016)]. Of course, it does not imply that PureSVD is always an optimal choice. However, it should be considered by beginner practitioners as a good starting model.

The *truncated SVD* can be computed with the help of an iterative *Lanczos procedure* [Golub and Van Loan (2012)] which invokes a Krylov subspace method and internally uses efficient bidiagonalization techniques supported by a Gram-Schmidt orthogonalization process. The key benefit of such approach is that in order to find r leading singular vectors and corresponding singular values it is only required to provide a rule of how to *multiply an interaction matrix by an arbitrary vector* from the right and from the left. More specifically, given the number of non-zero elements nnz_A of the matrix A that corresponds to the number of the observed interactions, the overall computational complexity of the SVD algorithm can be estimated as $O(nnz_A \cdot r) + O((M+N) \cdot r^2)$ [Halko *et al.* (2011)], where the first term corresponds to the complexity of a sparse matrix-vector product and the second term is related to an internal orthogonalization process.

2.3.2. *Biases and custom data transformation*

It has been already noted that user feedback is intrinsically subjective. One of the ways to partially address that subjectivity at least in the rating-based systems is to introduce the concept of the so called user and item bias. *User bias* captures a tendency of a user to systematically assign higher (or lower) ratings depending on how critical the user is in comparison with an average person. Likewise, *item bias* can be described as a tendency of items to receive higher (or lower) ratings. In practice, it turns out that the most part of an interaction "signal" (e.g. rating value) is accommodated by these biases. This allows for even non-personalized recommendation models, called *baseline predictors*, to demonstrate a fairly good prediction quality in the rating prediction task [Koren *et al.* (2009)].

These biases can be estimated with the help of simple statistics such as an average of user and item ratings calculated over the observed data sample. It is also possible to use more sophisticated estimation methods, e.g. averaging with value damping or even gradient-based optimization [Ekstrand *et al.* (2011)]. An overall bias b_{ij} (i.e. baseline predictor) can be expressed as a combination of all systematic biases[2]:

$$b_{ij} = \mu + t_i + f_j, \tag{2.8}$$

[2]The notation we use here slightly differs from what can be commonly seen in the literature — we assign different letters to user and item bias variables, as it helps to avoid an ambiguity in mathematical formulations which involve matrix-vector operations.

where μ is a global bias constant (e.g. global average rating); t_i denotes a user bias or a *tendency* to give higher or lower ratings; in turn f_j reflects an item bias, its *favouredness* or in some sense quality (based on the opinions of raters).

In the PureSVD model these biases can be used as a replacement for the missing data, which therefore reduces the distortion introduced by a straightforward zero-based imputation step and allows to partially address the subjectivity of user preferences. In the simplest scenario one could use Eq. (2.8) to replace the missing entries of A with the corresponding values of the baseline predictor. In this case the transformation T of the data is trivial. It is sufficient to simply subtract the bias values from the known entries of A. The unknown values can then be set to 0, preserving the same sparsity pattern as in A_0. After the preprocessing is done the standard PureSVD model is built on top of the centered data. When generating recommendations the *bias term should be added back* to the predicted scores of the model.

More elaborate data preprocessing techniques can also be supported without sacrificing the computational efficiency. As an example, consider the case when the missing elements of the rating matrix are first filled-in with the values of item average ratings $\boldsymbol{f} \in \mathbb{R}^N$ and then the resulting *complete* matrix is additionally normalized by subtracting user average ratings $\boldsymbol{t} \in \mathbb{R}^M$ [Sarwar *et al.* (2000)]. The elements of the obtained centered matrix $T(A) = \hat{A}$ can be expressed as:

$$\begin{cases} \hat{a}_{ij} = a_{ij} - t_i & \text{if } a_{ij} \text{ is known,} \\ \hat{a}_{ij} = f_j - t_i & \text{otherwise.} \end{cases}$$

By construction, the complete matrix $\hat{A}$ is likely to be *dense*. However, it can be split into the sum of a *sparse* matrix $\bar{A}$ with two rank-1 terms (outer products of vectors):

$$\hat{A} = \bar{A} - \boldsymbol{t}\boldsymbol{e}_N^T + \boldsymbol{e}_M\boldsymbol{f}^T, \tag{2.9}$$

where $\boldsymbol{e}_M, \boldsymbol{e}_N$ denote vectors of all ones of a conforming size and the elements of $\bar{A}$ are defined as follows:

$$\begin{cases} \bar{a}_{ij} = a_{ij} - f_j & \text{if } a_{ij} \text{ is known,} \\ \bar{a}_{ij} = 0 & \text{otherwise.} \end{cases}$$

Recall that in order to compute the truncated SVD it is only required to provide a matrix-vector multiplication rule. Multiplying Eq. (2.9) by an arbitrary vector v gives:

$$\hat{A}\boldsymbol{v} = \bar{A}\boldsymbol{v} - \boldsymbol{t}\langle \boldsymbol{e}_N, \boldsymbol{v}\rangle + \boldsymbol{e}_M\langle \boldsymbol{f}, \boldsymbol{v}\rangle, \tag{2.10}$$

where $\langle \cdot , \cdot \rangle$ stands for the scalar product of two vectors. Note, that the first term in Eq. (2.10) has the same computational complexity as in the original PureSVD approach as the matrix $\bar{A}$ by construction follows the sparsity pattern as A_0. The last 2 terms are linear with respect to the number of users and items, and therefore the *added complexity is only* $O(M + N)$, which is negligible as it is dominated by the complexity of the standard Lanczos procedure. Moreover, there is clearly *no need to explicitly form the dense matrix* $\hat{A}$ to compute SVD, which allows to avoid unnecessary memory overhead. This technique can be further used for an iterative variant of SVD [Kim and Yum (2005)] for achieving a much better performance in terms of the rating prediction task.

2.3.3. *Handling online updates*

Many recommendation services aim to provide an instant engagement for both known users and newcomers as well as quickly update the information about new items in the assortment. In the modern online world with its highly dynamic environment and an overwhelming amount of information this requires the ability to generate recommendations instantly. This, however, would be impossible for large scale recommender systems if the only way to accomplish that would be to recompute the whole model for every new (or unrecognized) user or a newly introduced item.

One common technique designed to support an instant service is called *folding-in* [Ekstrand *et al.* (2011)], which was initially proposed in the field of information retrieval for the semantic document-term analysis [Furnas *et al.* (1988)]. As long as at least one interaction with a new entity (i.e. user or item) is observed, it allows to approximately update the corresponding latent representation and quickly generate recommendations for this new entity without the need for the whole model recomputation. Note, that this setting is different from the so called *cold start* regime (see Chapter 8), where no interactions are available.

One of the greatest advantages of the SVD-based approach is an *analytical form of the folding-in*, which unlike many other MF methods does not require any additional optimization steps to calculate recommendations for a new user who is not a part of the model yet. For illustration purposes we will consider the new user scenario. New item scenario is trivially obtained by analogy.

Assuming that the model is expressive enough, a new user can be represented with high accuracy as a combination of previously seen users.

Therefore, the preference vector $\boldsymbol{a}$ of the new user (with imputed zeroes inplace of the unknowns) can be approximated as $\boldsymbol{a}^T \approx \boldsymbol{u}^T \Sigma V^T$, where $\boldsymbol{u}$ is unknown. Multiplying from the right the both parts of this equality by $V\Sigma^{-1}$ and using the orthonormality property $V^TV = I$ we obtain:

$$\boldsymbol{u}^T \approx \boldsymbol{a}^T V \Sigma^{-1}. \tag{2.11}$$

This represents an approximate embedding of a new user to the latent feature space. The formula can be further used to directly generate recommendations. By the virtue of Eq. (2.6) one could perform a reverse operation and restore the corresponding *new row* for the matrix R, which after transposing the result reads:

$$\boldsymbol{r} \approx VV^T \boldsymbol{a}, \tag{2.12}$$

where $\boldsymbol{r}$ is a vector of predicted relevance scores. Provided that there are k items in the preference vector $\boldsymbol{a}$, the overall complexity of generating recommendations for a single user is $O(Nkr)$, which is the result of the chain of matrix-vector multiplications.

Similarly to Eq. (2.7), recommendations for a new user can be generated by the orthogonal projection of the user's preferences onto the latent feature space. This suggests, that Eq. (2.12) can be used to *generate recommendations for both known and new users.* All it requires is a list of user preferences, even if it does not correspond to any particular known user. In the latter case it gives an estimate of possible user preferences, implicitly relying on the assumption that the learned model is expressive enough. In turn, for the known users it corresponds to the exact prediction formula.

As a precaution remark, the folding-in approach is only approximate and leads to the loss of orthogonality of factors. In the long run it accumulates an error and once in a while it is advised to fully recalculate the model, especially if a lot of new data is collected. Alternatively, *incremental update* techniques can be employed in order to avoid expensive recomputations [Berry *et al.* (1995); Zha and Zhang (2000); Brand (2002)].

2.3.4. *The family of eigendecomposition algorithms*

The PureSVD model can be viewed as a member of a broader family of eigendecomposition algorithms. Consider an SVD-based approximation $\tilde{A} \approx U\Sigma V^T$ for some complete matrix $\tilde{A}$ with standardized data. The corresponding correlation matrix $\tilde{A}^T\tilde{A} \approx V\Sigma^2V^T$ would represent the well known PCA with principal components given by $\tilde{A}V = U\Sigma$. The principal components can be then utilized to indicate similarity between users (or items in the transposed case) and build a neighborhood-based recommender system.

This path was initially explored by the authors of the Eigentaste model [Goldberg *et al.* (2001)] designed for the jokes recommendation system. The authors selected a subset (called the gauge set) of the observed data, where only items rated by all users were present. This has led to a *complete dense matrix* of ratings A. The only transformation $T(A)$ the authors used on top of it was the standardization of rating values, allowing to build a Pearson correlation matrix and apply classical PCA. The authors used the first 2 principal components and a clustering technique in this lower dimensional space in order to group like-minded users. In every group (or cluster) the rating for every *non-gauge* item was estimated as a mean value averaged across those users of the group who has provided rating for this item. As for the new users, they were requested to firstly provide ratings on the gauge items. After that the ratings were projected to the lower dimensional space allowing to assign the newcomers to the known clusters and generate averaging-based recommendations similarly to the known users.

Note, that rating predictions generated by Eigentaste are not fully personalized as they are assigned to a whole cluster of users at once. Moreover, the requirement of the dense gauge set can be fully satisfied only in specific environments with sufficiently large amount of user feedback and/or relatively small number of items to interact with (which is exactly the case with the jokes dataset used in the work). In many real-world settings with very high sparsity of the data these can be difficult or even impossible to guarantee. However, it turns out, that at least in the case of top-n recommendation task such restrictions can be alleviated. As shown by the authors of the EIGENREC model [Nikolakopoulos *et al.* (2015)], as long as the rating prediction is not one of the goals of a recommender system, one could build a more flexible and more general approach following the paradigm of PureSVD.

The authors make the following observation: *PureSVD can be viewed as an eigendecomposition of a scaled user-based or item-based cosine similarity matrix.* For instance, in an item-based case it solves an eigendecomposition problem for the following matrix cross-product:

$$A_0^T A_0 \equiv DCD \approx V\Sigma^2 V^T, \tag{2.13}$$

where the scaling matrix $D \in \mathbb{R}^{M\times M}$ is diagonal with diagonal elements $d_{ii} = \|\boldsymbol{a}_i\|_2$ and $\boldsymbol{a}_i$ denotes the ratings of the item i encoded within the i-th column of the matrix A_0. Each element c_{ij} of the symmetric matrix

$C \in \mathbb{R}^{M \times M}$ equals to the cosine similarity between item i and item j:

$$c_{ij} = cos(i, j) = \frac{\boldsymbol{a}_i^T \boldsymbol{a}_j}{d_{ii} d_{jj}}. \tag{2.14}$$

From here it follows, that by altering the scaling factors D and/or by replacing C with some other inter-item proximity or correlation matrix S:

$$DCD \rightarrow D^p \, S \, D^p,$$

one can obtain a new model with a different inner structure of the latent space. Here p is some real number (the authors used values in the range [-2, 2]) and S is a new proximity matrix, which can be based on Pearson correlation, Jaccard index or many other similarity measures. The authors emphasize, that in fact even *the choice of a scaling factor may have a significant impact* on the quality of recommendations. This scaling allows to control the sensitivity of the model to the popularity of items, and therefore to some extent mitigates the problem of unbalanced observation data present in the majority of recommender systems.

Similarly to PureSVD the authors use the Lanczos procedure in order to build an orthogonal basis. They propose their own parallel and highly efficient implementation of it. Therefore, the EIGENREC approach allows to preserve the benefits of PureSVD which include a good scalability and a quick way to generate recommendations according to Eq. (2.12) for both known and newly introduced users. The approach also gives more flexibility comparing to the standard PureSVD model and unlike the Eigentaste model allows to operate on the full assortment of items from the very beginning. It provides an instrument for a more intricate tuning, potentially making it suitable for a wider class of problems.

2.4. Weighted low-rank approximation

A straightforward data imputation is not the only way of dealing with missing values. Alternatively, one could try to avoid making any strict assumptions on the missing values and either ignore them completely or introduce some confidence-based description of it. Indeed, the fact that some interactions between users and items are unobserved does not immediately suggest that these interactions will never happen. For example, a user may never interact with an item simply due to inability to notice it among many other similar items in a large assortment. On the other hand, if a user consumes one item more often than another one, it may increase our confidence that the item is more relevant or more interesting for a user.

Hence, bringing the concept of a confidence-based weighting for both observed and unobserved interactions into a factorization model may help to create more accurate recommender systems. A common way to express the corresponding loss function reads:

$$\mathcal{L}\left(T(A), R\right) = \left\| W \circ \left(T(A) - R\right) \right\|_F^2, \tag{2.15}$$

where $W = \left[\sqrt{w_{ij}}\right]$ is a matrix of non-negative weights $w_{ij} \geq 0$ and $\circ$ denotes *Hadamard product*, i.e. an elementwise multiplication between two matrices. The weight values of W typically depend on the observed data $W = W(A)$. We take the square root of weights w_{ij} in order to conform with an equivalent elementwise formulation of the loss function:

$$\mathcal{L}\left(T(A), R\right) = \sum_{i,j} w_{ij}(a_{ij}^{(T)} - r_{ij})^2, \tag{2.16}$$

where $a_{ij}^{(T)}$ denotes an element of the matrix $T(A)$ at the intersection of the i-th row and j-th column.

One of the most popular choices of the weights is based on $\{0, 1\}$ values simply indicating the fact of interaction. The corresponding binary weight matrix W is then defined by:

$$\begin{cases} w_{ij} = 1 & \text{if } a_{ij} \text{ is known,} \\ w_{ij} = 0 & \text{otherwise.} \end{cases} \tag{2.17}$$

With this formulation, no data imputation is required as *all missing elements of the matrix A are simply ignored.* More elaborate weighting schemes are discussed in Sec. 2.4.3.

Typically, the number of users and items is very large while at the same time the number of observed interactions between them is very small. Therefore, the model obtained as a result of minimization of the loss function defined by Eq. (2.15) is likely to overfit and produce poor prediction quality on the unobserved part of the data. In order to prevent this overfitting additional constraints are typically imposed on the parameters of the model. Most commonly, a simple regularization is used for that purpose leading to the following regularized optimization objective:

$$\mathcal{J}(\Theta) = \mathcal{L}(\Theta) + \Omega(\Theta), \tag{2.18}$$

where $\mathcal{L}(\Theta)$ is defined by Eq. (2.15) (we omit the full notation of the input arguments for brevity) and $\Omega(\Theta)$ is some regularization function typically expressed in terms of some vector or matrix norm. Many factorization models use a simple quadratic term, allowing to penalize an undesired growth of the parameters' values:

$$\Omega(\Theta) = \lambda(\|P\|_F^2 + \|Q\|_F^2), \tag{2.19}$$

where $\lambda > 0$ is an additional model's hyper-parameter called regularization coefficient. In some cases a separate value is assigned to each factor matrix for more granular tuning of the model, which sometimes helps to achieve a better prediction quality. Altering the regularization function may also help to induce a specific structure on the resulting latent space, e.g. one could use l_1 norm to obtain sparse latent factors. In some other cases, when the data is strictly non-negative, imposing a non-negativity constraint on the factors helps to avoid meaningless predictions and may also improve generalization. In the case of binary input data it has a meaning of soft clustering or "fuzzy membership" [Takács *et al.* (2008)].

2.4.1. *Optimization techniques*

A recommendation model is learned as a solution to the corresponding optimization problem:

$$\Theta^* = \operatorname*{argmin}_{\Theta} \mathcal{J}(\Theta). \tag{2.20}$$

This can no longer be directly solved with the help of classical SVD and alternative optimization methods are required. Some of the most popular options are gradient-based methods, especially the *stochastic gradient descent* (SGD) [Bottou (2012)], and alternating minimization methods such as *alternating least squares* (ALS) [Zhou *et al.* (2008)] and *coordinate descent* (CD) [Yu *et al.* (2012)].

In general, these methods no longer guarantee global convergence and, therefore, the optimization requires careful initialization and hyper-parameters tuning. Nevertheless, the methods in practice exhibit fairly good convergence behavior which makes them the main building blocks for many recommender models. More advanced optimization techniques based on *Riemannian optimization* [Vandereycken (2013)] also seem promising in recommender systems settings, offering quick convergence and high scalability in low-rank approximation tasks [Yan *et al.* (2015)].

2.4.1.1. *Gradient-based techniques*

The main idea of the gradient-based approach (also called batch gradient) is to iteratively make steps in the direction that is opposite to the gradient of the optimization objective. Each iteration step in its naive implementation is based on the following sequential update rule for the model parameters:

$$\begin{cases} \boldsymbol{p}_i \leftarrow \boldsymbol{p}_i - \eta \dfrac{\partial \mathcal{J}}{\partial \boldsymbol{p}_i}, \\ \boldsymbol{q}_j \leftarrow \boldsymbol{q}_j - \eta \dfrac{\partial \mathcal{J}}{\partial \boldsymbol{q}_j}, \end{cases}$$

where η is a step size also called learning rate; its value can be a constant real number determined empirically with cross-validation or, in more advanced cases, depend on iterations.

The algorithm makes a full pass through all observations, called *epoch*, in order to perform a full update of matrices P and Q. Iterations continue until the maximum number of epochs is reached or a convergence criteria is met. Note that finding the gradient at each iteration can be quite computationally demanding and suffers from many redundant calculations. At large scale this may lead to both slow convergence and high memory load.

A more efficient implementation, which is the essence of SGD, is to approximate a full gradient with the gradient computed over a single observation or a small group of them (called mini-batch). Such smaller updates are easier to find at the cost of a less straightforward convergence. This allows to sweep through the entire dataset in a single pass for the full update of parameters and has a very low memory footprint. In the case of a single observation update, the update rules are as follows:

$$\begin{cases} \boldsymbol{p}_i \leftarrow \boldsymbol{p}_i + \eta(e_{ij}\boldsymbol{q}_j - \lambda \boldsymbol{p}_i), \\ \boldsymbol{q}_j \leftarrow \boldsymbol{q}_j + \eta(e_{ij}\boldsymbol{p}_i - \lambda \boldsymbol{q}_j), \end{cases} \tag{2.21}$$

where $e_{ij} = a_{ij}^{(T)} - r_{ij}$ measures how off is the prediction of the model at the current step from the ground-truth.

The method strongly depends on initialization of its parameters, performed at the beginning. A quite common practice is to use a normal distribution with zero mean and small deviation. It is also advised to shuffle the data prior to optimization in order to avoid unintended biases in the resulting model. The overall complexity of the approach is $O(nnz_A \cdot r)$.

Note that *SGD is inherently incremental*, which gives an "out-of-the-box" equivalent of the folding-in technique for the model updates. For example, in the case of a newly introduced user with at least a few known preferences one can simply iterate over these preferences with the first line of Eq. (2.21) until it converges. The other parameters related to items stay fixed in that case. New items can be handled in a similar fashion.

2.4.1.2. *Alternating minimization techniques*

In turn, the ALS-based methods decompose the optimization task into the sequence of the least squares problems. Note that while the optimization problem defined by Eq. (2.20) is non-convex, it is *bi-convex* with respect to its parameters. In other words, for fixed P it is convex in Q and for fixed Q it is convex in P. Moreover, the optimization problem can be solved independently for every row of P and Q. Therefore, one can iteratively minimize the objective function by switching between user and item factors and updating their rows as follows:

$$\begin{cases} \boldsymbol{p}_i \leftarrow \operatorname*{argmin}_{\boldsymbol{p}_i} \mathcal{J}(\Theta), \\ \boldsymbol{q}_j \leftarrow \operatorname*{argmin}_{\boldsymbol{q}_j} \mathcal{J}(\Theta). \end{cases} \tag{2.22}$$

After each iteration the objective function is guaranteed not to increase. However, unlike the unweighted case, there are no global guarantees for convergence in general. In practice, the algorithm is reported to require only around 10 or slightly more epochs to achieve a good approximation [Bell and Koren (2007); Hu *et al.* (2008)].

In order to find the update rules for Eq. (2.22), it is convenient to rewrite both $\mathcal{L}$ defined by Eq. (2.15) and Ω defined by Eq. (2.19) in the row-wise and column-wise forms, corresponding to $\boldsymbol{p}_i$ and $\boldsymbol{q}_j$ respectively. For example, in the user-wise case it reads:

$$\mathcal{J}(\Theta) = \sum_i (\boldsymbol{a}_i - Q\boldsymbol{p}_i)^T W^{(i)} (\boldsymbol{a}_i - Q\boldsymbol{p}_i) + \lambda \sum_i \boldsymbol{p}_i^T \boldsymbol{p}_i + \lambda \|Q\|_F^2, \tag{2.23}$$

where $W^{(i)} = \operatorname{diag}\{w_{i1}, w_{i2}, \ldots, w_{iN}\}$ is a diagonal matrix of weights and $\boldsymbol{a}_i$ is the *i-th row* of the matrix $T(A)$, i.e. it represents the *preference vector of user i with respect to all items.* After finding the derivative $\partial\mathcal{J}/\partial\boldsymbol{p}_i$ and setting it to zero one arrives at the following equation for $\boldsymbol{p}_i$:

$$\left(Q^T W^{(i)} Q + \lambda I\right) \boldsymbol{p}_i = Q^T W^{(i)} \boldsymbol{a}_i. \tag{2.24}$$

This gives a standard linear system of equations with the $r \times r$ symmetric positive definite matrix $\left(Q^T W^{(i)} Q + \lambda I\right)$. Direct solution of the system can be found in $O(r^3)$ time, for example, by the means of Cholesky decomposition. The resulting expression for the $\boldsymbol{p}_i$ update reads:

$$\boldsymbol{p}_i \leftarrow \left(Q^T W^{(i)} Q + \lambda I\right)^{-1} Q^T W^{(i)} \boldsymbol{a}_i. \tag{2.25}$$

Due to the symmetry of the objective function, in order to find an update rule for $\boldsymbol{q}_j$ one can simply replace $Q, W^{(i)}$ and $\boldsymbol{a}_i$ with their corresponding counterparts:

$$\boldsymbol{q}_j \leftarrow \left(P^T \bar{W}^{(j)} P + \lambda I\right)^{-1} P^T \bar{W}^{(j)} \bar{\boldsymbol{a}}_j, \tag{2.26}$$

where $\bar{W}^{(j)} = \text{diag}\{w_{1j}, w_{2j}, \ldots, w_{Mj}\}$ and $\bar{\boldsymbol{a}}_j$ denotes the j-*th column* of the matrix $T(A)$, i.e. the *preference vector of all users against item* j. At each epoch the algorithm updates all rows of the matrices P and Q, which can be done in parallel. As in the SGD case, the iteration process repeats until either the number of epochs exceeds some threshold value or the objective function ceases to decrease (with respect to a predefined tolerance). The overall complexity of the algorithm is estimated as $O(nnz_A \cdot r^2 + (M+N)r^3)$ [Pilászy *et al.* (2010)].

Note, that the same update rules can also be used to calculate approximate predictions for the *new entities*. Indeed, as every update is just the solution of the corresponding least squares problem, one can replace $\boldsymbol{a}_i$ or $\bar{\boldsymbol{a}}_j$ with the preference vector of a newly introduced user or item respectively. This technique is similar to the *folding-in* update used in PureSVD.

Another important consideration is that the time, required to solve Eq. (2.24), can be further reduced with additional computational tricks. For example, the straightforward application of the *Sherman-Woodbury-Morrison* formula gives an analytic expression for incremental calculations of the matrix inverse at each iteration. This, however, may not always provide a considerable speed-up and highly depends on the data sparsity [Pilászy *et al.* (2010)].

Alternatively, instead of the direct approach one could use iterative linear system solvers in order to find an approximate solution. A worth noting candidate is the *conjugate gradient* (CG) method [Golub and Van Loan (2012)], which is closely related to the Lanczos process and similarly requires only matrix-vector multiplications for performing the task. The method allows to reduce the complexity of the matrix inverse computation to $O(r)$ instead of $O(r^3)$ as in the original approach. It gives a decent trade-off between the accuracy of each individual update and the overall convergence speed [Takács *et al.* (2011)] and works quite well in practice[3].

[3]Its open-source implementation available at https://github.com/benfred/implicit is shown to provide a remarkable speedup almost without the drop in quality.

2.4.1.3. *Coordinate descent*

Another iterative approach for performing the optimization task is to employ the (block) coordinate descent method (CD) [Bertsekas (1999), Section 2.7]. In the context of the low-rank approximation of complete matrices it was explored in [Cichocki and Phan (2009)], where the authors additionally consider nonnegativity constraints. A few efficient variations of the method were also proposed for the missing value estimation in the recommender systems with explicit feedback [Bell *et al.* (2007); Pilászy *et al.* (2010); Yu *et al.* (2012)]. Recently, several efficient implementations were also proposed for the OCCF case [Yu *et al.* (2017); Bayer *et al.* (2017)].

Generally, instead of the bulk update of latent feature matrices performed in ALS, CD successively updates either blocks of variables (e.g. rows or columns of the factor matrices) or simply a single variable. Such formulation leads to a convex optimization and avoids computation of a matrix inverse. For example, by declaring the result of an update in the variable p_{ik} as θ one arrives at the following optimization subproblem:

$$f(\theta) = \sum_{ij} w_{ij} \left(a_{ij} - \left(\boldsymbol{p}_i^T \boldsymbol{q}_j - p_{ik} q_{jk}\right) - \theta q_{jk}\right)^2 + \lambda \theta^2, \tag{2.27}$$

where $f(\theta)$ is a univariate quadratic function. Its optimum value is then given by:

$$\theta^* = \frac{\sum_j w_{ij} \left(a_{ij} - \boldsymbol{p}_i^T \boldsymbol{q}_j + p_{ik} q_{jk}\right) q_{jk}}{\lambda + \sum_j w_{ij} q_{jk}^2}. \tag{2.28}$$

Similar expression can be obtained for updates in the matrix Q. This approach also offers a trade-off. The algorithm may require more epochs to converge, however, each iteration within every epoch becomes much cheaper. Despite being less popular than ALS and SGD, CD offers a competitive quality of recommendations with a number of computational advantages [Yu *et al.* (2012)].

2.4.2. *Biased matrix factorization*

As was already noted in Sec. 2.3.2, the rating prediction quality can be improved with the concept of biases, which absorb a significant part of the feedback signal. A similar data transformation procedure with manually crafted biases can be applied for the weighted MF problem as well. However, unlike the SVD-based case, the weighted formulation of the problem is more flexible and allows to declare bias variables as additional model parameters [Paterek (2007); Koren *et al.* (2009)].

With this approach, the predicted value of the rating r_{ij} assigned by user i to item j is modelled as follows:

$$r_{ij} = \mu + t_i + f_j + \boldsymbol{p}_i^T \boldsymbol{q}_j, \qquad (2.29)$$

where all bias variables $\{t_i\}$ and $\{f_j\}$ are *learned along with other model parameters*, i.e. $\Theta = \{\boldsymbol{t}, \boldsymbol{f}, P, Q\}$. The global average μ is usually pre-estimated based on the known values of ratings. One may conveniently rewrite the prediction formula in a compact matrix form, following the outer product rule similarly to Eq. (2.9):

$$R = \mu E + \bar{P}\bar{Q}^T,$$

where the block matrices $\bar{P} = [\boldsymbol{t}\, \boldsymbol{e}_M\, P]$ and $\bar{Q} = [\boldsymbol{e}_N\, \boldsymbol{f}\, Q]$ have a particular form of the first two columns comprised by the bias vectors and vectors of all ones; $E = \boldsymbol{e}_M \boldsymbol{e}_N^T$ is an $M \times N$ matrix of all ones.

Clearly, shifting the data values by μ would give similar to Eq. (2.2) form. However, the bias terms increase an overall rank of the solution by 2. Moreover, the result does not correspond to an arbitrary unbiased MF model of rank $r + 2$ due to a certain structure of the first 2 columns in the factor matrices. In some sense biases can be viewed as a specific constraint on the factors, which is used to reflect the core assumption about the underlying rating mechanism.

The SGD-based variation of this matrix factorization approach became popular after it was published in the famous blog post[4] by Simon Funk, when he attended the Netflix Prize competition. Due to that, sometimes this algorithm is also called *FunkSVD*. It has become an internal part of many other MF algorithms. The full update rule, including additional bias updates, reads:

$$\begin{cases} \boldsymbol{p}_i \leftarrow \boldsymbol{p}_i + \eta(e_{ij}\boldsymbol{q}_j - \lambda \boldsymbol{p}_i), \\ \boldsymbol{q}_j \leftarrow \boldsymbol{q}_j + \eta(e_{ij}\boldsymbol{p}_i - \lambda \boldsymbol{q}_j), \\ t_i \leftarrow t_i + \eta(e_{ij} - \lambda t_i), \\ f_j \leftarrow f_j + \eta(e_{ij} - \lambda f_j). \end{cases}$$

As a practical remark, such a representation via the bias terms also allows to quickly estimate the rating values for previously unobserved items or users with no associated ratings. In that case it falls back to the baseline value contained in the corresponding bias term and there is no contribution of the factorization part. This estimate can be further improved after at least one rating value is provided into the system with the incremental approach similarly to the unbiased MF case.

[4]http://sifter.org/simon/journal/20061211.html

2.4.3. *Confidence-based models*

Another important example of the weighted matrix factorization approach is based on a more flexible treatment of both observed and unobserved interactions. Consider the case, where users exhibit different behavior depending on how much they like a particular item. For example, when a user plays a particular sound track several times while skipping some other track just after listening to the first 10 seconds, this would be a clear indication that the first track is more interesting for the user. In other words, our *confidence* that the user enjoys the music is higher in the case of the first track. Likewise, the fact that user has never played some track does not immediately suggest that the track is not interesting — the user may be simply unaware of it. However, our confidence in the relevance of the track is lower in this case.

In order to account for such an uncertainty it seems reasonable to associate some confidence measure with every possible interaction. Instead of simply ignoring the missing data and assigning constant weights to the known interactions as in Eq. (2.17) we would like to change the weights of interactions depending on various conditions and to treat both observed and unobserved data in a more thorough way. The general form of a confidence-based loss function is slightly different from Eq. (2.15):

$$\mathcal{L}\left(T(A), R\right) = \|W(A) \circ (S - R)\|_F^2. \tag{2.30}$$

where the binary matrix $S \in \mathbb{B}^{M \times N}$ with elements

$$\begin{cases} s_{ij} = 1, & \text{if } a_{ij} \text{ is known,} \\ s_{ij} = 0, & \text{otherwise} \end{cases} \tag{2.31}$$

indicates whether a particular interaction has occurred. The weights matrix $W = [\sqrt{w_{ij}}]$ encodes a confidence in the observed feedback and *directly depends on the values of* A.

Note that this approach is not designed to predict an exact rating value. It rather focuses on the prediction of a probability of a certain event taking into account an additional information, be it a rating value, a browsing behavior or any other form of an explicit or implicit feedback that allows to quantify the corresponding confidence level. A particular choice of the confidence measure may significantly impact the performance of a recommendation model. A few different techniques were proposed independently by several research groups [Hu *et al.* (2008); Pan *et al.* (2008)]. The substantial difference in the proposed models is in the way the weighting is applied.

The authors of the so called *Weighted Regularized Matrix Factorization* model (WRMF) [Hu *et al.* (2008)], sometimes also called *implicit ALS* or *iALS*, propose to assign constant weight of 1 to the unobserved interactions, and increase the weight for any observed interaction proportionally to a satisfaction of a user with an item estimated from the expressed feedback:

$$w_{ij} = 1 + \alpha g\left(a_{ij}^{(T)}\right),$$

where α is an empirically determined coefficient of proportionality. The estimation function g is the most subjective part of the model and may vary depending on the domain of application and the type of available data. The authors give a few examples of it based on linear $g(x) = x$ and logarithmic $g(x) = \log(1 + \frac{x}{\epsilon})$ approximation (with an extra tuning parameter ϵ), which work well in practice.

The authors propose to use ALS optimization as it allows to efficiently handle computations with complete matrices S and W. The general form of the solution stays the same as in standard ALS:

$$\begin{cases} \boldsymbol{p}_i \leftarrow \left(Q^T W^{(i)} Q + \lambda I\right)^{-1} Q^T W^{(i)} \boldsymbol{s}_i, \\ \boldsymbol{q}_j \leftarrow \left(P^T \bar{W}^{(j)} P + \lambda I\right)^{-1} P^T \bar{W}^{(j)} \bar{\boldsymbol{s}}_j, \end{cases} \tag{2.32}$$

where $W^{(i)} = \text{diag}\{w_{i1}, w_{i2}, \ldots, w_{iN}\}$, $\bar{W}^{(j)} = \text{diag}\{w_{1j}, w_{2j}, \ldots, w_{Mj}\}$; $\boldsymbol{s}_i$ is a *binary* preference vector of user i with respect to all items, and $\bar{\boldsymbol{s}}_j$ is a *binary* preference vector of all users against item j. The authors came up with an elegant computational trick which allows to avoid redundant computations making the algorithm highly scalable.

Alternatively, the authors of the second approach, referred to as *weighted ALS* or *wALS* [Pan *et al.* (2008)], propose to assign the constant weight value of 1 for all known observations and in contrast to WRMF use alternate weighting schemes for the unobserved part. The weighting scheme can be based either on small constant values in the range [0, 1] or on some data aggregation which takes into account popularity effects. For example, in the user oriented approach the weights for the *unobserved* data are proportional to the number of ratings provided by user. The rationale behind is that the higher is the number of ratings provided by user, the more likely it is that the remaining non-rated items are irrelevant for that user. Likewise, in the item oriented case the lower popularity of an item would increase the corresponding weights for negative (unobserved) interactions.

2.4.4. *Combined latent representations*

A high level intuition behind latent features is often provided in terms of the ability to capture intrinsic item properties as well as user motivation and interests. Latent features are often treated as indicators of some user tastes and items' affinity to them. However, in practice, it can be quite difficult to directly map a single real feature to its latent representation [Takács *et al.* (2007)]. It is more likely that each latent feature will instead characterize some tangled combination of various aspects.

Moreover, the number of these aspects can be large and they may have a complicated, multifaceted nature making it hard to interpret them by a virtue of standard parametrization. On the other hand, this information may play an important role in the decision making process. Ignoring it may not only limit the expressiveness of a recommendation model, but also hinder its ability to uncover valuable implicit relations within the observed data.

One of the ways to improve sensitivity of a model to a multi-aspect input is to explicitly impose an aspect-based structure on the latent representation of users and items. As an example, consider an online retail shop where customers tend to purchase only a few items and rarely provide an explicit feedback. This would lead to a very sparse interaction matrix and make the decision making process obscure for a recommendation model.

Meanwhile, it is typically possible to collect additional information such as what pages users visit during their search for a product, what information they look for, what products they consider together, etc. Including such information into a model allows to increase an understanding of user interests, and therefore help to create a better prediction model. With the flexibility of a weighted matrix factorization this can be achieved directly by adjusting the optimization objective.

One of the earliest examples of such approach is the *NSVD* model [Paterek (2007)], where every user is characterized by a combination of items he or he interacted with. It can be especially helpful in the case of extreme sparsity and the lack of any side information, giving a more "smooth" representation of the data. The author of the model proposed 2 variations of such representation: based on binary vectors (simply denoting the fact of interaction) and based on latent features of items. In the latter case the solution can be sought in the following form:

$$R = SQQ^T \tag{2.33}$$

where S is a sparse matrix of aggregation coefficients with binary elements defined simialrly to Eq. (2.31). Here we omit biases as they can be trivially added.

The matrix product SQ in Eq. (2.33) gives an aggregated representation of every user via the latent features of consumed items. As a result, the contribution into the prediction score is defined by all the actions taken by the user, independently of the user-assigned rating values. Note that the model has a reduced number of parameters which can be especially suitable in the systems with very large amount of users and may potentially help to avoid a certain redundancy. It also has inspired further research in this direction and has led to more elaborate models such as *SVD++* and *Asymmetric-SVD* [Koren (2008)]. Later it was shown to be a special case of the more general models, namely *SVDFeature* [Chen *et al.* (2011)] and *Factorization Machines* [Rendle (2010)].

In the SVD++ model, which turned out to provide results superior to Asymmetric-SVD, the latent features of users are not replaced, but rather are augmented with an additional information about multiple aspects of user-item interactions in the following way:

$$R = (P + \bar{S}L)Q^T, \tag{2.34}$$

where L represents an independent of Q latent subspace, which is used to build neighborhoods of items rated together by the same user. The sparse aggregation matrix $\bar{S}$ has the same sparsity pattern as S. In contrast to NSVD, its values are not binary and are row-normalized, so that the norm of every row would be equal to 1, i.e. $\bar{S} = D^{-1}S$, where $D = \mathrm{diag}\{\|\boldsymbol{s}_1\|_2, \|\boldsymbol{s}_2\|_2, \ldots, \|\boldsymbol{s}_M\|_2\}$ and $\boldsymbol{s}_i$ is an i-th row of S. Such normalization prevents susceptibility of the model to popularity of items and to contribution of very active raters.

Note that multiple types of feedback can be easily incorporated into the model simply by adding more aggregation terms, i.e. $P + \bar{S}_1 L_1 + \bar{S}_2 L_2 + \ldots$, corresponding to different types of feedback (e.g. purchase activity, browsing history, etc.). The key drawback of such approach is an increased number of parameters, which makes the model more difficult to train and prone to overfitting.

Equation (2.34) can be reformulated as $R = (X\bar{P})Q^T$, where $X = [I\,\bar{S}]$ and $\bar{P}^T = [P^T L^T]$ are block matrices of aggregation coefficients and joint latent features respectively. Up until now we have used coefficients matrix X to combine items rated by the same user. However, it can also be used to reflect any sort of additional information which helps better describe the observed interactions. For example, instead of (or along with) indicating the rated-together items, it can be used to encode relevant user attributes and group users with respect to these attributes. The matrix $\bar{P}$ will be extended with the corresponding embeddings of these attributes onto the latent feature space similarly to how it was performed for items with the matrix L.

The same reasoning can be applied with respect to the matrix Q which can be replaced with an aggregated view on different item properties and the item-related interaction aspects. The most general formulation of such representation can be compactly described as:

$$R = XP(YQ)^T, \tag{2.35}$$

where the block matrices $P^T = [P_1^T\, P_2^T \ldots]$ and $Q^T = [Q_1^T\, Q_2^T \ldots]$ now represent various user-based, item-based and mutual aspects of the observed interactions. Sparse coefficient matrices $X = [X_1\, X_2 \ldots]$ and $Y = [Y_1\, Y_2 \ldots]$ with the corresponding block structure allow to aggregate various latent vectors to represent every interaction from a multi-aspect perspective. This aggregated model is known as *SVDFeature* [Chen *et al.* (2011)]. Due to its ability to take side information into account it can be considered as a representative of the so called *hybrid approach* (see Sec. 2.5.2 and also Chapter 4).

There is one nuance that is worth noting here. The authors of the model propose to represent the global bias as a weighted sum of global biases calculated with respect to different aspects. It is more convenient to demonstrate it with an equivalent to Eq. (2.35) elementwise formulation, now including all bias terms:

$$r_{ij} = b_0 + \boldsymbol{t}^T \boldsymbol{x}_i + \boldsymbol{f}^T \boldsymbol{y}_j + \boldsymbol{x}_i^T P Q^T \boldsymbol{y}_j, \tag{2.36}$$

where $b_0 = \sum_{g \in G} \gamma_g \mu_g$ is a global bias aggregated over the group of aspects denoted by G with individual weight coefficients γ_g and bias values μ_g.

Note that the bilinear form of Eq. (2.36) can be viewed as a special case of a polynomial expansion:

$$r(\boldsymbol{z}) = b_0 + \boldsymbol{b}^T \boldsymbol{z} + \boldsymbol{z}^T H \boldsymbol{z} + \ldots \tag{2.37}$$

The connection to the SVDFeature model can be seen with the following substitution: $\boldsymbol{b}^T = [\boldsymbol{t}^T\, \boldsymbol{f}^T]$ and $\boldsymbol{z}^T = [\boldsymbol{x}^T\, \boldsymbol{y}^T]$, where $\boldsymbol{x}^T$ and $\boldsymbol{y}^T$ are some rows of the matrices X and Y. The coefficients vector $\boldsymbol{z}$ now encodes the full information about an interaction between some user and some item with respect to all related aspects[5], as was discussed previously. Hence,

[5]In the case of categorical data, e.g. user or item id, user gender, movie genre, etc., this method of building a sparse representation of the multidimensional input data is called *one hot encoding*.

the quadratic term $\boldsymbol{z}^T H \boldsymbol{z}$ with symmetric positive semi-definite matrix H allows to account for an interplay between any entities and any aspects in their contribution to the final prediction score. Note that H subsumes matrices P and Q in a certain way and the parameters of the model are described as $\Theta = \{b_0, \boldsymbol{b}, H\}$.

Such a generalization leads to the next hybrid approach and a popular machine learning algorithm, namely *Factorization Machines* (FM) [Rendle (2010)], which has been proven to perform well in recommender systems. The author of the model notes that the matrix H should have a low-rank structure in order to deal with the sparsity problem and increase the expressiveness of the model:

$$H = VV^T$$

where V embeds all users, items and the corresponding side information onto the lower dimensional latent feature space. In addition to that, all self-influence terms (i.e. x_i^2) are excluded and the symmetry of the model (i.e. the equivalent contribution of both $x_i x_j$ and $x_j x_i$ interplay terms) is taken into account, which produces the following relevance score function:

$$r(\boldsymbol{z}) = b_0 + \sum_i b_i z_i + \sum_i \sum_{j=i+1} \langle \boldsymbol{v}_i, \boldsymbol{v}_j \rangle z_i z_j. \qquad (2.38)$$

The task of generating recommendations, therefore, boils down to solving the polynomial regression problem given the observation data.

Note that unlike SVDFeature or SVD++ the model allows to take into account additional interaction factors, e.g. it allows to include a "within-class" influence — an influence of entities and aspects of the same type on each other within a single observation. Indeed, indices i, j in $z_i z_j$ term may belong to 2 different items or 2 different features describing the same item. Depending on the problem, such extra interactions can be meaningless or undesirable. In order to control which interactions are allowed in the model one can replace $z_i z_j$ with $\delta_{ij} z_i z_j$, where binary variable δ_{ij} would indicate whether the corresponding interaction is allowed. Clearly, FM can be reduced to any of the previously discussed models by a proper choice of the model parameters and indicator coefficients. A popular variation of FM that uses this technique to separate the latent space for various groups of features is called Field-Aware FM (FFM) [Juan *et al.* (2016)].

FM models also have a close connection to a higher order approach based on Pairwise Interaction Tensor Factorization (PITF) model [Rendle and Schmidt-Thieme (2010)]. Unlike matrix-based models, PITF uses an array with 3 dimensions, called a 3rd order tensor, to encode pairwise relations between users, items and additional interaction aspects (tags). The model uses 2 independent latent feature spaces for tags: one for user-tag and another one for item-tag relations respectively. The PITF model per se is a member of a broader family of tensor-based methods, which allow to model *n-ary* relations (ternary, quaternary, etc.) not only in a pairwise but in a mutual way.

The topic of tensor methods in recommender systems deserves a separate discussion and we refer the reader to [Frolov and Oseledets (2017)] for a comprehensive overview. Worth noting here that tensor-based methods are often used for *context-aware* recommender systems, covered in details in Chapter 5. There are also several direct extensions of the FM idea to higher order cases, e.g. Tensor Machines [Yang and Gittens (2015)], Higher Order FM [Blondel *et al.* (2016)], Exponential Machines [Novikov *et al.* (2016)].

2.4.5. *Remark on connection with SVD*

As can be seen, there are some matrix factorization methods that have SVD acronym in their names. This may lead to a certain confusion, that should be avoided. Strictly speaking, most of these methods, like FunkSVD, SVD++, SVDFeature and their derivatives have very little in common with a mathematical formulation of SVD. Unlike conventional SVD, these methods do not build a space of singular vectors and do not compute singular values. Most of them do not preserve the orthogonality property. Weighted matrix factorization approach is designed specifically to work with incomplete matrices, often ignoring unknown entries or treating them not in the same way as in PureSVD. They form a separate family of methods with different optimization objectives and more flexible tuning. However, due to historical reasons, they are still sometimes are referenced as SVD-based methods.

As a matter of fact, it is, of course, possible to orthogonalize latent factors in Eq. (2.2) and get an equivalent to Eq. (2.6) form with orthonormal basis. This can be achieved by the virtue of the QR decomposition applied to both P and Q matrices (in order to get singular values as well one would have to additionally apply SVD to the product of the low dimensional upper triangular matrices resulted from the QR decomposition). Nevertheless, whenever the optimization objective Eq. (2.18) includes specific constraints other than simple quadratic regularization and the loss function is considerably different from Eq. (2.15), performing orthogonalization potentially leads to a loss of structure in the latent feature space imposed by those special conditions.

Also note that a simple regularization constraint similar to Eq. (2.19) can be added for SVD factors as well. Optimization of the corresponding loss function defined by Eq. (2.5) with this added constraint can be performed without the need to switch to general matrix factorization framework. The solution to such optimization problem, known as *quadratically regularized PCA*, has the same analytical form as the standard SVD and preserves its properties [Udell *et al.* (2016)]. There is also a connection of the latter to an iterative SVD-based approach called *softImpute* suitable for the rating prediction task [Hastie *et al.* (2015)].

2.5. Advanced methods

There is an overwhelming amount of factorization models that implement sophisticated modifications to standard MF formulation in order to address various problems. To name a few, the models may include the concept of metric learning [Hsieh *et al.* (2017)] for better latent representation, impose additional locality constraints [Chen *et al.* (2017)], rely on a more flexible probabilistic inference [Mnih and Salakhutdinov (2008); Salakhutdinov and Mnih (2008)], use kernel methods to capture non-linear effects [Rendle and Schmidt-Thieme (2008)], etc. A complete analysis of such a variety of methods falls beyond the scope of this chapter. We, however, briefly describe a few particular examples to illustrate how one can modify the components of standard problem formulation, given by Eq. (2.18), in order to obtain a new model with desired properties.

2.5.1. *Learning to rank*

One of the main concerns with the standard formulation of matrix factorization problems is that it is especially suitable for the rating prediction task, however, one can argue that this may not be the best choice for top-n recommendations, where the correct ranking of recommended items is more important than any particular prediction score. It turns out that there is a formal way to address this issue with the help of the *learning to rank* approach [Liu *et al.* (2009)]. Covering this broad topic would probably require a separate chapter in the book, so we will focus just on a few representative techniques that fit well into the general problem formulation of MF. In order to do that, let us consider three general categories of optimization objectives, which lead to different ranking mechanisms in recommender systems: *pointwise*, *pairwise* and *listwise* [Chapelle and Wu (2010)].

Pointwise objective directly depends on a pointwise loss function between the observations and the predicted values. This is the simplest case, which corresponds to previously discussed optimization problems, e.g. Eq. (2.5) or Eq. (2.15), and is not designed for the ranking task. Nevertheless, like in the case with PureSVD, which is formulated as a matrix completion problem and yet can be tuned to provide reasonably good precision-recall scores, it is also possible to empirically find a set of model hyper-parameters, which improve the ranking of recommendations. However, it is unlikely to get a significant improvement in this case.

Pairwise objective depends on a pairwise comparison of the predicted values and penalizes those cases where their ordering does not correspond to the ordering of observations. The total loss in that case may take the following (or similar) form:

$$\mathcal{L}(A, R) = \sum_{i} \sum_{j,j':a_{ij} > a_{ij'}} l(r_{ij} - r_{ij'}),$$

where $l(x_1 - x_2)$ is a pairwise loss function that decreases with the increase of the difference $x_1 - x_2$ (e.g. sigmoid function) and $r_i j$ is the predicted score. It allows to smoothly approximate an indicator function $\mathbb{I}(x_1 > x_2)$.

One of the most popular examples of the pairwise optimization is Bayesian Personalized Ranking (BPR) technique [Rendle *et al.* (2009)], which optimizes a smooth version of AUC with the help of SGD. Another variation of the pairwise approach is Weighted Approximate-Rank Pairwise (WARP) [Weston *et al.* (2011); Hong *et al.* (2013)], which implements an efficient iterative sampling procedure for negative examples. Alternatively, the authors of RankALS [Takács and Tikk (2012)] propose a modification of the WRMF model for the pairwise objective and propose an ALS-based optimization procedure.

Listwise objective optimizes the predicted ordering over entire lists of items at once. The corresponding listwise loss function can be schematically expressed as $l(\{a_{ij}\}, \{r_{ij}\})$. It penalizes the deviation of the predicted ranking of a given list of items from the ground truth ranking based on observations. This approach is considered to be the most suitable for the top-n recommendation task as it allows to directly optimize listwise metrics, e.g. mean average precision (MAP), normalized discounted cumulative gain (NDCG) or mean reciprocal rank (MRR) (see Chapter 9). The listwise approach follows a similar trick of a smooth approximation of the ranking metrics. For example, the reciprocal rank RR_{ij} of an item j recommended to a user i can be approximated by:

$$\mathrm{RR}_{ij} \approx \frac{1}{1+e^{-r_{ij}}}.$$

A few remarkable examples of this approach are CoFiRank [Weimer *et al.* (2008)], which implements a convex upper bound approximation of NDCG, and CLiMF [Shi *et al.* (2012b)], which instead optimizes a lower bound of a smooth reciprocal rank.

Worth noting here, that although both pairwise and listwise algorithms are likely to improve the quality of predicted ranking of elements, they are typically harder to implement and may require additional heuristics to reduce the computational complexity [Shi *et al.* (2012a)].

2.5.2. *Hybrid factorization models*

For a detailed overview of hybrid recommender systems we refer the reader to Chapter 4. Here we consider a particular category of hybrid systems, which extend standard MF approach with the ability to incorporate additional content data, i.e. side information about users and/or items. We have already described two methods from this category, namely FM and SVDFeature. Below are a few more examples based on a customization of the objective function.

One of the most straightforward ways to incorporate side knowledge is to impose additional constraints on the latent feature space via regularization. This can be effectively achieved with the help of *collective MF* [Singh and Gordon (2008)]. In its simplest variant, also known as *coupled MF*, the corresponding loss function can be formulated as follows [Fang and Si (2011)]:

$$\mathcal{L}(A,\Theta) = \left\|W \circ \left(A - PQ^T\right)\right\|_F^2 + \alpha\left(\|F - PW_F\|_F^2 + \|G - QW_G\|_F^2\right),$$

where $\Theta = \{P, Q, W_F, W_G\}$ represents model parameters, matrices F and G encode side information, i.e. user attributes and item characteristics, and α controls the contribution of the side information into the resulting model.

There are also several variations of the coupled factorization technique, where parametrization of the content matrices F, G is replaced with parametrization of the content-based similarity between users and items

Table 2.1. Comparison of low-rank approximation algorithms for explicit feedback data.

Algorithm	Overall complexity	Update complexity	Sensitivity	Optimality
SVD[a]	$O\left(nnz_A \cdot r + (M+N)r^2\right)$	$O\left(nnz_a \cdot r\right)$	Stable	Global
ALS	$O\left(nnz_A \cdot r^2 + (M+N)r^3\right)$	$O\left(nnz_a \cdot r + r^3\right)$	Stable	Local
CD	$O\left(nnz_A \cdot r\right)$	$O\left(nnz_a \cdot r\right)$	Stable	Local
SGD	$O\left(nnz_A \cdot r\right)$	$O\left(nnz_a \cdot r\right)$	Sensitive	Local

[a]For both standard and randomized implementations [Halko *et al.* (2011)].

[Shi *et al.* (2010); Barjasteh *et al.* (2015)]. The authors of the Local Collective Embeddings (LCE) model [Saveski and Mantrach (2014)] propose to add a locality constraint, so that the entities, which are close to each other in terms of real features, remain close to each other in the latent feature space. The authors use the model to specifically address the cold-start problem.

There are many more techniques that either directly embed content information into the latent space [Pilászy and Tikk (2009); Roy and Guntuku (2016)] or add new regularization terms to enforce feature-based proximity properties [Nguyen and Zhu (2013)]. It is also possible to use similar approach to incorporate *contextual information* [Liang *et al.* (2016)], which leads to context-aware models (see Chapter 5).

2.6. Practical aspects

There are many practical aspects that make particular algorithms more suitable in certain environments depending on the desired balance between technical and business requirements. For example, achieving the highest quality of recommendations with a state-of-the-art method may require a lot of computational resources or depend on a complex setup which is hard to maintain and support in production. In such cases a simpler approach with a more straightforward configuration and flexible tuning may become more favorable and help to find a better trade-off between a solution's complexity and the recommendations quality. The latter point is especially crucial when latent factors are used to build neighborhood-based models. In large scale setting an exact search of neighbors may take a prohibitively long time and has to be replaced with approximate solutions (see [Bachrach *et al.* (2014)] and also Chapter 11).

A thorough technical analysis of different algorithms is a non-trivial task and depends on various aspects. One of the most crucial ones is the scalability question, which includes an overall time complexity, memory and storage requirements, online updates support, parallelization efficiency in shared- and distributed-memory environments. Other aspects include stability of an algorithm and its convergence guarantees. General differences between the main algorithms discussed in this chapter are provided in Table 2.1. Note that unlike ALS and SVD, standard implementations of SGD and CD are inapplicable for OCCF problems, as their complexity becomes proportional to the total size of the rating matrix.

From the parallelization viewpoint, multi-core shared-memory systems are typically more preferred than distributed shared-nothing environments with multiple computational nodes. This allows to avoid the between-node communication and system state synchronization costs induced by hardware I/O capabilities and specific software implementations.

Moreover, a wide class of large-scale problems can be tackled in the shared-memory settings with the help of an up-to-date hardware [Hidasi and Tikk (2016)] and appropriate data preprocessing (e.g. cleansing, subsampling). For example, modern cloud computing services provide instances for memory-intensive applications with *several terabytes* of physical memory onboard, which may help to cover the needs in many practical cases. Therefore, as a rule of thumb, distributed setups should be avoided unless the data and/or model parameters do not fit into a single machine's memory [Amatriain and Agarwal (2016)].

2.6.1. *Parallel SGD*

As has been noted, the SGD algorithm is inherently sequential. Model's parameters are updated after every single learning step and parallelization of the algorithm becomes a challenging task. Within such computational environment various architectural choices on the data and model sharing, on data-accessing and data-passing strategies may have a dramatic impact on the algorithm's performance [Zhang and Ré (2014); Sallinen *et al.* (2016)].

A straightforward implementation of SGD in shared-memory environment directly leads to overwriting conflicts when, for example, several parallel workers operate on the ratings of the same user/item and, therefore, modify the same vector of user/item latent features. As a result, some of the standard techniques for SGD parallelization, such as *Hogwild!* [Recht *et al.* (2011)], may not be suitable for the matrix factorization case, unless the data is extremely sparse. As a lock-free algorithm, Hogwild! does not restrict parallel overwrites and it's performance is highly influenced by the data imbalance. Latent features of very popular items or very active users are likely to be updated and recomputed more frequently than latent features of entities with fewer ratings. In practice, this may result in a slower convergence and a degradation of an overall performance of the approach.

In turn, in the distributed case the synchronization of updated parameters of SGD between computational nodes may easily become the main bottleneck of computations. The described problems has led to many different approaches, which achieve certain trade-off's between effective communications and state synchronization, use various data partitioning techniques, implement elaborate locking strategies and rely on aggressive caching. Some of the approaches are only suitable for shared-memory systems [Zhuang *et al.* (2013); Pan *et al.* (2016)], others are designed for distributed systems [Gemulla *et al.* (2011); Schelter *et al.* (2014)] and some support both regimes [Yun *et al.* (2014)]. More details on technical implementation and analysis are provided in Chapter 11.

2.6.2. *Parallel ALS*

In contrast to SGD, parallel implementation of the standard ALS algorithm for weighted matrix factorization is much more straightforward as latent feature vectors for any user or item can be updated independently at each epoch. It is often said that the algorithm is *embarrassingly parallel*. However, at large scale and in the distributed settings the task may become more involved and IO intensive [Yu *et al.* (2012)], requiring elaborate data partitioning and parameters' synchronization [Das *et al.* (2017); Schelter *et al.* (2013)].

As an example, if factor matrices are too large to fit into a single machine's memory than one have to distribute rows of factor matrices P and Q across nodes and properly coordinate the between-node communications to ensure a consistent global state. The interactions data may also be distributed so that all interactions related to a single item or to a single user belong to the same computational node [Zhou *et al.* (2008)]. As this would require switching between columns and rows of the ratings matrix, which is typically stored in the compressed sparse row/column formats (CSR/CSC), two distributed copies of the data are used: a column-wise copy for item-related interactions and a row-wise copy for user-related interactions. This allows to avoid redundant computations and reduce the intensity of data transfer.

Nevertheless, communication overhead of the ALS in that case can still be considerable due to random access to the latent feature matrices and may not play well with widely accepted distributed data processing paradigms, such as map-reduce [Hidasi and Tikk (2016)]. Alternatively, in the shared-memory settings both ALS and iALS can be implemented very efficiently. iALS — high-performant computing in shared-memory [Gates *et al.* (2015)]

2.6.3. *Parallel CD*

The CD method can be considered as an attempt to combine the advantages of both ALS and SGD methods. It performs alternating optimization similar to ALS and consists of a more lightweight iteration steps. The authors of cyclic CD approach (CCD++) [Yu *et al.* (2012)] demonstrate that the method can be relatively easy adapted for both multi-core and distributed environments.

Similarly to [Bell *et al.* (2007); Cichocki and Phan (2009)], in CCD++ the standard row-wise updating scheme is replaced with the column-wise scheme, where the same component of the latent space is updated for all users or items at once. Basically, the CCD++ approach transforms the optimization problem into a sequence of local rank-1 subproblems, where the columns of latent feature matrices are alternatively updated and each alternating step is distributed across several parallel workers. The authors also note that repeating several alternating update cycles within a single subproblem allows to achieve better results.

Implementation of the algorithm is straightforward in the shared-memory settings. In distributed environment it requires additional synchronization of column factors after a complete rank-1 update. However, the authors estimate an overall communication overhead of the approach to be not significantly larger than in the case of popular distributed SGD algorithm [Gemulla *et al.* (2011)]. Moreover, it provides a more stable convergence. As demonstrated by the authors the approach shows favorable performance and scales well in both distributed and shared-memory environments.

2.6.4. *Parallel SVD*

As has been previously discussed, computation of SVD relies on the Lanczos procedure, which requires only matrix-vector products and can be made very efficient with the help of broadly available linear algebra kernels, such as Intel MKL or ARPACK. Internally, the computations are performed by calling highly optimized BLAS/LAPACK routines, tuned for a better utilization of hardware capabilities on the shared memory devices.

Implementation of the algorithm in the distributed setup is typically achieved via the distribution of the Gram matrix-vector product (assuming A is a tall matrix):

$$(A^T A)\boldsymbol{v} = \sum_i \boldsymbol{a_i}(\boldsymbol{a_i}^T \boldsymbol{v}).$$

After gathering the result one can use a linear algebra kernel locally to compute the top leading right singular vectors V by the virtue of an eigendecomposition of the Gram matrix. After that the matrix-matrix product AV can also be obtained in a distributed manner. The result is then collected and fed into the standard SVD to finally get the leading left singular vectors U and the corresponding singular values Σ. The main bottlenecks of the process are parallel data reads and communication overheads incurred by the matrix products. As has been demonstrated in [Gittens *et al.* (2016)] the scaling is very sensitive to implementation details of the distributed calculations and requires a careful investigation to achieve a better scalability.

One of the ways to achieve a better performance in both shared-memory and distributed setups is to replace exact SVD with its approximate variant, such as *Randomized SVD* [Halko *et al.* (2011)]. A higher computational efficiency of the algorithm comes at the cost of a less accurate result. This, however, is not a stopper as the exact rating prediction is not the main focus of the majority of recommender systems. Moreover, the quality of approximation can be improved by a higher rank. We refer the reader to Chapter 11 for further details.

2.6.5. *Hyper-parameters tuning*

Due to differences in convergence properties, the methods discussed above require various levels of involvement during the model selection process. Some of the methods are less demanding with respect to the hyper-parameters' choice, others exhibit more sensitive behavior (see Sensitivity column in Table 2.1).

Apparently, SVD can be treated as the most convenient method in this regard. Indeed, it only requires to tune a single parameter — the rank of the decomposition [Cremonesi *et al.* (2010)]. Moreover, due to the optimality of the algorithm, once the PureSVD model is computed for some rank value r, one can immediately obtain a model of any rank $r' < r$ simply by truncating the factor matrices to the first r' components and without any extra computations.

Both CD and ALS depend on at least one extra parameter related to regularization. However, the choice of its value in some reasonable range does not significantly affect the quality of the resulting model and the initialization may play a more important role due to potential abundance of local minima. However, WRMF methods introduce additional parameters related to the weighting scheme and require a careful tuning.

Lastly, SGD-based methods are the most sensitive to both initialization and hyper-parameters tuning. This is especially true for the learning rate [Zhuang *et al.* (2013)] and many practical implementations employ additional adaptive techniques [Bottou (2012)] to automatically select a more appropriate value depending on the convergence pattern and the distance from a minimum.

2.7. Conclusion

This chapter gives an overview of the most popular and widely spread matrix factorization techniques used in collaborative filtering models. It provides the key concepts related to the problem formulation, learning methods, tuning of models and their practical applications. Due to an outstanding composability of the MF approach it allows to address a great number of problems and challenges arising in recommender systems that go far beyond simple rating prediction task.

MF methods allow to naturally incorporate additional sources of information and impose specific constraints on the latent feature space, offering more meaningful interpretations and a better quality of recommendations. The algorithms offer various trade-offs between simplicity, computational efficiency, flexibility in tuning, online scenarios support and quality of recommendations. This remains up to a practitioner to validate the choice of a particular model based on a domain of application, available infrastructure and business requirements.

Even though we have tried to cover the major aspects of building MF models, the topic itself is very broad and evolving. New factorization models as well as more efficient modifications of the old ones are constantly being developed and successfully implemented in various application domains. Therefore, the material presented here should be treated more as a starting point for further research and exploration, rather than as a universal collection of techniques and methods, suitable for solving any recommendation problem. The next chapters of the book provide a greater level of details on many particular CF topics, which also include valuable examples for the MF approach.

References

Adomavicius, G. and Tuzhilin, A. (2005). Toward the next generation of recommender systems: A survey of the state-of-the-art and possible extensions, *IEEE transactions on knowledge and data engineering* **17**, 6, pp. 734–749.

Amatriain, X. and Agarwal, D. (2016). Tutorial: Lessons learned from building real-life recommender systems, in *Proceedings of the 10th ACM Conference on Recommender Systems* (ACM), pp. 433–433.

Amatriain, X., Pujol, J. M. and Oliver, N. (2009a). I like it... i like it not: Evaluating user ratings noise in recommender systems, in *International Conference on User Modeling, Adaptation, and Personalization* (Springer), pp. 247–258.

Amatriain, X., Pujol, J. M. and Oliver, N. (2009b). I like it... i like it not: Evaluating user ratings noise in recommender systems, in *International Conference on User Modeling, Adaptation, and Personalization* (Springer), pp. 247–258.

Bachrach, Y., Finkelstein, Y., Gilad-Bachrach, R., Katzir, L., Koenigstein, N., Nice, N. and Paquet, U. (2014). Speeding up the xbox recommender system using a euclidean transformation for inner-product spaces, in *Proceedings of the 8th ACM Conference on Recommender systems* (ACM), pp. 257–264.

Barjasteh, I., Forsati, R., Masrour, F., Esfahanian, A.-H. and Radha, H. (2015). Cold-start item and user recommendation with decoupled completion and transduction, in *Proceedings of the 9th ACM Conference on Recommender Systems* (ACM), pp. 91–98.

Bayer, I., He, X., Kanagal, B. and Rendle, S. (2017). A generic coordinate descent framework for learning from implicit feedback, in *Proceedings of the 26th International Conference on World Wide Web* (International World Wide Web Conferences Steering Committee), pp. 1341–1350.

Bell, R., Koren, Y. and Volinsky, C. (2007). Modeling relationships at multiple scales to improve accuracy of large recommender systems, in *Proceedings of the 13th ACM SIGKDD international conference on Knowledge discovery and data mining* (ACM), pp. 95–104.

Bell, R. M. and Koren, Y. (2007). Scalable collaborative filtering with jointly derived neighborhood interpolation weights, in *Data Mining, 2007. ICDM 2007. Seventh IEEE International Conference on* (IEEE), pp. 43–52.

Berry, M. W., Dumais, S. T. and O'Brien, G. W. (1995). Using linear algebra for intelligent information retrieval, *SIAM review* **37**, 4, pp. 573–595.

Bertsekas, D. P. (1999). *Nonlinear programming* (Athena scientific Belmont).

Billsus, D. and Pazzani, M. J. (1998). Learning collaborative information filters, in *Icml*, Vol. 98, pp. 46–54.

Blondel, M., Fujino, A., Ueda, N. and Ishihata, M. (2016). Higher-order factorization machines, in *Advances in Neural Information Processing Systems*, pp. 3351–3359.

Bottou, L. (2012). Stochastic gradient descent tricks, in *Neural networks: Tricks of the trade* (Springer), pp. 421–436.

Brand, M. (2002). Incremental singular value decomposition of uncertain data with missing values, *Computer Vision ECCV 2002*, pp. 707–720.

Chapelle, O. and Wu, M. (2010). Gradient descent optimization of smoothed information retrieval metrics, *Information retrieval* **13**, 3, pp. 216–235.

Chen, T., Zheng, Z., Lu, Q., Zhang, W. and Yu, Y. (2011). Feature-based matrix factorization, *arXiv preprint arXiv:1109.2271*.

Chen, Y., Zhao, X. and de Rijke, M. (2017). Top-n recommendation with high-dimensional side information via locality preserving projection, in *Proceedings of the 40th International ACM SIGIR Conference on Research and Development in Information Retrieval* (ACM), pp. 985–988.

Cichocki, A. and Phan, A.-H. (2009). Fast local algorithms for large scale nonnegative matrix and tensor factorizations, *IEICE transactions on fundamentals of electronics, communications and computer sciences* **92**, 3, pp. 708–721.

Cremonesi, P., Koren, Y. and Turrin, R. (2010). Performance of recommender algorithms on top-n recommendation tasks, in *Proceedings of the fourth ACM conference on Recommender systems*, pp. 39–46.

Das, A., Upadhyaya, I., Meng, X. and Talwalkar, A. (2017). Collaborative filtering as a case-study for model parallelism on bulk synchronous systems, in *Proceedings of the 2017 ACM on Conference on Information and Knowledge Management*, CIKM '17 (ACM, New York, NY, USA), pp. 969–977.

Deerwester, S., Dumais, S. T., Furnas, G. W., Landauer, T. K. and Harshman, R. (1990). Indexing by latent semantic analysis, *Journal of the American society for information science* **41**, 6, p. 391.

Desrosiers, C. and Karypis, G. (2011). A comprehensive survey of neighborhood-based recommendation methods, in *Recommender systems handbook* (Springer), pp. 107–144.

Eckart, C. and Young, G. (1936). The approximation of one matrix by another of lower rank, *Psychometrika* **1**, 3, pp. 211–218.

Ekstrand, M. D., Riedl, J. T., Konstan, J. A. *et al.* (2011). Collaborative filtering recommender systems, *Foundations and Trends® in Human–Computer Interaction* **4**, 2, pp. 81–173.

Fang, Y. and Si, L. (2011). Matrix co-factorization for recommendation with rich side information and implicit feedback, in *Proceedings of the 2nd International Workshop on Information Heterogeneity and Fusion in Recommender Systems* (ACM), pp. 65–69.

Frolov, E. and Oseledets, I. (2017). Tensor methods and recommender systems, *Wiley Interdisciplinary Reviews: Data Mining and Knowledge Discovery* **7**, 3.

Furnas, G. W., Deerwester, S., Dumais, S. T., Landauer, T. K., Harshman, R. A., Streeter, L. A. and Lochbaum, K. E. (1988). Information retrieval using a singular value decomposition model of latent semantic structure, in *Proceedings of the 11th annual international ACM SIGIR conference on Research and development in information retrieval* (ACM), pp. 465–480.

Gates, M., Anzt, H., Kurzak, J. and Dongarra, J. (2015). Accelerating collaborative filtering using concepts from high performance computing, in *Big Data (Big Data), 2015 IEEE International Conference on* (IEEE), pp. 667–676.

Gemulla, R., Nijkamp, E., Haas, P. J. and Sismanis, Y. (2011). Large-scale matrix factorization with distributed stochastic gradient descent, in *Proceedings of the 17th ACM SIGKDD international conference on Knowledge discovery and data mining* (ACM), pp. 69–77.

Gittens, A., Devarakonda, A., Racah, E., Ringenburg, M., Gerhardt, L., Kottalam, J., Liu, J., Maschhoff, K., Canon, S., Chhugani, J. *et al.* (2016). Matrix factorizations at scale: A comparison of scientific data analytics in spark and c+ mpi using three case studies, in *Big Data (Big Data), 2016 IEEE International Conference on* (IEEE), pp. 204–213.

Goldberg, K., Roeder, T., Gupta, D. and Perkins, C. (2001). Eigentaste: A constant time collaborative filtering algorithm, *Information Retrieval* **4**, 2, pp. 133–151.

Golub, G. H. and Van Loan, C. F. (2012). *Matrix computations*, 4th edn. (The Johns Hopkins University Press).

Halko, N., Martinsson, P.-G. and Tropp, J. A. (2011). Finding structure with randomness: Probabilistic algorithms for constructing approximate matrix decompositions, *SIAM review* **53**, 2, pp. 217–288.

Hastie, T., Mazumder, R., Lee, J. D. and Zadeh, R. (2015). Matrix completion and low-rank svd via fast alternating least squares, *Journal of Machine Learning Research* **16**, pp. 3367–3402.

Hidasi, B. and Tikk, D. (2016). Speeding up als learning via approximate methods for context-aware recommendations, *Knowledge and Information Systems* **47**, 1, pp. 131–155.

Hong, L., Doumith, A. S. and Davison, B. D. (2013). Co-factorization machines: modeling user interests and predicting individual decisions in twitter, in *Proceedings of the sixth ACM international conference on Web search and data mining* (ACM), pp. 557–566.

Hsieh, C.-K., Yang, L., Cui, Y., Lin, T.-Y., Belongie, S. and Estrin, D. (2017). Collaborative metric learning, in *Proceedings of the 26th International Conference on World Wide Web* (International World Wide Web Conferences Steering Committee), pp. 193–201.

Hu, Y., Koren, Y. and Volinsky, C. (2008). Collaborative filtering for implicit feedback datasets, in *Data Mining, 2008. ICDM'08. Eighth IEEE International Conference on* (IEEE), pp. 263–272.

Juan, Y., Zhuang, Y., Chin, W.-S. and Lin, C.-J. (2016). Field-aware factorization machines for ctr prediction, in *Proceedings of the 10th ACM Conference on Recommender Systems* (ACM), pp. 43–50.

Kim, D. and Yum, B.-J. (2005). Collaborative filtering based on iterative principal component analysis, *Expert Systems with Applications* **28**, 4, pp. 823–830.

Konstan, J. A. and Riedl, J. (2012). Recommender systems: from algorithms to user experience, *User modeling and user-adapted interaction* **22**, 1-2, pp. 101–123.

Koren, Y. (2008). Factorization meets the neighborhood: a multifaceted collaborative filtering model, in *Proceedings of the 14th ACM SIGKDD international conference on Knowledge discovery and data mining* (ACM), pp. 426–434.

Koren, Y., Bell, R. and Volinsky, C. (2009). Matrix factorization techniques for recommender systems, *Computer* **42**, 8.

Lee, J., Lee, D., Lee, Y.-C., Hwang, W.-S. and Kim, S.-W. (2016). Improving the accuracy of top-n recommendation using a preference model, *Information Sciences* **348**, pp. 290–304.

Liang, D., Altosaar, J., Charlin, L. and Blei, D. M. (2016). Factorization meets the item embedding: Regularizing matrix factorization with item co-occurrence, in *Proceedings of the 10th ACM conference on recommender systems* (ACM), pp. 59–66.

Liu, T.-Y. *et al.* (2009). Learning to rank for information retrieval, *Foundations and Trends® in Information Retrieval* **3**, 3, pp. 225–331.

Mnih, A. and Salakhutdinov, R. R. (2008). Probabilistic matrix factorization, in *Advances in neural information processing systems*, pp. 1257–1264.

Nguyen, J. and Zhu, M. (2013). Content-boosted matrix factorization techniques for recommender systems, *Statistical Analysis and Data Mining* **6**, 4, pp. 286–301.

Nikolakopoulos, A. N., Kalantzis, V. and Garofalakis, J. D. (2015). Eigenrec: An efficient and scalable latent factor family for top-n recommendation, *arXiv preprint arXiv:1511.06033*.

Novikov, A., Trofimov, M. and Oseledets, I. (2016). Exponential machines, *arXiv preprint arXiv:1605.03795*.

Pan, R., Zhou, Y., Cao, B., Liu, N. N., Lukose, R., Scholz, M. and Yang, Q. (2008). One-class collaborative filtering, in *Data Mining, 2008. ICDM'08. Eighth IEEE International Conference on* (IEEE), pp. 502–511.

Pan, X., Lam, M., Tu, S., Papailiopoulos, D., Zhang, C., Jordan, M. I., Ramchandran, K. and Ré, C. (2016). Cyclades: Conflict-free asynchronous machine learning, in *Advances in Neural Information Processing Systems*, pp. 2568–2576.

Paterek, A. (2007). Improving regularized singular value decomposition for collaborative filtering, in *Proceedings of KDD cup and workshop*, Vol. 2007, pp. 5–8.

Pilászy, I. and Tikk, D. (2009). Recommending new movies: even a few ratings are more valuable than metadata, in *Proceedings of the third ACM conference on Recommender systems* (ACM), pp. 93–100.

Pilászy, I., Zibriczky, D. and Tikk, D. (2010). Fast als-based matrix factorization for explicit and implicit feedback datasets, in *Proceedings of the fourth ACM conference on Recommender systems* (ACM), pp. 71–78.

Recht, B., Re, C., Wright, S. and Niu, F. (2011). Hogwild: A lock-free approach to parallelizing stochastic gradient descent, in *Advances in neural information processing systems*, pp. 693–701.

Rendle, S. (2010). Factorization machines, in *Data Mining (ICDM), 2010 IEEE 10th International Conference on* (IEEE), pp. 995–1000.

Rendle, S., Freudenthaler, C., Gantner, Z. and Schmidt-Thieme, L. (2009). Bpr: Bayesian personalized ranking from implicit feedback, in *Proceedings of the twenty-fifth conference on uncertainty in artificial intelligence* (AUAI Press), pp. 452–461.

Rendle, S. and Schmidt-Thieme, L. (2008). Online-updating regularized kernel matrix factorization models for large-scale recommender systems, in *Proceedings of the 2008 ACM conference on Recommender systems* (ACM), pp. 251–258.

Rendle, S. and Schmidt-Thieme, L. (2010). Pairwise interaction tensor factorization for personalized tag recommendation, in *Proceedings of the third ACM international conference on Web search and data mining* (ACM), pp. 81–90.

Roy, S. and Guntuku, S. C. (2016). Latent factor representations for cold-start video recommendation, in *Proceedings of the 10th ACM Conference on Recommender Systems* (ACM), pp. 99–106.

Salakhutdinov, R. and Mnih, A. (2008). Bayesian probabilistic matrix factorization using markov chain monte carlo, in *Proceedings of the 25th international conference on Machine learning* (ACM), pp. 880–887.

Sallinen, S., Satish, N., Smelyanskiy, M., Sury, S. S. and Ré, C. (2016). High performance parallel stochastic gradient descent in shared memory, in *Parallel and Distributed Processing Symposium, 2016 IEEE International* (IEEE), pp. 873–882.

Sarwar, B., Karypis, G., Konstan, J. and Riedl, J. (2000). Application of dimensionality reduction in recommender system-a case study, Tech. rep., Minnesota Univ Minneapolis Dept of Computer Science.

Saveski, M. and Mantrach, A. (2014). Item cold-start recommendations: learning local collective embeddings, in *Proceedings of the 8th ACM Conference on Recommender systems* (ACM), pp. 89–96.

Schelter, S., Boden, C., Schenck, M., Alexandrov, A. and Markl, V. (2013). Distributed matrix factorization with mapreduce using a series of broadcast-joins, in *Proceedings of the 7th ACM conference on Recommender systems* (ACM), pp. 281–284.

Schelter, S., Satuluri, V. and Zadeh, R. (2014). Factorbird-a parameter server approach to distributed matrix factorization, *arXiv preprint arXiv:1411.0602*.

Schnabel, T., Swaminathan, A., Singh, A., Chandak, N. and Joachims, T. (2016). Recommendations as treatments: Debiasing learning and evaluation, *arXiv preprint arXiv:1602.05352*.

Shi, Y., Karatzoglou, A., Baltrunas, L., Larson, M., Hanjalic, A. and Oliver, N. (2012a). Tfmap: optimizing map for top-n context-aware recommendation, in *Proceedings of the 35th international ACM SIGIR conference on Research and development in information retrieval* (ACM), pp. 155–164.

Shi, Y., Karatzoglou, A., Baltrunas, L., Larson, M., Oliver, N. and Hanjalic, A. (2012b). Climf: learning to maximize reciprocal rank with collaborative less-is-more filtering, in *Proceedings of the sixth ACM conference on Recommender systems* (ACM), pp. 139–146.

Shi, Y., Larson, M. and Hanjalic, A. (2010). Mining mood-specific movie similarity with matrix factorization for context-aware recommendation, in *Proceedings of the workshop on context-aware movie recommendation* (ACM), pp. 34–40.

Singh, A. P. and Gordon, G. J. (2008). Relational learning via collective matrix factorization, in *Proceedings of the 14th ACM SIGKDD international conference on Knowledge discovery and data mining* (ACM), pp. 650–658.

Steck, H. (2010). Training and testing of recommender systems on data missing not at random, in *Proceedings of the 16th ACM SIGKDD international conference on Knowledge discovery and data mining* (ACM), pp. 713–722.

Takács, G., Pilászy, I., Németh, B. and Tikk, D. (2007). Major components of the gravity recommendation system, *ACM SIGKDD Explorations Newsletter* **9**, 2, pp. 80–83.

Takács, G., Pilászy, I., Németh, B. and Tikk, D. (2008). Investigation of various matrix factorization methods for large recommender systems, in *Proceedings of the 2nd KDD Workshop on Large-Scale Recommender Systems and the Netflix Prize Competition* (ACM), p. 6.

Takács, G., Pilászy, I. and Tikk, D. (2011). Applications of the conjugate gradient method for implicit feedback collaborative filtering, in *Proceedings of the fifth ACM conference on Recommender systems* (ACM), pp. 297–300.

Takács, G. and Tikk, D. (2012). Alternating least squares for personalized ranking, in *Proceedings of the sixth ACM conference on Recommender systems* (ACM), pp. 83–90.

Udell, M., Horn, C., Zadeh, R., Boyd, S. *et al.* (2016). Generalized low rank models, *Foundations and Trends® in Machine Learning* **9**, 1, pp. 1–118.

Vandereycken, B. (2013). Low-rank matrix completion by riemannian optimization, *SIAM Journal on Optimization* **23**, 2, pp. 1214–1236.

Verstrepen, K., Bhaduriy, K., Cule, B. and Goethals, B. (2017). Collaborative filtering for binary, positiveonly data, *ACM SIGKDD Explorations Newsletter* **19**, 1, pp. 1–21.

Weimer, M., Karatzoglou, A., Le, Q. V. and Smola, A. J. (2008). Cofi rank-maximum margin matrix factorization for collaborative ranking, in *Advances in neural information processing systems*, pp. 1593–1600.

Weston, J., Bengio, S. and Usunier, N. (2011). Wsabie: Scaling up to large vocabulary image annotation, in *IJCAI*, Vol. 11, pp. 2764–2770.

Yan, Y., Tan, M., Tsang, I. W., Yang, Y., Zhang, C. and Shi, Q. (2015). Scalable maximum margin matrix factorization by active riemannian subspace search, in *IJCAI*, pp. 3988–3994.

Yang, J. and Gittens, A. (2015). Tensor machines for learning target-specific polynomial features, *arXiv preprint arXiv:1504.01697*.

Yu, H.-F., Bilenko, M. and Lin, C.-J. (2017). Selection of negative samples for one-class matrix factorization, in *Proceedings of the 2017 SIAM International Conference on Data Mining* (SIAM), pp. 363–371.

Yu, H.-F., Hsieh, C.-J., Si, S. and Dhillon, I. (2012). Scalable coordinate descent approaches to parallel matrix factorization for recommender systems, in *Proceedings of the 2012 IEEE 12th International Conference on Data Mining*, ICDM '12 (IEEE), pp. 765–774.

Yun, H., Yu, H.-F., Hsieh, C.-J., Vishwanathan, S. and Dhillon, I. (2014). Nomad: Non-locking, stochastic multi-machine algorithm for asynchronous and decentralized matrix completion, *Proceedings of the VLDB Endowment* **7**, 11, pp. 975–986.

Zha, H. and Zhang, Z. (2000). Matrices with low-rank-plus-shift structure: partial svd and latent semantic indexing, *SIAM Journal on Matrix Analysis and Applications* **21**, 2, pp. 522–536.

Zhang, C. and Ré, C. (2014). Dimmwitted: A study of main-memory statistical analytics, *Proceedings of the VLDB Endowment* **7**, 12, pp. 1283–1294.

Zhou, Y., Wilkinson, D., Schreiber, R. and Pan, R. (2008). Large-scale parallel collaborative filtering for the netflix prize, *Lecture Notes in Computer Science* **5034**, pp. 337–348.

Zhuang, Y., Chin, W.-S., Juan, Y.-C. and Lin, C.-J. (2013). A fast parallel sgd for matrix factorization in shared memory systems, in *Proceedings of the 7th ACM conference on Recommender systems* (ACM), pp. 249–256.

Chapter 3

Cutting-Edge Collaborative Recommendation Algorithms: Deep Learning

Balázs Hidasi

Gravity R&D
Budapest, Hungary
balazs.hidasi@gravityrd.com

After achieving great performance on complex domains, such as computer vision, natural language processing, speech recognition, reinforcement learning, etc., recommender systems research also turned towards deep learning recently. Although the recsys community was a late adopter and has been only focusing on deep learning since late 2015 / early 2016, it has already became one of the most promising and thus highly popular topic within the research community. Deep learning has the potential to revolutionize the field, just as it did with other domains. The aim of this chapter is to overview recent advances in the intersection of deep learning and recommender systems.

This chapter gives a short introduction to deep learning. However, the in-depth discussion of deep learning and its recent advances is outside of the scope of this book. Therefore, we refer the interested readers to deep learning books, such as [Goodfellow *et al.* (2016)]. The chapter also contains a short review on the history of deep learning methods in recommenders and then discusses the four main lines of current research in deep learning based recommender algorithms. As the field is evolving rapidly, the chapter focuses on the main research directions and seminal works instead of listing lots of algorithms. We mostly focus on algorithms working on implicit feedback, due to its high relevancy to practitioners. Finally, the chapter ends with a short guide for practitioner who would like to delve into this topic.

3.1. Introduction to deep learning

Deep learning is a subclass of machine learning algorithms in which data is processed through several non-linear layers, organized in complex and usually deep structures, where each layer learns a representation of the data; building on the representation learned by layers lower in the hierarchy. Lower layers learn basic, higher levels more abstract representations. For example, with image input lower layers learn to detect lines and edges; and higher levels recognize objects or object groups. Deep models are generally neural networks and deep learning is often referred to as the second resurgence (or third golden era) of neural networks. Deep models are modular in nature, therefore they can easily deal with heterogenous data.

3.1.1. *Neuron and neural networks*

The basic building block of neural networks is the artificial neuron. This unit is inspired by neurons of the human brain, yet it is but a rough abstraction of that. An artificial neuron has multiple inputs and one output. If the weighted sum of its inputs is over a certain threshold, the neuron fires, i.e. it emits a positive value. Otherwise it is deactivated (emits zero or a negative value). The neuron is capable of executing a very basic form of pattern recognition: it is activated for certain inputs and it is inactive for others. The input weights determine for which input the neuron activates. The inputs are weighted by these weights. In order to keep the unit differentiable, thresholding is replaced by an activation function. There are different activation functions. Originally sigmoid or sigmoid like functions were used. These later were replaced by functions that suit deep networks better. The output of a neuron can be expressed as:

$$y = f\left(\sum_{i=1}^{n} w_i x_i + b\right). \tag{3.1}$$

Neural networks are neurons connected to each other. Neurons can be connected to each other in many different ways. The most common architectures are the feedforward, recurrent and convolutional networks. Here we introduce the feedforward network. Recurrent neural networks will be introduced in Section 3.6.1. Convolutional networks are not discussed in detail, but are mentioned throughout the chapter.

In feedforward neural networks, neurons are organized in subsequent feedforward layers. The output of every neuron from the k-th layer is connected to the the input of every neuron in the subsequent, $(k+1)$-th layer. There are no connections between the neurons of the same layer. The first layer of the network is the input layer. Here the values are set to whatever we want the network to process. The final layer is the output

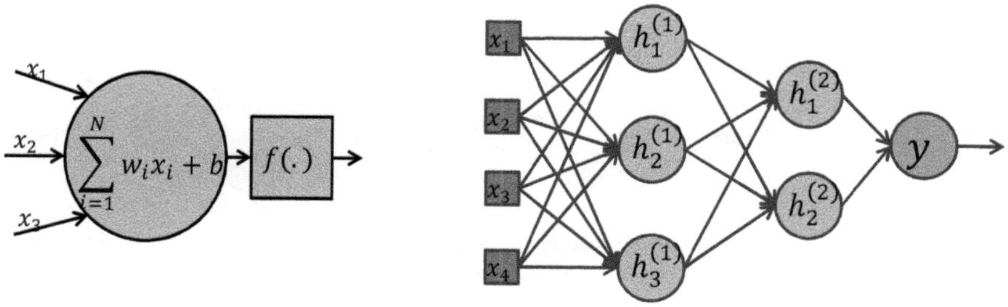

Fig. 3.1. Artificial neuron (left) and feedforward network composed of artificial neurons (right).

layer. The values produced by this layer are the final output of the network. Layers between the input and the output layers are called hidden layers.

The input is propagated through the layers of the network until it reaches the last layer. This is called the forward pass. A neural network with n hidden layer can be described as follows (using vector notation).

$$h^{(k)} = \begin{cases} x & k = 0 \text{ //input layer} \\ f_k\left(W^{(k)}h^{(k-1)} + b\right) = f_k\left(s^{(k)}\right) & k > 0 \end{cases} \tag{3.2}$$
$$y = h^{(n+1)} \text{ //output layer}$$

where x is the input of the network, $h^{(k)}$ is the output of the k-th hidden layer, f_k is the activation function of layer k (applied elementwise to each neuron in the layer) and $W^{(k)}$ is the weight matrix between the $(k-1)$-th and k-th layers. y is the output of the network.

The training of the network is done by setting the weights. There are multiple ways to do this, with gradient descent being the most widely used approach. For learning with gradient descent, we need to define a loss between the output of the network and its target. The gradient of this loss w.r.t. every weight is computed and weights are updated in the negative gradient direction to lower the loss. The update can be done efficiently by propagating the error back through the network as:

$$\frac{\partial L}{\partial W^{(k)}} = \left(h^{(k-1)}\right)^T \left(d^{(k)} \circ f_k'\left(s^{(k)}\right)\right)$$
$$d^{(k)} = \begin{cases} \frac{\partial L}{\partial y} & k = n+1 \\ d^{(k)} = \left(W^{(k+1)}\right)^T \left(d^{(k+1)} \circ f_{k+1}'\left(s^{(k+1)}\right)\right) & k \le n \end{cases} \tag{3.3}$$

where L is the loss and $d^{(k)}$ is the propagated error in the k-th layer.

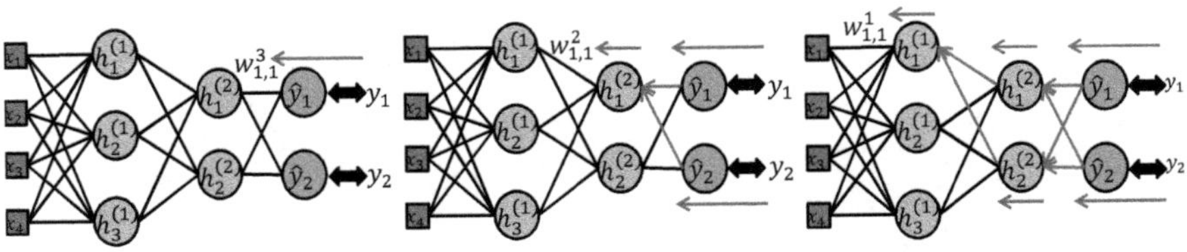

Fig. 3.2. Backpropagation: propagating the error backwards through the network to compute the gradient of the loss w.r.t. each weight.

Feedforward networks with at least one hidden layer are universal approximators [Hornik *et al.* (1989)], meaning that if they are large enough they can approximate any function with arbitrarily small error. Unfortunately, generally no tight bounds are known for what it means to be large enough. However, it was shown that the number of neurons required for good approximations is much lower if more hidden layers are used, i.e. if the network is deep[1]. Deep networks are much more expressive than shallow ones, thus they can be used more effectively for machine learning problems in practice.

3.1.2. *Techniques for easier training of deep networks*

The idea of training deep neural networks has been around since the 1980s. But their widespread application started only in the 2010s. There are several reasons for this. Besides being computationally expensive and thus requiring modern hardware, and only showing their advantages when trained on large enough databases — both of which became available only recently — training deep networks was considered to be hard. However, thanks to research breakthroughs from the early 2010s, most of the typical problems were alleviated, thus training of deep networks became more robust. Therefore a wider audience of researchers and engineers are now able to apply deep learning for various tasks. Below we highlight some of the most important techniques.

Activation functions: Although, there are other methods as well, the training of neural networks is usually done by stochastic gradient descent. The gradient of the loss w.r.t. the weights of the network is computed and weights are modified in the directions of the negative gradient to lower the loss. This can be done efficiently by propagating the prediction error back

[1]More accurately: for certain families of functions there exists a K number, so that if the network has at least K layers, the number of required neurons is polynomial in the size input and exponential otherwise.

Table 3.1. Properties of common activation functions.

Activation	$f(x)$	$f'(x)$	Parameters
Sigmoid	$f(x) = \frac{1}{1+e^{-x}}$	$f'(x) = f(x)(1 - f(x))$	–
ReLU	$f(x) = \max(x, 0)$	$f'(x) = \begin{cases} 1, & x \geq 0 \\ 0, & x < 0 \end{cases}$	–
Leaky ReLU	$f(x) = \begin{cases} x, & x \geq 0 \\ \alpha x, & x < 0 \end{cases}$	$f'(x) = \begin{cases} 1, & x \geq 0 \\ \alpha, & x < 0 \end{cases}$	$0 < \alpha < 1$
ELU	$f(x) = \begin{cases} x, & x \geq 0 \\ \alpha(e^x - 1), & x < 0 \end{cases}$	$f'(x) = \begin{cases} 1, & x \geq 0 \\ f(x) + \alpha, & x < 0 \end{cases}$	$\alpha > 0$
SELU	$f(x) = \lambda \begin{cases} x, & x \geq 0 \\ \alpha(e^x - 1), & x < 0 \end{cases}$	$f'(x) = \lambda \begin{cases} 1, & x \geq 0 \\ f(x) + \alpha, & x < 0 \end{cases}$	$\alpha > 0$ $\lambda > 1$

through the network. During this propagation each unit that is inbetween the output and the weight multiplies the gradient by the derivative of its activation function. The classic activation functions were sigmoid-like functions. The derivative of the sigmoid is $f(x)(1 - f(x))$, which is (a) close to zero if the absolute value of x is large; (b) and its maximum (without scaling) is below 1. Multiplying the error with small numbers coming from the derivative of the activation function makes the gradient disappear after passing through a few layers, thus weights of the lower layers will not be updated and become useless. This is called the vanishing gradient problem. This issue is solved with the introduction of new activations functions. The Rectified Linear Unit (ReLU) [Nair and Hinton (2010)] is a piecewise linear function which returns x if $x \geq 0$ and 0 otherwise. While ReLU solves the vanishing gradient problem, it can burn out, i.e. it can stuck in its zero state. The Leaky ReLU [Maas *et al.* (2013)] solves this issue, by returning αx if $x < 0$, where $0 < \alpha < 1$ is a parameter, but this comes at the cost of having a non noise-robust deactivation state. This problem is alleviated by the Exponential Linear Unit (ELU) [Clevert *et al.* (2015)]. This activation function can be enhanced further by scaling (Scaled ELU, SELU [Klambauer *et al.* (2017)]), which also has self-normalizing properties. The properties of the aforementioned activation functions is summarized in Table 3.1, activations and their derivatives are shown on Figure 3.3.

Dropout regularization: Deep neural network models have large capacity, which makes them susceptible for overfitting. Therefore it is crucial to use some form of regularization in neural models. Unfortunately,

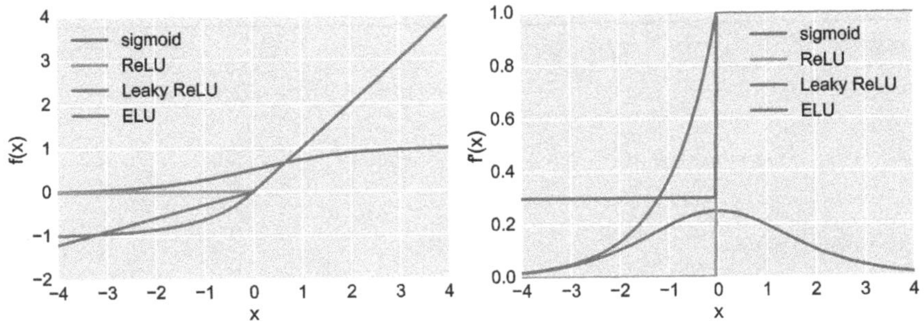

Fig. 3.3. Activation functions (left) and their derivatives (right).

the classic ℓ_2 regularization turned out to be not that efficient, thus other heuristics, such as limiting the activations of neurons were needed. Regularizing neural networks was tedious and was considered hard. Dropout [Srivastava *et al.* (2014)] is a simple regularization technique that works really well in practice. During training, in each pass — i.e. forward and backward propagation — a fraction of the neurons is randomly disabled. The selected neurons are completely dropped from the network for the duration of the whole pass. In order for the average activations to be unchanged, the activation of the remaining units are scaled up. (E.g. if we randomly drop 50% of the neurons, activations are multiplied by 2 for the remaining units.) Using dropout results in training slightly different networks that share their weights in each update step. The trained model is thus basically an ensemble of several networks. It is also assumed that due to randomly dropped units, each individual neuron learns a useful representation of the data, because it can not rely on other neurons being there. This effectively lowers model capacity and forces neurons to ignore observational noise in the training data.

Mini-batch training: As we already mentioned, the most common training method of neural networks is gradient descent, i.e. updating weights in the direction of the negative gradient of the loss. The two extreme ends of gradient descent is batch- and stochastic gradient descent (SGD). With batch gradient descent, all data points are propagated forward in the network first, then weights are updated based on the gradient of the average loss. This results in less frequent updates, but the error gradients are representative of the training data and computations are easily parallelizable. On the other hand, with the stochastic gradient method, the weight update is based on the gradient of the loss of only one (randomly selected) data point. This way, the updates are very frequent, but the gradients are noisy (not representative of the full training set) and parallelization of the training is not trivial. SGD usually converges faster than batch gradient. Mini-batch training is in between these two extremes and combines the advantages of both worlds. In mini-batch training, each update is based on the gradient of the average loss of several — few tens to few hundreds — (randomly selected) data points. The resulting gradients are less noisy than those of SGD and their computations can be parellelized, but updates are more frequent than with batch gradient. While it is somewhat counterintuitive, noisy gradients are also not necessarily bad for training. Small amount of noise on the updates serves as a regularizer during training, lowering the overfitting of the model. Without update noise and large enough model capacity, the network can be trained to fit the training data perfectly (including the observational noise) and does not generalize well. Of course, too much update noise slows down the learning and results in a suboptimal model. The size of the mini-batch can be used to control the amount of update noise indirectly. Smaller mini-batches tend to introduce more noise to the updates. Smaller mini-batches (few tens of examples) perform better for most tasks, but the optimal mini-batch size is primarily dependent on the data.

Adaptive learning rates: Gradient descent methods have their shortcomings. They can easily get stuck in narrow valleys or saddle points of the error surface and they are very sensitive to the choice of the learning rate parameter. Getting stuck can be handled by using momentum with SGD, i.e. having the direction (and size) of the previous update influence the current update. Selecting a good learning rate is more problematic. Large learning rates can make the training diverge. On the other hand small steps are slowing it down. Moreover, larger steps are beneficial during the beginning of the training, while the network is still far from the optimal configuration; but as it gets closer to the optimum later during training, smaller learning rates are more appropriate. The standard way of handling this problem was to use heuristics based learning rate schedules, e.g. halving the learning rate after every N updates; or utilizing a separate validation set and lowering the learning rate when the validation error not decreases. These heuristics are still sensitive to the choice of the initial learning rate and do not consider that certain parameters are not updated with the same frequency. If a weight is updated more frequently than another, it is effectively ahead in the training process and probably requires smaller updates. Adaptive learning rate methods aim to solve the aforementioned issues. They collect aggregated information on the updates of every model parameter and use those to scale the learning rate *per weight* according to what is optimal, based on different aspects of the estimated curvature. Adagrad [Duchi *et al.* (2011)] is the simplest method, scaling back learning rates with the square root of the summed square of gradients in the training so far. This results in constantly decreasing learning rates, but the schedule is different per model parameter. RMSProp [Tieleman and Hinton (2012)] is a similar method, but with a decay parameter. Adadelta [Zeiler (2012)] and Adam [Kingma and Ba (2014)] are more complex, yet popular adaptive learning methods based on estimation of the curvature. Table 3.2 shows how these methods compute the step size (η is the learning rate parameter, L_t is the loss during the t-th update). Adaptive learning methods are not too sensitive to the initial learning rate and make the training more robust. The downside is that additional memory is required for the accumulated gradient information whose size is in the order of the number of weights. Recently it was also shown [Wilson *et al.* (2017)] that on certain (artificial) dataset that popular adaptive learning methods converge to a point with arbitrarily high error. It is to be seen how this affects training on real data, but for the time being, adaptive learning methods still seem to perform better than standard SGD in practice.

Table 3.2. Adaptive learning methods (η is the learning rate parameter, L_t is the loss during the t-th update).

Method	Accumulators	Step size	Parameters
Adagrad	$G_t = G_{t-1} + (\nabla L)^2$	$\frac{\eta}{\sqrt{G_t+\epsilon}}$	$0 < \eta$
RMSProp	$G_t = \gamma G_{t-1} + (1-\gamma)(\nabla L)^2$	$\frac{\eta}{\sqrt{G_t+\epsilon}}$	$0 < \eta$ $0 < \gamma < 1$
Adadelta	$G_t = \gamma G_{t-1} + (1-\gamma)(\nabla L)^2$ $\Delta_t = \gamma \Delta_{t-1} + (1-\gamma)\left(\frac{\sqrt{\Delta_{t-1}+\epsilon}}{\sqrt{G_{t-1}+\epsilon}} \nabla L_t\right)^2$	$\frac{\sqrt{\Delta_{t-1}+\epsilon}}{\sqrt{G_t+\epsilon}}$	$0 < \gamma < 1$
Adam	$M_t = \beta_1 M_{t-1} + (1-\beta_1)\nabla L_t$ $V_t = \beta_2 V_{t-1} + (1-\beta_2)(\nabla L_t)^2$	$\frac{\eta \frac{M_t}{1-\beta_1^t}}{\sqrt{\frac{V_t}{1-\beta_2^t}+\epsilon}}$	$0 < \beta_1 < 1$ $0 < \beta_2 < 1$ $0 < \eta$

3.2. Deep learning for recommender systems

3.2.1. *Brief history of deep learning in recommender systems*

Neural networks were used for the rating prediction task during the Netflix Prize. [Salakhutdinov *et al.* (2007)] utilized Restricted Boltzmann Machines for the task and [Paterek (2007)] formalized asymmetric matrix factorization as a simple neural network. Between 2009 and 2015, neural models for recommendations were few and far between, even though deep learning in general started to gain a lot of attention even in 2012. Near the end of 2015, few papers applying deep learning for recommendations were published, signaling the beginning of the deep learning era for recommenders. The deep learning boom in RecSys started in 2016 with several papers continuing the exploration of the topic. The Deep Learning for Recommender Systems (DLRS) workshop series [Karatzoglou *et al.* (2016); Hidasi *et al.* (2017)] also started in that year, giving a boost to the research of the field. By the end of 2016, four main research directions have formed, which will be discussed in this chapter:

- Learning item embeddings,
- Deep collaborative filtering,
- Using content features in recommenders,
- Session-based recommendations with recurrent neural networks.

The trend continued in 2017, and now having multiple deep learning recommender system papers at top tier conferences is not rare. Since the this field is very young — most of the papers were published after 2015 — we expect significant contributions to come in the near future as well. This chapter discusses the state of the field and concentrates on papers published before September 2017; with the exception of Section 3.6.4, which is about the very recently formed research direction on deep generative models for recommendations. We focus on the practically important top-N recommendation task and networks that address this task. Since there is now a forming consensus in the recsys community to move away from rating prediction — due to its uselessness for the practical case of top-N recommendations [Cremonesi *et al.* (2010)] — deep learning models for rating prediction won't be discussed in this chapter.

3.2.2. *Advantages and drawbacks*

Deep learning has a lot of potential for recommender systems.

- **Feature extraction:** Deep learning methods are great feature extractors and can easily deal with unstructured data, such as image, audio, video or text. Therefore it is possible to use content directly in hybrid recommender algorithms instead of metadata.
- **Modularity:** Deep networks are modular, the output of a network can be easily put onto the input of another. This allows not just for replacing older feature extractors by the newest state-of-the-art ones, but it also makes it easy to deal with heterogeneous and multimodal data.
- **Sequence processing:** Certain networks — such as RNNs and certain types of CNNs — can deal with sequential data very well. This is especially useful in recommender systems, where the main source of information is the user history. It is also easier to do dynamic behaviour modeling and to follow the changes in the users' interest over time.
- **Complex models:** The complex models of deep neural networks can capture the nuances in user behaviour better than simple models, such as matrix factorization. Therefore they can produce both better user and item representations and better recommendations overall.

However, using deep learning comes at a cost. The complex models require lot of computational power. Training times can be expected to increase even with GPU support and parallelization. These models are completely black box, therefore explainability is low, but this is also true for matrix factorization models. Due to modularity, inexperienced engineers/researchers can quickly put together a very complex network, which can have subtle bugs, that might not become apparent immediately. Also, these very big networks can become unscalable if one is not careful enough during the planning phase. While these issues can be averted by careful planning, seeing that scalability is often dismissed or misunderstood in recommender systems research (not just in the deep learning branch), one should be cautious, before implementing a recommender network found in the literature. Due to writing very efficient code requires lot of work and strong programming skills, deep learning algorithms are generally written in deep learning frameworks. While these frameworks are really well written, they can still impose bottlenecks on the performance of the algorithm. Also, the productivization of such research code and its integration into a live recommender system is currently not straightforward.

3.3. Learning item embeddings

Item embeddings (or representations) are better known for recommender system experts as (latent) item feature vectors. Irrespective of the name, it is a vector assigned to each item in a latent feature space in such a way that the vectors of similar items are close to each other, and those of dissimilar items are further away. Item embeddings in themselves can be used to compute item similarities and to perform item-to-item recommendations (i.e. related products). They can be also used as input of other algorithms.

This task is by no means new to the recommendation domain. For example, matrix factorization also learns item and user representations/embeddings in a latent space. However, we will discuss a different approach that is based on the word2vec algorithm, which was originally used in natural language processing (NLP) for learning word embeddings.

3.3.1. *Word2vec*

Word2vec [Mikolov *et al.* (2013a)] is simple shallow neural model for learning word embeddings. Even though the model is shallow it is considered to be part of the "deep learning" toolkit. There are two different word2vec models, the Continuous Bag Of Words (CBOW) and the skip-gram model. Both of these models build upon the same data, but use it in different ways. The data is generated from natural text. A sliding window is used to process the text. In the middle of the sliding window is the actual or target word, while the previous and next T words are called context (words).

At the beginning of the training, each word is assigned with a random embedding vector. These vectors are modified during training in order to minimize the loss of the network.

CBOW model: The CBOW model uses the context words to predict the target word. The embedding vectors of the context words are averaged and the resulting context vector is put into a classifier whose desired output is the ID of the target word. The classifier uses the softmax transformation on its scores and cross-entropy to compute the loss. See the architecture on Figure 3.4.

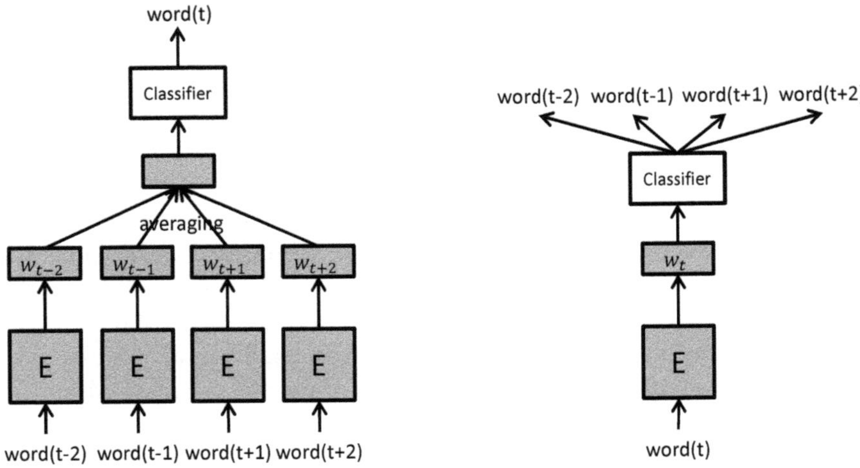

Fig. 3.4. The CBOW (left) and skip-gram (right) word2vec models.

Skip-gram model: The skip-gram model works in the opposite direction as the CBOW. It uses the embedding of the target word as an input of a multilabel classifier to predict the words of the context (see Figure 3.4). The classifier has the same structure as the one in the CBOW model. A variation of the skip-gram model moves away from the multilabel classification and uses a randomly selected word from the context as the desired output.

If the vocabulary is big, learning (for both CBOW and skip-gram) can become slow, because the algorithm requires to compute scores for all words in every step. This problem can be alleviated by using hierarchical softmax instead of the softmax transformation, that does not require the computation of all scores; or by sampling a few negative words to each target word and using the softmax on this sample only [Mikolov *et al.* (2013b)].

3.3.1.1. *Paragraph2vec*

The CBOW model was extended to learn paragraph or document embeddings besides word embeddings. Here, the ID of the document is also part of the context (besides the context words). The documents are also assigned with embedding vectors that is averaged with the embeddings of the context words to compute the input of the classifier (see Figure 3.5).

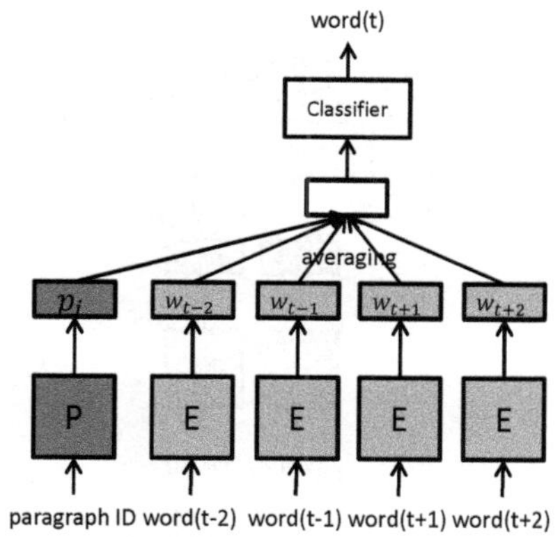

Fig. 3.5. The paragraph2vec model.

3.3.2. *Prod2vec*

Models from NLP are often adapted for the recommendation task. This is no different for word2vec. [Grbovic *et al.* (2015)] proposed to use the skip-gram word2vec model to learn item embeddings and coined the method prod2vec[2]. Words are replaced by items and documents with user histories. The underlying algorithm remains the same, but we learn item embeddings instead of word embeddings. The same authors also propose to use paragraph2vec to learn user representations simultaneously, by replacing the paragraph ID with the user ID.

The prod2vec model was later extended to handle multimodal information in order to learn item embeddings of higher quality. [Vasile *et al.* (2016)] utilizes item metadata besides the item ID. The input of the network is an item ID and the accompanying metadata. The output is the context items and metadata (i.e. other items from the user history and their accompanying metadata). The loss of the network is the combination of losses of different prediction tasks, performed simultaneously (see Figure 3.6):

- From the actual item predict the item context.
- From the actual item predict the metadata context.
- From the actual metadata predict the actual item.
- From the actual metadata predict the item context.
- From the actual metadata predict the metadata context.

[2]Later, the same idea was also published by different authors under the name of item2vec [Barkan and Koenigstein (2016)].

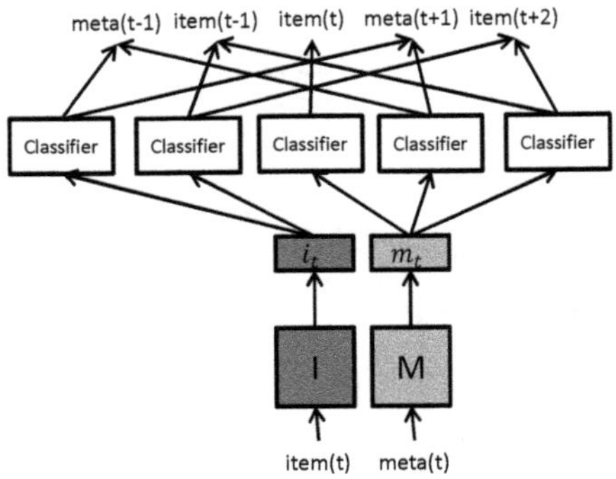

Fig. 3.6. The meta-prod2vec model.

The model was later further extended to consider other aspects of the item, such as the product image in [Nedelec *et al.* (2017)].

3.4. Deep collaborative filtering

It is straightforward to apply deep learning for collaborative filtering. Even the popular matrix factorization model can be represented as a very simple neural network. The input of this simple network is the one-hot encoded user ID[3]. The input-to-hidden weight matrix is the user feature matrix, thus the hidden layer is the user feature vector. The activation function of the hidden layer is the identity function. The hidden-to-output weight matrix is the item feature matrix and the output is the preference of the user over the items.

Several methods approach modeling from the perspective of making this simple model more neural network like. The MV-DNN model keeps the dot product model between user and item features, but runs both the user and item features through several non-linear layers. The MV in the name stands for multi-view, so the model is capable of handling multiple domains at once, by having a separate subnets for items from all domain (Figure 3.7, left).

The implicit CF-NADE model [Zheng *et al.* (2016)] adds non-linearities after both the hidden layer and the output. Following the classic work of [Hu *et al.* (2008)], implicit feedback is separated into a preference value (1 if the user has any events on the items, 0 otherwise) and a confidence/weight

[3]Since the model is symmetrical, this can be done the other way around, i.e. inputting the item ID and getting the preference of all users on this item on the output.

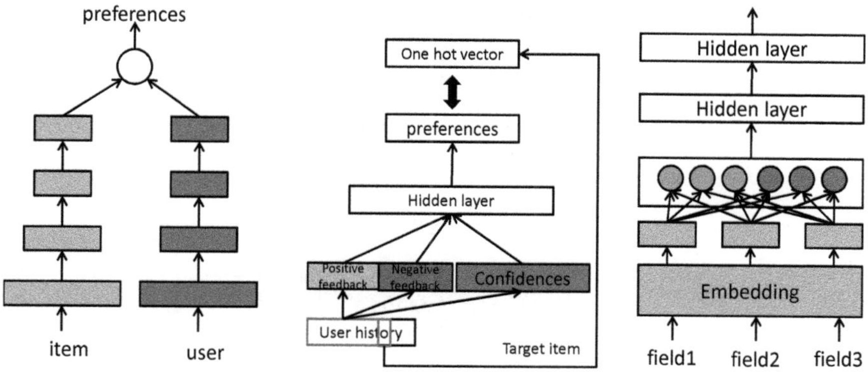

Fig. 3.7. **Left:** MV-DNN, **center:** CF-NADE, **right:** PNN.

value ($\alpha \text{supp}_{(u,i)} + 1$, $\alpha \gg 1$). The confidence is used to upweight positive feedback during the computation of the hidden state, and to increase the importance of the loss on items corresponding to positive feedback. The network takes a subset of the user's item on its input and predicts one other item from the user's history (Figure 3.7, center).

The Product Neural Network (PNN) [Qu *et al.* (2016)] is a deep version of the well-known Factorization Machine (FM) [Rendle (2012)]. First, the categorical features are embedded and the cross-products of these embeddings are computed. This cross product is fed to additional non-linear layers. The network is used for CTR prediction, thus its output is a single number, but modifying it for preference prediction would also be trivial (Figure 3.7, right).

There are many other methods (e.g. models that also consider the evolution of the user and/or item features, such as [Song *et al.* (2016); Dai *et al.* (2016)]), as deep collaborative filtering was one of the most popular research lines in the last few years. These are not discussed in detail here. Instead we introduce an important subclass of deep CF models that build on denoising autoencoders.

3.4.1. *Autoencoder based approaches*

The autoencoder is a neural model with one hidden layer. The expected output of the network is the same as its input. The autoencoder learns a (compressed) representation of the data in its hidden layer. This in itself can be already used for lossy compression. However, when compression is not the main task and thus learning the identity mapping is less useful,

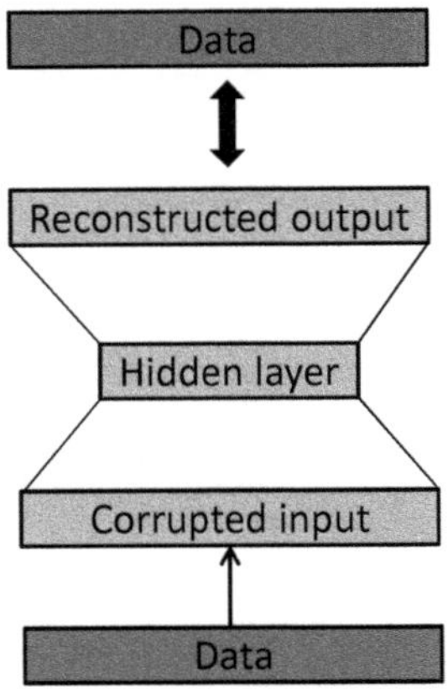

Fig. 3.8. Denoising autoencoder.

the generalization capabilities of the network can be increased by feeding it with noisy data and expecting it to restore the original data on its output (see Figure 3.8). This network is called the Denoising Autoencoder (DAE) [Vincent *et al.* (2008)]. During training, the network is fed with corrupted data. Corrupting the data can take different forms, e.g. white noise can be added to it, or randomly selected coordinates can be set to zero, etc. The task of the network is to denoise this data and reconstruct the uncorrupted data on its output. Both DAE and standard autoencoders can be made deep, by stacking multiple networks on the top of each other. DAE works better on real life data, due to its generalization and noise filtering capabilities.

Recommenders based on the denoising autoencoder model are fed with corrupted user histories during training. The user's history is transformed into a vector of ones and zeros for items the user did and did not consume, respectively. Then some of the ones are randomly set to zero. The network is requested to restore the original vector. For implicit data, the two most well known DAE based recommenders are Collaborative Deep Learning (CDL) [Wang *et al.* (2015)] and Collaborative Deep Autoencoder (CDAE) [Wu *et al.* (2016)].

CDL uses tags and item metadata instead of the item ID. Working on the metadata space has the advantage of being resistant to the item cold-start problem. These vectors are also denser than item based ones, which suits the training process of the DAE better. The underlying network is the Bayesian variation of the stacked denoising autoencoder.

CDAE works on item IDs and uses the standard DAE, but extends the input with a user node and the hidden layer with a bias node.

3.5. Direct use of content

Even though generally collaborative filtering (CF) has the best performance for recommendations ([Pilászy and Tikk (2009)]) it also has its weaknesses. The most well known is the cold-start problem, as CF can not recommend for new users or recommend new items. Therefore, practical recommenders usually use hybrid algorithms. These have a CF core and auxiliary parts based on other data for when CF is not applicable. The most common approach is to use the combination of CF and content-based filtering (CBF).

There are different ways to extend CF algorithms with CBF (also see Chapter 4). Common techniques include:

- **Weighting:** The CF and CBF algorithms predict preferences separately and their predictions are then weighted to produce the final scores. This is a very basic solution and often suboptimal, because the scales and distributions of the scores of the two methods differ significantly.
- **Cascading:** The CF algorithm is used for prediction, when it can give (sensible) predictions and the CBF is used otherwise. Even though this is a simple approach, it is surprisingly effective in practice.
- **Initializing:** The features extracted from the content can be used to initialize item features. E.g. factorizing the item–metadata matrix and using the item features in a standard user–item matrix factorization as starting values of the item matrix. See e.g. [Hidasi and Tikk (2013)].
- **Joint training:** There is a single item representation that is part of two jointly trained optimization problems. E.g. factorize the user–item and the item–metadata matrices simultaneously and restrict item features to be the same in both factorizations, see [Singh and Gordon (2008)].
- **Regularizing:** Add a term to the loss function on the difference between the content features and the item feature vectors computed by the CF part. Content features are either pretrained or learned simultaneously. This way the item features are regularized to be close to their content-based counterparts, thus the model can work

fairly well with new items, that only have content features. Can be viewed as the relaxed version of joint training, because content a behaviour based item features do not have to be the same, but should be close. See e.g. [Hu *et al.* (2012)].

- **Mapping content:** Build a separate model that maps the content to latent features produced by CF. New items then can be instantly assigned with a feature vector and can be recommended. See e.g. [Van den Oord *et al.* (2013)].

Despite of its name, CBF traditionally does not work with the content directly, but rather utilizes the item metadata. The reason for not using content is that automatically extracting features from images, video, audio and sometimes even from text is challenging, while manual feature selection is tedious, domain specific and often suboptimal.

Deep learning methods greatly improved upon how we can handle unstructured data, such as content. Convolutional Neural Networks (CNN) can extract high level features from images and videos or any image-like data automatically; Recurrent Neural Networks (RNN) can summarize natural text and can deal with any kind of sequential data. All of the approaches mentioned above can work without any modifications with pretrained content features instead of metadata based features. Moreover, due to the modularity of deep models, it is very easy to use heterogenous, multimodal data, i.e. different aspects of the items, in the same model. Content features can be pretrained (with optional fine tuning during training) or the entire network can be trained end-to-end.

In the rest of this section we describe one model for using image, music and text data, respectively.

Product images: [He and McAuley (2016)] uses visual features from a pretrained CNN to extend the behaviour based item feature vector (i.e. the visual and behaviour based item feature vectors are concatenated). Beforehand, the high dimensionality of the visual feature vector is reduced by multiplying it with a separate weight matrix. The preference model consists of the dot product of the (full) item feature vector and the user feature vector, an item bias and visual biases. All but the visual features are learned. Learning is performed using the BPR algorithm [Rendle *et al.* (2009)]. See Figure 3.9.

Music recommenders: In order to overcome the item cold-start problem, [Van den Oord *et al.* (2013)] uses deep convolutional networks to predict latent item features. First, matrix factorization is performed on non

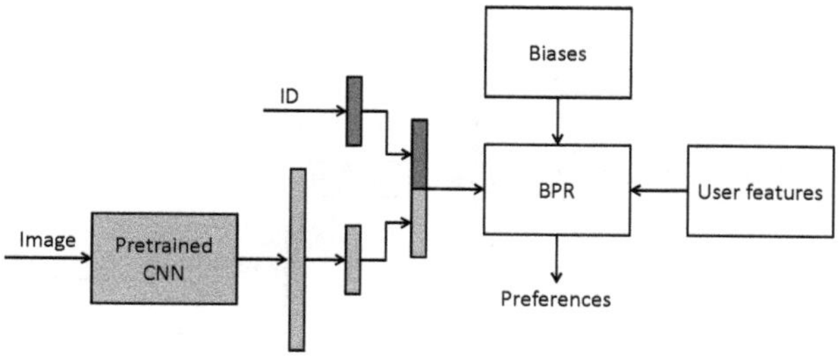

Fig. 3.9. The visual BPR model using pretrained visual features for fashion recommendation.

cold-start items. Then a CNN is trained on the time-frequency representation of the songs. The target of this network is the item feature vectors from the matrix factorization. This way, when a new song becomes available in the system, it can be immediately assigned with a feature vector based on its content; and this feature vector is in the same latent space as the end result of the factorization. See Figure 3.10.

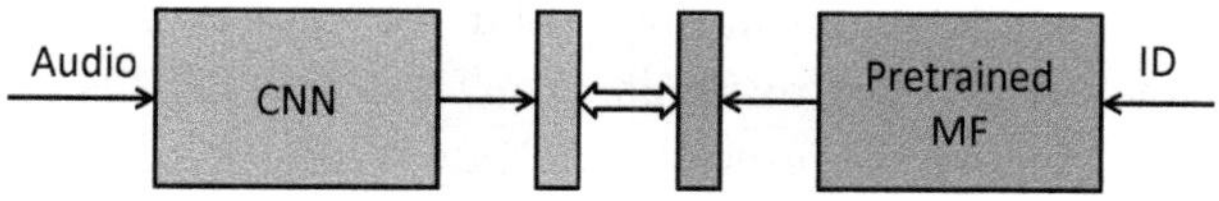

Fig. 3.10. Matching content to pretrained item features to overcome item cold-start.

Textual content: The content of scientific articles is processed word-by-word by a two layered recurrent neural network in [Bansal *et al.* (2016)]. The hidden states of the RNN are averaged and result in the feature vector of the processed text. The item feature vector is the sum of the text vector and and a behaviour based item feature vector (i.e. embedding of the ID). The item feature vector then is used to predict which users would like the item as well as what tags would suite the item well. The whole model is trained end-to-end, i.e. the parameters of the text processing part are learned at the same time as other parts of the model. See Figure 3.11

3.6. Session-based recommendations

The traditional recommendation setup assumes easily identifiable users who have long histories of interactions with the items; and this history encodes the static or slowly changing preferences of the user. This is the classic item-to-user recommendation scenario, where collaborative filtering methods,

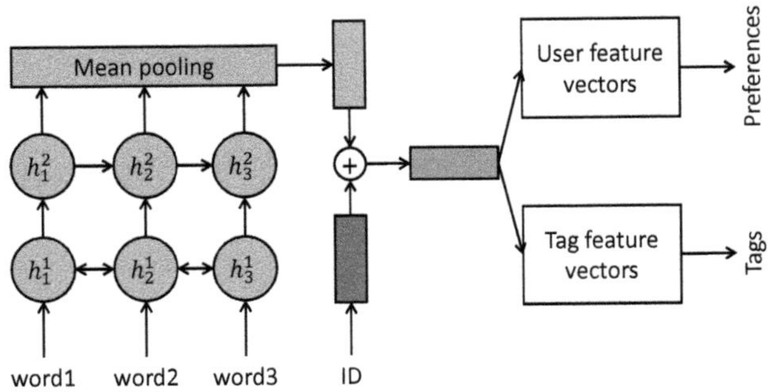

Fig. 3.11. Text processing with recurrent model to enhance recommendations.

such as matrix factorization, shine. This setup is mostly accurate for domains, such as movie or book recommendations, where the basis of consuming an item is heavily preference based and where the identification of the users can be done in a sensible way. However, the domains of recommender systems are much more diverse and there exist several application areas where these assumptions are not valid. In some domains, user identification is not trivial. For example, YouTube like OTT video services generally do not require the users to log in. While there are other ways for user identification, the lack of well defined user accounts makes it less reliable, because the user can use multiple devices to access the service and the same device can be used by multiple users. There are domains where the majority of the users actively try to avoid identification, e.g. on adult content sites.

Even if we have a more or less reliable way of user identification, treating all of a user's events as a consistent history that signals global preferences of the user is oftentimes incorrect. This is especially true when the service has an item catalog of very diverse items accommodating for different uses, and if either (a) consuming an item does not entail huge commitment; or (b) choosing an item is primarily need-based as opposed to preference-based. For example, OTT video services are often used in different ways by the same user in different sessions: e.g. listening to music during work, watching cartoons with their kids on weekends, watching videos in a certain topic that is actually interesting for the user, etc. Users of e-commerce stores of electronic devices, household appliances, etc. are usually looking for a certain kind of product that they need at that time. If a user is looking for a fridge, they won't be interested in televisions and the next time they come back looking for televisions, their past preferences over fridges will be of little importance to determine the best product to recommend[4]. The same holds for classified sites and online marketplaces. The common denominator of these three examples is that the user history is composed of distinct, very loosely connected sub histories. This results in a permanent user cold-start problem, where the recommender can not rely on past information when the user starts a new session. Most of the applications of recommenders fall into these "non-traditional" categories and not into domains where long term consistent user histories are available. But even in the classical domains, user-based approaches have their shortcomings. (1) They do not account for user intent, e.g. if the user wants to buy something for himself, or wants to buy someone else a present or just browses. (2) The user history is only consistent w.r.t. interactions signaling higher commitment (e.g. purchase, watching a full movie), but the amount of other interactions (e.g. viewing a product page) is much larger and much more erratic and it is often the only type of information available for lots of users.

To accommodate for the aforementioned challenges, the recommender system needs to quickly adapt to the actual intent of the user, thus it should be able to update its internal model after each user action and refine the recommendation content as the session progresses. The topic of session-based recommenders revels around this problem. Session-based recommendation is not just a single task, as there are multiple ways to approach the problem and the task should be selected to match the actual needs of the domain.

The two most prevalent tasks of session-based recommendations are next click prediction and intent prediction. The former aims to predict the next item in the session given the previous one, while the goal of the latter is to guess the user's intent in the actual session, given the session (e.g. the user wants to buy something, or just browses). A session is a sequence of events in close proximity, produced by the same user. If no explicit session identifier is available, it is common practice to assume the session's end if the time between two consecutive events of the user is above a threshold. Many session-based algorithms only use the items of the events — similarly to most user-based collaborative filtering algorithms — but there are algorithms capable of utilizing additional information of the events (e.g. time, context) or the items (e.g. metadata).

[4]Some information on the user's preferences can be derived from past sessions on global preference parameters — from more trivial things like price range to hidden connections unrevealed by transfer learning — but this is also beyond the scope of classic methods.

Session-based recommendation was addressed even before the advent of deep learning, but papers on this topics were few and far between and methods did not work well in the pure session-based scenario, due to the complexity of the topic and classic methods being unable to deal with sequences appropriately. Therefore, in practice the de facto solution for this problem was the traditional item-to-item recommendation [Linden *et al.* (2003)], in which items are recommended based on their similarity to or co-occurrences with the item of the user's last event while the rest of the session is ignored. While this approach works fairly well, this solution is heavily non-personalized (each user gets the same items on a given item page). It is also less accurate, since it does not consider earlier events from the session. Deep learning revitalized session-based recommendations by applying Recurrent Neural Networks (RNNs) — the de facto neural network family for dealing with sequences — for session data.

3.6.1. *Recurrent Neural Networks*

RNNs are for modeling sequential data. The key difference between a recurrent and a feedforward layer is the existence of an internal hidden state, which serves as the summary of the processed sequence. The hidden state is the function of the actual input and the previous value of the hidden state. The hidden state in the vanilla RNN is computed as:

$$h_t = f\left(Wx_t + Uh_{t-1} + b\right) \tag{3.4}$$

here $f(\cdot)$ is the activation function (usually tanh or logistic sigmoid). Recurrent layers are trained using backpropagation through time (BPTT) [Williams and Zipser (1995)]: a sequence is propagated forward through the RNN for T steps and the gradient w.r.t. the weights is computed considering all T steps. Another way to think about this is that the RNN is unrolled for T steps and the resulting network is trained with backpropagation, but the weights between the layers are shared (see Figure 3.12). The RNN layer by itself is a deep architecture, but multiple RNN layers can be used after one another to process the data.

RNNs have a vast and quickly expanding literature and discussing all aspects is beyond the scope of this book. Here we only briefly discuss the gated RNNs and some of the common architectures.

The traditional RNNs suffer from the vanishing and exploding gradient problem. The reason of vanishing/exploding gradients in RNNs is due to the recursion in their structure. Thus it is present for non-saturating

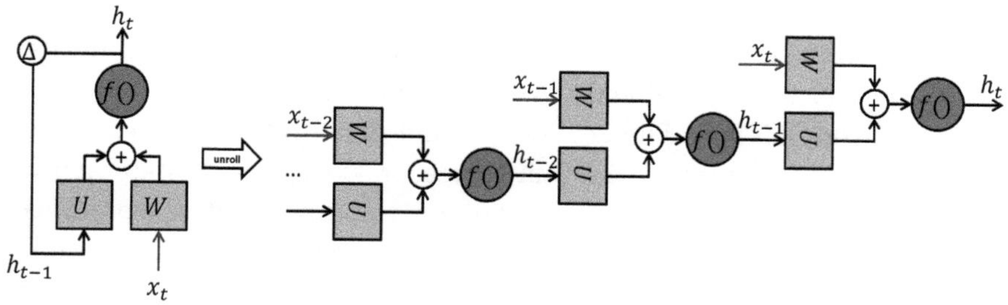

Fig. 3.12. Unrolling an RNN unit.

activation functions as well, unlike the vanishing gradient problem of feed-forward networks mentioned in Section 3.1.2.

The gradient w.r.t. the first input of the sequence (x_1) can be computed as[5]:

$$\frac{\partial h_t}{\partial x_1} = \frac{\partial h_t}{\partial h_{t-1}} \cdot \frac{\partial h_{t-1}}{\partial h_{t-2}} \cdots \frac{\partial h_2}{\partial h_1}\frac{\partial h_1}{\partial x_1} = \left(U^{t-1}W\right)^T \tag{3.5}$$

Depending on the spectral norm of U:

- $||U||_2 < 1 \rightarrow$ The effect of x_1 on h_t is negligible if $t \gg 1$. The network quickly forgets and the hidden state is determined by only the few most recent inputs (vanishing gradients).
- $||U||_2 > 1 \rightarrow$ Minor changes (e.g. noise) in x_1 significantly changes h_t if $t \gg 1$. The weight updates will be large and the network becomes unstable (exploding gradients).

There are different ways to overcome this problem:

- **Gradient clipping:** Define a threshold and if the gradient exceeds this anywhere, scale back the gradient so that its maximal value equals to the threshold value. This solves the exploding gradient problem, but dos not address the vanishing gradient, so the network still forgets quickly.
- **Unitary RNNs:** Make sure that the spectral norm of U is 1, so gradients will never vanish nor explode. This can be done by using unitary matrices for U. However, weight updates ruin the unitary property and restoring it after each update is costly and can also slow down convergence. Therefore it is better to compose U as

[5]For the sake of clarity it is assumed that f is the identity function. The same can be shown for other activation functions.

the product of parameterized simple unitary matrices and update these parameters directly [Arjovsky *et al.* (2016)].

- **Gated RNNs:** The key idea behind gated RNNs is to compose the hidden state as the sum of the previous hidden state and a difference (delta) which depends on the input and the previous hidden state (as opposed to computing the new value directly): $h_t = h_{t-1} + \Delta h_t = h_{t-1} + f(x_t, h_{t-1})$. This in itself solves the vanishing gradient, but makes the network unable to forget. To overcome this and to solve the exploding gradient, gates are introduced to compute the information flow. There are different variants of this idea with the *Long Short Term Memory* (LSTM) [Hochreiter and Schmidhuber (1997)] and the *Gated Recurrent Unit* (GRU) [Cho *et al.* (2014)] being the two most well known.

3.6.1.1. *Long Short Term Memory*

The LSTM [Hochreiter and Schmidhuber (1997)] is the first and most well known of the gated RNNs. Here, the hidden state is decomposed into two: (1) a memory cell (c_t), which serves as an intermediate representation (state updates are done here); and (2) a hidden state. The hidden state's value is read from the memory cell, masked by the output gate. This decouples the computations of the gates from the internal state representation (i.e. the output gate can break the information flow, without having to reset the state representation). The LSTM has three gates. Each gate is the function of the actual input and the previous hidden state and their value is always between 0 and 1.

- **Input gate:** Controls what portion of the delta is added to the memory cell. If it is zero, the current input will not contribute to the memory cell (and to the hidden state).
- **Forget gate:** Controls how much of the previous memory cell is kept. If it is zero, the network completely forgets the previously accumulated knowledge of the sequence.
- **Output gate:** Controls, how much of the memory cell's value is written into the hidden state.

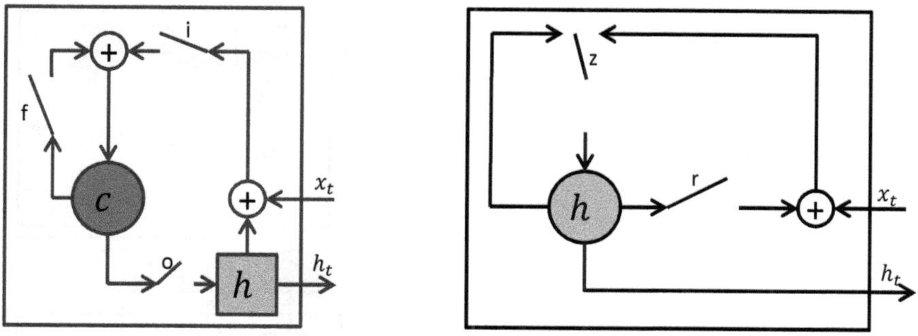

Fig. 3.13. Gated units: LSTM (left) and GRU (right).

The LSTM model (also see Figure 3.13):

$$\begin{aligned}
i_t &= \sigma\left(W_i x_t + U_i h_{t-1} + b_i\right) \\
f_t &= \sigma\left(W_f x_t + U_f h_{t-1} + b_f\right) \\
o_t &= \sigma\left(W_o x_t + U_o h_{t-1} + b_o\right) \\
c_t &= f_t \circ c_{t-1} + i_t \circ \tanh\left(W_c x_t + U_c h_{t-1} + b_c\right) \\
h_t &= o_t \circ \tanh\left(c_t\right)
\end{aligned} \tag{3.6}$$

3.6.1.2. *Gated Recurrent Unit*

GRU [Cho *et al.* (2014)] is a simpler gated RNN, in which there is no memory cell, the input and forget gates are merged into an update gate (z_t), and the information flow is broken by a reset gate (r_t) (instead of the output gate and separating the state representation from the hidden state). It is reported to perform similarly to the LSTM. The GRU model (also see Figure 3.13):

$$\begin{aligned}
r_t &= \sigma\left(W_r x_t + U_r h_{t-1} + b_r\right) \\
z_t &= \sigma\left(W_z x_t + U_z h_{t-1} + b_z\right) \\
h_t &= z_t \circ h_{t-1} + (1 - z_t) \circ \tanh\left(W_h x_t + U_h (r_t \circ h_{t-1}) + b_h\right)
\end{aligned} \tag{3.7}$$

3.6.1.3. *Common architectures with RNNs*

The RNN layer can be used in different architectures and in different ways, depending on the task. The most basic approach is to put a feedforward layer on the top of the RNN layer to perform prediction (e.g. classification) from the value of the hidden state. But even in this simple case we can decide the timing of the prediction (e.g. do we want to predict in each step

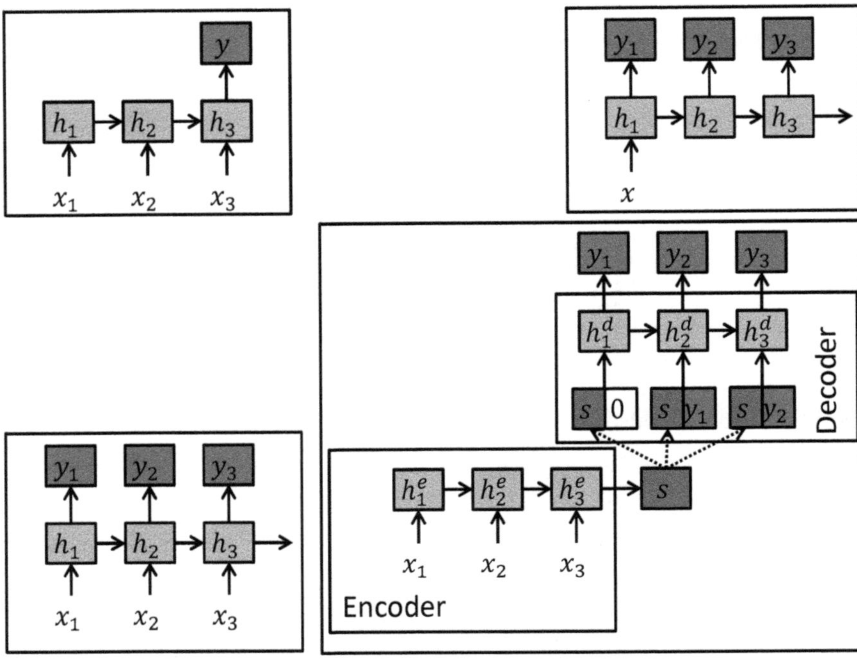

Fig. 3.14. RNN architectures corresponding to common tasks: sequence to value (top left), sequence generation (top right), next symbol prediction (bottom left), sequence to sequence learning (bottom right).

or only at the end of the sequence, etc.). See the most common tasks and the appropriate RNN architecture below and on Figure 3.14:

- **Sequence to value:** Assign a value/label to a sequence after it is processed in its entirety. E.g. time series classification, document classification.
- **Sequence generation:** Starting from a single input generate a sequence, i.e. decode the input into a sequence. E.g. image caption generation.
- **Next symbol prediction:** Predict the next symbol/value in the sequence after processing the actual input. E.g. next video frame prediction, next click prediction.
- **Sequence to sequence learning:** Given a sequence, generate another sequence. The difference to next symbol prediction is that the output sequence is generated after the input sequence is fully processed and not simultaneously. While it is possible to do this

with one RNN, it is more efficient to use an encoder-decoder architecture [Sutskever *et al.* (2014)] in which one RNN encodes the input sequence by processing it and another RNN uses this code to seed the output generation. The performance of such architectures can be improved by using attention mechanisms [Mnih *et al.* (2014)] which allows the decoder to look at the hidden state's value of the input network at all time steps at once. E.g. machine translation.

3.6.2. *The GRU4Rec algorithm*

The GRU4Rec algorithm [Hidasi *et al.* (2015); Hidasi and Karatzoglou (2017)] is a GRU based network adapted to the recommendation domain[6]. Sessions here are represented as sequences of item IDs. The input of the network is the actual item ID in the session and the task is to predict the next item of the session (next click prediction). Items IDs are represented by one-hot vectors. The network architecture consists of one or more GRU layers preceded by an optional embedding layer and followed by a feedforward layer that predicts the likelihood of each item to be the next in the session (see Figure 3.15).

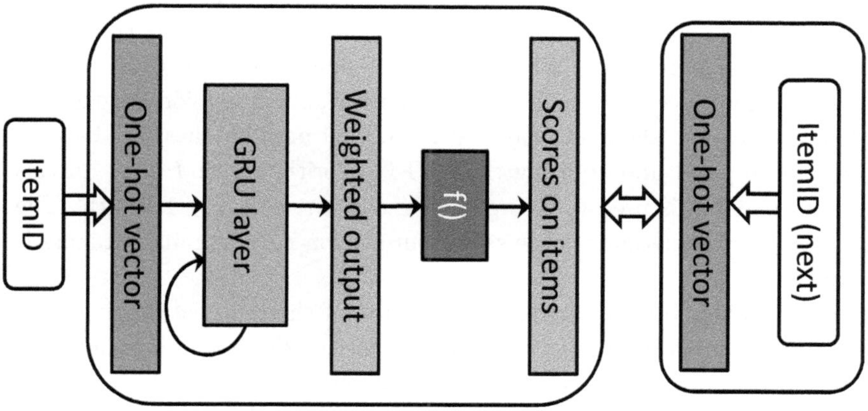

Fig. 3.15. Schematic of GRU4Rec.

While GRUs are excellent when it comes to dealing with sequences, the recommendation domain has several properties which need to be accounted for when designing an algorithm. One such speciality is session length: it both varies greatly (from 2 to several hundreds) and short sessions are very

[6]Code available: https://github.com/hidasib/GRU4Rec

common. In fact, sessions with 1 or 2 events are the most common in practice in many domains. This makes the use of sequence by sequence training and the BPTT algorithm hard. Also, the number of items (which correspond to the number of inputs and outputs of the network) can easily be in the millions (or at least several hundreds of thousands). Predicting for all items in each training step takes a lot of time; and due to another speciality of the domain, it is unacceptable for recommender algorithms to train for days, because they need to be retrained frequently to be able to handle new items and users.

Session-parallel mini-batches: Instead of using the BPTT algorithm on each session separately, GRU4Rec does one step updates on mini-batches assembled from multiple sessions. Since multiple steps are computed at once (one step for each mini-batch), multiple hidden states are needed to follow each example. At the start of the training the mini-batch is composed from the items of the first events of the first M sessions (M is the mini-batch size) and the desired outputs are the items of the next events of the same sessions. After the gradient update, the mini-batch is now composed of the items of the second events of the sessions and the desired outputs are tied to the third events. This continues until one or more of the sessions has no more events. When this happens, the corresponding hidden state is reset to zero and the next available session is put in the place of the completed session in the mini-batch, starting from its first event (see Figure 3.16). This training procedure fits data with high variance in session length well. Alternatively, the algorithm can learn from 2D mini-batches, i.e. batches of sequences, and update using the BPTT algorithm with $T > 1$. In this version, padding is used to fill up shorter sequences to $T + 1$ length. The accuracy of this methods is the same, but session-parallel mini-batching is faster.

Mini-batch based sampling: The size of the item catalog in a recommender system in practice is usually in the hundreds of thousands, and it is not uncommon to see item catalogs of several millions of items. During training, GRU4Rec does a prediction step in the order of the number of events in the training data. If it were to predict for all items in each step, the training would scale with the product of the number of events and items, which would result in poor scalability and slow training times in practical systems. To overcome this issue, GRU4Rec does not predict for all items, but only for the target items and for several other items (negative samples). This significantly speeds up training process at the cost of only approximating the loss. However, this is a widespread approach for many algorithms

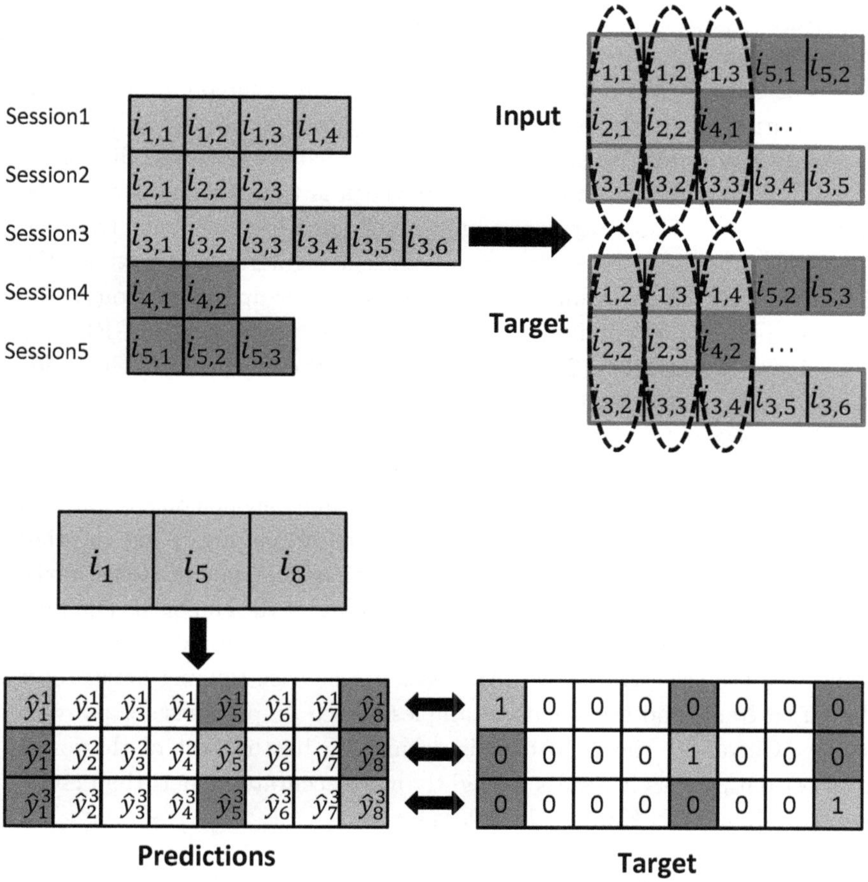

Fig. 3.16. **Top:** Session parallel mini-batches. The circled boxes form a mini-batch. **Bottom:** Mini-batch based sampling. Predictions are computed for the non-white items only.

in both recommenders and other domains. GRU4Rec introduces a very efficient sampling mechanism by using the items of the desired outputs of the other mini-batch examples as the negative items for the given mini-batch example[7] (see Figure 3.16). This step can be implemented very efficiently on GPU and it doesn't require additional sampling. These two properties make this step very fast. Another advantageous property is that the sampling

[7]E.g. $M = 3$ and the desired outputs in a step are item 4, 7 and 9. The target for the first example is item 4 with negative examples item 7 and 9; the target for the second example is item 7, with item 4 and 9 as negative examples and so on.

procedure samples negative items proportionally to their popularity. Popularity-based negative sampling is often better than uniform sampling due to (a) learning the item representations of more common items quicker; (b) capturing less of the popularity bias.

Additional sampling: Even though mini-batch based sampling is efficient in terms of computations, its drawback is tying sample size to the mini-batch size. Generally, lower mini-batch sizes are better for training, due to the gradient noise serving as an efficient regularization method. On the other hand, larger sample size allows for better approximation of the loss and better gradient updates. Therefore the v2 version of GRU4Rec [Hidasi and Karatzoglou (2017)] introduced additional sampling that adds shared negative samples to each example of the mini-batch. These additional samples are sampled in proportion to their support on the power of α ($0 \leq \alpha \leq 1$). By setting α, sampling can be balanced between uniform and popularity-based sampling. The optimal value depends on the dataset, but is usually around 0.5[8]. Mini-batch and additional sampling together can both quickly learn good item representations for popular items in the beginning of the training and fit for long tail items at the end of the training.

Loss functions: Different loss functions were proposed for the algorithm, see the selection below. All losses are listwise ranking losses (over the target items and the negative samples) based on pointwise or pairwise scores. Currently, the BPR-max is deemed to be the best of the losses for most datasets as it results in slightly more accurate models than cross-entropy [Hidasi and Karatzoglou (2017)].

- *Cross-entropy:* Standard categorical cross-entropy between the predicted distribution and the target one-hot vector, preceded by the softmax transformation of the scores. While cross-entropy is pointwise in itself, the combination with the softmax transformation results in a listwise ranking. Originally reported to be unstable, but was fixed in later versions. Performs similarly to BPR and TOP1 when only a few negative samples are used, but is superior to them when additional samples are used as well.

$$L_{\text{xe}} = -\log s_i = -\log \frac{e^{r_i}}{\sum_{j=1}^{N} e^{r_j}} \tag{3.8}$$

- *BPR:* The average BPR loss [Rendle *et al.* (2009)] between the target item and the negative samples. On some datasets performs

[8]If the mini-batch samples are ignored, the optimal value goes up to around 1 and the overall accuracy drops.

better than the TOP1 loss.

$$L_{\text{bpr}} = -\frac{1}{N_S}\sum_{j=1}^{N_S} \log \sigma(r_i - r_j) \tag{3.9}$$

- *TOP1:* A heuristic loss of two parts: the first part pushes the score of the target item above that of the negative samples; the second part regularizes the score of the negative items.

$$L_{\text{top1}} = \frac{1}{N_S}\sum_{j=1}^{N_S} \sigma(r_j - r_i) + \sigma(r_j^2) \tag{3.10}$$

- *BPR-max:* Introduced in the second version of the algorithm, BPR-max combines the benefits of pairwise losses, the softmax transformation and score regularization. [Hidasi and Karatzoglou (2017)] concluded that the BPR and TOP1 losses are unsuitable when the number of negative examples is high, due to the gradient vanishing when the score of the target item is around the score of the top $\sim 10\%$ of the negative samples. With many samples, this results in the training stopping before the target item is put to the top of the list. The solution is to weight the BPR losses with the softmax score of the negative samples, thus ignoring samples whose score is much lower than that of the targets, which solves the vanishing gradient problem. The loss also includes a score regularization term similarly to the TOP1 loss, as it was found to be beneficial during training.

$$L_{\text{bpr-max}} = -\log \sum_{j=1}^{N_S} s_j \sigma(r_i - r_j) + \lambda \sum_{j=1}^{N_S} s_j r_j^2 \tag{3.11}$$

- *TOP1-max:* The softmax weighted version of the TOP1 loss. Inferior to the BPR-max loss.

$$L_{\text{top1-max}} = \sum_{j=1}^{N_S} s_j \left(\sigma(r_j - r_i) + \lambda\sigma(r_j^2)\right) \tag{3.12}$$

Constrained embeddings: Adding an embedding layer before the first GRU layer has been reported to slightly decrease the accuracy of the network. This is only counterintuitive for the first sight. The reason — yet again — is the large number of items, of which some are very rarely visited during training. Both the embedding and the output weight matrix are item representations, but they are learned separately from each other. Thus input and output representations can become slightly incompatible. The easiest solution is to have a shared weight matrix instead of two separate ones, or in other words: constrain the embedding matrix on the output weight matrix (see Figure 3.17). This results in unified item representations, which makes easier for the network to learn proper representations quicker and thus converge better.

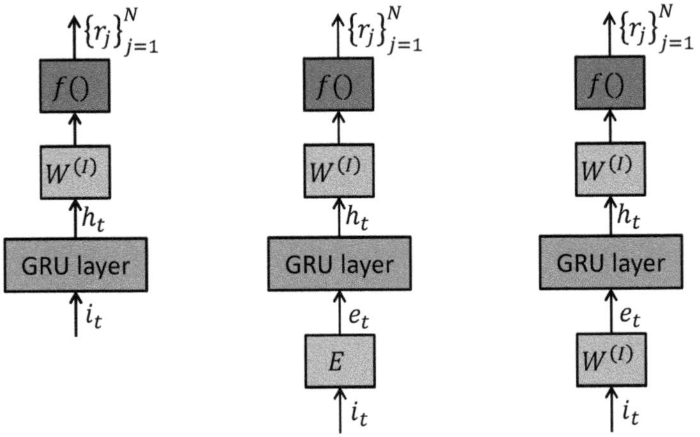

Fig. 3.17. GRU4Rec without embedding (left), with embedding (center), and with constrained embedding (right).

3.6.3. *Extending the model*

Since its inception, the original idea was extended in different directions to accommodate for various data and recommendation tasks. Below we shortly introduce the most notable extensions.

p-RNN for multimodal data: One way to improve the accuracy of the model is to use multimodal data. Instead of just relying to the item ID, we can make use of its image (e.g. thumbnail of a video) and textual item description. The parallel RNN (p-RNN) [Hidasi *et al.* (2016)] is an architecture for such multimodal data, where each view of the input is processed by a separate subnet. The subnets are then merged on their hidden states (see Figure 3.18). Training is done via alternating training, i.e. each subnet trains while the others are fixed.

Contextualization: Contextual information can help increasing the accuracy and adaptivity of recommender systems, therefore extending the model to be able to handle context is an important step. However, the naive

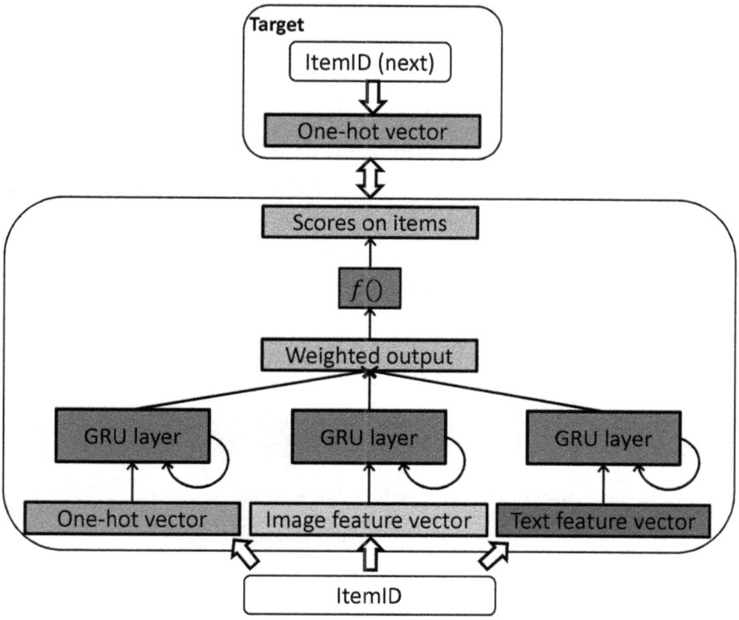

Fig. 3.18. The parallel RNN architecture.

inclusion of the context in GRU does not benefit the model. [Smirnova and Vasile (2017)] proposed different ways to include the context and they created an architecture that performs really well. It uses the context of the actual item on the input by concatenating the context embedding with the item embedding. The context of the next item[9] is also used in the model: the hidden state of the network is multiplied by the context embedding in elementwise fashion. The authors also introduce a structure coined "context wrapper" which is used to allow context-aware transitions of the hidden state in the network by replacing GRU equations (3.7) with the following:

$$\begin{aligned} r_t &= \sigma\left((W_r x_t + U_r h_{t-1}) \circ (V_r c_t) + b_r\right) \\ z_t &= \sigma\left((W_z x_t + U_z h_{t-1}) \circ (V_z c_t) + b_z\right) \\ h_t &= z_t \circ h_{t-1} + (1 - z_t) \circ \tanh\left((W_h x_t + U_h(r_t \circ h_{t-1})) \circ (V_h c_t) + b_h\right) \end{aligned} \tag{3.13}$$

Personalization: Even though the sessions of the users are disjoint, in certain domains, earlier sessions may contain information on the subsequent sessions of the user. E.g. two users with similar lifestyle will generate similar

[9]This is known during the training and can be set during inference time according to the actual context.

sessions in the same order. [Quadrana *et al.* (2017)] proposed to personalize session-based recommendations by predicting the initial hidden state of the session based RNN based on previous sessions of the user. The proposed architecture is a hierarchical RNN, in which session-based predictions are done by the standard GRU4Rec and another RNN is used on the sequence of sessions of the user to predict the initial hidden state of the session-based RNN (see Figure 3.19). By doing so, recommendations become personalized and adapted to the user from the start of the session, while this usually takes a few steps for the original algorithm.

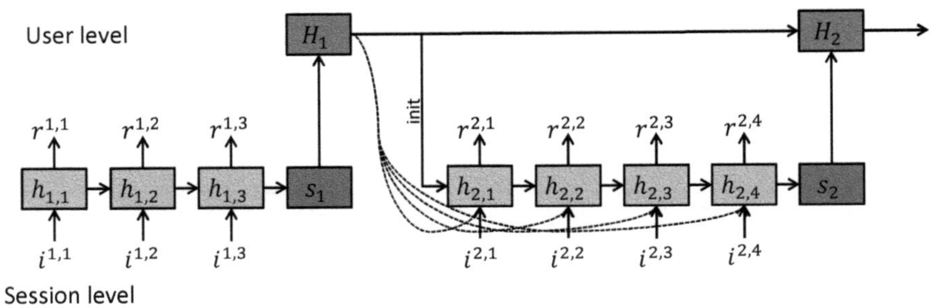

Fig. 3.19. Hierarchical RNN.

Intent prediction: The focus of this line of research so far has been next click prediction. This is appropriate for certain domains where the goal is to keep the user there by recommending more relevant content — like news or video recommendations. In other domains, the end goal is to generate revenue, therefore it is important for the algorithm to distinguish between users who are just browsing and those who are also likely to purchase something. This is the intent prediction task: based on (partial) sessions, we want to predict the intent (end goal) of the user. [Loyola *et al.* (2017)] is a preliminary work in this direction. The proposed architecture is an encoder-decoder based sequence-to-sequence learner for session data. The advantage of this architecture is that the output of the encoder can be used with different decoders simultaneously. One of the decoders can predict the rest of the session (next click prediction), while the other can predict the user's intent.

3.6.4. *Brief introduction to generative models*

So far this chapter discussed discriminative models only. Even though deep generative models — such as the Generative Adversarial Network (GAN) or the Variational Autoencoder (VAE) — are in the focus of the mainstream deep learning research, their preliminary application for recommenders happened only recently. This section gives a brief introduction to two of the most popular classes of deep generative models, Generative Adversarial Networks and Variational Autoencoders and discusses their preliminary applications in the RecSys domain.

Generative Adversarial Networks: The GAN [Goodfellow *et al.* (2014)] consists of two neural networks, the Generator (G) and the Discriminator (D) (see Figure 3.20). The goal of the generator is to capture the underlying distribution of the data and be able to generate new data points according to this distribution. In order to achieve this objective it plays a competitive game with the discriminator whose objective is to discriminate between samples coming from the training data and samples generated by the generator. The two networks are trained simultaneously and as D improves in discriminating real and generated samples, so does the ability of G to generate more realistic samples. In theory, this min-max game is played until the networks reach the Nash equilibrium and can not improve further. In practice, GANs are known to be fairly unstable and require careful hyperparameter tuning. They also suffer from mode collapse. This means that the generator learns to exploit one or few of the modes of the training data instead of capturing the entirety of the underlying distribution. A vast body of literature is available on solutions for this problem, including modified training and different loss functions for D and G [Arjovsky *et al.* (2017); Mao *et al.* (2017); Bellemare *et al.* (2017); Salimans *et al.* (2016)]. Modeling continuous data (e.g. images, sounds) with GAN is trivial, modeling discrete data (e.g. sequences of symbols) requires significant modifications [Kusner and Hernández-Lobato (2016); Hjelm *et al.* (2017); Yu *et al.* (2017)]. GANs are well known for their application on images, such as realistic image generation [Radford *et al.* (2015)], image super resolution [Ledig *et al.* (2016)], image-to-image translation [Isola *et al.* (2017)] (e.g. creating photorealistic images based on doodles) and many more; but they have also been used in other areas, e.g. for speech enhancement [Pascual *et al.* (2017)].

IRGAN: In order to unify the generative and discriminative models of information retrieval in a single framework [Wang *et al.* (2017)] proposed to adapt GANs for the information retrieval domain and coined this framework IRGAN. IRGAN is a conditional GAN as the generator is conditioned on the query (e.g. search query, recommendation request). Instead of generating a new document (item), the generator of IRGAN selects documents form the

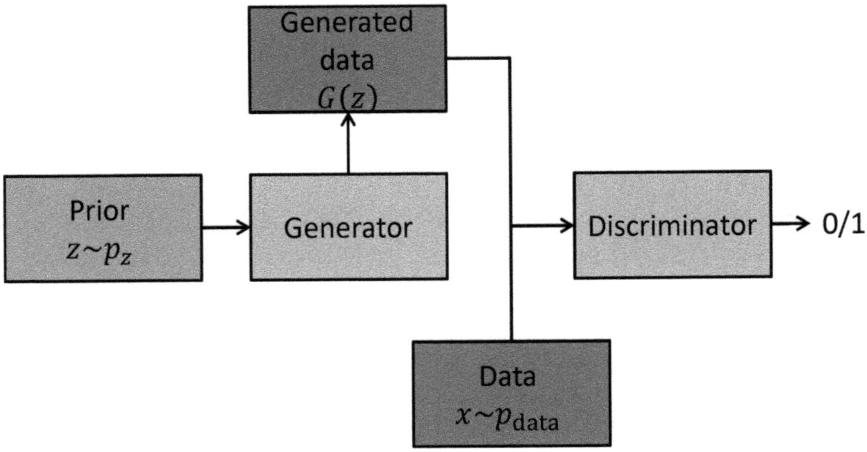

Fig. 3.20. High level schematic of the GAN.

available documents. The documents are selected independently from each other. The discriminator rates the choices of G and tries to distinguish them from documents that are known to be relevant for the given query. A different way to look at this framework is that G is an adaptive negative sample generator for D, which is a relevance predictor. IRGAN is also extended for the pairwise case. In this mode, G generates a pair of items where the first item is preferred to the second one for the given query. This is compared to the item pairs sampled from the training data by the discriminator. In practice, a real pair is sampled first and G only replaces the preferred item within it. By doing this, the pairwise IRGAN becomes a novel sampling strategy on its own, where the negative samples sought are already more relevant than the less preferred item of a pairwise feedback. IRGAN is demonstrated to work on the recommendation task (along with web search and question answering) in preliminary offline experiments, but it is unknown how it compares to various negative sampling strategies.

Variational Autoencoder: Similarly to autoencoders and denoising autoencoders (Section 3.4.1), a variational autoencoder [Kingma and Welling (2013)] also consist of an encoder and a decoder network. The key difference is that the encoder of an autoencoder encodes the input into a latent feature vector (code); meanwhile the encoder of VAE encodes it into a parameterized distribution in the latent space, called variational distribution (see Figure 3.21). In practice, the variational distribution is often a multivariate, fully factorized Gaussian distribution (i.e. Gaussian with

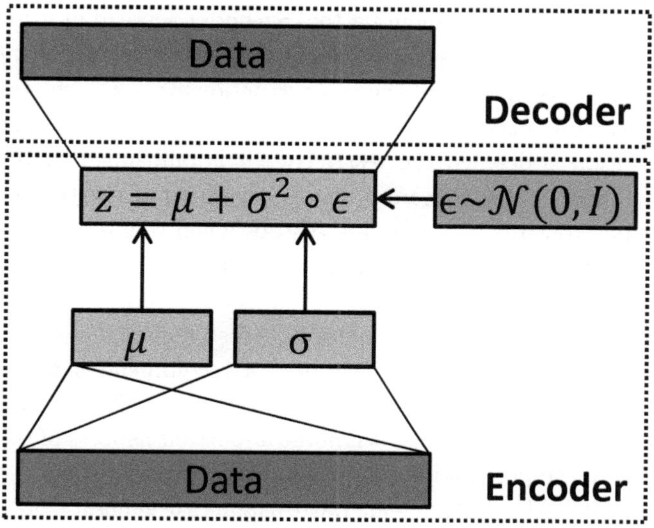

Fig. 3.21. High level schematic of the VAE with the reparameterization trick.

diagonal covariance matrix), thus the encoder needs to compute the mean vector and the diagonal of the covariance. The decoder takes a sample from this distribution and tries to reconstruct the original input. Since sampling does not work with SGD directly, it is common to use the reparameterization trick [Rezende *et al.* (2014)] instead. This means that the sample is computed as the sum of the mean vector and the diagonal of the covariance matrix multiplied by a sample from $\mathcal{N}(0, I)$ (white noise). This way, sampling is decoupled from the parameters of the encoder, which enables the use of SGD for training. The objective of the VAE is the negative reconstruction loss minus the KL divergence between the variational distribution and the prior distribution[10].

VAE for collaborative filtering: Variational Autoencoders for Collaborative Filtering (VAE-CF) [Liang *et al.* (2018)] improves upon the autoencoder based recommenders (Section 3.4.1) by using a variational autoencoder instead of a denoising autoencoder. In order to adapt VAE to the recommendation domain, two modifications are made by the authors. Firstly, multinomial conditional likelihood is used instead of Gaussian as it

[10]While this description of the VAE is easy to understand and allows for comparisons with autoencoders, it is not the usual formulation of the model and misses some details. We refer the reader to [Kingma and Welling (2013); Rezende *et al.* (2014)] for the full description of the VAE.

fits the implicit feedback problem better. Secondly, instead of using the evidence lower bound (i.e. the standard loss of the VAE) for optimization, the loss is interpreted as the sum of the reconstruction error and the regularization (i.e. constraining the variational distribution on the prior distribution). As regularization is deemed too restrictive, it is multiplied by $0 \leq \beta \leq 1$ to soften its effect. β starts from 0 at the beginning of the training and is slowly annealed towards 1. Increasing β stops when ranking metrics on the validation set start decreasing. The results with VAE-CF are promising, especially on sparser data.

VAE for slate recommendations: Most recommender algorithms only do relevance estimation and the set of recommended items is generated greedily by returning the most relevant items. While this strategy works well enough in most cases, it ignores important aspects of recommendations, such as the positioning of recommended items or their interdependencies. Slate-CVAE [Jiang *et al.* (2018)] aims to generate optimal sets of recommended items (slates), where optimality is defined as maximizing user interaction with the slate. The training data for this task consists of slates and user interactions with slates. The input of the encoder is the concatenated item embeddings of a slate. The condition contains the number of interactions each item of the slate received. It is important to note, that while the condition vector can use other information as well, it is not mandatory to add anything else to it. This means that the algorithm is capable of non-personalized, personalized, context-aware and context-driven [Pagano *et al.* (2016)] recommendations, depending on the information used in the context vector. The decoder tries to reconstruct the slate from the sampled code and the condition vector. The final layer of the decoder is a soft nearest neighbor layer, so it is able to return one of the slates of the training data. Slate-CVAE is definitely an interesting direction, but it is only evaluated via proxies: simulated data and sessions interpreted as slates. Even though more thorough evaluation is needed, these proxy evaluations suggest that the method has potential.

3.7. Practical guide

This section gives a brief overview of deep learning frameworks and discusses the most important best practices for researching and developing deep learning algorithms.

3.7.1. *Frameworks*

Creating efficient implementations of deep models from scratch can be a challenging and time consuming task. Fast execution requires highly optimized code and low level optimization for the hardware. Fortunately, there are several open source frameworks available in which this low level optimization is already available. Thus researchers and practitioners can focus on designing the model, which then will be executed quickly on either CPU or GPU. These frameworks also provide other kinds of supports to shorten development time. One of these is the automatic computations of gradients. This feature is immensely helpful for the gradient descent training as it eliminates the (often tedious) manual computation of gradients, which can easily be the cause of errors. The core of these frameworks are tensor operations extended with other operators that are common in deep models (e.g. convolutions). Below we detail some of the popular frameworks.

Theano[11] [The Theano Development Team (2016)] is an open source python framework — developed by the MILA lab of the University of Montreal — for defining and evaluating mathematical expressions efficiently. First a mathematical expression needs to be defined using symbolic variables and operations. These variables and operations are organized into a directed computational graph. This graph is then compiled into highly optimized code. This code can be executed several times with different input values (e.g. the actual data points). The framework has both a CPU and a GPU backend. While finding the optimal operators for efficient Theano code can be tricky, optimized Theano programs often run faster than those in other frameworks. Theano is a low level framework in the sense, that the focus is on efficiently evaluating any mathematical expression, thus any kind of model can be defined easily and previously defined models can be modified quickly. On the other hand — while there is clearly additional support for deep learning models built into the framework — this also means that predefined models/layers are not part of the core package. Those either have to be assembled by the user or imported from high level packages built on Theano. These properties make it ideal for research, but less so for production. It is one of the older frameworks with its development started in 2007. In September 2017, the cease of development was announced after the next release with support and bug fixes being provided for another year.

[11] `http://deeplearning.net/software/theano`

TensorFlow[12] [Abadi *et al.* (2016)] is Google's machine learning framework that was opensourced at the end of 2015. The ideas behind the framework are very similar to Theano: machine learning models are represented as computational graphs, which are then compiled into efficient code and then executed on CPU, GPU or custom hardware. The framework quickly gained popularity and has become the de facto deep learning framework. Besides the quick development, clear focus on machine learning and Google's support, this is also due to it being not just a standalone framework for research but part of a deep learning ecosystem, tailored towards production. Other parts of this ecosystem have been made public as well. For example, TensorBoard visualizes learning of the networks, tf.layers provides high level predefined and optimized layers as building blocks, TensorFlow Serving helps with running models in production, TensorFlow models can be easily run on smart phones, etc. While TensorFlow is generally slower than Theano, it is much easier to put together complex models in it thanks to the high level functionality available in the framework.

pyTorch[13] is developed by Facebook and follows a different approach. It provides quick tensor computations on GPU and supports deep networks by having an automatic differentiation library. Since there is no need to predefine expressions and compile the computational graph, pyTorch is more similar to traditional programming and thus can be seen as a more lightweight framework from the user's point of view. Since computational graphs are defined on the fly, implementing dynamic neural models, such as RNNs, is easier. The developers of the framework put a lot of emphasis on efficient execution both in terms of speed and memory footprint. Even though the framework was released only in 2017, it quickly gained popularity, due to the aforementioned features making it an ideal research framework.

Keras[14] is a high level framework that provides layers and networks as building blocks of complex models. It has backends for Theano, TensorFlow and CNTK, so it can be installed over any of these low level frameworks.

There are many other deep learning frameworks, e.g. the Microsoft CNTK or the Apache MXNet, which is the default machine learning environment on Amazon AWS. The major frameworks all support both CPU and GPU execution and distributed (e.g. multi GPU) training. There is

[12] `https://www.tensorflow.org/`
[13] `http://pytorch.org/`
[14] `https://keras.io/`

some difference is speed, but as the frameworks are constantly developed and optimized, the title of fastest framework (for a given network) often changes hands. The key difference is mostly in syntax and the available selection of high level features. The optimal choice thus is mostly up to personal preference and whether it is to be used for research or production.

3.7.2. *Best practices*

Hyperparameter optimization: Neural networks have several hyperparameters that influence their performance. Unfortunately the effect of these can be significant, so hyperparameter tuning is often required. It is best to have a separate validation set for this optimization. Training neural networks requires a lot of resources (compared to simpler models) and the space of hyperparameters is large, thus optimization takes a long time. Optimizing the code for speed before the parameter search can significantly decrease the total time required. Note however, that even with the use of a deep learning framework, optimizing for GPU and CPU can differ slightly. Sometimes the frameworks themselves can be the bottleneck, e.g. a frequently used operator is implemented inefficiently. In this case, using custom code for the part in question can be the answer, but the trade-off between implementation and execution time should be kept in mind. Fortunately, not all hyperparameters have huge effects on the performance and interaction between the parameters is limited. With limited resources, finding just the optimal network size and learning rate can give reasonably good results.

Scalability & training times: Scalability is a really important property of recommender algorithms in practice. While researchers started to take scalability into account recently, it is still not addressed every time and is often misinterpreted. Scalability is not the same as having an acceptable training time on a dataset. Scalability is how much more computation will be required if you train it on a bigger dataset. For example, if an algorithm scales with the product of the number of users and items, it scales poorly, even if it can be trained on the research dataset in an hour. Deep models are complex and require significant amount of computational power, but can scale both well and poorly, depending on the model design. The general rule of thumb is that the algorithm should scale linearly with the number of training events/users/items in order to be robust enough for use in practice. Unfortunately, it is not just scalability that matters. Recommendation models are retrained frequently in order to accommodate for new items/users and to follow the trends in the data. If training takes several days, the model is already outdated when it is freshly trained. This is in contrast with other application domains of deep learning, where the model can even train for months and then can be used for a longer time period without any retraining. It is very easy to go overboard and design a deep model for recommendation that takes lot of time to train. The threshold for acceptable training time depends on the domain, as well as the ability of the algorithm of handling new users/items without retraining. A trade-off should be sought between model complexity (more accurate, but slower training model) and model recency.

Reproducibility: Reproducibility is of key importance for research, yet it is usually undermined by pragmatic considerations. Fortunately, the deep learning research community holds reproducibility in high regards. It is common that a published paper about a new model or technique is accompanied by a public implementation, including the algorithm and the experimental setup. Experiments are often carried out on public benchmark datasets that help both reproducibility and the comparison of methods. Recommender systems process user generated behavior data, which can tell a lot about the service that generated the data and the users themselves, thus companies are often reluctant to share their data. In the meantime, their research focuses on improving their recommendations on their own data and care less about public benchmarks (not to mention that the offline evaluation of recommenders is just a proxy for their actual performance). However, it is still encouraged to include measurements on related public datasets, even if they are not the primary targets. The choice of dataset should be appropriate for the problem set out to be solved by the proposed algorithm. Making code public is also strongly encouraged: the solution can generate higher interest and can be more easily used as a baseline when publicly available.

3.8. Summary

Although using deep learning in recommender systems is a very recent, it still boasts of good results and even more potential. The direct use of content has a huge possibility for hybrid systems, recurrent neural networks have revitalized session-based recommendations and the modularity of neural network can be easily exploited to handle heterogeneous data. Due to it being a novel research direction, there is much research to be done, but some robust methods have already been developed. Working with deep learning within this domain is challenging due to having to adhere to practical constraints, but is also rewarding.

References

Abadi, M., Agarwal, A., Barham, P., Brevdo, E., Chen, Z., Citro, C., Corrado, G. S., Davis, A., Dean, J., Devin, M. *et al.* (2016). Tensorflow: Large-scale machine learning on heterogeneous distributed systems, *arXiv preprint arXiv:1603.04467*.

Arjovsky, M., Chintala, S. and Bottou, L. (2017). Wasserstein gan, *arXiv preprint arXiv:1701.07875*.

Arjovsky, M., Shah, A. and Bengio, Y. (2016). Unitary evolution recurrent neural networks, in *International Conference on Machine Learning*, pp. 1120–1128.

Bansal, T., Belanger, D. and McCallum, A. (2016). Ask the GRU: Multi-task learning for deep text recommendations, in *Proceedings of the 10th ACM Conference on Recommender Systems* (ACM), pp. 107–114.

Barkan, O. and Koenigstein, N. (2016). Item2vec: neural item embedding for collaborative filtering, in *Machine Learning for Signal Processing (MLSP), 2016 IEEE 26th International Workshop on* (IEEE), pp. 1–6.

Bellemare, M. G., Danihelka, I., Dabney, W., Mohamed, S., Lakshminarayanan, B., Hoyer, S. and Munos, R. (2017). The cramer distance as a solution to biased wasserstein gradients, *arXiv preprint arXiv:1705.10743*.

Cho, K., Van Merriënboer, B., Bahdanau, D. and Bengio, Y. (2014). On the properties of neural machine translation: Encoder-decoder approaches, *arXiv preprint arXiv:1409.1259*.

Clevert, D.-A., Unterthiner, T. and Hochreiter, S. (2015). Fast and accurate deep network learning by exponential linear units (elus), *arXiv preprint arXiv:1511.07289*.

Cremonesi, P., Koren, Y. and Turrin, R. (2010). Performance of recommender algorithms on top-n recommendation tasks, in *Proceedings of the fourth ACM conference on Recommender systems* (ACM), pp. 39–46.

Dai, H., Wang, Y., Trivedi, R. and Song, L. (2016). Recurrent coevolutionary latent feature processes for continuous-time recommendation, in *Proceedings of the 1st Workshop on Deep Learning for Recommender Systems* (ACM), pp. 29–34.

Duchi, J., Hazan, E. and Singer, Y. (2011). Adaptive subgradient methods for online learning and stochastic optimization, *Journal of Machine Learning Research* **12**, Jul, pp. 2121–2159.

Goodfellow, I., Bengio, Y. and Courville, A. (2016). *Deep learning*, MIT Press.

Goodfellow, I., Pouget-Abadie, J., Mirza, M., Xu, B., Warde-Farley, D., Ozair, S., Courville, A. and Bengio, Y. (2014). Generative adversarial nets, in *Advances in neural information processing systems*, pp. 2672–2680.

Grbovic, M., Radosavljevic, V., Djuric, N., Bhamidipati, N., Savla, J., Bhagwan, V. and Sharp, D. (2015). E-commerce in your inbox: Product recommendations at scale, in *Proceedings of the 21th ACM SIGKDD International*

Conference on Knowledge Discovery and Data Mining (ACM), pp. 1809–1818.

He, R. and McAuley, J. (2016). Vbpr: Visual bayesian personalized ranking from implicit feedback, in *AAAI*, pp. 144–150.

Hidasi, B. and Karatzoglou, A. (2017). Recurrent neural networks with top-k gains for session-based recommendations, *arXiv preprint arXiv:1706.03847* `http://arxiv.org/abs/1706.03847`.

Hidasi, B., Karatzoglou, A., Baltrunas, L. and Tikk, D. (2015). Session-based recommendations with recurrent neural networks, *International Conference on Learning Representations (ICLR 2016)* `http://arxiv.org/abs/1511.06939`.

Hidasi, B., Karatzoglou, A., Sar-Shalom, O., Dieleman, S., Shapira, B. and Tikk, D. (2017). DLRS 2017: Second workshop on deep learning for recommender systems, in *Proceedings of the Eleventh ACM Conference on Recommender Systems*, RecSys '17 (ACM, New York, NY, USA), ISBN 978-1-4503-4652-8, pp. 370–371, doi:10.1145/3109859.3109953, `http://doi.acm.org/10.1145/3109859.3109953`.

Hidasi, B., Quadrana, M., Karatzoglou, A. and Tikk, D. (2016). Parallel recurrent neural network architectures for feature-rich session-based recommendations, in *Proceedings of the 10th ACM Conference on Recommender Systems*, RecSys '16 (ACM, New York, NY, USA), ISBN 978-1-4503-4035-9, pp. 241–248, doi:10.1145/2959100.2959167, `http://doi.acm.org/10.1145/2959100.2959167`.

Hidasi, B. and Tikk, D. (2013). Initializing matrix factorization methods on implicit feedback databases, *J. UCS* **19**, 12, pp. 1834–1853.

Hjelm, R. D., Jacob, A. P., Che, T., Trischler, A., Cho, K. and Bengio, Y. (2017). Boundary-seeking generative adversarial networks, *arXiv preprint arXiv:1702.08431*.

Hochreiter, S. and Schmidhuber, J. (1997). Long short-term memory, *Neural computation* **9**, 8, pp. 1735–1780.

Hornik, K., Stinchcombe, M. and White, H. (1989). Multilayer feedforward networks are universal approximators, *Neural networks* **2**, 5, pp. 359–366.

Hu, F., Chen, T., Liu, N. N., Yang, Q. and Yu, Y. (2012). Discriminative factor alignment across heterogeneous feature space, in *Joint European Conference on Machine Learning and Knowledge Discovery in Databases* (Springer), pp. 757–772.

Hu, Y., Koren, Y. and Volinsky, C. (2008). Collaborative filtering for implicit feedback datasets, in *Data Mining, 2008. ICDM'08. Eighth IEEE International Conference on* (IEEE), pp. 263–272.

Isola, P., Zhu, J.-Y., Zhou, T. and Efros, A. A. (2017). Image-to-image translation with conditional adversarial networks, *arXiv preprint*.

Jiang, R., Gowal, S., Mann, T. A. and Rezende, D. J. (2018). Optimizing slate recommendations via slate-cvae, *arXiv preprint arXiv:1803.01682*.

Karatzoglou, A., Hidasi, B., Tikk, D., Sar-Shalom, O., Roitman, H. and Shapira, B. (2016). Recsys'16 workshop on deep learning for recommender systems (DLRS), in *Proceedings of the 10th ACM Conference on Recommender Sys-*

tems, RecSys '16 (ACM, New York, NY, USA), ISBN 978-1-4503-4035-9, pp. 415–416, doi:10.1145/2959100.2959202, http://doi.acm.org/10.1145/2959100.2959202.

Kingma, D. and Ba, J. (2014). Adam: A method for stochastic optimization, *arXiv preprint arXiv:1412.6980.*

Kingma, D. P. and Welling, M. (2013). Auto-encoding variational bayes, *arXiv preprint arXiv:1312.6114.*

Klambauer, G., Unterthiner, T., Mayr, A. and Hochreiter, S. (2017). Self-normalizing neural networks, *arXiv preprint arXiv:1706.02515.*

Kusner, M. J. and Hernández-Lobato, J. M. (2016). Gans for sequences of discrete elements with the gumbel-softmax distribution, *arXiv preprint arXiv:1611.04051.*

Ledig, C., Theis, L., Huszár, F., Caballero, J., Cunningham, A., Acosta, A., Aitken, A., Tejani, A., Totz, J., Wang, Z. *et al.* (2016). Photo-realistic single image super-resolution using a generative adversarial network, *arXiv preprint.*

Liang, D., Krishnan, R. G., Hoffman, M. D. and Jebara, T. (2018). Variational autoencoders for collaborative filtering, *arXiv preprint arXiv:1802.05814.*

Linden, G., Smith, B. and York, J. (2003). Amazon.com recommendations: Item-to-item collaborative filtering, *IEEE Internet computing* **7**, 1, pp. 76–80.

Loyola, P., Liu, C. and Hirate, Y. (2017). Modeling user session and intent with an attention-based encoder-decoder architecture, in *Proceedings of the Eleventh ACM Conference on Recommender Systems*, RecSys '17 (ACM, New York, NY, USA), ISBN 978-1-4503-4652-8, pp. 147–151, doi:10.1145/3109859.3109917, http://doi.acm.org/10.1145/3109859.3109917.

Maas, A. L., Hannun, A. Y. and Ng, A. Y. (2013). Rectifier nonlinearities improve neural network acoustic models, in *Proc. ICML*, Vol. 30.

Mao, X., Li, Q., Xie, H., Lau, R. Y., Wang, Z. and Smolley, S. P. (2017). Least squares generative adversarial networks, in *2017 IEEE International Conference on Computer Vision (ICCV)* (IEEE), pp. 2813–2821.

Mikolov, T., Chen, K., Corrado, G. and Dean, J. (2013a). Efficient estimation of word representations in vector space, *arXiv preprint arXiv:1301.3781.*

Mikolov, T., Sutskever, I., Chen, K., Corrado, G. S. and Dean, J. (2013b). Distributed representations of words and phrases and their compositionality, in *Advances in neural information processing systems*, pp. 3111–3119.

Mnih, V., Heess, N., Graves, A. *et al.* (2014). Recurrent models of visual attention, in *Advances in neural information processing systems*, pp. 2204–2212.

Nair, V. and Hinton, G. E. (2010). Rectified linear units improve restricted Boltzmann machines, in *Proceedings of the 27th international conference on machine learning (ICML-10)*, pp. 807–814.

Nedelec, T., Smirnova, E. and Vasile, F. (2017). Specializing joint representations for the task of product recommendation, in *Proceedings of the 2nd Workshop on Deep Learning for Recommender Systems* (ACM), pp. 10–18.

Pagano, R., Cremonesi, P., Larson, M., Hidasi, B., Tikk, D., Karatzoglou, A. and Quadrana, M. (2016). The contextual turn: From context-aware to context-driven recommender systems, in *Proceedings of the 10th ACM conference on recommender systems* (ACM), pp. 249–252.

Pascual, S., Bonafonte, A. and Serra, J. (2017). Segan: Speech enhancement generative adversarial network, *arXiv preprint arXiv:1703.09452*.

Paterek, A. (2007). Improving regularized singular value decomposition for collaborative filtering, in *Proceedings of KDD cup and workshop*, Vol. 2007, pp. 5–8.

Pilászy, I. and Tikk, D. (2009). Recommending new movies: even a few ratings are more valuable than metadata, in *Proceedings of the third ACM conference on Recommender systems* (ACM), pp. 93–100.

Qu, Y., Cai, H., Ren, K., Zhang, W., Yu, Y., Wen, Y. and Wang, J. (2016). Product-based neural networks for user response prediction, in *Data Mining (ICDM), 2016 IEEE 16th International Conference on* (IEEE), pp. 1149–1154.

Quadrana, M., Karatzoglou, A., Hidasi, B. and Cremonesi, P. (2017). Personalizing session-based recommendations with hierarchical recurrent neural networks, in *Proceedings of the Eleventh ACM Conference on Recommender Systems*, RecSys '17 (ACM, New York, NY, USA), ISBN 978-1-4503-4652-8, pp. 130–137, doi:10.1145/3109859.3109896, `http://doi.acm.org/10.1145/3109859.3109896`.

Radford, A., Metz, L. and Chintala, S. (2015). Unsupervised representation learning with deep convolutional generative adversarial networks, *arXiv preprint arXiv:1511.06434*.

Rendle, S. (2012). Factorization machines with libfm, *ACM Transactions on Intelligent Systems and Technology (TIST)* **3**, 3, p. 57.

Rendle, S., Freudenthaler, C., Gantner, Z. and Schmidt-Thieme, L. (2009). Bpr: Bayesian personalized ranking from implicit feedback, in *Proceedings of the twenty-fifth conference on uncertainty in artificial intelligence* (AUAI Press), pp. 452–461.

Rezende, D. J., Mohamed, S. and Wierstra, D. (2014). Stochastic backpropagation and approximate inference in deep generative models, *arXiv preprint arXiv:1401.4082*.

Salakhutdinov, R., Mnih, A. and Hinton, G. (2007). Restricted boltzmann machines for collaborative filtering, in *Proceedings of the 24th international conference on Machine learning* (ACM), pp. 791–798.

Salimans, T., Goodfellow, I., Zaremba, W., Cheung, V., Radford, A. and Chen, X. (2016). Improved techniques for training gans, in *Advances in Neural Information Processing Systems*, pp. 2234–2242.

Singh, A. P. and Gordon, G. J. (2008). Relational learning via collective matrix factorization, in *Proceedings of the 14th ACM SIGKDD international conference on Knowledge discovery and data mining* (ACM), pp. 650–658.

Smirnova, E. and Vasile, F. (2017). Contextual sequence modeling for recommendation with recurrent neural networks, in *Proceedings of the 2nd Workshop on Deep Learning for Recommender Systems*, DLRS 2017, ISBN 978-1-4503-5353-3, pp. 2–9, doi:10.1145/3125486.3125488, `http://doi.acm.org/10.1145/3125486.3125488`.

Song, Y., Elkahky, A. M. and He, X. (2016). Multi-rate deep learning for temporal recommendation, in *Proceedings of the 39th International ACM SIGIR conference on Research and Development in Information Retrieval* (ACM), pp. 909–912.

Srivastava, N., Hinton, G. E., Krizhevsky, A., Sutskever, I. and Salakhutdinov, R. (2014). Dropout: a simple way to prevent neural networks from overfitting, *Journal of machine learning research* **15**, 1, pp. 1929–1958.

Sutskever, I., Vinyals, O. and Le, Q. V. (2014). Sequence to sequence learning with neural networks, in *Advances in neural information processing systems*, pp. 3104–3112.

The Theano Development Team (2016). Theano: A python framework for fast computation of mathematical expressions, *CoRR* **abs/1605.02688**, arXiv:1605.02688, `http://arxiv.org/abs/1605.02688`.

Tieleman, T. and Hinton, G. (2012). Lecture 6.5-rmsprop: Divide the gradient by a running average of its recent magnitude, *COURSERA: Neural networks for machine learning* **4**, 2, pp. 26–31.

Van den Oord, A., Dieleman, S. and Schrauwen, B. (2013). Deep content-based music recommendation, in *Advances in neural information processing systems*, pp. 2643–2651.

Vasile, F., Smirnova, E. and Conneau, A. (2016). Meta-prod2vec: Product embeddings using side-information for recommendation, in *Proceedings of the 10th ACM Conference on Recommender Systems* (ACM), pp. 225–232.

Vincent, P., Larochelle, H., Bengio, Y. and Manzagol, P.-A. (2008). Extracting and composing robust features with denoising autoencoders, in *Proceedings of the 25th international conference on Machine learning* (ACM), pp. 1096–1103.

Wang, H., Wang, N. and Yeung, D.-Y. (2015). Collaborative deep learning for recommender systems, in *Proceedings of the 21th ACM SIGKDD International Conference on Knowledge Discovery and Data Mining* (ACM), pp. 1235–1244.

Wang, J., Yu, L., Zhang, W., Gong, Y., Xu, Y., Wang, B., Zhang, P. and Zhang, D. (2017). Irgan: A minimax game for unifying generative and discriminative information retrieval models, in *Proceedings of the 40th International ACM SIGIR conference on Research and Development in Information Retrieval* (ACM), pp. 515–524.

Williams, R. J. and Zipser, D. (1995). Gradient-based learning algorithms for recurrent networks and their computational complexity, *Backpropagation: Theory, architectures, and applications* **1**, pp. 433–486.

Wilson, A. C., Roelofs, R., Stern, M., Srebro, N. and Recht, B. (2017). The marginal value of adaptive gradient methods in machine learning, *arXiv preprint arXiv:1705.08292*.

Wu, Y., DuBois, C., Zheng, A. X. and Ester, M. (2016). Collaborative denoising auto-encoders for top-n recommender systems, in *Proceedings of the Ninth ACM International Conference on Web Search and Data Mining* (ACM), pp. 153–162.

Yu, L., Zhang, W., Wang, J. and Yu, Y. (2017). Seqgan: Sequence generative adversarial nets with policy gradient, in *AAAI*, pp. 2852–2858.

Zeiler, M. D. (2012). Adadelta: an adaptive learning rate method, *arXiv preprint arXiv:1212.5701*.

Zheng, Y., Liu, C., Tang, B. and Zhou, H. (2016). Neural autoregressive collaborative filtering for implicit feedback, in *Proceedings of the 1st Workshop on Deep Learning for Recommender Systems* (ACM), pp. 2–6.

Chapter 4

Hybrid Collaborative Recommendations: Practical Considerations and Tools to Develop a Recommender

Michal Kompan, Peter Gašpar and Maria Bielikova
Slovak University of Technology in Bratislava
Faculty of Informatics and Information Technologies
Ilkovičova 2, 842 16, Bratislava, Slovak Republic
{name.surname}@stuba.sk

Hybrid collaborative recommender systems were developed to increase performance of single recommenders similarly to the ensemble machine learning methods. There are several types of hybridization that can be utilized in various scenarios. In this chapter, we summarize the overview of hybrid collaborative recommender systems. We focus on how they are beneficial with respect to the standard collaborative filtering techniques and how they can be utilized to solve the main issues of the collaborative filtering techniques. Examples of frameworks and libraries that implement existing recommendation approaches are complemented with the comprehensive list of datasets available (including state-of-the-art results reported on these datasets). Evaluation process of recommender systems is discussed followed by the examples of the evaluation frameworks. We summarize pros and cons of several hybrid approaches and conclude with suggestions for practical implementations.

4.1. Introduction

The diversity as a concept has been proved by the evolution to be very successful. The researches follow this idea in many modern approaches. The ensemble methods in the machine learning often outperform standalone methods. In the context of the recommender systems, similar idea is used as so-called hybrid recommenders.

As a rule, two standard recommendation techniques became established [Balabanovic *et al.*, 1997]: *collaborative filtering* and *content-based recommendation*. In a collaborative filtering a recommender relies on the user evaluation of the items and based on that it calculates the user and item similarity [Kim *et al.*, 2009]. Such a recommender is able to find users that have a similar taste to a given user or the items that would be similarly rated by the other users [Balabanovic *et al.*, 1997]. Content-based recommenders utilize content representation of the items (i.e., a set of attributes that characterize an item) and predict, whether a user would like items that are similar to those he/she preferred in the past.

It is clear that all these techniques have their advantages and disadvantages, which researchers try to solve in many different ways. Luckily, characteristics of these approaches are often disjointed and thus their reasonable combination can result in better performing approach.

Therefore, we adopt the following definition. *Hybrid recommender systems* are based on the combination of two or more monolithic recommendation techniques, such that the advantages of one recommender are utilized in order to solve the disadvantages of the other recommender [Burke *et al.*, 2011]. Hybrid recommendation is sometimes referred as another recommendation technique (alongside to content-based and collaborative filtering) [Adomavicius *et al.*, 2005].

Hybrid recommenders were introduced to increase an accuracy of existing monolithic techniques and also to reduce or eliminate their major drawbacks (such as the cold-start problem, sparsity, or user outliers). The most common technique is to employ collaborative filtering with the combination of other technique (e.g., content-based or knowledge-based recommender) [Burke, 2002]. However, there are also other combinations of techniques that are suitable depending on the studied problem and sometimes even a domain.

Hybrid recommender systems also attracted many companies and they became employed as a part of their products. In the domain of news, Google developed [Liu *et al.*, 2010] a hybrid recommender that was aimed to combine both user interests and trends in Google News (a combination of content-based recommendation and collaborative filtering). As they reported, proposed hybrid improved Click-Through-Rate by 30% (in comparison to the baseline collaborative approach).

In 2006 a movie streaming company Netflix announced a competition, where the goal was to recommend movies such that the proposed algorithm made an improvement over 10% in comparison to the Netflix baseline recommender. Several solutions were proposed (such as [Koren, 2009]) and many of them were actually the hybrid approaches based on the collaborative filtering recommendation (including the winning ones). Netflix also uses hybrid recommenders nowadays [Gomez-Uribe *et al.*, 2015], for example in the search results, where it combines user movie playbacks, search data and metadata.

The recommender system domain historically connects academia with the business. As a result, plenty of libraries and frameworks have been proposed. These cover various programing languages and as a rule implement state-of-the-art approaches. To give a short overview, we present a comparison of several recommendations libraries and frameworks with emphasis on the hybrid recommenders.

The typical evaluation of a recommenders starts with offline experiments. To obtain a reliable and comparable results, appropriate methodology and dataset is a necessity. In this chapter, we also compare typical datasets used for the evaluation of recommenders with metrics reported by several authors.

The chapter systematically covers the following topics:

- Overview of hybrid recommender techniques with emphasis on pros and cons (Sec. 4.2)
- Comparison of recommender system libraries and frameworks (Sec. 4.3)
- Practical hints for the evaluation of recommenders (Sec. 4.4)
- Comparison of datasets and reported metrics (Sec. 4.4.4)

4.2. Hybrid recommender systems — pros and cons

Hybrid recommender systems were proposed to improve existing monolithic recommendation techniques. By using and combining these techniques, hybrids aim to improve recommendation performance (from several points of view) and thus enhance overall user experience. Correspondingly, there are problems that monolithic techniques suffer from and hybrid recommenders are able to solve.

4.2.1. *Types of combinations*

When picking a hybrid recommendation technique, we have to select, which combination will be utilized. In other words, which monolithic recommendation techniques will be employed and also how they will be combined.

Following the Burke's taxonomy [Burke, 2002] we distinguish between seven basic types of hybrid combinations: weighted, switching, mixed, feature combination, cascade, meta-level, and feature augmentation. We would like to emphasize that these combinations types define how two or more approaches are combined (no restrictions for specific recommender type).

In *weighted hybrid*, the underlying recommenders calculate a score for an item and these scores are then combined to produce a final (single) score for the recommended item. [Hornung *et al.*, 2013] built a weighted music recommender that combined collaborative recommender (for track similarity) and two content-based recommenders (for tag and time similarity). To enrich the final list of recommendations, they also generated the additional serendipitous music tracks by considering a similarity of the users. Moreover, the famous Netflix prize winner algorithm combined 24 monolithic predictors in order to provide final estimate. The gradient boosted decision trees were used to combine single models covering neighborhood, matrix factorization or regression models [Koren, 2009].

Switching hybrid specifies a condition, which determines which recommendation technique will be selected and used for the recommendation (depending on the situation). A switching hybrid was proposed in [Ghazanfar *et al.*, 2014], where the authors utilized clustering approach to detect the gray-sheep users. These gray-sheep users then received recommendations generated by the separate content-based recommender.

In *mixed hybrid*, each underlying recommender generates a list of recommendations that are combined to produce a final recommendation. In other words, both lists are presented to the user. A TV recommendation using mixed hybrid was proposed in [Barragans-Martínez *et al.*, 2010], where the collaborative filtering was combined with the content-based recommendation. During the merging strategy, they used an average rating of TV shows (calculated by the recommenders). Kaššák *et al.* proposed a mixed hybrid recommender that aggregated the content-based and collaborative filtering candidates within the group of people [Kaššák *et al.*, 2016].

Feature combination uses multiple types of features that are combined to learn a single recommender model. For instance, the ratings of users combined with the content features of the specific item [Basu *et al.*, 1998]. [Zanker *et al.*, 2009] utilized a single collaborative filtering recommender that combined various features (called *rating domains*), such as the navigation actions, viewed items, items added to the shopping basket, or the user context.

Cascade hybrid is based on the idea of refinements, where the first-level recommender generates recommended items. The role of the second recommender is to adjust the items returned by the first recommender, but here the focus is only on those items that need refinements. Lampropoulos [Lampropoulos *et al.*, 2012] presented a cascade hybrid that employed a two-step solution. Firstly, a content-based recommender was used as a one-class classifier that identified the items suitable for a particular user. Then, a second-level collaborative filtering recommender assigned ratings to the items identified by the content-based recommender.

In *meta-level hybrid*, the first recommender learns a model, which is used as an input to the second recommender. By analyzing the rule-based preferences from historical user interactions, a collaborative filtering model was learned and used as an input to the knowledge-based recommender in [Zanker, 2008].

Similar idea is applied for the *feature augmentation hybrid*, where the *result* (not a model) of the first recommender prediction is used as a feature to the second recommender. [Campos *et al.*, 2010] created a hybrid recommender that used weights produced during the content-based recommendation as an input to the collaborative filtering recommender.

As noted by [Burke, 2002], some combinations (e.g., switching or mixed hybrid) require an initial effort that must be done before we may employ the hybrid strategy. For example, in case of the switching hybrid, we must define the criteria to switch between the recommendation techniques beforehand. The weighted hybrid requires setting the weights that apply for the results of particular hybrids.

For the feature augmentation, the cascade and the meta-level hybrid, a dependency may cause issues, if the second-level recommender relies on the results of the first-level recommender (Table 4.1 summarizes the pros and cons of these combinations).

Another perspective for the hybrid recommender classification was proposed by Aggarwal [Aggarwal, 2016]. He recognizes three high-level types:

- *ensemble design* — analogy to ensemble methods in machine learning. Several algorithms are combined into a single output (switching, weighted, cascade, feature augmentation),
- *monolithic design* — refers to a recommender combining several data sources (feature combination, meta-level),
- *mixed systems* — combines both ensemble and monolithic design.

This taxonomy offers a valuable (from the machine learning perspective) view which addresses the nature of Burke's insight.

4.2.2. *Hybrids as a solution for recommendation issues*

There are several issues that standard recommendation techniques suffer from. In the worst-case scenario, it results in an inability to recommend any items. Most of these issues are related to how recommender systems work.

We further examine the problems related to the collaborative filtering recommendations (the problems collaborative filtering is either suffering from or is able to help to deal with):

- *cold-start problem* (a problem of a new user/item, or a new context in case of the context-aware recommender systems),
- *over-specialization* (inability to recommend items outside-the-box),
- *sparsity* (of a user-item matrix),
- *extremes* (gray and black sheep),
- *lack of diversity*.

Hybrid recommenders are capable of reducing these problems by hybridization of collaborative filtering with the other recommendation technique.

Table 4.1. Pros and cons of hybrid combinations (based on [Burke, 2002]).

Type	*Pros*	*Cons*
Weighted	Possible to adjust weights of hybrids. Can be used in datasets with implicit feedback.	Value of the particular recommendation techniques should be uniform across the algorithms. All the techniques apply the weights to each item, which may be redundant.
Switching	System is more sensitive to strengths and weaknesses of the particular recommenders.	Switching criteria must be defined.
Mixed	Suitable where it is possible to make a large number of recommendations simultaneously. Allows to recommend both popular and new items.	Combination technique must be employed. Rules for solving conflicting situations must be also defined.
Feature combination	Combines features from several algorithms which results to improved similarities.	May require feature selection in content-based recommender [Basu *et al.*, 1998].
Feature augmentation	Allows to improve an accuracy of a system without modifying it.	A quality of second recommender may depend on the recommendations of the first (augmenting) recommender.
Cascade	Allows to employ second recommender to only relevant items (results of the first recommender). More efficient than weighted.	Quality of second-level recommender may depend on the recommendations of the first-level recommender.
Meta-level	Learned model is a compressed representation of the user-item preferences.	A quality of the second recommender may depend on the quality of the representation of the first recommender.

4.2.2.1. Cold-start problem

One of the most notable problems occurs when a *new user* or a *new item* is introduced to a recommender. This problem is also referred as a *cold-start problem*. Here, the recommender fails to generate appropriate recommendations since it does not have enough knowledge about the user preferences (see Chapter 8).

When a new user appears, a low number of user-item interactions causes that the recommender is unable to unmask user preferences. This problem is usually present in both content-based and collaborative recommenders. Specifically, for the collaborative recommenders, a cold-start problem occurs also when a new item appears. Since it is not rated by any users, it is not possible to score how appropriate would be to recommend such an item [Schein *et al.*, 2002].

There are several domains, which suffer from the new item cold-start problem more as others. In some domains recommended items are relevant for only a short time period (e.g., news, discounts) and thus the value of the recommended item decreases exponentially over the time.

The cold-start problem is not usually an isolated state of the system, but it is a process (its effect decay over the time, i.e., user activity). It is clear, that there is no specific line (e.g., an amount of user ratings) to be recognized as the "no cold-start". In [Visnovsky *et al.*, 2014], authors analyzed the influence of the amount of user rating to the quality of user similarity search (cluster quality). As we can see (Fig. 4.1), the increasing number of the user ratings logarithmically improves the cluster quality. For the MovieLens dataset approx. 50 ratings are required to obtain similar clusters as considering all the user ratings.

However, a new item problem does not affect the content-based recommenders, hence the content-based recommender can extract item properties without any user ratings[1]. Therefore, the content-based recommenders can be used to reduce the cold-start problem of collaborative filtering [Ronen *et al.*, 2013]. Moreover, several approaches aim at addressing not only important content-based features, but also important features selection [Cella *et al.*, 2017].

---

[1]This assumes that content (and similarity search) can be processed and computed immediately.

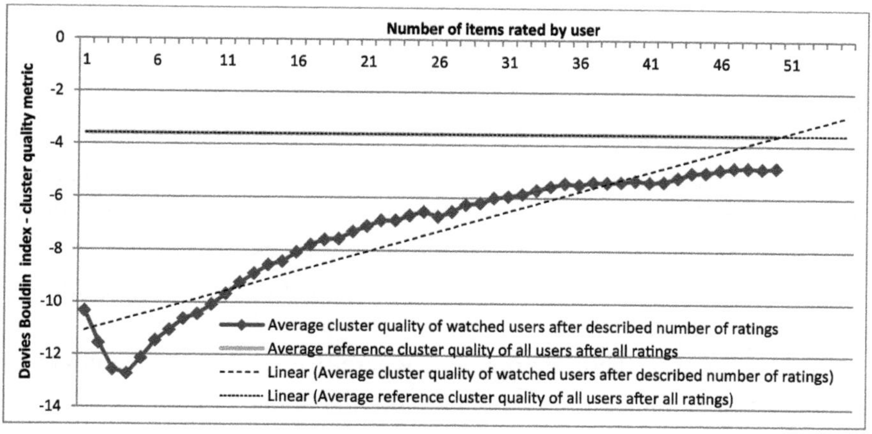

Fig. 4.1. The influence of the amount of user ratings to the cluster quality (similar user search task) in MovieLens dataset [Višňovský *et al.*, 2014].

Hybrid recommendation is able to solve the cold-start problem for both a new user and a new item. One example is a work of Schein [Schein *et al.*, 2002], where they fit a model using content and collaborative information. They present a two-way aspect model and Naïve Bayes recommender that uses content features in order to predict the ratings for the non-rated items.

The cold-start problem was further explored in [Braunhofer *et al.*, 2014], where the authors applied hybrid recommender to solve a problem of *a new context*. A cold-start problem of a new context occurs when an existing user is exposed to a new contextual situation. They proposed a switching hybrid recommender that combined a demographic-based context-aware recommender and demographics-based context-aware recommender.

However, as they outlined, evaluating such a hybrid recommender that was in addition extended by the contextual feature was a demanding task since there was a lack of large datasets suitable for this task. There are several datasets that can be used for this task: STS [Elahi *et al.*, 2013], CoMoDa [Odic *et al.*, 2013], and Music [Baltrunas *et al.*, 2011].

4.2.2.2. *Over-specialization*

One of the shortcomings of the content-based recommenders is that they are not able to recommend the *outside-the-box* items, also referred as a problem of *over-specialization* [Shardanand *et al.*, 1995]. Since the content-based recommender relies on the content descriptions during the user preferences analysis, it is limited to find the similar items to only those that user previously liked.

For example, if a user watches movie from the comedy and adventure genre, content-based recommender learns this information and builds a user model that is used to recommend only movies from these genres. Therefore, it may fail to recommend, for example horror movies even if a user would appreciate some.

On the contrary, there are domains where it is useless to recommend similar items. If someone bought an expensive camera, he/she probably won't buy another (within some reasonable time period).

Here, the hybridization can be beneficial if we combine both collaborative filtering and content-based recommendation. In such a hybrid environment, collaborative recommendation can be helpful in recommending the items outside-the-box. Moreover, the hybridization may eliminate the trade-off between recommendation accuracy and diversity of recommended items [Yoshii *et al.*, 2008].

On the contrary, the specific settings and domain characteristics may bring the over-specialization problem to the collaborative recommenders as well. As the collaborative filtering usually uses the most similar users, if these are highly consistent (and recommender is not designed to bring diversity), only highly specific items will be recommended (similarly to the content-based over-specialization problem).

The over-specialization problem refers to recommending highly tailored items to user past preferences. This often results to the problem of diversity lack. These are, however, two separate concepts. We may lack the diversity of recommended items without over-specialization problem (e.g., user likes adventure and receives sci-fi recommendations).

4.2.2.3. Sparsity

Real-world web applications contain tremendous amount of content and users. This is unfortunately a problem for the collaborative recommender approaches, which often use a user-item matrix (Fig. 4.2). In fact, such a

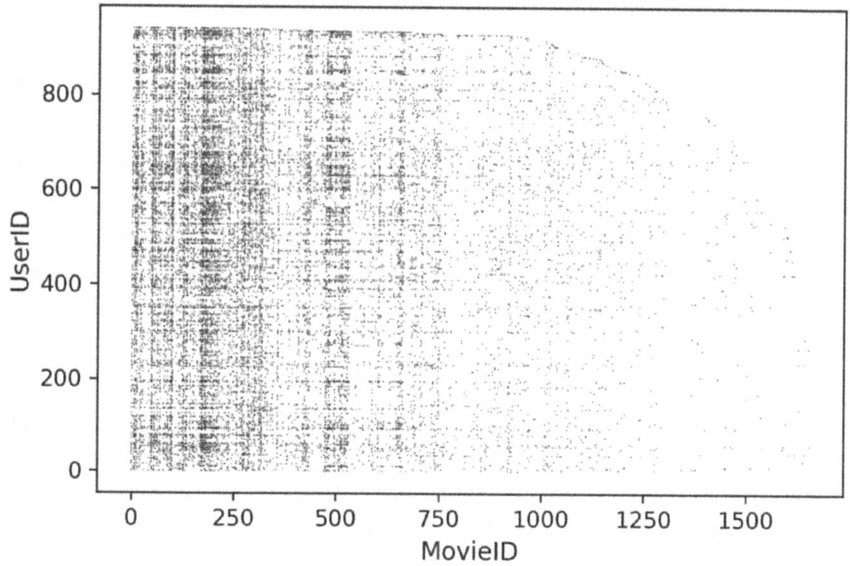

Fig. 4.2. User-item interaction matrix from the MovieLens 100k dataset.

matrix is extremely sparse in an average system. This is usually a result of the fact that many users interact with only few items.

One example is the MovieLens 20M dataset [Harper *et al.*, 2015], which contains 27 000 items (movies) and 138 000 users. An upper bound for the maximum number of ratings is therefore $3{,}726 * 10^9$, however the dataset contains only $2 * 10^7$ ratings.

Fortunately, *sparsity* is yet another issue of the recommender systems that can be reduced by the hybrid recommenders. By utilizing a hybrid model, the missing items from the matrix can be calculated, which solves the problem of sparsity. Several hybridization types are helpful, e.g., the feature combination. By combining several recommender sources (e.g., content and collaborative), we reduce the rating matrix sparsity. An example hybrid recommender was proposed in [Kim *et al.*, 2012], where the authors used a social network and trust scores between users to reduce data sparsity.

4.2.2.4. Gray and black sheep

Another problem of the standard recommendation techniques is the specific users (*extremes*), for which a particular approach can be not sufficient enough. Here we distinguish between two basic extremes: *gray sheep* and *black sheep*.

Gray sheep users do not have consistent opinions and thus do not clearly fall into any of the groups of people sharing the same opinion [Claypool *et al.*, 1999]. This problem occurs namely in the small and medium community of users.

Also, as noted in [Claypool *et al.*, 1999], unlike in the cold-start problem, even by gathering more ratings from such users, a recommender is unable to produce precise predictions. Depending on the dataset and a number of gray sheep users [Ghazanfar *et al.*, 2014], a presence of the gray sheep users may affect the quality of the recommendation for the whole community.

On the other hand, *black sheep* users [McCrae *et al.*, 2004] have no or few people that they correlate with. Therefore, recommendation approaches relying on the user-to-user correlations are unable to generate any predictions. Su *et al.* pointed out that although this is clearly a failure of the recommender system, non-electronic recommenders are unable to properly recommend items to black sheep users as well [Su *et al.*, 2009]. Therefore, we may consider such a failure to be acceptable.

Both gray and black sheep users cannot benefit from the collaborative recommendation. This is a consequence of the inability of the recommender to find a relationship between such a user and other users in the community. Similarly, a demographic recommender may have the same issue, since it uses demographic information about the users to categorize them into groups. However, here the solution of the problem is a hybridization where a collaborative or demographic recommender can be combined with a content-based recommendation.

[Ghazanfar *et al.*, 2014] utilized K-means clustering to identify the gray sheep users and proposed a switching hybrid recommender that was able to decrease the recommendation error rate by switching between the collaborative filtering and the content-based recommender.

4.2.3. *Drawbacks of the hybrid recommenders*

One reason to employ a hybrid recommendation is to *improve the performance* of individual — *monolithic* recommenders, such that the hybrid recommender performs better than any underlying recommender.

However, this requires that the underlying recommenders should be also well-tuned such that they are able to recommend items with satisfying accuracy. If the underlying recommender performs poorly, a hybrid recommender may fail in improving the accuracy and it may end up with the drop, indeed.

We need to choose which recommendation techniques we need to employ and optimize its parameters. Moreover, these recommender techniques need to be properly evaluated. For this step, it is required to have a good knowledge of the underlying recommendation techniques, but also, we need to understand the domain.

Here we should take into consideration the basic *domain characteristics* [Burke *et al.*, 2011]: heterogeneity (of items in the domain), degree of risk (for a user accepting a recommendation), degree of churn (whether a recommender face a continual stream of new items), preferences (stable or unstable), interaction style (implicit or explicit), and scrutability (whether an explanation of recommendation is required by the recommender). Analysis of the domain allows us to choose an appropriate recommendation technique and consider the conditions within which it would run. For instance, in the news domain, where the degree of churn is relatively high, we need to consider the scalability of a hybrid approach.

Explanation of recommendations is a still an open research problem in monolithic recommenders [Herlocker *et al.*, 2000]. In case of hybrid techniques, the problem grows even further, hence we need to properly present an information about the source of the recommendation. By using for instance, a weighed hybrid, it could become cumbersome to determine which recommender contributes the most to the result and even more how this should be presented to a user.

Recently, there have been attempts to solve the issues with the explanation of a hybrid recommendation. For example, Bostandjiev [Bostandjiev *et al.*, 2012] used visual interactive interface that was intended to explain recommendation process and elicit additional user preferences.

This is related to another issue with the hybridization. Not only a particular recommendation technique may need some training phase, but also a hybrid recommender need to be trained in order to handle such particular recommenders. In other words, a hybrid recommender itself adds another parameter that need to be tuned [Campos *et al.*, 2010]. A cross-validation may be employed in order to set parameters (weights), such that the combination of recommendation techniques would fit the problem the best (e.g., which recommenders should be picked for the switching hybrid).

This is a case especially in a weighed hybrid, where the weights need to be estimated. Here, some heuristics may be applied, or these weights can be set with the machine learning. Moreover, these weights can be also personalized, which requires not even more time to train recommender, but also more training instances.

Finally, a hybrid recommender usually requires an additional computation complexity (as more methods are used), which results in worse performance than the monolithic approaches [Cremonesi *et al.*, 2011]. The issues of scalability and distributed approaches is deeply discussed in Chapter 11.

4.3. Practical implementation considerations

The concept of combining several recommenders to overcome notorious shortcomings is widely accepted. Most of studies in the recommender systems field pointing improved results when used hybrid recommenders. Thanks to this "agreement", there are plenty of libraries and frameworks implementing (or supporting) hybrid recommender approaches. In this section, we will briefly analyze the most important features of these (Table 4.2).

Table 4.2. Comparison of libraries and frameworks supporting hybrid recommendation.

Name	*Language*	*License*	*Type of combination*	*Evaluation*	*Note*
Mrec	Python	BSD	Weighted, Cascade	yes	-
Matchbox	AzureML	Microsoft online services	Switching	yes	-
Surprise	Python	BSD-3 Clause	–	yes	Custom hybrid implementation is required
LightFM	Python	Apache v2	Feature combination	yes	-
Librec	Java	GNU GPL	Weighted	yes	-
LensKit	Java	LGPL v2.1	Weighted	yes	-
MyMediaLite	.NET	GNU GPL v3	Weighted	yes	-
Easyrec	Java	GNU GPL	–	no	Custom hybrid implementation is required
PredictionIO	Scala	Apache Licence v2.0	Multiple	yes	-
FluRS	Python	MIT	–	yes	Custom hybrid implementation is required
Seldon	Python	Apache Licence v2.0	Cascade	yes	-
Recommenderlab	R	GNU GPL v2	Weighted	yes	-
Prea	Java	Free BSD	–	yes	Custom hybrid implementation is required
Duine	Java	LGPL v3	Switching	yes	-

4.3.1. *Mrec[2] recommender system library*

Mrec is a Python recommender and evaluation library developed at Mendeley [Mendeley, 2017]. As a part of it, there are several algorithms implemented, which can be used either standalone or as a part of the recommender. The library provides an implementation for:

- SLIM item similarity,
- Weighted matrix factorization WRMF,
- Weighted approximately ranked pairwise ranking loss (WARP),
- Hybrid model which optimizes WARP based on user-item matrix and content features,
- various evaluation metrics (such as Precision, Recall, or Mean Reciprocal Rank).

For a fast development, a command-line interface is available. In addition, the library supports parallelization using IPython. The input for the hybrid recommender consists of the user-item matrix and the content features. A core approach for the library is the WARP algorithm, which reached promising results on the well-established image dataset ImageNet[3] — in the mean of the speed, memory usage, and the performance as well [Weston *et al.*, 2010].

4.3.2. *Matchbox[4] recommender*

Azure machine learning is getting more and more attention in the last years. The Matchbox recommender, which is available as a part of this machine learning platform, is a large-scale recommender system. It includes both collaborative and content-based approach. These are combined based on the Bayesian probabilistic model.

The main idea is to use the content-based approach first (when a user is relatively new to the system and has only few ratings). Next, the smooth transition to the collaborative filtering is performed as more and more ratings for the user are available.

---

[2]https://mendeley.github.io/mrec
[3]http://www.image-net.org
[4]https://msdn.microsoft.com/en-us/library/azure/dn905987.aspx

Two types of content-based features are supported — item and user content features (characteristics). The framework also supports three types of feedback [Stern *et al.*, 2009]: (a) explicit user ratings of items, (b) binary preferences (likes and dislikes), (c) ordinal ratings on a user-specific scale. One of the major shortcomings is the lack of an online training (model has to be retrained periodically).

Model optimal parameters search is offered through the Tune Hyperparameter Module and Cross Validation Module. Also, several metrics to evaluate the performance are available (e.g., MAE, RSME, Precision, AUC).

As the experiments showed [Stern *et al.*, 2009], the content-based features are especially important in the cold-start phase. Together as a hybrid approach, the Matchbox reflects the state-of-the-art performance.

4.3.3. *Surprise*[5] *library*

SciPy provides a collection of packages for scientific computation. The Surprise library is a Scikit (SciPy toolkit) library for building and analyzing recommenders [Hug, 2017]. Although it is intended for an easy implementation of custom recommenders, it also provides a range of popular algorithms. The core functionality covers:

- dataset handling (MovieLens and Jester included),
- prediction algorithms — neighborhood methods (kNN), matrix factorization (SVD, SVD++, PMF, NMF), and similarity measures (cosine, Pearson, MSD),
- evaluation support (cross-validation), parameter optimization.

The library itself does not implement any of the hybrid approaches. The ecosystem allows to create custom recommenders, though. In this way, we are able to create a variety of recommenders on the level of a rating prediction or rank reordering.

The performance of the algorithms is evaluated based on the RMSE, MAE, or FCP metrics. One of the important characteristics is the documentation, which provides relevant information and a plethora of examples.

---

[5]http://surpriselib.com

4.3.4. *LightFM[6] library*

Yet another Python implementation. The name is derived from "factorization machines" and combines the content and collaborative ideas [Kula, 2015]. The users and items are represented as the latent vectors, which are defined by the linear combinations of embeddings of the content features (users and items).

Implemented model reflects the data available for the training. If there are no content features provided, it acts as a pure collaborative filtering approach. When the content features are available, these are considered in the optimization process (also useful for the cold-start problem reduction). In total, four loss functions are implemented:

- Logistic,
- Bayesian probabilistic rating,
- Weighted approximate-rank pairwise,
- k-OS Weighted approximate-rank pairwise (k^{th} positive example as a bias).

A model performance evaluation is supported by the implementation of the standard metrics: Precision, Recall, AUC, and Reciprocal rank. Moreover, LightFM allows to easily obtain the MovieLens 100k dataset[7] and use it for the fast experiments.

The LightFM is also available as a Docker container. The documentation provides several examples over various scenarios.

4.3.5. *Librec[8]*

Librec is a Java library, which includes plenty (over 70) of algorithms implementations. The library consists of several modules, which cover the whole process of recommendation (Fig. 4.3).

The library implements a weighted hybrid recommender, which uses a linear combination of HeatS and ProbS algorithms (derived from the heat and probability spreading) [Zhou *et al.*, 2010]. In total, six types of

---

[6]https://github.com/lyst/lightfm
[7]http://grouplens.org/datasets/movielens/100k/
[8]https://www.librec.net

recommenders are included, while each of them consists of several algorithm implementations:

- Abstract Recommender — provides a set of basic algorithms (e.g., most popular, collaborative, association rules, global average, hybrid),
- Probabilistic Graphical Recommender (e.g., clustering, LDA, PLSA, BUCM),
- Matrix Factorization Recommender (e.g., SVD, BPR, WRMF, RBM),
- Factorization Machines Recommender (e.g., FMALS, FMSGS),
- Social Recommender (e.g., TrustMF, TrustSVD, SOREG, RSTE),
- Tensor Recommender (e.g., BPTF, PITF).

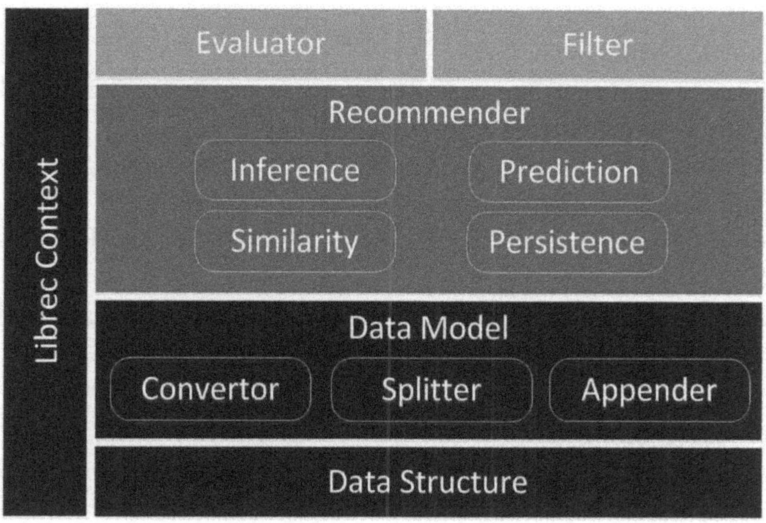

Fig. 4.3. Librec modules overview. The final algorithm is a combination of these components. A set of interfaces allows a flexible implementation of any new algorithms[8].

Several metrics for the performance evaluation are also included, e.g., AUC, nDCG, Precision, Recall, MAE, MPE, and RMSE. Also, a FilmTrust dataset was extracted and included. The documentation provides details for the library usage, with references to the active blogs and discussion forums.

4.3.6. *LensKit*[9]

Lenskit is an open-source toolkit for building and researching recommender systems, created at University of Minnesota by the GroupLens research group [Ekstrand *et al.*, 2011]. The toolkit was used in over 40 research papers and is also a part of the MovieLens project.

LensKit consists of the several modules focused on the similarity calculation, recommendation, and evaluation of the performance. Four basic algorithms are implemented:

- item-based collaborative filtering,
- user-based collaborative filtering,
- matrix factorization (FunkSVD),
- slope-one rating prediction.

The linear weighted hybrid recommender mechanism is also provided, which allows to combine two recommender lists.

To evaluate the performance of build algorithms, two groups of metrics are supported: prediction accuracy metrics (e.g., RMSE, MAE, Coverage) and top-n (or ranking) metrics (e.g., MAP, MRR, Precision, Recall, and nDCG).

Since the toolkit is supported by the one of major recommenders research group, the community is highly active.

4.3.7. *MyMediaLite*[10]

The library was created and currently is maintained by the research group at University of Hildesheim [Gantner *et al.*, 2011]. Thanks to its academic background, it has been utilized in over 20 research papers. There are several algorithms implemented in the library, while in addition, own approaches are supported as well. Two basic scenarios are feasible — the rating prediction and the item prediction:

- Item recommenders (e.g., Random, Most popular, Incremental),
- Rating prediction (e.g., SlopeOne, BPSO, Latent-feature log linear, Matrix factorization with factor-wise learning).

---

[9]http://lenskit.org
[10]http://www.mymedialite.net

For the hyperparameter optimization, a grid search and the Nealder-Mead algorithm is used. As a part of the library, a weighted hybrid recommendation is also provided.

Evaluation module includes the cross-validation and the online evaluation for the several standard metrics (e.g., MAE, RMSE, AUC, nDCG, Precision). MyMediaLite also supports the real-time incremental updates for the selected recommenders.

4.3.8. *Easyrec*[11]

Easyrec service is made available for the public usage by using the instance provided by the Smart Agent Technologies of the Research Studios Austria. However, the source code is accessible and allows to run an own instance. Easyrec also supports the third-party plugins to integrate with the popular web-based applications (e.g., Drupal, Mediawiki) via the RESTful Web services.

Several non-personalized and personalized algorithms are already implemented within the service:

- Bought together,
- Popular,
- SlopeOne,
- Association rule miner.

The service is designed such that there is no need to implement any recommenders, which partially limits its possibilities, though. Also, there is no evaluation support provided within the service.

4.3.9. *PredictionIO*[12]

PredictionIO is currently an incubating project of Apache covering the predictive engines for various machine learning tasks. The platform consists of three parts (Fig. 4.4):

- core machine learning stack (intended for building, evaluating and deploying algorithms),

---

[11]http://easyrec.org

[12]http://predictionio.incubator.apache.org

- event server (unifying the events from multiple platforms),
- a template gallery (a storage of algorithm implementations).

The recommenders template gallery contains a number of implementations aiming at the specific tasks (e.g., in the domain of e-shops):

- Collaborative filtering: user and item based,
- Content based: products similarity,
- Association rules, Frequent pattern,
- Complimentary purchases,
- Personalized ranking,
- Hybrid recommendations.

Among the recommender templates, also a classification, regression, clustering, and NLP tasks are supported. The hybrid idea is supported in each of these tasks, while various combining (ensemble) mechanism can be utilized.

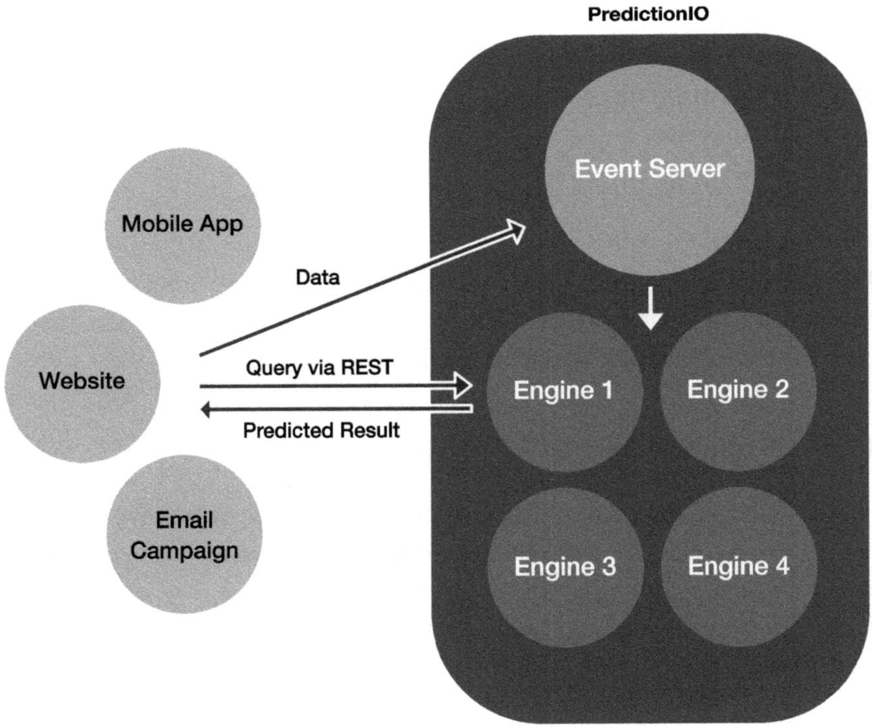

Fig. 4.4. PredictionIO core components[12].

An evaluation of the performance is provided by the Tuning and Evaluation module, which supports the optimal parameters search and standard evaluation metrics (e.g., Precision, Recall, Accuracy).

PredicionIO is a highly scalable platform as it bases on Apache Hadoop, HBase, Spark and ElasticSearch (also available as Docker container). The project benefits from the extensive documentation with a plenty of examples and highly active community of the developers.

4.3.10. *FluRS*[13]

Build in Python, FluRS is a small open-source project for an online item recommendation for Python. Its main idea is to provide the "fluent", i.e., incremental recommendation algorithms. Several algorithms are implemented, such as:

- Incremental collaborative filtering (based on the kNN),
- Incremental Matrix factorization and Matrix factorization with BPR optimization,
- Incremental Factorization machines.

A native support for the hybrid recommenders is not provided, on the contrary several metrics for the performance evaluation are available (e.g., MAP, MRR, Precision, Recall). As the project is relatively small and new, the documentation is still evolving.

4.3.11. *Seldon*[14]

Seldon is a platform supporting machine learning tasks intended for the deploy in the production. It runs within a Kubernetes Clusters and supports several model-building tools. Two basic endpoints are available:

- Prediction (several Python pipelines support),
- Recommendation (similar users, latent factor models, association rules, content based, collaborative, hybrid).

Seldon is capable of the weighted and cascading combination of several algorithms and also allows to implement an own hybridization approach.

---

[13]https://github.com/takuti/flurs

[14]https://www.seldon.io

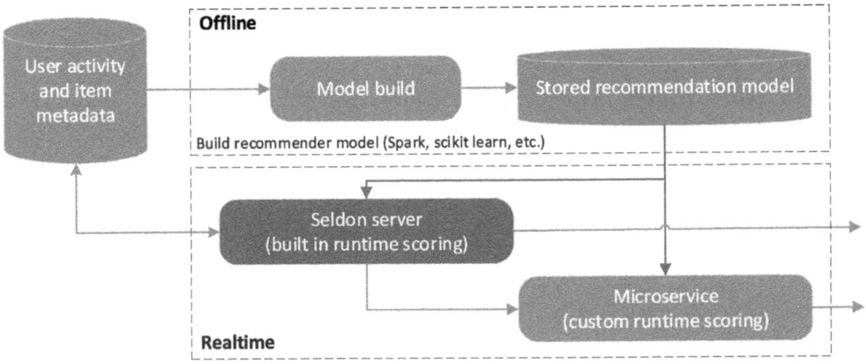

Fig. 4.5. Seldon recommendation components[14].

The process of generating recommendations is divided into the offline and the real-time part (Fig. 4.5).

The platform supports extensive monitoring and analytics via the third-party applications. Also, a paid support for commercial projects is available.

Seldon community is highly active, which results in a rich documentation with many examples.

4.3.12. *Recommenderlab*[15]

Framework Recommenderlab aims at providing general research infrastructure rather than to create recommender applications. Authors focus on optimizing the process of experiments covering efficient data handling, easy incorporation of algorithms and evaluation [Hahsler, 2017]. Several algorithms are available within the framework:

- Collaborative filtering: user and item based,
- Association rules,
- Most popular, Random,
- Hybrid recommenders: weighting scheme.

The framework supports standard evaluation metrics (e.g., MAE, RMSE, ROC, Precision, Recall). Similarly, the cross-validation, bootstrap sampling serves for the model performance comparison. As a standard for

[15]http://lyle.smu.edu/IDA/recommenderlab

the R libraries, Recommenderlab is well documented and a variety of examples is included in the documentation.

4.3.13. *Prea*[16]

Toolkit Prea is focused on the collaborative filtering approaches [Lee *et al.*, 2012]. Natively, there is no hybridization technique available, but the support for the own recommenders is provided. Comparing to the Mahout or MyMedia, Prea also implements several Matrix factorization approaches. Moreover, many additional algorithms are implemented:

- Random, Constant, Average,
- User and Item based, Slope-One,
- Matrix factorization (SVD, NMF, PMF).

For the evaluation of the performance, the toolkit has built-in some basic metrics (e.g., RMSE, MAE, nDCG). Surprisingly, Precision and Recall are not supported. For the data split, a cross-validation is supported. In [Lee *et al.*, 2012], the authors compared the implementations of the algorithm to the MyMedia framework resulting in the very similar values for MAE and RMSE metrics respectively.

4.3.14. *Duine*[17]

Duine is rather a smaller framework mainly focused on the predictive tasks for the recommendation. Its idea is based on a concept of plugins (e.g., profile models, feedback processors). From the hybrid recommenders perspective, the switching mechanism is utilized. Interesting feature of Duine is an Explanation API, which helps with creating user-friendly explanations for the end-users. Several prediction techniques are implemented:

- Average,
- User-based collaborative filtering,
- Content-based,
- Case-based reasoning.

---

[16]http://prea.gatech.edu

[17]http://www.duineframework.org

The framework provides a variety of practical examples; however, the last update comes from 2009. Despite this, it can be used as a solid starting point for the own implementations.

4.3.15. *Summarization*

Whether attempting to create a new recommender, or using an existing one, recommendation libraries can be useful during the whole development process. However, it is sometimes cumbersome to choose a proper one.

Firstly, it is important to know, which task we aim to do — a rating prediction or an item recommendation (i.e., a Top-N item recommendation). In fact, nearly every library that is capable of the rating prediction can be used for the item recommendation task as well (items can be sorted by their predicted ratings and returned in a Top-N list). However, some libraries are adapted to the item recommendation task and also utilize various techniques in order to improve the order of the recommended items (e.g., learning to rank) with respect to the ranking metrics (such as precision, or recall). One notable example is a LightFM library that uses various loss functions that are meant to optimize the ranking of the items (e.g., Weighted approximate-rank pairwise).

In the training and evaluation phase, the dataset (about users, items, and their interactions) plays an important role. One source is to use the existing recommendation datasets (such as MovieLens; we further describe several available datasets later) or to use the custom ones. The main advantage of the existing datasets is that we are able to compare our results with the research community without the need of implementing the state-of-the-art techniques. However, when creating a recommender for the production environment, there is also a need to evaluate the performance of the recommender on custom data. Here, we may utilize the pre-implemented state-of-the-art recommendation methods.

Some of the libraries support data preprocessing, e.g., the normalization. For instance, (e.g., Surprise) offer tools to handle data preprocessing, or machine learning libraries can be utilized (such as Scikit-learn[18] for Python).

Finally, an important criterion is whether we plan to deploy our recommendation technique to a production environment. This can be usually achieved using every library listed above, however, several of them are more suitable for this task and support the whole pipeline of the process of the recommendation: data preprocessing and storage, model training and storage, evaluation, and serving recommendation. Also, there are other relevant factors:

- Do we need to scale to many users (i.e., the process of serving recommendation need to be fast)?
- Do we need to keep the recommendation model up-to-date (i.e., method of recommendation should be capable of a frequent model training without any significant impact on the performance)?
- Is it possible to perform an incremental model update?

Examples of the libraries that are suitable for the production environment are PredictionIO and Seldon.

4.4. Practical evaluation considerations

There are several best practices that should be followed when evaluating a recommender system. We list some common principles that are applicable to any technique and add those specific for the hybrid recommendation techniques.

4.4.1. *Defining an experiment*

One of the first important things that need to be considered during the evaluation is the dataset. Here a correct splitting criterion should be chosen. A common approach is to split the whole dataset into the training and the test set. Usually, it is also suggested to interchange these sets and conduct several successive experiments, such that the train and test set is always different (e.g., k-fold cross validation).

---

[18]http://scikit-learn.org

Besides the train and tests sets, also a validation set is important. Especially, in a case of hybrid recommenders, where we need to set and test various parameters, weights, and combinations, the experiments conducted on a validation set are a necessity. The validation set need to be different from the train and test sets.

In the domain of recommenders, there are two basic types of experiments: online and offline (see Chapter 9). Offline experiments are conducted using the pre-collected datasets [Ricci *et al.*, 2015] (for details see Sec. 4.4.4). These experiments are usually performed as soon as the first prototype of the recommender is available and are intended to well-tune the parameters and to explore recommender basic characteristics.

The offline experiments should be not interpreted as the exact measure of the recommender performance. They provide rather the worst-case scenario estimate (as the users do not have a chance to see recommendations and to interact with them).

In an online experiment scenario, we are able to measure how the recommender system influences user behavior, while he/she interacts with the presented items [Ricci *et al.*, 2015]. Clearly, the advantage is that the experiment is conducted with the real users performing the real tasks. Usually, users are split into the groups, where each group experience a different recommendation setup (A/B testing). There are also some drawbacks, such as the risk that the users will be faced a non-relevant recommendation and they leave the experiment too early. There, an online study is done after the offline experiments since the recommender is supposed to be well-tuned and several parameters are already set.

4.4.2. *Evaluation in hybrid recommenders*

Since the hybrid recommender may consist of different underlying recommender techniques, this need to be considered during the *evaluation*. This also depends on the used combination strategy. Therefore, there are several approaches that were adopted by the researches and were used during their evaluation phase.

The most common practice is to compare the performance of the hybrid recommender (as a whole) to the other state-of-the-art techniques. The state-of-the-art techniques can be either baseline monolithic techniques (such as collaborative filtering) or other hybrid recommenders.

Problem with such a simple approach is that the hybrid recommender is a black-box and we do not know how the underlying recommenders perform. Therefore, many authors also examine different setups of hybrids with respect to the choice of underlying recommenders, used features, and parameters.

Specifically, for the cascade hybrid combination [Lampropoulos *et al.*, 2012], it is suitable to focus on the improvement of the second-level recommender with respect to the first-level recommender (i.e., whether a second-level recommender is viable to *improve* the performance of the first-level recommender). However, as we have discussed before, it also important to have the first-level recommender well-tuned and evaluated, as well.

In the case of hybrids that combine results of two or more monolithic recommenders, it is more appropriate to evaluate these monolithic approaches separately and then as a whole. For instance, different weights should be set in order to tune the weighted recommender. We can go even further and analyze all the underlying recommenders to compare whether their predictions are similar and how close they are to each other [Hornung *et al.*, 2013].

For instance, in feature combination hybrids, various feature configurations can be investigated [Zanker *et al.*, 2009]. Here, not only different inputs should be considered, but also a different weighing of these inputs. Moreover, if we keen to bring a more personalized experience, a relevance scores of the inputs can be measured for the user (of group of users) separately.

An initial motivation to create hybrid recommenders was not only to improve performance, but also to address the issues associated with the monolithic approaches. However, here we need to focus on the data that we analyze and also for the metrics we choose.

To verify a cold-start problem, a performance for the new user, new item, or new context need to be investigated [Braunhofer *et al.*, 2014]. Based on the combination strategy, several comparisons need to be utilized for each scenario. To illustrate the cold-start problem, there should be also suitable data containing users (or items, or contexts) with the low number of ratings. In some scenarios, these cold-start users (or items, or contexts) may be evaluated separately. If the dataset does not contain such data, we may simulate the cold-start problem by removing the selected number of interactions.

Similarly, for the gray and black sheep users, we first need to identify those users and measure how the recommendation quality differs based on the approach. However, it is usually necessary to verify whether a recommender is able to provide relevant items to non-extreme (i.e., other than gray and black sheep) users.

[Miranda *et al.*, 1999] investigate whether gray sheep users would benefit from using hybrid recommendation rather than collaborative or content-based only. Evaluation procedure was conducted using an online experiment, where these users were actually a part of the study. [Ghazanfar *et al.*, 2014] utilized state-of-the-art datasets and performed offline evaluation, where gray sheep users were separated from the whole dataset and the performance of monolithic approaches was compared to the performance of the switching hybrid.

An issue of diversity is measured as a metric itself — we can measure how diverse the resulting recommendations are or, more precisely, how diverse their *properties* are. For instance, [Burke *et al.*, 2014] analyzed user-based diversity (how users differ within group) and tag-based diversity (how different item tags are) and compared them between three different hybrid approaches.

4.4.3. *Evaluation frameworks*

Usually, a framework which supports prediction tasks for the recommendation also supports some kind of the performance evaluation. There are, however, frameworks designed specifically for the evaluation of recommender algorithms.

4.4.3.1. WrapRec[19]

WrapRec is a configuration based open-source project (under the MIT license) written in C#. Its idea is to support a fast evaluation of the custom algorithms or the algorithms adopted from the other frameworks [Loni and Said, 2014]. WrapRec architecture is split to three parts (Fig. 4.6).

First one is a module which brings a native support to MyMediaLite and LibFM frameworks. Moreover, also the third-party libraries and custom build recommenders can be evaluated by extending WrapRec. The second module is the Split, which defines the way the data are handled from the train and test evaluation perspective. After the data is loaded through Data reader, they are stored in Data container. By extending the split class, a split can be fully customized. Finally, the metrics used for the evaluation are defined in the Evaluation context module. Multiple evaluators are supported.

The WrapRec is a part of the research project CrowdRec based on the Delft University of Technology. It comes with an extensive documentation including several examples.

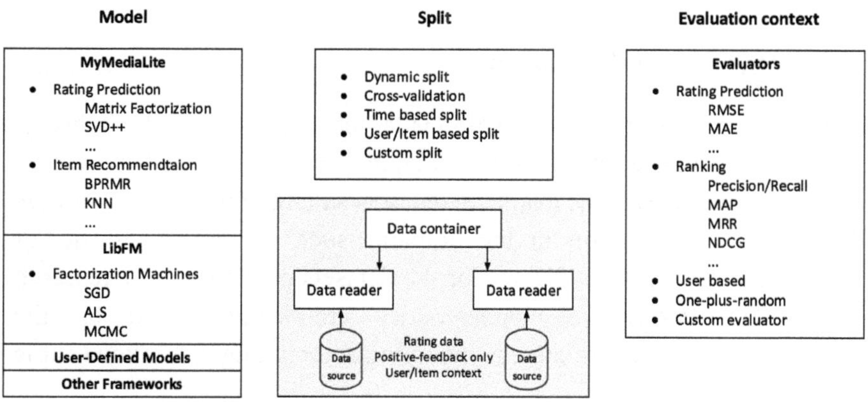

Fig. 4.6. WrapRec evaluation framework architecture[19].

[19]http://babakx.github.io/WrapRec

4.4.3.2. *Rival*[20]

Another open-source toolkit designed especially for the recommender evaluation is Rival, which is written in Java under LGPL v2.1 license. As it is quite new toolkit, its documentation needs to be improved in the future. Rival consists of four basic modules (important from our point of view).

The evaluation module implements several metrics and strategies for the evaluation. Error metrics include MAE and RMSE, while Precision, Recall, nDCG and MAP as ranking metrics are included.

A recommendation module integrates algorithms from the LensKit and Apache Mahout, which provides a complement to their evaluation tools. The split module is, as expected, responsible for the train and test data splitting. Standard Cross validation is supported. Moreover, a random split and temporal split (considering a timestamp of instances) is available.

Last, but not least, the example module provides examples of the toolkit usage on a real-case scenario. The toolkit is available via the Maven repository.

4.4.4. *Overview of the datasets*

Since the domain of recommendation is largely connected to the industry, there are many real-world datasets available that are being used in research papers. We summarize the available datasets suitable for the collaborative filtering recommendation in the domains such as movies, music, or e-commerce (Table 4.3). We opt for the datasets covered by the research papers. The descriptive characteristics are supplemented by the comparison of the performance in the rating and ranking prediction (Table 4.4).

A majority of the listed datasets is intended to be used for the evaluation of the explicit feedback, namely to predict the ratings of the items. However, in many scenarios, these ratings can be also used as an implicit feedback, as well and can be utilized to evaluate the Top-N collaborative recommendations. In addition, many selected datasets

---

[20]http://rival.recommenders.net

contain not only the interactions between users and items, but also the characteristics of either the users (demography) or the items (content characteristics) that can be utilized in the hybrid collaborative recommendation.

Table 4.3. Overview of available recommendation datasets.

Dataset	*Released*	*Ratings*	*Users*	*Items*	*Sparsity (%)*
MovieLens 100k	10/2016	100K	1K	9K	98.889
MovieLens 1M	2/2003	1M	6K	4K	95.833
MovieLens 10M	1/2009	10M	72K	10K	98.611
MovieLens 20M	10/2016	20M	138K	27K	99.463
MovieLens 26M	8/2017	26M	270K	45K	99.786
DouBan	2011	16.8M	129K	59K	99.779
Flixter	9/2010	8.2M	1M	49K	99.983
Last.fm 1K	5/2010	19.1M	992	1.5M	98.716
Last.fm 360K	3/2010	17.6M	359K	294K	99.983
Yahoo Music R1	3/2004	11.5M	-	98K	-
Yahoo Music R2	2006	717M	1.8M	136K	99.707
Yahoo Music R3	2006	300K	15K	1K	98.000
Yahoo Music KDD Cup Track 1	2011	262M	1M	625K	99.958
The Million Playlist Dataset	1/2018	1 000 000 playlists with 5-250 of tracks			
Epinions (product ratings)	10/2007	664K	50K	140K	99.990
Epinions (trust ratings)	10/2007	487K	-	-	-
Amazon Product Data	2015	142M	-	-	-
Yelp	2018	5.2M	-	174K	-
Book-Crossing	9/2004	1.1M	279K	271K	99.999
Jester Dataset 1	2003	4.1M	73K	100	43.836
Jester Dataset 2	2012	2.2M	79K	150	81.435

Table 4.4. Results of the evaluation of recommendation approaches using the selected datasets.

Dataset	*RMSE*	*MAE*	*Precision*	*Recall*	*F-score*	*AUC*
MovieLens 100k	0.8906[A]		0.3526[A]			
MovieLens 1M	0.8333[B]		-			
MovieLens 10M	0.7764[C]		-			
MovieLens 20M	0.7762[C]		-			
Flixter	1.0954[D]		0.046[E]			
DouBan	0.6988[D]		0.082[E]			
Last.fm 1K	-			0.301[F]		
Yahoo Music T1	21.2634[G]					
Yahoo Music T1	21.879[H]					
Yelp	1.0072[I]	0.7920[I]			0.63[J]	0.85[J]
Yelp					0.0152[K]	
Epinions		0.9321[L]				
Epinions		0.825[M]				
Amazon (clothing)						0.7961 / 0.7317[N]
Amazon (home)						0.7155 / 0.6396[N]
Amazon					0.033[K]	
Book-Crossing (item-based)			0.0364[O]	0.0732[O]		
Book-Crossing (user-based)			0.0369[O]	0.0576[O]		
Jester Dataset 1	4.1229[P]	3.1606[P]				

A http://www.mymedialite.net
B [Lee *et al.*, 2013]
C [Strub *et al.*, 2016]
D [Ma *et al.*, 2011]
E [Wu *et al.*, 2017]
F [Yang *et al.*, 2012]
G [Zheng *et al.*, 2012]
H [Koenigstein *et al.*, 2011]
I [Hu *et al.*, 2014]
J [Tsai, 2016]
K results are reported at F-Score@5; [Chen *et al.*, 2016]
L detailed results are reported in the paper; [Ma *et al.*, 2008]
M [Pham *et al.*, 2011]
N first result is a warm-start setting, second result is a cold-start setting; [Liu *et al.*, 2017]
O [Ziegler *et al.*, 2005]
P [Takacs *et al.*, 2009]

Movies has been one of the most dominant domains in collaborative filtering recommendation for many years. Its popularity is ascribed namely to the Netflix Prize competition, however, also to a great availability of the datasets. The MovieLens datasets[21] published by the GroupLens contain movie ratings and tagging activity of the users on the online portal MovieLens.org. Basic content information is also available, but this can be easily extended through the mapping to the external sources: IMDB[22] and TMDB[23]. Douban is the Chinese social network allowing users to rate, review and recommend movies, music, and books [Ma *et al.*, 2011]. Ma *et al.* crawled the movie section of the portal and besides the ratings they obtained also 1.7M user-to-user relationships (friends links). Similarly, user-to-user relationships are also available for the Flixster dataset [Jamali and Ester, 2010]. From totally 1M users, the user-movie ratings are available for only 150K of them. Although the number of ratings is considerably lower than in the case of MovieLens or DouBan, Flixster dataset stands out in much greater number of social interactions (26.7M). An example of TV and movie recommender is presented in Chapter 14.

Yahoo published several recommendation datasets in the domain of *music* with varying amount of information[24]. Yahoo Music R1 contains the rating activity of users over artists (i.e., the items are artists). Yahoo Music R2, a much bigger dataset, contains the rating activity of users on the songs. In addition, there are also metadata available (such as artist, album, genres), though these metadata are represented only using the anonymous identifiers. R2 dataset was used in an evaluation of a data-parallel low-rank matrix factorization [Schelter *et al.*, 2013], where authors applied MapReduce technique to improve the performance of the computations in order to better match production environment requirements. The third dataset — Yahoo Music R3 contains a sample of ratings for the songs that were collected from the users' interactions and from the online survey conducted by Yahoo Research. An example of music recommender is presented in Chapter 15.

---

[21]https://grouplens.org/datasets/movielens/
[22]https://www.imdb.com/
[23]https://themoviedb.org/
[24]https://webscope.sandbox.yahoo.com/catalog.php?datatype=r

As a part of the KDD Cup 2011 competition[25], another music ratings dataset was released that contained 10-years (1999–2009) rating activity on four types of items: tracks, albums, artists, and genres. There are also four different versions of this dataset available varying by the amount of rating activity, where one version is focused on the learning-to-rank problem rather than rating prediction.

Last.fm is a popular online service intended to provide music recommendations based on the music that users listen on their devices. In 2010 there were two datasets released [Celma, 2010]: Last.fm 1K[26] and Last.fm 360K[27]. Last.fm 1K contains full listening history of 992 users represented by the tuple user-artist-track. Last.fm 360K provides history for 359,347 users, however, this dataset contains only information about the playcount of the artists. Both datasets contain users' demography (gender, age, country, sign-up date) and MusicBrainz[28] ID for artists and tracks (if available), which allows to extend the dataset for additional metadata.

For the task of the playlist continuation prediction The Million Playlist Dataset (MPD) was published as a part of the Recommender Systems Challenge 2018[29]. Provided dataset contains 1 million playlists created by the U.S. users in the online music streaming service Spotify[30]. When the dataset was generated, there were several criteria for picking up the representative playlists (e.g., minimum number of artists in a playlist, minimum number of albums in a playlist). In comparison to other datasets available in this domain, one disadvantage is the missing information about the playlist authors (i.e., user ids). Since at the time of writing this book the challenge was just announced, there were no published papers using this dataset.

---

[25]http://www.kdd.org/kdd2011/kddcup.shtml

[26]http://www.dtic.upf.edu/~ocelma/MusicRecommendationDataset/lastfm-1K.html

[27]http://www.dtic.upf.edu/~ocelma/MusicRecommendationDataset/lastfm-360K.html

[28]https://musicbrainz.org

[29]https://recsys-challenge.spotify.com/

[30]https://www.spotify.com

Amazon[31] is one of the most famous online retailers with roughly millions of daily users. In 2015 Julian McAuley released[32] the Amazon *product reviews* dataset [McAuley *et al.*, 2015], which contained the reviews and product metadata from the May 1996–July 2014 period. Reviews consist of numerical ratings, textual reviews, and helpfulness votes. Product metadata include descriptions, categories, price, brand, related products, sales rank, and visual features extracted from the product images (4096-length feature vector extracted using the deep convolutional neural network). Reviews can be retrieved as a whole or as the subsets distinguished by the product categories. In addition, the author provides the raw ratings (a tuple user-item-rating-timestamp) that can be suitable for the training using some of the recommender systems libraries (e.g., MyMediaLite).

Epinions.com was an online service (run by eBay, now discounted) more focused on the opinions of the users on various items (e.g., cars, books, movies, software) [Massa and Avesani, 2007]. Moreover, users were able to "rate" other reviews and express their trust to the reviews of these users (i.e., which reviews they find valuable and which they found offensive, inaccurate, or not valuable at all). Massa *et al.* crawled Epinions.com website in order to evaluate a trust-aware collaborative filtering recommender. The dataset[33] that they released contained both the ratings of the users on the products, but also the ratings of users on other users' reviews (i.e., the users' trust statements).

Another dataset containing reviews is from Yelp[34], where the businesses are being reviewed. Here, the dataset is published directly by Yelp and is updated periodically, since it is a part of an ongoing research challenge. It contains more than 5 million reviews of 174K businesses supplemented by 200K pictures. Moreover, there are 1.1M tips by 1.3M users available and also aggregated hourly check-ins for each business. Besides the reviews and check-ins, there are several metadata about the businesses (such as address, average rating, whether it is a take-out

---

[31] https://www.amazon.com

[32] http://jmcauley.ucsd.edu/data/amazon/

[33] http://www.trustlet.org/downloaded_epinions.html

[34] https://www.yelp.com/dataset

restaurant, availability of parking, business categories, etc.) and the users (such as list of friends, votes given by the user, opinions received by other users, etc.).

In a *book* domain, Book-Crossing dataset[35] was released, which contains implicit and explicit ratings (on a scale 1–10) of people on books from Book-Crossing.com [Ziegler *et al.*, 2005]. Book-Crossing[36] is an online community portal, where the users may exchange books between themselves. There are also demographic data available (users' location and age) and books metadata (title, author, year of publication, publisher, cover images).

Another popular dataset among the research community is from Jester Online Recommender System[37]. Dataset contains users' ratings on *jokes* on a rating scale from −10 to 10 [Goldberg *et al.*, 2001]. What is specific for this dataset is that the number of items is very low (100 and 150), thus resulting in low sparsity of the rating matrix.

4.4.5. *Summarization*

Generally, an evaluation of any recommender system is a systematic process where a researcher need to take into account several aspects: experiment setup, used data, examined metrics, and desired outcomes. Especially, if we think of hybrid recommender systems, there also other factors, such as whether we need to evaluate any monolithic recommenders, or whether we need to evaluate various combinations of attributes in data.

Existing recommendation libraries may simplify this process by providing pre-implemented tools. There are also libraries that are focused specifically on the recommender systems evaluations. In case of data, there are several datasets from a couple of domains publicly available, which allows not only to easily evaluate new recommendation approaches but also to be more transparent while comparing the results (of evaluation) with the research community.

---

[35]http://www2.informatik.uni-freiburg.de/~cziegler/BX/
[36]https://www.bookcrossing.com
[37]http://eigentaste.berkeley.edu/

4.5. Conclusions

Recommender systems have been proved beneficial in many domains. There are several types of recommendation techniques, such as content-based, collaborative, or demographic that are utilized with respect to the domain, user characteristics, item characteristics, and a goal. However, many of these techniques still suffer from various issues that have their roots namely in a lack of data or a specific user behavior.

Hybrid recommender systems were designed to combine the advantages of these techniques in order to solve their major issues and to improve an overall recommendation performance. There are several hybrid combinations that we may choose from and can be used as a template of how to combine two or more recommendation techniques.

Each combination may be suitable for a different scenario and in a different domain. Moreover, we should take into consideration their advantages and disadvantages. It is also important to note that these combinations are not tightened to any particular recommendation techniques. In addition, researchers are not limited to basic combinations, but there are many parameters that can be tuned (e.g., weights, switching criteria). Allowing to tune the parameters brings a possibility to create another level of personalization.

There are several issues related to collaborative filtering that can be reduced by employing a hybrid recommender. Most notable are the cold-start problem, over-specialization, lack of diversity, user extremes, and sparsity. Collaborative filtering may either suffer from these problems or can be used a solution to this problems with the combination of another technique. A major advantage of hybridization is that it enables system to recommend items in some scenarios that collaborative filtering may fail in (such as a new item or a user extreme).

Hybrid recommenders have become a part of many frameworks and libraries in the domain of recommendation. These libraries are also available in various programming languages (such as R, Python, Java, Scala) and support different hybrid combinations. Moreover, many of these libraries are also extendable (e.g., Surprise, Easyrec, FluRS), which allows researchers to easily create their own combinations and yield more valuable research results. They also consist of many other algorithms that are related to the recommendation, such as similarity measures, plenty of matrix factorization approaches, or prediction algorithms.

In most cases, the libraries also provide tools to evaluate performance of the recommendation and measure some basic and also advanced metrics. For this purpose, several state-of-the-art datasets were created and are publicly available as well.

Evaluation of hybrid recommenders inherits many specifics of basic recommendation techniques. Firstly, it is important to pick a suitable dataset and split the data into train and test sets. While designing an experiment, we may opt into online or offline evaluation. Choice of the evaluation metrics should be performed with the respect to the recommendation task.

There are also specifics that are related to the evaluation of the hybrid recommenders. Here we should not only evaluate hybrid recommender as a whole, but also consider evaluating particular underlying recommenders, as well as their parameters. Moreover, if a hybrid recommender was designed to solve the recommendation issues, it is necessary to investigate whether (and how) the hybrid was able to tackle them.

There are also some evaluation frameworks that were designed specifically for this purpose: WrapRec and Rival. Both of them allow to use existing libraries and algorithms. They also offer evaluation measures that are related to the recommendation tasks. Moreover, they are open-source, which allows researchers to do a further development.

Besides all the advantages of the hybrid recommenders, the hybridization may bring up some issues. Hybrids suffer from the problem of the scalability since the training phase may require additional time cost. Another related issue is that we need more data in order to well-tune parameters of hybrid combinations, which is not needed in monolithic approaches.

Yet there are still some unexplored hybrid combinations that were not studied and verified in current literature. Also, many state-of-the-art approaches are only compared to monolithic techniques, with missing comparison of hybrids between each other. There are also many domains, where the hybrid recommenders were not applied yet.

Acknowledgement

This chapter is based on work and was partially supported by the grants No. APVV-15-0508, VG 1/0667/18 and KEGA 028STU-4/2017.

References

Adomavicius, G. and Tuzhilin, A. (2005). Toward the next generation of recommender systems: a survey of the state-of-the-art and possible extensions. IEEE Transactions on Knowledge and Data Engineering, 17(6), 734-749.

Aggarwal, C. C. (2016). Recommender Systems: The Textbook. Cham: Springer. ISBN: 978-3-319-29657-9.

Balabanovic, M. and Shoham, Y. (1997). Fab: Content-based, collaborative recommendation. Commun. ACM, 40(3), 66-72.

Baltrunas, L., Kaminskas, M., Ludwig, B., Moling, O., Ricci, F., Aydin, A., Luke, K.-H., and Schwaiger, R. (2011). InCarMusic: Context-Aware Music Recommendations in a Car, pp. 89-100. Springer Berlin Heidelberg, Berlin, Heidelberg.

Barragans-Martinez, A. B., Costa-Montenegro, E., Burguillo, J. C., Rey-Lopez, M., Mikic-Fonte, F. A., and Peleteiro, A. (2010). A hybrid content-based and item-based collaborative filtering approach to recommend TV programs enhanced with singular value decomposition. Information Sciences, 180(22), pp. 4290-4311.

Basu, C., Hirsh, H., and Cohen, W. (1998). Recommendation as classification: Using social and content-based information in recommendation. In Proceedings of the Fifteenth National/Tenth Conference on Artificial Intelligence/Innovative Applications of Artificial Intelligence, AAAI '98/IAAI '98, pp. 714-720, Menlo Park, CA, USA. American Association for Artificial Intelligence.

Braunhofer, M., Codina, V., and Ricci, F. (2014). Switching hybrid for cold-starting context-aware recommender systems. In Proceedings of the 8th ACM Conference on Recommender Systems, RecSys '14, pp. 349-352, New York, NY, USA. ACM.

Bostandjiev, S., O'Donovan, J., and Hollerer, T. (2012). Tasteweights: A visual interactive hybrid recommender system. In Proceedings of the Sixth ACM Conference on Recommender Systems, RecSys '12, pp. 35-42, New York, NY, USA. ACM.

Burke, R. (2002). Hybrid recommender systems: Survey and experiments. User Modeling and User-Adapted Interaction, 12(4), pp. 331-370.

Burke, R. and Ramezani, M. (2011). Matching Recommendation Technologies and Domains, pp. 367-386. Springer US, Boston, MA.

Burke, R., Vahedian, F., and Mobasher, B. (2014). Hybrid Recommendation in Heterogeneous Networks, pages 49-60. Springer International Publishing, Cham.

Celma, O. (2010). Music Recommendation, pp. 43–85. Springer Berlin Heidelberg, Berlin, Heidelberg.

Chen, X., Qin, Z., Zhang, Y., and Xu, T. (2016). Learning to rank features for recommendation over multiple categories. In Proceedings of the 39th International ACM SIGIR Conference on Research and Development in Information Retrieval, SIGIR '16, pp. 305–314, New York, NY, USA. ACM.

Cremonesi, P., Turrin, R., and Airoldi, F. (2011). Hybrid algorithms for recommending new items. In Proceedings of the 2nd International Workshop on Information Heterogeneity and Fusion in Recommender Systems, HetRec '11, pp. 33-40, New York, NY, USA. ACM.

de Campos, L. M., Fernandez-Luna, J. M., Huete, J. F., and Rueda-Morales, M. A. (2010). Combining content-based and collaborative recommendations: A hybrid approach based on bayesian networks. International Journal of Approximate Reasoning, 51(7): pp. 785-799.

Cella, L., Cereda, S., Quadrana, M., and Cremonesi, P. (2017) Deriving Item Features Relevance from Past User Interactions. Proceedings of the 25th Conference on User Modeling, Adaptation and Personalization, ACM, pp. 275-279.

Claypool, M., Gokhale, A., Miranda, T., Murnikov, P., Netes, D., and Sartin, M. (1999). Combing content-based and collaborative filters in an online newspaper.

Ekstrand, M., D., Ludwig, M., Konstan, J., A., and Riedl, J., T. (2011). Rethinking the recommender research ecosystem: reproducibility, openness, and LensKit. In Proceedings of the fifth ACM conference on Recommender systems (RecSys '11). ACM, New York, NY, USA, pp. 133-140.

Elahi, M., Braunhofer, M., Ricci, F., and Tkalcic, M. (2013). Personality-based active learning for collaborative filtering recommender systems. In Proceeding of the XIIIth International Conference on AI*IA 2013: Advances in Artificial Intelligence - Volume 8249, pp. 360-371, New York, NY, USA. Springer-Verlag New York, Inc.

Gantner, Z., Rendle, S., Freudenthaler, Ch. and Schmidt-Thieme, L. (2011). MyMediaLite: a free recommender system library. In Proceedings of the fifth ACM conference on Recommender systems (RecSys '11). ACM, New York, NY, USA, pp. 305-308.

Ghazanfar, M. A. and Prugel-Bennett, A. (2014). Leveraging clustering approaches to solve the gray-sheep users problem in recommender systems. Expert Syst. Appl., 41(7), pp. 3261-3275.

Goldberg, K., Roeder, T., Gupta, D., and Perkins, C. (2001). Eigen-taste: A constant time collaborative filtering algorithm. Information Retrieval, 4(2):133–151.

Gomez-Uribe, C. A. and Hunt, N. (2015). The Netflix Recommender System: Algorithms, Business Value, and Innovation. ACM Trans. Manage. Inf. Syst., 6(4), pp. 13:1-13:19.

Hahsler, M. (2017). recommenderlab: A Framework for Developing and Testing Recommendation Algorithms. Technical report.

Harper, F. M. and Konstan, J. A. (2015). The movielens datasets: History and context. ACM Trans. Interact. Intell. Syst., 5(4), pp. 19:1-19:19.

Herlocker, J. L., Konstan, J. A., and Riedl, J. (2000). Explaining collaborative filtering recommendations. In Proceedings of the 2000 ACM Conference on Computer Supported Cooperative Work, CSCW '00, pp. 241-250, New York, NY, USA. ACM.

Hornung, T., Ziegler, C.-N., Franz, S., Przyjaciel-Zablocki, M., Schatzle, A., and Lausen, G. (2013). Evaluating hybrid music recommender systems. In Proceedings of the 2013 IEEE/WIC/ACM International Joint Conferences on Web Intelligence (WI) and Intelligent Agent Technologies (IAT) - Volume 01, WI-IAT '13, pp. 57-64, Washington, DC, USA. IEEE Computer Society.

Hu, L., Sun, A., and Liu, Y. (2014). Your neighbors affect your ratings: On geographical neighborhood influence to rating prediction. In Proceedings of the 37th International ACM SIGIR Conference on Research & Development in Information Retrieval, SIGIR '14, pp. 345–354, New York, NY, USA. ACM.

Hug, N. (2017). Surprise, a Python library for recommender systems. http://surpriselib.com

Jamali, M. and Ester, M. (2010). A matrix factorization technique with trust propagation for recommendation in social networks. In Proceedings of the Fourth ACM Conference on Recommender Systems, RecSys '10, pp. 135–142, New York, NY, USA. ACM.

Kaššák, O., Kompan, M., and Bieliková, M. (2016). Personalized hybrid recommendation for group of users: Top-n multimedia recommender. Information Processing Management, 52(3):459–477.

Kim, J. K., Jang, M. K., Kim, H. K., and Cho, Y. H. (2009). A hybrid recommendation procedure for new items using preference boundary. In Proceedings of the 11th International Conference on Electronic Commerce, ICEC '09, pp. 289-295, New York, NY, USA. ACM.

Kim, S.-C., Park, C.-S., and Kim, S. K. (2012). A Hybrid Recommendation System Using Trust Scores in a Social Network, pp. 107–112. Springer Netherlands, Dordrecht.

Koenigstein, N., Dror, G., and Koren, Y. (2011). Yahoo! music recommendations: Modeling music ratings with temporal dynamics and item taxonomy. In Proceedings of the Fifth ACM Conference on Recommender Systems, RecSys '11, pp. 165–172, New York, NY, USA. ACM.

Koren, Y. (2009). The bellkor solution to the netflix grand prize.

Kula, M. (2015). Metadata Embeddings for User and Item Cold-start Recommendations. In Proceedings of the Workshop on New Trends in Content-Based Recommender Systems. Ceur-WS, pp. 14-21.

Lampropoulos, A. S., Sotiropoulos, D. N., and Tsihrintzis, G. A. (2012). Evaluation of a cascade hybrid recommendation as a combination of one-class classification and collaborative filtering. In 2012 IEEE 24th International Conference on Tools with Artificial Intelligence, volume 1, pp. 674-681.

Lee, J., Kim, S., Lebanon, G., and Singer, Y. (2013). Local low-rank matrix approximation. In Dasgupta, S. and McAllester, D., editors, Proceedings of the 30th International Conference on Machine Learning, volume 28 of Proceedings of Machine Learning Research, pp. 82–90, Atlanta, Georgia, USA.

Lee, J., Sun, M., and Lebanon, G. (2012). PREA: personalized recommendation algorithms toolkit. J. Mach. Learn. Res. 13, 1 (September 2012), pp. 2699-2703.

Liu, J., Dolan, P., and Pedersen, E. R. (2010). Personalized news recommendation based on click behavior. In Proceedings of the 15th International Conference on Intelligent User Interfaces, IUI '10, pp. 31-40, New York, NY, USA. ACM.

Liu, Q., Wu, S., and Wang, L. (2017). Deepstyle: Learning user preferences for visual recommendation. In Proceedings of the 40th International ACM SIGIR Conference on Research and Development in Information Retrieval, SIGIR '17, pp. 841–844, New York, NY, USA. ACM.

Loni, B., and Said, A. (2014). WrapRec: an easy extension of recommender system libraries. In Proceedings of the 8th ACM Conference on Recommender systems (RecSys '14). ACM, New York, NY, USA, pp. 377-378.

Ma, H., Zhou, D., Liu, C., Lyu, M. R., and King, I. (2011). Recommender systems with social regularization. In Proceedings of the Fourth ACM International Conference on Web Search and Data Mining, WSDM '11, pp. 287–296, New York, NY, USA. ACM.

Ma, H., Yang, H., Lyu, M. R., and King, I. (2008). Sorec: Social recommendation using probabilistic matrix factorization. In Proceedings of the 17th ACM Conference on Information and Knowledge Management, CIKM '08, pp. 931–940, New York, NY, USA. ACM.

Massa, P. and Avesani, P. (2007). Trust-aware recommender systems. In Proceedings of the 2007 ACM Conference on Recommender Systems, RecSys'07, pp. 17–24, New York, NY, USA. ACM.

McAuley, J., Targett, C., Shi, Q., and van den Hengel, A. (2015). Image-based recommendations on styles and substitutes. In Proceedings of the 38th International ACM SIGIR Conference on Research and Development in Information Retrieval, SIGIR '15, pp. 43–52, New York, NY, USA. ACM.

McCrae, J., Piatek, A., and Langley, A. (2004). Collaborative filtering.

Miranda, T., Claypool, M., Gokhale, A., Mir, T., Murnikov, P., Netes, D., & Sartin, M. (1999). Combining content-based and collaborative filters in an online newspaper. In Proceedings of ACM SIGIR Workshop on Recommender Systems.

Mendeley. (2017). Mrec library documentation. http://mendeley.github.io/mrec

Odic, A., Tkalcic, M., Tasic, J. F., and Kosir, A. (2013). Predicting and detecting the relevant contextual information in a movie-recommender system. Interact-ing with Computers, 25(1), pp. 74-90.

Pham, M. C., Cao, Y., Klamma, R., and Jarke, M. (2011). A clustering approach for collaborative filtering recommendation using social network analysis. 17(4):583–604.

Ricci, F., Rokach, L., and Shapira, B. (2015). Recommender Systems Handbook. Springer US, New York, NY, USA, 2nd edition.

Ronen, R., Koenigstein, N., Ziklik, E., and Nice, N. (2013). Selecting content-based features for collaborative filtering recommenders. In Proceedings of the 7th ACM

Conference on Recommender Systems, RecSys '13, pp. 407-410, New York, NY, USA. ACM.

Schein, A. I., Popescul, A., Ungar, L. H., and Pennock, D. M. (2002). Methods and metrics for cold-start recommendations. In Proceedings of the 25th Annual International ACM SIGIR Conference on Research and Development in Information Retrieval, SIGIR '02, pp. 253-260, New York, NY, USA. ACM.

Schelter, S., Boden, C., Schenck, M., Alexandrov, A., and Markl, V. (2013). Distributed matrix factorization with mapreduce using a series of broadcast-joins. In Proceedings of the 7th ACM Conference on Recommender Systems, RecSys '13, pp. 281–284, New York, NY, USA. ACM.

Shardanand, U. and Maes, P. (1995). Social information filtering: Algorithms for automating "word of mouth". In Proceedings of the SIGCHI Conference on Human Factors in Computing Systems, CHI'95, pp. 210-217, New York, NY, USA. ACM Press/Addison-Wesley Pub. Co.

Stern, D., H., Herbrich, R. and Graepel, T. (2009). Matchbox: Large Scale Online Bayesian Recommendations. In Proceedings of the 18th international conference on World wide web (WWW '09). ACM, New York, NY, USA, pp. 111-120.

Strub, F., Gaudel, R., and Mary, J. (2016). Hybrid recommender system based on autoencoders. In Proceedings of the 1st Workshop on Deep Learning for Recommender Systems, DLRS 2016, pp. 11–16, New York, NY, USA. ACM.

Su, X. and Khoshgoftaar, T. M. (2009). A survey of collaborative filtering techniques. Advances in Artificial Intelligence, 2009, pp. 4:2–4:2.

Takacs, G., Pilaszy, I., Nemeth, B., and Tikk, D. (2009). Scalable collaborative filtering approaches for large recommender systems. J. Mach. Learn. Res., 10:623–656.

Tsai, C.-H. (2016). A fuzzy-based personalized recommender system for local businesses. In Proceedings of the 27th ACM Conference on Hypertext and Social Media, HT '16, pp. 297–302, New York, NY, USA. ACM.

Višnovský, J., Kaššák, O., Kompan, M. and Bieliková, M. (2014) The Cold Start: Minimal User's Rating Activity Estimation. In 1st Workshop on Recommender Systems for Television and online Video (RecSysTV) in conjunction with 8th ACM Conference on Recommender Systems, Foster City, USA, p. 4.

Weston, J., Bengio, S., and Usunier, N. (2010). Large scale image annotation: learning to rank with joint word-image embeddings. Mach. Learn. 81, 1, pp. 21-35.

Wu, Q., Liu, S., and Miao, C. (2017). Modeling uncertainty driven curiosity for social recommendation. In Proceedings of the International Conference on Web Intelligence, WI '17, pp. 790–798, New York, NY, USA. ACM.

Yang, D., Chen, T., Zhang, W., Lu, Q., and Yu, Y. (2012). Local implicit feedback mining for music recommendation. In Proceedings of the Sixth ACM Conference on Recommender Systems, RecSys '12, pp. 91–98, New York, NY, USA. ACM.

Yoshii, K., Goto, M., Komatani, K., Ogata, T., and Okuno, H. G. (2008). An efficient hybrid music recommender system using an incrementally trainable probabilistic generative model. IEEE Transactions on Audio, Speech, and Language Processing, 16(2), pp. 435-447.

Zanker, M. (2008). A collaborative constraint-based meta-level recommender. In Proceedings of the 2008 ACM Conference on Recommender Systems, RecSys'08, pp. 139-146, New York, NY, USA. ACM.

Zanker, M. and Jessenitschnig, M. (2009). Collaborative feature-combination recommender exploiting explicit and implicit user feedback. In 2009 IEEE Conference on Commerce and Enterprise Computing, pp. 49-56.

Zheng, Z., Chen, T., Liu, N., Yang, Q., and Yu, Y. (2012). Rating prediction with informative ensemble of multi-resolution dynamic models. In Dror, G., Koren, Y., and Weimer, M., editors, Proceedings of KDD Cup 2011, volume 18 of Proceedings of Machine Learning Research, pp. 75–97. PMLR.

Zhou, T., Kuscsik, Z., Liu, J., Medo, M., Wakeling, J., and Zhang, Y. (2010). Solving the apparent diversity-accuracy dilemma of recommender systems. Proceedings of the National Academy of Sciences, 107, 4511–4515.

Ziegler, C.-N., McNee, S. M., Konstan, J. A., and Lausen, G. (2005). Improving recommendation lists through topic diversification. In Proceedings of the 14th International Conference on World Wide Web, WWW '05, pp. 22–32, New York, NY, USA. ACM.

Chapter 5

Context-Aware Recommendations

Yong Zheng[1] and Bamshad Mobasher[2]

[1]*Illinois Institute of Technology, Chicago, Illinois, 60616, USA*
[2]*DePaul University, Chicago, Illinois, 60604, USA*
yong.zheng@iit.edu, mobasher@cs.depaul.edu

With increasingly more complex information spaces on the Internet, recommender systems have emerged as critical tools to alleviate information overload and to assist decision making. Although such systems generate personalized recommendations based on users' learned preferences, most recommenders do not consider the fact that a user's decision or preferences may vary across contexts and situations. Context-aware recommender systems, on the other hand, not only adapt to user preferences, but also consider the contextual situations. This chapter provides a broad introduction to Context-Aware Recommender Systems (CARS). We especially focus on algorithmic approaches for integrating context into the recommendation framework, including approaches based on context selection and context-aware collaborative filtering. We also discuss evaluation strategies for CARS; available data sets and open-source libraries; and lessons learnt from the practical applications.

5.1. Introduction

Recommender Systems have been effective tools in alleviating the information overload problem. They can assist users in decision making by recommending item or resources that are tailored to those user tastes or preferences. The underlying principle behind recommender systems is to make predictions on users' preferences for items by inferring or learning from their past preferences or activity. The "item" refers to the information entity in a specific domain that can be selected or consumed by a user. For example, the item could be a video on YouTube, a movie on Netflix, a playlist on Pandora, a restaurant on Yelp, a hotel in TripAdvisor, or a book sold on Amazon.com. User preferences on items are expressed through either explicit or implicit feedback. Examples of explicit feedback are user ratings or thumb up/down behaviors. Implicit feedback may be obtained through textual reviews, click-throughs, purchase behavior, etc. Several recommendation algorithms have been developed for personalized item recommendation by exploiting users' preference histories. Common approaches include collaborative filtering [Resnick *et al.* (1994); Koren *et al.* (2009)] in Chapter 1, content-based recommenders [Lops *et al.* (2011)], and hybrid recommenders [Burke (2002)] in Chapter 4.

Traditional recommender systems, however, ignore the impact of contextual factors (e.g., time, location, companion, weather) which may affect the user preferences on items. Time-aware recommender systems [Koren (2010); Campos *et al.* (2014)] made the first attempt to take a specific contextual factor, namely time, into account. The assumption behind such a system is that a user's preferences may change over time. For example, a user may prefer cartoons when he is a child, but he may like other types of movies when he grows up. But, there are many other contextual factors, such as location, companion, occasion, weather, mood, and others that may have an impact on user preferences or on the utility of an item for a user in a given situation. For example, a user may choose a different type of movie when he or she is going to watch the movie with different "companions" such as *with a partner* rather than *with kids*.[1] Also, users may choose a more formal restaurant for *a business dinner*, but a fast food restaurant may be more appropriate for a *quick lunch* alone.[2] Research in context-Aware Recommender Systems (CARS) has been focused on addressing the issue of integrating contextual factors into the recommendation process.

Many context-aware recommendation algorithms have been developed in a variety of application domains. It has been demonstrated that considering context in recommendation can result in significant improvements in the effectiveness of recommendation in various domains such as movies [Košir *et al.* (2011); Zheng *et al.* (2013b)], music [Baltrunas *et al.* (2010, 2011a)], tourism [Zheng *et al.* (2012); Braunhofer *et al.* (2013)], restaurants [Ono *et al.* (2009); Ramirez-Garcia and Garca-Valdez (2014)], mobile apps [Baltrunas *et al.* (2015)], and others.

This chapter provides an overview of CARS as well as practical introduction to algorithms and applications of context-aware recommendation. We first introduce the notion of "context" in recommendation and discuss how to identify most influential context information. Next, we review a variety of context-aware collaborative filtering algorithms and discuss their practical implementation. We also discuss evaluation methods in CARS, available data sets and open source libraries that can help in conducting experiments. Finally, we also introduce the lessons learnt from our experiences, followed by the discussion of some open issues and challenges.

[1] *Partner* and *kids* are viewed as two contextual conditions for "companion".

[2] *Business dinner* and *quick lunch* are viewed as two different occasions.

5.2. Background and Preliminaries

5.2.1. *Context Definition and Categorization*

The notion of context and its role in our daily interactions with our environment has been studied in psychology, linguistics, artificial intelligence, information retrieval, pervasive/ubiquitous computing, mobile computing, and more recently, recommender systems. While little agreement exists among researchers as to what constitutes context, the importance of context is undisputed. In psychology, a change in context during learning has been shown to have an impact on recall [Smith (1979); Bartlett and Santrock (1979)], suggesting a key role played by context in structuring of and retrieval from human memory. Research into linguistics has shown that context plays the important role of a disambiguation function, that is, it reduces the possible interpretations of a message that exists in abstraction from its context [Leech (1981)].

In more computational settings, the definition of context varies from domain to domain, which results in different understandings and applications of context in each area. The most commonly used definition for context in ubiquitous computing was attributed to Abowd *et al.* in 1999 [Abowd *et al.* (1999)]: "context is any information that can be used to characterize the situation of an entity. An entity is a person, place, or object that is considered relevant to the interaction between a user and an application, including the user and applications themselves."

In information retrieval, the notion of context has been used to better capture a user's information need and intent which may not be available from an ambiguous query alone [Brown and Jones (2001)]. For example, time and location could be useful contextual information that can help the system provide more relevant results given a user query. Google may return different results for the same query depending on user's location. Context can be defined as "all cognitive and social factors as well as the user's aims and intentions during a search session [Belkin *et al.* (2004)]". In other words, the user's intent inferred by the topics in the query could be also considered context [Bai *et al.* (2007)]. A query "apple" could mean a fruit, a tech company, a type of computer, or a reference to New York City ("the Big Apple"). The intent of the user, however, may need to be inferred based on the topics in corpus and possibly based on the user's learned profile.

In recommender systems, any variable influencing users' decision making could be viewed as context. Contextual variables may include time, location, emotional state, weather conditions, type of activity, and other variables considered relevant in a given domain. The earliest work in CARS [Van Setten *et al.* (2004)] dates back to more than a decade ago. Despite the considerable amount of work since then, the research community still used competing definitions of context. For example, there are no clear guidelines for selecting appropriate contextual variables when designing recommender systems. In some academic publications, researchers have blended user characteristics (e.g., gender) and item features (e.g., genre) into the scope of contexts. This blurs the line between the context-aware and content-based recommender systems.

Adomavicius *et al.* [Adomavicius *et al.* (2011)] introduce an analysis to categorize context factors resulting in six possible general classes for context factors, as shown in Figure 5.1.

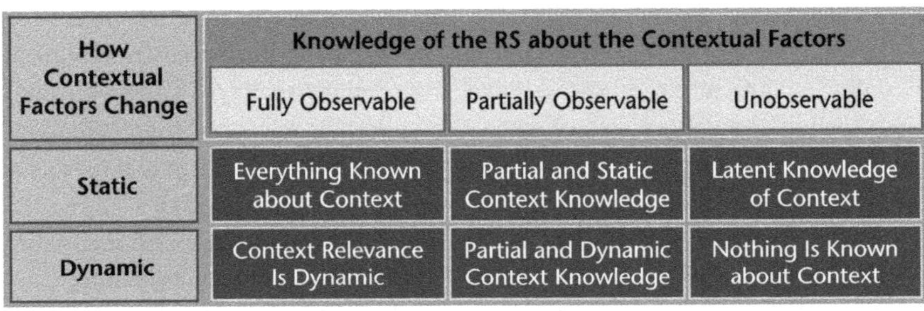

Fig. 5.1. Classification of Contextual Factors.

More specifically, they introduce a two-part classification of contextual information based on two considerations: what a recommender system knows about contextual factors (i.e., contextual factors may be fully observable, partially observable or unobservable) and how the contextual factors change over time (i.e., the environment is static or dynamic). This classification enables the understanding of context under different scenarios when designing recommender systems. But, even in a situation when there are a set of static and full observable contextual factors, it may still be difficult to determine which subset of factors should be used to characterize context.

Zheng [Zheng (2015b)] revisited the notion of context and defined context as the set of attributes which are related to the experience of users on items. For example, the time and location may be different each time someone watches a movie. Similarly, changes in a user's emotional state may influence the choice of music a user may be streaming online. In short, we find that most contextual features are the attributes of the activity itself, such as time, location and companion. Certain user characteristics can also be considered as contextual factors, such as user's emotional states or demographic features. By contrast, item features, such as movie genre, music album, news category, are usually viewed as describing the content of items in recommender systems.

Dourish [Dourish (2004)] distinguished between two views of context: the representational contexts and the interactional contexts. The representational view, dominant in ubiquitous computing and recommender systems, assumes context is a form of delineable information that can be described using a set of "appropriate" observable attributes. Furthermore, these attributes do not change and are clearly distinguishable from features describing the underlying activity undertaken by the user within the context. In the interactional view [Hosseinzadeh Aghdam *et al.* (2015); Hariri *et al.* (2015)], the scope of contextual features is defined dynamically and occasioned rather than static. Rather than assuming that context defines the situation within which an activity occurs, Dourish suggests a cyclical relationship between context and activity, where the activity gives rise to context. While context itself is may not be characterized using a set of observable variables, the behavior arising from context is observable and may be explained by a set of latent factors.

This chapter is focused on CARS with representational contexts — generally, we assume there are a set of fully observed contextual variables available at hand, and we assume these variables are static and their contextual influence is relatively stable for a specific domain. For example, *time*, *location* and *companion* can be three relevant contextual variables in the movie domain and the system can adapt to users' preferences in different contextual situations based on combinations of values for these observed variables.

Table 5.1. Contextual Ratings on Movies.

User	Item	Rating	Time	Location	Companion
U1	T1	3	weekend	home	alone
U1	T1	5	weekend	cinema	girlfriend
U1	T1	?	weekday	home	family

As an example, consider the situation depicted in the table above: there is one user $U1$, one item $T1$, and three contextual dimensions — Time (weekend or weekday), Location (at home or cinema) and Companion (alone, girlfriend, family).

Generally, we use the phrase *context dimension* to denote a contextual variable, e.g. "Location". The term *context condition* refers to a specific value in a dimension, e.g. "home" and "cinema" are two contextual conditions for "Location". A *context* or *context situation* is, therefore, a set of contextual conditions, e.g. $\{weekend, home, family\}$.

5.2.2. *Context Selection*

Once the context variables are identified from the data, context selection becomes the next task, since not all the contextual variables are influential in the recommendation task. It is similar to the feature selection process in machine learning tasks such as classification. It is an essential step in CARS because irrelevant contextual variables may not only increase computational costs but also result in noise that could negatively affect the effectiveness of recommendations. Furthermore, too many context variables may result in the "*cold-start context*" problem which makes it difficult to predict user's taste in contexts in which they do not have any preference histories.

Odic *et al.* [Odic *et al.* (2012)] identified two approaches to context selection: online survey and statistical detection for context relevancy assessment. In the user survey approach, subjects are asked to explicitly give information about how important or influential each context is from their perspective. For example, in [Baltrunas *et al.* (2012)] users' opinions on contextual information in tourism domain were acquired by asking users to imagine a given situation. Statistical detection is a more objective way which utilize a statistical test (e.g, t-test) to evaluate whether users gave significant different ratings in different contextual conditions. This strategy was adopted in the context-aware splitting approaches [Baltrunas and Ricci (2014); Zheng *et al.* (2014a)].

There are some drawbacks in these two context selection approaches. Using surveys is straightforward but it will require additional user effort which is undesirable from user perspective in practice. In addition, some research [Asoh *et al.* (2010)] has pointed out that users' opinions by imagining a situation is usually not reliable. The results from statistical detection may also not be reliable if there is not a sufficient number rating profiles in different contextual conditions. However, statistical detection is a widely used approach in practice since it does not require extra user effort.

5.3. Context-Aware Collaborative Filtering

In the past decade, several context-aware recommendation algorithms have been developed in order to adapt the recommendations to different contextual situations. According to Adomavicius [Adomavicius and Tuzhilin (2011)], there are three ways to incorporate contexts into recommender systems which can be described by the Figure 5.2.

Contextual pre-filtering uses contexts as filters to filter out irrelevant rating profiles and then applies a standard recommender to generate the recommendation

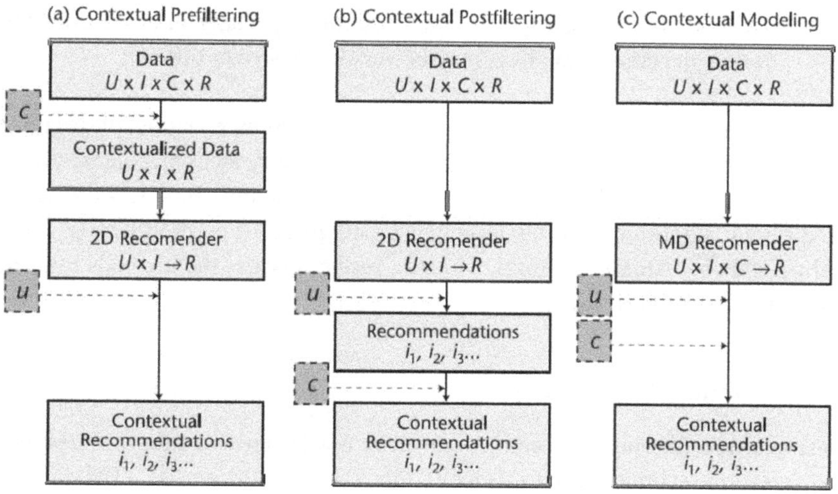

Fig. 5.2. Three ways to incorporate contexts into recommendation algorithms.

list. Contextual post-filtering applies a standard recommender first and then uses contexts as filters to filter out irrelevant recommendations or re-rank the item recommendations. In contextual modeling approaches, contexts are utilized as parts of the learning process to develop the predictive models based on which recommendations are generated.

In this section, we will focus on collaborative filtering (CF) approach to recommendation and discuss how context information can be incorporated into the CF algorithms. CF is one of the most popular recommendation approaches and has several advantages over other types of recommendation models:

- CF can provide recommendations based on user ratings and it does not require additional information about users or items, such as user demographic information or item features. This reduces the complexity of data collection and alleviates user privacy problems since the pure CF technique does not rely on content information.
- CF techniques are simple and straightforward. They are easy-implemented and capable for online learning or real-time recommendations, especially the matrix factorization based approaches in Chapter 2.
- The nature of CF allows researchers to further explore underlying patterns or understand the insights of recommendations. Many applications utilize CF to exploit explanations of item recommendations [Herlocker *et al.* (2000)], such as Amazon.com.

- CF approaches are applicable and useful, especially in some domains where the content analysis may be very expensive or difficult.

For more details about the collaborative filtering recommendation, please refer to the first and second chapter of this book. We specifically focus on context-aware recommendation algorithms based on matrix factorization. Below, we categorize different algorithms into the aforementioned groups (i.e., pre-filtering, post-filtering and contextual modeling), and then further discuss the methods that are based on matrix factorization.

5.3.1. *Contextual Pre-Filtering*

Contextual pre-filtering uses contexts to filter out irrelevant rating profiles and then applies a recommender to generate the recommendation list. The earliest approach falling into this category was the *reduction-based approach* [Adomavicius *et al.* (2005a)] developed by Adomavicius *et al.* in 2005 which uses only the rating profiles that pertain to the context of the user-specified criteria in which a recommendation is made. *DaVI* (Dimensions as Virtual Items) [Domingues *et al.* (2011)] is another approach developed by Domingues *et al.* in 2009. This approach treats contextual conditions as new items (i.e., virtual item), and create an item-item similarity matrix to discover the relationships between items and contexts (i.e., virtual items). The top similar items will be recommended to a user within a given context by inferring from this similarity matrix. Semantic pre-filtering (SPF) [Codina *et al.* (2013, 2015)] is another approach which tries to learn similarity between context conditions based on a user-context or item-context matrix, and then aggregates the similarity of two contextual situations to further filter out ratings placed in dissimilar contexts.

Differential context modeling is another category of algorithms which utilize contexts as filters. These approaches can be viewed as hybrid models that combine pre-filtering and contextual modeling. There are two types of algorithms falling into this category: differential context relaxation (DCR) [Zheng *et al.* (2012)] and differential context weighting (DCW) [Zheng *et al.* (2013a)]. DCR uses different context dimensions as filters for various components in the recommendation algorithm. Take user-based collaborative filtering for example, the algorithm may need to calculate user-user similarities, aggregate neighborhood ratings to make predictions. These different steps or parts are considered as algorithm components. Zheng *et al.* [Zheng *et al.* (2012)] used particle swarm optimization to identify the optimal context filters to maximize the contextual contribution of each algorithm components. DCR follows the notion of context matching. The *context matching* refers to the process of seeking matched contexts from the rating profiles. For example, if we are going to recommend movies to a user for him or her to watch at weekend with kids, an exact context matching will look for ratings that were left in the same situation (i.e., at weekend with kids) in the preference history. DCR does not require an exact match in two contextual situations, which helps alleviate the sparsity problem. DCW is an improvement over the DCR model, where DCW measures the similarity between two contexts by learning the weights for each context dimension. Only the ratings with a similar context situation will be incorporated in the prediction model. Both DCR and DCW are general approaches to be applied on any traditional recommendation models.

Another popular approach is *Item splitting* [Baltrunas and Ricci (2009)] proposed by Baltrunas *et al.* in 2009. It assumes that the nature of an item, from the user's point of view, may change in different contextual conditions, hence it may be useful to consider it as two different items. Once the influential contextual conditions are identified for each item, this item can be split into different new item entities, one for each context. In this way, the context information is fused into the new items, which converts the original multidimensional rating data to a traditional two-dimension data which only contains ratings associated with users-item pairs. Similarly, we can split users rather than items. Baltrunas *et al.* (2009) developed *micro-profiling* [Baltrunas and Amatriain (2009)], which was the first attempt to examine user splitting. Said *et al.* made a similar proposal called *contextual user profiles* [Said *et al.* (2011)]. It is also possible to split users and items at the same time, which results in the *UISplitting* approach [Zheng *et al.* (2014a)].

Due to the simplicity and flexibility in these context-aware splitting approaches, they have become some of the most popular pre-filtering algorithms for context-aware recommendation. In this section, we specifically introduce how item splitting works. User splitting and UISplitting can be performed in a similar way. Take the original rating matrix in Table 5.2 as an example. We have three rating profiles associated with a same user $U1$ and item $T1$. Item splitting iterates over all contextual conditions in each context dimension and evaluates the splits based on the impurity criteria. It finds the best split for each item in the rating matrix and then items are split into two new ones, where contexts are eliminated from the original matrix. Thus, it transforms the original multi-dimensional rating matrix to a 2D matrix. Assume that the best contextual condition to split item T1 in Table 5.2 is "Location = *home and not home*", T1 can be split into T11 (movie T1 being seen at home) and T12 (movie T1 being seen not at home). Once the best split has been identified, the rating matrix can be transformed as shown by Table 5.3(a). Accordingly, if we plan to perform user splitting only, the result could be shown as Table 5.3(b) if we assume "Time = *Weekday or not*" as the best split.

Table 5.2. Original Rating Matrix.

User	Item	Rating	Time	Location	Companion
U1	T1	3	Weekend	Home	Friend
U1	T1	5	Weekend	Cinema	Girlfriend
U1	T1	?	Weekday	Home	Family

Table 5.3. Transformed Rating Matrix.

(a) by Item Splitting

User	Item	Rating
U1	T11	3
U1	T12	5
U1	T11	?

(b) by User Splitting

User	Item	Rating
U12	T1	3
U12	T1	5
U11	T1	?

This example shows a *simple split*, in which a single contextual condition is used to split the item. It is also possible to perform a *complex split* using multiple conditions across multiple context dimensions. However, as discussed in [Baltrunas and Ricci (2009)], there are significant costs of sparsity and potential overfitting when using multiple conditions. Once the splitting operations are completed, we can apply any traditional recommendation algorithms, such as matrix factorization, to produce item recommendation based on the transformed two-dimensional rating matrix.

The main challenge is how to find the best context conditions to split the user or the item. Impurity criteria [Baltrunas and Ricci (2009)] has been used to determine whether and how much the items are rated differently in binary context conditions. For example, t-test can be applied on two list of ratings associated with the binary context conditions, such as "Time = *Weekday or not*". The p-value based on the t-test can tell us whether there is a significant difference in the ratings. And the t-value can be used to compare which context condition is the best split if there are multiple significant split options.

Based on our experience, it is difficult to determine whether item splitting or user splitting is the best option for a particular domain. We suggest using UISplitting directly which splits users and items at the same time. The risk behind UISplitting is that it may increase the sparsity of the rating matrix since the number of users and items will be enlarged. A balance between the sparsity and the splits should be taken into consideration to find the best parameter or the best splitting approach by given a data set. Furthermore, techniques like matrix factorization help manage the sparsity problem in certain situations.

5.3.2. *Contextual Post-Filtering*

Compared to the other two categories, contextual post-filtering has not been as heavily investigated. The idea behind post-filtering is straightforward — the recommender system produces the predicted ratings or the list of top-*N* items without considering contexts using a traditional recommendation algorithm. Next, items that are irrelevant to the target context are removed, or the recommendation list of the items is reranked to favor items more appropriate to the target context. Panniello *et al.* [Panniello *et al.* (2009)] proposed the first post-filtering method which can be described by Equation 5.1.

$$\widehat{R}(u,t,c) = \begin{cases} \widehat{R}(u,t) & Pr(u,t,c) \geq p \\ 0 & Pr(u,t,c) < p \end{cases} \tag{5.1}$$

$\widehat{R}(u,t,c)$ refers to the predicted rating for the user u on item t within context situation c, while $\widehat{R}(u,t)$ is the predicted rating without considering contexts by a traditional recommendation algorithm. They additionally calculate a probability with which the user will choose a certain type of item in a given context which is denoted by $Pr(u,t,c)$. This probability is computed as the number of neighbors (i.e., users similar to u) who purchased or consumed the same item t in contexts c divided by the number of the total number of neighbors. We set $\widehat{R}(u,t,c)$ as zero if this probability is smaller than a threshold p. In other words, the item t is not qualified to be recommended. Otherwise, the model will use $\widehat{R}(u,t)$ to represent $\widehat{R}(u,t,c)$. We call this method post-filtering based on user neighborhood and denote it by "*PoF_Ngbr*". The notion of user neighborhood comes from the neighborhood-based collaborative filtering [Resnick *et al.* (1994)], where the neighborhood can be identified by measuring user-user similarities by corresponding metrics, such as cosine similarity or Pearson correlations. This method is only valid for evaluating the top-*N* recommendations, since they mark $\widehat{R}(u,t,c)$ as zero if item t is not going to be recommended.

Ramirez *et al.* [Ramirez-Garcia and Garca-Valdez (2014)] proposed a post-filtering method by adjusting the predicted ratings. We refer to this method as "*PoF_Adj*". The predicted rating can be obtained by Equation 5.2. $\overline{R}(t,c)$ denotes the average value of the ratings that are placed on the item t within context c. The predicted contextual rating, therefore, is composed by this average rating and the predicted rating without considering context (i.e., $\widehat{R}(u,t)$). They set a ratio β (0¡β¡1) to control the contributions of each part.

$$\widehat{R}(u,t,c) = \beta \times \widehat{R}(u,t) + (1-\beta) \times \overline{R}(t,c) \tag{5.2}$$

Inspired by these two approaches, Zheng [Zheng (2018)] proposed a new approach which utilizes the rating deviations between a contextual rating and the rating without contexts. This rating deviation is used to estimate whether the user likes a specific item in context c. More specifically, the prediction method can be described by Equations 5.3 and 5.4.

$$\widehat{R}(u,t,c) = \begin{cases} \widehat{R}(u,t) & Dev(u,t,c) > 0 \\ 0 & Dev(u,t,c) \leq 0 \end{cases} \tag{5.3}$$

$$Dev(u,t,c) = \frac{\sum_{a \varepsilon N}(R(a,t,c) - R(a,t)) \times sim(a,u)}{\sum_{a \varepsilon N} sim(a,u)} \tag{5.4}$$

$Dev(u,t,c)$ is used to estimate the rating deviation of u's rating on item t with and without considering context c. It tells that it is appropriate to recommend the item t to u in context c if the deviation is positive. Otherwise, we set the predicted rating as zero to remove this item from the recommendation list. To compute $Dev(u,t,c)$, we utilize Equation 5.4. N denotes the top-K nearest neighbors for user u who rated item t in our knowledge base, and user a is a user neighbor in set N. The function $sim(a,u)$ is used as a weight to aggregate the contributions by these neighbors. This similarity can be calculated by the popular user-user similarity metrics, e.g., cosine similarity. $R(a,t)$ denotes neighbor a's rating on item t, while $R(a,t,c)$ tells a's rating on item t in contexts c.

The model by Equation 5.3 is similar to *PoF_Ngbr* as shown in Equation 5.1, where we set the predicted rating value as zero to remove an inappropriate item from the recommendation list. From another perspective, we can also use a similar method in the *PoF_Adj* model to adjust the predicted ratings. It can be shown in Equation 5.5. Due to the fact that the value of $Dev(u,t,c)$ could be a positive or negative one, it implies that we should apply a bonus or penalty to the user u's rating on item t if the context information c is taken into account.

$$\widehat{R}(u,t,c) = \widehat{R}(u,t) + Dev(u,t,c) \tag{5.5}$$

As a summary, these post-filtering techniques are simple and straightforward. They use a traditional recommendation algorithm to produce the predicted rating without considering contexts (i.e., $\widehat{R}(u,t)$), then try to contextualize this predicted rating by removing irrelevant items associated with the contexts or adjusting the predicted rating. Apparently, the key challenge is how to contextualize the predicted rating or recommendations without considering contexts.

5.3.3. *Contextual Modeling*

In contextual modeling, contextual factors are directly considered as part of learning the prediction function. One such approach is based on tensor factorization (TF) [Karatzoglou *et al.* (2010); Frolov and Oseledets (2017)]. In TF, each context variable is viewed as an individual dimension in the multidimensional rating space in addition to the traditional user and item dimensions. TF directly takes advantage of the multidimensional data representation, and it assumes each dimension is independent from others. There are also other context-dependent recommendation models which are demonstrated to be more effective and easier to interpret [Zheng (2017c)]. CARS^2 [Shi *et al.* (2014)] is an approach which learns context representations. More specifically, it uses a separate representation for users and items under a specific context to capture the interactions among users, items and contexts. In this section, we specifically discuss two classes of context-dependent algorithms: deviation-based contextual modeling and similarity-based contextual modeling.

5.3.3.1. *Deviation-Based Models*

The deviation-based contextual modeling is a learning algorithm that attempts to minimize the squared rating prediction errors by learning the rating deviations between two contextual situations. Table 5.4 illustrates this concept through an example.

Table 5.4. Example of Rating Deviations.

Context	**D1: Time**	**D2: Location**
c_1	Weekend	Home
c_2	Weekday	Cinema
Dev(Di)	0.5	-0.1

In this example, there are two context dimensions: Time and Location. Each context situation is constructed by the context conditions in these two variables. There are two contexts c_1:{Weekend, Home} and c_2:{Weekday, Cinema}. The last row in Table 5.4 depicts the rating deviation from c_1 to c_2 in each context variable. For example, Dev(D1) represents the rating deviation in the variable "Time" from c_1 to c_2. More specifically, it indicates that a user's rating in weekday is generally higher than his or her rating at weekend by 0.5. Accordingly, Dev(D2) is -0.1, which tells that a user's rating in cinema is generally lower than his or her rating at home by 0.1.

Therefore, we can predict a user's rating on a specific item within contexts c_2 if we know his or her rating on the same item within contexts c_1. For example, if user u rated item t in context c_1 as a four-star, his or her rating on t in context c_2 can be simply estimated as the four star plus the aggregated rating deviations in each context variable. Namely, the predicted rating will be 4.4 (i.e., 4 + 0.5 - 0.1).

Theoretically, we can learn the rating deviations between every pair of context conditions within the same context variable. However, that approach may introduce sparsity problems if there are many context conditions in a single context variable. A simple solution to alleviate this problem is to use a single virtual baseline context to which all other contexts are compared. Take Table 5.5 for example, we introduce a special context situation c_0, where the context conditions in all the context variables are "N/A" (i.e., not available). The ratings in c_0 can be interpreted as a user's ratings on the items without considering contextual situations.

Table 5.5. Example of Rating Deviations.

Context	D1: Time	D2: Location
c_0	N/A	N/A
c_2	Weekday	Cinema
Dev(Di)	0.5	-0.1

Therefore, the predictive function for user's rating on an item within a context can be described by Equation 5.6.

$$F(u,t,c) = P(u,t) + \sum_{1}^{N} Dev(D_i,c) \tag{5.6}$$

where $F(u,t,c)$ is the prediction function to estimate user u's rating on item t within context c. $P(u,t)$ is the predicted rating given by u on t without considering any context situations. $Dev(D_i,c)$ tells the rating deviation at the ith context variable from c_0 to c, where N is the number of context dimensions in the data set.

$P(u,t)$, as the predicted rating given by u on t, can be replaced by any predictive function in the traditional recommendation algorithms. For example, it could be the prediction function in user-based collaborative filtering, or the function by matrix factorization. What we are going to learn are the rating deviations in each context variable from c_0 to c, i.e., the rating deviation between two context conditions in each context variable.

In Equation 5.6, we simply assume the $Dev(D_i,c)$ is the same for all the users and the items. A finer-grained model may assume $Dev(D_i,c)$ may vary from users to users, from items to items. For example, a user-specific model could be described by Equation 5.7, where we assign a $Dev(D_i,c,u)$ to each user by assuming that different users may have personalized values in $Dev(D_i,c)$. According, an item-specific model can be developed too.

$$F(u,t,c) = P(u,t) + \sum_{1}^{N} Dev(D_i,c,u) \tag{5.7}$$

The abovementioned framework provides a high-level picture of how the rating deviations in different contexts can be incorporated into the recommendation model. Context-aware matrix factorization (CAMF) [Baltrunas *et al.* (2011b)] is one approach that falls into the category of deviation-based contextual modeling approaches. CAMF replaces $P(u,t)$ in the above formula with the predictive function in matrix factorization, as shown in the Equation 5.8:

$$\hat{r}_{uic_{k,1}c_{k,2}\ldots c_{k,L}} = \mu + b_u + \sum_{j=1}^{L} B_{ijc_{k,j}} + \overrightarrow{p_u} \cdot \overrightarrow{q_i} \tag{5.8}$$

In this model, μ represents the average rating, and b_u indicates the user bias in the rating. Furthermore, $\overrightarrow{p_u}$ is used to denote a user feature vector, and $\overrightarrow{q_i}$ the item feature vector. Assuming that there are L contextual dimensions in total, $c_k = \{c_{k,1}c_{k,2}\ldots c_{k,L}\}$ a contextual situation, where $c_{k,j}$ denotes the contextual condition in the jth context dimension. Therefore, $B_{ijc_{k,j}}$ indicates the contextual rating deviation associated with item i and the contextual condition in the jth dimension. Deviation-based contextual sparse linear method (CSLIM) [Zheng *et al.* (2014b)] is another example which utilizes the prediction function in sparse linear method [Ning and Karypis (2011)] as the component $P(u,t)$.

5.3.3.2. *Similarity-Based Models*

Instead of learning ratings deviations across context conditions, similarity-based contextual modeling tries to learn the similarities between pairs of context situations. In this setting, the terms "context similarity" or "similarity of contexts" refer to the similarity of a user's rating behavior in two contexts.

Table 5.6. Example of Context Similarity.

Context	D1: Time	D2: Location
c_0	N/A	N/A
c_2	Weekday	Cinema
Sim(Di)	0.5	0.1

Table 5.6 gives an example of context similarity. The table is similar to the one that is used to represent rating deviations in contexts. The last row in Table 5.6 provides the similarity values of the contexts c_0 and c_2 in each context variable.

The predicted rating or ranking score can be described as:

$$F(u,t,c) = P(u,t) \times Sim(c_0,c) \tag{5.9}$$

Again, any predictive function in the traditional recommender systems can be used to replace the $P(u,t)$. The challenge becomes how to measure the similarity between c_0 and c.

Zheng *et al.* [Zheng *et al.* (2015c,d,b)] proposed three methods to represent the similarity of contexts. Independent Context Similarity (ICS) assumes the similarity between two contexts is the product of the similarities between pairs of context conditions in each context variable. In this case, the model will learn the similarity between every two context conditions in the same context variable. Latent Context Similarity (LCS) is an improved method based on the ICS where each context condition is represented by a latent vector, and therefore the similarity between two context conditions can be estimated by the dot product of the two corresponding vectors. LCS is used to alleviate the cold-start context problem in ICS. Multidimensional Context Similarity (MCS) is the most effective but also the most complex model. It assigns a weight to each context condition, where a context situation can be viewed as a point in the multidimensional space. As a result, the dissimilarity of two contexts is represented as the distance between points in this space. Readers interested in a more detailed description of these approaches may refer to our previous work [Zheng *et al.* (2015c,d)].

5.4. Experimentation in CARS: Library, Data and Evaluations

5.4.1. *Recommendation Library: CARSKit*

In the area of recommender systems, many recommendation algorithms have been developed. Many of these have been incorporated in to recommendation software libraries and frameworks such as Mahout[3], Duine[4], Cofi[5], EasyRec[6], GraphLab Create[7], LensKit[8], LibRec[9], MyMediaLite[10].

However, until recently no libraries have included implementations of context-aware recommendation algorithms due to the challenges associated with data formats and usage of contexts in the recommendation algorithms. We developed CARSKit [Zheng *et al.* (2015a)] as a tool for facilitating CARS research and to support recommender system community. It is the 1st open-source Java-based context-aware recommendation library. The library is hosted on the Github[11].

CARSKit provides a flexible architecture so that it is easy to expand the scope of context-aware recommendation algorithms and provides a framework based on

[3]Mahout, http://mahout.apache.org/
[4]Duine, http://www.duineframework.org/
[5]Cofi, http://www.nongnu.org/cofi/
[6]EasyRec, http://easyrec.org/
[7]GraphLab Create, https://dato.com/products/create/
[8]LensKit, http://lenskit.org/
[9]LibRec, http://www.librec.net/
[10]MyMediaLite, http://www.mymedialite.net/
[11]CARSKit, https://github.com/irecsys/CARSKit

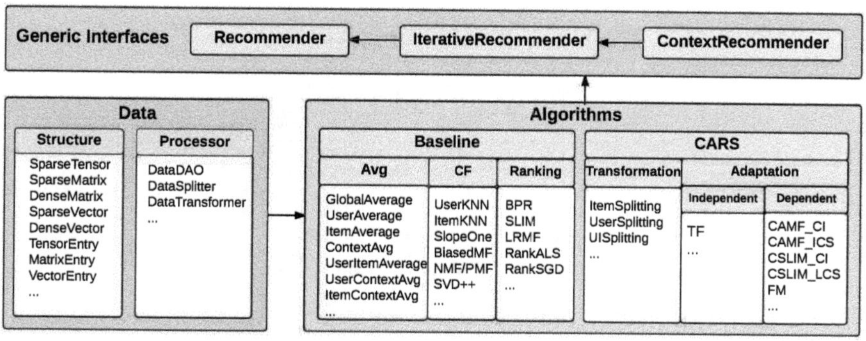

Fig. 5.3. The Architecture and Design in CARSKit.

which new algorithms can be developed in the future. The design of the library is depicted in the Figure 5.3.

The workflow is straightforward: different recommendation algorithms are the implementations and extensions of the generic interfaces where the shared and common functions are defined, such as functions for rating or score prediction in a specific context. Evaluation metrics for rating predictions and top-N recommendations are embedded into the *Recommender*.

Due to the special characteristics of the data format and processing methods in CARS, there were several challenges in the design of the library. The key features in CARSKit are described below.

5.4.1.1. *Data Transformer*

Before discussing data structures, we introduce the data transformer component in CARSKit. Usually, the contextual rating data can be stored in two formats: *loose* format and *compact* format, as shown in tables below. The loose format assumes that there is only one rating for each <user, item> pair in associated contexts, where the compact format allows to store multiple ratings to a same <user, item> pair in different contextual situations. Take the example shown in the two tables illustrating formats. The first two rows in loose format actually represent a single rating by U1 for T1 within contexts {Weekend, Work}. In the compact format, each row represents a single contextual rating profile, that is, there are only two contextual rating profiles in the loose format but four rating profiles in the compact format in this example.

Most contextual information is in the form of categorical data. In this case, both the loose and compact formats will increase storage and computational costs. In CARSKit, we store contextual ratings in *binary* format as shown in

Table 5.7. Loose Format.

UserID	ItemID	Rating	Context	Condition
U1	T1	3	Time	Weekend
U1	T1	3	Location	Work
U2	T2	4	Time	Weekday
U2	T2	4	Location	Home

Table 5.8. Compact Format.

UserID	ItemID	Rating	Time	Location
U1	T1	3	Weekend	Work
U2	T2	4	Weekday	Home
U1	T1	4	Weekend	Home
U2	T2	2	Weekday	Work

Table 5.9. Binary Format.

UserID	ItemID	Rating	Time:Weekend	Time:Weekday	Location:Home	Location:Work
U1	T1	3	1	0	0	1
U2	T2	4	0	1	1	0
U1	T1	4	1	0	1	0
U2	T2	2	0	1	0	1

Table 5.9, which is able to significantly boost the running performance. To assist the end users to prepare the rating data, we provide two methods *TransformationFromLooseToBinary* and *TransformationFromCompactToBinary* as the data transformer in our toolkit.

5.4.1.2. *Data Structures*

Since contexts are considered additional inputs beyond users and items, the data structure becomes the most important part and has a direct impact on the flexibility and running performance of the models.

There are two factors involved in designing the data structure: data storage and data operations. Intuitively, the context-aware data can be represented in N-dimensional space as tensors where each context dimension is considered as an individual dimension in the rating space. In this case, we build *SparseTensor* and *TensorEntry* to record the indices of each user, item, context dimension, and the associated rating. This structure is useful for the context-aware recommendation algorithms using N-dimensional operations, such as the multiverse recommendation algorithm described in [Karatzoglou *et al.* (2010)].

There are also other contextual recommendation algorithms that consider the dependencies between contexts and user/item dimensions, where two-dimensional operation is still the most frequent one adopted in those algorithms. In such cases, the modules such as *SparseMatrix* and *SparseVector* can be useful. They well-recognized and efficient data representations in existing recommendation libraries, such as LensKit and LibRec. SparseMatrix uses the compressed row and column storage[12] which was also demonstrated to boost the running efficiency in the design of LibRec [Guo *et al.* (2015)].

5.4.1.3. *Recommendation Algorithms*

As mentioned before, CARSKit is the 1st library specifically designed for CARS. As shown in Figure 5.3, we divide the contextual algorithms into two categories: transformation algorithms and adaptation algorithms.

The transformation algorithms try to pre-process the data and convert the contextual data set to a 2-dimensional rating matrix which only contains users, items and ratings, so that any traditional recommendation algorithms can be applied to. One of the most effective classes of techniques falling into this category is the context-aware splitting approaches [Zheng *et al.* (2014a)].

The adaptation algorithms directly incorporate contexts into the prediction function. There are two subcategories: *independent modeling* (e.g., TF [Karatzoglou *et al.* (2010)]) which assumes contexts are independent with users (and items); and *dependent modeling* which exploits the dependencies among users, items and contexts, such as CAMF [Baltrunas *et al.* (2011b)] and contextual sparse linear method (CSLIM) [Zheng *et al.* (2014b,c)]. Dependent modeling can be built in two ways: by modeling contextual rating deviations [Baltrunas *et al.* (2011b); Zheng *et al.* (2014c)] and by learning context similarities [Zheng *et al.* (2015c,d)]. Factorization machines (FM) [Rendle *et al.* (2011)] is a finer-grained algorithm which exploits pairwise relationships in its learning process. Among those algorithms, TF and CAMF are two popular ones which have been recognized as the standard baselines in CARS.

In addition to those state-of-the-art contextual recommendation algorithms, we also included some traditional recommendation algorithms in the package *baseline*. We did not re-compile those algorithms and directly reuse the classical recommenders provided by LibRec. There are two main purposes to include those traditional recommendation algorithms — on one hand, those algorithms can be applied after the data transformation (e.g., splitting operations), which is an essential step in the context-aware transformation algorithms. On the other hand, it is usually common to compete a contextual recommendation algorithm with non-contextual algorithms to judge whether the contextual effect is significant, or a context-aware recommendation algorithm is necessary or not.

[12]Sparse Matrix Storage, http://netlib.org/linalg/html_templates/node90.html

5.4.1.4. *Configuration and Evaluations*

For evaluation purpose, we provide *DataSplitter* which enables the users to adopt either train-testing evaluation or the N-folds cross validations.

Most of the recommendation algorithms embedded in CARSKit are able to perform two recommendation tasks: *rating prediction* and *item recommendation*, except those specifically designed for top-N recommendation, such as CSLIM. But the evaluation is different from traditional approaches since contexts are additional inputs in the evaluation process. Typically, the rating prediction can be evaluated by different prediction errors, such as mean absolute error (MAE), root mean square error (RMSE) and mean prediction error (MPE). The item recommendation can be evaluated through *relevance* metrics, such as precision and recall, and *ranking* metrics, such as mean average precision (MAP), normalized discounted cumulative gain (NDCG) and mean reciprocal rank (MMR).

Moreover, CARSKit provides flexible configurations by a single file which includes both *algorithm configuration* (e.g., algorithm parameters) and *experimental configuration*, e.g., input and output, evaluation strategy and metrics, etc. For more details about the configurations, please visit the Github page of CARSKit to review the latest manual published on the page.

5.4.2. *Context-Aware Data*

Context-aware recommendation is still an emerging area of research. Thus, the number of context-aware data sets available for research is limited due to the difficulty of context collections as well as the potential risks to user privacy. The data sets used in this research are described by Table 5.10.

Table 5.10. List of Available Context-aware Data Sets.

	Food	Restaurant	AdomMovie	CoMoDa	Music	STS	Frappe
# of users	212	50	116	121	42	325	957
# of items	20	40	226	1232	139	249	4082
# of context dimensions	2	2	4	8	5	11	3
# of context conditions	8	7	23	37	21	53	14
Rating Type and Scale	Rating (1 - 5)	Rating (1 - 5)	Rating (1 - 13)	Rating (1 - 5)	Rating (1 - 5)	Rating (1 - 5)	Logged Frequency (0 - 4.46)
# of ratings	6360	2309	1717	2292	3251	2354	87580
Density (i, c)	75%	71.4%	36.2%	26.2%	42.1%	10.2%	31.3%

More details about these data sets are provided below.

- The *Food* data [Ono *et al.* (2009)] was collected from surveys and it includes subjects' ratings on the Japan food menus in two contextual dimensions: degree of hungriness in real situations, and degree of hungriness in assumed or imagined situations. Typical context conditions in these two dimensions are full, hungry, normal. There are 212 subjects, and every subject gave 30 ratings on their selected 5 food menus in 6 different contextual situations. Generally, this is a good data set for exploring contextual preferences, since each user gave multiple ratings on a same item in different contexts.

- The *Restaurant* data [Ramirez-Garcia and Garca-Valdez (2014)] is also a data set collected from survey. Subjects gave ratings to the popular restaurants in Tijuana, Mexico by considering two contextual variables: time and location.
- The *Adom* data [Adomavicius *et al.* (2005b)] could be the earliest available context-aware movie data set. Subjects were asked to leave ratings on movies in different situations: time, location and companion. This data set is very sparse in ratings and very few users rated the same movie for multiple times in different contexts, so contextual effect may be very weak or difficult to be captured in this data set.
- The *CoMoDa* data [Košir *et al.* (2011)] is a publicly available context-aware movie data collected from surveys. There are 12 context dimensions which captured users' various situations, including mood, weather, time, location, companion, etc.
- The *South Tyrol Suggests (STS)* data [Braunhofer *et al.* (2013)] was collected from a mobile app which provides context-aware suggestions for attractions, events, public services, restaurants, and much more for South Tyrol. There are 14 contextual dimensions in total, such as budget, companion, daytime, mood, season, weather, etc.
- The *Music* data [Baltrunas *et al.* (2011a)] was collected from InCarMusic which is a mobile application (Android) offering music recommendations to the passengers of a car. Users are requested to enter ratings for some items using a web application. Here is the list of contextual variables included in the data: driving style, road type, landscape, sleepiness, traffic conditions, mood, weather, natural phenomena.
- The *Frappe* data [Baltrunas *et al.* (2015)] comes from the mobile usage in the app named as Frappe which is a context-aware app discovery tool that will recommend the right apps for the right moment. We used 3 context dimensions for experimental evaluations, including time of the day, day of the week and location. This data captures the frequencies of an app used by each user within 2 months.

All of the context-aware data sets listed above are real-world data, most of which were collected from surveys. The music, STS, and Frappe data sets were obtained from the real-usage data. A problem with survey data is that biases may be included from subjective opinions. On the other hand, there is only limited context information in real-world usage data. But survey data is able to collect user's tastes in more context situations which are difficult to collect from practice, such as emotional states. You can find most of these data sets available at the CARSKit repository on the Github[13].

5.4.3. *Evaluation Protocols*

The quality of context-aware recommendations can be evaluated similarly as in the traditional recommender systems — by *rating prediction* or *top-N recommendations*. Due to the fact that context information will be utilized as parts of the inputs to the system, the evaluation metrics may be calculated in a different way.

The task in rating prediction evaluations is relatively easy and straightforward — given a user and one item, along with the specific contextual situations where the user is going to consume or enjoy the item, the system will predict the rating for the tuple <user, item, contexts>. Afterwards, traditional evaluation metrics in the rating prediction task, such as mean absolute error (MAE), root mean squared error (RMSE), or mean squared error (MSE), can be produced in the same way. The example of MAE can be shown as follows.

$$MAE = \frac{1}{|T|} \sum_{(a,t)\varepsilon T} abs(P_{a,t,c} - R_{a,t,c}) \tag{5.10}$$

T represents the test set, where $|T|$ denotes the total number of ratings in the test set. $R_{a,t,c}$ is the actual rating given by user a on item t within contexts c. (a,t,c) is the <user, item, contexts> tuple in the test set. $P_{a,t,c}$ is the predicted rating by the context-aware recommendation model. The "abs" function is able to return the absolute value of the prediction error.

When it comes to the top-N recommendation task, we can still use the relevance metrics (such as precision and recall) and ranking metrics (such as Mean Reciprocal Rank (MRR) and Normalized Discounted Cumulative Gain (NDCG)) as the evaluation protocols. However, the calculation should be different since we are recommending a list of items to a user within a given contextual situation. Take precision at 10 for example, we produce a ranked list of top-10 items that user a may like in the context c. Precision measures what is the ratio of the relevant items in these ten recommendations. The ground truth should be the items

[13]Context-aware Data Sets, https://github.com/irecsys/CARSKit/tree/master/context-aware_data_sets

user a likes within the same context c in the test set. Therefore, precision should be computed as the average of precision values over all of the <user, contexts> pairs in the test set.

In the CARSKit library, we further group the <user, contexts> pairs by each unique user, calculate the precision by each user and finally obtain the average value as the result for precision. This operation allows us to compare precision values based on user-basis. Accordingly, other top-N evaluation metrics, such as recall, MRR, NDCG, can be adjusted in a similar way.

5.5. Discussion

5.5.1. *Lessons Learnt*

Based on our practical experience in the area of context-aware recommendation, we would like to share some lessons we learnt.

- We believe the process of context selection is important and necessary. When there are many context dimensions or conditions, sparsity becomes a problem and the computational costs, not to mention that irrelevant contexts may negatively affect recommendation performance.
- Contextual pre-filtering techniques may work well in smaller datasets. However, one cannot generalize the results since the sparsity of contextual ratings may introduce biases into the recommendation models. It is highly recommended that the context-aware recommendation algorithms should be evaluated over larger and denser data sets if possible.
- There are different CARS algorithms which can be built upon the neighborhood-based collaborative filtering approaches, the matrix factorization techniques, or the sparse linear methods. Those based on neighborhood-based collaborative filtering may consume a lot of memory and spend more time to produce recommendations due to the expensive process of the similarity calculations. In contrast to the algorithms based on the matrix factorization, fewer learning iterations may be required for the SLIM-based recommendation approaches, but they are sensitive to the algorithm parameters, such as the initial values, learning rates, etc. It may take a while to find the optimal parameters for these algorithms. Context-aware recommendation algorithms based on matrix factorization may not be the best choice in terms of recommendation accuracy, but they are easily implemented and extended, and it is also easy to tune up the learning parameters. That is also the reason why we use

matrix factorization based context-aware recommendation algorithms on the case studies in this chapter.

- Last but not least, A/B testing or user studies could be the best way to evaluate different context-aware recommendation models. The process of decision making in specific contexts could be very subjective. It is not guaranteed that one algorithm that has a good offline performance will also perform well in real setting. It is always suggested to use A/B tests or user studies before reaching final conclusions.

5.5.2. *Open Issues*

We believe that there are several aspects of context-aware recommendation that are worth exploring as future research.

- **User-Centric Evaluations**
  Most CARS research has focused on simulation-based evaluation of recommendation effectiveness. However, we believe it is important to conduct user studies and A/B tests to evaluate how different context-aware recommendation algorithms perform in real practice. We expect to build real-world applications and examine the corresponding algorithms based on user-centric evaluations in our future work.
- **Post-filtering Recommendation Algorithms**
  There is very limited research on contextual post-filtering algorithms. One of the reasons is the data sparsity problem which may result in bad performance in post-filtering. In our experimental evaluations, we found that SLIM works well in the top-N recommendation, even if it does not take contexts into consideration. We expect to fuse contextual post-filtering with the recommendations produced by SLIM to improve context-aware recommendations.
- **Deep Learning for CARS**
  The popular deep learning techniques have been incorporated into the area of recommender systems in Chapter 3, but there are few of applications in CARS.
- **Sparsity and Cold-start Problems in Chapter 8**
  The sparsity problem is always a challenge in context-aware recommendation. Braunhofer *et al.* [Braunhofer (2014); Braunhofer *et al.* (2014)] have explored hybrid methods as one of solutions. We believe there is still a long way to go on dealing with sparsity problems in context-aware recommendations. For example, we may utilize cross-domain

knowledge to alleviate the sparsity problems. A "domain" can be a different or related application domain, or another context, or information extracted from different devices.

- **Numeric v.s. Categorical Contexts**
In current development of CARS, we mainly use categorical context information, but ignore the real valued contextual variables. In the earlier development of time-aware recommender systems, the usage of time information was based on discrete time information, such as time in seconds, weeks or years. By contrast, we segment this information to different categories (e.g., morning, evening, weekend, weekday). It is worth exploring the combination of numerical and categorical representations of time information in time-aware recommender systems. Moreover, other contextual variables may also be represented with real numbers, such as temperature, degree of happiness or other emotional status, and so forth. Dealing with these numerical variables in addition to segmenting them into different categories is still an area of investigation.
- **Context Suggestions**
Both traditional recommender systems and CARS produce item recommendations, while the context-awareness may bring new recommendation opportunities, such as context suggestion. Context suggestion [Baltrunas *et al.* (2010); Zheng (2015a); Zheng *et al.* (2016); Zheng (2017b,a)] aims to recommend appropriate context situations to users in order for them to better consume or enjoy the items. Not only appropriate item recommendations but also the suitable contexts can guarantee good user experience in the real-world applications.
- **Recommendation Explanations by Contexts**
Recommendation explanation is always an important problem in recommender systems research. It helps us to better understand the recommendations and help end users trust the recommendations and be more engaged. Many types of new recommender systems try to provide users with recommendation explanations. For example, Netflix found that taking social connections into account in their movie recommender systems may not have an effect unless the system explicitly explains them [Amatriain and Basilico (2015)]. Accordingly, we believe contexts also plays an important role in recommendation explanations [Zheng (2017d)] and we will explore it in future work.
- **Privacy in CARS**
There are multiple privacy concerns in the recommender systems, such as user privacy and RecSys privacy, as mentioned in Chapter 13. Context

privacy is another concern, e.g., whether and how should we utilize user locations, and so forth.

- **New User Interfaces and Interactions**
  Both academia and industry have realized the importance of context in the recommendation process. To design better context-aware recommender systems or use context for recommendation explanation, it is important to develop new user interfaces and new ways for users to interact with the application in different contexts.

References

Abowd, G. D., Dey, A. K., Brown, P. J., Davies, N., Smith, M. and Steggles, P. (1999). Towards a better understanding of context and context-awareness, in *Handheld and ubiquitous computing*, pp. 304–307.

Adomavicius, G., Mobasher, B., Ricci, F. and Tuzhilin, A. (2011). Context-aware recommender systems, *AI Magazine* **32**, 3, pp. 67–80.

Adomavicius, G., Sankaranarayanan, R., Sen, S. and Tuzhilin, A. (2005a). Incorporating contextual information in recommender systems using a multidimensional approach, *ACM Transactions on Information Systems (TOIS)* **23**, 1, pp. 103–145.

Adomavicius, G., Sankaranarayanan, R., Sen, S. and Tuzhilin, A. (2005b). Incorporating contextual information in recommender systems using a multidimensional approach, *ACM Transactions on Information Systems (TOIS)* **23**, 1, pp. 103–145.

Adomavicius, G. and Tuzhilin, A. (2011). Context-aware recommender systems, in *Recommender systems handbook* (Springer), pp. 217–253.

Amatriain, X. and Basilico, J. (2015). Recommender systems in industry: A netflix case study, in *Recommender Systems Handbook* (Springer), pp. 385–419.

Asoh, H., Motomura, Y. and Ono, C. (2010). An analysis of differences between preferences in real and supposed contexts, in *ACM RecSys' 10, Proceedings of the 2nd International Workshop on Context-Aware Recommender Systems.*

Bai, J., Nie, J.-Y., Cao, G. and Bouchard, H. (2007). Using query contexts in information retrieval, in *Proceedings of the 30th annual international ACM SIGIR conference on Research and development in information retrieval* (ACM), pp. 15–22.

Baltrunas, L. and Amatriain, X. (2009). Towards time-dependant recommendation based on implicit feedback, in *ACM RecSys, the 4th Workshop on Context-Aware Recommender Systems.*

Baltrunas, L., Church, K., Karatzoglou, A. and Oliver, N. (2015). Frappe: Understanding the usage and perception of mobile app recommendations in-the-wild, *CoRR* **abs/1505.03014**, `http://arxiv.org/abs/1505.03014`.

Baltrunas, L., Kaminskas, M., Ludwig, B., Moling, O., Ricci, F., Aydin, A., Lüke, K.-H. and Schwaiger, R. (2011a). Incarmusic: Context-aware music recommendations in a car, in *E-Commerce and Web Technologies* (Springer), pp. 89–100.

Baltrunas, L., Kaminskas, M., Ricci, F., Rokach, L., Shapira, B. and Luke, K.-H. (2010). Best usage context prediction for music tracks, in *Proceedings of the 2nd Workshop on Context Aware Recommender Systems.*

Baltrunas, L., Ludwig, B., Peer, S. and Ricci, F. (2012). Context relevance assessment and exploitation in mobile recommender systems, *Personal and Ubiquitous Computing* **16**, 5, pp. 507–526.

Baltrunas, L., Ludwig, B. and Ricci, F. (2011b). Matrix factorization techniques for context aware recommendation, in *Proceedings of the fifth ACM conference on Recommender systems* (ACM), pp. 301–304.

Baltrunas, L. and Ricci, F. (2009). Context-based splitting of item ratings in collaborative filtering, in *Proceedings of ACM conference on Recommender systems*, pp. 245–248.

Baltrunas, L. and Ricci, F. (2014). Experimental evaluation of context-dependent collaborative filtering using item splitting, *User Modeling and User-Adapted Interaction* **24**, 1-2, pp. 7–34.

Bartlett, J. C. and Santrock, J. (1979). Affect-depedent episodic memory in young children, *Child Development* **5**, pp. 513–518.

Belkin, N., Muresan, G. and Zhang, X. (2004). Using users context for ir personalization, in *Proceedings of the ACM/SIGIR Workshop on Information Retrieval in Context* (Citeseer).

Braunhofer, M. (2014). Hybrid solution of the cold-start problem in context-aware recommender systems, in *User Modeling, Adaptation, and Personalization* (Springer), pp. 484–489.

Braunhofer, M., Codina, V. and Ricci, F. (2014). Switching hybrid for cold-starting context-aware recommender systems, in *Proceedings of the 8th ACM Conference on Recommender systems* (ACM), pp. 349–352.

Braunhofer, M., Elahi, M., Ricci, F. and Schievenin, T. (2013). Context-aware points of interest suggestion with dynamic weather data management, in *Information and Communication Technologies in Tourism 2014* (Springer), pp. 87–100.

Brown, P. J. and Jones, G. J. (2001). Context-aware retrieval: Exploring a new environment for information retrieval and information filtering, *Personal and Ubiquitous Computing* **5**, 4, pp. 253–263.

Burke, R. (2002). Hybrid recommender systems: Survey and experiments, *User Modeling and User-Adapted Interaction* **12**, 4, pp. 331–370.

Campos, P. G., Díez, F. and Cantador, I. (2014). Time-aware recommender systems: a comprehensive survey and analysis of existing evaluation protocols, *User Modeling and User-Adapted Interaction* **24**, 1-2, pp. 67–119.

Codina, V., Ricci, F. and Ceccaroni, L. (2013). Exploiting the semantic similarity of contextual situations for pre-filtering recommendation, in *User Modeling, Adaptation, and Personalization* (Springer), pp. 165–177.

Codina, V., Ricci, F. and Ceccaroni, L. (2015). Distributional semantic pre-filtering in context-aware recommender systems, *User Modeling and User-Adapted Interaction*, pp. 1–32.

Domingues, M. A., Jorge, A. M. and Soares, C. (2011). Using contextual information as virtual items on top-n recommender systems, *arXiv preprint arXiv:1111.2948.*

Dourish, P. (2004). What do we talk about when we talk about context, *Personal and Ubiquitous Computing* **8**, 1, pp. 19–30.

Frolov, E. and Oseledets, I. (2017). Tensor methods and recommender systems, *Wiley Interdisciplinary Reviews: Data Mining and Knowledge Discovery* **7**, 3.

Guo, G., Zhang, J., Sun, Z. and Yorke-Smith, N. (2015). Librec: A java library for recommender systems, in *Posters, Demos, Late-breaking Results and Workshop Proceedings of the 23rd International Conference on User Modeling, Adaptation and Personalization*.

Hariri, N., Mobasher, B. and Burke, R. (2015). Adapting to user preference changes in interactive recommendation, in *IJCAI*, Vol. 15, pp. 4268–4274.

Herlocker, J. L., Konstan, J. A. and Riedl, J. (2000). Explaining collaborative filtering recommendations, in *Proceedings of the 2000 ACM conference on Computer supported cooperative work* (ACM), pp. 241–250.

Hosseinzadeh Aghdam, M., Hariri, N., Mobasher, B. and Burke, R. (2015). Adapting recommendations to contextual changes using hierarchical hidden markov models, in *Proceedings of the 9th ACM Conference on Recommender Systems* (ACM), pp. 241–244.

Karatzoglou, A., Amatriain, X., Baltrunas, L. and Oliver, N. (2010). Multiverse recommendation: n-dimensional tensor factorization for context-aware collaborative filtering, in *Proceedings of the fourth ACM conference on Recommender systems* (ACM), pp. 79–86.

Koren, Y. (2010). Collaborative filtering with temporal dynamics, *Communications of the ACM* **53**, 4, pp. 89–97.

Koren, Y., Bell, R. and Volinsky, C. (2009). Matrix factorization techniques for recommender systems, *IEEE Computer* **42**, 8, pp. 30–37.

Košir, A., Odic, A., Kunaver, M., Tkalcic, M. and Tasic, J. F. (2011). Database for contextual personalization, *ELEKTROTEHNISKI VESTNIK* **78**, 5, pp. 270–274.

Leech, G. (1981). *Semantics: The Study of Meaning*, 2nd edn. (Penguin).

Lops, P., De Gemmis, M. and Semeraro, G. (2011). Content-based recommender systems: State of the art and trends, in *Recommender systems handbook* (Springer), pp. 73–105.

Ning, X. and Karypis, G. (2011). SLIM: Sparse linear methods for top-n recommender systems, in *2011 IEEE 11th International Conference on Data Mining* (IEEE), pp. 497–506.

Odic, A., Tkalcic, M., Tasic, J. F. and Košir, A. (2012). Relevant context in a movie recommender system: Users opinion vs. statistical detection, in *ACM RecSys' 12, Proceedings of the 4th International Workshop on Context-Aware Recommender Systems*.

Ono, C., Takishima, Y., Motomura, Y. and Asoh, H. (2009). Context-aware preference model based on a study of difference between real and supposed situation data, pp. 102–113.

Panniello, U., Tuzhilin, A., Gorgoglione, M., Palmisano, C. and Pedone, A. (2009). Experimental comparison of pre-vs. post-filtering approaches in context-aware recommender systems, in *Proceedings of the third ACM conference on Recommender systems* (ACM), pp. 265–268.

Ramirez-Garcia, X. and Garca-Valdez, M. (2014). Post-filtering for a restaurant context-aware recommender system, in *Recent Advances on Hybrid Approaches for Designing Intelligent Systems*, Vol. 547 (Springer), ISBN 978-3-319-05169-7, pp. 695–707.

Rendle, S., Gantner, Z., Freudenthaler, C. and Schmidt-Thieme, L. (2011). Fast context-aware recommendations with factorization machines, in *Proceedings of the 34th international ACM SIGIR conference on Research and development in Information Retrieval* (ACM), pp. 635–644.

Resnick, P., Iacovou, N., Suchak, M., Bergstrom, P. and Riedl, J. (1994). Grouplens: an open architecture for collaborative filtering of netnews, in *Proceedings of the 1994 ACM conference on Computer supported cooperative work* (ACM), pp. 175–186.

Said, A., De Luca, E. W. and Albayrak, S. (2011). Inferring contextual user profiles – improving recommender performance, in *ACM RecSys, the 4th Workshop on Context-Aware Recommender Systems*.

Shi, Y., Karatzoglou, A., Baltrunas, L., Larson, M. and Hanjalic, A. (2014). Cars2: Learning context-aware representations for context-aware recommendations, in *Proceedings of the 23rd ACM International Conference on Conference on Information and Knowledge Management* (ACM), pp. 291–300.

Smith, S. M. (1979). Remembering in and out of context, *Journal of Experimental Psychology: Human Learning and Memory* **5**, pp. 460–471.

Van Setten, M., Pokraev, S. and Koolwaaij, J. (2004). Context-aware recommendations in the mobile tourist application COMPASS, in *Adaptive hypermedia and adaptive web-based systems* (Springer), pp. 235–244.

Zheng, Y. (2015a). Context suggestion: Solutions and challenges, in *Proceedings of the 15th IEEE International Conference on Data Mining Workshops* (IEEE), pp. 1602–1603.

Zheng, Y. (2015b). A revisit to the identification of contexts in recommender systems, in *Proceedings of the 20th ACM Conference on Intelligent User Interfaces Companion* (ACM), pp. 133–136, doi:10.1145/2732158.2732167.

Zheng, Y. (2017a). Context suggestion: empirical evaluations vs user studies, in *Proceedings of the International Conference on Web Intelligence* (ACM), pp. 753–760.

Zheng, Y. (2017b). Indirect context suggestion, in *Proceedings of the 25th Conference on User Modeling, Adaptation and Personalization* (ACM), pp. 399–400.

Zheng, Y. (2017c). Interpreting contextual effects by contextual modeling in recommender systems, in *CIKM' 17, Proceedings of the Workshop on Interpretable Data Mining (IDM) Bridging the Gap between Shallow and Deep Models* (ACM).

Zheng, Y. (2017d). Interpreting contextual effects by contextual modeling in recommender systems, *arXiv preprint arXiv:1710.08516*.

Zheng, Y. (2018). Context-aware mobile recommendations by a novel post-filtering approach, in *The 31st International Florida Artificial Intelligence Research Society Conference*.

Zheng, Y., Burke, R. and Mobasher, B. (2012). Differential context relaxation for context-aware travel recommendation, in *E-Commerce and Web Technologies*, Lecture Notes in Business Information Processing (Springer Berlin Heidelberg), pp. 88–99.

Zheng, Y., Burke, R. and Mobasher, B. (2013a). Recommendation with differential context weighting, in *User Modeling, Adaptation, and Personalization*, Lecture Notes in Computer Science (Springer Berlin Heidelberg), pp. 152–164.

Zheng, Y., Burke, R. and Mobasher, B. (2013b). The role of emotions in context-aware recommendation, in *ACM RecSys' 13, Proceedings of the 3rd International Workshop on Human Decision Making in Recommender Systems* (ACM), pp. 21–28.

Zheng, Y., Burke, R. and Mobasher, B. (2014a). Splitting approaches for context-aware recommendation: An empirical study, in *Proceedings of the 29th Annual ACM Symposium on Applied Computing* (ACM), pp. 274–279.

Zheng, Y., Mobasher, B. and Burke, R. (2014b). CSLIM: Contextual SLIM recommendation algorithms, in *Proceedings of the 8th ACM Conference on Recommender Systems* (ACM), pp. 301–304, doi:10.1145/2645710.2645756.

Zheng, Y., Mobasher, B. and Burke, R. (2014c). Deviation-based contextual SLIM recommenders, in *Proceedings of the 23rd ACM Conference on Information and Knowledge Management* (ACM), pp. 271–280, doi:10.1145/2661829.2661987.

Zheng, Y., Mobasher, B. and Burke, R. (2015a). CARSKit: A java-based context-aware recommendation engine, in *Proceedings of the 15th IEEE International Conference on Data Mining Workshops* (IEEE).

Zheng, Y., Mobasher, B. and Burke, R. (2015b). Incorporating context correlation into context-aware matrix factorization, in *Proceedings of the 2015 International Conference on Constraints and Preferences for Configuration and Recommendation and Intelligent Techniques for Web Personalization-Volume 1440* (CEUR-WS.org), pp. 21–27.

Zheng, Y., Mobasher, B. and Burke, R. (2015c). Integrating context similarity with sparse linear recommendation model, in *User Modeling, Adaptation, and Personalization, Lecture Notes in Computer Science*, Vol. 9146 (Springer Berlin Heidelberg), ISBN 978-3-319-20266-2, pp. 370–376, doi:10.1007/978-3-319-20267-933.

Zheng, Y., Mobasher, B. and Burke, R. (2015d). Similarity-based context-aware recommendation, in *Web Information Systems Engineering*, Lecture Notes in Computer Science (Springer Berlin Heidelberg).

Zheng, Y., Mobasher, B. and Burke, R. (2016). User-oriented context suggestion, in *Proceedings of the 24th ACM Conference on User Modeling, Adaptation, and Personalization* (ACM).

Chapter 6

Group Recommendations

Ludovico Boratto[a] and Alexander Felfernig[b]

[a]*Eurecat, Centre Tecnològic de Catalunya,*
[a]*Graz University of Technology*
Email: ludovico.boratto@acm.org, alexander.felfernig@ist.tugraz.at

Group recommender systems have been developed to produce suggestions in contexts in which more than one person is involved in the recommendation process. As it happens with the systems that produce recommendation for single users, collaborative algorithms have been employed for most of the group-based solutions presented in the literature. Producing group recommendations, however, is not trivial and a system has to deal with additional aspects, such as the combination of the individual preferences into group preferences, or the prediction of the ratings straight for a group. In this chapter, we explore collaborative group recommender systems, analyzing the different families of approaches and architectural solutions available to build them, the algorithms that can be developed to produce the recommendations, the existing solutions in the literature, and the current open challenges in this area.

6.1. Introduction

While recommender systems suggest items that single users might like, group recommender systems are designed to *provide suggestions in scenarios where a group of users is engaged* [Jameson and Smyth (2007); Felfernig *et al.* (2018b)]. Group recommendation can be naturally adopted in application scenarios that involve groups (e.g., suggest a restaurant to a group of people who want to dine together).

Group recommendation has been highlighted as a challenging research area, with the first survey on the topic [Jameson and Smyth (2007)] being placed in the *Challenges* section of the widely-known book "The Adaptive Web", and research indicating it as a future direction in recommender systems, since it presents numerous open issues and challenges [Ricci (2014)].

Indeed, with respect to classic recommender systems, those that operate with groups present several additional aspects that characterize them and cannot be dealt with recommender systems for single users. Examples of these challenging aspects are the following:

(1) *Preference acquisition.* A group recommender system might acquire the preferences by considering only those expressed by the individual group members, or by allowing the groups to express them. It is also known that the interactions between the members of a group can help refining the individual preferences [Delic *et al.* (2016)].
(2) *Group building.* At the moment, no public dataset that contains both the groups' structure and the individual preferences exists, so there is a need to detect and/or synthesize groups that have the same characteristics of those handled by the system.
(3) *Group modeling* is the process adopted to combine the individual preferences in a unique model that represents the group.
(4) *Rating prediction* is the most characterizing aspect in all the types of recommender systems, and also plays an important role when working with groups, since the ratings might be predicted for single users or specifically for groups.
(5) *Help group members to achieve consensus.* This task is adopted in order to find an agreement on what should be proposed to the group.
(6) *Explanation of the recommendations*, i.e., the task performed by some of the systems to justify why an item has been suggested to the group.

To this day, 300+ research papers can be counted in the literature on this topic. As for single user recommender systems, also the vast majority of the research on group-based ones has been devoted to collaborative algorithms[1], exploring both memory-based and model-based algorithms previously presented in Chapter 1. Hence, in this chapter we focus on collaborative group recommender systems. In particular, we will consider the architectural and algorithmic solutions that can be employed to build these systems, thus focusing on the second, third, and fourth aspect of the previous list (i.e., *group building*, *group modeling*, and *rating prediction* tasks). While *preference acquisition*, *helping the members to achieve a consensus*, and *generating explanations* are important aspects to consider to build effective systems, these topics deviate from the collaborative focus

[1]For a discussion of group recommendation algorithms based on content-based filtering, constraint-based, critiquing-based, and hybrid recommendation we refer to [Felfernig *et al.* (2018b)].

of this book, and readers are referred to [Jameson and Smyth (2007)] for additional details. In this chapter, we assume that users provide individual ratings for the items, that the system recommends to the group the top-n items in the list, without involving group members to reach a consensus, and without providing any explanations. Since it would be impossible to cover the whole literature in collaborative group recommender systems in this chapter, Table 6.1 presents a summary of the most relevant approaches (considered as the ones that contain a collaborative algorithm and with over fifty citations at the time of the writing of this chapter), in order to give readers a summary of representative literature.

The remainder of this chapter is organized as follows: Section 6.2 illustrates different families of approaches that exist to produce group recommendations, showing the different high-level architectural representations of a system; Section 6.3 presents the strategies adopted in the literature to define groups; Section 6.4 focuses on the group modeling task, aimed at creating a unique representation of the preferences of a group; Section 6.5 illustrates how collaborative rating prediction algorithms are employed in group recommender systems; Section 6.6 presents a case-study that compares the different families of approaches by employing a user-based collaborative approach; Section 6.7 introduces future directions that the definition of collaborative group recommender systems can take; Section 6.8 contains concluding remarks.

6.2. Families of Approaches: Architectural Solutions

Given a set of individual preferences, expressed in the form of ratings for the items that a user evaluated, group preferences can be generated by using one of the following three families of approaches [Jameson and Smyth (2007)]:

- constructing group preference models and then predicting missing ratings for each group by using such models;
- predicting ratings for the items not rated by each user and merging the individual recommendations made for the members of a group;
- aggregating predictions built for every user into a group preference.

We will now describe in detail each of them, by also presenting the architecture of the system (both, the details and the figures have been adapted from the ones originally presented in [Boratto and Carta (2015); Boratto *et al.* (2017)]).

Table 6.1. Most cited collaborative group recommender systems.

Author(s)	*Domain*	*Group building*	*Prediction algorithm*	*Group modeling*	*Dataset(s)*	*Evaluation type*	*Evaluation metric(s)*
[O'Connor *et al.* (2001)]	Movies	Predefined groups	User-based CF	Least misery	MovieLens (actual system)	User study	Surveys
[Chen *et al.* (2008)]	Movies	[Not specified]	Item-based CF	Weighted Average	MovieLens (not specified)	Offline	MAE, Precision
[Amer-Yahia *et al.* (2009)]	Movies	Based on similarity between users	User-based CF	Least misery	MovieLens-10M	Offline	DCG
[Boratto *et al.* (2009a)]	Movies	Commu-nity Detection (Louvain)	Item-based CF	Weighted Average	MovieLens-1M	Offline	RMSE
[Campos *et al.* (2009)]	Movies	Based on similarity between users	Bayesian Networks	Average, Least Misery, Most Pleasure, Approval Voting	MovieLens-100k	Offline	Percentage of success, MAE
[Recio-García *et al.* (2009)]	Movies	Predefined groups	User-based CF	Weighted Average	MovieLens	User study	Defined by the authors
[Baltrunas *et al.* (2010)]	Movies	Based on similarity between users	SVD	Average Least misery, Borda	MovieLens-100k	Offline	nDCG
[Berkovsky and Freyne (2010)]	Food	Predefined groups	User-based CF	Weighted Average	Wellbeing Diet book	Offline	F1, MAE, Precision@k Coverage
[Kim *et al.* (2010)]	Books	Predefined groups	User-based CF	Popularity	[Not available]	User study	Precision
[Quijano-Sánchez *et al.* (2011)]	Movies	People explicitly joining social events	User-based CF	Average	Facebook and Tuenti	User study	Precision@3
[Ntoutsi *et al.* (2012)]	Movies	Based on similarity between users	User-based CF	Least Misery, Average, Most Pleasure	MovieLens-100k	Offline	Defined by the authors
[Liu *et al.* (2012)]	POIs	Predefined groups	Probabilistic Latent Topic Models	Weighted Average	whrrl.com, Meetup	Offline	Recall
[Ye *et al.* (2012)]	Music	Predefined groups	Model-based CF	Weighted Average	lastfm.com, whrrl.com	Offline	Relative Ranking

6.2.1. *Constructing Group Preference Models*

This approach detects groups of users according to the constraints imposed by the system (e.g., demographic features, or homogeneous preferences), builds a group model by using the preferences expressed by each user, and via that model predicts a rating for the items not rated by the group. Algorithm 1 presents how the approach works.

Algorithm 1 Approach that constructs group preference models.

```
Detect a set of groups G
Construct a model m_g for each group g ∈ G, which represents the preferences of the whole group
for all items i not rated by the group g do
  Use m_g to predict a group rating p_gi
end for
Calculate an aggregate rating r_gi from the ratings of the members of the group, either expressed (r_ui) or predicted (p_ui)
```

The architecture of a system that uses this approach is reported in Fig. 6.1. As mentioned earlier, the system has first to detect the existing groups in the dataset (TASK 1), considering the available information about the users (like the ratings they gave to the items or their demographic information, INPUT 1) and the constraints that the system has (e.g., the number of groups to detect). In order to build the predictions for a group, the system has to produce a group model that contains the group's preferences (TASK 2). The task that builds the model receives as input the ratings for the items evaluated by each user (INPUT 1) and the groups detected by the previous task. Each group model is used to predict the ratings for the group (TASK 3). After the system has predicted the ratings, group recommendations are selected.

6.2.2. *Merging Recommendations Made for Single Users*

This approach merges the items with the highest predicted ratings for each group member. Algorithm 2 presents how the approach works.

The architecture of a system that uses this approach is reported in Fig. 6.2. The system first detects the groups (TASK 1), considering both the user information (INPUT 1) and the existing constraints when detecting the groups (INPUT 2). The ratings for the items evaluated by each user

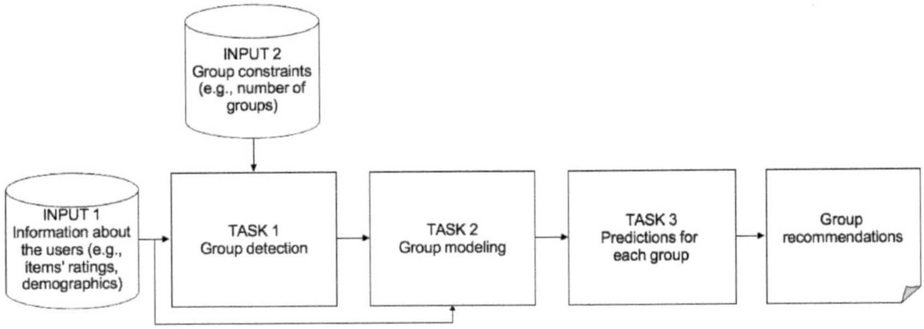

Fig. 6.1. Architecture of a system that uses group preference models to build the predictions.

Algorithm 2 Approach that merges the recommendations made for single users.

Detect a set of groups G
for all members u of a group $g \in G$ **do**
  for all items i not rated by the user u **do**
    Predict a rating p_{ui}
  end for
end for
Select the set C_i of items with the highest predicted ratings p_{ui} for each user u

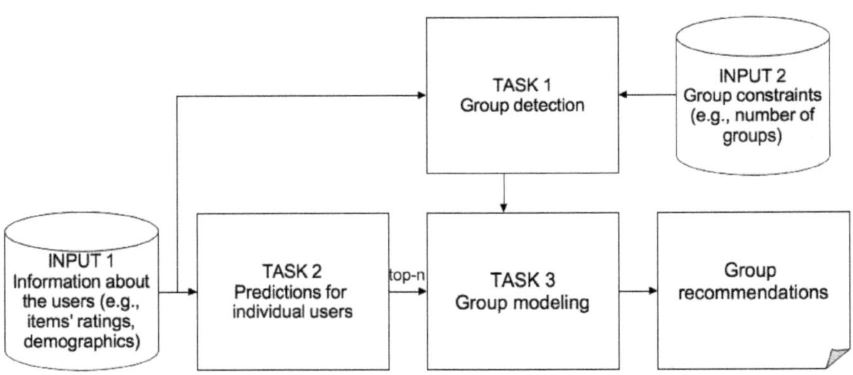

Fig. 6.2. Architecture of a system that merges the recommendations produced for each user.

(INPUT 1) are then used to predict the missing ratings for the items (TASK 2). The output of this task (the top-n predictions, which represent the recommendations for a user) is given as input to the task that merges the recommendations, usually with a union (group modeling, TASK 3), along with the composition of each group. After the system has merged the individual recommendations, group recommendations are selected.

6.2.3. *Aggregating Individual Predictions*

This approach first detects the groups, then predicts individual preferences for all the items not rated by each user, and aggregates the individual preferences for an item to derive a group preference. While the previous approach merged the individual recommendations (top-n items) through a union, this approach considers all the individual predictions and aggregates them in a unique group score. Algorithm 3 presents how the approach works.

Algorithm 3 Approach that aggregates the predictions made for single users.

Detect a set of groups G
for all items i **do**
  for all members u of a group $g \in G$ who did not rate i **do**
    predict a rating p_{ui}
  end for
end for
Calculate an aggregate rating r_{gi} from the ratings of the members of the group, either expressed (r_{ui}) or predicted (p_{ui})

The architecture of a system that uses this approach is shown in Fig. 6.3. This system applies an approach that is similar to the one described above. First, it detects the groups (TASK 1), considering both the user information (INPUT 1) and the existing constraints when detecting the groups (INPUT 2). The ratings for the items evaluated by each user (INPUT 1) are used as input to predict the missing ratings for each user's items (TASK 2). The output (i.e., all the calculated predictions) is given as input, along with the ratings given by the users for the items (INPUT 1) and each group's composition, to the task that models the group preferences (TASK 3). After the system has modeled each group's preferences, group recommendations are selected.

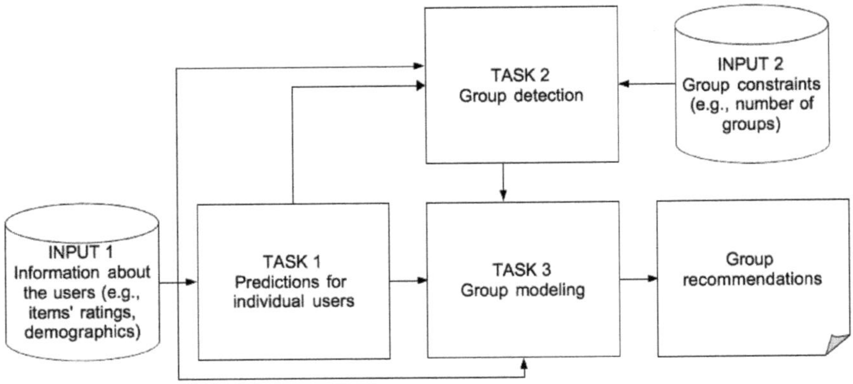

Fig. 6.3. Architecture of a system that aggregates individual predictions.

A variant of this approach, which switches the first two tasks, was introduced in [Boratto *et al.* (2017)]. As the architecture in Fig. 6.4 shows, a system can predict the ratings for the single users before detecting the groups, in order to use the predicted ratings as additional information in the group building task. This allows to avoid the data sparsity that usually characterizes the ratings, detect more cohesive groups in terms of their preferences, and produce more accurate group recommendations.

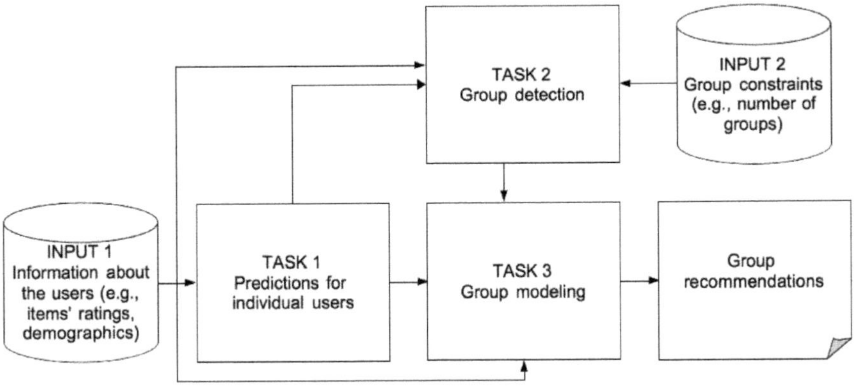

Fig. 6.4. Architecture of a system that aggregates individual predictions, which are also used to detect the groups.

6.3. Group Building

Since no dataset with information about both the structure of groups and the ratings of users exist, it is necessary to process datasets that do not contain group features as, for example, the household or group membership. The most common approach to induce explicit group features is to use rating data, and the similarity between two users can be computed in different ways. Baltrunas *et al.* [Baltrunas *et al.* (2010)] compute the Pearson's correlation between two users, and forms groups of two, three, four, and eight users, whose similarity is over 0.27. Other approaches employ the k-means clustering algorithm on the ratings matrix, to detect groups of users with similar preferences [Boratto *et al.* (2017)].

Group construction might also involve features specifying certain demographic information about the user. Using demographic information, synthetic groups can be created, for example, using location, interest, age, gender, etc.

Another example of utilizing non-group features in order to generate groups is to make use of the users' social networks. Especially in social network graphs, where connections are asymmetric (follower/followee), strong social ties (two users are both followers/followees of each other) represent a strong group connection between two users. Is is also possible to identify larger groups where all users are each others followers and followees. Depending on the nature of the social network, these strongly connected social graphs can either represent similarities in taste, interests, or actual groups in real world [Mislove *et al.* (2010)].

6.4. Group Modeling

In order to manage and provide information related to a group, it is necessary to first *model* the group [Masthoff (2015)]. A group is composed of single users that get together for a particular aim. So, the first aspect to consider when modeling a group is an individual user model, made of interests for a set of items. A group model can be considered as a "synthesis" of the user models, built by combining the preferences of the users in a group.

In group recommendation, building a group model is strongly related to the idea of collective choice, i.e., making a choice for a group, by taking into account the opinions of the users that belong to it. The aggregation of individual preferences is made by using a particular strategy and the usefulness of a strategy has to be evaluated in the environment in which the modeling is done.

6.4.1. *Existing Strategies*

Existing group modeling strategies are now presented. Each strategy is described with an example of how individual ratings are combined, by considering three users (u_1, u_2, and u_3) that rate ten items (identified by $i_1, ..., i_{10}$) with a rating from 1 to 10.

Additive Utilitarian. Individual ratings for each item are summed and a list of the group ratings is produced (the higher the sum is, the earlier the item appears in the list). The ranked group list of items is exactly the same that would be produced when averaging the individual ratings, so this strategy is also called 'Average strategy'. Table 6.2 shows an example of how the strategy works. The strategy has shown to be very effective in several contexts, especially those in which the users have the same relevance in the group [Boratto and Carta (2014)]. However, the function causes problems in the context of larger groups since the opinions of individuals count less.

Table 6.2. Additive Utilitarian strategy.

	i_1	i_2	i_3	i_4	i_5	i_6	i_7	i_8	i_9	i_{10}
u_1	8	10	7	10	9	8	10	6	3	6
u_2	7	10	6	9	8	10	9	4	4	7
u_3	5	1	8	6	9	10	3	5	7	10
Group	20	21	21	25	26	28	22	15	14	23

Pocket RestaurantFinder [McCarthy (2002)] recommends restaurants to a group of people, by averaging the individual preferences of the users in the group on different types of features (e.g., location, cost, cuisine). In [Pessemier *et al.* (2013)], the authors illustrate that modeling users with an average is the best way to aggregate individual preferences in different contexts; so, this strategy is employed in their experiments, along with the *Average Without Misery* strategy, which will be presented later.

Multiplicative Utilitarian. For each item, the ratings given by the users are multiplied and a ranked list of items is produced (the higher the product is, the earlier the item appears in the list). Table 6.3 shows an example of how the strategy works.

In [Christensen and Schiaffino (2011)], this strategy is adopted in order to produce music recommendations.

Table 6.3. Multiplicative Utilitarian strategy.

	i_1	i_2	i_3	i_4	i_5	i_6	i_7	i_8	i_9	i_{10}
u_1	8	10	7	10	9	8	10	6	3	6
u_2	7	10	6	9	8	10	9	4	4	7
u_3	5	1	8	6	9	10	3	5	7	10
Group	280	100	336	540	648	800	270	120	84	420

Borda Count. Each item gets a number of points, according to the position in the list of each user. The least favorite item gets 0 points and a point is added each time the next item in the list is considered. If a user gave the same rating to more items, points are distributed. In the example in Table 6.4, items i_8 and i_9 were rated by user u_2 with the lowest rating and should "share" the lowest positions with 0 and 1 points, so both the items get (0+1)/2=0.5 points. A group preference is obtained by adding the individual points of an item.

This strategy was implemented in [Baltrunas *et al.* (2010)].

Table 6.4. Borda Count strategy.

	i_1	i_2	i_3	i_4	i_5	i_6	i_7	i_8	i_9	i_{10}
u_1	4.5	8	3	8	6	4.5	8	1.5	0	1.5
u_2	3.5	8.5	2	6.5	5	8.5	6.5	0.5	0.5	3.5
u_3	2.5	0	6	4	7	8.5	1	2.5	5	8.5
Group	10.5	16.5	11	18.5	18	21.5	15.5	4.5	5.5	13.5

Copeland Rule. It is a form of majority voting that sorts the items according to their *Copeland index*, which is calculated as the number of times in which an alternative beats the others, minus the number of times it loses against the other alternatives. In the example in Table 6.5, item i_2 beats item i_1, since both u_1 and u_2 gave it a higher rating.

The approach proposed in [Felfernig *et al.* (2012)] proved that a form of majority voting is the most successful in a *requirements negotiation* context, in which the system resolves the existing conflicts between requirements and decides the ones that are implemented.

Plurality Voting. Each user votes for her/his favorite option. The alternative that receives the highest number of votes wins. If more than one

Table 6.5. Copeland Rule strategy.

	i_1	i_2	i_3	i_4	i_5	i_6	i_7	i_8	i_9	i_{10}
i_1	0	+	-	+	+	+	+	-	-	0
i_2	-	0	-	0	-	0	0	-	-	-
i_3	+	+	0	+	+	+	+	-	-	+
i_4	-	0	-	0	-	+	-	-	-	-
i_5	-	+	-	+	0	+	+	-	-	-
i_6	-	0	-	-	-	0	-	-	-	-
i_7	-	0	-	+	-	+	0	-	-	-
i_8	+	+	+	+	+	+	+	0	0	+
i_9	+	+	+	+	+	+	+	0	0	+
i_{10}	0	+	-	+	+	+	+	-	-	0
Index	-2	+6	-5	+6	+1	+8	+4	-8	-8	-2

Table 6.6. Plurality Voting strategy.

	1	2	3	4	5	6
u_1	i_2, i_4, i_7	i_4, i_7	i_5	i_1	i_3	i_8
u_2	i_2, i_6	i_4, i_7	i_5	i_1, i_{10}	i_3	i_8, i_9
u_3	i_6, i_{10}	i_{10}	i_{10}	i_{10}	i_3	i_9
Group	i_2, i_6	i_4, i_7	i_5	i_1, i_{10}	i_3	i_8, i_9

alternative needs to be selected, the options that received the highest number of votes are selected. Table 6.6 presents an example of how the strategy works.

This strategy was implemented and tested by [Senot *et al.* (2010, 2011)] in the TV domain.

Approval Voting. Each user can vote for as many items as she/he wants and a point is assigned to all the items a user likes. To show how the strategy works, we are going to suppose that each user votes for all the items with a rating above a certain threshold (in the example in Table 6.7, the threshold rating is 5). A group preference is obtained by adding the individual points of an item.

When choosing the pages to recommend to a group, *Let's Browse* [Lieberman *et al.* (1999)] evaluates if the page currently considered by the system matches with the user profile above a certain threshold and recommends the one that gets the highest score.

Table 6.7. Approval Voting strategy.

	i_1	i_2	i_3	i_4	i_5	i_6	i_7	i_8	i_9	i_{10}
u_1	1	1	1	1	1	1	1	1		1
u_2	1	1	1	1	1	1	1			1
u_3			1	1	1	1			1	1
Group	2	2	3	3	3	3	2	1	1	3

Least Misery. The rating assigned to an item for a group is the lowest rating expressed for that item by any of the members of the group. This strategy is usually employed to model small groups, to make sure that every group member is satisfied. A drawback of this strategy is that if the majority of the group members really likes something, but one person does not, the item will not be recommended to the group. This is what happens in the example in Table 6.8 for items i_2 and i_7.

This strategy is used by *PolyLens* [O'Connor *et al.* (2001)], in order to produce movie recommendations that satisfy the small groups handled by the system.

Table 6.8. Least Misery strategy.

	i_1	i_2	i_3	i_4	i_5	i_6	i_7	i_8	i_9	i_{10}
u_1	8	10	7	10	9	8	10	6	3	6
u_2	7	10	6	9	8	10	9	4	4	7
u_3	5	1	8	6	9	10	3	5	7	10
Group	5	1	6	6	8	8	3	4	3	6

Most Pleasure. The rating assigned to an item for a group is the highest rating expressed for that item by a member of the group. Table 6.9 presents an example of how the strategy works.

This strategy is adopted by [Quijano-Sánchez *et al.* (2012)] in a case-based group recommender system, proposed as a solution to the cold start problem.

Average Without Misery. The rating assigned to an item for a group is the average of the ratings assigned by each user for that item. All the items that were evaluated by a user with a rating under a certain threshold are not considered in the group model (in the example in Table 6.10, the threshold rating is 4).

Table 6.9. Most Pleasure strategy.

	i_1	i_2	i_3	i_4	i_5	i_6	i_7	i_8	i_9	i_{10}
u_1	8	10	7	10	9	8	10	6	3	6
u_2	7	10	6	9	8	10	9	4	4	7
u_3	5	1	8	6	9	10	3	5	7	10
Group	8	10	8	10	9	10	10	6	7	10

Table 6.10. Average Without Misery strategy.

	i_1	i_2	i_3	i_4	i_5	i_6	i_7	i_8	i_9	i_{10}
u_1	8	10	7	10	9	8	10	6	3	6
u_2	7	10	6	9	8	10	9	4	4	7
u_3	5	1	8	6	9	10	3	5	7	10
Group	20	-	21	25	26	28	-	15	-	23

In order to model the preferences of a group for each genre of music that can be played in a gym, *MusicFX* [McCarthy and Anagnost (1998)] sums the individual ratings expressed by each user, discarding the ones under a minimum degree of satisfaction. As previously mentioned, in [Pessemier *et al.* (2013)], the authors illustrate that an average is the best way to represent individual preferences in a group model in different contexts and employ this strategy in their study.

Fairness. This strategy is based on the idea that users can be recommended something they do not like, as long as they also get recommended something they like. This is done by allowing each user to choose her/his favorite item. If two items have the same rating, the choice is based on the other users' preferences. This is done until everyone made a choice. Next, everyone chooses a second item, starting from the person who chose last the first time.

If in the example, if we suppose that user u_1 chose first, she/he would consider i_2, i_4, and i_7, and would choose i_4, because it has the highest average considering the other users' ratings. Next, u_2 would choose between i_2 and i_6 and would select i_6 for the same reason. Then, u_3 would choose item i_{10}. Since everyone chose an item, it would be u_3's turn again and i_5 would be chosen. User u_2 would choose i_2, which has the highest rating along with i_6 (which was already chosen). Then, u_1 would choose i_7, which is the one with the highest rating and was not chosen yet. The final sequence of items that models the group would be: i_4, i_6, i_{10}, i_5, i_2, i_7, i_1, i_3, i_9, i_8.

This strategy was adopted in the music recommender system proposed in [Christensen and Schiaffino (2011)].

Most Respected Person (Dictatorship). This strategy selects the items according to the preferences of the most respected person, using the preferences of the other users just in case more than one item received the same evaluation. The idea behind this strategy is that there are scenarios is which a group is guided/dominated by a person. In the example in Table 6.11, it is supposed that u_1 is the most respected person.

Table 6.11. Most Respected Person strategy (Dictatorship).

	i_1	i_2	i_3	i_4	i_5	i_6	i_7	i_8	i_9	i_{10}
u_1	8	10	7	10	9	8	10	6	3	6
u_2	7	10	6	9	8	10	9	4	4	7
u_3	5	1	8	6	9	10	3	5	7	10
Group	8	10	7	10	9	8	10	6	3	6

This strategy is used by *INTRIGUE* [Ardissono *et al.* (2005)] that, in order to build a group model, advantages the preferences of a subset of users with particular needs. The studies performed on the *G.A.I.N.* [Pizzutilo *et al.* (2005)] system show that when people interact, a user or a small portion of the group influences the choices of the whole group. In [Jung (2012)], Jung presented an approach to identify long tail users, i.e., users who can be considered as an expert group on a certain attribute. So, the ratings given by the long tail user groups are considered in order to provide a relevant recommendation to the non-expert user group, called short head group.

6.4.2. *Discussion*

As highlighted in [Pizzutilo *et al.* (2005)], there is no strategy useful in every context independently from the environment, and the choice of the strategy that best models a group should be made after a deep analysis of the context in which the group is modeled.

However, in the current group recommendation literature, Additive Utilitarian is the strategy that is most-widely employed. Indeed, as previously highlighted, it has been proven to be effective in several scenarios [Pessemier *et al.* (2013)]. This could also be due to the fact, in practice, groups are usually built from existing datasets. Hence, as shown in [Boratto *et al.* (2016)], where groups are built with the k-means clustering algorithm:

- an average, which is a single value that is meant to typify a set of different values, allows to weigh the preferences of all the users in equal ways;
- creating a group model with an average of the individual values for each item is like re-creating the centroid of the cluster, i.e., a super-user that connects every user in the group.

Figure 6.5 and Table 6.12 illustrate a comparison of the different group modeling strategies in the collaborative group recommender system previously mentioned [Boratto *et al.* (2016)][2]. The results present the RMSE obtained in the MovieLens-1M dataset (whose details are presented in Section 6.6), using the different group modeling strategies for a varying number of groups. As these results show, Additive Utilitarian obtains the highest accuracy (lowest RMSE) for all the groups.

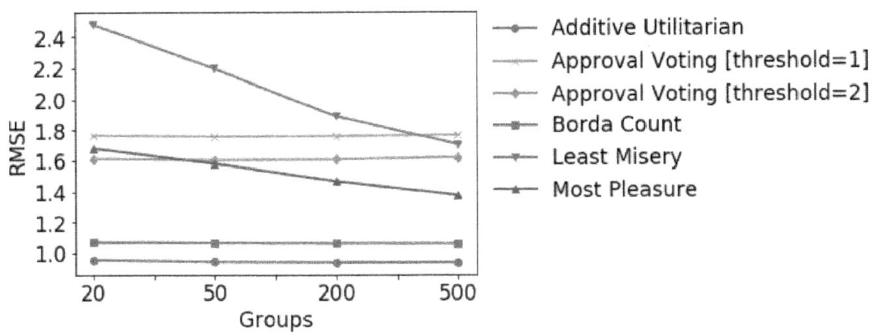

Fig. 6.5. RMSE values obtained by using the different group modeling strategies.

Table 6.12. RMSE values obtained by using the different group modeling strategies.

	20 groups	**50 groups**	**200 groups**	**500 groups**
AU	0.9554	0.9435	0.9395	0.9385
AV [threshold=1]	1.7634	1.7558	1.7573	1.7629
AV [threshold=2]	1.6112	1.6025	1.6057	1.6193
BC	1.0667	1.0624	1.0596	1.0570
LM	2.4782	2.1972	1.8868	1.7024
MP	1.6786	1.5796	1.4648	1.3735

[2] The paper compared only the strategies that produce an actual rating for a group and some of them were discarded for different reasons. Readers can refer to the original paper for more details.

6.5. Rating Prediction

Rating prediction is a key task to build effective group recommender systems. Given that the algorithms have been presented in Chapter 1, in this section we will analyze how the different classes of prediction algorithms that have been previously presented (i.e., memory-based and model-based) have been employed in the group recommendation context. One aspect that emerges is that there is not a one-fits-all solution, and the choice of the algorithm employed by the system strongly depends on the studied domain.

In this section, we will provide relevant examples of how the different algorithms have been employed for the different families of approaches.

Sometimes, a comparison between the different classes of algorithms is even necessary, in order to explore what is the most appropriate solution. For example, in [Pessemier *et al.* (2013)], a model-based approach (SVD) was compared to two memory-based algorithms (user- and item-based) and to content-based and hybrid systems; results showed that the item-based algorithm outperformed the user-based one (thus it was employed also in the hybrid solution) and that SVD outperformed the rest of the algorithms. Castro *et al.* [Castro *et al.* (2017)] compared user-based and item-based algorithms in a noise handling context, finding out that the user-based approach was the most effective.

6.5.1. *Memory-based Algorithms*

Here, we will analyze how memory-based algorithms have been employed in group recommender systems, dividing the analysis into user- and item-based approaches.

6.5.1.1. *User-based Algorithms*

This class of algorithms is by far the most widely-employed in the group recommendation literature. As previously mentioned, sometimes new domains need to explore different solutions. Even though it might be clear that user-based filtering is the most appropriate solution, the family of approaches in which the algorithm should be embedded needs to be studied. Berkovsky and Freyne [Berkovsky and Freyne (2010)] considered the case of recipe recommendation to the members of a family and employed both an approach based on group preference models and one based on aggregating individual predictions, finding out that the first one was more effective in this context.

User-based collaborative filtering has been employed in several group recommender systems that built the predictions using a group preference model. The approach presented in [Anand (2013)], provides movie recommendations to groups characterized by similar content-based features, and [Chen *et al.* (2008)] that derived subgroup ratings by considering the interactions between the members of a group; these subgroup ratings are employed by the rating prediction task.

When the family of approaches that merges single user recommendations is considered, the most famous example is PolyLens [O'Connor *et al.* (2001)], which recommends movies to small groups of users.

Examples of systems that aggregate all the individual predictions are that by Quijano-Sánchez *et al.* [Quijano-Sánchez *et al.* (2011)], in which the predictions are combined in a model that also considers the group personality composition and the social connections between the single users, and e-Tourism [Sebastia *et al.* (2009)], which combines the predictions produced by a user-based algorithm with that produced with content-based, demographic, and knowledge-based filterings.

6.5.1.2. *Item-based Algorithms*

Item-based collaborative filtering has not been widely-employed in the group recommendation literature. Examples of systems that employ it in a system based on group preference models are [Boratto *et al.* (2009b)] and [Boratto and Carta (2014)], which use item-based algorithms to produce recommendations for large groups, respectively detected with community-detection and clustering algorithms.

Some systems aggregate individual predictions with item-based algorithms when considering the disagreement between the group members [Amer-Yahia *et al.* (2009)] and when considering fairness in package recommendations to a group [Serbos *et al.* (2017)].

6.5.2. *Model-based algorithms*

Some exploratory studies have been conducted to compare the use of model-based algorithms on different families of approaches, as we saw with user-based algorithms. Ortega *et al.* [Ortega *et al.* (2016)] compared the use of Matrix Factorization algorithms on systems that used group preference models and aggregated individual predictions, while Hu *et al.* [Hu *et al.* (2011)] compared SVD on the same two families of approaches. The first study found out that the use of group preference models is better with large datasets and medium/large groups and individual predictions should be preferred with small groups, while the second study found out that the use of group preference models outperforms the aggregation of individual predictions in the Moviepilot dataset.

Different types of model-based algorithms are employed for the different families of approaches. Regarding group preference models, Yuan *et al.* [Yuan *et al.* (2014)] presented a Latent Dirichlet Allocation (LDA) based generative model on groups, while Purushotham *et al.* [Purushotham *et al.* (2014)] used classic Matrix Factorization to build group-activity recommendations in Location-Based Social Networks.

Model-based algorithms have also been employed when aggregating individual predictions. The approach by Baltrunas *et al.* [Baltrunas *et al.* (2010)] built individual predictions with SVD to produce a ranked list of items to recommend to a group. Christensen and Schiaffino [Christensen and Schiaffino (2014)] built individual predictions with a classic Matrix Factorization algorithm, which are combined with information such as the trusted relationships, the social similarity, and the social centrality of the users, in order to build group recommendations.

6.6. Case-study: Comparing Families of Approaches

This section presents a case-study, in which we compare the different families of approaches introduced in this chapter (readers can refer to [Boratto *et al.* (2017)] for more details on the study).

For each of the four architectures presented in Section 6.2, we built a group recommender system. More specifically, the association between the systems and the families of approaches presented in Section 6.2 is shown in Table 6.13.

Table 6.13. Systems developed according to each family of approaches and architecture.

Name	Family of approaches	Architecture
ModelBased (MB)	Group preference models (Section 6.2.1)	Fig. 6.1
MergeRecommendations (MR)	Merging recommendations (Section 6.2.2)	Fig. 6.2
Predict&Cluster (PC)	Merging predictions (Section 6.2.3)	Fig. 6.3
Cluster&Predict (CP)	Merging predictions (Section 6.2.3)	Fig. 6.4

6.6.1. *Experimental setup*

The datasets used to perform the study are MovieLens-1M (ML–1M)[3] and a subset of Yahoo! Webscope (R4)[4]. Details of the datasets' structure are provided in Table 6.14.

Table 6.14. Datasets details.

	ML–1M	R4
Users	6040	5070
Items	1682	1647
Ratings	100,000,000	153,461

The accuracy of the predicted ratings was measured through the Root Mean Squared Error (RMSE) metric. It compares each rating r_{ui}, given by a user u for an item i in the test set, with the rating p_{gi}, predicted for the item i for the group g under which user u is subsumed. The related formula is shown below:

$$RMSE = \sqrt{\frac{\sum_{i=0}^{n}(r_{ui} - p_{gi})^2}{n}}$$

n is the number of ratings available in the test set. Measuring the effectiveness of the group recommendations as an average of the individual accuracies is necessary since, as we will highlight in Section 6.7, no metric specific for group recommendation exists. This way of measuring the accuracy might have some drawbacks, since it does not take into account if some users have a different role/importance/reputation in the group or the size of the group itself. However, since our groups are detected via a clustering algorithm (hence, no user has a different role), this approach seems fair to evaluate how accurate are the group recommendations are for the individual group members.

The details of the algorithms used to implement each of the tasks considered in this study are the following:

Group building. Groups are detected with the k-means clustering algorithm. In order to study different group recommendation scenarios, we detected 20, 50, 200, and 500 groups. In addition, we will also present two settings that will serve as baselines, in which we create a unique group with all the users (named "1 group") and we do not group the users, thus creating individual recommendations for each user (named "P", which stands for "Personalized").

[3]http://www.grouplens.org/
[4]https://webscope.sandbox.yahoo.com/catalog.php?datatype=r

Group modeling. Group preferences have been modeled with the Additive Utilitarian strategy previously presented, which was shown to be the most effective to model groups detected with the k-means clustering algorithm [Boratto and Carta (2014)].

Rating prediction. In the families of approaches that merge recommendations for single users and aggregate individual predictions, ratings have been predicted for individual users with the user-based algorithm presented in Chapter 1; the number of neighbors chosen to run the algorithm is 100. In the family of approaches that constructs group preference models, the predictions for the groups have been built with with the item-based algorithm presented in Chapter 1; the number of neighbors chosen is 20 in the MovieLens-1M dataset, and 10 in the Webscope (R4) dataset (the difference in the number of neighbors is due to a parameter setting, done via experiments that can be found in [Boratto and Carta (2015)]).

6.6.2. *Results*

Figure 6.6 and Table 6.15 report the results obtained by each system with the MovieLens-1M dataset, while Fig. 6.7 and Table 6.16 report the results for the Webscope (R4) dataset.

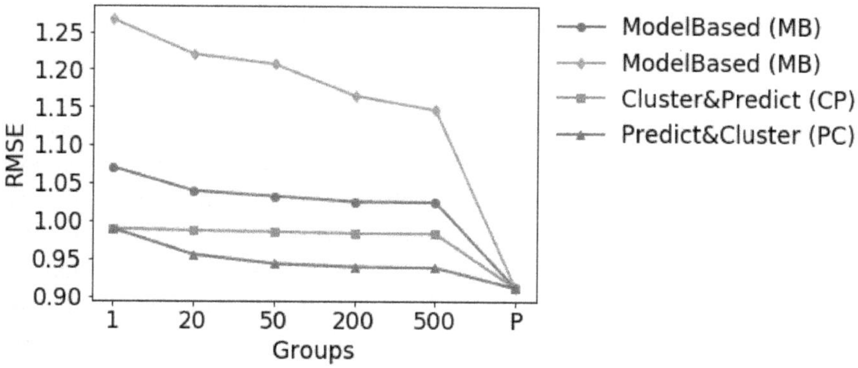

Fig. 6.6. RMSE obtained by the each system in the MovieLens-1M dataset.

Table 6.15. RMSE obtained by the each system in the MovieLens-1M dataset.

	1 group	**20 groups**	**50 groups**	**200 groups**	**500 groups**	**Personalized**
MB	1.0705	1.0398	1.0327	1.0257	1.0249	0.9120
MR	1.2667	1.2207	1.2075	1.1653	1.1461	0.9120
CP	0.9895	0.9872	0.9857	0.9837	0.9832	0.9120
PC	0.9895	0.9554	0.9435	0.9395	0.9385	0.9120

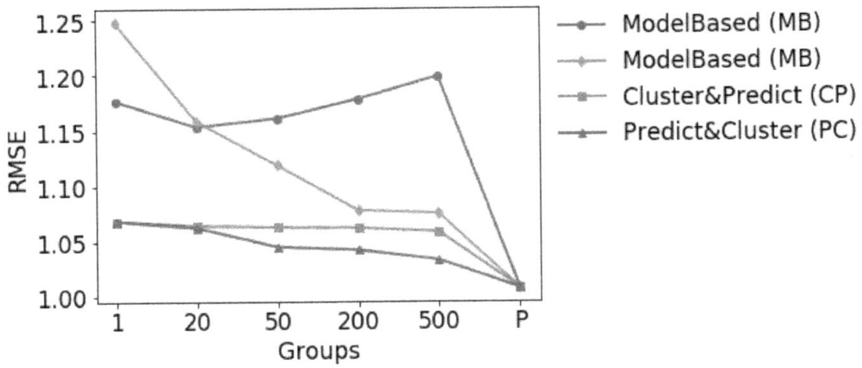

Fig. 6.7. RMSE obtained by the each system in the Webscope (R4) dataset.

Table 6.16. RMSE obtained by the each system in the Webscope (R4) dataset.

	1 group	**20 groups**	**50 groups**	**200 groups**	**500 groups**	**Personalized**
MB	1.1767	1.1534	1.1610	1.1780	1.1985	1.0074
MR	1.2477	1.1585	1.1187	1.0780	1.0751	1.0074
CP	1.0686	1.0644	1.0631	1.0622	1.0587	1.0074
PC	1.0686	1.0626	1.0454	1.0421	1.0330	1.0074

As it can be observed, as the number of groups increases, the RMSE values decrease. In other words, as we expected, a system can perform better when more recommendations can be produced, since groups get smaller and it is easier for the system to produce more effective recommendations when the preferences of less users have to be met. This is not true for the *ModelBased* system in the Webscope (R4) dataset (which is the smaller and sparser one), which cannot provide accurate predictions for small groups (the 200 and 500 groups settings). This apparently negative feedback demonstrates that a model-based group recommender system is not accurate for small groups. More specifically, given a small amount of information about the preferences of the users for the items, a group model cannot be effectively employed to generate group recommendations.

As it can be noticed, the three approaches to produce group recommendations are clearly separated in the MovieLens-1M results. Webscope (R4) shows a different behavior for the *ModelBased* system due to the latter's inability to provide accurate predictions for small groups. Conversely, the other three systems (i.e., *MergeRecommendations*, *Predict&Cluster*, and *Cluster&Predict*) behave as the MovieLens-1M dataset.

In both cases, however, the approach that merges individual recommendations (i.e., *MergeRecommendations*) and that based on a group model (i.e., *ModelBased*) produced the worst results. This means that if we employ an approach that merges individual recommendations, only a small amount of preferences per user is available and a group recommender system is not able to properly satisfy the users. As for the approach based on group preference models, these results confirm the need of relying on preferences expressed by single users, in order to build accurate group recommendations.

The systems that merge individual preferences (i.e., *Cluster&Predict* and *Predict&Cluster*) are the ones that achieve the best results. As it can also be noticed, the accuracy of *Predict&Cluster* is much higher than that of *Cluster&Predict*, and this proves that enhancing clustering with individual predictions leads to great improvements in the quality of the predicted results.

6.7. Future Directions: Open Issues and Challenges

This section presents the current open issues and challenges in the development of collaborative group recommender systems.

Explanation with model-based algorithms. A current open challenge when working with model-based collaborative filtering algorithms, such as Matrix Factorization, is the lack of explanations on *why* and item has been recommended to the group, due to the fact that the recommendation is based on latent features. The problem is amplified in the group recommendation scenario, since the individual preferences are usually combined into a group model, which makes it even more difficult to connect the produced recommendations to the individual preferences of the users. An initial overview of approaches to explain recommendation to groups is given in [Felfernig *et al.* (2018c)].

Understanding and employing group dynamics. It is known that group members are influenced in their evaluations by the composition of the group [Gartrell *et al.* (2010)] and by the interactions between the members [Delic *et al.* (2016)]. Integrating the evolution of the individual preferences that happens because of the group dynamics into collaborative algorithms is still an open issue. An initial overview of approaches to take into account group dynamics in group recommendations scenarios is given in [Tkalčič *et al.* (2018)].

Adapting to group constraints. At the moment, users are not allowed to express specific constraints in the context of a group. While a step in this direction has been made in the INTRIGUE system [Ardissono *et al.* (2005)], which gives a different weight to the preferences of users with specific needs (e.g., children or disabled people), moving from generic needs to individual constraints to produce more tailored group recommendations still represents a challenge. An initial overview of related approaches is given in [Felfernig *et al.* (2018a)].

Lack of existing datasets. The fact that in practice we need to detect artificial groups in order to build group recommender systems, makes it hard to evaluate their effectiveness in real-world scenarios, in which users might be connected because of unexpected reasons or might interact among themselves, thus providing relevant information for the building of effective systems.

Evaluation metrics. As [Ricci (2014)] highlights, the evaluation of group recommender systems is a challenging aspect. Indeed, there is no metric to evaluate the satisfaction of a group as a whole, and the metrics currently used to evaluate single user recommender systems only measure the individual satisfactions with the group recommendations.

Lack of software frameworks. Chapter 9 surveyed the existing software frameworks for recommender systems. Unfortunately, none of them directly allows to generate group recommendations. Hence, when adopting a regular (single user) recommender system for group recommendation purposes, some form of pre- or post-processing filter function is needed, that is aware of the given group structure in the dataset. In the case when the system works with group profiles, the pre-filtering function would merge the individual preferences into one group profile (i.e., the function would treat the group as a super-user). In case the system treats the user profiles separately to calculate lists of recommendations, a post-processing filter function would be needed that is able to aggregate the individual recommendation lists into a unique group recommendation list.

6.8. Conclusions

This chapter introduced group recommender systems from the collaborative point of view. Group recommendation is characterized by several peculiar and challenging aspects that do not characterize the systems that produce recommendations for single users. Hence, starting from these aspects, we analyzed how to build a system. We considered the existing families of approaches and presented the high-level architecture of the systems. For each of the tasks in the architectures, we presented the algorithms and techniques that are employed to build them. A case-study compared the different families of approaches on different datasets, showing which family of approaches is more effective when groups are detected with a clustering algorithm. Despite the big efforts made to develop effective group recommender systems, several open issues still exist, which represent interesting perspectives and directions to do exciting research in this area.

References

Amer-Yahia, S., Roy, S. B., Chawla, A., Das, G. and Yu, C. (2009). Group recommendation: Semantics and efficiency, *Proceedings of the VLDB Endowment* **2**, 1, pp. 754–765.

Anand, D. (2013). Group movie recommendations via content based feature preferences, *International Journal of Scientific & Engineering Research* **4**, 2.

Ardissono, L., Goy, A., Petrone, G. and Segnan, M. (2005). A multi-agent infrastructure for developing personalized web-based systems, *ACM Transactions on Internet Technology* **5**, 1, pp. 47–69.

Baltrunas, L., Makcinskas, T. and Ricci, F. (2010). Group recommendations with rank aggregation and collaborative filtering, in X. Amatriain, M. Torrens, P. Resnick and M. Zanker (eds.), *Proceedings of the 2010 ACM Conference on Recommender Systems, RecSys 2010, Barcelona, Spain, September 26-30, 2010* (ACM), pp. 119–126, doi:10.1145/1864708.1864733, `http://doi.acm.org/10.1145/1864708.1864733`.

Berkovsky, S. and Freyne, J. (2010). Group-based recipe recommendations: analysis of data aggregation strategies, in X. Amatriain, M. Torrens, P. Resnick and M. Zanker (eds.), *Proceedings of the 2010 ACM Conference on Recommender Systems, RecSys 2010, Barcelona, Spain, September 26-30, 2010* (ACM), ISBN 978-1-60558-906-0, pp. 111–118.

Boratto, L. and Carta, S. (2014). Modeling the preferences of a group of users detected by clustering: a group recommendation case-study, in R. Akerkar, N. Bassiliades, J. Davies and V. Ermolayev (eds.), *4th International Conference on Web Intelligence, Mining and Semantics (WIMS 14), WIMS '14, Thessaloniki, Greece, June 2-4, 2014* (ACM), pp. 16:1–16:7, doi:10.1145/2611040.2611073, `http://doi.acm.org/10.1145/2611040.2611073`.

Boratto, L. and Carta, S. (2015). The rating prediction task in a group recommender system that automatically detects groups: architectures, algorithms, and performance evaluation, *J. Intell. Inf. Syst.* **45**, 2, pp. 221–245, doi:10.1007/s10844-014-0346-z, `https://doi.org/10.1007/s10844-014-0346-z`.

Boratto, L., Carta, S., Chessa, A., Agelli, M. and Clemente, M. L. (2009a). Group recommendation with automatic identification of users communities, in *Proceedings of the 2009 IEEE/WIC/ACM International Joint Conference on Web Intelligence and Intelligent Agent Technology - Volume 03*, WI-IAT '09 (IEEE Computer Society, Washington, DC, USA), ISBN 978-0-7695-3801-3, pp. 547–550, doi:10.1109/WI-IAT.2009.346, `http://dx.doi.org/10.1109/WI-IAT.2009.346`.

Boratto, L., Carta, S., Chessa, A., Agelli, M. and Clemente, M. L. (2009b). Group recommendation with automatic identification of users communities, in *Proceedings of the 2009 IEEE/WIC/ACM International Conference on Web Intelligence and International Conference on Intelligent Agent Technology - Workshops, Milan, Italy, 15-18 September 2009* (IEEE Computer Society), pp. 547–550, doi:10.1109/WI-IAT.2009.346, `https://doi.org/10.1109/WI-IAT.2009.346`.

Boratto, L., Carta, S. and Fenu, G. (2016). Discovery and representation of the preferences of automatically detected groups, *Future Gener. Comput. Syst.* **64**, C, pp. 165–174, doi:10.1016/j.future.2015.10.007, `http://dx.doi.org/10.1016/j.future.2015.10.007`.

Boratto, L., Carta, S. and Fenu, G. (2017). Investigating the role of the rating prediction task in granularity-based group recommender systems and big data scenarios, *Inf. Sci.* **378**, pp. 424–443, doi:10.1016/j.ins.2016.07.060, `https://doi.org/10.1016/j.ins.2016.07.060`.

Campos, L. M., Fernández-Luna, J. M., Huete, J. F. and Rueda-Morales, M. A. (2009). Managing uncertainty in group recommending processes, *User Modeling and User-Adapted Interaction* **19**, 3, pp. 207–242, doi:10.1007/s11257-008-9061-1, `http://dx.doi.org/10.1007/s11257-008-9061-1`.

Castro, J., Toledo, R. Y. and Martínez, L. (2017). An empirical study of natural noise management in group recommendation systems, *Decision Support Systems* **94**, pp. 1–11, doi:10.1016/j.dss.2016.09.020, `http://dx.doi.org/10.1016/j.dss.2016.09.020`.

Chen, Y.-L., Cheng, L.-C. and Chuang, C.-N. (2008). A group recommendation system with consideration of interactions among group members, *Expert Systems with Applications* **34**, 3, pp. 2082–2090.

Christensen, I. A. and Schiaffino, S. N. (2011). Entertainment recommender systems for group of users, *Expert Systems with Applications* **38**, 11, pp. 14127–14135.

Christensen, I. A. and Schiaffino, S. N. (2014). Social influence in group recommender systems, *Online Information Review* **38**, 4, pp. 524–542, doi:10.1108/OIR-08-2013-0187, `http://dx.doi.org/10.1108/OIR-08-2013-0187`.

Delic, A., Neidhardt, J., Nguyen, T. N., Ricci, F., Rook, L., Werthner, H. and Zanker, M. (2016). Observing group decision making processes, in S. Sen, W. Geyer, J. Freyne and P. Castells (eds.), *Proceedings of the 10th ACM Conference on Recommender Systems, Boston, MA, USA, September 15-19, 2016* (ACM), pp. 147–150, doi:10.1145/2959100.2959168, `http://doi.acm.org/10.1145/2959100.2959168`.

Felfernig, A., Atas, M., Helic, D., Tran, T. N. T., Stettinger, M. and Samer, R. (2018a). Algorithms for group recommendation, in *Group Recommender Systems: An Introduction* (Springer International Publishing, Cham), ISBN 978-3-319-75067-5, pp. 27–58, doi:10.1007/978-3-319-75067-5_2, https://doi.org/10.1007/978-3-319-75067-5_2.

Felfernig, A., Boratto, L., Stettinger, M. and Tkalčič, M. (2018b). *Group Recommender Systems: An Introduction* (Springer).

Felfernig, A., Tintarev, N., Tran, T. N. T. and Stettinger, M. (2018c). Explanations for groups, in *Group Recommender Systems: An Introduction* (Springer International Publishing, Cham), ISBN 978-3-319-75067-5, pp. 105–126, doi:10.1007/978-3-319-75067-5_6, https://doi.org/10.1007/978-3-319-75067-5_6.

Felfernig, A., Zehentner, C., Ninaus, G., Grabner, H., Maalej, W., Pagano, D., Weninger, L. and Reinfrank, F. (2012). Group decision support for requirements negotiation, in L. Ardissono and T. Kuflik (eds.), *Advances in User Modeling - UMAP 2011 Workshops, Girona, Spain, July 11-15, 2011, Revised Selected Papers, Lecture Notes in Computer Science*, Vol. 7138 (Springer), ISBN 978-3-642-28508-0, pp. 105–116.

Gartrell, M., Xing, X., Lv, Q., Beach, A., Han, R., Mishra, S. and Seada, K. (2010). Enhancing group recommendation by incorporating social relationship interactions, in W. G. Lutters, D. H. Sonnenwald, T. Gross and M. Reddy (eds.), *Proceedings of the 2010 International ACM SIGGROUP Conference on Supporting Group Work, GROUP 2010, Sanibel Island, Florida, USA, November 6-10, 2010* (ACM), pp. 97–106, doi:10.1145/1880071.1880087, http://doi.acm.org/10.1145/1880071.1880087.

Hu, X., Meng, X. and Wang, L. (2011). Svd-based group recommendation approaches: an experimental study of moviepilot, in *Proceedings of the 2nd Challenge on Context-Aware Movie Recommendation*, CAMRa '11 (ACM, New York, NY, USA), ISBN 978-1-4503-0825-0, pp. 23–28, doi:10.1145/2096112.2096117, http://doi.acm.org/10.1145/2096112.2096117.

Jameson, A. and Smyth, B. (2007). Recommendation to groups, in *The Adaptive Web, Methods and Strategies of Web Personalization, Lecture Notes in Computer Science*, Vol. 4321 (Springer, Berlin), ISBN 978-3-540-72078-2, pp. 596–627.

Jung, J. J. (2012). Attribute selection-based recommendation framework for short-head user group: An empirical study by movielens and imdb, *Expert Systems with Applications* **39**, 4, pp. 4049–4054, doi:10.1016/j.eswa.2011.09.096, http://dx.doi.org/10.1016/j.eswa.2011.09.096.

Kim, J. K., Kim, H. K., Oh, H. Y. and Ryu, Y. U. (2010). A group recommendation system for online communities, *International Journal of Information Management* **30**, 3, pp. 212–219, doi:10.1016/j.ijinfomgt.2009.09.006, http://dx.doi.org/10.1016/j.ijinfomgt.2009.09.006.

Lieberman, H., Dyke, N. W. V. and Vivacqua, A. S. (1999). Let's browse: A collaborative web browsing agent, in *IUI*, pp. 65–68.

Liu, X., Tian, Y., Ye, M. and Lee, W.-C. (2012). Exploring personal impact for group recommendation, in X. wen Chen, G. Lebanon, H. Wang and M. J. Zaki (eds.), *21st ACM International Conference on Information and Knowledge Management, CIKM'12, Maui, HI, USA, October 29 - November 02, 2012* (ACM), ISBN 978-1-4503-1156-4, pp. 674–683.

Masthoff, J. (2015). Group recommender systems: Aggregation, satisfaction and group attributes, in F. Ricci, L. Rokach and B. Shapira (eds.), *Recommender Systems Handbook* (Springer), pp. 743–776, doi:10.1007/978-1-4899-7637-6_22, https://doi.org/10.1007/978-1-4899-7637-6_22.

McCarthy, J. (2002). Pocket RestaurantFinder: A situated recommender system for groups, in *Workshop on Mobile Ad-Hoc Communication at the 2002 ACM Conference on Human Factors in Computer Systems*, http://interrelativity.com/joe/publications/PocketRestaurantFinder-CHI2002ws-AdHoc.pdf.

McCarthy, J. F. and Anagnost, T. D. (1998). Musicfx: An arbiter of group preferences for computer supported collaborative workouts, in S. E. Poltrock and J. Grudin (eds.), *CSCW '98, Proceedings of the ACM 1998 Conference on Computer Supported Cooperative Work, Seattle, WA, USA, November 14-18, 1998* (ACM), ISBN 1-58113-009-0, pp. 363–372.

Mislove, A., Viswanath, B., Gummadi, P. K. and Druschel, P. (2010). You are who you know: inferring user profiles in online social networks, in B. D. Davison, T. Suel, N. Craswell and B. Liu (eds.), *Proceedings of the Third International Conference on Web Search and Web Data Mining, WSDM 2010, New York, NY, USA, February 4-6, 2010* (ACM), pp. 251–260, doi:10.1145/1718487.1718519, http://doi.acm.org/10.1145/1718487.1718519.

Ntoutsi, E., Stefanidis, K., Nørvåg, K. and Kriegel, H.-P. (2012). Fast group recommendations by applying user clustering, in P. Atzeni, D. W. Cheung and S. Ram (eds.), *Conceptual Modeling - 31st International Conference ER 2012, Florence, Italy, October 15-18, 2012. Proceedings*, *Lecture Notes in Computer Science*, Vol. 7532 (Springer), ISBN 978-3-642-34001-7, pp. 126–140.

O'Connor, M., Cosley, D., Konstan, J. A. and Riedl, J. (2001). Polylens: A recommender system for groups of users, in W. Prinz, M. Jarke, Y. Rogers, K. Schmidt and V. Wulf (eds.), *Proceedings of the Seventh European Conference on Computer Supported Cooperative Work, 16-20 September 2001, Bonn, Germany* (Kluwer), pp. 199–218.

Ortega, F., Hernando, A., Bobadilla, J. and Kang, J. H. (2016). Recommending items to group of users using matrix factorization based collaborative filtering, *Inf. Sci.* **345**, C, pp. 313–324, doi:10.1016/j.ins.2016.01.083, http://dx.doi.org/10.1016/j.ins.2016.01.083.

Pessemier, T., Dooms, S. and Martens, L. (2013). Comparison of group recommendation algorithms, *Multimedia Tools and Applications*, pp. 1–45, doi:10.1007/s11042-013-1563-0, http://dx.doi.org/10.1007/s11042-013-1563-0.

Pizzutilo, S., Carolis, B. D., Cozzolongo, G. and Ambruoso, F. (2005). Group modeling in a public space: Methods, techniques and experiences, in *Proceedings of WSEAS AIC 05* (ACM, Malta).

Purushotham, S., Kuo, C. J., Shahabdeen, J. and Nachman, L. (2014). Collaborative group-activity recommendation in location-based social networks, in R. A. de By and C. Wenk (eds.), *Proceedings of the 3rd ACM SIGSPATIAL International Workshop on Crowdsourced and Volunteered Geographic Information, GeoCrowd '14, Dallas, Texas, USA, November 4, 2014* (ACM), pp. 8–15, doi:10.1145/2676440.2676442, `http://doi.acm.org/10.1145/2676440.2676442`.

Quijano-Sánchez, L., Bridge, D. G., Díaz-Agudo, B. and Recio-García, J. A. (2012). A case-based solution to the cold-start problem in group recommenders, in B. Díaz-Agudo and I. Watson (eds.), *Case-Based Reasoning Research and Development - 20th International Conference, ICCBR 2012, Lyon, France, September 3-6, 2012. Proceedings*, *Lecture Notes in Computer Science*, Vol. 7466 (Springer), ISBN 978-3-642-32985-2, pp. 342–356.

Quijano-Sánchez, L., Recio-García, J. A., Díaz-Agudo, B. and Jiménez-Díaz, G. (2011). Social factors in group recommender systems, *ACM Transactions on Intelligent Systems and Technology*.

Recio-García, J. A., Jiménez-Díaz, G., Sánchez-Ruiz-Granados, A. A. and Díaz-Agudo, B. (2009). Personality aware recommendations to groups, in L. D. Bergman, A. Tuzhilin, R. D. Burke, A. Felfernig and L. Schmidt-Thieme (eds.), *Proceedings of the 2009 ACM Conference on Recommender Systems, RecSys 2009, New York, NY, USA, October 23-25, 2009* (ACM), ISBN 978-1-60558-435-5, pp. 325–328.

Ricci, F. (2014). Recommender systems: Models and techniques, in *Encyclopedia of Social Network Analysis and Mining* (Springer, New York), pp. 1511–1522.

Sebastia, L., Garcia, I., Onaindia, E. and Guzman, C. (2009). *E-Tourism*: a tourist recommendation and planning application, *International Journal on Artificial Intelligence Tools* **18**, 5, pp. 717–738, doi:10.1142/S0218213009000378, `http://dx.doi.org/10.1142/S0218213009000378`.

Senot, C., Kostadinov, D., Bouzid, M., Picault, J. and Aghasaryan, A. (2011). Evaluation of group profiling strategies, in T. Walsh (ed.), *IJCAI 2011, Proceedings of the 22nd International Joint Conference on Artificial Intelligence, Barcelona, Catalonia, Spain, July 16-22, 2011* (IJCAI/AAAI), ISBN 978-1-57735-516-8, pp. 2728–2733.

Senot, C., Kostadinov, D., Bouzid, M., Picault, J., Aghasaryan, A. and Bernier, C. (2010). Analysis of strategies for building group profiles, in P. D. Bra, A. Kobsa and D. N. Chin (eds.), *User Modeling, Adaptation, and Personalization, 18th International Conference, UMAP 2010, Big Island, HI, USA, June 20-24, 2010. Proceedings*, *Lecture Notes in Computer Science*, Vol. 6075 (Springer), ISBN 978-3-642-13469-2, pp. 40–51.

Serbos, D., Qi, S., Mamoulis, N., Pitoura, E. and Tsaparas, P. (2017). Fairness in package-to-group recommendations, in *Proceedings of the 26th International Conference on World Wide Web*, WWW '17 (International World Wide Web Conferences Steering Committee, Republic and Canton of Geneva, Switzerland), ISBN 978-1-4503-4913-0, pp. 371–379, doi:10.1145/3038912.3052612, https://doi.org/10.1145/3038912.3052612.

Tkalčič, M., Delic, A. and Felfernig, A. (2018). Personality, emotions, and group dynamics, in *Group Recommender Systems: An Introduction* (Springer International Publishing, Cham), ISBN 978-3-319-75067-5, pp. 157–167, doi:10.1007/978-3-319-75067-5_9, https://doi.org/10.1007/978-3-319-75067-5_9.

Ye, M., Liu, X. and Lee, W.-C. (2012). Exploring social influence for recommendation: a generative model approach, in *Proceedings of the 35th international ACM SIGIR conference on Research and development in information retrieval*, SIGIR '12 (ACM, New York, NY, USA), ISBN 978-1-4503-1472-5, pp. 671–680, doi:10.1145/2348283.2348373, http://doi.acm.org/10.1145/2348283.2348373.

Yuan, Q., Cong, G. and Lin, C. (2014). COM: a generative model for group recommendation, in S. A. Macskassy, C. Perlich, J. Leskovec, W. Wang and R. Ghani (eds.), *The 20th ACM SIGKDD International Conference on Knowledge Discovery and Data Mining, KDD '14, New York, NY, USA - August 24-27, 2014* (ACM), pp. 163–172, doi:10.1145/2623330.2623616, http://doi.acm.org/10.1145/2623330.2623616.

Chapter 7

User Preference Sources: Explicit vs. Implicit Feedback

Paolo Cremonesi, Franca Garzotto and Maurizio Ferrari Dacrema
Politecnico di Milano, Department of Electronic, Information and Bioengineering, Via Ponzio 34/5, 20133 Milano, Italy
paolo.cremonesi@polimi.it, franca.garzotto@polimi.it, maurizio.ferrari@polimi.it

The process of collecting preferences from users is fundamental during the normal operational life of a recommender system. The preference elicitation strategy can affect both the "user utility" (how well the system can make good recommendations to the new user who is undergoing the elicitation process) and the "system utility" (how well the system can provide good recommendations to all users, given what it learns from the new users). Not only do recommender systems need to gather information from users; they also need this information to be reliable and noiseless, as inconsistencies in user preferences limit prediction accuracy.

This chapter provides an overview of several design criteria for making the elicitation process more effective. The chapter analyzes two sources of preferences: (i) implicit feedbacks, inferred from the observable user activity, such as purchases or clicks; and (ii) preferences the user has explicitly stated for particular items.

Keywords: preference elicitation, system design, implicit feedback, explicit feedback.

7.1. Introduction

Whenever a user joins a recommender systems, the system needs to learn something from the user in order to provide personalized recommendations [Kluver *et al.* (2012); Sparling and Sen (2011); Ekstrand *et al.* (2011)].

These information needs come up under the guise of the *cold-start problem*. The cold-start problem concerns the issue of providing recommendations when data are not yet available on which to base those predictions. There are three types of cold-start problems: (i) *new-user*, whenever a new user joins the system but their preferences are not yet known [Cosley *et al.* (2003)], (ii) *new-item*, whenever new items are added to the system (e.g., when a new movie or book is released) but these have not yet received enough preferences to be recommendable [Schein *et al.* (2002)], and (iii) *system bootstrap*, where a system has no information about any user preferences (an extreme intersection of both new-user and new-item problems).

The *elicitation* process is fundamental not only during cold-start but also during the normal operational life of the system, as user's preferences evolve with time [Drenner *et al.* (2008); Konstan and Riedl (2012); Rashid *et al.* (2002); Cremonesi *et al.* (2012a, 2013)].

The preference elicitation strategy can affect both the "user utility" (how well the system can make good recommendations to the new user who is undergoing the elicitation process) and the "system utility" (how well the system can provide good recommendations to all users, given what it learns from the new users) [Drenner *et al.* (2008); Konstan and Riedl (2012); Rashid *et al.* (2002)].

Not only do recommender systems need to gather information from users; they also need this information to be reliable and noiseless, as inconsistencies in user preferences limit prediction accuracy [Kluver *et al.* (2012)], a requirement that is referred to as the "magic barrier" [Said *et al.* (2012b,a); Amatriain *et al.* (2009a)].

Preference is not a function of the item only: contextual factors such as time, mood, and social environment can influence a user's preference.

Consumption ratings are provided when the user has just consumed (or is consuming) the item in question. These ratings are particularly common in online streaming media environments, such as the Spotify music recommender or NetFlix instant streaming.

Memory ratings are provided based on the user's memory of experiencing the item. When a user eats at a restaurant then rates it on TripAdvisor, they are providing a memory preference. In some ways, consumption ratings are the most reliable, as the item is fresh in the user's mind. Memory ratings are also based on experience, so the user has a fuller set of data on which to base their rating than in the expectation case, but their impression of the item may not be accurately remembered or may be influenced by various external factors.

Therefore the elicited data may contain information about the context in which the user experienced the item and, in the case of memory preferences, intervening events or changes of mind as well as their preference for the item itself.

Maximizing both utility and ease of use are somehow conflicting requirements in the design of an elicitation process. The system needs to learn from users and to collect enough "good" information to generate satisfying recommendations. Not gathering enough information can result in a poor user model, which may lead to limited accuracy of recommendations and, in turn, may negatively affect the quality of the user interaction with the recommender system. Still, requiring users to spend too much time and energy with the system to express their opinions can be annoying, and may cause users either to stop using the system or to provide random information (or no information at all); this in turn will result in poor recommendations, increasing the risk of dissatisfying the user. Hence, elicitation strategies must face a potential design tension: to raise utility by increasing the amount of information gathered from users, and to make the elicitation process smooth from a user interaction perspective, limiting complexity and user effort during sign-up tasks [Cremonesi *et al.* (2012b)].

The chapter is structured as follows. Section 7.2 outlines the different design choices available when eliciting users' preferences. Section 7.3 investigates the impact of feedback noise on the quality of recommendations. Sections 7.4 and 7.5 investigate the trade-off between quality and effort during explicit elicitation. Section 7.6 investigates the impact of the user interface on explicit elicitation. Section 7.7 analyzes strategies and problems related to implicit feedback collection. Section 7.8 outlines future research aimed at improving the quality of the elicitation process.

7.2. Design dimensions

A number of design criteria have been identified for making the elicitation process more effective in terms of both utility and quality of the user interaction with the recommender system. Such criteria can be organized along different design dimensions.

7.2.1. *Demographics vs. Preferences*

A first design dimension is the type of information the system needs to collect in order to provide recommendations.

Historically, the first recommender systems focused on collecting *demographic* attributes from users and providing recommendations based on demographic classes [Rich (1979); Al-Shamri (2016)]. Age, gender, occupation, income, nationality, are examples of demographic data used by these systems. Today, demographics-based recommender systems raise some concerns due to security and privacy issues and the difficulty to obtain reliable demographic data from the users.

Modern systems provide personalized recommendations based on same users' *preferences* for items in a domain (e.g., ratings of movies). Even a few users' preferences on some of the items are more valuable than any demographic data in terms of quality of recommendations [Pilászy and Tikk (2009)].

In this chapter we deal only with *preference-based* elicitation.

7.2.2. *Sources of Preferences*

A second design dimension is the source of preferences. Preference data comes from two primary sources:

- *implicit* feedbacks are inferred by the system from the observable user activity, such as purchases or clicks;
- *explicit* feedbacks are preferences the user has explicitly stated for particular items.

System designers must be careful when using implicit data to understand the mapping between user interaction and actual user preferences, needs and intentions. Explicit feedbacks avoid the potentially difficult inference problems for user preferences that affect the collection of implicit ratings. However, explicit feedback suffers from the drawback that there can be a discrepancy between what the users say and what they do [O'Mahony *et al.* (2006)].

In this chapter we analyze both explicit (Sections 7.3–7.5) and implicit (Section 7.7) feedback.

7.2.3. *Categorical vs. Numerical Explicit Preferences*

An *explicit* preference elicitation process means that the system learns from specific facts provided by users about their preferences. When dealing with explicit preferences, a third design dimension concerns the type of feedback collected from the user. Explicit preferences can be divided into rating-based (i.e., numerical) feedbacks and categorical feedbacks [Awang *et al.* (2016)].

With *rating-based* elicitation, user's preferences are mapped to numerical values (ratings). Rating scales are arbitrary and vary widely in granularity.

Categorical preferences have no order or structure. Examples of categorical preferences are tagging items [Drenner *et al.* (2008)] and preferences on product attributes (e.g., favorite actors) [Adomavicius and Tuzhilin (2005)].

Categorical preferences are not analyzed in this chapter.

7.2.4. *Rating Scale*

Many recommender systems represent a user's opinion about an item as a single number on a rating scale.

The choice of rating scales is a major concern in recommender systems [Friedman and Amoo (1999); Amoo and Friedman (2001); Garland (1991); Cosley *et al.* (2003)]. The rating scale should be high enough so that users can create the correct number of categories for them to distinguish between levels of liking, but not so high that users require too much effort in order to make appropriate judgments among the categories. The scale should also allow the system to make accurate predictions.

Rating scales are analyzed in more details in Section 7.4.

7.2.5. *Human vs. System Controlled Elicitation*

When designing the interaction process for explicit elicitation, two possible strategies are available: (i) *human-controlled* — the user selects autonomously the items to evaluate, and (ii) *system-controlled* — the system creates the list of items for the user to evaluate [Mobasher (2007)].

One limitation of the human-controlled elicitation approach is that the users may not be able to identify the items to evaluate that maximize utility (either for the user or for the system).

System-controlled elicitation involves the issue of which items the user should rate and in which order, in order to maximize user or system utility. The most common approaches for system-controlled elicitation are based on *active learning*, which selectively chooses the items to be rated by the users [Elahi *et al.* (2016)].

This chapter assumes that a human-controlled elicitation is in place whenever we are dealing with explicit feedback, although most of the considerations apply to system-controlled elicitation as well. The benefits of active learning with respect to human-controlled elicitation are analyzed with more details in Chapter 11.

7.2.6. *Number of Ratings*

One of the difficult issues when designing a recommender system is the number of ratings — i.e., the profile length — which should be collected from a user before providing recommendations [Cremonesi *et al.* (2012b)].

A design tension exists which is induced by two conflicting requirements. On the one hand, the system must collect "enough" ratings from the users in order to learn their preferences and improve the accuracy of recommendations. On the other hand, gathering more ratings adds a burden on the users, which may negatively affect the quality of their experience.

Several research works investigated the effects of profile length from both a subjective (user-centric) point of view and an objective (accuracy-based) perspective [Drenner *et al.* (2008); Konstan and Riedl (2012); Rashid *et al.* (2002)]. These studies identify a potentially optimal profile length for an explicit, rating based, and human controlled elicitation strategy [Cremonesi *et al.* (2012b)].

The effects of profile length on recommendations are analyzed in Section 7.5.

7.3. Noise

Most recommender systems assume user ratings accurately represent user preferences. However, several research works show that user ratings are imperfect and noisy. Moreover, this noise limits the predictive power of any recommender system [Kluver *et al.* (2012)]. This limit has been named the *magic barrier* [Said *et al.* (2012a); Bellogín *et al.* (2014); Amatriain *et al.* (2009a)]. Recent work has begun to explore the quality of the ratings themselves [Cosley *et al.* (2003); O'Mahony *et al.* (2004); Sparling and Sen (2011)].

Rating errors (the difference between users' ratings and their true preferences) can be divided into two categories: *natural noise*, where user ratings differ from true preference due to errors in the elicitation process, and *malicious noise*, which refers to discrepant ratings introduced for the purpose of manipulating the system's recommendations or predictions [O'Mahony *et al.* (2006)]. In this chapter, we are only concerned with natural noise. Malicious noise is analyzed in Chapter 17.

Natural noise can result from normal human errors, interface design (such as the order in which items are presented for rating), the context in which an item was consumed (e.g., eating in a restaurant at lunch or dinner times), the user's bias, and other confounding factors [Amatriain *et al.* (2009b)].

One additional potential source of noise is the time gap between when the user consumed the item and when they rated it. *Consumption ratings* are provided when the user has just consumed (or is consuming) the item in question. These ratings are particularly common in online streaming media environments, such as the Spotify and NetFlix streaming services. *Memory ratings* are provided based on the user's memory of experiencing the item. When a user eats at a restaurant and later they rate it on TripAdvisor, they are providing a memory rating. In some ways, consumption ratings are the most reliable, as the item is fresh in the user's mind. Memory ratings are also based on experience but their impression of the item may not be accurately remembered or may be later influenced by various external factors.

Two non-exclusive approaches are possible to mitigate the negative impact of rating noise: (i) reducing the noise at the elicitation level, or (ii) designing noise-aware recommender algorithms.

Detecting and compensating for noise in the users input ratings potentially results in better recommender systems. From the academic point of view, natural noise in ratings can be detected by asking users to re-rate items [Amatriain *et al.* (2009a); Cosley *et al.* (2003)]. This approach is almost impractical in commercial systems, as asking users to re-rate item would likely annoy users and negatively affect their experience. Proper choices during the rating elicitation process — such as rating scale, profile length, interface designs — can be adopted more effectively to reduce natural noise. However, re-rating can be used in experimental setups to help defining the best rating scale, as explained in Section 7.4.1.

Recommender systems that understand and adapt to noise in user ratings may be able to mitigate the magic barrier problem and generate more accurate recommendations [Toledo *et al.* (2015)]. Some probabilistic models assumes that user-provided ratings are drawn from a normal distribution whose mean is the user's true preference. Analyzing ratings with these models in a naive Bayesian framework allows the system to compensate for noise in individual ratings [Babas *et al.* (2013); Adomavicius and Tuzhilin (2005)]. User biases can be accounted with pre-normalization of the ratings, as with global effects normalization [Bell and Koren (2007)]. Context noise can be accounted with ad-hoc context-aware recommender systems (see Chapter 5).

7.4. Rating scale

Many recommender systems represent a user's opinion about an item as a single number on a rating scale.

Scales vary widely in their granularity [Cosley *et al.* (2003)]. These scales are appealing to the system designer, as they provide an integral or real-valued rating that is easy to use in computations. They also benefit from the body of research about Likert scales that is available in the area of survey design [Amoo and Friedman (2001); Friedman and Amoo (1999); Garland (1991)].

When choosing a ratings scale, a number of design options are available:

- the number of categories (or *granularity*) of the scale (e.g., 2 values, 5 values, 10 values, ...);
- the labels attached to each category (e.g., stars, thumbs up/down, ...);
- the presence of a neutral mid point in the scale.

The impact of these choices on the accuracy of recommendations is analyzed in the following subsections.

7.4.1. *Granularity*

Mapping user preference to a rating scale is not an easy task, as different users may have different preferences with respect to the rating scales [Gena *et al.* (2011)].

Ideally, a rating scale should allow users to express their opinions in a meaningful way without too much effort. The rating scale should be high enough so that users can create the correct number of categories for them to distinguish between levels of liking, but not so high that users require too much time and effort to make judgments between the categories. System designers may face a fundamental trade-off between a coarse rating scale that is relatively quick and noisy, and a granular rating scale that requires more effort with less noise.

There is not a clear consensus on the optimal granularity that balances quality and effort. A number of psychological studies use information theory to measure how much information different rating scales transmit. One of the fundamental measures of information is called *entropy*, which measures how hard a random variable is to predict [Shannon (2001)]. By studying the entropy of ratings, results suggest that a less granular rating scale increases rating noise, although this effect is limited [Garner (1960)]. For instance, users seem to give borderline items the benefit of the doubt when forced to rate on a coarse scale, i.e., they rate borderline items higher than with a finer rating scale [Cosley *et al.* (2003)]. However, quality of recommendations does not increase indefinitely with the "rating effort": we cannot gain more information by arbitrarily increasing the resolution of the rating scale [Kluver *et al.* (2012)].

On the contrary, other studies comparing scales with varying numbers of response categories show that users rate items consistently even when using different scales, and as few as two response categories may be adequate in practice [Matell and Jacoby (1971, 1972)].

Other studies show that the effectiveness of rating scales depends on both the human-computer interface (e.g., mobile, desktop) and the application domain [Cosley *et al.* (2003)].

To directly address the quality/effort trade-off [Kluver *et al.* (2012)] measure and compare three quantities: preference bits per rating, mean time per raring and preference bits per second.

- *Preference bits per rating* (*pbpr*) measures how much an elicited rating is able to reduce the entropy of a new rating, and it is assessed either by asking users to re-rate an item or by comparing elicited and predicted ratings. Pbpr are strongly correlated with the accuracy of recommendations.
- The mean time required to rate an item is used as a measure of the user's mental effort [Sparling and Sen (2011)].
- *Preference bits per second* (*pbps*) measures the ratio between the amount of information collected from a user divided by the effort for that user. For a rating scale, *pbps* is defined as the *pbpr* for that rating scale divided by the number of seconds per rating with that scale. Preferences bits per second can be considered as a measure of the *speed* at which we can elicit preference information from a user.

Table 7.1 reports the values of the three metrics (pbpr, mean time, pbps) collected on two different experiments with two different movie datasets (values re-elaborated from [Kluver *et al.* (2012)]). The results suggest that, despite the faster ratings, the binary scale gives us slower information than other scales. There is evidence that a rating scale with 5 or 10 values matches how users think about movies better than a 2 values rating scale. The results show also that there is not a strong difference between the 5 and 10 values rating scale, in terms of pbps.

Table 7.1. Effort and quality for each scale and dataset.

	Preference bits per rating (pbpr)			Mean rating time per item (sec)			Preference bits per second (pbps)		
Scale	2	5	10	2	5	10	2	5	10
ML	0.15	0.270	0.270	3.91	4.09	4.33	0.038	0.66	0.65
Jester	0.12	0.210	0.240	15.47	16.39	16.55	0.008	0.13	0.17

7.4.2. *Labels*

Labels associated with each choice in a rating scale can affect the distribution of responses [Friedman and Amoo (1999)]. For instance, changing a scale's range without changing the granularity (e.g., from -5 ...5 to 0 ...10) influences users' responses [Amoo and Friedman (2001)].

When using a Likert scale, users specify their level of preference for an item with a bipolar like-dislike scale, measuring either positive or negative preferences for an item [Croasmun and Ostrom (2011)].

Several studies report that users with both like and dislike options provide more ratings than users with only "like" options [Sen *et al.* (2007)]. For this reason, rating scales should exhibit "symmetry". Symmetry means that a scale contains an equal numbers of positive and negative positions, or labels, whose distances are symmetric about the "neutral" mid value (whether or not that value is present in the scale).

The labels for a five-level rating scale, for example, could be: 1-Strongly Dislike, 2-Dislike, 3-Neither Like nor Dislike, 4-Dislike, 5-Strongly Like. If labels are missing and only numerical values are presented to the users, most users will still assume a balanced and symmetrical numerical scale. However, some users could have a biased mapping between numbers and labels. Referring to the previous example, some users might map the neutral "Neither Like or Dislike" label with 4, other users with 2. This discrepancy from the intended, implicit labeling of a numerical-only rating scale and the user interpretation introduce *biases* in the rating behavior of users.

The user bias is a form of rating noise and should be considered when building a recommender algorithm (see Section 7.3).

7.4.3. *Mid-point*

The use of a mid-point in ratings is still debated. Using a mid-point value has been shown to affect the data [Croasmun and Ostrom (2011)].

Since they have no neutral point, even-numbered rating scales force the respondent to commit to a certain polarized position, even if the respondent may not have a definite opinion.

Odd-numbered scales provide an option for indecision or neutrality. By giving responders a neutral response option, they are not required to decide one way or the other on an issue. This may reduce the chance of response biases, which is the tendency to favor either positive or negative ratings. Respondents do not feel forced to have an opinion if they do not have one.

Some researches suggest that excluding a middle choice can reduce respondents' bias toward providing positive replies, but that doing so can increase the rating effort and reduce the number of collected ratings [Garland (1991)].

7.5. Number of ratings

One issue when designing a recommender system is the number of ratings — i.e., the profile length — which should be collected from a new user before providing recommendations [Drenner *et al.* (2008)].

On the one hand, the system must collect enough ratings from the users in order to learn their preferences and improve the accuracy of recommendations. On the other hand, gathering more ratings adds a burden on the user, which may negatively affect the user experience [Cremonesi *et al.* (2012b)].

Several works confirm that the profile length of a users is positively correlated to the accuracy of recommendations [Konstan and Riedl (2012); Rashid *et al.* (2002)]. However, Figure 7.1 shows that the accuracy (recall) of two state-of-art CF algorithms — PureSVD and AsySVD — measured on the Netflix dataset does not increases indefinitely with the profile length [Cremonesi *et al.* (2012b)].

Other works suggest to minimize the number of elicited preferences in the profile initialization [Drenner *et al.* (2008); Golbandi *et al.* (2010); Rashid *et al.* (2002)]. According to behavioral decision theories [Häubl and Trifts (2000)], users are likely to settle on the immediate benefit of saving effort over the delayed gratification of higher accuracy. A number of studies find that the optimal number of ratings in the user profile is between 5 and 20 ratings (more likely 10 ratings). Eliciting more ratings does not significantly improve the quality of recommendations, while it adds an useless burden on the user, with the risk of increasing noise [Cremonesi *et al.* (2012b); Kluver and Konstan (2014)].

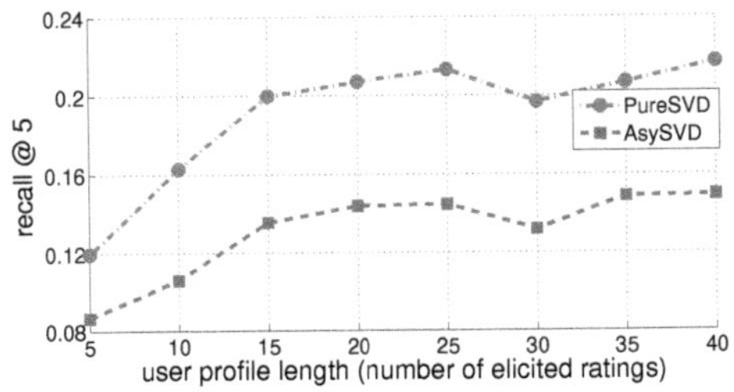

Fig. 7.1. Accuracy of recommendations vs. number of elicited ratings for two matrix-factorization based CF algorithms.

7.6. Interface

Designers of recommender systems interfaces try to help users form consistent and useful preferences by helping them to remember the item, for example by presenting descriptions and attributes of it.

In the course of eliciting ratings, the system interface may affect users' opinions of the rated items. When a person is asked to evaluate an item, the evaluation of it may depend on the information provided in the interface [Nguyen *et al.* (2013)].

For instance, if the interface displays some kind of predefined opinion on the items the users are asked to rate, (e.g., average rating, or predicted rating), the system is actually influencing their beliefs, persuading them to rate items consistently with the opinions shown by the interface [Cosley *et al.* (2003)]. If opinions are influenced by the interface, they might be less valuable for making recommendations [Cosley *et al.* (2003)].

7.7. Implicit feedback

Preferences can also be inferred from user behavior [Oard *et al.* (1998)]. Many e-commerce systems often use purchase decisions as a proxy for ratings, resulting in either a unary scale (bought items are "liked", others unknown) or a binary scale (bought items "liked", unbought items "disliked").

User actions such as time spent reading the description of an item, saving an item in the favorite list, buying an item, and any other user "*signals*" potentially contribute to the building of an implicit user profile [Morita and Shinoda (1994); Tan *et al.* (2016); Quadrana *et al.* (2017)].

Observed or inferred preference information are expected to be noisier than explicitly provided data. However, one problem with explicit ratings is that the point in time when users provide a rating can be quite different from the point in time when they consumed or purchased an item (e.g., when a user rates a bunch of movies while registering for a movie recommendation service), which in turn creates some natural noise in the ratings and might easily mislead the recommender system. For this reason, in this situations it may happen that implicit feedback can build a more accurate long-term profile of a user's taste then users themselves can articulate with explicit ratings [O'Mahony *et al.* (2006)].

When implicit ratings are collected, the observed user events must be translated into computationally useful estimators of preference. The most common setup is to map implicit feedback into explicit ratings, and to later use traditional CF techniques to provide recommendations [Oard *et al.* (1998)]. For instance, if a user skips a track while listening to music from an online streaming service, this signal could be considered as a negative feedback. On the contrary, if a user listens to the same track again and again, this signal could be considered as a positive feedback. Other forms of data, such as the time spent on a page, also provide useful signals for inferring user preferences. There is a substantial correlation between the time a user spend reading the description of an item and his/her preference for that item [Morita and Shinoda (1994); Konstan *et al.* (1997)].

While this approach supports the usage of traditional CF algorithms, it also oversimplifies the underlying problem potentially.

There are also other interesting patterns that can be derived from interaction logs but are not easily tied to a positive or negative feedback. Based on the ordering and the timestamps of the user actions, we can for example detect interest drifts of individual users over time, or detect short-term popularity trends in the community that can be exploited by recommendation algorithms. These patterns are exploitable only with the design of ad-hoc CF algorithms tuned for implicit feedback, as it happens in *session-based* recommender systems [Tan *et al.* (2016); Quadrana *et al.* (2017)].

The main input to a *session-based* recommender system is a — usually time-stamped — list of past user actions. Each action can be associated with one user of the system or be an anonymous action. Each action can furthermore be associated with one of the recommendable items. Finally, each action can be of one of several pre-defined types (e.g., item-view, item-purchase, add-to-cart, etc.) and each action, user, and item may have a number of additional attributes. In a traditional setup, all implicit feedback are associated to one of the known users and items, and a mapping is designed between each user action and an explicit rating. However, anonymous user actions are not uncommon, e.g., in the e-commerce domain, where users are often not logged in. Moreover, not all actions are related to an item, since, for example, relevant information can be extracted from the users search or navigation behavior as well [Quadrana *et al.* (2018)].

In session-based recommenders, recommendations can be provided based solely on the interactions in the current user session (e.g., when the users are anonymous). But there are also cases where a user might be logged-in or some form of user identifier might be present (e.g., by effect of cookies or other identifiers). In these cases it is reasonable to assume that the user behavior in past sessions might offer valuable information for providing recommendations in the next session [Quadrana *et al.* (2017)].

In the literature, the difference between explicit and implicit feedback recommender systems only lies in the type of the available preference signals. Many research researches on implicit feedback assume implicit feedbacks as positive opinions of users and map these to a positive rating. These approaches do not consider multiple interactions over time and do not consider signals that express negative feedback. In particular, in session-based recommendation scenarios, the users' short-term intents, which can be estimated from their very last actions, can represent a crucial form of context information that should be taken into account when recommending [Jannach *et al.* (2015)].

Explicit and implicit rating data are not mutually exclusive [Koren (2010)]. Recommender systems can incorporate both into the recommendation process. Whenever a new user join a recommender system, the system can first collect some implicit ratings (e.g., through page views). As the user is persuaded to provide explicit ratings, hybrid data becomes available to the system. Many commercial recommender systems make use of both explicit and implicit data in providing their various recommendations (e.g., Amazon, Netflix).

7.8. Future directions

The wide set of studies on preference elicitation provides a number of answers to different questions, but also highlights the discrepancy between what we know and what we need to know, pinpointing directions for current and future research. In this respect, the following issues are particularly challenging.

(1) *Investigation of pairwise preference elicitation.* Since ratings represent evaluations measured against an absolute benchmark, it could be difficult for the user to consistently rate items. For instance, if a user rates an item with the highest value in the rating scale and successively finds an even better item, then there is no way to express such a preference. Another way to articulate preferences is based on pairwise comparisons: one item is preferred with respect to another item. Although some existing research shows that pairwise-based recommendation technology can lead to a better quality of recommendations [Blėdaitė and Ricci (2015)], there is still no consensus whether it is easier to decide which item is preferred among two, rather than rating them in some predefined scale [Nobarany *et al.* (2012)]. In that respect, further work is needed in the development of user interfaces and algorithms that make use of pairwise preferences.
(2) *Investigation of session-based recommender systems.* The development of new effective session-based recommenders is still an open research issue. In e-commerce domains, implicit feedback in the form of page views and item purchases is the primary source of preference data for recommendation [Linden *et al.* (2003)]. System designers and analysts must be careful when using implicit data to understand the mapping between system-visible user identities (frequently accounts) and actual people.
(3) *Investigation of mixed feedback.* One of the most common sources of natural noise in collected feedbacks (either explicit or implicit) is that users of e-commerce systems will often share their account with other users. For instance, a family may have a single Amazon.com account, thus providing the system with ill-defined aggregate information about several people's preferences. Moreover, users will often purchase items as gifts for others. Those purchases and related activities do not necessarily communicate anything about the user's tastes. A similar scenario happens for TV recommender systems: the same TV set is used by different members of a family (either individually or in group), making difficult for the recommender system to guess who is in front of the TV screen. A lot of research effort has been focused on group recommender systems but little work has been done to identify single users from group feedbacks.

References

Adomavicius, G. and Tuzhilin, A. (2005). Toward the next generation of recommender systems: A survey of the state-of-the-art and possible extensions, *IEEE transactions on knowledge and data engineering* **17**, 6, pp. 734–749.

Al-Shamri, M. Y. H. (2016). User profiling approaches for demographic recommender systems, *Know.-Based Syst.* **100**, C, pp. 175–187, doi:10.1016/j.knosys.2016.03.006, https://doi.org/10.1016/j.knosys.2016.03.006.

Amatriain, X., Pujol, J. M. and Oliver, N. (2009a). I like it... i like it not: Evaluating user ratings noise in recommender systems, in *Proceedings of the 17th International Conference on User Modeling, Adaptation, and Personalization: Formerly UM and AH*, UMAP '09 (Springer-Verlag, Berlin, Heidelberg), ISBN 978-3-642-02246-3, pp. 247–258, doi:10.1007/978-3-642-02247-0_24, http://dx.doi.org/10.1007/978-3-642-02247-0_24.

Amatriain, X., Pujol, J. M., Tintarev, N. and Oliver, N. (2009b). Rate it again: Increasing recommendation accuracy by user re-rating, in *Proceedings of the Third ACM Conference on Recommender Systems*, RecSys '09 (ACM, New York, NY, USA), ISBN 978-1-60558-435-5, pp. 173–180, doi:10.1145/1639714.1639744, http://doi.acm.org/10.1145/1639714.1639744.

Amoo, T. and Friedman, H. H. (2001). Do numeric values influence subjects responses to rating scales?

Awang, Z., Afthanorhan, A. and Mamat, M. (2016). The likert scale analysis using parametric based structural equation modeling (sem), *Computational Methods in Social Sciences* **4**, 1, p. 13.

Babas, K., Chalkiadakis, G. and Tripolitakis, E. (2013). You are what you consume: A bayesian method for personalized recommendations, in *Proceedings of the 7th ACM Conference on Recommender Systems*, RecSys '13 (ACM, New York, NY, USA), ISBN 978-1-4503-2409-0, pp. 221–228, doi:10.1145/2507157.2507158, http://doi.acm.org/10.1145/2507157.2507158.

Bell, R. M. and Koren, Y. (2007). Lessons from the netflix prize challenge, *Acm Sigkdd Explorations Newsletter* **9**, 2, pp. 75–79.

Bellogín, A., Said, A. and de Vries, A. P. (2014). The magic barrier of recommender systems–no magic, just ratings, in *International Conference on User Modeling, Adaptation, and Personalization* (Springer), pp. 25–36.

Blédaité, L. and Ricci, F. (2015). Pairwise preferences elicitation and exploitation for conversational collaborative filtering, in *Proceedings of the 26th ACM Conference on Hypertext & Social Media*, HT '15 (ACM, New York, NY, USA), ISBN 978-1-4503-3395-5, pp. 231–236, doi:10.1145/2700171.2791049, http://doi.acm.org/10.1145/2700171.2791049.

Cosley, D., Lam, S. K., Albert, I., Konstan, J. A. and Riedl, J. (2003). Is seeing believing?: How recommender system interfaces affect users' opinions, in *Proceedings of the SIGCHI Conference on Human Factors in Computing Systems*, CHI '03 (ACM, New York, NY, USA), ISBN 1-58113-630-7, pp. 585–592, doi:10.1145/642611.642713, http://doi.acm.org/10.1145/642611.642713.

Cremonesi, P., Garzotto, F. and Turrin, R. (2012a). Investigating the persuasion potential of recommender systems from a quality perspective: An empirical study, *ACM Trans. Interact. Intell. Syst.* **2**, 2, pp. 11:1–11:41, doi:10.1145/2209310.2209314, `http://doi.acm.org/10.1145/2209310.2209314`.

Cremonesi, P., Garzotto, F. and Turrin, R. (2013). User-centric vs. system-centric evaluation of recommender systems, in P. Kotzé, G. Marsden, G. Lindgaard, J. Wesson and M. Winckler (eds.), *Human-Computer Interaction – INTERACT 2013* (Springer Berlin Heidelberg, Berlin, Heidelberg), ISBN 978-3-642-40477-1, pp. 334–351.

Cremonesi, P., Garzottto, F. and Turrin, R. (2012b). User effort vs. accuracy in rating-based elicitation, in *Proceedings of the Sixth ACM Conference on Recommender Systems*, RecSys '12 (ACM, New York, NY, USA), ISBN 978-1-4503-1270-7, pp. 27–34, doi:10.1145/2365952.2365963, `http://doi.acm.org/10.1145/2365952.2365963`.

Croasmun, J. T. and Ostrom, L. (2011). Using likert-type scales in the social sciences, *Journal of Adult Education* **40**, 1, p. 19.

Drenner, S., Sen, S. and Terveen, L. (2008). Crafting the initial user experience to achieve community goals, in *Proceedings of the 2008 ACM Conference on Recommender Systems*, RecSys '08 (ACM, New York, NY, USA), ISBN 978-1-60558-093-7, pp. 187–194, doi:10.1145/1454008.1454039, `http://doi.acm.org/10.1145/1454008.1454039`.

Ekstrand, M. D., Riedl, J. T., Konstan, J. A. *et al.* (2011). Collaborative filtering recommender systems, *Foundations and Trends® in Human–Computer Interaction* **4**, 2, pp. 81–173.

Elahi, M., Ricci, F. and Rubens, N. (2016). A survey of active learning in collaborative filtering recommender systems, *Comput. Sci. Rev.* **20**, C, pp. 29–50, doi:10.1016/j.cosrev.2016.05.002, `http://dx.doi.org/10.1016/j.cosrev.2016.05.002`.

Friedman, H. H. and Amoo, T. (1999). Rating the rating scales.

Garland, R. (1991). The mid-point on a rating scale: Is it desirable, *Marketing bulletin* **2**, 1, pp. 66–70.

Garner, W. R. (1960). Rating scales, discriminability, and information transmission, *Psychological review* **67**, 6, p. 343.

Gena, C., Brogi, R., Cena, F. and Vernero, F. (2011). The impact of rating scales on user's rating behavior, in *Proceedings of the 19th International Conference on User Modeling, Adaption, and Personalization*, UMAP'11 (Springer-Verlag, Berlin, Heidelberg), ISBN 978-3-642-22361-7, pp. 123–134, `http://dl.acm.org/citation.cfm?id=2021855.2021867`.

Golbandi, N., Koren, Y. and Lempel, R. (2010). On bootstrapping recommender systems, in *Proceedings of the 19th ACM International Conference on Information and Knowledge Management*, CIKM '10 (ACM, New York, NY, USA), ISBN 978-1-4503-0099-5, pp. 1805–1808, doi:10.1145/1871437.1871734, `http://doi.acm.org/10.1145/1871437.1871734`.

Häubl, G. and Trifts, V. (2000). Consumer decision making in online shopping environments: The effects of interactive decision aids, *Marketing science* **19**, 1, pp. 4–21.

Jannach, D., Lerche, L. and Jugovac, M. (2015). Adaptation and evaluation of recommendations for short-term shopping goals, in *Proceedings of the 9th ACM Conference on Recommender Systems*, RecSys '15 (ACM, New York, NY, USA), ISBN 978-1-4503-3692-5, pp. 211–218, doi:10.1145/2792838.2800176, `http://doi.acm.org/10.1145/2792838.2800176`.

Kluver, D. and Konstan, J. A. (2014). Evaluating recommender behavior for new users, in *Proceedings of the 8th ACM Conference on Recommender Systems*, RecSys '14 (ACM, New York, NY, USA), ISBN 978-1-4503-2668-1, pp. 121–128, doi:10.1145/2645710.2645742, `http://doi.acm.org/10.1145/2645710.2645742`.

Kluver, D., Nguyen, T. T., Ekstrand, M., Sen, S. and Riedl, J. (2012). How many bits per rating? in *Proceedings of the Sixth ACM Conference on Recommender Systems*, RecSys '12 (ACM, New York, NY, USA), ISBN 978-1-4503-1270-7, pp. 99–106, doi:10.1145/2365952.2365974, `http://doi.acm.org/10.1145/2365952.2365974`.

Konstan, J. A., Miller, B. N., Maltz, D., Herlocker, J. L., Gordon, L. R. and Riedl, J. (1997). Grouplens: Applying collaborative filtering to usenet news, *Commun. ACM* **40**, 3, pp. 77–87, doi:10.1145/245108.245126, `http://doi.acm.org/10.1145/245108.245126`.

Konstan, J. A. and Riedl, J. (2012). Recommender systems: From algorithms to user experience, *User Modeling and User-Adapted Interaction* **22**, 1-2, pp. 101–123, doi:10.1007/s11257-011-9112-x, `http://dx.doi.org/10.1007/s11257-011-9112-x`.

Koren, Y. (2010). Factor in the neighbors: Scalable and accurate collaborative filtering, *ACM Trans. Knowl. Discov. Data* **4**, 1, pp. 1:1–1:24, doi:10.1145/1644873.1644874, `http://doi.acm.org/10.1145/1644873.1644874`.

Linden, G., Smith, B. and York, J. (2003). Amazon.com recommendations: Item-to-item collaborative filtering, *IEEE Internet computing* **7**, 1, pp. 76–80.

Matell, M. S. and Jacoby, J. (1971). Is there an optimal number of alternatives for likert scale items? study i: Reliability and validity, *Educational and psychological measurement* **31**, 3, pp. 657–674.

Matell, M. S. and Jacoby, J. (1972). Is there an optimal number of alternatives for likert-scale items? effects of testing time and scale properties, *Journal of Applied Psychology* **56**, 6, p. 506.

Mobasher, B. (2007). The adaptive web, chap. Data Mining for Web Personalization (Springer-Verlag, Berlin, Heidelberg), ISBN 978-3-540-72078-2, pp. 90–135, `http://dl.acm.org/citation.cfm?id=1768197.1768201`.

Morita, M. and Shinoda, Y. (1994). Information filtering based on user behavior analysis and best match text retrieval, in *Proceedings of the 17th Annual International ACM SIGIR Conference on Research and Development in Information Retrieval*, SIGIR '94 (Springer-Verlag New York, Inc., New York, NY, USA), ISBN 0-387-19889-X, pp. 272–281, `http://dl.acm.org/citation.cfm?id=188490.188583`.

Nguyen, T. T., Kluver, D., Wang, T.-Y., Hui, P.-M., Ekstrand, M. D., Willemsen, M. C. and Riedl, J. (2013). Rating support interfaces to improve user experience and recommender accuracy, in *Proceedings of the 7th ACM*

Conference on Recommender Systems, RecSys '13 (ACM, New York, NY, USA), ISBN 978-1-4503-2409-0, pp. 149–156, doi:10.1145/2507157.2507188, http://doi.acm.org/10.1145/2507157.2507188.

Nobarany, S., Oram, L., Rajendran, V. K., Chen, C.-H., McGrenere, J. and Munzner, T. (2012). The design space of opinion measurement interfaces: Exploring recall support for rating and ranking, in *Proceedings of the SIGCHI Conference on Human Factors in Computing Systems*, CHI '12 (ACM, New York, NY, USA), ISBN 978-1-4503-1015-4, pp. 2035–2044, doi:10.1145/2207676.2208351, http://doi.acm.org/10.1145/2207676.2208351.

Oard, D. W., Kim, J. *et al.* (1998). Implicit feedback for recommender systems, in *Proceedings of the AAAI workshop on recommender systems* (Menlo Park, CA: AAAI Press), pp. 81–83.

O'Mahony, M., Hurley, N., Kushmerick, N. and Silvestre, G. (2004). Collaborative recommendation: A robustness analysis, *ACM Trans. Internet Technol.* **4**, 4, pp. 344–377, doi:10.1145/1031114.1031116, http://doi.acm.org/10.1145/1031114.1031116.

O'Mahony, M. P., Hurley, N. J. and Silvestre, G. C. (2006). Detecting noise in recommender system databases, in *Proceedings of the 11th International Conference on Intelligent User Interfaces*, IUI '06 (ACM, New York, NY, USA), ISBN 1-59593-287-9, pp. 109–115, doi:10.1145/1111449.1111477, http://doi.acm.org/10.1145/1111449.1111477.

Pilászy, I. and Tikk, D. (2009). Recommending new movies: Even a few ratings are more valuable than metadata, in *Proceedings of the Third ACM Conference on Recommender Systems*, RecSys '09 (ACM, New York, NY, USA), ISBN 978-1-60558-435-5, pp. 93–100, doi:10.1145/1639714.1639731, http://doi.acm.org/10.1145/1639714.1639731.

Quadrana, M., Jannach, D. and Cremonesi, P. (2018). Sequence-aware recommender systems, *ACM Computing Surveys*.

Quadrana, M., Karatzoglou, A., Hidasi, B. and Cremonesi, P. (2017). Personalizing session-based recommendations with hierarchical recurrent neural networks, in *Proceedings of the Eleventh ACM Conference on Recommender Systems*, RecSys '17 (ACM, New York, NY, USA), ISBN 978-1-4503-4652-8, pp. 130–137, doi:10.1145/3109859.3109896, http://doi.acm.org/10.1145/3109859.3109896.

Rashid, A. M., Albert, I., Cosley, D., Lam, S. K., McNee, S. M., Konstan, J. A. and Riedl, J. (2002). Getting to know you: Learning new user preferences in recommender systems, in *Proceedings of the 7th International Conference on Intelligent User Interfaces*, IUI '02 (ACM, New York, NY, USA), ISBN 1-58113-459-2, pp. 127–134, doi:10.1145/502716.502737, http://doi.acm.org/10.1145/502716.502737.

Rich, E. (1979). User modeling via stereotypes, *Cognitive science* **3**, 4, pp. 329–354.

Said, A., Jain, B. J., Narr, S. and Plumbaum, T. (2012a). Users and noise: The magic barrier of recommender systems, in *Proceedings of the 20th International Conference on User Modeling, Adaptation, and Personalization*, UMAP'12 (Springer-Verlag, Berlin, Heidelberg), ISBN 978-3-642-31453-7, pp. 237–248, doi:10.1007/978-3-642-31454-4_20, http://dx.doi.org/10.1007/978-3-642-31454-4_20.

Said, A., Jain, B. J., Narr, S., Plumbaum, T., Albayrak, S. and Scheel, C. (2012b). Estimating the magic barrier of recommender systems: A user study, in *Proceedings of the 35th International ACM SIGIR Conference on Research and Development in Information Retrieval*, SIGIR '12 (ACM, New York, NY, USA), ISBN 978-1-4503-1472-5, pp. 1061–1062, doi:10.1145/2348283.2348469, http://doi.acm.org/10.1145/2348283.2348469.

Schein, A. I., Popescul, A., Ungar, L. H. and Pennock, D. M. (2002). Methods and metrics for cold-start recommendations, in *Proceedings of the 25th Annual International ACM SIGIR Conference on Research and Development in Information Retrieval*, SIGIR '02 (ACM, New York, NY, USA), ISBN 1-58113-561-0, pp. 253–260, doi:10.1145/564376.564421, http://doi.acm.org/10.1145/564376.564421.

Sen, S., Harper, F. M., LaPitz, A. and Riedl, J. (2007). The quest for quality tags, in *Proceedings of the 2007 International ACM Conference on Supporting Group Work*, GROUP '07 (ACM, New York, NY, USA), ISBN 978-1-59593-845-9, pp. 361–370, doi:10.1145/1316624.1316678, http://doi.acm.org/10.1145/1316624.1316678.

Shannon, C. E. (2001). A mathematical theory of communication, *SIGMOBILE Mob. Comput. Commun. Rev.* **5**, 1, pp. 3–55, doi:10.1145/584091.584093, http://doi.acm.org/10.1145/584091.584093.

Sparling, E. I. and Sen, S. (2011). Rating: How difficult is it? in *Proceedings of the Fifth ACM Conference on Recommender Systems*, RecSys '11 (ACM, New York, NY, USA), ISBN 978-1-4503-0683-6, pp. 149–156, doi:10.1145/2043932.2043961, http://doi.acm.org/10.1145/2043932.2043961.

Tan, Y. K., Xu, X. and Liu, Y. (2016). Improved recurrent neural networks for session-based recommendations, in *Proceedings of the 1st Workshop on Deep Learning for Recommender Systems*, DLRS 2016 (ACM, New York, NY, USA), ISBN 978-1-4503-4795-2, pp. 17–22, doi:10.1145/2988450.2988452, http://doi.acm.org/10.1145/2988450.2988452.

Toledo, R. Y., Mota, Y. C. and Martínez, L. (2015). Correcting noisy ratings in collaborative recommender systems, *Knowledge-Based Systems* **76**, pp. 96–108.

Chapter 8

User Preference Elicitation, Rating Sparsity and Cold Start

Mehdi Elahi, Matthias Braunhofer, Tural Gurbanov and Francesco Ricci

Free University of Bozen-Bolzano,
Piazza Domenicani, 3
Italy - 39100, Bozen-Bolzano
{meelahi,mbraunhofer,tgurbanov,fricci}@unibz.it

A prerequisite for implementing collaborative filtering recommender systems is the availability of users' preferences data. This data, typically in the form of ratings, is exploited to learn the tastes of the users and to serve them with personalized recommendations. However, there may be a lack of preference data, especially at the initial stage of the operations of a recommender system, i.e., in the *Cold Start* phase. In particular, when a new user has not yet rated any item, the system would be incapable of generating relevant recommendations for this user. Or, when a new item is added to the system catalogue and no user has rated it, the system cannot recommend this item to any user.

This chapter discusses the cold start problem and provides a comprehensive description of techniques that have been proposed to address this problem. It surveys algorithmic solutions and provides a summary of their performance comparison. Moreover, it lists publicly available resources (e.g., libraries and datasets) and offers a set of practical guidelines that can be adopted by researchers and practitioners.

8.1. Introduction

The research on Recommender Systems (RSs) has seen a continuous growth, starting from the mid-90s when early works have been published [Resnick *et al.* (1994); Shardanand and Maes (1995)]. One of the most popular approaches in recommender systems is Collaborative Filtering (CF), which is the main focus of this chapter. In collaborative filtering a set of preferences provided by a community of users, the relationships and similarities among the user preferences, are exploited to generate personalized recommendations for a target user. A variety of collaborative filtering techniques, which are typically classified as user-based [Herlocker *et al.* (1999)], item-based [Linden *et al.* (2003)], or Matrix Factorization (MF) [Koren and Bell (2011)], have been proposed and evaluated in both industry and academia [Koren and Bell (2015); Shi *et al.* (2014)].

Moreover, various types of preference data can be exploited by collaborative filtering recommender systems, and they can roughly divided in two wide categories: *Explicit Feedback* and *Implicit Feedback* (see Chapter 7). Explicit feedback, which is typically considered as a more informative signal of user's preference, refers to item evaluations that the user explicitly reports, e.g., five star ratings for movies in Netflix, or like/dislikes for posts in Facebook [Koren and Bell (2015); Stern *et al.* (2009); Agarwal and Chen (2009)]. Despite their usefulness, eliciting explicit feedback requires some user effort [Elahi *et al.* (2014)] and still might not completely reveal actual users' needs [Neidhardt *et al.* (2014)].

Hence, implicit feedback, i.e., actions performed by the users on the items, such as viewing an item, have been used to infer preference information [Oard *et al.* (1998); Hu *et al.* (2008a); Gurbanov and Ricci (2017)]. This data is much more abundant and simpler to collect than explicit feedback. For example, the purchase or browsing history of a user in Amazon.com can be used by the system to predict additional interests of the user and ultimately generate recommendations for her. A user who frequently purchases books of a specific author will likely be interested in that author.

It is worth noting that despite the clear usefulness of the implicit feedback, the usage of this type of data has some limitations: for instance, it is easier to infer a "positive" feedback from a user action rather than a "negative" one. Most importantly, the system can only guess the actual user interests from the tracked behaviour of a user. In fact, purchasing an item may not necessarily indicate that the user was satisfied about it, as the user may have regret purchasing that item. But, repeated actions of the users on items may increase the confidence in such an inferred preference [Hu *et al.* (2008a)].

Preference data can be also classified into *Absolute* or *Relative*. The former type of preference is the most popular; these are expressed in the form of absolute evaluations in a predefined scale, e.g., the five star ratings used by Netflix.com [Linden *et al.* (2003); Sarwar *et al.* (2001); Koren and Bell (2015)]. However, this type of preference data have limitations. For example, a user who loves two items, but prefers one over another, might be pushed to rate them alike, or to penalize one, due to the limited number of points in the rating scale. This, and other problems, could be resolved by eliciting relative preferences, i.e., pairwise comparisons: "I prefer Blade Runner to Indiana Jones". One can also express how much an item is preferred (or considered inferior) to another, hence generating pair scores (positive or negative). The larger the score the more one item is preferred to the other item in the pair [Kalloori *et al.* (2016); Blėdaitė and Ricci (2015); Jones *et al.* (2011); Rendle *et al.* (2009)].

Regardless of the type of the data used, collaborative filtering recommender systems commonly suffer from a challenging problem, i.e., *Cold Start.* This problem occurs when the system has not yet acquired sufficient preference data (e.g., ratings or pairwise scores) to generate relevant recommendations for users. The most common example of the cold start problem occurs when the system is not capable of properly recommending any item to a new user (*New User* problem) or recommending a new item (not yet evaluated by any user) to a user (*New Item* problem) [Adomavicius and Tuzhilin (2005); Schein *et al.* (2002)]. In extreme cases, both problems may take place, e.g., when a recommender system has been recently launched and the system database contains very limited or no information about the preferences of users on items [Su and Khoshgoftaar (2009)].

In addition to these problems, preference data *Sparsity* may be observed. This occurs when the system has only collected a small percentage of all the potential preferences, which, for instance, in a collaborative filtering system is the full set of ratings of all the users for all the items. In fact, even in a mature system the users have rated a small percentage of the available items. Sparsity becomes a significant problem when the number of available ratings of each user is extremely smaller than the overall number of items [Adomavicius and Tuzhilin (2005)]. In general, preference data sparsity makes it very challenging for the recommender to generate accurate recommendations.

The cold start problem is even more challenging in Context-Aware Recommender Systems, known as CARSs (see Chapter 5). These systems generate recommendations which are adapted not only to the user's preferences but also to the contextual situation [Adomavicius and Tuzhilin (2011)]. Here, there is a cold start problem when the system is requested to generate recommendations for contextual situations under which the observed users did not express preferences (*New Context* problem) [Braunhofer (2015)]. Indeed, CARS needs to collect preference data that are augmented with the indication of the contextual situation of the user when experiencing the evaluated item. For that reason, in CARS, it is not only important to have ratings, but also to acquire them in several different contextual situations, so that the system can learn the users' contextually dependent preferences. Hence, the cold start problem in CARS occurs more frequently, compared to traditional recommender systems; there is often a large number of possible alternative contextual situations that may be observed in realistic scenarios [Braunhofer *et al.* (2014)].

Various approaches for addressing the cold start problem and improving collaborative filtering have been proposed in the literature. Here we briefly introduce them, and in the other sections of this chapter, we discuss them in details.

A popular approach to cope with the cold start problem consists of implementing *hybrid* recommendation techniques (see Chapter 4), e.g., to combine collaborative and content-based filtering [Nicholas and Nicholas (1999); Burke (2002); Ge *et al.* (2015); Adomavicius and Tuzhilin (2005); Ricci *et al.* (2015); Vartak *et al.* (2017); Kim *et al.* (2016)]. It is also possible to address the cold-start problem with cross-domain recommendation techniques. These techniques aim at improving the recommendations in a target domain by making use of information about the user preferences in an auxiliary domain [Fernández-Tobías *et al.* (2016); Cantador and Cremonesi (2014)]. In this case, knowledge of the preferences of the user is transferred from an auxiliary domain to the target domain. More complete and accurate user models and item recommendations in the target domain can then be built even when not much preference data is available in this domain. For example, by using the knowledge of ratings and tags assigned by the users to items in a movie domain (auxiliary) it is possible to better learn the user preferences in a book domain (target) [Enrich *et al.* (2013)].

Another approach to cold-start consists of complementing the rating data with other sources of information about the items. For example, in multimedia recommender systems, it has been shown that audio-visual features, e.g., variation of colour, camera and object motion, and lighting, can be automatically extracted from movies in order to solve the cold start problem and to effectively generate relevant recommendations [Elahi *et al.* (2017); Deldjoo *et al.* (2016, 2018)]. As an additional example, in a food recommender, it has been shown that one can exploit the information brought by tags assignments of the users to recipes to improve the recommendation performance of a system that is using only ratings for recipes [Ge *et al.* (2015); Massimo *et al.* (2017)]. Or, in a restaurant RS, it is possible to use information about the restaurant cuisine or location [Burke (2000); Adomavicius and Tuzhilin (2005); Ricci *et al.* (2015)].

In an alternative group of approaches the cold start problem is tackled by better profiling the users with additional information about them, such as their personality [Pu *et al.* (2012)]. In fact, studies conducted on user personality characteristics have shown that it is useful to exploit this information in collaborative filtering [Hu and Pu (2011, 2009); Tkalcic *et al.* (2013); Schedl *et al.* (2018); Elahi *et al.* (2013)]. Another example in this group of approaches, is discussed in [Trevisiol *et al.* (2014)]. The authors propose a graph-based method to solve the cold start problem in news recommendation. This method uses referral link of the new user as well as the currently visited page in order to generate recommendations for new users.

The cold-start problem can also attacked by directly enlarge the size of the preference data set by programming the system to selectively acquire new users' preferences with *Active Learning* [Elahi *et al.* (2016)]. In fact, Active learning tackles the cold start problem at the root, by identifying high quality data that better represents a user's preferences and improves the performance of the system. This can be done in various forms, e.g., by requesting the user to assess items one-by-one [Rashid *et al.* (2008a); Elahi *et al.* (2014)] or alternatively to evaluate a set of them altogether [Ekstrand *et al.* (2014); Loepp *et al.* (2014)].

In one seminal paper on this subject [Rashid *et al.* (2002)], the authors present several active learning techniques that have been tested on the well-known movie recommender system MovieLens. The goal was to create an effective sign up procedure that can help the system to build the initial user profile by acquiring specific ratings. Hence, when a new user registers to the system, the user is requested to rate movie items that are estimated to be more informative. Experimental results have shown the effectiveness of the approach.

Active learning has been used in a Points of Interest (POI) CARS named South Tyrol Suggests (STS) [Braunhofer (2015)]. In the early stages of the deployment, STS was in an extreme cold-start situation where only a few hundred of contextual ratings were provided by users for nearly 27,000 POIs stored in the system database. To solve this problem, the system acquired the users' personality to identify which items they could have visited and hence requested to provide their contextual ratings for those items [Elahi *et al.* (2013)].

In addition to the above-mentioned methods, it is also possible to use stereotypes as a mechanism for generating recommendations for users in the cold start situation. In this case, the system recommends items that matches a generic profile of that user, built upon the available data for a group the target user belongs to [Adomavicius and Tuzhilin (2005)]. For example, the systems attempts to recommend items that are popular within a certain age group or gender [Braunhofer *et al.* (2015a)]. In the extreme cold start scenario, if absolutely no data is available, the system may recommend popular items or items with the highest average ratings. Although the recommendations are non-personalized, still a satisfactory level of recommendation quality might be achieved [Fernandez Tobias *et al.* (2016)].

The rest of the chapter is structured as the following. Section 8.2 reviews the algorithmic solutions to tackle with the cold start and sparsity problem. Section 8.3 introduces the available resources such as libraries and datasets applicable in the research on cold start. Section 8.4 compares the discussed solutions and provides an overview of their performances. In Section 8.5, a set of practical guidelines are provided that can be used by researchers and practitioners. Finally, Section 8.6 concludes the chapter and discusses the directions of the future works.

8.2. Algorithmic solutions

As we have discussed in the introduction, various and different approaches to alleviate the cold-start problem have been proposed in the literature. We present them by grouping them into two broad classes according to which kind of knowledge is mainly used. The approaches in the first and more important class exploit in the recommendation process additional knowledge sources about either the users, the items or the contextual situations. The five approaches identified in this class are: active learning, cross-domain recommendation, recommendation based on implicit feedback, content-based recommendation and demographic-based recommendation. On the other hand, the approaches in the second class try to better process and leverage existing knowledge rather than acquiring new one.

8.2.1. *Active Learning*

Active learning (AL) applied to recommender systems refers to technologies aimed at eliciting the most informative preference data. When the system is using ratings, AL tries to elicit from the user those ratings that best reveal the user's interests, and hence should better improve the quality of subsequent recommendations [Rubens *et al.* (2015)]. [Rashid *et al.* (2002)] discusses six different AL rating elicitation strategies:

Entropy where items with the largest rating entropy are asked to be rated by the user;

Random which randomly selects items to be rated;

Popularity which measures the number of already acquired ratings for the items, and requests the user to rate the most frequently rated ones;

Popularity * Entropy where items that are both popular and have diverse ratings are requested to be rated by the user;

Log(popularity) * Entropy which is similar to the previous strategy except that it considers the log of the number of ratings before computing popularity;

Item-Item Personalized where the items are selected randomly until a first rating is acquired. Then a recommender is used to predict the items that the user is likely to have seen; these items are requested to the user to rate.

The results of off line and on line experiments conducted on the MovieLens dataset demonstrated that *log(popularity) * entropy* improves more the performance of the recommender system in terms of rating prediction accuracy.

In [Rashid *et al.* (2008a)], the authors extend their early work [Rashid *et al.* (2002)] by proposing additional AL strategies, namely:

Entropy0 which differs from the entropy strategy, as described above, in that it treats missing ratings as a separate category, in addition to the acquired ratings on the 1-5 scale;

HELF Harmonic mean of Entropy and Logarithm of Frequency selects the items with the largest harmonic mean of the logarithm of the rating frequency and the entropy;

IGCN Information Gain through Clustered Neighbours constructs a decision tree where the nodes are the items to rate and a node's branches according to the different ratings that a user can give to the item.

The authors evaluated these strategies in a preliminary off line experiment. Then, in an on line experiment, they used the MovieLens web-based movie recommender and every new user, during the sign-up process, was asked to rate 20 movies selected by one of the AL strategies (training set). After the sign-up process the new users were asked to complete a user satisfaction survey and to provide some more movie ratings, which were used to check the prediction accuracy (testing set). Based on the obtained results, the authors concluded that, overall, IGCN and entropy0 are the best-performing strategies.

Additional AL strategies can be found in [Elahi *et al.* (2014, 2011)], including:

Binary Prediction which first transforms the original rating matrix into a binary matrix, by mapping null entries to 0 and not null entries to 1, and then learns a predictive model that identifies the items the user is likely to have experienced and thus is able to rate;

Highest Predicted where the items with the highest predicted ratings are asked to be rated by the user;

Lowest Predicted which, differently from Highest Predicted, requests the user to rate the items with the lowest predicted ratings;

Highest and Lowest Predicted where items with either extreme low or extreme high predicted ratings are selected;

Voting which combines together the lists of top candidate items for rating elicitation from different strategies to produce a single list.

The authors have evaluated the considered strategies for their system-wide effectiveness implementing a simulation loop that models the gradual process of rating elicitation and rating dataset growth. The procedure is initiated by splitting the matrix of ratings into three different matrices with the equal number of rows and columns [Elahi *et al.* (2014)]:

- K: contains the ratings which are known to the recommender system at a certain point of time;
- X: contains the ratings which are known by the users but not by the recommender system. These ratings are iteratively acquired, i.e., they are transferred into K if the recommender system requests the (simulated) users to rate them;
- T: contains a subset of the ratings that are known by the users but are withheld from X for the evaluation purpose.

In every *iteration* of the experiment, the system carefully selects a set of items from the item catalog, according to a specific active learning strategy. The selected items are checked to find those with ratings available in X. The ratings of these items are assigned to the corresponding entries in K. Then the assigned ratings are removed from the X. Finally, the evaluation metrics are measured on the ratings in T, and the prediction model of the system is trained again using the new set of ratings in K. This process is repeated for 170 iterations, till almost all the ratings are elicited and the system performance gets stabilized.

Figure 8.1 illustrates the comparison of the rating prediction accuracy (in terms of Mean Absolute Error) of the described active learning strategies on MovieLens dataset. As it can be seen, there are two separable groups of active learning strategies [Elahi *et al.* (2014)]:

(1) Monotone error reduction strategies which include lowest-highest predicted, lowest predicted, voting and random strategies.
(2) Non-monotone error reduction strategies that include binary predicted, highest predicted, popularity, log(popularity)*entropy and variance strategies.

The strategies in the first group have a better performance, specially in the middle phase of the rating elicitation process. However, at the very beginning and at the end, the strategies in the second group outperform the first ones. Indeed, at the very beginning, the binary-predicted and the

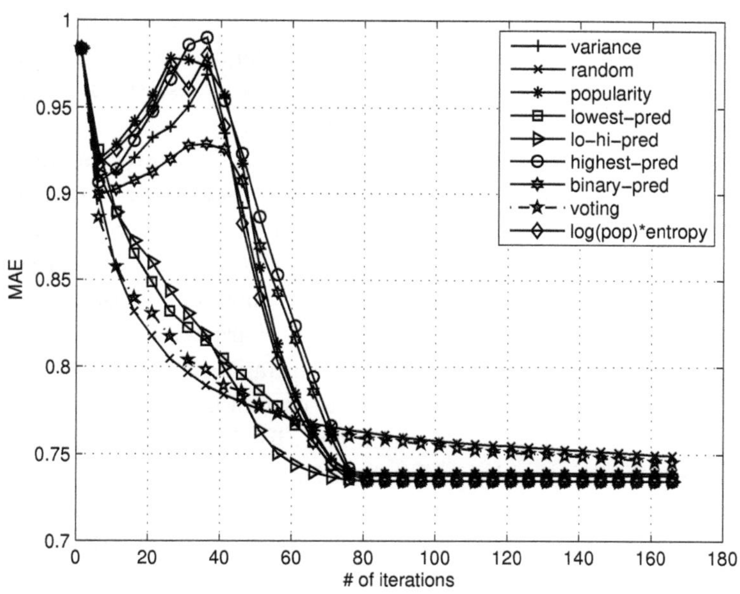

Fig. 8.1. Performance comparison of some well known Active Learning strategies in terms of prediction accuracy [Elahi *et al.* (2014)].

voting strategies (both from the first group) perform the best. Then, the random and the lowest-highest-predicted strategies (both from the second group) have good performance. Finally, at the ending phase, the MAE stabilizes for most of the strategies since they are not able to elicit anymore ratings.

The non-monotone behavior of the strategies in the second group occurs since they have a strong selection bias. The highest predicted strategy is perhaps the best known since it is the default strategy in recommender systems (people typically provide ratings to the recommendations). At the beginning of the rating elicitation process, this strategy elicits primarily items that have received high ratings. This leads to acquire significantly more high ratings than low ratings and ultimately biasing the recommender towards overestimating the predicted ratings. This is the reason why the error increases in the middle stage and drops afterward, as the items with high ratings are finished and the strategy begins to select also items with lower ratings.

Overall, all the conducted experiments have shown that each strategy has its own strengths and weaknesses. Prediction-based strategies are ineffective to deal with new users or new items, whereas voting, popularity and log(popularity) * entropy can select items for new users, though not for new items. Moreover, some strategies, such as, highest predicted and lowest predicted may bring a system-wide bias and increase the system error as they try to add only ratings with certain values.

It is worth noting that all the aforementioned AL strategies have been designed to work on traditional, two-dimensional user-item rating matrices, hence they are only suitable for attacking the new user and new item problems. In case of multidimensional user-item-context matrices, it is necessary to modify these strategies so that they output not only a list of items to be rated but also the contextual factors which characterize the situation of the items' consumption. Identifying and acquiring exactly those contextual information that matter ensures that (i) the user effort in specifying contextual information is kept to a minimum, and (ii) the system's performance is not negatively impacted by irrelevant information.

A solution to that problem can be found in [Baltrunas *et al.* (2012)], where the authors present a survey-based approach to identify the most useful contextual factors and conditions, as well as, to capture data about how the context influences user ratings. In their approach, they first estimated the dependency of the user preferences from an initial candidate set of contextual factors. This was achieved through a web tool, in which users were requested to evaluate if a particular contextual condition (e.g.,"it is a cold day") has a positive, negative or no influence on the user's rating of a particular type of POI (e.g., spa). Using the obtained data, they were able to establish the most important contextual factors for different types of POI.

The same problem mentioned above, i.e., identifying the contextual factors that are truly relevant for a particular recommender system, was addressed by Odić *et al.* [Odić *et al.* (2012)]. They developed two approaches: the first one is called "assessment" and it is based on surveying the users, while the second is called "detection" of the context relevance and is performed by mining the rating data. These two approaches are very different in terms of when each one could be used (e.g., assessment can be used before rating acquisition, whereas detection cannot), what information is needed (e.g., detection requires a substantial number of in-context ratings, while assessment does not) and whether they rely on real situations (as in detection) or hypothetical situations (as in assessment). In order to determine which of the two is better, the authors used real rating data and a survey data set to construct two (possibly different) lists of relevant and irrelevant contextual factors. Then, they considered all the contextual information one by one as input to a contextualized matrix factorization model, and finally counted for both approaches the number of times a contextual factor estimated as relevant led the system to obtain a higher prediction accuracy than using a factor estimated as irrelevant. Based on the obtained results, they concluded that the detection method performs better than the assessment one for identifying the contextual factors that should be exploited in the rating prediction model.

In a related paper [Odić *et al.* (2013)], the same authors investigate in more detail the "detection" approach and provide several statistical measures for relevant-context detection, i.e., unalikeability, entropy, variance, χ^2 test and Freeman-Halton test. Among these measures, they identify the Freeman-Halton test as the most useful and flexible measure to distinguish relevant from irrelevant contextual factors. Moreover, the authors show that the rating prediction performance was significantly better when using contextual factors detected as relevant than when using contextual factors detected as irrelevant.

Another example of selecting the most relevant contextual factors can be found in [Vargas-Govea *et al.* (2011)]. In this paper, the authors focus on a context-aware recommender system for restaurants, and show that its efficiency and predictive accuracy can be improved by using a reduced subset of contextual factors. To select contextual factors, the Las Vegas Filter (LVF) algorithm was chosen. LVF repeatedly generates random subsets of factors, computes their evaluation measure based on an inconsistency criterion, which tests the extent to which a reduced subset can still predict the rating values, and finally returns the subset yielding the best evaluation measure.

Finally, in [Braunhofer *et al.* (2015b); Braunhofer and Ricci (2016)], a context acquisition algorithm is proposed, that given a user-item pair, parsimoniously and adaptively identifies the most useful contextual factors, i.e., those that when elicited together with the user ratings improve more the quality of future recommendations, both for that user and for other users of the system. "Parsimonious" means that it selectively requests and possibly elicits only the most relevant contextual factors, whereas "adaptive" means that it personalizes the selection of the most relevant contextual factors to each individual user and item.

8.2.2. *Cross-Domain Recommendation*

Another option for collecting preference data in the form of ratings comes from collecting them in an auxiliary and possibly better known domain with the plan to transfer the knowledge brought by this preference data to the target domain [Cremonesi *et al.* (2011)]. Cross-domain approaches are about that, and are classified into two classes: those that aggregate knowledge from various auxiliary domains to perform recommendations in a target domain, and those that link and transfer knowledge between domains to support recommendations.

An example of knowledge aggregation for cross-domain recommendation is described in [Berkovsky *et al.* (2007)]. Here, the authors present four aggregation-based methods: (i) centralized prediction — the aggregated knowledge consists of user preferences (i.e., ratings), (ii) distributed peer identification — the aggregated knowledge is composed of user similarities, (iii) distributed neighbourhood formation — the aggregated knowledge is made up of user neighbourhoods, and (iv) distributed prediction — the aggregated knowledge is composed of single-domain recommendations. The authors show that the proposed methods can improve the accuracy of target domain recommendations in case of data sparsity, however, these methods can only be applied if the involved recommenders share some users or items and secondly, they must use the same recommendation technique.

One example of a knowledge transfer approach for cross-domain recommendations is presented in [Enrich *et al.* (2013)]. This paper introduces three novel cross-domain rating prediction models (UserItemTags, UserItemRelTags and ItemRelTags), which are based on the SVD matrix factorization model [Koren *et al.* (2009)]. They leverage the preference knowledge contained in tag assignments in order to improve the rating prediction task. The underlying hypothesis is that the information about how users tag items in an auxiliary domain can be exploited to improve the rating prediction accuracy in a different target domain, provided that there is an overlap between the set of tags used in the two domains. All the proposed models add tag factor vectors to the latent item vectors, which are then combined with the latent user features to compute rating predictions. The difference between these models lies in the set of tags used for rating prediction. UserItemTags and UserItemRelTags predict a target user rating by exploiting the tags this user has attached to the target item, whereas, ItemRelTags considers all the tags assigned by any user on the target item to predict ratings. To evaluate whether the exploitation of user tags collected in one domain could be useful to improve the rating prediction accuracy in a completely different domain, the authors carried out a series of off line experiments using the MovieLens and LibraryThing datasets. The obtained results indicate that the usage of tags is beneficial and their proposed models outperform the traditional matrix factorization model which only uses ratings [Koren *et al.* (2009)].

In another paper, Fernández-Tobías *et al.* [Fernández-Tobías *et al.* (2016)] have studied the quality of cross-domain recommendations in terms of accuracy, diversity and catalog coverage, by evaluating a number of algorithms (i.e., popular items, user-based nearest neighbours, item-based nearest neighbours, matrix factorization for positive-only feedback) on two datasets with positive-only feedback. Their results show an increased ranking accuracy in cold-start situations when cross-domain information is used. The results for diversity varied across the two datasets, and hence the authors conclude that in general the results depend on the target domain. Finally, the results also indicate that a greater item diversity on the source profile can harm the perform in the target domain.

8.2.3. *Recommendation Based on Implicit Feedback*

In many real-world scenarios explicit feedback can be difficult to obtain or unavailable (e.g., news portals). In these cases, recommender systems can use implicit feedback. Indeed, they can monitor the users' behaviour, such as noting which items they browse or purchase, the duration of time spent viewing an item and the search terms they use, and use these observations in order to infer the user preferences regardless of the user's willingness to actively provide explicit evaluations, such as, ratings. In this scenario, recommenders have been built either by leveraging implicit feedback data only [Hu *et al.* (2008a); Rendle *et al.* (2009); Gurbanov *et al.* (2016)] or by extending recommender models based on explicit feedback. A notable example of this last category of approaches is SVD++ [Koren (2008)], which extends the standard matrix factorization model by adding a new set of item factor vectors in order to leverage preference information coming from the items for which users provided implicit feedback. In particular, in this model each user is characterized not only by a user factor vector, but with additional factor vectors which represent the contribution of the implicit feedback to the model of the user in the factors space.

We conclude this section by pointing out that approaches based on implicit and explicit feedback are only applicable if feedback information from users is available, which is not the case for newly registered users. Hence, the applicability of these approaches in the new user scenario is in both cases severely restricted.

8.2.4. *Content-Based Recommendation*

A classical approach for addressing the cold-start problem, and in particular the new item problem suffered from standard collaborative filtering-based recommenders, is to rely on preference information brought by the features associated with the liked items (content). For example, if a user has positively rated a POI of type "hiking trails" then the system can predict that other POIs of the same type too will be liked.

Among the many systems where content information is used, we mention the work of Manzato [Manzato (2013)]. The proposed gSVD++ model shows how content metadata can be directly incorporated into matrix factorization and in particular into the SVD++ model proposed by Koren [Koren (2008)]. This is achieved by introducing a new set of latent factor vectors for content metadata (e.g., the type of a POI, the actors or the genre of a movie) that are combined with the item's latent factor vector. To evaluate gSVD++, the author used the MovieLens 100k dataset, and have found that it has lower MAE and RMSE than SVD++ and a previous model proposed in [Manzato (2012)], which also performs factorization on item's metadata.

Fernández-Tobias and Cantador [Fernández-Tobías and Cantador (2014)] adapted gSVD++ by enhancing the items' latent factor vectors with additional latent factor vectors for the tags that were applied to them. Likewise, the user's latent factor vector is extended with additional latent factor vectors representing the tags of the user, to better capture these interests. These enhancements tackle the new item problem and the new user problem. Moreover, it suffices that a tag is available for an item, but not necessarily a rating, in order to make the item recommendable. This happens in many situations, for instance, in the social bookmarking website Delicious[1], in which users can apply tags to their bookmarks, but are not asked to rate the bookmarked website.

[Shi *et al.* (2014)] provides a comprehensive overview of collaborative filtering recommender systems that integrate rich side information about items and users as well. The authors mostly focused on social networks and user-contributed information, such as user tags, geotags, multimedia content and reviews/comments, which has become widely available since the introduction of Web 2.0 technologies. Based on this focus, they have identified two types of challenges for recommender system research: 1) new conditions and tasks (e.g., social recommendation, group recommendation, long tail recommendation, cross-domain collaborative filtering); and 2) challenges due to the interaction between recommender systems and other areas of research (e.g., search, interaction and economics).

To conclude, by incorporating content information, a recommender system is capable of recommending items not yet rated by any user and hence can overcome the new item problem. Nevertheless, it does not solve the new user problem, since still enough ratings have to be collected before the system can really understand the user preferences and provide accurate recommendations.

8.2.5. *Leveraging User Descriptions*

A notable way to solve the new user problem is to utilize user attributes (e.g., demographic data and personality characteristics) in the recommendation process. In [Vozalis and Margaritis (2004)], the authors enhance traditional collaborative filtering by integrating in to the model demographic information (i.e., age, gender, job and zip code). Specifically, when generating rating predictions, the authors propose to use not only the ratings-based correlation, but also the demographic correlation between the active user

[1]https://del.icio.us/

and a neighbour one. Demographic correlation is calculated by representing each user with a vector of 27 demographic features and applying vector similarity on the generated demographic vectors.

The same idea is used also in the approach presented in [Koren *et al.* (2009)]. Here, the authors extend their original SVD++ algorithm [Koren (2008)] by incorporating known user attributes. In practice, they consider Boolean attributes and a user u is described by the subset of their attributes $A(u)$. Attributes can, for instance, describe the gender, age group, zip code, income level, and so on. Then, the proposed algorithm computes a distinct factor vector $y_a \in \mathbb{R}^f$ for each attribute of the user.

User personality has also been used in collaborative filtering [Fernandez Tobias *et al.* (2016)]. In this article, the authors investigate three different approaches: (a) personality-based collaborative filtering, which directly improves the recommendation prediction model by incorporating user personality information; (b) personality-based active learning, which exploits personality information to first identify useful user preference data to be elicited in a target domain, and then improve the recommendation prediction model; and (c) personality-based cross-domain recommendation, which utilizes personality information to enhance the usage of preference data from auxiliary domains in order to compensate the lack of preference data in the target domain. The results of their experiments on the myPersonality dataset [Bachrach *et al.* (2012)] show that the proposed approaches are viable methods for improving collaborative filtering to tackle the new user problem.

To conclude, we note that the exploitation of user-related information can help to generate better recommendations for new users, i.e., for which none or a limited number of ratings is available. However, in general, recommendations based only on user features are less accurate than those based on user preferences. In fact, for instance, demographic factors account only for a little part of the variance of the rating behaviour.

8.2.6. *Better Using Existing Preference Knowledge*

Instead of acquiring new information about either the user or the application domain, as in the approaches illustrated in the previous sections, one can simply try to better use the available information.

A popular solution moving in this direction is offered by hybrid recommender systems. These systems combine two or more recommendation techniques in order to improve their performance and to overcome the limitations of the combined techniques [Burke (2007)]. For instance, the new item problem, which is an issue for pure Collaborative Filtering (CF), can be overcome by hybridizing CF with a content-based component. According to [Burke (2002)], it is possible to identify seven hybridization techniques:

Weighted The scores of different recommendation components are combined into a single weighted score.

Switching The system selects and applies a single recommender from among its constituents based on the current recommendation situation.

Mixed Recommendations generated by different recommendation components are presented side-by-side in a combined list.

Feature Combination Features derived from different recommenders are combined and injected into a single recommendation algorithm.

Cascade Recommendation components are strictly hierarchically ordered from the strongest to the weakest, with the ties observed by using a technique are broken by the following one.

Feature Augmentation A feature or set of features computed by one recommendation technique are passed as input to the next technique.

The idea of better using existing preference data is especially important in context-aware recommenders. In fact, since these systems need significantly more preference data than context-free ones, for CARS all the available preference data should be exploited as much as possible. Hence, some techniques have been proposed in the CARS literature aiming at better exploiting the (usually sparse) set of contextually-tagged ratings. An approach worth to be mentioned is the generalized pre-filtering approach proposed by Adomavicius *et al.* [Adomavicius and Tuzhilin (2015)]. It exploits hierarchical relationships between different contexts in order to generalize a target context when the number of ratings acquired in that particular context is small. For example, if we would like to identify which POIs to recommend on a Monday, we may use not only the ratings for POIs collected on Mondays, but also those obtained on other weekdays (but not in the weekend). This results in a significantly better usage of the existing rating dataset, and allows to build more robust and general rating prediction models.

A similar idea to deal with the cold-start problem in context-aware recommenders is used by Zheng *et al.* [Zheng *et al.* (2012)]. They also try to increase the number of ratings used to generate predictions by defining a relaxed notion of match between the target context and the contexts associated with ratings.

The main limitation of the two aforementioned approaches is that it is necessary to choose the right level of generalization for contexts. This might be easy for datasets with relatively dense in-context ratings, but for datasets where only a small amount of in-context ratings is available, finding the proper level of generalization remains problematic.

One possible solution, is offered by the Differential Context Weighting (DCW) approach [Zheng *et al.* (2013)], which relies on the concept of weighting vectors and similarity of contexts in order to weight the contribution of individual contextual factors and ratings in the rating prediction algorithm, respectively. More specifically, weighting vectors are used to indicate the influential power of each contextual dimension in the various components involved in the recommendation process, whereas similarities are used to assess how much to weight a rating under a particular target context. This allows to consider all (weighted) ratings in the rating prediction process instead of completely filtering them out when they have non-matching contexts.

Similarly, also Codina *et al.* [Codina *et al.* (2013)] show that, in the recommendation process, instead of using only contextual ratings that exactly match the target context, it is possible to reuse also those that were provided in similar contexts. In their approach, two contexts are deemed as similar if they are influencing the items' (or users') ratings in a similar way.

Yet another example is the Contextual Sparse LInear Method (CSLIM) approach of [Zheng *et al.* (2014, 2015)]. It also reuses ratings collected in contexts that do not match the target context for rating prediction, i.e., by weighting them via the learned contextual rating deviation or similarity (correlation) terms.

8.3. Tools & Datasets

In this section, we discuss tools and datasets that can be used to build RSs that are resistant to the cold start and data sparsity problems. While it is impossible to provide an exhaustive survey of the ever-growing number of available resources, this selection should present a proper initial selection and where tools and datasets might be headed next.

8.3.1. *Tools*

In general, there are two ways of integrating a RS into a product, either to use a tool providing the RS as a service, i.e., Software as a Service (SaaS) RS, or to use a software framework providing functionality that can be incorporated into the source code of the product.

Yusp[2] (former *Gravity*) provides one of the most well-known SaaS solutions for RSs. The company offers to their customers a recommendation engine that generates recommendations by collecting and utilizing data of a customer's users through an API. Customers can manage the engine settings using a special dashboard. Similarly to many other SaaS products, this recommendation engine is closed-source. To solve the cold start and the data sparsity problems, Yusp collects knowledge from different resources[3] and uses hybridization techniques[4]. The available hybridization strategies, which are here called *logics*, can be selected from a list of the pre-defined ones or can be created from scratch using the dashboard tool. The weighting, switching or cascade strategies are available, but there are some more [Burke (2007)]. For example, the "optimized more like this" strategy attempts to make recommendations based on collaborative filtering, however, if there is not enough behavioural data (e.g., clicks or views) for an item, it "falls back" to content-based models. In addition, Yusp can distinguish the users who come to the service only to browse from those who know what they are looking for specific items[5]. Knowing the users' intent the recommendation method or the recommended items are further adapted. For instance, if someone clicks on everything from phone cases to real estate within a short period of time, the system will assume she is only there for browsing and will not use their click history for recommendations.

Other companies, such as *RetailRocket*[6], *nosto*[7] or *Relap*[8], provide recommendation functionality similar to those included in Yusp. However, little or nothing is known about the recommendation techniques and methods used by these companies. The target application domains for SaaS RSs are usually e-commerce, media and news.

[2]http://www.yusp.com/
[3]http://support.yusp.com/support/solutions/articles/5000717574-actions-what
[4]http://support.yusp.com/support/solutions/articles/5000712803-select-a-recommendation-logic
[5]http://www.yusp.com/blog/cold-start-problem-recommender-systems/
[6]https://retailrocket.net/
[7]http://www.nosto.com/
[8]https://relap.io/

In comparison to SaaS solutions, almost all RS software frameworks are open-source and well-documented, that makes them more attractive for developers and researchers. The frameworks provide generic functionality that can be further tailored to the application specific goals by additional code written on top of the framework. Some of the frameworks, such as rrecsys [Çoba and Zanker (2016)], SurPRISE[9], LibRec [Guo *et al.* (2015)], and Hi-Rec[10] [Elahi *et al.* (2017)] have been designed for rapid prototyping of recommendation algorithms. These frameworks contain only basic recommendation models. Others, such as *Turi*[11], *Apache Mahout*[12], *Mortar*[13], *Seldon*[14] and *Oryx*[15], can be used for building large scale fully-fledged recommendation engines that can cope with the cold start and the data sparsity problems. In this section, we will consider only frameworks from the second group.

To overcome the cold start and the data sparsity problems, frameworks, as well as SaaS RSs, combine data from different knowledge sources. This is done either by utilizing recommendation models that can work with multiple data sources or by employing hybridization techniques. For example, Apache Mahout generates recommendations by leveraging information about actions of different types performed by users while interacting with the product (clicks, views, purchases, etc.)[16]. The open-source recommender engine Mortar uses a graph-based approach that can blend collaborative filtering with content-based recommendations and other signals[17]. Finally, the Seldon recommender engine allows the product owners to specify in a configuration file the set of models that must be used, the order in which these models should be applied, and the priority of each model[18] (i.e., hybridize several recommendation models).

Table 8.1 summarizes the properties and features of the aforementioned frameworks. The *Specialization* column contains a brief description of the most specific feature of the frameworks. It can be noted that only Mortar has been developed specifically for solving RS tasks. All the other

[9] http://surpriselib.com
[10] https://fmoghaddam.github.io/Hi-Rec/
[11] https://turi.com/
[12] http://mahout.apache.org/
[13] https://github.com/mortardata/mortar-recsys
[14] https://www.seldon.io/
[15] http://oryx.io/
[16] https://mahout.apache.org/users/algorithms/intro-cooccurrence-spark.html
[17] http://help.mortardata.com/data_apps/recommendation_engine/recommendation_engine_basics
[18] http://docs.seldon.io/advanced-recommender-config.html

Table 8.1. Properties and features of the frameworks.

	Frameworks				
	Apache Mahout	Mortar	Oryx	Seldon	Turi
Specialization	Scalable algorithms	Personalized recom. at scale	Real-time large scale ML	Scalable ML with deployment	End-to-end large-scale data analysis and data product development
Components	Apache Hadoop, Apache Spark, A search engine	Mortar Development Framework	Apache Spark, Apache Kafka, Apache Hadoop, Apache Zookeper	Apache Spark, MySQL, Kubernetes	Graphlab Create, Turi Distributed framework
Minimum requirements	-	Mortar account, Github account, AWS account	Hadoop cluster	Kubernetes cluster	Turi license
Main language	Java	Pig	Java	Scala/Python	Python
RS models	Item-based CF, MF with ALS, IMF, SVD++, Content-based	Graph-based	MF with ALS, Popularity-based	MF with ALS, User-Clustered MF, Popularity-based	Item-based CF, Content-based, MF with ALS/SGD/adagrad, IMF, SVD++, Popularity-based
Actions of multiple types	Yes	Yes	No	No	No
Recommendations delivery	Hadoop command-line interface	Stores all the recommendations in Dynamo DB	HTTP API	Seldon command-line interface, HTTP API	Graphlab SDK, HTTP API
Anonymous user	No	No	Yes, user is defined by a set of items she interacted with	Yes, by creating a hybrid model	Yes, user is defined by a set of items she interacted with
Diversity	No	Yes	Yes	No	Yes
Explanations	No	Yes	Yes	No	No

frameworks, even though they provide a range of prediction tasks employed in RSs, they have been developed as scalable Machine Learning (ML) frameworks for general purposes. The *Components* row shows additional tools and frameworks that are used to support the target RS framework and let it run on a computer cluster. In fact, all the frameworks can be run on a computer cluster. To employ a framework some *Minimum Requirements* should be met. For example, Oryx can be deployed only on Hadoop cluster[19], while Seldon requires Kubernetes[20]. The recommendation techniques supported by the frameworks are listed in the *RS models* row. Turi provides the largest set of techniques, including, matrix factorization (MF) [Koren *et al.* (2009)], implicit feedback MF (IMF) [Hu *et al.* (2008a)], SVD++ [Koren (2008)], item-based CF [Deshpande and Karypis (2004)], content-based and popularity-based models.

All the frameworks, except Oryx, allow filtering items using *meta-parameters*, such as tags and content information (genres, categories, etc.). Mortar and Apache Mahout can leverage information about user actions of different types, but do not provide out-of-the-box HTTP API and cannot provide recommendations for anonymous users.

When using Turi and Oryx the request to the RS can specify the set of items that a user has observed or has interacted with[21,22].

These items are used by the framework to generate the user's profile on-the-fly and, therefore, to generate recommendations for *anonymous* users (often are cold users). Knowing the similarity between items, Oryx and Mortar build recommendations containing diverse items, that can be useful to acquire more information about users' preferences. Moreover, they provide functionality that can be used to explain the recommendations. One framework only supports hybridization (cascading), this is Seldon. And only one framework supports active learning, this is Oryx. It provides the "surprise me" functionality, which recommends items that a user least-likely will interact with.

To sum up, though numerous RS tools and frameworks is currently available, only few of them are equipped with functionality capable to mitigate the cold start and the data sparsity problems. The problems are usually treated by using additional knowledge sources about the users, the

[19]`http://hadoop.apache.org/`
[20]`https://kubernetes.io/`
[21]`http://oryx.io/apidocs/com/cloudera/oryx/app/serving/als/RecommendToAnonymous.html`
[22]`https://github.com/turi-code/userguide/blob/master/recommender/making-recommendations.md#making-recommendations-for-new-users`

items or the contextual situations. For example, Apache Mahout, Turi and Oryx use algorithmic solutions based on implicit feedback data, while Yusp and Mortar employ content-based filtering. Beyond that, Yusp and Seldon provide hybridization and basic active learning techniques. Yusp offers weighting, switching and cascading hybridization approaches, while Seldon supports the lowest predicted (a.k.a. "surprise me") AL strategy. Whilst these solutions allow reducing the problems we believe that the efficient implementation of more advanced hybridization and AL techniques described in this chapter can significantly improve the existing tools.

8.3.2. *Datasets*

There are many publicly available datasets that can be used to conduct research in the field of recommender systems; in particular, for analyzing, testing and comparing existing recommenders. In this section, we will mention datasets that can be employed to train models resistant to the cold start and the data sparsity problems. Such datasets contain information about interactions between users and items and satisfy at least one of the following conditions:

- the dataset contains user and item metadata;
- the dataset contains additional interaction information (i.e., user actions of multiple types);
- the dataset reflects the real-time (or near real-time) changes in the user behaviour.

The first two conditions are essential for building and testing systems which are based on hybridization or use additional knowledge sources, while the third one might be helpful in building an active learning, rating elicitation, process.

One of the most popular recommender system data sets is MovieLens [Harper and Konstan (2015)]. The dataset comprises 5-star ratings and free-text tags collected by a movie recommendation service. Using tags assigned by users to movies as well as movie genres one can build a content-based recommender system. Other data sets containing item and user metadata are Amazon product data [He and McAuley (2016)] and Yelp dataset[23]. The first one includes reviews (ratings, text, helpfulness votes), product metadata (descriptions, category information, price, brand, and image features), and links (also viewed/also bought graphs) from Amazon.

[23]`https://www.yelp.com/dataset/`

The second one is a subset of Yelp's businesses, reviews, and user data that incorporates over 1.2 million business attributes like hours, parking, availability, and ambiance. It also aggregates check-ins over time for each of the 174,000 businesses.

In addition to user and item metadata, datasets can include user (behavioral) data, such as information about personalty or actions of different types performed by the users while interacting with the system. For instance, My Personality dataset[24] and STS dataset[25] both contain the personality of the users together with other types of data. The RetailRocket[26] as another dataset that contains logs of users' actions, such as, clicks, add to carts, and transactions, that were collected from the e-commerce website over a period of 4.5 months. Another good example is XING's dataset [Abel *et al.* (2017)] which contains the logs of interactions performed by users on job postings (see Chapter 17). These interactions include clicks, bookmarks, replies, and deletes. It also includes reciprocal interactions, e.g., whether or not a recruiter has showed his interest into a user by clicking on a candidate user's profile. It is worth noting that the available XING dataset is a semi-synthetic sample of another data set, and it is enriched with artificial users whose presence contributes to the anonymization. Moreover, some noise was added to the data: not all the users' interactions are contained in the dataset while some of the interactions are artificial (have actually not been performed by the users). Four types of actions are contained in the dataset: clicks, bookmarks, replies and deletes.

Moreover, recently a set of datasets have been published, called Mise-en-scene [Deldjoo *et al.* (2016)][27] and MPEG-7 visual datasets[28] [Deldjoo *et al.* (2018)].

These datasets contain various types of visual features extracted from movie trailers. The idea is that, for the new movie items in the extreme cold start situation, where there is no data available, visual features can be used to generate personalized recommendation. These features are descriptive of the aesthetic elements (e.g., color, light, and motion), encapsulated within the movies. The features are automatically extracted by analyzing frame-by-frame of each movie and aggregating them over the entire movie.

---

[24]`https://mypersonality.org`

[25]`https://www.researchgate.net/publication/305682479_Context-Aware_Dataset_STS_-_South_Tyrol_Suggests_Mobile_App_Data`

[26]`https://www.kaggle.com/retailrocket/ecommerce-dataset`

[27]`https://www.researchgate.net/publication/305682388_Mise-en-Scene_Dataset_Stylistic_Visual_Features_of_Movie_Trailers_description`

[28]`https://goo.gl/ZYij61`

Finally, *Satori*[29] and *webhose.io*[30] are services that allow collecting data about social activity, e-commerce reviews, and news in near real-time. The data can be used to explore the influence of different events on the users' preferences. Using this knowledge one can separate the reviews which reflect more stable user preferences from those that were mostly influenced by some exogenous event.

8.4. Performance Comparison

In this section, we will summarize the experimental results contained in research studies that focused on collaborative filtering systems in cold start scenarios. However, preliminary, we would like to make a few important observations.

First of all, most of the studies conducted in this particular research area have adopted an off line evaluation setting and measured the rating prediction error (e.g., Mean Absolute Error — MAE, or Root Mean Squared Error — RMSE). Hence, we do not have any knowledge about how the surveyed techniques may perform with respect to other important metrics, such as those measuring ranking quality (e.g., Normalized Discounted Cumulative Gain — NDCG, or Mean Average Precision — MAP). Moreover, the majority of the surveyed studies have focused on the movie recommendation domain and used the well-known MovieLens and Netflix datasets.

Most importantly, in spite of the similarities among the adopted evaluation methods, still there are some substantial differences, and hence, it is almost impossible to draw any final conclusion, as well as to generalize the results obtained in the experiments.

As summarized in Table 8.2 and described in more detail in this section, the surveyed techniques for tackling the cold-start problem have their own advantages and drawbacks and have been evaluated on different datasets[31].

Please note that *HLU* refers to *Half Life Utility* and it measures the utility of a recommendation list for every user, assuming that the likelihood that she will notice and choose a recommended item decays exponentially with respect to the ranking of the item's [Breese *et al.* (1998); Pan *et al.*

---

[29] `https://www.satori.com/`

[30] `https://webhose.io/`

[31] MovieLens [Harper and Konstan (2016)], Netflix [Koren (2009)], STS [Braunhofer *et al.* (2014)], EachMovie [GroupLens (2018)], MyPersonality [Kosinski *et al.* (2015)], LibraryThing [Librarything (2018)], 7TV [Gurbanov *et al.* (2016)], Rossmann [Rendle *et al.* (2009)], Adom [Adomavicius *et al.* (2005)], CoMoDa [Košir *et al.* (2011)], InCarMusic [Baltrunas *et al.* (2011)], TripAdvisor [TripAdvisor (2018)], IMDB [IMDB (2018)].

Table 8.2. Summary of the performance of various cold-start solutions on different data sets.

Approach	Cold-start problem			Metric			Evaluation		Datasets
	New user	New item	New context	RMSE, MAE	Precision, MAP	NDCG, HLU, MPR	Offline	Online	
Active Learning	✓	✓	✓	✓	✓	✗	✓	✓	CoMoDa, MovieLens, Netflix, STS
Cross-domain recommendation	✓	✓	✓	✓	✓	✓	✓	✗	EachMovie, MyPersonality, MovieLens, LibraryThing
Implicit feedback	✗	✗	✗	✓	✓	✓	✓	✗	7TV, MyPersonality, Netflix, Rossmann
Content-based recommendation	✗	✓	✗	✓	✓	✓	✓	✗	MovieLens, LibraryThing
Demographic-based recommendation	✓	✗	✗	✓	✗	✗	✓	✗	MovieLens
Hybrid recommendation	✓	✓	✓	✓	✓	✗	✓	✗	IMDb, MovieLens, Netflix
Context-aware models optimized for data sparsity	✗	✗	✓	✓	✓	✓	✓	✗	Adom, CoMoDa, InCarMusic, TripAdvisor, STS

(2008); Schedl *et al.* (2018)]. Hence, greater value of *HLU* may correspond to a superior recommendation performance. Additionally, *MPR* refers to *Mean Percentile Ranking* and it is a metric that measures the user satisfaction of the items in a ranked list of recommendations, and it is computed as the average of the percentile ranking of the test items within the ranked list of recommendations for every test user [Hu *et al.* (2008b); Schedl *et al.* (2018)]. It is worth noting that the smaller *MPR*, the better recommendation performance.

First, let us consider Active Learning — it has been shown in offline and online experiments that it can be used to effectively tackle the new user, new item and new context problem by selecting informative items for users to rate. More specifically, experimental results obtained in previous research have provided evidence that a proper choice of an active learning elicitation strategy allows to improve the performance of a recommender system in terms of rating prediction and ranking accuracy for new users, new items and new contextual situations [Rashid *et al.* (2002, 2008a); Elahi *et al.* (2014); Odić *et al.* (2012); Braunhofer and Ricci (2016)]. However, users may perceive active learning as tedious or even unpleasant, as it takes time and effort for them to browse the proposed items and rate them.

Also cross-domain recommendation can be a compelling approach to solve the cold-start problem. It has been shown that by exploiting cross-domain recommendation techniques it is possible to improve the prediction accuracy [Berkovsky *et al.* (2007); Enrich *et al.* (2013)] as well as the ranking accuracy [Fernández-Tobías *et al.* (2016)] in the target domain in various cold-start situations. Cross-domain recommendation techniques, however, require the co-existence of users, items and contextual situations across the considered and different domains, which is not always the case. Otherwise, wrong conclusions about the user preferences might be transferred from the auxiliary domain to the target domain.

A similar comment holds for recommendation approaches that incorporate implicit feedback. Implicit feedback data is generally easier to collect than explicit ratings data, and it has been shown that by extending rating-based recommender models using implicit feedback it is possible to achieve a higher recommendation performance [Koren (2008)]. However, recommendation approaches based on implicit data fail on users, items and contextual situations which were just entered into the system and for which no implicit feedback is yet available.

The other techniques, i.e., content-based recommendation and demographic-based recommendation, deal with just one specific type of cold-start problems. For instance, content-based techniques can lead to a higher prediction and ranking accuracy for new items (new item problem) [Manzato (2013); Fernández-Tobías and Cantador (2014)], whereas demographic-based techniques can improve the prediction and ranking recommendation accuracy for new users (new user problem) [Koren *et al.* (2009); Fernandez Tobias *et al.* (2016)].

As we have already mentioned in Section 8.2.6, the observed advantages and drawbacks of these techniques motivated the development of hybrid approaches that exploit simultaneously more than one technique, and adaptively use them for recommendation depending on their strengths and weaknesses in a given (cold-start) situation.

8.4.1. *Caveats*

There are several important issues related to the evaluation of recommendation techniques to deal with the cold-start problem that should also discussed. We list these aspects in the rest of this section.

Direct comparison of cold-start solutions. Different cold-start solutions have been typically evaluated in isolation with each other, i.e., new active learning solutions are compared to state-of-the-art active learning solutions, newly proposed cross-domain recommendation algorithms are confronted with other cross-domain recommendation algorithms, and so on. It is still an open and important question which one is more effective, e.g., in terms of recommendation accuracy, effort required from the users or simplicity of implementation.

User-centric evaluation studies. Off line evaluations are the most common evaluation methods for cold-start solutions. Although they allow for low-cost simulation of the behaviour of users when interacting with different algorithms, they can not substitute on line evaluations. On line evaluations are more costly, however, as they are carried out with real users in almost real-life settings, they allow to draw more reliable conclusions about the true merits of an algorithm.

Definition of benchmark results, evaluation procedures and public large-scale datasets. The evaluation of cold-start solutions for recommender systems is complex for several reasons: (1) it requires large feedback datasets with user and item attributes; otherwise it is difficult or even impossible to test the effects of using user and item attributes in the prediction models; (2) the evaluation must cover multiple perspectives, e.g., it must consider new users, new items, new contextual situations, mixtures of elementary cold-start cases, different degrees of coldness and different types of user and item attributes available; and finally (3) the evaluations and results of the cold-start solutions proposed so far are difficult to compare as there exists no standard or reference evaluation procedure as well as metrics.

8.5. Guidelines

In this section, we suggest a number of practical guidelines that can be adopted for the design and development of an operational collaborative filtering system, resistant to cold start problem.

Algorithms should adapt to the time evolution of user/item profiles: one of the important factors that could impact on the behavior of collaborative filtering systems, is the temporal dynamics of the user preferences which could be dependent on the application domain. As an example, the ratings that the users provide to TV programmes or books are more stable than the ones provided to news items, as the news items change more rapidly [Picault *et al.* (2011)]. This is important to take into account when designing the algorithms that can better adapt to the dynamic nature of the reality of collaborative filtering systems. Adaptive algorithms can tackle this issue by learning the temporal evolution of the user/item profile recognizing the different evolution patterns, e.g., stable profiles, progressive profiles, fast changing profiles, etc. [Picault and Ribiere (2008); Hawalah and Fasli (2015)].

Scalability of the recommender algorithm should be taken into account: the computation load performed by collaborative filtering algorithms can increase dramatically with the growth of the dimensions of the rating matrix, i.e., the number of users and items in the dataset [Park *et al.* (2012)]. It will also increase the sparsity of the data and hence create severe cold start problem. Consequently, a recommendation algorithm that could operate with a limited volume of users and items may fail to generate recommendations in a reasonable time if a considerable number of new users and new items are added to the dataset. Hence, in real world recommender systems, with huge databases (big data), it is crucial to adopt algorithms that are capable of scaling up (see Chapters 11, 17, 21, and 14). For instance, collaborative filtering algorithms based on Dimensionality Reduction methods (e.g., Singular Value Decomposition — SVD) are able to significantly reduce the computational cost of a recommendation and still to generate highly accurate recommendations [Koren and Bell (2015); Isinkaye *et al.* (2015)].

Ratings on recommended items can cause serious system bias: recommender systems typically obtain ratings on items with the highest predicted ratings (recommendations). This is due to the fact that it is more likely that the users assess and rate recommended items which are supposed to be interesting to them. However, it has been shown [Elahi *et al.* (2014)] that ratings given for such items may inject in the data set a large number of high ratings, which will ultimately cause a serious bias of the prediction algorithm for high ratings (see Chapter 10). Therefore, it is also important that the system adopts certain strategies in order to obtain ratings on items that the users may dislike and would rate low.

Natural acquisition of ratings can strongly impact the system performance: in operational recommender systems the ratings are added to the system database through two different processes [McNee *et al.* (2003); Carenini *et al.* (2003); Rashid *et al.* (2008b); Elahi *et al.* (2014)]: (i) the system explicitly requests the user to rate a set of selected items (active learning), or (ii) the user voluntary explores the item catalog and provides some ratings (natural acquisition of ratings). The majority of the research works have studied the performance of collaborative filtering systems in cold start settings, under the assumption that the new ratings are added only as a response to the system requests. However, in reality, user ratings are generated by both processes with diverse impact on the system performance. Hence, it is important to consider a *mixed initiative* scenario [Rashid *et al.* (2008b); Pu *et al.* (2012)], by considering both rating elicitation sources. This may provide a better prediction of the temporal evolution of the system performance and a more realistic scenario [Elahi *et al.* (2014, 2012)].

User ratings on random items could be beneficial: supporting users in exploring items and providing ratings on random items has shown to be useful in reducing the system bias. This could be more effective particularly when the system contains very popular items that receive almost all the ratings from the users. For systems that include rating elicitation in the sign-up process, a small portion of randomly selected items can be included in the lists the users are proposed to rate [Zhao *et al.* (2013); Christakopoulou *et al.* (2016)].

Conversational interaction models could improve user profiling: the standard preference elicitation model of current collaborative filtering systems, is mainly supporting the generation of the initial user profiling, during the sign-up process; the system builds the user profile by requesting the user to rate a set of items [Carenini *et al.* (2003)]). However, the user should be able to update her profile (by ratings more items) whenever she likes. This approach is analysed in [Carenini *et al.* (2003); Sun *et al.* (2013)]. Here the user rates items (selected by the system) in the sign-up process, and later on, she still gets notifications from the system motivating her to rate more items, which are selected by the system. This process includes explanations which clarify the benefits of providing more ratings. Therefore, the control is still in the user's hands, while a sense of co-operation between the user and the system is formed.

Elicitation of contextual ratings are necessary for CARS: in context aware recommender systems (CARS), the context of the user plays a significant role in recommendation accuracy. While there are several types of contextual information that can be automatically obtained from sensors (e.g., weather, temperature, location, daytime, season, and weekday), there are also contextual information that have to be specified by the user (e.g., budget, companion, mood, and transport mean). However, not all the contextual factors are equally useful for the system to improve the recommendation accuracy. Hence, actively selecting the contextual factors that are the most informative to explain the rating given by the user to an item is important and can potentially improve the system accuracy more while being easier for the user to provide them. So, the application of active learning in context-aware recommender systems is in order [Baltrunas (2011)].

Hybridization can result in robustness: while collaborative filtering approaches are considered powerful and effective in generating relevant recommendations, they have several limitations, which are impacting on their performance in cold start situations. For instance, if two items have synonymous names, e.g., espresso machine and coffee maker, a collaborative filtering systems may not consider these items similar until they are similarly rated by the users [Isinkaye *et al.* (2015)]. Another example is recommendation in the news domain where the items (news articles) may get outdated quickly and the users may not be willing to read them any more. As a consequence, there would be low overlap among the ratings of users, and the recommendation quality may drop considerably. In both the above-mentioned problems, hybridizing collaborative filtering with content-based could provide an effective solution [Burke (2007)]. This is probably the reason why almost all real-world recommender systems employ hybridization techniques in order to increase system robustness.

8.6. Conclusions and Future Directions

As was mentioned previously, collaborative filtering suffer from the cold start problem which occurs when the system is not capable of identifying items that could be recommended to a new user or users that may receive a recommendation for a new item. Moreover, when the system has collected only a small percentage of all the potential ratings, the rating sparsity problem may be observed. These problems are even worse in Context-Aware Recommender Systems (CARSs) where the recommendations can be requested for contextual situations under which the users did not rated any item.

Various approaches for addressing the cold start and data sparsity problems and improving collaborative filtering have been proposed. These approaches can be split into two major groups. Solutions in the first group exploit additional knowledge sources about the users, the items or the contextual situations, and incorporates active learning (AL), cross-domain, implicit feedback, content-based and demographic-based approaches. While solutions in the second group leverage hybrid approaches that try to make a better use of the available information by combining two or more recommendation techniques.

However, no solution is fully solving the considered problem and each of them has drawbacks. For instance, one cannot leverage implicit feedback data in the new user scenario, because even this type of user/item interactions are missing. Similar limitations can be found for the content-based approach where a sufficient number of ratings have to be collected before the system can understand the user preferences and provide accurate recommendations. The demographic-based approaches are easier to implement but less accurate than personalized recommendations, and the cross-domain approaches require the co-existence of users, items and contextual situations across different domains, which is not always the case. Meanwhile, users do not like to enter ratings requested by the system, as this is tedious and takes time and effort. Finally, only basic active learning and hybridization techniques are present in the existing tools and frameworks for building RSs. Thus, by taking into account all these aspects, we would like to close this chapter by briefly indicating possible future directions for addressing the cold start and data sparsity problems.

As IT-technology gets more advanced, it becomes possible to collect and process more information from different sources. For example, "Internet-of-Things" devices, are capable of sensing their environment and collect users' contextual and behavioural information [Tu *et al.* (2016)]. The availability of data of different types (collected from various sources) entails the creation of more advanced hybrid approaches.

Another direction is related to the recent development of artificial intelligence and its application in dialog systems (e.g., chatbots or smart assistants). This type of technologies can transform conversational recommender systems from a "bothersome process" to an enjoyable one, satisfying both the system objectives and the user at the same time [Rubens *et al.* (2015)].

Finally, the efficient implementation of the advanced hybridization and active learning techniques in the existing systems tools and frameworks may stimulate the research community to find even better solutions to solve to the cold start and data sparsity problems.

References

Abel, F., Deldjoo, Y., Elahi, M. and Kohlsdorf, D. (2017). Recsys challenge 2017: Offline and online evaluation, in *Proceedings of the Eleventh ACM Conference on Recommender Systems* (ACM), pp. 372–373.

Adomavicius, G., Sankaranarayanan, R., Sen, S. and Tuzhilin, A. (2005). Incorporating contextual information in recommender systems using a multidimensional approach, *ACM Transactions on Information Systems (TOIS)* **23**, 1, pp. 103–145.

Adomavicius, G. and Tuzhilin, A. (2005). Toward the next generation of recommender systems: A survey of the state-of-the-art and possible extensions, *IEEE Trans. on Knowl. and Data Eng.* **17**, 6, pp. 734–749, doi: 10.1109/TKDE.2005.99.

Adomavicius, G. and Tuzhilin, A. (2011). Context-aware recommender systems, in *Recommender Systems Handbook* (Springer), pp. 217–253.

Adomavicius, G. and Tuzhilin, A. (2015). Context-aware recommender systems, in *Recommender Systems Handbook* (Springer), pp. 191–226, doi:10.1007/978-1-4899-7637-6_6.

Agarwal, D. and Chen, B.-C. (2009). Regression-based latent factor models, in *Proceedings of the 15th ACM SIGKDD International Conference on Knowledge Discovery and Data Mining*, KDD '09 (ACM, New York, NY, USA), ISBN 978-1-60558-495-9, pp. 19–28, doi:10.1145/1557019.1557029, `http://doi.acm.org/10.1145/1557019.1557029`.

Bachrach, Y., Kosinski, M., Graepel, T., Kohli, P. and Stillwell, D. (2012). Personality and patterns of facebook usage, in *Proceedings of the 4th Annual ACM Web Science Conference*, WebSci '12 (ACM, New York, NY, USA), ISBN 978-1-4503-1228-8, pp. 24–32, doi:10.1145/2380718.2380722, `http://doi.acm.org/10.1145/2380718.2380722`.

Baltrunas, L. (2011). *Context-Aware Collaborative Filtering Recommender Systems*, Ph.D. thesis, The Department of Computer Science, Free University of Bozen-Bolzano.

Baltrunas, L., Kaminskas, M., Ludwig, B., Moling, O., Ricci, F., Aydin, A., Lüke, K.-H. and Schwaiger, R. (2011). Incarmusic: Context-aware music recommendations in a car, in *International Conference on Electronic Commerce and Web Technologies* (Springer), pp. 89–100.

Baltrunas, L., Ludwig, B., Peer, S. and Ricci, F. (2012). Context relevance assessment and exploitation in mobile recommender systems, *Personal Ubiquitous Comput.* **16**, 5, pp. 507–526, doi:10.1007/s00779-011-0417-x, `http://dx.doi.org/10.1007/s00779-011-0417-x`.

Berkovsky, S., Kuflik, T. and Ricci, F. (2007). Distributed collaborative filtering with domain specialization, in *Proceedings of the 2007 ACM Conference on Recommender Systems*, RecSys '07 (ACM, New York, NY, USA), ISBN 978-1-59593-730–8, pp. 33–40, doi:10.1145/1297231.1297238, `http://doi.acm.org/10.1145/1297231.1297238`.

Blédaité, L. and Ricci, F. (2015). Pairwise preferences elicitation and exploitation for conversational collaborative filtering, in *Proceedings of the 26th ACM Conference on Hypertext & Social Media*, HT '15 (ACM, New York, NY, USA), ISBN 978-1-4503-3395-5, pp. 231–236, doi:10.1145/2700171.2791049, `http://doi.acm.org/10.1145/2700171.2791049`.

Braunhofer, M. (2015). *Techniques for Context-Aware and Cold-Start Recommendations*, Ph.D. thesis, Free University of Bozen-Bolzano.

Braunhofer, M., Elahi, M. and Ricci, F. (2014). Techniques for cold-starting context-aware mobile recommender systems for tourism, *Intelligenza Artificiale* **8**, 2, pp. 129–143, doi:10.3233/IA-140069.

Braunhofer, M., Elahi, M. and Ricci, F. (2015a). User personality and the new user problem in a context-aware point of interest recommender system, in *Information and Communication Technologies in Tourism 2015* (Springer), pp. 537–549.

Braunhofer, M., Fernández-Tobías, I. and Ricci, F. (2015b). Parsimonious and adaptive contextual information acquisition in recommender systems, in *IntRS@RecSys, CEUR Workshop Proceedings*, Vol. 1438 (CEUR-WS.org), pp. 2–8.

Braunhofer, M. and Ricci, F. (2016). Contextual information elicitation in travel recommender systems, in *Information and Communication Technologies in Tourism 2016* (Springer), pp. 579–592.

Breese, J. S., Heckerman, D. and Kadie, C. (1998). Empirical analysis of predictive algorithms for collaborative filtering, in *Proceedings of the 14th Conference on Uncertainty in Artificial Intelligence*, (Morgan Kaufmann Publishers Inc), pp. 43–52.

Burke, R. (2000). Knowledge-based recommender systems, in *Encyclopedia of library and information science*, 32 (CRC Press), pp. 181–201.

Burke, R. (2002). Hybrid recommender systems: Survey and experiments, *User Modeling and User-Adapted Interaction* **12**, 4, pp. 331–370, doi:10.1023/A:1021240730564, `http://dx.doi.org/10.1023/A:1021240730564`.

Burke, R. (2007). The adaptive web, in P. Brusilovsky, A. Kobsa and W. Nejdl (eds.), *The Adaptive Web: Methods and Strategies of Web Personalization*, chap. Hybrid Web Recommender Systems (Springer-Verlag, Berlin, Heidelberg), ISBN 978-3-540-72078-2, pp. 377–408, `http://dl.acm.org/citation.cfm?id=1768197.1768211`.

Cantador, I. and Cremonesi, P. (2014). Tutorial on cross-domain recommender systems, in *Proceedings of the 8th ACM Conference on Recommender Systems*, RecSys'14 (ACM, New York, NY, USA), ISBN 978-1-4503-2668-1, pp. 401–402, doi:10.1145/2645710.2645777.

Carenini, G., Smith, J. and Poole, D. (2003). Towards more conversational and collaborative recommender systems, in *Proceedings of the 8th international conference on Intelligent user interfaces*, IUI '03 (ACM, New York, NY, USA), ISBN 1-58113-586-6, pp. 12–18, doi:10.1145/604045.604052.

Christakopoulou, K., Radlinski, F. and Hofmann, K. (2016). Towards conversational recommender systems, in *Proceedings of the 22Nd ACM SIGKDD International Conference on Knowledge Discovery and Data Mining*, KDD '16 (ACM, New York, NY, USA), ISBN 978-1-4503-4232-2, pp. 815–824, doi:10.1145/2939672.2939746, `http://doi.acm.org/10.1145/2939672.2939746`.

Çoba, L. and Zanker, M. (2016). rrecsys: An r-package for prototyping recommendation algorithms, in *RecSys Posters*.

Codina, V., Ricci, F. and Ceccaroni, L. (2013). Exploiting the semantic similarity of contextual situations for pre-filtering recommendation, in *User Modeling, Adaptation, and Personalization* (Springer), pp. 165–177.

Cremonesi, P., Tripodi, A. and Turrin, R. (2011). Cross-domain recommender systems, in *Data Mining Workshops (ICDMW), 2011 IEEE 11th International Conference on* (IEEE), pp. 496–503.

Deldjoo, Y., Elahi, M., Cremonesi, P., Garzotto, F., Piazzolla, P. and Quadrana, M. (2016). Content-based video recommendation system based on stylistic visual features, *Journal on Data Semantics*, pp. 1–15, doi:10.1007/s13740-016-0060-9.

Deldjoo, Y., Elahi, M., Quadrana, M. and Cremonesi, P. (2018). Using visual features based on mpeg-7 and deep learning for movie recommendation, *International Journal of Multimedia Information Retrieval*.

Deshpande, M. and Karypis, G. (2004). Item-based top-n recommendation algorithms, *ACM Trans. Inf. Syst.* **22**, 1, pp. 143–177, doi:10.1145/963770.963776, `http://doi.acm.org/10.1145/963770.963776`.

Ekstrand, M. D., Harper, F. M., Willemsen, M. C. and Konstan, J. A. (2014). User perception of differences in recommender algorithms, in *Proceedings of the 8th ACM Conference on Recommender systems* (ACM), pp. 161–168.

Elahi, M., Braunhofer, M., Ricci, F. and Tkalcic, M. (2013). Personality-based active learning for collaborative filtering recommender systems, in *Congress of the Italian Association for Artificial Intelligence* (Springer), pp. 360–371.

Elahi, M., Deldjoo, Y., Bakhshandegan Moghaddam, F., Cella, L., Cereda, S. and Cremonesi, P. (2017). Exploring the semantic gap for movie recommendations, in *Proceedings of the Eleventh ACM Conference on Recommender Systems*, RecSys '17 (ACM, New York, NY, USA), ISBN 978-1-

4503-4652-8, pp. 326–330, doi:10.1145/3109859.3109908, http://doi.acm.org/10.1145/3109859.3109908.

Elahi, M., Ricci, F. and Repsys, V. (2011). System-wide effectiveness of active learning in collaborative filtering, in *Proceedings of the International Workshop on Social Web Mining, Co-located with IJCAI, Barcelona, Spain (July 2011)*.

Elahi, M., Ricci, F. and Rubens, N. (2012). Adapting to natural rating acquisition with combined active learning strategies, in *ISMIS'12: Proceedings of the 20th international conference on Foundations of Intelligent Systems* (Springer-Verlag, Berlin, Heidelberg), ISBN 978-3-642-34623-1, pp. 254–263, doi:http://dx.doi.org/10.1007/978-3-642-34624-8_30.

Elahi, M., Ricci, F. and Rubens, N. (2014). Active learning strategies for rating elicitation in collaborative filtering: A system-wide perspective, *ACM Transactions on Intelligent Systems and Technology* **5**, 1, pp. 13:1–13:33, doi:10.1145/2542182.2542195.

Elahi, M., Ricci, F. and Rubens, N. (2016). A survey of active learning in collaborative filtering recommender systems, *Comput. Sci. Rev.* **20**, C, pp. 29–50, doi:10.1016/j.cosrev.2016.05.002, http://dx.doi.org/10.1016/j.cosrev.2016.05.002.

Enrich, M., Braunhofer, M. and Ricci, F. (2013). Cold-start management with cross-domain collaborative filtering and tags, in *E-Commerce and Web Technologies - 14th International Conference, EC-Web 2013, Prague, Czech Republic, August 27-28, 2013. Proceedings*, pp. 101–112.

Fernandez Tobias, I., Braunhofer, M., Elahi, M., Ricci, F. and Ivan, C. (2016). Alleviating the new user problem in collaborative filtering by exploiting personality information, *User Modeling and User-Adapted Interaction (UMUAI)* **26**, Personality in Personalized Systems, doi:10.1007/s11257-016-9172-z.

Fernández-Tobías, I. and Cantador, I. (2014). Exploiting social tags in matrix factorization models for cross-domain collaborative filtering, in *CBRecSys@RecSys*, pp. 34–41.

Fernández-Tobías, I., Tomeo, P., Cantador, I., Di Noia, T. and Di Sciascio, E. (2016). Accuracy and diversity in cross-domain recommendations for cold-start users with positive-only feedback, in *Proceedings of the 10th ACM Conference on Recommender Systems*, RecSys '16 (ACM, New York, NY, USA), ISBN 978-1-4503-4035-9, pp. 119–122, doi:10.1145/2959100.2959175, http://doi.acm.org/10.1145/2959100.2959175.

Ge, M., Elahi, M., Fernaández-Tobías, I., Ricci, F. and Massimo, D. (2015). Using tags and latent factors in a food recommender system, in *Proceedings of the 5th International Conference on Digital Health 2015*, DH '15 (ACM, New York, NY, USA), ISBN 978-1-4503-3492-1, pp. 105–112, doi:10.1145/2750511.2750528.

GroupLens (2018). Eachmovie, https://grouplens.org/datasets/eachmovie/.

Guo, G., Zhang, J., Sun, Z. and Yorke-Smith, N. (2015). Librec: A java library for recommender systems, in *UMAP Workshops*.

Gurbanov, T. and Ricci, F. (2017). Action prediction models for recommender systems based on collaborative filtering and sequence mining hybridization, in *Proceedings of the Symposium on Applied Computing*, SAC '17 (ACM, New York, NY, USA), ISBN 978-1-4503-4486-9, pp. 1655–1661, doi:10.1145/3019612.3019759, http://doi.acm.org/10.1145/3019612.3019759.

Gurbanov, T., Ricci, F. and Ploner, M. (2016). Modeling and predicting user actions in recommender systems, in *Proceedings of the 2016 Conference on User Modeling Adaptation and Personalization*, UMAP '16 (ACM, New York, NY, USA), ISBN 978-1-4503-4368-8, pp. 151–155, doi:10.1145/2930238.2930284, http://doi.acm.org/10.1145/2930238.2930284.

Harper, F. M. and Konstan, J. A. (2015). The movielens datasets: History and context, *ACM Trans. Interact. Intell. Syst.* **5**, 4, pp. 19:1–19:19, doi:10.1145/2827872, http://doi.acm.org/10.1145/2827872.

Harper, F. M. and Konstan, J. A. (2016). The movielens datasets: History and context, *ACM Transactions on Interactive Intelligent Systems (TiiS)* **5**, 4, p. 19.

Hawalah, A. and Fasli, M. (2015). Dynamic user profiles for web personalisation, *Expert Syst. Appl.* **42**, 5, pp. 2547–2569, doi:10.1016/j.eswa.2014.10.032, http://dx.doi.org/10.1016/j.eswa.2014.10.032.

He, R. and McAuley, J. (2016). Ups and downs: Modeling the visual evolution of fashion trends with one-class collaborative filtering, in *Proceedings of the 25th International Conference on World Wide Web*, WWW '16 (International World Wide Web Conferences Steering Committee, Republic and Canton of Geneva, Switzerland), ISBN 978-1-4503-4143-1, pp. 507–517, doi:10.1145/2872427.2883037, https://doi.org/10.1145/2872427.2883037.

Herlocker, J. L., Konstan, J. A., Borchers, A. and Riedl, J. (1999). An algorithmic framework for performing collaborative filtering, in *Proceedings of the 22nd Annual International ACM SIGIR Conference on Research and Development in Information Retrieval*, SIGIR '99 (ACM, New York, NY, USA), ISBN 1-58113-096-1, pp. 230–237, doi:10.1145/312624.312682, http://doi.acm.org/10.1145/312624.312682.

Hu, R. and Pu, P. (2009). A comparative user study on rating vs. personality quiz based preference elicitation methods, in *Proceedings of the 14th international conference on Intelligent user interfaces*, IUI '09 (ACM, New York, NY, USA), ISBN 978-1-60558-168-2, pp. 367–372, doi:10.1145/1502650.1502702.

Hu, R. and Pu, P. (2011). Enhancing collaborative filtering systems with personality information, in *Proceedings of the fifth ACM conference on Recommender systems*, RecSys '11 (ACM, New York, NY, USA), ISBN 978-1-4503-0683-6, pp. 197–204, doi:10.1145/2043932.2043969.

Hu, Y., Koren, Y. and Volinsky, C. (2008a). Collaborative filtering for implicit feedback datasets, in *Proceedings of the 2008 Eighth IEEE International Conference on Data Mining*, ICDM '08 (IEEE Computer Society, Washington, DC, USA), ISBN 978-0-7695-3502-9, pp. 263–272, doi:10.1109/ICDM.2008.22, http://dx.doi.org/10.1109/ICDM.2008.22.

Hu, Y., Koren, Y. and Volinsky, C. (2008b). Collaborative filtering for implicit feedback datasets, in *Proceedings of the 8th IEEE International Conference on Data Mining* (IEEE), pp. 263–272.

IMDB (2018). Internet movie database: Movies, tv and celebrities, http://www.imdb.com/.

Isinkaye, F., Folajimi, Y. and Ojokoh, B. (2015). Recommendation systems: Principles, methods and evaluation, *Egyptian Informatics Journal* **16**, 3, pp. 261–273.

Jones, N., Brun, A. and Boyer, A. (2011). Comparisons instead of ratings: Towards more stable preferences, in *Proceedings of the 2011 IEEE/WIC/ACM International Conferences on Web Intelligence and Intelligent Agent Technology - Volume 01*, WI-IAT '11 (IEEE Computer Society, Washington, DC, USA), ISBN 978-0-7695-4513-4, pp. 451–456, doi:10.1109/WI-IAT.2011.13, http://dx.doi.org/10.1109/WI-IAT.2011.13.

Kalloori, S., Ricci, F. and Tkalcic, M. (2016). Pairwise preferences based matrix factorization and nearest neighbor recommendation techniques, in *Proceedings of the 10th ACM Conference on Recommender Systems*, RecSys '16 (ACM, New York, NY, USA), ISBN 978-1-4503-4035-9, pp. 143–146, doi:10.1145/2959100.2959142, http://doi.acm.org/10.1145/2959100.2959142.

Kim, D., Park, C., Oh, J., Lee, S. and Yu, H. (2016). Convolutional matrix factorization for document context-aware recommendation, in *Proceedings of the 10th ACM Conference on Recommender Systems*, RecSys '16 (ACM, New York, NY, USA), ISBN 978-1-4503-4035-9, pp. 233–240, doi:10.1145/2959100.2959165, http://doi.acm.org/10.1145/2959100.2959165.

Koren, Y. (2008). Factorization meets the neighborhood: A multifaceted collaborative filtering model, in *Proceedings of the 14th ACM SIGKDD International Conference on Knowledge Discovery and Data Mining*, KDD '08 (ACM, New York, NY, USA), ISBN 978-1-60558-193-4, pp. 426–434, doi:10.1145/1401890.1401944, http://doi.acm.org/10.1145/1401890.1401944.

Koren, Y. (2009). The bellkor solution to the netflix grand prize, *Netflix prize documentation* **81**, pp. 1–10.

Koren, Y. and Bell, R. (2011). *Advances in Collaborative Filtering* (Springer US, Boston, MA), ISBN 978-0-387-85820-3, pp. 145–186, doi:10.1007/978-0-387-85820-3_5, https://doi.org/10.1007/978-0-387-85820-3_5.

Koren, Y. and Bell, R. (2015). Advances in collaborative filtering, in *Recommender Systems Handbook* (Springer), pp. 77–118, doi:10.1007/978-1-4899-7637-6_3.

Koren, Y., Bell, R. and Volinsky, C. (2009). Matrix factorization techniques for recommender systems, *Computer* **42**, 8, pp. 30–37.

Kosinski, M., Matz, S. C., Gosling, S. D., Popov, V. and Stillwell, D. (2015). Facebook as a research tool for the social sciences: Opportunities, challenges, ethical considerations, and practical guidelines, *American Psychologist* **70**, 6, p. 543.

Košir, A., Odic, A., Kunaver, M., Tkalcic, M. and Tasic, J. F. (2011). Database for contextual personalization, *Elektrotehniški vestnik* **78**, 5, pp. 270–274.

Librarything (2018). Catalog your books online 2018, `https://www.librarything.com/`.
Linden, G., Smith, B. and York, J. (2003). Amazon.com recommendations: Item-to-item collaborative filtering, *IEEE Internet Computing* **7**, 1, pp. 76–80, doi:10.1109/MIC.2003.1167344.
Loepp, B., Hussein, T. and Ziegler, J. (2014). Choice-based preference elicitation for collaborative filtering recommender systems, in *Proceedings of the SIGCHI Conference on Human Factors in Computing Systems* (ACM), pp. 3085–3094.
Manzato, M. G. (2012). Discovering latent factors from movies genres for enhanced recommendation, in *Proceedings of the Sixth ACM Conference on Recommender Systems*, RecSys '12 (ACM, New York, NY, USA), ISBN 978-1-4503-1270-7, pp. 249–252, doi:10.1145/2365952.2366006, `http://doi.acm.org/10.1145/2365952.2366006`.
Manzato, M. G. (2013). gsvd++: Supporting implicit feedback on recommender systems with metadata awareness, in *Proceedings of the 28th Annual ACM Symposium on Applied Computing*, SAC '13 (ACM, New York, NY, USA), ISBN 978-1-4503-1656-9, pp. 908–913, doi:10.1145/2480362.2480536, `http://doi.acm.org/10.1145/2480362.2480536`.
Massimo, D., Elahi, M., Ge, M. and Ricci, F. (2017). Item contents good, user tags better: Empirical evaluation of a food recommender system, in *Proceedings of the 25th Conference on User Modeling, Adaptation and Personalization* (ACM), pp. 373–374.
McNee, S. M., Lam, S. K., Konstan, J. A. and Riedl, J. (2003). Interfaces for eliciting new user preferences in recommender systems, in *Proceedings of the 9th international conference on User modeling*, UM '03 (Springer-Verlag, Berlin, Heidelberg), ISBN 3-540-40381-7, pp. 178–187, `http://dl.acm.org/citation.cfm?id=1759957.1759988`.
Neidhardt, J., Schuster, R., Seyfang, L. and Werthner, H. (2014). Eliciting the users' unknown preferences, in *Proceedings of the 8th ACM Conference on Recommender systems* (ACM), pp. 309–312.
Nicholas, I. S. C. and Nicholas, C. K. (1999). Combining content and collaboration in text filtering, in *In Proceedings of the IJCAI'99 Workshop on Machine Learning for Information Filtering*, pp. 86–91.
Oard, D. W., Kim, J. *et al.* (1998). Implicit feedback for recommender systems, in *Proceedings of the AAAI workshop on recommender systems*, pp. 81–83.
Odić, A., Tkalčič, M., Tasič, J. F. and Košir, A. (2012). Relevant context in a movie recommender system: Users' opinion vs. statistical detection, *ACM RecSys 2012, Proceedings of the 4th International Workshop on Context-Aware Recommender Systems (CARS 2012)*.
Odić, A., Tkalčič, M., Tasič, J. F. and Košir, A. (2013). Predicting and detecting the relevant contextual information in a movie-recommender system, *Interacting with Computers*, p. iws003.
Pan, R., Zhou, Y., Cao, B., Liu, N. N., Lukose, R., Scholz, M. and Yang, Q. (2008). One-class collaborative filtering, in *Proceedings of the 8th IEEE International Conference on Data Mining* (IEEE), pp. 502–511.

Park, D. H., Kim, H. K., Choi, I. Y. and Kim, J. K. (2012). A literature review and classification of recommender systems research, *Expert Systems with Applications* **39**, 11, pp. 10059–10072.

Picault, J. and Ribiere, M. (2008). An empirical user profile adaptation mechanism that reacts to shifts of interests, in *Proceedings of the European conference in artificial intelligence (ECAI)*.

Picault, J., Ribiere, M., Bonnefoy, D. and Mercer, K. (2011). How to get the recommender out of the lab? in *Recommender Systems Handbook* (Springer), pp. 333–365.

Pu, P., Chen, L. and Hu, R. (2012). Evaluating recommender systems from the user's perspective: survey of the state of the art, *User Modeling and User-Adapted Interaction* **22**, 4-5, pp. 317–355, doi:10.1007/s11257-011-9115-7.

Rashid, A. M., Albert, I., Cosley, D., Lam, S. K., McNee, S. M., Konstan, J. A. and Riedl, J. (2002). Getting to know you: Learning new user preferences in recommender systems, in *Proceedings of the 7th International Conference on Intelligent User Interfaces*, IUI '02 (ACM, New York, NY, USA), ISBN 1-58113-459-2, pp. 127–134, doi:10.1145/502716.502737, `http://doi.acm.org/10.1145/502716.502737`.

Rashid, A. M., Karypis, G. and Riedl, J. (2008a). Learning preferences of new users in recommender systems: An information theoretic approach, *SIGKDD Explor. Newsl.* **10**, 2, pp. 90–100, doi:10.1145/1540276.1540302, `http://doi.acm.org/10.1145/1540276.1540302`.

Rashid, A. M., Karypis, G. and Riedl, J. (2008b). Learning preferences of new users in recommender systems: an information theoretic approach, *SIGKDD Explor. Newsl.* **10**, pp. 90–100, doi:10.1145/1540276.1540302.

Rendle, S., Freudenthaler, C., Gantner, Z. and Schmidt-Thieme, L. (2009). Bpr: Bayesian personalized ranking from implicit feedback, in *Proceedings of the twenty-fifth conference on uncertainty in artificial intelligence* (AUAI Press), pp. 452–461.

Resnick, P., Iacovou, N., Suchak, M., Bergstrom, P. and Riedl, J. (1994). Grouplens: An open architecture for collaborative filtering of netnews, in *Proceedings of the 1994 ACM Conference on Computer Supported Cooperative Work*, CSCW '94 (ACM, New York, NY, USA), ISBN 0-89791-689-1, pp. 175–186, doi:10.1145/192844.192905, `http://doi.acm.org/10.1145/192844.192905`.

Ricci, F., Rokach, L. and Shapira, B. (2015). Recommender systems: Introduction and challenges, in *Recommender Systems Handbook* (Springer US), pp. 1–34, doi:10.1007/978-1-4899-7637-6_1.

Rubens, N., Elahi, M., Sugiyama, M. and Kaplan, D. (2015). Active learning in recommender systems, in *Recommender Systems Handbook - chapter 24: Recommending Active Learning* (Springer US), pp. 809–846, doi:10.1007/978-1-4899-7637-6_24.

Sarwar, B., Karypis, G., Konstan, J. and Riedl, J. (2001). Item-based collaborative filtering recommendation algorithms, in *Proceedings of the 10th international conference on World Wide Web* (ACM), pp. 285–295.

Schedl, M., Zamani, H., Chen, C.-W., Deldjoo, Y. and Elahi, M. (2018). Current challenges and visions in music recommender systems research, *International Journal of Multimedia Information Retrieval* doi:10.1007/s13735-018-0154-2, https://doi.org/10.1007/s13735-018-0154-2.

Schein, A. I., Popescul, A., Ungar, L. H. and Pennock, D. M. (2002). Methods and metrics for cold-start recommendations, in *SIGIR '02: Proceedings of the 25th annual international ACM SIGIR conference on Research and development in information retrieval* (ACM, New York, NY, USA), ISBN 1-58113-561-0, pp. 253–260, doi:10.1145/564376.564421.

Shardanand, U. and Maes, P. (1995). Social information filtering: Algorithms for automating “word of mouth”, in *Proceedings of the SIGCHI Conference on Human Factors in Computing Systems*, CHI '95 (ACM Press/Addison-Wesley Publishing Co., New York, NY, USA), ISBN 0-201-84705-1, pp. 210–217, doi:10.1145/223904.223931, http://dx.doi.org/10.1145/223904.223931.

Shi, Y., Larson, M. and Hanjalic, A. (2014). Collaborative filtering beyond the user-item matrix: A survey of the state of the art and future challenges, *ACM Comput. Surv.* **47**, 1, pp. 3:1–3:45, doi:10.1145/2556270, http://doi.acm.org/10.1145/2556270.

Stern, D. H., Herbrich, R. and Graepel, T. (2009). Matchbox: Large scale online bayesian recommendations, in *Proceedings of the 18th International Conference on World Wide Web*, WWW '09 (ACM, New York, NY, USA), ISBN 978-1-60558-487-4, pp. 111–120, doi:10.1145/1526709.1526725, http://doi.acm.org/10.1145/1526709.1526725.

Su, X. and Khoshgoftaar, T. M. (2009). A survey of collaborative filtering techniques, *Adv. in Artif. Intell.* **2009**, pp. 4:2–4:2, doi:10.1155/2009/421425, http://dx.doi.org/10.1155/2009/421425.

Sun, M., Li, F., Lee, J., Zhou, K., Lebanon, G. and Zha, H. (2013). Learning multiple-question decision trees for cold-start recommendation, in *Proceedings of the Sixth ACM International Conference on Web Search and Data Mining*, WSDM '13 (ACM, New York, NY, USA), ISBN 978-1-4503-1869-3, pp. 445–454, doi:10.1145/2433396.2433451.

Tkalcic, M., Kosir, A. and Tasic, J. (2013). The ldos-peraff-1 corpus of facial-expression video clips with affective, personality and user-interaction metadata, *Journal on Multimodal User Interfaces* **7**, 1-2, pp. 143–155, doi:10.1007/s12193-012-0107-7.

Trevisiol, M., Aiello, L. M., Schifanella, R. and Jaimes, A. (2014). Cold-start news recommendation with domain-dependent browse graph, in *Proceedings of the 8th ACM Conference on Recommender Systems*, RecSys '14 (ACM, New York, NY, USA), ISBN 978-1-4503-2668-1, pp. 81–88, doi:10.1145/2645710.2645726.

TripAdvisor (2018). Read reviews, compare prices and book, 2018, https://www.tripadvisor.com/.

Tu, M., Chang, Y.-K. and Chen, Y.-T. (2016). A context-aware recommender system framework for iot based interactive digital signage in urban space, in *Proceedings of the Second International Conference on IoT in Urban*

Space, Urb-IoT '16 (ACM, New York, NY, USA), ISBN 978-1-4503-4204-9, pp. 39–42, doi:10.1145/2962735.2962736, http://doi.acm.org/10.1145/2962735.2962736.

Vargas-Govea, B., González-Serna, G. and Ponce-Medellın, R. (2011). Effects of relevant contextual features in the performance of a restaurant recommender system, *ACM RecSys* **11**.

Vartak, M., Thiagarajan, A., Miranda, C., Bratman, J. and Larochelle, H. (2017). A meta-learning perspective on cold-start recommendations for items, in *Advances in Neural Information Processing Systems*, pp. 6907–6917.

Vozalis, M. and Margaritis, K. G. (2004). Collaborative filtering enhanced by demographic correlation, in *in Proceedings of the AIAI Symposium on Professional Practice in AI, part of the 18th World Computer Congress*, pp. 293–402.

Zhao, X., Zhang, W. and Wang, J. (2013). Interactive collaborative filtering, in *Proceedings of the 22nd ACM international conference on Conference on information & knowledge management*, CIKM '13 (ACM, New York, NY, USA), ISBN 978-1-4503-2263-8, pp. 1411–1420, doi:10.1145/2505515.2505690, http://doi.acm.org/10.1145/2505515.2505690.

Zheng, Y., Burke, R. and Mobasher, B. (2012). Differential context relaxation for context-aware travel recommendation, in *E-Commerce and Web Technologies* (Springer), pp. 88–99.

Zheng, Y., Burke, R. and Mobasher, B. (2013). Recommendation with differential context weighting, in *User Modeling, Adaptation, and Personalization* (Springer), pp. 152–164.

Zheng, Y., Mobasher, B. and Burke, R. (2014). Cslim: Contextual slim recommendation algorithms, in *Proceedings of the 8th ACM Conference on Recommender Systems*, RecSys '14 (ACM, New York, NY, USA), ISBN 978-1-4503-2668-1, pp. 301–304, doi:10.1145/2645710.2645756, http://doi.acm.org/10.1145/2645710.2645756.

Zheng, Y., Mobasher, B. and Burke, R. (2015). Integrating context similarity with sparse linear recommendation model, in *User Modeling, Adaptation and Personalization* (Springer), pp. 370–376.

Chapter 9

Offline and Online Evaluation of Recommendations

Alejandro Bellogín[†] and Alan Said[‡]
[†] *Universidad Autónoma de Madrid, Madrid, Spain*
alejandro.bellogin@uam.es
[‡] *University of Skövde, Skövde, Sweden*
alansaid@acm.org

Recommender Systems have been a popular research topic within personalized systems and information retrieval since the mid nineties. Throughout this time, various models of recommendation have been developed, e.g., approaches using collaborative filtering for purposes such as retrieval of ranked lists of items for consumption, or for the rating prediction task (which was made very popular through the Netflix prize). Today, the use of recommender systems has spread to a very wide area of topics, including personalized healthcare, online news portals, food, social networks, exercise, jobs, investment, transportation, shopping, etc. Given the various situations recommendations can be applied to, it follows that evaluation of these systems needs to be tailored to the specific setting, domain, user-base, context, etc. This chapter aims to give an overview of some of the more commonly used evaluation methods and metrics used for various types of recommendation techniques. We also provide a summary of the available resources in this topic, in addition to some practical considerations, experimental results, and future directions about evaluation in recommendation.

9.1. Introduction

The evaluation of Recommender Systems (RS) has been, and still is, the subject of active research in the field, where open questions remain [Herlocker *et al.* (2004); Gunawardana and Shani (2015)]. Since the advent of the first RS, recommendation performance has been usually equated to the accuracy of rating prediction, that is, estimated ratings are compared against actual ratings, and differences between them are computed by means of error-based metrics such as the Mean Absolute Error (MAE) or Root Mean Squared Error (RMSE). In terms of the effective utility of recommendations for users, there is however an increasing realization that the quality (precision) of a ranking of recommended items can be more important than the accuracy in predicting specific rating values [Knijnenburg and Willemsen (2015)]. As a result, precision-oriented metrics are being increasingly considered in the field, and a large amount of recent work has focused on evaluating top-N ranked recommendation lists with the above type of metrics. Precision in this context can be interpreted as the relevance of the recommended item, i.e. the likelihood of the item being watched, liked, or otherwise consumed by the user.

Nonetheless, the recent realization that high prediction accuracy might not translate to a higher perceived appreciation from the users has brought a plethora of novel metrics and methods, focusing on other aspects of recommendation [Said *et al.* (2013a); Castells *et al.* (2015)]. More specifically, other dimensions apart from accuracy — such as coverage, diversity, novelty, and serendipity — have been recently taken into account and analyzed when considered what makes a good recommendation [Said *et al.* (2014b); Cremonesi *et al.* (2011); McNee *et al.* (2006); Bollen *et al.* (2010); Mesas and Bellogín (2017)].

With this in mind comes the understanding that evaluation is the key to identifying how well an algorithm or a system works. Deploying a new algorithm in a new system will have an effect on the overall performance of the system — in terms of accuracy and other types of metrics. Both prior deploying the algorithm, and after the deployment, it is important to evaluate the system performance. However, the evaluation strategies, metrics, and methodologies need to consider that the use of RSs has spread to a very wide area of topics, including personalized healthcare [Elsweiler *et al.* (2015); Luo *et al.* (2016)], online news portals [Said *et al.* (2014a)], food [Elahi *et al.* (2015, 2014)], social networks [Guy (2015)], exercise [Berkovsky *et al.* (2012)], jobs [Abel (2015)], investment [Zhao *et al.* (2015)], transportation [Bistaffa *et al.* (2015)], shopping [Jannach *et al.* (2015)], etc. and adapt specifically to each use-case.

The rest of the chapter is organized as follows. Section 9.1.1 defines basic concepts used when evaluating recommender systems, then, Sec. 9.1.2 and Sec. 9.1.3 present in more detail the specifics of offline and online evaluation, and Sec. 9.2 provides some algorithmic solutions specifically devoted for evaluation in the areas described in the first part of this book. Then, Sec. 9.3 shows resources — such as datasets and libraries — currently available to perform RS evaluation. Section 9.4 presents experimental results where we show how the different evaluation methodologies and metrics compare against each other. And, finally, in Sec. 9.5 and Sec. 9.6 we discuss some practical considerations about the problem of RS evaluation, and include future directions worth of further involvement from the community.

9.1.1. *Basic Concepts in Evaluation*

The evaluation of RSs has been a major object of study in the field since its earliest days, and it is still a topic of ongoing research, where open questions remain [Herlocker *et al.* (2004); Gunawardana and Shani (2015)]. It is acknowledged that the evaluation of RSs should take into account the goal of the system itself. For example, [Herlocker *et al.* (2004)] identify two main user tasks: *annotation in context* and *find good items.* In these tasks the users only care about errors in the item rank order provided by the system, not the predicted rating value itself. Based on this consideration, researchers have started to use precision-based metrics to evaluate recommendations — as we shall see later — although most works also still report error-based metrics for comparison with state of the art approaches. Moreover, other authors [Herlocker *et al.* (2004); Gunawardana and Shani (2015)] encourage considering alternative performance criteria, like the novelty of the suggested items and the item coverage of a recommendation method. Throughout this chapter we describe these types of evaluation metrics.

However, not all the evaluation metrics can be computed under any experimental settings. Different evaluation protocols exist and they impose constraints on the type of data that can be measured and analyzed. Two main evaluation protocols are usually considered: offline and online evaluation. These protocols present a clear tradeoff between effort (time, users, etc.) and usefulness/trustworthiness of the results, which will be discussed in the subsequent sections. We will also briefly introduce how user studies fit in this context, a protocol with some advantages and disadvantages of the two previously mentioned evaluation protocols.

9.1.2. *Offline evaluation*

Offline evaluation allows to compare a wide range of candidate algorithms at a low cost, it is easy to conduct and does not require any interaction with real users. However, user studies and online experiments are more trustworthy — since the system is used by real users and interacted with in real time — but care must be taken to consider biases in the experimental design [Gunawardana and Shani (2015)].

An important decision in the experimental configuration of RS evaluation is the dataset partitioning strategy. How the datasets are partitioned into training and test sets may have a considerable impact on the final performance results, and may cause some recommenders to obtain better or worse results depending on how the partitioning process is configured.

There are several choices to be made when considering offline evaluation, first, we should choose whether or not to take time into account [Gunawardana and Shani (2015)]. Time-based approaches require that user interaction records have timestamps. A simple approach is to select a time point in the available interaction data timeline, and to separate the data at that point. The resulting data subsets can then be used as training data (all interaction records prior to the time point) and test data (all interaction dated after the split time point), a simple example is shown in Figure 9.1 (right). The split point can be set so as to, for instance, have a desired training/test ratio in the experiment or defining a specific time window for the training and test splits [Campos *et al.* (2014)]. The ratio between training and test data can be global, with a single common split point for all users, or user-specific, to ensure the same ratio per user. Time-based approaches have the advantage of more realistically matching working application scenarios, where ' "future"' user likes (which would translate to positive response to recommendations by the system) are to be predicted based on past evidence.

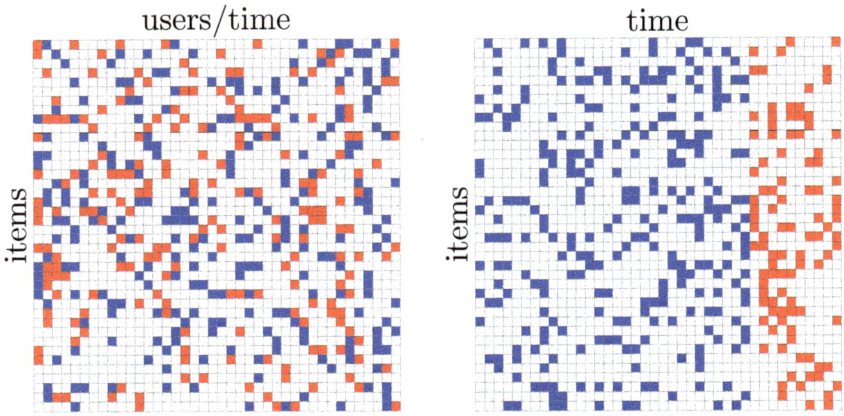

Fig. 9.1. How the dataset would be split into training (blue) and test (red) sets, for a random (left) and temporal (right) split. White cells denote unknown values in the user-item matrix.

In the case when time is not a factor, there are at least the following three strategies to select which items are selected into the training data and which are selected into the test data correspondingly. These are: a) sample a fixed number (different) for each user; b) sample a fixed (but the same for all) number for each user, also known as *given n* or *all but n* protocols; c) sample a percentage of all the interactions using cross-validation. Commonly, the last protocol is used [Goldberg *et al.* (2001)], although several authors have also used the *all but n* protocol [Breese *et al.* (1998)]. Figure 9.1 (left) shows an example of a random dataset partition not taking time into consideration.

Recently, some researchers have started to prune the datasets by creating p-cores, where every user and item has at least p interactions [Jäschke *et al.* (2007)], in order to reduce the inherent sparsity existing in RS datasets. Nonetheless, independently from the dataset partitioning, it is recognized that the goals for which an evaluation is performed may be different for each situation, and thus, a different setting (and partitioning protocol) should be developed [Herlocker *et al.* (2004); Gunawardana and Shani (2015)]. If that is not the case, the results obtained in a particular setting would be biased and difficult to use in further experiments, for instance, in an online experiment.

Regarding the actual evaluation process, there is a relation between the evaluation protocol and the evaluation metrics that can be computed. Error metrics require explicit ground truth values for every evaluated user-item pair — that is, only items in the test set of each user will be considered. Ranked recommendations (using the metrics mentioned before), on the other hand, require for a target user u to select two sets of items, namely relevant and not relevant items. The following candidate generation strategies, where L_u denotes the set of target items the recommender ranks (candidate items), have been proposed (we follow the notation presented in [Said and Bellogín (2014)]):

UserTest (UT) This strategy takes the same target item sets as standard error-based evaluation: for each user u, the list L_u consists of items rated by u in the test set. The smallest set of target items for each user is selected, including no unrated items. A relevance threshold is used to indicate which of the items in the user's test are considered relevant. Threshold variations can be static for all users [Jambor and Wang (2010)], or per-user [Basu *et al.* (1998)].

TrainItems (TI) Every rated item in the system is selected — except those rated by the target user. This strategy is useful when simulating a real system where no test is available, i.e. no need to look into the test set to generate the rankings [Bellogín *et al.* (2011)]. The relevant items for each user consist of those included in her test set; the use of a threshold to consider only highly rated items is optional although recommended.

RelPlusN (RPN) For each user, a set of highly relevant items is selected from the test set. Then, a set of non-relevant items is created by randomly selecting N additional items. In [Cremonesi *et al.* (2010)], N is set to $1,000$ stating that the non-relevant items are selected from items in the test set not rated by u. Finally, for each highly relevant item i, the recommender produces a ranking of the union between this item and the non-relevant items.

As it was observed in [Bellogín *et al.* (2011)] and [Said and Bellogín (2014)], each of these candidate generation strategies may produce a different ranking of recommendation performance — TrainItems and RelPlusN produce consistent results (although with different absolute values) whereas UserTest obtains results closer to those from error-based metrics. Hence, we should pay attention to the strategy used when ranking metrics are computed, since the amount of relevant items considered can drastically change the output of the experiment. In contrast to other fields such Information Retrieval (IR), in RS we have to define training and test sets, whereas in IR, we would have the whole dataset available, first, for the indexing task, and then, for the retrieval and evaluation tasks. In RS, we need to separate the data into training and test; the more training available, the better the algorithm will learn the users' preferences. However, the smaller the test set, the smaller the confidence on the obtained results. This sparsity in the ground truth dimension may produce biases in the evaluation results, as observed in [Bellogín *et al.* (2017)].

Once the splitting and evaluation methodology are decided, we can estimate the performance of the system by computing different evaluation metrics on the results reported by the recommendation algorithm. As mentioned before, depending on the selected methodology, it might not be possible to compute some metrics. More importantly, depending on the algorithm being tested some of the evaluation metrics that we are going to present now cannot be applied.

In the classical formulation of the recommendation problem, user preferences for items are represented as numeric ratings, and the goal of a recommendation algorithm consists of predicting unknown ratings based on known ratings and, in some cases, additional information about users, items, and the context. In this scenario, the accuracy of recommendations has been commonly evaluated by measuring the error between predicted and known ratings, using **error metrics**. Traditionally, the most popular metrics to measure the accuracy of a RS have been the Mean Absolute Error (MAE), and the Root Mean Squared Error (RMSE):

$$\mathrm{MAE} = \frac{1}{|\mathrm{Te}|} \sum_{(u,i)\in \mathrm{Te}} |\tilde{r}(u,i) - r(u,i)| \tag{9.1}$$

$$\mathrm{RMSE} = \sqrt{\frac{1}{|\mathrm{Te}|} \sum_{(u,i)\in \mathrm{Te}} (\tilde{r}(u,i) - r(u,i))^2} \tag{9.2}$$

where $\tilde{r}$ and r denote the predicted and real rating, respectively, and Te corresponds to the test set. The RMSE metric is usually preferred to MAE because it penalizes larger errors.

A critical limitation of these metrics is that they do not make any distinction between the errors made on the top items predicted by a system, and the errors made for the rest of the items. Furthermore, they can only be applied when the recommender predicts a score in the allowed range of rating values. Because of that, implicit and some content-based and probabilistic recommenders cannot be evaluated in this way, since $\tilde{r}(u,i)$ would represent a probability or, in general, a preference score, not a rating.

Although dominant in the literature, some authors have argued that the error-based evaluation methodology is detrimental to the field since the recommendations obtained in this way are not the most useful for users [McNee *et al.* (2006)]. Acknowledging this, recent work has evaluated top-N ranked recommendation lists with **precision-oriented metrics** [Cremonesi *et al.* (2010); McLaughlin and Herlocker (2004); Bellogín *et al.* (2011)], drawing from well studied evaluation methodologies in the IR field.

Among the wide range of precision-oriented metrics based on rankings, the most typical ones are precision, recall, normalized discounted cumulative gain, mean average precision, and mean reciprocal rank. Each of these metrics captures the quality of a ranking from a slightly different angle. More specifically, precision accounts for the fraction of recommended items that are relevant, whereas recall is the fraction of the relevant items that has been recommended. Both metrics are inversely related, since an improvement in recall typically produces a decrease in precision. They are typically computed up to a ranking position or cutoff k, being denoted as $P@k$ and $R@k$ [Baeza-Yates and Ribeiro-Neto (2011)]. Note that recall has also been referred to as hit-rate in [Deshpande and Karypis (2004)]. Hit-rate has also been defined as the percentage of users with at least one correct recommendation [Bellogín *et al.* (2013)], corresponding to the success metric (or first relevant score), as defined by TREC [Tomlinson (2012)].

The mean average precision (MAP) metric provides a single summary of the user's ranking by averaging the precision figures obtained after each new relevant item is obtained [Baeza-Yates and Ribeiro-Neto (2011)]. Normalized discounted cumulative gain (nDCG) uses graded relevance that is accumulated starting at the top of the ranking and may be reduced, or discounted, at lower ranks [Järvelin and Kekäläinen (2002)]. Using a different discount function, the rank score or half-life utility metric [Breese *et al.* (1998); Herlocker *et al.* (2004)] can be obtained from the nDCG formulation. Mean reciprocal rank (MRR) favors rankings whose first correct result occurs near the top ranking results [Baeza-Yates and Ribeiro-Neto (2011)]. This metric is similar to the average rank of correct recommendation (ARC) proposed in [Burke (2004)] and to the average reciprocal hit-rank (ARHR) defined in [Deshpande and Karypis (2004)].

In a more modern formulation of the recommendation problem, the ratings are no longer important. Instead, the consumption of recommended items by users is key, i.e., whether a recommended movie will be seen or a music track listened to. In this context, it is important to ask oneself what the recommender system should bring the user. If a recommendation algorithm suggests an item that the user is already aware of, what is the value of the system? Will this recommendation result in the consumption of the item? The common assumption is that a recommender system, in this context, should bring the user something she might not yet be familiar with, i.e., something novel, unexpected, or serendipitous. Still, the novel, unexpected, or serendipitous recommendations need to fulfill the requirement of the items being of actual interest to the user. These, so-called, non-accuracy metrics focus on the variety, popularity, novelty and similar aspects of the items, or lists of items, that are recommended [Castells *et al.* (2015)].

Due to the nature of non-accuracy metrics, there are often various definitions of them, each tailored towards the context they are used in. Furthermore, it is difficult to create a ground truth dataset to use with these metrics. Hence, recommendation algorithms which are specifically tailored towards non-accuracy metrics will often perform badly in terms of accuracy-based metrics [Said *et al.* (2013a)]; and symmetrically, if an algorithm is tailored towards accuracy metrics, it will often perform badly in terms of non-accuracy metrics.

Perhaps the most well-known non-accuracy metric, serendipity, attempts to model what is often referenced to as a *pleasant and unexpected surprise*. Serendipity is a compound metric of, among others, novelty and diversity. Novelty expresses how new a recommended item is (for a user). The underlying motivation for novelty being an interesting aspect of recommendation is that items which are old, or rather not new, can already have been seen by the user. If this is the case, recommending items which are already known by the user might not be of much value, since the recommendation does not actually present the user with something she could not find herself. Novelty can be directly measured in online experiments by asking users whether they are familiar with the recommended item [Celma and Herrera (2008)]. However, it is also interesting to measure novelty in an offline experiment, so as not to restrict its evaluation to costly and hard to reproduce online experiments. While novelty often can have a negative effect on accuracy, diversity can often be increased without necessarily sacrificing accuracy. Diversity expresses, as implied, the variety of the recommended items. There are multiple ways of measuring diversity in a set of recommended items [Castells *et al.* (2015)], commonly this is done by measuring the intra-list diversity (ILD) [Smyth and McClave (2001)] which is defined as

$$\text{ILD} = \frac{1}{|R|(|R|-1)} \sum_{i \in R} \sum_{j \in R} (1 - s(i,j)) \tag{9.3}$$

where R is the list of recommended items and $s(i,j)$ is a similarity measure reporting on the similarity of items i and j given some predefined set of item features. In essence, what ILD calculates is the aggregate diversity of the list of recommended items, i.e. the more similar (opposite of diverse) the items in the list are (given the selected similarity measure), the lower diversity will the list have. When using ILD as a measure, the goal is to generate a list of recommended items that contains items that are both accurate and diverse.

Furthermore, an often forgotten dimension of evaluation, at least in academic research, are those metrics related to the development, and maintenance of the recommender system itself, such as CPU cost per recommendation, storage cost, cost of re-training the model, etc. The coverage of the list of recommended items — i.e., how well does the list correspond to what is currently in stock or elseways available — also fits in this dimension, and it can be applied not only to the catalog of available items but to the users, hence, an algorithm which can recommend very accurate items, but only for a small portion of users might not be as "good" as a slightly less accurate algorithm with a higher user coverage. In [Gunawardana and Shani (2015)] two metrics are proposed for measuring item coverage: one based on the Gini index, and another based on Shannon's entropy. In [Ge *et al.* (2010)] the authors propose simple ratio quantities to measure such metrics, and to discriminate between the percentage of the items for which the system is able to generate a recommendation (prediction coverage), and the percentage of the available items that are effectively ever recommended (catalog coverage).

Finally, a last step once the results from the evaluation metrics have been obtained is to perform some type of statistical testing. There exist different options to check for significance on the results or on the difference between a method and a baseline: paired/unpaired tests, effect size, confidence intervals, etc. [Sakai (2014)]. It is important to be as specific as possible regarding which procedure was followed and the method used; additionally, to facilitate the interpretation of the results, related statistics such as the mean, variance, and population size of the samples should also be reported.

More importantly, when performing any type of statistical testing method, the data on which the method was computed must be specified, since, as with other aspects of the recommendation process, there is no standard procedure yet, especially when running cross-validated splits, where more than one test split is used and, hence, more than one performance measurement is obtained, which could lead to inconsistent conclusions about the significance of the results if performed in an split basis, e.g., a significant difference is found in some folds but not in others. On the other hand, if the results from each split (on a user basis, as it is done in IR for queries) are concatenated one after the other, may distort the test, because the data points are not independent (the same user appears more than once) and the number of points increase substantially [Bouckaert (2003); Kosir *et al.* (2013)].

9.1.3. *Online evaluation and User studies*

Online evaluation, as opposed to traditional offline evaluation is performed through direct involvement of a system's users in order to establish a *qualitative* assessment of the system's quality *as perceived by the end users.* To illustrate one of the key differences between offline evaluation and online evaluation, consider this top-N recommendation scenario: We have a user-item interaction matrix, as shown in Table 9.1. The table shows a matrix of 5 users and 6 items and their interactions, e.g., a 1 represents an interaction (rating, purchase, etc.), a 0 the lack of such. The training/test split is illustrated by the dashed line. In this case, an offline evaluation will only recognize item i_5 as a true positive recommendation for user u_3 and items i_5 and i_6 for user u_4. Users u_1, u_2 and u_5 will not have any true positive recommendations since they have not interacted with any of the items. The evaluation does not consider that the items might actually be liked by the user, if recommended in a real-world situation. Similarly, the fact that u_3 has interacted with i_5 does not need to imply that the item is a good recommendation.

Table 9.1. A user-item matrix divided into a training set (above the dashed line) and a test set (below the dashed line).

	u_1	u_2	u_3	u_4	u_5
i_1	1	1	0	0	1
i_2	1	0	1	1	1
i_3	0	0	0	1	0
i_4	1	0	1	0	1
i_5	0	0	1	1	0
i_6	0	0	0	1	0

In order to overcome this deficiency, online evaluation attempts to capture the quality of the recommendation as perceived by the users by analyzing their interaction patterns with the system together with explicitly asking questions [Gunawardana and Shani (2015); Pu *et al.* (2012); Kohavi *et al.* (2009)]. Online evaluation sometimes involves a user study. Users can be made aware or encouraged to participate, or participate unknowingly. In real-life systems, the concept of *A/B testing* is readily used to estimate different algorithms' qualities [Kohavi *et al.* (2009)]. A/B testing involves assigning a subset of a system's users to the algorithm under evaluation. In studies of real life systems, users are usually not made aware of their participation in tests [Kohavi *et al.* (2009)]. The interactions of the users are then analyzed and compared to a baseline.

More elaborate user studies — including questionnaires and other explicitly collected information — serve as an alternative to A/B testing. This type of studies commonly involve asking the users questions throughout, or after, their interaction with the system. In studies like this, the participants are naturally aware of their participation in the study. In order to be able to analyze the results quantitatively, the users are asked to agree or disagree with a question in the form of a statement, or we can collect data that is later analyzed at different granularity levels [Knijnenburg and Willemsen (2015)].

There is no default, quality-related, set of questions to ask when performing a recommender systems user study, instead questions are based on the type of quality that is sought for; whether relating to the concepts mentioned above or to rather technical qualities, e.g., time of recommendation, number of items recommended, etc. This type of user studies need to be meticulously planned and executed. If poorly executed, there is a risk of changing the users' opinions, e.g., through suggestive questions, or excessive workload or time involved in answering the questions. Workload and time-related issues can be mitigated by creating an incentive for the users to fulfill the survey, e.g., raffling off vouchers, prizes, etc. If no incentive is given, the time involved in answering the survey creates a decaying effect on the fraction of users who complete the study. When the users are given an incentive, there is a risk that some users will answer the questions quickly (at random) in order to be eligible for the award. In order to mitigate these effects, the number of questions and work load should be kept relatively low.

9.2. Algorithmic Solutions

9.2.1. *Evaluating collaborative filtering algorithms*

The process of evaluating traditional collaborative filtering recommender systems follows the processes described in Sec. 9.1.1. Historically speaking, collaborative filtering recommender algorithms have been the de facto standard for recommendation. In this context, it is only natural that most standard evaluation methods and strategies have been built with those in mind.

The process of evaluating collaborative filtering recommendation algorithms is inherently tied to the setting of the recommendation, e.g. whether the recommendation algorithm is intended to predict ratings or whether it is intended to identify relevant items for the users. Knowing this allows for further specification of evaluation settings, i.e. selecting whether to perform the evaluation in an online or offline manner (as discussed in Sec. 9.1.1). However, regardless of whether online or offline, the most central selection to make in the evaluation is the objective function, i.e. RMSE or MAE in a rating prediction scenario; accuracy or non-accuracy metric in the top-N recommendation scenario, or more complex values such as dwell-time or churn rate in online evaluation settings.

9.2.2. *Evaluating social recommendation algorithms*

Social recommendation algorithms take the end user's social graph or other social information into consideration. As such, the evaluation of social recommendation algorithms can be performed identically to the evaluation of traditional collaborative filtering algorithms (as discussed above), unless the recommendation algorithms tailors to a non-trivial social context, such as mutual recommendation of users (e.g. in dating apps). Let us focus on the mutual recommendation scenario. Consider the social graph exemplified in Figure 9.2. If a recommender were to recommend social connections in this graph not knowing the edges seen in the figure. Identifying node 1 as a recommendation to node 6 could be deemed a false positive knowing that the social connection between the two nodes is unidirectional, i.e. node 1 is connected to node 6 and not the other way around. Instead, if we consider the case of nodes 3 and 4, a correct recommendation would be to recommend both nodes to each other. However a recommendation algorithm might only identify node 3 as a recommendation to node 4 (and not the analog inverse). All of these cases need to be covered by the objective function selected for the purpose of evaluating the social recommendations in this graph. Whereas in the case of evaluating collaborative filtering recommendation algorithms, traditional metrics such as precision and recall could be used verbatim, in the case of evaluation of social recommendation, the metrics need to be adapted (or aggregated) to fit the recommendation context. For a more in-depth overview of this topic, see Chap. 16.

9.2.3. *Evaluating group recommendation algorithms*

Technically, the evaluation methods and metrics for single user item recommendations mentioned in Sec. 9.1 can be employed for evaluating group

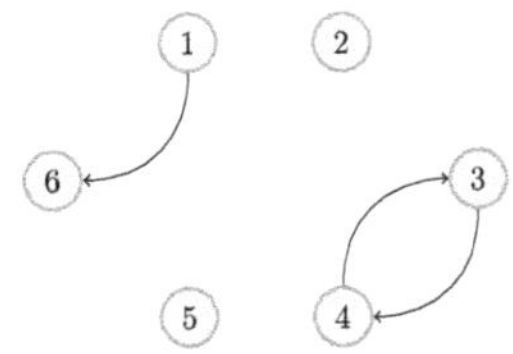

Fig. 9.2. Unidirectional (nodes 1–6) and bidirectional (nodes 3–4) recommendations.

recommendation algorithms. However, two fundamentally different strategies are common for group recommender evaluation: (*i*) evaluation of recommendations based on individual user profiles with aggregation of items recommended to each member of a group and (*ii*) recommendation based on aggregated group profiles. Any common evaluation metric can be used as long as the group is accounted for, by evaluating individual users first, and averaging the evaluation scores for all users in a group, for instance. An example of such an adaptation for the $RMSE$ measure is given in Equation 9.4 where $RMSE_G$ is the $RMSE$ value for the set of users belonging to a group G, where $|G|$ denotes the size of the group.

$$RMSE_G = \frac{1}{|G|} \sum_{u \in G} RMSE_u \tag{9.4}$$

More details about group recommendation can be seen in Chap. 6.

9.2.4. *Evaluating context-aware algorithms*

Evaluation of context-aware algorithms requires either a tailored objective function that is able to take into consideration the context of the recommendation or a completely separate evaluation process for each possible instance of our contextual variable. However, again, it is possible to use a traditional evaluation method if certain preparations are done in terms of the dataset. For instance, dividing the users or items into separate contextual profiles, i.e. consider that a user interacts with items on week days and on the weekend. This user could be split up into two separate contextual users, one corresponding to the actions the user has taken on week days, and one corresponding to the weekend activities as exemplified in Table 9.2. Thus, when recommending an item for the weekend profile of said user, only items in the users weekend catalog would be considered true positive recommendations, and, hence, they would affect metrics such as precision or recall.

Table 9.2. An example of how a user-item rating matrix can be contextualized by, e.g. dividing the users into separate profiles for ratings given during weekdays (wd), and ratings given during weekends (we).

(a) Without context.

	u_1	u_2	u_3
i_1	4	4	2
i_2	3	2	
i_3			3

(b) With context.

	u_1^{wd}	u_1^{we}	u_2^{wd}	u_2^{we}	u_3^{wd}	u_3^{we}
i_1	4			4	2	
i_2		3		4		
i_3						3

9.3. Available Resources

In this section, we present some resources related to RS evaluation that are publicly available, including APIs and libraries (Sec. 9.3.1), datasets (Sec. 9.3.2), and competitions (Sec. 9.3.3). We do not consider general tools or environments such as Amazon Mechanical Turk[1] or Crowdflower[2] because they are not specifically tailored to recommender systems experimentation, but they are means to prepare experiments and obtain data from. We do consider, however, frameworks developed in similar areas such as Machine Learning or Information Retrieval where, either explicitly or implicitly, recommendation algorithms or evaluation metrics can be computed or generated.

9.3.1. *APIs and libraries for evaluation*

In general, there are no known APIs that are used to evaluate provided results, mostly because it would involve knowing everything about the recommendation system (users, items, features, etc.). There are, indeed, companies based on serving recommendations as-a-service, but those are also out of the scope of this chapter because the techniques used are generally not disclosed (some examples include BrainSins[3], Criteo[4], and YOOCHOOSE[5]). There are some services based on APIs open to researchers such as plista[6], but since a challenge was organized based on this data, the following section will address this service in detail.

Regarding the software frameworks and libraries, Table 9.3 presents

[1]`https://www.mturk.com/mturk`, accessed October 2017.
[2]`https://www.crowdflower.com`, accessed October 2017.
[3]`https://www.brainsins.com`, accessed October 2017.
[4]`https://www.criteo.com`, accessed October 2017.
[5]`https://yoochoose.com`, accessed October 2017.
[6]`https://www.plista.com`, accessed October 2017.

Table 9.3. An overview of some of the most common open source frameworks used for recommender systems.

Name	License	Language	Updated	Link
CARSKit	GPL v2	Java	2017	Source
CofiRank	MPL	C++	2013	Source
Crab	BSD	Python	2012	Website
EasyRec	GPL v2	Java	2016	Website
FastFM	BSD 3	Python	2017	Website
Hi-Rec	MIT	Java	2018	Website
Implicit	MIT	Python	2018	Source
Lenskit	LGPL v2.1	Java	2018	Website
LibFM	GPL v 3	C++	2018	Website
LibRec	GPL v3	Java	2018	Website
LightFM	Apache 2.0	Python	2018	Source
mrec	BSD 3	Python	2016	Source
MyMediaLite	GPL	C# & Java	2017	Website
PREA	BSD	Java	2014	Source
Predictor	MIT	Ruby	2015	Source
Python-recsys	N/A	Python	2014	Source
RankSys	MPL 2.0	Java	2017	Source
RapidMiner	AGPL	Java	2017	Website
RecDB	BSD	PostGreSQL	2018	Source
Recommendable	MIT	Ruby	2018	Source
Recommender 101	Custom	Java	2015	Website
Recommenderlab	GPL v2	R	2017	Website
RecSys.jl	MIT	Julia	2016	Source
RiVal	Apache 2.0	Java	2017	Website
rrecsys	GPL v3	R	2018	Source
SLIM	Other	C	2012	Website
Surprise	BSD 3	Python	2018	Website
SVDFeature	Apache 2.0	C++	2014	Source
TagRec	AGPL 3.0	Java	2018	Source
trec_eval	Other	C	2016	Source
Turi	Apache 2.0	C++	2018	Website
Waffles	LGPL	C++	2018	Source
WrapRec	MIT	C#	2018	Website
QMF	Apache 2.0	C++	2017	Source

a list of the most common open source frameworks that can be used at different stages of the recommendation pipeline. Some of these frameworks/libraries are focused on other fields (Machine Learning, Natural Language Processing, or Information Retrieval) but they provide resources that can be used for recommendation, such as clustering, nearest-neighbors, matrix factorization, etc.

Depending on the application the framework was conceived for, the evaluation techniques and tools could be more or less applicable to recommendation, in particular, when a specific task is aimed. For instance, Machine Learning libraries typically implement error-based metrics such as MAE (see Sec. 9.1.2), whereas Information Retrieval libraries are mostly focused on precision-oriented metrics.

9.3.2. *Datasets for evaluation*

Public datasets including interactions between users and items are of paramount importance in recommender systems research. They serve as input for recommendation algorithms, as simulation data, or for evaluation purposes. Despite their importance, publicly accessible datasets with ratings, clicks, transactions, and so on, have not been abundantly available until very recently, leaving researchers not many options besides classical datasets such as MovieLens, focused on the movie domain.

Table 9.4 shows an heterogeneous list of datasets, including a large number of movie datasets. This might be attributed to historical reasons, since the first public datasets were based on the MovieLens[7] system [Harper and Konstan (2016)] and most of the research developed since has been focused on the movie rating prediction task, neglecting other tasks or domains until recently, when researchers have had access to other, more diverse data sources.

9.3.3. *Competitions about evaluation*

In 2006, one initiative, the Netflix Prize[8], created a focus on recommender systems and contributed to major advancements in the field during its three year run. Similar initiatives have led to great improvements in related fields — e.g., the Text Retrieval Conference[9] in Information Retrieval, and the KDD Cup[10] in Machine Learning and Data Mining communities. Following the success of these, different competitions related to recommendation have appeared, one of them — the RecSys Challenge[11] — organized in conjunction with the ACM Conference on Recommender Systems [Said (2016)].

Throughout the duration of the Netflix Prize, significant advancements were made in the RS research field, e.g., establishing matrix factorization methods such as SVD as state-of-the-art in recommendation. At the 2010 ACM RecSys conference, the seed for what would become the RecSys Challenge was organized as the Challenge on Context-aware Movie Recommendation (CAMRa) [Adomavicius *et al.* (2010)]. CAMRa attracted a moderate number of participants, but contributed to establishing the RecSys Challenge series.

---

[7] `https://movielens.org`, accessed October 2017.

[8] `http://www.netflixprize.com`

[9] `http://trec.nist.gov`, accessed October 2017.

[10] `http://www.kdd.org/kdd-cup`, accessed October 2017.

[11] `http://www.recsyschallenge.com`

Table 9.4. Datasets commonly used for recommender system evaluation, summarizing the number of users, items, and events (ratings, clicks, or interactions in general) included in the dataset.

Name	Domain	Events	Users	Items	Density	Source
Book-crossing	Books	$1.1 \cdot 10^6$	$2.8 \cdot 10^5$	$2.7 \cdot 10^5$	0.001%	Website
LibimSeTi	Dating	$1.7 \cdot 10^7$	$1.4 \cdot 10^5$	$1.7 \cdot 10^5$	0.076%	Website
Xing (2016)	Jobs	$8.8 \cdot 10^6$	$1.4 \cdot 10^6$	$1.4 \cdot 10^6$	$5 \cdot 10^{-6}$%	Website
Jester	Jokes	$4.1 \cdot 10^6$	$7.3 \cdot 10^4$	10^2	56.34%	Website
Moviepilot (2010)	Movies	$4.5 \cdot 10^6$	$1.1 \cdot 10^5$	$2.5 \cdot 10^4$	0.002%	Website
CAMRa2011 (Moviepilot)		$4.4 \cdot 10^6$	$1.7 \cdot 10^5$	$3.0 \cdot 10^4$	0.001%	Website
CAMRa2010 (Filmtipset time)		$5.8 \cdot 10^6$	$3.5 \cdot 10^4$	$5.4 \cdot 10^4$	0.003%	Website
CAMRa2010 (Filmtipset social)		$3.1 \cdot 10^6$	$1.7 \cdot 10^4$	$2.4 \cdot 10^4$	0.008%	Website
Mise-en-scène		$1.3 \cdot 10^7$	$1.8 \cdot 10^5$	$1.3 \cdot 10^4$	$6 \cdot 10^{-9}$%	Website
MovieLens 100k		$1.9 \cdot 10^5$	10^3	$1.7 \cdot 10^3$	6.30%	Website
MovieLens 1M		10^6	$6.0 \cdot 10^3$	$3.9 \cdot 10^3$	4.25%	Website
MovieLens 10M		10^7	$7.2 \cdot 10^4$	$1.1 \cdot 10^4$	1.31%	Website
MovieLens 20M		$2 \cdot 10^7$	$1.4 \cdot 10^5$	$2.7 \cdot 10^4$	0.05%	Website
MovieLens HetRec		$8.6 \cdot 10^5$	$2.1 \cdot 10^3$	10^4	0.04%	Website
MovieTweetings		$7.0 \cdot 10^5$	$5.3 \cdot 10^4$	$2.6 \cdot 10^4$	0.04%	Website
Netflix		10^8	$4.8 \cdot 10^5$	$1.8 \cdot 10^4$	1.17%	Website
Last.fm 1K	Music	$1.9 \cdot 10^7$	$9.9 \cdot 10^3$	$1.8 \cdot 10^5$	10.91%	Website
Last.fm 360K		$1.7 \cdot 10^7$	$3.6 \cdot 10^5$	$2.9 \cdot 10^5$	0.016%	Website
Last.fm HetRec		$9.3 \cdot 10^4$	$1.9 \cdot 10^3$	$1.8 \cdot 10^4$	0.003%	Website
Yahoo Music (KDD-cup'11)		$2.6 \cdot 10^8$	10^6	$6.2 \cdot 10^5$	0.042%	Website
South Tyrol Suggests	Point of interest	$2.5 \cdot 10^4$	$3.3 \cdot 10^3$	$2.5 \cdot 10^3$	3.1%	Website
Delicious	Tags	$4.2 \cdot 10^8$	10^3	$1.3 \cdot 10^8$	$3 \cdot 10^{-6}$%	Website
Delicious HetRec		$4.4 \cdot 10^5$	$1.8 \cdot 10^3$	$6.9 \cdot 10^4$	0.003%	Website
YOOCHOOSE	Retail	$3.3 \cdot 10^7$	$9.2 \cdot 10^6$	$5.3 \cdot 10^4$	$7 \cdot 10^{-5}$%	Website

Table 9.5. Events and initiatives related to recommender system evaluation.

Event	**Occurrence** (first time)	**Link**
ECMLPKDD Discovery Challenge	Not fixed (2008)	Website
Kaggle competitions	Not fixed (2012)	Website
KDD Cup	Not fixed (2007)	Website
NewsREEL	Not fixed (2013)	Website
Netflix Prize	Once (2006)	Website
RecSys Challenge	Yearly (2010)	Website
TREC Contextual Suggestion	Yearly (2012)	Website
WSDM challenge	Not fixed (2018)	Website

The RecSys Challenge has followed a similar structure since its inception: *i)* a dataset and problem are presented, *ii)* teams sign up and participate, *iii)* participants submit their solutions in time for a deadline, *iv)* participants submit papers outlining their approaches, *v)* during a workshop at the ACM RecSys conference participants present their approaches and winners are announced. One of the main features of the challenge is to make available a new real world dataset. This structure is common to other competitions related to RS, such as those presented in Table 9.5.

The RecSys challenge has been constantly evolving to adapt to different trends in the RS community. It has incorporated tasks different to rating prediction and integrating diverse domains — from identifying which groups of users to recommend certain ad campaigns to user engagement prediction in Twitter or e-commerce. Over the years, the challenge has established itself as a benchmarking event for current recommender system research. It has attracted participants from academia and industry, allowing researchers and practitioners to learn from, and cooperate with each other, in a community-driven event. Each yearly instance takes on new research challenges based on ongoing trends in industry and academia, which is also evidenced in some of the related events organized in parallel in different venues, such as KDD or WSDM conferences.

Current and future research in RS acknowledge that certain recommender system settings require online evaluation, i.e., an instantaneous feedback loop between the users of the system and the algorithm. In some instantiations of these competitions, a second stage (if available) usually bring real interactions with users from the system. A special mention deserves the NewsREEL initiative, which allows researchers to receive real recommendation requests by news providers, in the context of the plista recommendation-as-a-service environment. This event started in 2013 as a challenge associated to an ACM RecSys workshop on news recommendation, but has been organized independently since then.

9.4. Experimental Results

A large amount of recommender systems research is based on comparisons of recommendation algorithm's predictive accuracy: the better the evaluation metrics (higher accuracy scores or lower predictive errors), the better the recommender system. However, it is usually difficult to put in context and compare these results, mostly because too many alternatives exist when designing and implementing an evaluation strategy (more on this on Sec. 9.5), and the actual implementation of a recommendation algorithm sometimes diverges considerably from the well-known ideal formulation, frequently due to manual tuning and modifications observed to work better in some situations.

In [Said and Bellogín (2014)], a thorough experimental comparison of different evaluation techniques and recommendation algorithms was presented. In order to achieve comparable evaluation protocols when using different recommendation frameworks and datasets, the authors had complete control of the evaluation dimensions being benchmarked: data splitting, evaluation strategies, and evaluation metrics. The main result found was that there is a large difference in recommendation accuracy across frameworks and strategies, specifically, the same baseline method may perform orders of magnitude better or worse depending on the framework. We include here more details about these results and their corresponding discussion.

Figure 9.3 shows the catalog coverage and nDCG metrics computed using a controlled evaluation protocol. These figures show a wide combination of strategies for data splitting, recommendation and candidate items generation. We notice that, except for the UserTest candidate items strategy, MyMediaLite outperforms Mahout and LensKit in terms of nDCG@10, when using a matrix factorization algorithm (SVD in the figure) and a user-based with Pearson similarity (UB Pea). This high precision comes at the expense of lower coverage, specifically of the catalog (item) coverage. As a consequence, MyMediaLite seems to be able to recommend at least one item per user, but far from the complete set of items, in particular compared to the other frameworks. In terms of nDCG@10, the best results are obtained with the UserTest strategy, with noticeable differences between

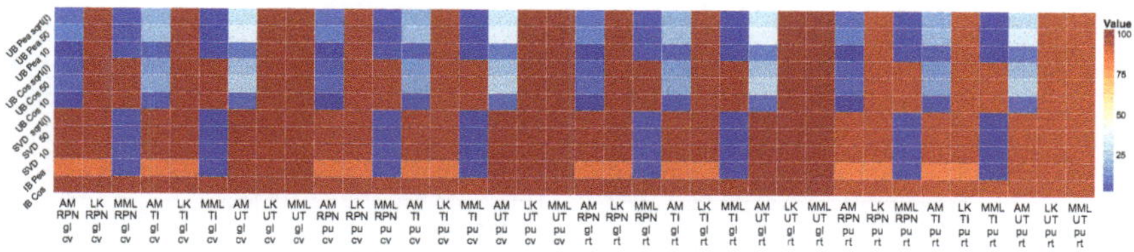

(a) Catalog coverage (in percent).

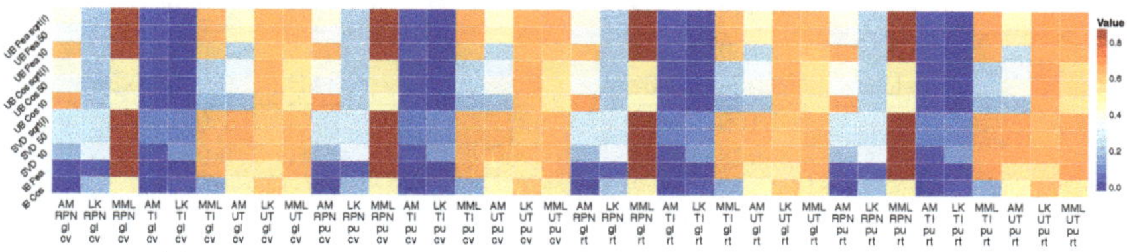

(b) Normalized discounted cumulative gain at 10 (nDCG@10).

Fig. 9.3. Catalog coverage and nDCG for the controlled evaluation. RPN, TI and UT refer to RelPlusN, TrainItems and UserTest strategies, IB and UB refer to item- and user-based respectively; Pea and Cos to Pearson and Cosine; gl and pu to global and per user; AM (Mahout), LK (LensKit), MML (MyMediaLite) to the frameworks; and cv, rt to cross validation and ratio.

recommender types, i.e. IB performs poorly, SVD performs well, UB in between, in accordance with, e.g., [Koren and Bell (2015)]. The splitting strategy has little effect on the results in this setting.

Additionally, the results of a framework-dependent evaluation are shown in Table 9.6. For this, we use each framework's internal evaluation classes and report the results obtained. Table 9.6a shows evaluation results in terms of nDCG, generated by Mahout and LensKit, whereas Table 9.6b shows the RMSE values from LensKit and MyMediaLite. We start by studying the results presented in Table 9.6a, where it seems that LensKit outperforms Mahout at several orders of magnitude. The highest nDCG obtained by Mahout (0.2868) is less than one third of the lowest value obtained by LensKit (0.9422). This should be taken in the context of each framework's evaluator. Note that Mahout's evaluator will only consider users with a certain minimum number of preferences (two times the level of recall). Our level of recall was set to 50, meaning only users with at least 100 preferences are evaluated (corresponding only to circa 33% of the users in this dataset).

Looking at the RMSE results obtained by LensKit and MyMediaLite in Table 9.6b, the difference between the frameworks is not as large as in the previous case. All RMSE results are between 7.5% (UBCos50) and

Table 9.6. Results using the internal evaluation methods of each framework.

(a) nDCG for AM and LK.

Alg.	F.W.	nDCG
IBCos	AM	0.000414780
	LK	0.942192050
IBPea	AM	0.005169231
	LK	0.924546132
SVD50	AM	0.105427298
	LK	0.943464094
UBCos50	AM	0.169295451
	LK	0.948413562
UBPea50	AM	0.169295451
	LK	0.948413562

(b) RMSE values for LK and MML.

Alg.	F.W.	RMSE
IBCos	LK	1.01390931
	MML	0.92476162
IBPea	LK	1.05018614
	MML	0.92933246
SVD50	LK	1.01209290
	MML	0.93074012
UBCos50	LK	1.02545490
	MML	0.95358984
UBPea50	LK	1.02545490
	MML	0.93419026

11.7% (UBCos10) of each other. In this case, both frameworks created five instances of training/tests splits with an 80%-20% ratio. The splits are randomly seeded, meaning that even though the frameworks only create five training/test datasets, they are not the same between the two frameworks.

A further observation to be made is how the framework's internal evaluation compares to our controlled evaluation. We see that the frameworks perform better in the controlled environment than in the internal ditto. In the case of nDCG (Table 9.6a), we see that Mahout's values fluctuate more in both versions of the controlled evaluation (RPN and UT) than in Mahout's own evaluation. The internal evaluation results consistently remain lower than the corresponding values in the controlled setting — although the RPN values are closer to Mahout's own results. The UT (UserTest) values obtained in the controlled evaluation are several orders of magnitude higher than in the internal setting, even though the setting resembles Mahout's own evaluation closer than the RPN setting. LensKit's internal evaluation consistently shows better results than the controlled setting. We believe this could be an effect related to how the final averaged nDCG metric is calculated, or how the training/test splitting is performed by each framework, where some users might be ignored because of a low number of preferences.

Given the widely differing results, it seems pertinent that in order to perform an inter-framework benchmarking, the evaluation process needs to be clearly defined and both recommendations and evaluations performed in a controlled and transparent environment.

9.5. Practical Considerations

In the light of the previous section, it stands clear that even though different recommendation frameworks implement algorithms in a similar fashion, the results are not comparable, i.e., the performance of an algorithm implemented in one cannot be compared to the performance of the same algorithm in another. Not only do there exist differences in algorithmic implementations, but also in the evaluation methods themselves.

There are no *de facto* rules or standards on how to evaluate a recommendation algorithm. This also applies to how recommendation algorithms of a certain type should be realized (e.g., default parameter values, use of backup recommendation algorithms, and other *ad-hoc* implementations). However, this should perhaps not be seen as something negative *per se*. Yet, when it comes to performance comparison of recommendation algorithms, a standardized (or controlled) evaluation is crucial [Konstan and Adomavicius (2013)]. Without which, the relative performance of two or more algorithms evaluated under different conditions becomes essentially meaningless. In order to objectively and definitively characterize the performance of an algorithm, a controlled evaluation, with a defined evaluation protocol is a prerequisite.

Next, we discuss three main issues that should be considered when implementing and evaluating recommendation algorithms, besides performing a controlled evaluation. First, in Sec. 9.5.1 we describe some typical practical considerations, mostly related to technical aspects and design issues. Then, Sec. 9.5.2 summarizes some issues that have an impact on the reproducibility of RS experimental results.

9.5.1. *Design issues to evaluate recommender systems*

One of the first and most obvious design issues that researchers typically assume when evaluating RS is that a user will never consume an item more than once. This results in test splits with no overlap with the training split — as common practice in Machine Learning. Such hypothesis was probably inherited from the most common domain used in the first years by the community: the movie domain; here, the datasets would only contain one interaction (i.e., a rating) once for each user-item pair. This is in contrast with other domains such as music or e-commerce [Schedl *et al.* (2015); Linden *et al.* (2003)], where repeated consumption is paramount and encouraged by the system.

Another constraint (somewhat artificially) imposed in many works when evaluating RS is that of ignoring those users or items with few interactions. This is a well-known problem known as *cold start*, and it is inherent to the task of a real recommender system, where there will always exist recently created items or users. Nonetheless, when those cases are ignored, it should be made crystal clear and explicitly stated like that in the paper, otherwise the validity of this comparison would not be fair. At the same time, it is interesting to note that cold-start tailored evaluation methodologies have been proposed [Kluver and Konstan (2014)], aiming to provide unbiased measurements when evaluating users with a limited number of training and test groundtruth information. Evaluation in such a scarce situation is not easy, and hence, more studies should be devoted to properly understand any deficiency that may arise in this context.

Additionally, multi-criteria recommendation poses the problem of how to combine various metrics into something that can be optimized [Adomavicius and Kwon (2015)]. Even though there is a significant body of work on multi-criteria RS optimization, there is no clear guideline on how to perform this type of evaluation [Jambor and Wang (2010); Ge *et al.* (2010); Said *et al.* (2013b)]. State of the art methods in multi-criteria evaluation require a trade-off between the different metrics to be evaluated, this can then be applied to offline evaluation. Online evaluation using several criteria requires elaborate qualitative analyses of long term results of recommendation approaches.

Finally, in very sparse datasets such as those used for music recommendation or in location-based RS, it is possible to evaluate by surrogates, which means that a recommendation is assessed as correct if some attribute of the item matches the target item, instead of the actual item, due to the task being extremely difficult and to find the correct item in large catalogs. As stated in [Schedl *et al.* (2015)], evaluation in music RS has been carried out for a long time using genre as proxy and modeling a genre prediction task. Similarly, in location-aware recommendation, the item category has been used as surrogate, assuming that a recommendation would be received by the user in the same way if a museum was recommended, even though that was not the actual museum she visited according to the data [Li *et al.* (2012); Kumar *et al.* (2017)].

These examples evidence that other, more complex environments and evaluation situations can, and should, be defined when evaluating recommendation systems. Either new evaluation metrics, experimental methodologies, or data splitting techniques could be explored to bring closer the RS evaluation to the actual tasks that want to be modeled.

Additional issues related to offline evaluation focus on aspects of the underlying data which is used for both training and evaluation of RSs. The concept known as the *magic barrier* of RS [Herlocker *et al.* (2004)] concerns the fact that user interactions are not concise, i.e., they contain noise and other irregularities which make the data not fully trustworthy. The effect of this is that when optimizing towards a certain metric, there is an upper level, a threshold beyond which optimization is useless [Said *et al.* (2012)]. This threshold is unknown; at best, it can be estimated assuming there are enough interactions provided by each user [Said and Bellogín (2018)].

9.5.2. *Reproducibility issues on evaluation*

Evaluating recommender systems is not an easy task. In all fairness, it is not *a* task, it is a set of several interconnected and standalone tasks that, when viewed together, should result in one (or several) measure(s) which then state the quality of the recommender. The challenge becomes, for instance, what metrics to use when evaluating, or how many metrics to use.

The results discussed in Sec. 9.4 highlight the differences in recommendation accuracy between implementations of the same algorithms on different frameworks, distinct levels of accuracy and variations of evaluation results on the same dataset and in the same framework when employing various evaluation strategies. It is important to note, thus, that inter-framework comparisons of recommendation quality can potentially point to incorrect results and conclusions, unless performed with great caution and in a controlled, framework independent, environment.

Furthermore, there are several aspects when designing and implementing a recommender system that may affect its final results and hinder a potential replication of the reported settings and insights, from using different model parameters to how specific parts of the system are implemented. For instance, the data splitting strategy may have a deep impact on the final results being reported, making this an important aspect to take into account when reporting details about an experiment. Moreover, there are several aspects in recommendation algorithms that are not standard in the community which, eventually, turn out to be implementation-dependent. For example, the term kNN (to specify collaborative filtering algorithms based on k-nearest neighbors) is usually linked to the user-based (UB) variation of these methods, even though item-based (IB) algorithms are also available and, in some situations, work better than user-based [Sarwar *et al.* (2001)]. Furthermore, there exist several different formulations for the user-based kNN algorithm, none of which is regarded as the "standard" one.

Additionally, there are also different alternatives in the literature when computing neighbors. Although the most typical one consists of taking the top-k closest users to the target user, there are other techniques based on thresholds that should be properly reported and specified when used in experiments because of their unpopularity [Ning *et al.* (2015)]. Besides that, a very important detail that is usually not reported is when the neighbors are actually computed: at training or test (i.e., evaluation) time.

Finally, it is not difficult to find implementations where a capping is applied after the score is predicted — probably inherited by the rating prediction task, the most popular recommendation task for a long time — so such score is bounded by the rating range. However, this may lead to further problems when producing a ranking, since ties are more likely to occur and, hence, tie-breaking strategies have to be implemented and reported, something we have seldom found explicitly in the analyzed frameworks and public implementations. At the same time, it is very important to report what happens when a recommender cannot predict a score. This decision has an impact on the coverage of the system — if a baseline recommender is used instead, every algorithm will always have complete coverage —, but also on how performance metrics are evaluated in those cases.

In summary, the issues of replication — obtaining the exact same results in the same setting — and reproducibility — obtaining comparable results using a different setting — are very difficult challenges at the moment. They force researchers to reimplement the baseline algorithm they want to compare their approach against, or to pay extra attention to every algorithmic and evaluation detail, ignoring if the observed discrepancies with respect to what already was published come from omitted details from the original papers (parameters, methodologies, protocols, etc.) or a wrong interpretation of any of these intermediate steps.

9.6. Future directions

The empirical evaluation of RS is acknowledged to be an open problem in the field, with open issues yet to be addressed [Gunawardana and Shani (2015)]. Many experimental approaches and metrics have been developed over the years, which the community is well acquainted with, but key aspects and details in the design and application of available methodologies are open to configuration and interpretation, where even apparently subtle details may create a considerable difference. This results in a significant divergence in experimental practice, hindering the comparison and proper assessment of contributions and advances to the field.

The discussion and definition of the basic elements of the experimental conditions (and their requirements) are critical to support continuous innovation in any discipline. The offline evaluation of RS requires an implementation of the algorithm or technique to be evaluated, a set of quality measures for comparative evaluation, and an experimental protocol establishing how to handle the data and compute metrics in detail. Online evaluation similarly requires an algorithm implementation and a population of users to survey (by means of an A/B test, for instance). Here again, perhaps even more importantly than in offline evaluation, an experimental protocol needs to be established and adhered to.

In order to seek reproducibility and replication, several strategies can be considered, such as source code sharing, standardization of agreed evaluation metrics and protocols, or releasing public experimental design software, all of which have difficulties of their own. Furthermore, for online evaluation, an extensive analysis of the population of test users should be provided. While the problem of reproducibility and replication has been recognized in the community, the need for a solution remains largely unmet.

Another open problem when evaluating RS is the relation between online and offline experiments. Several authors have explored this issue in some domains but no conclusive results have been obtained [Garcin *et al.* (2014); Beel *et al.* (2013); de Souza Pereira Moreira *et al.* (2015)]. The main question is how to align the offline evaluation to the online (usually, with A/B tests) results. Some possibilities include designing a good evaluation methodology or using a sensible evaluation metric to the problem at hand. An interesting output that could be produced by a better understanding of this issue is that offline evaluation would predict which methods, parameters, or configurations will work better when integrated and tested in an online evaluation. Recent approaches apply counterfactual reasoning to compute offline estimators of business metrics, by computing the expected reward of a new method based on logs collected on a technique running in production [Gilotte *et al.* (2018); Joachims and Swaminathan (2016)]; in this way, we could answer *what if?* questions for a range of algorithms not tested by the users with some level of certainty.

Additionally, there should be an increased effort in making metrics meaningful, especially tailored to specific problems in RS, such as time-aware recommendation, cross-domain, cold-start users and items, repeated consumption, and so on. We should also understand their inherent biases, where a more in-depth analysis and some solutions are surveyed in Chap. 10. Furthermore, several algorithms exist for temporal contexts, however, experimental evaluation methodologies have not been analyzed until recently [Campos *et al.* (2014)], where evaluation dimensions such as novelty and diversity remained largely unexplored, even though there exist obvious relations between these concepts and time-aware recommendations [Lathia *et al.* (2010)]. Similarly, there exist several approaches aiming to exploit cross-domain patterns for recommendation [Cantador *et al.* (2015)], however the evaluation of these systems has been limited to analyze the sensitivity to the amount of user or item overlap, target domain density, and user profile size, neglecting other properties such as the dependence on the groundtruth sparsity levels on either the source or the target domains, or even defining new metrics specifically tailored to this problem, e.g., where the amount of knowledge that is transferred from the source domain is considered inside the metric.

References

Abel, F. (2015). We know where you should work next summer: Job recommendations, in *Proceedings of the 9th ACM Conference on Recommender Systems, RecSys 2015, Vienna, Austria, September 16-20, 2015* (ACM), p. 230.

Adomavicius, G. and Kwon, Y. (2015). Multi-criteria recommender systems, in *Recommender Systems Handbook* (Springer), pp. 847–880.

Adomavicius, G., Tuzhilin, A., Berkovsky, S., De Luca, E. W. and Said, A. (2010). Context-awareness in recommender systems: research workshop and movie recommendation challenge, in *Proceedings of the 2010 ACM Conference on Recommender Systems, RecSys 2010, Barcelona, Spain, September 26-30, 2010* (ACM), pp. 385–386.

Baeza-Yates, R. A. and Ribeiro-Neto, B. A. (2011). *Modern Information Retrieval - the concepts and technology behind search, Second edition* (Pearson Education Ltd., Harlow, England), ISBN 978-0-321-41691-9.

Basu, C., Hirsh, H. and Cohen, W. W. (1998). Recommendation as classification: Using social and content-based information in recommendation, in *Proceedings of the Fifteenth National Conference on Artificial Intelligence and Tenth Innovative Applications of Artificial Intelligence Conference, AAAI 98, IAAI 98, July 26-30, 1998, Madison, Wisconsin, USA.* (AAAI Press / The MIT Press), pp. 714–720.

Beel, J., Genzmehr, M., Langer, S., Nürnberger, A. and Gipp, B. (2013). A comparative analysis of offline and online evaluations and discussion of research paper recommender system evaluation, in *Proceedings of the*

International Workshop on Reproducibility and Replication in Recommender Systems Evaluation, RepSys 2013, Hong Kong, China, October 12, 2013 (ACM), pp. 7–14.

Bellogín, A., Cantador, I., Díez, F., Castells, P. and Chavarriaga, E. (2013). An empirical comparison of social, collaborative filtering, and hybrid recommenders, *ACM TIST* **4**, 1, p. 14.

Bellogín, A., Castells, P. and Cantador, I. (2011). Precision-oriented evaluation of recommender systems: an algorithmic comparison, in *Proceedings of the 2011 ACM Conference on Recommender Systems, RecSys 2011, Chicago, IL, USA, October 23-27, 2011* (ACM), pp. 333–336.

Bellogín, A., Castells, P. and Cantador, I. (2017). Statistical biases in information retrieval metrics for recommender systems, *Information Retrieval Journal.*

Berkovsky, S., Freyne, J. and Coombe, M. (2012). Physical activity motivating games: Be active and get your own reward, *ACM Trans. Comput.-Hum. Interact.* **19**, 4, pp. 32:1–32:41.

Bistaffa, F., Filippo, A., Chalkiadakis, G. and Ramchurn, S. D. (2015). Recommending fair payments for large-scale social ridesharing, in *Proceedings of the 9th ACM Conference on Recommender Systems, RecSys 2015, Vienna, Austria, September 16-20, 2015* (ACM), pp. 139–146.

Bollen, D. G. F. M., Knijnenburg, B. P., Willemsen, M. C. and Graus, M. P. (2010). Understanding choice overload in recommender systems, in *Proceedings of the 2010 ACM Conference on Recommender Systems, RecSys 2010, Barcelona, Spain, September 26-30, 2010* (ACM), pp. 63–70.

Bouckaert, R. R. (2003). Choosing between two learning algorithms based on calibrated tests, in T. Fawcett and N. Mishra (eds.), *Machine Learning, Proceedings of the Twentieth International Conference (ICML 2003), August 21-24, 2003, Washington, DC, USA* (AAAI Press), pp. 51–58.

Breese, J. S., Heckerman, D. and Kadie, C. M. (1998). Empirical analysis of predictive algorithms for collaborative filtering, in *UAI '98: Proceedings of the Fourteenth Conference on Uncertainty in Artificial Intelligence, University of Wisconsin Business School, Madison, Wisconsin, USA, July 24-26, 1998* (Morgan Kaufmann), pp. 43–52.

Burke, R. D. (2004). Hybrid recommender systems with case-based components, in *Advances in Case-Based Reasoning, 7th European Conference, ECCBR 2004, Madrid, Spain, August 30 - September 2, 2004, Proceedings, Lecture Notes in Computer Science*, Vol. 3155 (Springer), pp. 91–105.

Campos, P. G., Díez, F. and Cantador, I. (2014). Time-aware recommender systems: a comprehensive survey and analysis of existing evaluation protocols, *User Model. User-Adapt. Interact.* **24**, 1-2, pp. 67–119.

Cantador, I., Fernández-Tobías, I., Berkovsky, S. and Cremonesi, P. (2015). Cross-domain recommender systems, in *Recommender Systems Handbook* (Springer), pp. 919–959.

Castells, P., Hurley, N. J. and Vargas, S. (2015). Novelty and diversity in recommender systems, in *Recommender Systems Handbook* (Springer), pp. 881–918.

Celma, Ò. and Herrera, P. (2008). A new approach to evaluating novel recommendations, in *Proceedings of the 2008 ACM Conference on Recommender Systems, RecSys 2008, Lausanne, Switzerland, October 23-25, 2008* (ACM), pp. 179–186.

Cremonesi, P., Garzotto, F., Negro, S., Papadopoulos, A. V. and Turrin, R. (2011). Comparative evaluation of recommender system quality, in *Proceedings of the International Conference on Human Factors in Computing Systems, CHI 2011, Extended Abstracts Volume, Vancouver, BC, Canada, May 7-12, 2011* (ACM), pp. 1927–1932.

Cremonesi, P., Koren, Y. and Turrin, R. (2010). Performance of recommender algorithms on top-n recommendation tasks, in *Proceedings of the 2010 ACM Conference on Recommender Systems, RecSys 2010, Barcelona, Spain, September 26-30, 2010* (ACM), pp. 39–46.

de Souza Pereira Moreira, G., de Souza, G. A. and da Cunha, A. M. (2015). Comparing offline and online recommender system evaluations on long-tail distributions, in *Poster Proceedings of the 9th ACM Conference on Recommender Systems, RecSys 2015, Vienna, Austria, September 16, 2015, CEUR Workshop Proceedings*, Vol. 1441 (CEUR-WS.org).

Deshpande, M. and Karypis, G. (2004). Item-based top-N recommendation algorithms, *ACM Trans. Inf. Syst.* **22**, 1, pp. 143–177.

Elahi, M., Ge, M., Ricci, F., Fernández-Tobías, I., Berkovsky, S. and Massimo, D. (2015). Interaction design in a mobile food recommender system, in *Proceedings of the Joint Workshop on Interfaces and Human Decision Making for Recommender Systems, IntRS 2015, co-located with ACM Conference on Recommender Systems (RecSys 2015), Vienna, Austria, September 19, 2015, CEUR Workshop Proceedings*, Vol. 1438 (CEUR-WS.org), pp. 49–52.

Elahi, M., Ge, M., Ricci, F., Massimo, D. and Berkovsky, S. (2014). Interactive food recommendation for groups, in *Poster Proceedings of the 8th ACM Conference on Recommender Systems, RecSys 2014, Foster City, Silicon Valley, CA, USA, October 6-10, 2014, CEUR Workshop Proceedings*, Vol. 1247 (CEUR-WS.org).

Elsweiler, D., Harvey, M., Ludwig, B. and Said, A. (2015). Bringing the "healthy" into food recommenders, in *Proceedings of the 2nd International Workshop on Decision Making and Recommender Systems, Bolzano, Italy, October 22-23, 2015, CEUR Workshop Proceedings*, Vol. 1533 (CEUR-WS.org), pp. 33–36.

Garcin, F., Faltings, B., Donatsch, O., Alazzawi, A., Bruttin, C. and Huber, A. (2014). Offline and online evaluation of news recommender systems at swissinfo.ch, in *Eighth ACM Conference on Recommender Systems, RecSys '14, Foster City, Silicon Valley, CA, USA - October 06 - 10, 2014* (ACM), pp. 169–176.

Ge, M., Delgado-Battenfeld, C. and Jannach, D. (2010). Beyond accuracy: evaluating recommender systems by coverage and serendipity, in *Proceedings of the 2010 ACM Conference on Recommender Systems, RecSys 2010, Barcelona, Spain, September 26-30, 2010* (ACM), pp. 257–260.

Gilotte, A., Calauzènes, C., Nedelec, T., Abraham, A. and Dollé, S. (2018). Offline A/B testing for recommender systems, in Y. Chang, C. Zhai, Y. Liu and Y. Maarek (eds.), *Proceedings of the Eleventh ACM International Conference on Web Search and Data Mining, WSDM 2018, Marina Del Rey, CA, USA, February 5-9, 2018* (ACM), pp. 198–206.

Goldberg, K. Y., Roeder, T., Gupta, D. and Perkins, C. (2001). Eigentaste: A constant time collaborative filtering algorithm, *Inf. Retr.* **4**, 2, pp. 133–151.

Gunawardana, A. and Shani, G. (2015). Evaluating recommender systems, in *Recommender Systems Handbook* (Springer), pp. 265–308.

Guy, I. (2015). Social recommender systems, in *Recommender Systems Handbook* (Springer), pp. 511–543.

Harper, F. M. and Konstan, J. A. (2016). The movielens datasets: History and context, *TiiS* **5**, 4, pp. 19:1–19:19.

Herlocker, J. L., Konstan, J. A., Terveen, L. G. and Riedl, J. (2004). Evaluating collaborative filtering recommender systems, *ACM Trans. Inf. Syst.* **22**, 1, pp. 5–53.

Jambor, T. and Wang, J. (2010). Optimizing multiple objectives in collaborative filtering, in *Proceedings of the 2010 ACM Conference on Recommender Systems, RecSys 2010, Barcelona, Spain, September 26-30, 2010* (ACM), pp. 55–62.

Jannach, D., Lerche, L. and Jugovac, M. (2015). Adaptation and evaluation of recommendations for short-term shopping goals, in *Proceedings of the 9th ACM Conference on Recommender Systems, RecSys 2015, Vienna, Austria, September 16-20, 2015* (ACM), pp. 211–218.

Järvelin, K. and Kekäläinen, J. (2002). Cumulated gain-based evaluation of IR techniques, *ACM Trans. Inf. Syst.* **20**, 4, pp. 422–446.

Jäschke, R., Marinho, L. B., Hotho, A., Schmidt-Thieme, L. and Stumme, G. (2007). Tag recommendations in folksonomies, in *Knowledge Discovery in Databases: PKDD 2007, 11th European Conference on Principles and Practice of Knowledge Discovery in Databases, Warsaw, Poland, September 17-21, 2007, Proceedings*, *Lecture Notes in Computer Science*, Vol. 4702 (Springer), pp. 506–514.

Joachims, T. and Swaminathan, A. (2016). Counterfactual evaluation and learning for search, recommendation and ad placement, in R. Perego, F. Sebastiani, J. A. Aslam, I. Ruthven and J. Zobel (eds.), *Proceedings of the 39th International ACM SIGIR conference on Research and Development in Information Retrieval, SIGIR 2016, Pisa, Italy, July 17-21, 2016* (ACM), pp. 1199–1201.

Kluver, D. and Konstan, J. A. (2014). Evaluating recommender behavior for new users, in *Eighth ACM Conference on Recommender Systems, RecSys '14, Foster City, Silicon Valley, CA, USA - October 06-10, 2014* (ACM), pp. 121–128.

Knijnenburg, B. P. and Willemsen, M. C. (2015). *Evaluating Recommender Systems with User Experiments* (Springer US, Boston, MA), ISBN 978-1-4899-7637-6, pp. 309–352.

Kohavi, R., Longbotham, R., Sommerfield, D. and Henne, R. M. (2009). Controlled experiments on the web: survey and practical guide, *Data Min. Knowl. Discov.* **18**, 1, pp. 140–181.

Konstan, J. A. and Adomavicius, G. (2013). Toward identification and adoption of best practices in algorithmic recommender systems research, in *Proceedings of the International Workshop on Reproducibility and Replication in Recommender Systems Evaluation, RepSys 2013, Hong Kong, China, October 12, 2013* (ACM), pp. 23–28.

Koren, Y. and Bell, R. M. (2015). Advances in collaborative filtering, in *Recommender Systems Handbook* (Springer), pp. 77–118.

Kosir, A., Odic, A. and Tkalcic, M. (2013). How to improve the statistical power of the 10-fold cross validation scheme in recommender systems, in A. Bellogín, P. Castells, A. Said and D. Tikk (eds.), *Proceedings of the International Workshop on Reproducibility and Replication in Recommender Systems Evaluation, RepSys 2013, Hong Kong, China, October 12, 2013* (ACM), pp. 3–6.

Kumar, G., Jerbi, H. and O'Mahony, M. P. (2017). Towards the recommendation of personalised activity sequences in the tourism domain, in *Proceedings of the 2nd Workshop on Recommenders in Tourism co-located with 11th ACM Conference on Recommender Systems (RecSys 2017), Como, Italy, August 27, 2017, CEUR Workshop Proceedings*, Vol. 1906 (CEUR-WS.org), pp. 26–30.

Lathia, N., Hailes, S., Capra, L. and Amatriain, X. (2010). Temporal diversity in recommender systems, in *Proceeding of the 33rd International ACM SIGIR Conference on Research and Development in Information Retrieval, SIGIR 2010, Geneva, Switzerland, July 19-23, 2010* (ACM), pp. 210–217.

Li, W., Eickhoff, C. and de Vries, A. P. (2012). Want a coffee?: predicting users' trails, in *The 35th International ACM SIGIR conference on research and development in Information Retrieval, SIGIR '12, Portland, OR, USA, August 12-16, 2012* (ACM), pp. 1171–1172.

Linden, G., Smith, B. and York, J. (2003). Amazon.com recommendations: Item-to-item collaborative filtering, *IEEE Internet Computing* **7**, 1, pp. 76–80.

Luo, L., Li, B., Berkovsky, S., Koprinska, I. and Chen, F. (2016). Who will be affected by supermarket health programs? tracking customer behavior changes via preference modeling, in *Advances in Knowledge Discovery and Data Mining - 20th Pacific-Asia Conference, PAKDD 2016, Auckland, New Zealand, April 19-22, 2016, Proceedings, Part I, Lecture Notes in Computer Science*, Vol. 9651 (Springer), pp. 527–539.

McLaughlin, M. R. and Herlocker, J. L. (2004). A collaborative filtering algorithm and evaluation metric that accurately model the user experience, in *SIGIR 2004: Proceedings of the 27th Annual International ACM SIGIR Conference on Research and Development in Information Retrieval, Sheffield, UK, July 25-29, 2004* (ACM), pp. 329–336.

McNee, S. M., Riedl, J. and Konstan, J. A. (2006). Being accurate is not enough: how accuracy metrics have hurt recommender systems, in *Extended Abstracts Proceedings of the 2006 Conference on Human Factors in Computing*

Systems, CHI 2006, Montréal, Québec, Canada, April 22-27, 2006 (ACM), pp. 1097–1101.

Mesas, R. M. and Bellogín, A. (2017). Evaluating decision-aware recommender systems, in *Proceedings of the Eleventh ACM Conference on Recommender Systems, RecSys 2017, Como, Italy, August 27-31, 2017* (ACM), pp. 74–78.

Ning, X., Desrosiers, C. and Karypis, G. (2015). A comprehensive survey of neighborhood-based recommendation methods, in *Recommender Systems Handbook* (Springer), pp. 37–76.

Pu, P., Chen, L. and Hu, R. (2012). Evaluating recommender systems from the user's perspective: survey of the state of the art, *User Model. User-Adapt. Interact.* **22**, 4-5, pp. 317–355.

Said, A. (2016). A short history of the recsys challenge, *AI Magazine* **37**, 4, pp. 102–104.

Said, A. and Bellogín, A. (2014). Comparative recommender system evaluation: benchmarking recommendation frameworks, in *Eighth ACM Conference on Recommender Systems, RecSys '14, Foster City, Silicon Valley, CA, USA - October 06-10, 2014* (ACM), pp. 129–136.

Said, A. and Bellogín, A. (2018). Coherence and inconsistencies in rating behavior: estimating the magic barrier of recommender systems, *User Modeling and User-Adapted Interaction.*

Said, A., Bellogín, A., Lin, J. J. and de Vries, A. P. (2014a). Do recommendations matter?: news recommendation in real life, in *Computer Supported Cooperative Work, CSCW '14, Baltimore, MD, USA, February 15-19, 2014, Companion Volume* (ACM), pp. 237–240.

Said, A., Fields, B., Jain, B. J. and Albayrak, S. (2013a). User-centric evaluation of a k-furthest neighbor collaborative filtering recommender algorithm, in *Computer Supported Cooperative Work, CSCW 2013, San Antonio, TX, USA, February 23-27, 2013* (ACM), pp. 1399–1408.

Said, A., Jain, B. J. and Albayrak, S. (2013b). A 3d approach to recommender system evaluation, in *Computer Supported Cooperative Work, CSCW 2013, San Antonio, TX, USA, February 23-27, 2013, Companion Volume* (ACM), pp. 263–266.

Said, A., Jain, B. J., Narr, S. and Plumbaum, T. (2012). Users and noise: The magic barrier of recommender systems, in *User Modeling, Adaptation, and Personalization - 20th International Conference, UMAP 2012, Montreal, Canada, July 16-20, 2012. Proceedings, Lecture Notes in Computer Science*, Vol. 7379 (Springer), pp. 237–248.

Said, A., Tikk, D. and Cremonesi, P. (2014b). Benchmarking - A methodology for ensuring the relative quality of recommendation systems in software engineering, in *Recommendation Systems in Software Engineering* (Springer), pp. 275–300.

Sakai, T. (2014). Statistical reform in information retrieval? *SIGIR Forum* **48**, 1, pp. 3–12.

Sarwar, B. M., Karypis, G., Konstan, J. A. and Riedl, J. (2001). Item-based collaborative filtering recommendation algorithms, in *Proceedings of the Tenth International World Wide Web Conference, WWW 10, Hong Kong, China, May 1-5, 2001* (ACM), pp. 285–295.

Schedl, M., Knees, P., McFee, B., Bogdanov, D. and Kaminskas, M. (2015). Music recommender systems, in *Recommender Systems Handbook* (Springer), pp. 453–492.

Smyth, B. and McClave, P. (2001). Similarity vs. diversity, in *Case-Based Reasoning Research and Development, 4th International Conference on Case-Based Reasoning, ICCBR 2001, Vancouver, BC, Canada, July 30 - August 2, 2001, Proceedings, Lecture Notes in Computer Science*, Vol. 2080 (Springer), pp. 347–361.

Tomlinson, S. (2012). Measuring robustness with first relevant score in the TREC 2012 microblog track, in *Proceedings of The Twenty-First Text REtrieval Conference, TREC 2012, Gaithersburg, Maryland, USA, November 6-9, 2012*, Vol. Special Publication 500-298, (National Institute of Standards and Technology (NIST)).

Zhao, X., Zhang, W. and Wang, J. (2015). Risk-hedged venture capital investment recommendation, in *Proceedings of the 9th ACM Conference on Recommender Systems, RecSys 2015, Vienna, Austria, September 16-20, 2015* (ACM), pp. 75–82.

Chapter 10

Recommendations Biases and Beyond-Accuracy Objectives in Collaborative Filtering

Pasquale Lops, Fedelucio Narducci, Cataldo Musto, Marco de Gemmis, Marco Polignano and Giovanni Semeraro

Department of Computer Science, University of Bari Aldo Moro, Via E. Orabona 4 - I70126 Bari, Italy, firstname.lastname@uniba.it

Recommender systems research, traditionally focused on accuracy, is currently paying more and more attention to additional factors for evaluating the perceived quality and usefulness of recommendation lists. In this paper, we present a survey of the most important dimensions, other than accuracy, usually taken into account in the collaborative filtering literature. We survey beyond-accuracy objectives, i.e. novelty, diversity and serendipity, and the main techniques for increasing them. Moreover, we discuss possible undesired biases occurring in collaborative filtering algorithms, and how to effectively deal with them.

10.1. Introduction

Even though the primary goal of recommender systems is to find items that best match the model of user preferences, in the last years there has been a huge attention to the analysis of what recommenders recommend, i.e. what they include in the top-N recommendation lists [Jannach *et al.* (2015)]. This allows to take into account additional factors, other than accuracy, which contribute to the perceived quality and usefulness of recommendations, and at the same time this allows to assess if recommendation lists exhibit possibly undesired biases in order to adopt specific countermeasures to deal with them.

The importance of taking into account additional factors, other than accuracy, is acknowledged and emphasized in several studies [Herlocker *et al.* (2004); McNee *et al.* (2006); Ge *et al.* (2010); Hurley and Zhang (2011); Zhang *et al.* (2012); de Gemmis *et al.* (2015); Castells *et al.* (2015); Kaminskas and Bridge (2016)]. Among the possible factors, most studies emphasize the importance of *novelty*, i.e. how different a recommendation is with respect to what has been previously seen or experienced by a user [Castells *et al.* (2015)], *serendipity* [Iaquinta *et al.* (2008); Onuma *et al.* (2009); Kawamae (2010); Zhang *et al.* (2012); de Gemmis *et al.* (2015)], i.e. the experience of receiving unexpected suggestions helping the user to find surprisingly interesting items she might not have otherwise discovered, and *diversity* [Hurley and Zhang (2011); Adomavicius and Kwon (2012); Castells *et al.* (2015)], at the level of *individual* users [Ziegler *et al.* (2005)], or *aggregate* across all users [Adomavicius and Kwon (2012)].

Those additional factors, such as the aforementioned novelty, serendipity or diversity, have been recognized as a goal that often conflicts with accuracy [Fleder and Hosanagar (2009)], therefore it is important that systems were designed and evaluated by taking into account the need of properly balancing the different factors [Jugovac *et al.* (2017)]. Several optimization strategies have been proposed for *beyond accuracy* objectives. Most of them focus on one specific objective, even though recently the attention has been posed on assessing how optimization strategies affect the different objectives, i.e. how different objectives relate one to each other [Kaminskas and Bridge (2016)].

In this paper, we focus on the wider perspective of recommender systems aiming at generating more valuable and useful recommendations, beyond the simple prediction accuracy. In particular, we envisage a chapter describing the new dimensions of evaluation for collaborative-based recommender systems. In Section 10.2, we discuss the main additional dimensions, other than accuracy, usually taken into account in the Collaborative Filtering (CF) literature as well as some possible observed biases. Then, strategies to adapt or extend CF recommenders to take into account the new dimensions of evaluations or to deal with possible undesired biases are presented in Section 10.3. Finally, we describe some libraries supporting the evaluation in terms of both accuracy and beyond-accuracy metrics (Section 10.4), and close the chapter with a summary and discussion of the main challenges and directions for future research (Section 10.5).

10.2. Popularity Bias and Beyond Accuracy Metrics

In this section, we discuss the popularity bias of CF systems and the beyond-accuracy metrics usually taken into account in the evaluation, such as diversity, novelty and serendipity. As regards CF techniques, we focus on:

- Neighborhood-based methods, both *user-based* [Ning *et al.* (2015)] and *item-based* [Koren and Bell (2015)], whose key idea is to recommend a user with items preferred in the past by like-minded users.

- Matrix Factorization (MF), which trains a model by performing dimensionality reduction on the user-item rating matrix, in order to represent users and items into a subspace of latent factors, hopefully capturing implicit properties of both users and items.

10.2.1. *Popularity Bias*

CF systems generate recommendations for the target user by exploiting information about items that other users with similar tastes liked or bought in the past. Thus, the similarity in taste is calculated based on the similarity of users' rating history [Ning *et al.* (2015)].

The main drawback of this approach is that users are more likely to provide feedback on *popular* items than on niche ones, and this introduces a bias towards popular items in the observed user feedback compared to the user's true interests [Steck (2011)]. Popularity follows Zipf's Law, a power law that represents the discrete form of the Pareto distribution, from which the well-known Pareto principle, also known as the 80-20 rule, derives. For example, in the MovieLens1M dataset [Harper and Konstan (2016)] the top 20% most popular items cumulatively have many more ratings than the 80% items in the long tail. This leads to the creation of a *rich-get-richer effect* for popular items and vice-versa for unpopular ones. This problem is called *popularity bias*, and structural reasons for that bias are also identified in [Cañamares and Castells (2017)], upon a probabilistic reformulation of memory-based CF.

In offline evaluations, recommending popular items represents a very strong baseline with respect to accuracy measures such as precision and recall [Cremonesi *et al.* (2010); Steck (2011); Bellogín *et al.* (2017)], although this could lead towards recommendation of items interesting neither to users — possibly disappointed by the recommender system — nor to content providers — often interested in digital markets to boost sales of items in the long tail, i.e. less known or niche items.

It has been shown by real-world studies that the recommendation of popular items might be misleading since it does not always lead to the desired sales or persuasion effects [Jannach and Hegelich (2009)], rather it leads to undesired blockbuster effects [Fleder and Hosanagar (2009)]. Indeed, recommendations based on sales or ratings prevent recommendations of items with limited historical data, even if they would be rated favorably. Moreover, popularity-based recommendation could fail in the so-called *bounded domains*, i.e. those where the same product or service can be consumed by a limited number of users (e.g., hotel rooms) [Cremonesi *et al.* (2013)]. Hence, it becomes important that recommender systems achieve a good balance between popular and less-popular items [Abdollahpouri *et al.* (2017)].

In the literature, popularity is usually measured by counting the number of ratings received by an item or, depending on specific application scenarios, by the number of sales of a product, or the number of listens of a song, etc. This way of measuring popularity does not take into account whether ratings assigned to the items are positive or negative. This means that an item may be popular even though it is disliked by most of the users. Hence, another popularity indicator is the item's average rating, even if this measure is not able to describe if the item is liked or disliked by a large or small number of users.

Popularity is usually analyzed through other measures, e.g. *aggregate diversity* (see Section 10.2.2), to assess the level of personalization provided by a recommender system. As an example, a popularity-based recommendation algorithm suggesting the same most popular items to all the users across the system will have a very low aggregate diversity (low level of personalization), whereas a recommendation algorithm suggesting different top-N items to each user will have a high aggregate diversity (high level of personalization). For this reason, in the literature the analysis of popularity bias is often performed in combination with aggregate diversity and concentration bias, i.e. the actual distribution of recommended items, in order to capture inequalities with respect to how frequently certain items are recommended to users. To this purpose, the Gini index [Gini (1921)] is often adopted, which takes values between 0 and 1, where 0 represents an equal distribution of frequencies, i.e. all the items are recommended with the same frequency, while 1 corresponds to maximal inequality, i.e. all the recommendations are concentrated on a single item.

However, as discussed in literature and well summarized in [Bellogín *et al.* (2017)], the offline evaluation settings for recommender systems, usually based on Information Retrieval methodologies, involve subtle differences, which result in substantial biases to the effectiveness measurements that may distort the empirical observations and hinder comparison across systems and experiments. For example, one of the main problem discussed in [Steck (2011)], is that ratings are *missing not at random* (MNAR), and subject to biases affecting both the input for algorithms and the data for evaluation.

10.2.2. *Diversity*

Diversity is one of the qualities considered particularly useful for improving the user experience with a recommender system. Diversity is rooted in the Information Retrieval (IR) research area [Kaminskas and Bridge (2016)]. The motivations behind the diversification of the results of an IR system is determined by the need of effectively matching the user information need. For example, if the user query contains an ambiguous word like *apple* and it is not possible to disambiguate the user intention, a solution consists in producing a list of documents related to the different meanings of the ambiguous query term [Clarke *et al.* (2008)]. In order to accomplish this task, a measure for defining the differences between the retrieved documents is thus required. This measure is actually a function that computes how different two documents are and it can be based on the features the retrieved documents share, such as metadata, topics, document types [Carbonell and Goldstein (1998); Clarke *et al.* (2008); Wang and Zhu (2009); Agrawal *et al.* (2009)]. This measure is the key concept in the diversity definition.

Similarly to the IR field, the goal of a recommender system is to produce a set of recommendations that maximizes the utility of the target user. Some authors express the diversity in terms of *rarity* of the items in the recommendation list [Patil and Taillie (1982)], others define the diversity in terms of dissimilarity of all pairs of items in a given set [Hurley and Zhang (2011)].

Inspired by the formulation in [Smyth and McClave (2001)], proposed for computing the diversity between cases in case-based recommender systems, Kaminskas and Bridge [Kaminskas and Bridge (2016)] generalize the formula as follows:

$$Diversity(R) = \frac{1}{|R|(|R|-1)} \sum_{i \in R} \sum_{j \in R, j \neq i} dist(i,j) \qquad (10.1)$$

where R denotes the list of the recommended items, $|R|$ denotes the number of items in the list R ($|R| > 1$), and *dist* is a diversity measure (one of those previously introduced). Equation (10.1) computes the *average diversity*, also known as *intra-list diversity* or *individual diversity* [Ziegler *et al.* (2005)] since it considers diversity from an individual user perspective [Adomavicius and Kwon (2012)].

The distance between two items can be computed in several ways, and it is usually defined as the complement of similarity [Smyth and McClave (2001)]. It can be based on some item features [Ziegler *et al.* (2005)], or vector representation of the items, e.g. in [Kelly and Bridge (2006)] the Hamming distance computed on the binary rating vectors of the items is used as distance metric. In [Vargas and Castells (2014)], a metric based on the genre diversification of items is defined. The authors consider the random selection of items as the optimal strategy for generating pure genre diversity. The proposed metric, called *binomial diversity*, computes the distance of the genre distribution in the list of recommendations with the distribution obtained by a random selection of the items. A similar idea is proposed in [Di Noia *et al.* (2017)]. Here, a re-ranking methodology based on multi-attribute diversification is based on the user propensity to diversity on a specific attribute (e.g. genre, director, starring in the movie domain).

Bellogín *et al.* [Bellogín *et al.* (2013)] have adapted the metric α-*nDCG* proposed by Clarke *et al.* in IR field [Clarke *et al.* (2008)] to the recommendation scenario. α-*nDCG* accounts for relevance and diversity together and it is a variant of the *nDGC* which penalizes the documents sharing features with documents ranked higher in the list. The α parameter is used to balance the emphasis between relevance and diversity.

An example of three recommendation lists with different intra-list diversity is reported in Fig. 10.1.

Fig. 10.1. Example of recommendation lists with different degree of diversity: the 1st list has very low diversity since all the movies have the same actor (Javier Bardem) and the same genre (drama, romance); in the 2nd list the actor is the same but there are different genres; the 3rd list as higher diversity, in terms of both actor and genre.

Opposite to individual diversity, *aggregate diversity* takes into account the recommendation diversity across all users [Brynjolfsson *et al.* (2011); Adomavicius and Kwon (2012)], and it is measured as the total number of distinct items recommended across all users. Aggregate diversity is also called *catalog coverage* since it measures the capability of a recommender system to cover the item catalog. For top-N recommendations, aggregate diversity is usually measured as the total number of distinct items recommended across all users based on the top-N recommended items lists. It is called *DiversityInTopN* and is formally defined as follows:

$$DiversityInTopN = \left| \bigcup_{u \in U} R_N(u) \right| \tag{10.2}$$

where $R_N(u)$ is the list of the most highly ranked N items for the target user u.

It is worth to note that a high individual diversity does not imply a high aggregate diversity. Indeed, a recommender that suggests the same list of *diverse* items to all the users will have a high intra-list diversity, but the lowest aggregate diversity.

Aggregate diversity represents an important quality dimension from both a business and a user perspective since an increase of the catalog coverage and of the distribution of the items across the users may increase both sales and user satisfaction [Vargas and Castells (2014)].

There is now a large consensus that users are more satisfied when they receive diversified recommendations, even though this causes a physiological loss in terms of accuracy [Ziegler *et al.* (2005); Shi *et al.* (2012); Vargas *et al.* (2014)]. For example, a higher accuracy can be achieved by recommending the most popular items, but this approach generates less personalized recommendations and, consequently, a reduction in diversity (in particular aggregate diversity). Conversely, if the recommender tries to improve the diversity, it generally recommends not popular items for which the prediction of the user liking is a hard task. In [Adomavicius and Kwon (2012)] the authors carried out a simple experiment: they compared the results in terms of accuracy and diversity in two extremely different settings: in the first one, only popular items were recommended to users, while in the second one long-tail items were recommended. The experiment was performed on the MovieLens dataset, where popularity-based recommendations show a *precision-in-top-1* of 0.82 by recommending 49 distinct items of the whole catalog, whereas the long-tail recommendations achieve an accuracy of 0.68 with 695 distinct items. This very simple experiment demonstrates that it is possible to easily increase the diversity of the system by recommending less popular items. However, the goal of a recommender system should be to improve diversity without dramatically worsening accuracy. To sum up, the intuition is that a recommendation algorithm should increase diversity without including too many items that are not relevant to the user.

10.2.3. *Novelty*

The novelty of a piece of information generally refers to how different it is with respect to what has been previously seen or experienced by a person [Castells *et al.* (2015)].

Basically, there is a subtle connection between the concept of novelty and that of diversity, since the former refers to the difference between the information a person is currently enjoying and the information she enjoyed in the past, while the latter is atemporal and rather related to the differences between the components of something a person is currently experiencing (e.g., the differences between item pairs in a recommendation list).

Similarly to diversity, the definition of novelty finds its roots in IR research. The problem of providing users with novel information was first raised by Baeza-Yates and Ribeiro-Neto [Baeza-Yates *et al.* (1999)], who proposed to calculate the novelty of a set of retrieved documents as the fraction of relevant documents which are unknown to the user. The higher the number of unknown (and relevant) documents which are retrieved, the higher the novelty. A similar idea was proposed by Zhang *et al.* [Zhang *et al.* (2002)], who defined the novelty of a document as the opposite of its redundancy, which can be in turn calculated as the average distance between the current document and those the user previously consumed. This vision was further investigated by Allan *et al.* [Allan *et al.* (2003)], who extended the concept of redundancy at a sentence level, by considering the amount of non-redundant information contained in a single sentence of a document.

Recommender Systems literature inherited and extended these definitions, since a novel recommendation is typically a recommendation which is unknown to the user and different from the items she has consumed before. The main problem arising from this definition lies in the fact that it is not easy to precisely frame what is unknown for the current user. The easiest way to tackle this problem is to explicitly ask the users whether each item in her recommendation list is known or not. Such a process requires that a huge (and costly) user study is arranged every time a new recommendation list or a new recommendation algorithm has to be evaluated, and this is not feasible for almost all scenarios.

As a consequence, most of the literature tried to handle this problem by approximating the concept of novelty, that is to say, by defining some metrics whose goal is to estimate whether the user knows the item or not, without explicitly asking. Basically, two different visions have been proposed in literature [Vargas and

Castells (2011)]: an identity-based novelty and a global novelty. In the first case, novelty is calculated on the ground of previous behavior of the user (e.g., which items she consumed). In the latter, the global behavior of the whole community of the users is taken into account.

Most of the metrics proposed in literature fall into the first class: as an example, in [Nakatsuji *et al.* (2010a)], Nakatsuji et al. calculate the novelty as the distance between the present interests of the user and the items she received as recommendations. Basically, the distance between two items a and b can be calculated as the opposite of their similarity $sim(a,b)$, which can be in turn calculated as the number of overlapping properties describing the items (*content-based similarity*) or as the difference between the ratings the users give them (*rating-based similarity*).

By following a similar insight, Zhang [Zhang (2013)] points out that a novel recommendation should have three characteristics: being unknown, being relevant, being dissimilar from the previous recommendations. Kapoor [Kapoor *et al.* (2015)] further extended this definition by also including the concept of temporal novelty, which refers to those items which are known but forgotten, a very typical scenario for domains with frequent and repeated item consumption as music recommendation.

In all these cases, the concept of novelty is focused on the experience of the target user, since a novel item is an item which contains information different from that she previously consumed. The opposite vision, which is typically referred to as global novelty, relates the novelty of a document to its popularity among all the users [Celma and Herrera (2008)]. In this case, only the items that are in the so-called long-tail, i.e. the part of the item catalog seen (rated or purchased) by a small part of the user community, are labeled as novel. According to this definition, the novelty can be simply calculated as the opposite of the popularity. In other terms, the less popular an item is, the more likely it is to be unknown to the users. Clearly, as stated by Celma *et al.* [Celma and Herrera (2008)], the unpopularity of an items is not always a good indication of its novelty. Indeed, a user familiar with rare items would probably know other similar rare items, but this simple calculation provides a good approximation for measuring the novelty without arranging a long and costly user study. More recently, other heuristics to approximate the popularity of an item have been proposed: as an example, Jambor and Wang [Jambor and Wang (2010)] used rating variance in the dataset, while Oh *et al.* [Oh *et al.* (2011)] exploited exogenous information sources, such as box office earnings for movies, with the simple insight that the smaller the earnings, the lower the popularity of a movie.

Recently, Vargas and Castells [Vargas and Castells (2011)] made an effort to define a solid theoretical framework to model the concept of novelty. In their work, several variants of novelty metrics were proposed. Specifically they defined Mean Popularity Complement (MPC) as the opposite of popularity, and Mean Inverse User Frequency (MIUF) as the logarithm of the average number of users who consumed a specific item. Such a slight variation aims at giving more importance to very rare items.

Formally, given a set of recommendations R and given a function $p(i)$ which returns the popularity of an item i, MPC and MIUF are defined as follows:

$$MPC(R) = \frac{1}{|R|} \sum_{i \in R} 1 - p(i) \tag{10.3}$$

$$MIUF(R) = \frac{1}{|R|} \sum_{i \in R} -log_2 p(i) \tag{10.4}$$

In order to evaluate recommendation techniques with respect to novelty, in both cases the overall novelty is obtained by averaging the values of novelty over all the recommendations generated by the algorithm.

10.2.4. *Serendipity*

Overspecialization is a typical problem of recommender systems which stems from the fact that they aim at finding items that best match the model of user preferences in order to improve accuracy, regardless of the actual usefulness of the suggestions [McNee *et al.* (2006)]. For example, if the target user likes the movie *Star Trek into Darkness*, user-based collaborative algorithms will suggest movies liked by other people who liked that movie. Most of the target user's neighbors will probably be science-fiction fans, hence the user will be provided with items within her existing range of interests and her tendency towards a certain behavior is reinforced by creating a self-referential loop. A possible solution to this problem is to design recommendation algorithms which provide *serendipitous* suggestions helping the users find surprisingly interesting items they might not have otherwise discovered, or that would have been really hard to discover [Herlocker *et al.* (2004)].

Serendipity is a factor which contributes to improve the perceived quality of recommendations [Zhang *et al.* (2012)]. According to McNee *et al.* [McNee *et al.* (2006)], serendipity is the experience of receiving an unexpected and fortuitous item recommendation, while other authors associate serendipity to a positive emotional response of the user about novel items, which involves also surprise [Shani and Gunawardana (2011); Gunawardana and Shani (2015); de Gemmis *et al.* (2015)]. According to these definitions, serendipity in recommender systems is achieved by providing *unexpected* suggestions, while maintaining *high relevance*. While relevance is usually determined in terms of closeness to the user profile, the assessment of unexpectedness of recommendations is not immediate. Previous studies agreed on defining unexpectedness as the deviation from a benchmark model or primitive prediction method that generates expected recommendations [Murakami *et al.* (2008); Ge *et al.* (2010)]. For instance, in case of a movie recommender system, expected recommendations could be blockbusters seen by many people, or movies related to those already seen by the user, such as sequels, or those with same genre and director. A limitation of this approach is its sensitivity to the choice of the primitive prediction method.

Furthermore, it is useful to point out the differences with related notions of *novelty* and *diversity*. Continuing with movie recommendations, if the system recommends a movie the user was not aware of, the movie will be novel, but not serendipitous. On the other hand, a movie by a young, not very popular director is more likely to be serendipitous (and also novel). Although diversity is very different from serendipity, a relationship between the two notions exists, in the sense that providing the user with a diverse list can facilitate unexpectedness [Adamopoulos and Tuzhilin (2011)]. We can reasonably assume that users could be surprised to some extent when a romantic movie is recommended within a list of science-fiction movies. However, the diversification of recommendations does not necessarily imply serendipity because diverse items could all fall into the range of user preferences.

Serendipity has been recognized as a goal that often conflicts with accuracy: as an extreme case, let us consider random recommendations, which improve serendipity but cause a drastic loss in accuracy, making the system actually ineffective [Fleder and Hosanagar (2009)]. Therefore, it is important that systems were designed and evaluated by taking into account the need of properly balancing these two factors.

10.3. Algorithmic solutions

In this section, we review the techniques to deal with the popularity bias of CF systems, and to increase diversity, novelty and serendipity. Algorithmic solutions and the corresponding references are summarized in Table 10.1.

Table 10.1. Overview of algorithmic solutions for dealing with the popularity bias and for increasing diversity, novelty and serendipity.

Goal	Algorithmic solutions	Main references	Comment
Dealing with Popularity Bias	Application of a post-processing scheme	Ranking-based techniques to promote long-tail (less popular) items, with a small level of accuracy loss [Adomavicius and Kwon (2012)]	Easy way to deal with different biases of different recommendation algorithms
		Personalized Bias Adjustment (PBA) of relevance scores to deal with user and recommendation lists biases regarding popular items [Jannach *et al.* (2015)]	
	Tuning hyperparameters of recommendation algorithms	More iterations of gradient descent to reduce popularity and concentration bias of Funk-SVD [Jannach *et al.* (2015)]	No need to modify or having in-depth knowledge of inner working of recommendation algorithms
		Reducing regularization for Factorization Machines with ALS reduces the concentration bias [Jannach *et al.* (2015)]	
		More factors for MF increase coverage, diversity and novelty [Kaminskas and Bridge (2016)]	
	Proper adaptation of recommendation algorithms	Probabilistic neighbor-selection (k-PN) for selecting similar but diverse neighbors for k-NN methods [Adamopoulos and Tuzhilin (2014)]	Possible not straightforward adaptation of recommendation algorithms to avoid the bias
		Biasing sampling process of implicit feedback statements in BPR algorithm [Jannach *et al.* (2015)]	
		Adaptive BPR: learning also from feedback coming from items browsed by users in BPR [Pan *et al.* (2015)]	
		Fairness-aware regularization for RankALS [Abdollahpouri *et al.* (2017)]	
		Recommendation neutrality [Kamishima *et al.* (2014)]	

Table 10.1. (*Continued*).

Goal	Algorithmic solutions	Main references	Comment
Increasing Diversity	Re-ranking techniques	Re-ranking algorithms using an objective function which combines relevance and diversity [Kaminskas and Bridge (2016)] [Smyth and McClave (2001)] [Ziegler *et al.* (2005)] [Vargas and Castells (2014)]	Techniques that can be used with any recommendation algorithm
		Pareto frontier to find the best balance of recommendation algorithms maximizing different qualities, such as accuracy, diversity and novelty [Ribeiro *et al.* (2012)]	
	Modeling techniques	Use of an objective function to balance predicted relevance and variance of the recommendation list, based on latent factors coming from MF [Shi *et al.* (2012)]	Recommendation algorithms need to be properly extended
		Definition of the Set Diversity Bias, working at item-set level rather than at single-item level. Prediction of an item score based on both relevance and diversity of the item set [Su *et al.* (2013)]	
		Incorporate diversity into a personalized ranking objective [Hurley (2013)]	
		Clustering based on priority-modeoids [Boim *et al.* (2011)] and multidimensional clustering [Li and Murata (2012)]	

Table 10.1. (*Continued*).

Goal	Algorithmic solutions	Main references	Comment
Increasing Novelty	Re-ranking techniques	Use of an objective function to balance relevance and novelty of the recommendation [Kaminskas and Bridge (2016)]	Techniques that can be used with any recommendation algorithm
		Use of a distance measure to re-rank recommendations by promoting items far from those previously consumed [Rao *et al.* (2013)]	
	Modeling techniques	Clustering long-tail items and calculating prediction for such items by leveraging ratings in the clusters [Park and Tuzhilin (2008)]	Recommendation algorithms need to be properly extended
		Bipartite graph-based data model with a spreading strategy to promote items voted by less users [Zhou *et al.* (2010)], or with the assignment of a lower cost to edges connecting less popular items [Shi (2013)]	
Increasing Serendipity	Finding unexpected recommendations out of the filter bubble	AURALIST defines the bubble as the cluster of preferred artists and recommends those outside cluster groups [Zhang *et al.* (2012)]	
		Outside-The-Box defines the bubble as region of interests and recommends items not falling into those regions [Abbassi *et al.* (2009)]	
	Analysis of data related to users or items	Analysis of purchase logs of innovators, i.e. users with similar preferences with the active user who purchase the same items earlier [Kawamae (2010)]	
		Knowledge infusion approach based on content analysis of items able to discover hidden and non-obvious correlations not detectable using standard similarity scores [de Gemmis *et al.* (2015)]	

Table 10.1. (*Continued*).

Goal	Algorithmic solutions	Main references	Comment
	Analysis of relatedness among users	K-NN recommendation algorithm based on k-furthest neighbors, i.e. users dissimilar to the target user [Said *et al.* (2013)], or based on relatedness among users computed through random walk with restarts on a user similarity graph [Nakatsuji *et al.* (2010b)]	
		TANGENT leverages likeminded users well connected to familiar items and to unrelated others as well [Onuma *et al.* (2009)]	

10.3.1. *Dealing with Popularity Bias*

As reported in [Jannach *et al.* (2015)], recommendation algorithms tend to exhibit different undesired biases, even though they implement very similar techniques. For example, it has been shown that two factorization machines variants, i.e. Alternating Least Squares (ALS) and Markov Chain Monte Carlo optimization (MCMC) [Rendle (2012)], can lead to quite different results in terms of the concentration bias even though they solely differ in terms of the optimization procedure.

In the next sections, we present possible approaches to deal with popularity and concentration biases, specifically devised for CF methods.

10.3.1.1. *Application of a post-processing scheme to deal with the undesired bias*

The easiest way to deal with undesired biases is to apply a proper post-processing step to the results produced by state-of-the-art recommendation algorithms, aiming at reducing or avoiding the biases. This can be achieved by implementing very basic or more complex strategies. For example, in order to deal with the popularity bias, in the final recommendation list the predicted rating of very popular items could be lowered, while the ratings of unpopular items could be increased, in order to modify the ranking of items. Of course, the definition of the adjustment of predicted ratings is an issue since it could lead to the promotion of irrelevant items.

More general approaches apply proper post-processing methods on the recommendation lists generated using state-of-the-art recommendation algorithms, usually optimized for Root Mean Squared Error (RMSE), in order to remove or mitigate some biases. As described in Section 10.3.2, this approach has been already implemented to improve the aggregate diversity of recommender systems, using ranking-based techniques [Adomavicius and Kwon (2012)]: even though the main goal of the proposed techniques is to improve the trade-off between accuracy and diversity, it has been shown that ranking approaches are able to deal with the popularity bias since they recommend significantly more long-tail items with a small level of accuracy loss.

Recently, a post-processing scheme based on user-specific rating adjustments has been proposed in [Jannach *et al.* (2015)], in order to deal with the popularity bias. The technique, called *personalized bias adjustment (PBA)*, provides an adjustment of relevance scores of rating prediction-based algorithms which takes into account both the original rating bias in the user profile and that of the recommendation list. PBA assesses the 1) *user's bias*, i.e. tendency of the user to generally like or dislike popular items, by taking into account both the popularity of items rated in the user profile as well as the ratings assigned to those items; 2) the *recommendation list bias*, computed in the same way as the user bias, by considering the popularity of recommended items as well as their predicted rating. PBA tries to make these biases aligned, i.e. smaller, through a proper optimization function matching the list and user biases and a simple gradient descent approach to solve the minimization problem. Experiments on the MovieLens400k dataset, a subset of the MovieLens10M dataset used in [Jannach *et al.* (2015)], show that applying PBA post-processing to matrix factorization (MF) [Koren (2010)] and Factorization Machines with ALS does not lead to a change in the overall accuracy of the algorithms in terms of RMSE, precision and recall, but at the same time leads to a decrease of the concentration bias.

The technique could be easily applied with different rating prediction algorithms and could be applied to deal with different biases.

10.3.1.2. *Tuning the hyperparameter settings of the recommendation algorithm*

The same recommendation algorithm configured with different hyperparameters might generate different top-*N* recommendations and create undesired biases.

Funk-SVD algorithm [Funk (2006)], a state of the art MF technique using gradient descent as an optimization procedure, exhibits a strong concentration bias when the optimal number of iterations (training steps of the gradient descent) in terms of RMSE is used [Jannach *et al.* (2015)]. The analysis carried out on MovieLens shows that increasing the number of iterations leads on one hand to a small increase of RMSE due to the overfitting, and on the other hand to a reduction of the concentration bias, to a lower popularity bias and also to more diverse recommendations in terms of Intra-List Diversity (ILD) [Jannach *et al.* (2015)].

Changing the number of training steps does not influence factorization machines with ALS in terms of accuracy or other quality factors; on the contrary they are sensible to the variation of the regularization hyperparameter λ used to penalize model overfitting. Similarly to the results of the previous experiment, in correspondence to the value of λ where the best RMSE is observed, the algorithm shows a concentration bias. Reducing λ slightly increases the RMSE, but at the same time reduces the concentration bias and increases the recommendation diversity [Jannach *et al.* (2015)].

The influence of recommendation algorithms parameters on the difference performance metrics is also acknowledged in [Kaminskas and Bridge (2016)], where the authors performed an analysis on the MovieLens dataset, in order to assess the influence of the number of factors for a MF algorithm, and of the neighborhood size for CF algorithms, on different performance metrics, such as recall, coverage, novelty and diversity. For MF, increasing the number of factors leads to a loss of recall, and to an increase in terms of coverage, diversity and novelty, while for CF a higher number of neighbors leads to a decrease of coverage, diversity and novelty, and to an increase of recall.

The practical implication of this analysis is that a proper tuning of the hyperparameters may help to reduce potential concentration or popularity biases, by accepting a small loss of accuracy, without the need of modifying the recommendation algorithms or also having an in-depth knowledge of their inner working.

10.3.1.3. *Adaptation of the recommendation algorithm in order to avoid a certain bias*

In case the tuning of hyperparameters is not feasible in terms of complexity or does not report the expected outcomes, it could be useful to understand the inner logic of the recommendation algorithm, in particular which characteristics lead to the undesired bias, in order to design a solution to avoid that bias. The main problems with this approach are its requirement in terms of in-depth knowledge of the recommendation mechanisms, whose complexity depends on the algorithm, and the definition of countermeasures which can be algorithm-specific.

For classical k-NN methods, a possible countermeasure for avoiding the popularity and concentration bias is reported in [Adamopoulos and Tuzhilin (2014)], which describes an alternative neighbor selection scheme, called probabilistic neighbor selection (k-PN). The rating estimation is obtained by aggregating the ratings of the k probabilistically selected neighbors rather than those of the k most similar neighbors. The main intuition is that *similar but diverse neighbors should be used in neighborhood-based methods*, since leveraging the most similar neighbors might lead to obvious recommendations. The theoretical motivation behind this intuition is based on the phenomenon of *hubness* [Radovanovic *et al.* (2010)], where *hubs* are users (items) which appear in many more k-NN lists than other users (items), as well as to the *ensemble learning theory*, where it is known that the decrease of the generalization error of the model is obtained with estimators of the ensemble which are negatively correlated. The problem of hubness can be alleviated by a more equal distribution of the number of times each user (or item) is included in a neighborhood, while the need of negatively correlated estimators is obtained by considering neighbors as estimators, and by reducing their covariance [Adamopoulos and Tuzhilin (2013)]. Experiments on RecSys HetRec 2011 MovieLens [Cantador *et al.* (2011)], MovieTweetings [Dooms *et al.* (2013)] and Amazon [McAuley and Leskovec (2013)] datasets showed that systematically selecting diverse sets of neighbors allows to reduce the concentration bias while preserving a high accuracy.

For a more complex algorithm, such as Bayesian Personalized Ranking (BPR), which learns to rank items based on implicit feedback [Rendle *et al.* (2009)], a similar strategy based on the adaptation of the inner working of the algorithm is explained in [Jannach *et al.* (2015)]. It has been shown that the popularity bias in BPR is originated by the non-uniform distribution of implicit feedback statements used to optimize the model parameters through the gradient descent. Implicit feedback statements are in the form (u,i,j), which means that user u prefers item i more than j, and the bias is generated by the random sampling strategy, which leads to the selection of statements with high popular i and less popular j. A possible countermeasure is to bias the sampling process of items i and j, in order to sample statements where i is less popular and j is more popular, i.e. learning from feedback where the user liked an item i more than a generally popular item j, which is more informative in terms of user preferences. Several experiments have been performed on MovieLens400k and Yahoo!Movies datasets, using different distribution functions to determine the sampling strategy. Generally the method seems to be able to balance accuracy, popularity and concentration bias of BPR [Jannach *et al.* (2015)]. Other authors worked on the improvement of the inner logic of BPR, even though they did not take into account possible biases generated by the algorithm. In [Pan *et al.* (2015)], the authors extend implicit feedback coming from transactions, i.e. items bought, with that coming from examination records, i.e. items browsed (of high uncertainty w.r.t. the user's true preference), learn a confidence for each uncertain feedback in an adaptive manner, and propose a novel preference learning algorithm called Adaptive Bayesian Personalized Ranking (ABPR) able to leverage both the types of implicit feedback. Empirically, ABPR shows very promising recommendation results on MovieLens and Netflix in terms of ranking oriented evaluation metrics.

Other approaches propose more straightforward adaptations of recommendation algorithms. The method in [Abdollahpouri *et al.* (2017)] is based on the use of regularization to control the popularity bias of a recommender system based on a learning to rank setting, such as RankALS [Takács and Tikk (2012)]. The regularization term of the objective function is defined so that it will be minimized when the distribution of recommendations is *fair*, i.e. a list that achieves 50/50 balance between popular items and those in the long tail. The technique is called *fairness-aware* regularization, it could be applied to any learning algorithm in which the objective function is formulated as a pairwise function of items (e.g. BPR), and could be extended by defining a different distribution between popular and long tail items. Experiments on MovieLens1M and Epinions datasets show that the approach is able to provide a tunable mechanism for controlling the trade-off between accuracy and coverage.

Finally, another way to correct the popularity bias in CF is obtained by enhancing *recommendation neutrality* [Kamishima *et al.* (2014)], where neutrality is specified with respect to a specific viewpoint. If the viewpoint is that users have no interest in the popularity of items and wish to ignore this information, they can obtain recommendations that are neutral with respect to that aspect. The algorithm is based on a variant of the probabilistic MF model, where a constraint term is adopted to enhance neutrality by enforcing the statistical independence between the target viewpoint and a rating. Experiments on the Flixster dataset show that the method was able to correct the popularity bias with a not significantly worsening of prediction error in terms of MAE.

This analysis makes clear that the adaptation of the recommendation algorithm to avoid a certain bias might be very complex. Indeed, a preliminary analysis of the algorithm is needed to understand both the reasons for the bias and which part of the algorithm is responsible for that. Depending on the recommendation algorithm, the design of a variant of the algorithm to avoid the bias could be not straightforward.

10.3.2. *Increasing Diversity*

Similarly to the strategies that can be adopted to deal with the popularity bias of recommender systems, the techniques used for optimizing the diversity can be split in two main types: 1) techniques that do not modify the behavior of the recommendation algorithm, but adopt a post-processing step on the results (i.e. re-ranking), and 2) techniques that try to induce the diversity by directly generating diversified recommendations.

The main characteristic of the first type of techniques is that they can be applied to every algorithm and do not modify the recommendation process since the diversification is performed by a re-ranking step. The rank of the items is modified in order to avoid the occurrence of too similar items in the first positions of the list. The second type of techniques extends the recommendation algorithm for optimizing the diversity when the recommendations are generated. So, there is not a re-ranking step, and the list of recommended items is already diversified according to some criteria.

10.3.2.1. *Re-ranking Techniques*

Most of recommender systems rank the items to be shown to the user according to a relevance score (i.e., the predicted rating) computed by the recommendation algorithm. Indeed, a simple method is to compute the rank of an item as the reciprocal of its predicted rating. Therefore the top-N recommendation list will be composed of items with the highest predicted rating value. This ranking strategy is defined as *standard ranking* [Adomavicius and Kwon (2012)] and is shared also with IR systems that rank the retrieved documents in order of decreasing relevance score [Robertson (1977)].

More formally, the standard rank is computed as follow:

$$rank_{standard}(i) = \frac{1}{r(u,i)} \tag{10.5}$$

where $r(u,i)$ is the relevance score computed by the recommender system on the item i for the user u. Since the recommendation algorithm is generally programmed for optimizing recommendation accuracy, standard ranking does not consider diversification of the recommendations. In [Adomavicius and Kwon (2012)], standard ranking is compared with *popularity-based ranking* in order to evaluate the impact on the diversity of these two different ranking approaches. Popularity-based ranking is defined as follows:

$$rank_{popularity}(i) = |U(i)| \tag{10.6}$$

where $U(i)$ is the number of users who rated the item i. Accordingly, the larger the number of users who rated an item, the higher the position in the ranked recommendation list. In that comparison, Adomavicius and Kwon demonstrated that popularity-based ranking increased aggregate diversity by 3.6 times compared to the standard ranking approach, using the MovieLens dataset and item-based CF.

Hence, recommending the most popular items is a simple strategy to increase the *diversity*, even though they observed an accuracy loss of 20%. To overcome this problem, a parameterized version of the ranking function is proposed, where the parameter is a rating threshold which allows to consider only the items that were predicted above the predefined threshold to ensure an acceptable level of accuracy, as is typically done in recommender systems. If the user desires a high accuracy, the threshold could be set close to the maximum rating available, otherwise by decreasing the threshold more diversified recommendations could be obtained.

In [Kaminskas and Bridge (2016)], the authors describe a greedy re-ranking algorithm based on the idea of defining an objective function which is a combination of *relevance* and *diversity* of a given item. At each step the algorithm puts into the recommendation list the item which maximizes the objective function. This approach was originally proposed in the IR literature by Carbonell and Goldstein [Carbonell and Goldstein (1998)], who defined the Maximum Marginal Relevance (MMR) approach. MMR defined the objective function as a linear combination of the item's relevance and the negative of its maximum similarity to items already in the result list. Several recommendation algorithms [Smyth and McClave (2001); Ziegler *et al.* (2005)] defined the objective function as a linear combination of the relevance and average distance to the items in the recommendation list. More formally:

$$f_{objective}(i,R) = \alpha \cdot rel(i) + (1-\alpha) \cdot \frac{1}{|R|} \sum_{j \in R} dist(i,j) \tag{10.7}$$

where α is the weight that allows to control the impact of *relevance* and *distance* in the re-ranking process, similarly to the role of the above mentioned rating threshold; $rel(i)$ is the relevance score for the item i, and $dist(i,j)$ is the diversity function between two items. Smyth and McClave [Smyth and McClave (2001)] tested their algorithm on the Travel case base[1] and demonstrated that retrieval diversity can be improved without sacrificing retrieval similarity. Ziegler *et al.* [Ziegler *et al.* (2005)] conducted a large scale user survey and showed that the user's overall liking of recommendation lists goes beyond accuracy and involves other factors (e.g., the users' perceived list diversity). A different approach has been proposed in [Barraza-Urbina *et al.* (2015)], where the objective function computes the diversity between the recommended item and the items in the user profile. An experimental evaluation carried out on the MovieLens100k dataset demonstrated that the approach can be tuned towards more exploitative diverse results or more explorative diverse results with controlled sacrifice over relevance. A greedy strategy is also used in [Vargas and Castells (2014)]. Here a

[1] http://www.ai-cbr.org

greedy re-ranking approach optimizes the ranking by defining an objective function that linearly combines relevance and binomial diversity (see Section 10.2.2). Experimental results on the Netflix Prize dataset[2] and on the Million Song Dataset Challenge [McFee *et al.* (2012)] confirmed the effectiveness of their proposals. In [Zhang and Hurley (2008)], the authors do not apply a greedy approach but tackle the issue as a binary optimization problem, to control the diversity and matching quality of recommendation lists. The evaluation was performed on the MovieLens1M dataset and it has been shown that the method can increase the likelihood to recommend novel items, while maintaining good performance.

Finally, in [Ribeiro *et al.* (2012)], similarly to a hybrid recommendation approach, the recommendation list is composed by merging the items generated by different recommendation algorithms. More specifically, the recommendations are generated by user-based and item-based nearest neighbor approaches, popularity-based approach, content-based and demographic-based nearest-neighbor approaches. The Pareto frontier is used for finding the best balance of the three recommendation objectives: accuracy, diversity, and novelty. The idea is that each recommendation algorithm maximizes a recommendation quality, so the proposed model finds the best balance (i.e. the Pareto frontier) where none of the three qualities can be improved without penalizing the other two. Experiments conducted on the MovieLens and LastFM datasets showed that it is possible to combine different algorithms in order to produce better recommendations and to control the desired balance between accuracy, novelty and diversity.

10.3.2.2. *Modeling Techniques*

In the previous section we analyzed re-ranking techniques that can be used with any recommendation algorithm. In this section we present some approaches which extend the recommendation algorithms in order to generate diversified recommendations.

In [Shi *et al.* (2012)], a combination of MF and portfolio theory is proposed. The portfolio theory of IR [Wang and Zhu (2009)], inspired by the Modern Portfolio Theory which deals with investments on financial markets, argues that ranking under uncertainty does not consist only in retrieving individual relevant documents, but also in *choosing the right combination of relevant documents*. Accordingly, the list of the retrieved documents will satisfy an expected overall relevance and *variance*. An optimal ranking balances the overall relevance against the variance. Shi *et al.* [Shi *et al.* (2012)] defined an objective function that balances the predicted relevance and the variance of the recommendation list, where the

[2] http://www.netflixprize.com/

variance is based on latent factor vectors obtained from the MF approach. This model is able to adapt the diversification level to the user tastes. Indeed, if a user generally rates diverse items, the corresponding latent factors will have higher variance than a user who generally rates similar items. An experimental evaluation carried out on the MovieLens1M dataset demonstrated the effectiveness of the approach for adapting result diversification to the users' needs without accessing to explicit item properties. In addition, they also showed that the proposed approach is capable of effectively adjusting the trade-off between the relevance and the diversity of recommended items, and thus could further contribute to the overall recommendation quality. This idea to analyze the *tendencies of the individual users* based on their past behavior in order to adapt the diversification of the recommendations is also adopted in [Jugovac *et al.* (2017)], even though in that work a re-ranking approach is exploited.

A set-oriented CF model for diversified top-N recommendations is proposed in [Su *et al.* (2013)]. The authors integrate the diversity concept into a traditional MF model. This model works at item-set level rather than at single-item level by defining the new concept of *Set Diversity Bias*. The set-oriented CF model predicts the score on a given item by considering both relevance and diversity of the item set. It is possible to control the weight of diversity in the model through a parameter. The diversity of an item set is associated with the latent factors of the items in it. The experimental evaluation on the MovieLens1M dataset showed that the model outperforms traditional models in terms of personalized diversity and maintains good performance on relevance prediction.

A method to incorporate diversity into a personalized ranking objective in the context of ranking-based recommendation using implicit feedback is proposed in [Hurley (2013)]. The idea is to learn a prediction formula as the product of user and item feature vectors modified to weight the difference in ratings between two items by their dissimilarity. Promising results were obtained on the MovieLens1M dataset in terms of intra-list diversity with just a small loss in relevance.

Other techniques are based on clustering for diversifying the recommendations. The model proposed in [Boim *et al.* (2011)] is based on the notion of *priority-medoids*. The medoid of a given cluster is an element in the cluster for which the sum of the distances from it to the other items in the cluster is minimal. The priority-medoids adds the requirement that the representatives (i.e. the medoids) are the ones having highest rating in their corresponding clusters. Thus the clusters based on priority-medoids are different than the clusters formed when considering standard medoids. Experiments carried out on the Netflix Prize dataset demonstrated the effectiveness of the model.

In [Li and Murata (2012)] the authors propose a model based on multidimensional clustering and collaborative filtering. Multidimensional factors are used for ranking the recommendation list to substantially increase diversity while maintaining comparable levels of accuracy. The similarity between items is thus combined with multidimensional data clusters. This helps in the exploration of other clusters, which have similarity closer to the target user and provide him with good set of recommendations. The proposed approach performs better than traditional two-dimensional approaches in terms of improving quality of recommendations on the MovieLens100K dataset.

10.3.3. *Increasing Novelty*

Bellogin *et al.* [Bellogín *et al.* (2013)] showed that CF algorithms tend to have a lower novelty when compared with content-based counterparts. This behavior is also confirmed by Pampin *et al.* [Pampın *et al.* (2015)]: in their work they investigated beyond-accuracy metrics and they noted that the algorithm with the lowest novelty is the item-based k-NN. Conversely, in their experiments user-based k-NN shows a small improvement in terms of novelty, especially when the parameter k (number of neighbors) is small. However, this is a not a general behavior since other results [Kaminskas and Bridge (2016)] showed that item-based can overcome user-based k-NN on some datasets.

Regardless these little differences, the clear trend which emerges from all these empirical comparisons is that *CF algorithms generally have a lower novelty than other techniques*. This is confirmed also by a a large-scale evaluation conducted by Celma and Herrera [Celma and Herrera (2008)] on Last.fm, who showed that CF approach produces recommendations which are more familiar to the user. This issue is partially due to the *new item problem* which is typical of CF. Indeed, when an item receives just a few ratings it is very difficult to make it included in the recommendation lists, which are typically oriented towards more popular items (see Section 10.2.1).

As a consequence, several methodologies to generate more novel recommendations have been proposed in literature. The easiest (and trivial) way to provide users with novel recommendations is to filter out all the items she already rated or used. Obviously, such a trivial approach does not fulfill the goal, since in every recommendation scenario it is not realistic to ask the user to rate all the items she knows, thus it is necessary to think about other methodologies.

A very simple and empirical approach for promoting novel items has been proposed by Herlocker *et al.* [Herlocker *et al.* (2004)]. In their work the authors suggest to create a list of *obvious* recommendations and to remove such items from the recommendation lists generated by the algorithm. However, also this approach is not fully convincing, since the set of obvious items might be different for each user.

A more sophisticated technique was proposed by Hijikata *et al.* [Hijikata *et al.* (2009)], who designed a CF algorithm relying on two different rating profiles: the traditional one, based on the explicit ratings given by the users, and an *acquaintance* one, where the user express her familiarity (that is to say, known or unknown) with each item. Next, several hybridization strategies which combine both the profiles are proposed by the authors. However, even if the experiments confirmed the effectiveness of the approach, such a very simple technique has the important issue of doubling the cognitive load of the user, since two different profiles have to be constructed in order to get novel recommendations.

Generally speaking, the techniques to provide users with novel recommendations can be roughly split in two classes, which are the same classes we already introduced for the discussion of the diversity: *re-ranking techniques* and *model-based techniques*. In the first case, a set of candidate recommendations that have been selected on the basis of relevance are re-ranked in order to improve the novelty of the recommendations. In the latter, some shrewdness aiming at promoting novel items is directly injected in the recommendation model.

Re-ranking approaches for generating novel recommendation follow the same insight which is used to *diversify* a recommendation list. As previously shown, a very common way to generate *different* recommendations is to define a new objective function that combines the relevance of the item with other criteria. Formally, a simple variant of Equation (10.7) is proposed:

$$f_{objective}(i,R) = \alpha \cdot rel(i) + (1-\alpha) \cdot novelty(i) \tag{10.8}$$

As discussed in [Kaminskas and Bridge (2016)], $novelty(i)$ can be defined through the Mean Popularity Complement (MPC) and the Mean Inverse User Frequency (MIUF), as defined in Equations (10.3) and (10.4), respectively. In both cases, items which are both relevant and not popular among the users are provided with a higher overall score and are more likely to be recommended. As shown by the authors, even such a simple approach based on linear combination of different objective functions leads to good experimental results on both MovieLens and Last.fm data, since in all the cases the resulting novelty got a significant increase thanks to these formulas. Another attempt is proposed in [Rao *et al.* (2013)], where the authors exploit a distance measure based on a taxonomy of the concepts mentioned in a set of news articles in order to re-rank a recommendation list by promoting more novel items, i.e. far from those the user previously consumed.

On the other side, the main goal of *model-based techniques* was to adapt the behavior of recommendation algorithms in order to promote novel items. As an example, Park and Tuzhilin [Park and Tuzhilin (2008)] focused on long-tail items and proposed an algorithm to specifically improve the prediction accuracy on such items, in order to increase the likelihood of putting them in the recommendation lists. Their approach is based on the idea of clustering long-tail items and of calculating the prediction for such items by exploiting all the ratings in the clusters. Experiments on MovieLens and Bookcrossing datasets showed that such an approach can actually reduce the prediction error, but it cannot guarantee that novel items are introduced in the recommendation list.

An alternative approach was proposed by Zhou *et al.* [Zhou *et al.* (2010)]. In this work the authors define a bipartite graph-based data model representing users and items and introduce a weight spreading strategy specifically oriented at promoting novel recommendations by just giving more weight to the items voted by less users. A similar attempt was also proposed by Shi [Shi (2013)], who defined a graph-based data model and introduced the concept of *edge transition costs*, aiming at giving more importance to novel items by assigning a lower cost to the edges which connect less popular items. In both the works, the experiments carried out on MovieLens100k and Last.fm datasets, showed that these techniques are able to improve the overall novelty of the algorithms.

Finally, a very relevant attempt towards the generation of novel recommendations was proposed by Oh *et al.* [Oh *et al.* (2011)], who introduced the concept of *personal popularity tendency* in order to identify the subset of users that are more willing of receiving novel recommendations and further personalize their ranking by penalizing the items which are very popular in the community.

10.3.4. *Increasing Serendipity*

The idea of introducing serendipity in information seeking systems in an *operational* way was proposed in the early 2000s [Toms (2000); Campos and de Figueiredo (2001); Erdelez (2004); Foster and Ford (2003)].

One of the most common strategies of *programming for serendipity* is determining the *filter bubble* of the user, i.e. the *familiar* information that a personalized filter (such as a recommender system) creates for each of us, and then finding unexpected recommendations out of the bubble. For instance, the Auralist system finds musical bubbles, i.e. clusters of artists the user usually listens to, in order to recommend artists outside established cluster groups [Zhang *et al.* (2012)]. Evaluation performed over 360k Last.fm users and a proper user study show that Auralist's emphasis on serendipity indeed improves user satisfaction.

The *Outside-The-Box* system [Abbassi *et al.* (2009)] suggests items which do not fall into regions of interests (the bubble) the user is familiar with. Unfamiliarity can arise either when a user does not like items in that region and chooses not to rate them, or when the user has not been exposed enough to the region. In the latter case that region is likely to contain serendipitous items. Experiments on the MovieLens10M dataset show that, compared to a standard CF approach, most of the recommendations rank higher up in the returned list which makes them different and novel, hence, potentially useful.

Among collaborative approaches, in [Kawamae (2010)] the authors propose a strategy for suggesting surprisingly interesting items to a user by identifying purchase history logs of users who have similar preferences and a high degree of purchase precedence (i.e., purchasing the same items earlier) relative to that user. These users are called "innovators" since they become aware of items well before their release, and purchase them just after their release. The method assigns higher weights to innovators, and can rank these novel items first in the recommendation list. This should help to find items that match the latest user preferences, but also items she might not have otherwise discovered. Experiments on the Netflix Prize dataset and on two other datasets consisting of purchase histories obtained from real on-line music and video download services in Japan demonstrate that this algorithm offers a high degree of recommendation serendipitousness.

An item-item approach is described in [de Gemmis *et al.* (2015)], where the authors define a knowledge infusion process based on item content analysis able to discover hidden and non-obvious correlations between items which cannot be detected by using standard similarity scores, such as the number of users that co-rated the items. The approach computes a correlation score used to fill in an item-item correlation matrix that can be used by any recommendation method. Offline experiments on a subset of the HetRec2011-MovieLens-2K dataset demonstrate that the proposed algorithm produces more serendipitous suggestions than other collaborative or content-based recommendation algorithms, showing better balancing of relevance and unexpectedness. Furthermore, a preliminary user study was performed to assess both the acceptance and the actual perception of serendipity of recommendations, through the administration of questionnaires and the analysis of users' emotions, respectively. Results showed that the presence of positive emotions, such as happiness and surprise, could help to assess the actual perception of serendipity.

In [Said *et al.* (2013)], the authors propose a k-furthest neighbor recommendation algorithm, that overcomes the bias towards popular items of classical k-NN approaches by suggesting items coming from neighborhoods of users *dissimilar* to the target user. A similar approach is suggested by Nakatsuji *et al.*

[Nakatsuji *et al.* (2010b)], whose k-NN algorithm calculates ratings based on relatedness among users, inferred by running random walk with restarts on a user similarity graph. Another user-based approach is the one adopted by TANGENT [Onuma *et al.* (2009)], which detects groups of likeminded users and suggests items relevant to users from different groups. Likeminded users are grouped by selecting nodes in a graph connecting users with movies they like, giving high scores to nodes that are well connected to the older choices of the user, and at the same time well connected to unrelated choices, in order to broaden the user horizons. This strategy allows the recommendation of items close enough to a user's current interests, but also towards a new area that the user has not discovered yet.

A serendipity-oriented recommendation mechanism that models unexpectedness by combining the concepts of item rareness and dissimilarity is proposed in [Zheng *et al.* (2015)]. An unexpected item is assumed to show low popularity as well as low distance from a user's profile. PureSVD, (variation of singular value decomposition) is adopted to compute recommendations.

10.4. Available resources

Initially the typical evaluation of recommender systems has been focused on classical accuracy metrics, and for this reason the vast majority of libraries or resources to perform evaluations allowed to only consider these aspects, without taking into account the variety of beyond-accuracy metrics. Diversity, novelty, or serendipity are often tested individually, and there is a lack of systems which allow to perform comparative studies taking into account accuracy and beyond-accuracy metrics at the same time.

In this section, we describe the common and widely adopted libraries for implementing and testing recommender systems, with a specific focus on the beyond-accuracy evaluation, if supported.

- **MyMediaLite [Gantner *et al.* (2011)]**[3]. The framework is implemented in C#, is available to developers under the terms of the GNU General Public License (GPL), and can be easily included in programs written in Ruby, Python and partially in Java. It addresses the two most common scenarios in CF, i.e. rating prediction, and item prediction from positive-only feedback (e.g. from clicks, likes, or purchase actions). Among a variety of algorithms, for rating prediction it implements different variants of k-NN models, a method for MF and simple baseline methods, such as Slope-One and averages, while for item prediction it implements simple

[3]http://www.mymedialite.net/index.html

baselines as random and most popular items, and advanced MF models, such as BPRMF (Bayesian Personalized Ranking Matrix Factorization), and BPRSLIM (Sparse Linear Methods). Moreover, it includes evaluation routines for the most common evaluation metrics: Mean Absolute Error (MAE) and Root Mean Squared Error (RMSE) for the rating prediction task, and Area under the ROC-curve (AUC), Precision@N, Mean Average Precision (MAP), and normalized Discounted Cumulative Gain (nDCG) for item prediction. The framework allows to perform cross-validation and hyperparameter selection using grid search, but it does not support the evaluation of beyond-accuracy metrics.

- **LensKit [Ekstrand *et al.* (2011)]**[4]. It has been developed as a flexible platform for reproducible recommender systems research. It is implemented in Java, and it can be used through command line or included in an external project. It provides the most common recommendation algorithms, such as k-NN models, matrix factorization (e.g. Funk-SVD) and Slope-One. LensKit provides a flexible framework for offline evaluations. Currently, train-test evaluation of recommender prediction accuracy is supported, with several metrics for measuring accuracy, such as MAE, RMSE, nDCG, and coverage.
- **Mahout [Owen *et al.* (2012)]**[5]. It is a Java-based environment for implementing machine learning applications and Recommender Systems. It is supported by the Apache Software Foundation, and it is based on Apache Hadoop for managing data distribution for scalable applications. The Apache support makes the environment stable and efficient. It supports k-NN models, and Matrix factorization with Alternating Least Squares (also on implicit feedback), and provides routines for the evaluation in terms of MAE, RMSE, Precision, Recall and nDCG.
- **rrecSys [Çoba and Zanker (2016)]**[6]. It is an open source extension package in R for rapid prototyping and assessment of recommender system algorithms, specifically designed for education purposes. It supports both rating predictions and lists of recommended items, and includes average-based baselines, Slope-One, k-NN models, FunkSVD, BPR and weighted Alternating Least Squares. The algorithms can be evaluated in terms of MAE, RMSE, Precision, Recall, AUC, nDCG, and coverage measures.

[4]http://lenskit.org/
[5]https://mahout.apache.org
[6]https://cran.r-project.org/web/packages/rrecsys/index.html

- **RankSys [Castells *et al.* (2015)]**[7]. RankSys is a new framework for the implementation and evaluation of recommendation algorithms and techniques, specifically devised for the assessment and enhancement of novelty and diversity. It contains implementations of the novelty and diversity metrics and re-ranking strategies based on 1) a direct re-scoring of the scores provided by the original recommendation ranking, and 2) a greedy selection in which some set-wise magnitude is maximized by iteratively selecting those items that maximize it. The code is licensed under the Mozilla Public License (MPL) 2.0[8].

Even though the evaluation of the beyond-accuracy metrics is deemed crucial in the recommender systems community, at the moment RankSys is the only library which supports that kind of evaluation, while the other platforms still do not provide any support for it.

Besides the evaluation of accuracy and beyond-accuracy objectives, another aspect which is considered challenging in the recommender systems community concerns the fair evaluation of results, specifically in very complex experiments where different dimensions usually take place. This complexity makes difficult the comparison of results across different recommendation frameworks, since it is not unusual that the reported accuracy of an algorithm in one framework differs from the same algorithm in a different framework [Said and Bellogín (2014)]. To this purpose, it is possible to use benchmark frameworks which allow to leverage a common evaluation pipeline. Recently, *RiVal* [Said and Bellogín (2014)][9] emerged as a new toolkit for evaluation benchmark of recommender systems. It allows for fine-grained control of the evaluation methodology to obtain results comparable across different recommendation frameworks. It is open source, written in Java, and distributed by a Creative Commons Attribution 3.0 Unported license. It does not include recommendation algorithms, rather it provides wrappers for incorporating recommendations obtained with MyMediaLite, LensKit, and Apache Mahout. The evaluation pipeline proposed by *RiVal* is based on three phases: 1) dataset splitting, 2) selection of candidate items to evaluate, 3) evaluation of results. The dataset splitting allows to define the approaches for subdividing the dataset into a training and test set. Standard approaches based on time, cross-validation, random selection and percent ratio splitting are implemented. The data are then shared with the framework which generates the recommendations in order to obtain the predicted rating or the list of recommended items. The candidate

[7] http://ranksys.org/
[8] https://www.mozilla.org/en-US/MPL/2.0/
[9] http://rival.recommenders.net/

items obtained are then opportunely formatted, and information about the possible ground truth is extracted. Finally in the evaluation step, the framework uses the configured metrics (e.g., RMSE, MAE, nDCG, Precision, Recall, MAP) for evaluating the items and to provide results. There is no support for beyond-accuracy metrics.

10.5. Challenges and Future directions

In this chapter, we have reviewed the state-of-the-art research on beyond-accuracy objectives, with a specific focus on collaborative recommender systems. We have focused on how to measure and possibly increase diversity, novelty and serendipity, and on how to deal with possible undesired biases occurring in CF systems.

In our opinion, there are the following interesting challenges to address:

Adaptation of beyond-accuracy objectives to specific recommendation domains. An important challenge in beyond-accuracy research is to devise solutions tailored to specific recommendation domains. There are domains where users may be particularly interested in consuming items already consumed in the past [Mourão *et al.* (2017)]: this happens for example for music or restaurants, while this may be not useful for items such as books. In [Mourão *et al.* (2017)], subsets of items that were already consumed long ago are proposed as good candidates for reconsumption. Indeed, they are usually neglected by recommender systems, they could improve unexpectedness and serendipity since users do not expect to consume them presently, and finally, they may enhance the user experience by improving diversity. Experiments demonstrated that the long-term history of each user can be effectively exploited as a new source of unexpectedness and diversity.

Adaptation of beyond-accuracy objectives to individual users. Each specific beyond-accuracy objective could be different and tailored to each individual user, according to her needs and preferences. For example, different users could be inclined to diversifying only with respect to some specific item dimensions (e.g., director and year in the movie domain) and not be interested in diverse suggestions related to other ones (e.g. genre in the movie domain). As in [Di Noia *et al.* (2017)], this may allow to define an adaptive multi-attribute diversification approach able to customize the degree of individual diversity by taking into account the inclination of each user to diversifying over different content-based item dimensions. In general, the challenge is to look at the tendencies of the individual users based on their past behavior, in order to emulate these tendencies in the final recommendation lists [Jugovac *et al.* (2017)].

Finally, another aspect worth to be investigated is the development of proper software libraries for implementing and evaluating the optimization strategies for the beyond-accuracy objectives, and for properly balancing the trade-off between accuracy and beyond-accuracy metrics.

References

Abbassi, Z., Amer-Yahia, S., Lakshmanan, L. V. S., Vassilvitskii, S. and Yu, C. (2009). Getting Recommender Systems to Think Outside the Box, in *Proceedings of the ACM Conference on Recommender Systems, RecSys 2009, New York, USA* (ACM), ISBN 978-1-60558-435-5, pp. 285–288.

Abdollahpouri, H., Burke, R. and Mobasher, B. (2017). Controlling popularity bias in learning-to-rank recommendation, in P. Cremonesi, F. Ricci, S. Berkovsky and A. Tuzhilin (eds.), *Proceedings of the Eleventh ACM Conference on Recommender Systems, RecSys 2017, Como, Italy, August 27-31, 2017* (ACM), ISBN 978-1-4503-4652-8, pp. 42–46, doi:10.1145/3109859.3109912.

Adamopoulos, P. and Tuzhilin, A. (2011). On Unexpectedness in Recommender Systems: Or How to Expect the Unexpected, in P. Castells, J. Wang, R. Lara and D. Zhang (eds.), *Proceedings of the ACM RecSys 2011 Workshop on Novelty and Diversity in Recommender Systems (DiveRS), CEUR Workshop Proceedings*, Vol. 816 (CEUR-WS.org), pp. 11–18.

Adamopoulos, P. and Tuzhilin, A. (2013). Probabilistic neighborhood selection in collaborative filtering systems, Working paper: CBA-13-04, NYU, `http://hdl.handle.net/2451/31988`.

Adamopoulos, P. and Tuzhilin, A. (2014). On over-specialization and concentration bias of recommendations: probabilistic neighborhood selection in collaborative filtering systems, in A. Kobsa, M. X. Zhou, M. Ester and Y. Koren (eds.), *Eighth ACM Conference on Recommender Systems, RecSys '14, Foster City, Silicon Valley, CA, USA - October 06-10, 2014* (ACM), ISBN 978-1-4503-2668-1, pp. 153–160, doi:10.1145/2645710.2645752.

Adomavicius, G. and Kwon, Y. (2012). Improving aggregate recommendation diversity using ranking-based techniques, *IEEE Trans. Knowl. Data Eng.* **24**, 5, pp. 896–911, doi:10.1109/TKDE.2011.15, `https://doi.org/10.1109/TKDE.2011.15`.

Agrawal, R., Gollapudi, S., Halverson, A. and Ieong, S. (2009). Diversifying search results, in *Proceedings of the Second ACM International Conference on Web Search and Data Mining*, WSDM '09 (ACM, New York, NY, USA), ISBN 978-1-60558-390-7, pp. 5–14, doi:10.1145/1498759.1498766, `http://doi.acm.org/10.1145/1498759.1498766`.

Allan, J., Wade, C. and Bolivar, A. (2003). Retrieval and novelty detection at the sentence level, in *Proceedings of the 26th annual international ACM SIGIR conference on Research and development in informaion retrieval* (ACM), pp. 314–321.

Baeza-Yates, R., Ribeiro-Neto, B. *et al.* (1999). *Modern information retrieval*, Vol. 463 (ACM Press New York).

Barraza-Urbina, A., Heitmann, B., Hayes, C. and Ramos, A. C. (2015). Xplodiv: An exploitation-exploration aware diversification approach for recommender systems, in *FLAIRS Conference*, pp. 483–488.

Bellogín, A., Cantador, I. and Castells, P. (2013). A comparative study of heterogeneous item recommendations in social systems, *Inf. Sci.* **221**, pp. 142–169, doi:10.1016/j.ins.2012.09.039, `http://dx.doi.org/10.1016/j.ins.2012.09.039`.

Bellogín, A., Cantador, I., Díez, F., Castells, P. and Chavarriaga, E. (2013). An empirical comparison of social, collaborative filtering, and hybrid recommenders, *ACM Transactions on Intelligent Systems and Technology (TIST)* **4**, 1, p. 14.

Bellogín, A., Castells, P. and Cantador, I. (2017). Statistical biases in information retrieval metrics for recommender systems, *Inf. Retr. Journal* **20**, 6, pp. 606–634, doi:10.1007/s10791-017-9312-z.

Boim, R., Milo, T. and Novgorodov, S. (2011). Diversification and refinement in collaborative filtering recommender, in *Proceedings of the 20th ACM International Conference on Information and Knowledge Management*, CIKM '11 (ACM, New York, NY, USA), ISBN 978-1-4503-0717-8, pp. 739–744, doi:10.1145/2063576.2063684, `http://doi.acm.org/10.1145/2063576.2063684`.

Brynjolfsson, E., Hu, Y. and Simester, D. (2011). Goodbye pareto principle, hello long tail: The effect of search costs on the concentration of product sales, *Management Science* **57**, 8, pp. 1373–1386.

Campos, J. and de Figueiredo, A. D. (2001). Searching the Unsearchable: Inducing Serendipitous Insights, in *Proceedings of the Workshop Program at the Fourth International Conference on Case-Based Reasoning* (ICCBR), pp. 159–164.

Cañamares, R. and Castells, P. (2017). A probabilistic reformulation of memory-based collaborative filtering: Implications on popularity biases, in N. Kando, T. Sakai, H. Joho, H. Li, A. P. de Vries and R. W. White (eds.), *Proceedings of the 40th International ACM SIGIR Conference on Research and Development in Information Retrieval, Shinjuku, Tokyo, Japan, August 7-11, 2017* (ACM), ISBN 978-1-4503-5022-8, pp. 215–224.

Cantador, I., Brusilovsky, P. and Kuflik, T. (2011). 2nd workshop on information heterogeneity and fusion in recommender systems (hetrec 2011), in *Proceedings of the 5th ACM conference on Recommender systems*, RecSys 2011 (ACM, New York, NY, USA).

Carbonell, J. and Goldstein, J. (1998). The use of mmr, diversity-based reranking for reordering documents and producing summaries, in *Proceedings of the 21st Annual International ACM SIGIR Conference on Research and Development in Information Retrieval*, SIGIR '98 (ACM, New York, NY, USA), ISBN 1-58113-015-5, pp. 335–336, doi:10.1145/290941.291025, `http://doi.acm.org/10.1145/290941.291025`.

Castells, P., Hurley, N. J. and Vargas, S. (2015). Novelty and diversity in recommender systems, in F. Ricci, L. Rokach and B. Shapira (eds.), *Recommender Systems Handbook* (Springer), ISBN 978-1-4899-7636-9, pp. 881–918, doi:10.1007/978-1-4899-7637-6_26,
`https://doi.org/10.1007/978-1-4899-7637-6_26`.

Celma, Ò. and Herrera, P. (2008). A new approach to evaluating novel recommendations, in *Proceedings of the 2008 ACM conference on Recommender systems* (ACM), pp. 179–186.

Clarke, C. L., Kolla, M., Cormack, G. V., Vechtomova, O., Ashkan, A., Büttcher, S. and MacKinnon, I. (2008). Novelty and diversity in information retrieval evaluation, in *Proceedings of the 31st Annual International ACM SIGIR Conference on Research and Development in Information Retrieval*, SIGIR '08 (ACM, New York, NY, USA), ISBN 978-1-60558-164-4, pp. 659–666, doi:10.1145/1390334.1390446, http://doi.acm.org/10.1145/1390334.1390446.

Çoba, L. and Zanker, M. (2016). rrecsys: An R-package for Prototyping Recommendation Algorithms, in I. Guy and A. Sharma (eds.), *Proceedings of the Poster Track of the 10th ACM Conference on Recommender Systems (RecSys 2016), Boston, USA, September 17, 2016, CEUR Workshop Proceedings*, Vol. 1688 (CEUR-WS.org), http://ceur-ws.org/Vol-1688.

Cremonesi, P., Garzotto, F. and Quadrana, M. (2013). Evaluating top-n recommendations "when the best are gone", in Q. Yang, I. King, Q. Li, P. Pu and G. Karypis (eds.), *Seventh ACM Conference on Recommender Systems, RecSys '13, Hong Kong, China, October 12-16, 2013* (ACM), ISBN 978-1-4503-2409-0, pp. 339–342, doi:10.1145/2507157.2507225.

Cremonesi, P., Koren, Y. and Turrin, R. (2010). Performance of recommender algorithms on top-n recommendation tasks, in X. Amatriain, M. Torrens, P. Resnick and M. Zanker (eds.), *Proceedings of the 2010 ACM Conference on Recommender Systems, RecSys 2010, Barcelona, Spain, September 26-30, 2010* (ACM), ISBN 978-1-60558-906-0, pp. 39–46, doi:10.1145/1864708.1864721.

de Gemmis, M., Lops, P., Semeraro, G. and Musto, C. (2015). An investigation on the serendipity problem in recommender systems, *Inf. Process. Manage.* **51**, 5, pp. 695–717, doi:10.1016/j.ipm.2015.06.008, https://doi.org/10.1016/j.ipm.2015.06.008.

Di Noia, T., Rosati, J., Tomeo, P. and Di Sciascio, E. (2017). Adaptive multi-attribute diversity for recommender systems, *Information Sciences* **382**, pp. 234–253.

Dooms, S., Pessemier, T. D. and Martens, L. (2013). Movietweetings: a movie rating dataset collected from twitter, in *Workshop on Crowdsourcing and Human Computation for Recommender Systems*, RecSys 2013.

Ekstrand, M. D., Ludwig, M., Konstan, J. A. and Riedl, J. T. (2011). Rethinking the recommender research ecosystem: reproducibility, openness, and lenskit, in *Proceedings of the fifth ACM conference on Recommender systems* (ACM), pp. 133–140.

Erdelez, S. (2004). Investigation of Information Encountering in the Controlled Research Environment, *Information Processing and Management* **40**, 6, pp. 1013–1025.

Fleder, D. and Hosanagar, K. (2009). Blockbuster Culture's Next Rise or Fall: The Impact of Recommender Systems on Sales Diversity, *Management Science* **55**, 5, pp. 697–712.

Foster, A. and Ford, N. (2003). Serendipity and Information Seeking: an Empirical Study, *Journal of Documentation* **59**, 3, pp. 321–340.

Funk, S. (2006). Netflix Update: Try This At Home, URL: http://sifter.org/simon/journal/20061211.html.

Gantner, Z., Rendle, S., Freudenthaler, C. and Schmidt-Thieme, L. (2011). Mymedialite: A free recommender system library, in *Proceedings of the fifth ACM conference on Recommender systems* (ACM), pp. 305–308.

Ge, M., Delgado-Battenfeld, C. and Jannach, D. (2010). Beyond Accuracy: Evaluating Recommender Systems by Coverage and Serendipity, in X. Amatriain, M. Torrens, P. Resnick and M. Zanker (eds.), *Proceedings of the ACM Conference on Recommender Systems* (ACM), ISBN 978-1-60558-906-0, pp. 257–260.

Gini, C. (1921). Measurement of Inequality and Incomes, *The Economic Journal* **31**, pp. 124–126.

Gunawardana, A. and Shani, G. (2015). Evaluating recommender systems, in F. Ricci, L. Rokach and B. Shapira (eds.), *Recommender Systems Handbook* (Springer), pp. 265–308, doi:10.1007/978-1-4899-7637-6_8, https://doi.org/10.1007/978-1-4899-7637-6_8.

Harper, F. M. and Konstan, J. A. (2016). The MovieLens Datasets: History and Context, *TiiS* **5**, 4, pp. 19:1–19:19, doi:10.1145/2827872, http://doi.acm.org/10.1145/2827872.

Herlocker, L., Konstan, J. A., Terveen, L. G. and Riedl, J. T. (2004). Evaluating Collaborative Filtering Recommender Systems, *ACM Transactions on Information Systems* **22**, 1, pp. 5–53.

Hijikata, Y., Shimizu, T. and Nishida, S. (2009). Discovery-oriented collaborative filtering for improving user satisfaction, in *Proceedings of the 14th international conference on Intelligent user interfaces* (ACM), pp. 67–76.

Hurley, N. and Zhang, M. (2011). Novelty and Diversity in Top-N Recommendation - Analysis and Evaluation, *ACM Trans. Internet Techn.* **10**, 4, pp. 14:1–14:30, doi:10.1145/1944339.1944341, http://doi.acm.org/10.1145/1944339.1944341.

Hurley, N. J. (2013). Personalised ranking with diversity, in *Proceedings of the 7th ACM Conference on Recommender Systems*, RecSys '13 (ACM, New York, NY, USA), ISBN 978-1-4503-2409-0, pp. 379–382, doi:10.1145/2507157.2507226, http://doi.acm.org/10.1145/2507157.2507226.

Iaquinta, L., de Gemmis, M., Lops, P., Semeraro, G., Filannino, M. and Molino, P. (2008). Introducing Serendipity in a Content-Based Recommender System, in F. Xhafa, F. Herrera, A. Abraham, M. Köppen and J. M. Benítez (eds.), *8th International Conference on Hybrid Intelligent Systems* (IEEE Computer Society), ISBN 978-0-7695-3326-1, pp. 168–173.

Jambor, T. and Wang, J. (2010). Optimizing multiple objectives in collaborative filtering, in *Proceedings of the fourth ACM conference on Recommender systems* (ACM), pp. 55–62.

Jannach, D. and Hegelich, K. (2009). A case study on the effectiveness of recommendations in the mobile internet, in L. D. Bergman, A. Tuzhilin, R. D. Burke, A. Felfernig and L. Schmidt-Thieme (eds.), *Proceedings of the 2009 ACM Conference on Recommender Systems, RecSys 2009, New York, NY, USA, October 23-25, 2009* (ACM), ISBN 978-1-60558-435-5, pp. 205–208, doi:10.1145/1639714.1639749.

Jannach, D., Lerche, L., Kamehkhosh, I. and Jugovac, M. (2015). What recommenders recommend: an analysis of recommendation biases and possible countermeasures, *User Model. User-Adapt. Interact.* **25**, 5, pp. 427–491, doi:10.1007/s11257-015-9165-3, https://doi.org/10.1007/s11257-015-9165-3.

Jugovac, M., Jannach, D. and Lerche, L. (2017). Efficient optimization of multiple recommendation quality factors according to individual user tendencies, *Expert Syst. Appl.* **81**, C, pp. 321–331, doi:10.1016/j.eswa.2017.03.055, https://doi.org/10.1016/j.eswa.2017.03.055.

Kaminskas, M. and Bridge, D. (2016). Diversity, serendipity, novelty, and coverage: A survey and empirical analysis of beyond-accuracy objectives in recommender systems, *ACM Trans. Interact. Intell. Syst.* **7**, 1, pp. 2:1–2:42, doi:10.1145/2926720, `http://doi.acm.org/10.1145/2926720`.

Kamishima, T., Akaho, S., Asoh, H. and Sakuma, J. (2014). Correcting popularity bias by enhancing recommendation neutrality, in L. Chen and J. Mahmud (eds.), *Poster Proceedings of the 8th ACM Conference on Recommender Systems, RecSys 2014, Foster City, Silicon Valley, CA, USA, October 6-10, 2014, CEUR Workshop Proceedings*, Vol. 1247 (CEUR-WS.org), `http://ceur-ws.org/Vol-1247/recsys14_poster10.pdf`.

Kapoor, K., Kumar, V., Terveen, L., Konstan, J. A. and Schrater, P. (2015). I like to explore sometimes: Adapting to dynamic user novelty preferences, in *Proceedings of the 9th ACM Conference on Recommender Systems* (ACM), pp. 19–26.

Kawamae, N. (2010). Serendipitous Recommendations Via Innovators, in F. Crestani, S. Marchand-Maillet, H.-H. Chen, E. N. Efthimiadis and J. Savoy (eds.), *Proceedings of the 33rd International ACM SIGIR Conference on Research and Development in Information Retrieval* (ACM), ISBN 978-1-4503-0153-4, pp. 218–225.

Kelly, J. P. and Bridge, D. (2006). Enhancing the diversity of conversational collaborative recommendations: a comparison, *Artificial Intelligence Review* **25**, 1, pp. 79–95.

Koren, Y. (2010). Factor in the neighbors: Scalable and accurate collaborative filtering, *ACM Transactions on Knowledge Discovery from Data (TKDD)* **4**, 1, pp. 1:1–1:24, doi:10.1145/1644873.1644874, `http://doi.acm.org/10.1145/1644873.1644874`.

Koren, Y. and Bell, R. M. (2015). Advances in collaborative filtering, in F. Ricci, L. Rokach and B. Shapira (eds.), *Recommender Systems Handbook* (Springer), ISBN 978-1-4899-7636-9, pp. 77–118.

Li, X. and Murata, T. (2012). Multidimensional clustering based collaborative filtering approach for diversified recommendation, in *Computer Science & Education (ICCSE), 2012 7th International Conference on* (IEEE), pp. 905–910.

McAuley, J. J. and Leskovec, J. (2013). Hidden factors and hidden topics: understanding rating dimensions with review text, in Q. Yang, I. King, Q. Li, P. Pu and G. Karypis (eds.), *Seventh ACM Conference on Recommender Systems, RecSys '13, Hong Kong, China, October 12-16, 2013* (ACM), pp. 165–172, doi:10.1145/2507157.2507163.

McFee, B., Bertin-Mahieux, T., Ellis, D. P. and Lanckriet, G. R. (2012). The million song dataset challenge, in *Proceedings of the 21st International Conference on World Wide Web*, WWW '12 Companion (ACM, New York, NY, USA), ISBN 978-1-4503-1230-1, pp. 909–916, doi:10.1145/2187980.2188222, `http://doi.acm.org/10.1145/2187980.2188222`.

McNee, S. M., Riedl, J. and Konstan, J. A. (2006). Being Accurate is not Enough: How Accuracy Metrics have Hurt Recommender Systems, in G. M. Olson and R. Jeffries (eds.), *Extended Abstracts Proceedings of the 2006 Conference on Human Factors in Computing Systems* (ACM), pp. 1097–1101.

Mourão, F., da Rocha, L. C., Araujo, C. S., Jr., W. M. and Konstan, J. A. (2017). What surprises does your past have for you? *Inf. Syst.* **71**, pp. 137–151.

Murakami, T., Mori, K. and Orihara, R. (2008). Metrics for Evaluating the Serendipity of Recommendation Lists, in K. Satoh, A. Inokuchi, K. Nagao and T. Kawamura (eds.), *New Frontiers in Artificial Intelligence, Lecture Notes in Computer Science*, Vol. 4914 (Springer), ISBN 978-3-540-78196-7, pp. 40–46.

Nakatsuji, M., Fujiwara, Y., Tanaka, A., Uchiyama, T., Fujimura, K. and Ishida, T. (2010a). Classical music for rock fans?: novel recommendations for expanding user interests, in *Proceedings of the 19th ACM international conference on Information and knowledge management* (ACM), pp. 949–958.

Nakatsuji, M., Fujiwara, Y., Tanaka, A., Uchiyama, T., Fujimura, K. and Ishida, T. (2010b). Classical music for rock fans?: Novel recommendations for expanding user interests, in *Proceedings of the 19th ACM International Conference on Information and Knowledge Management*, CIKM '10 (ACM, New York, NY, USA), ISBN 978-1-4503-0099-5, pp. 949–958, doi:10.1145/1871437.1871558, `http://doi.acm.org/10.1145/1871437.1871558`.

Ning, X., Desrosiers, C. and Karypis, G. (2015). A comprehensive survey of neighborhood-based recommendation methods, in F. Ricci, L. Rokach and B. Shapira (eds.), *Recommender Systems Handbook* (Springer), ISBN 978-1-4899-7636-9, pp. 37–76, doi: 10.1007/978-1-4899-7637-6_2.

Oh, J., Park, S., Yu, H., Song, M. and Park, S.-T. (2011). Novel recommendation based on personal popularity tendency, in *Data Mining (ICDM), 2011 IEEE 11th International Conference on* (IEEE), pp. 507–516.

Onuma, K., Tong, H. and Faloutsos, C. (2009). TANGENT: A Novel, 'Surprise me', Recommendation Algorithm, in J. F. Elder IV, F. Fogelman-Soulié, P. A. Flach and M. J. Zaki (eds.), *Proceedings of the 15th ACM SIGKDD International Conference on Knowledge Discovery and Data Mining* (ACM), ISBN 978-1-60558-495-9, pp. 657–666.

Owen, S., Anil, R., Dunning, T. and Friedman, E. (2012). *Mahout in action* (Manning Shelter Island, NY).

Pampın, H. J. C., Jerbi, H. and O'Mahony, M. P. (2015). Evaluating the relative performance of collaborative filtering recommender systems, *Journal of Universal Computer Science* **21**, 13, pp. 1849–1868.

Pan, W., Zhong, H., Xu, C. and Ming, Z. (2015). Adaptive bayesian personalized ranking for heterogeneous implicit feedbacks, *Knowl.-Based Syst.* **73**, pp. 173–180, doi:10.1016/j.knosys.2014.09.013.

Park, Y.-J. and Tuzhilin, A. (2008). The long tail of recommender systems and how to leverage it, in *Proceedings of the 2008 ACM conference on Recommender systems* (ACM), pp. 11–18.

Patil, G. and Taillie, C. (1982). Diversity as a concept and its measurement, *Journal of the American statistical Association* **77**, 379, pp. 548–561.

Radovanovic, M., Nanopoulos, A. and Ivanovic, M. (2010). Hubs in space: Popular nearest neighbors in high-dimensional data, *Journal of Machine Learning Research* **11**, pp. 2487–2531, `http://portal.acm.org/citation.cfm?id=1953015`.

Rao, J., Jia, A., Feng, Y. and Zhao, D. (2013). Taxonomy based personalized news recommendation: Novelty and diversity, in *International Conference on Web Information Systems Engineering* (Springer), pp. 209–218.

Rendle, S. (2012). Factorization machines with libfm, *ACM TIST* **3**, 3, pp. 57:1–57:22, doi:10.1145/2168752.2168771, http://doi.acm.org/10.1145/2168752.2168771.

Rendle, S., Freudenthaler, C., Gantner, Z. and Schmidt-Thieme, L. (2009). BPR: bayesian personalized ranking from implicit feedback, in J. A. Bilmes and A. Y. Ng (eds.), *UAI 2009, Proceedings of the Twenty-Fifth Conference on Uncertainty in Artificial Intelligence, Montreal, QC, Canada, June 18-21, 2009* (AUAI Press), pp. 452–461.

Ribeiro, M. T., Lacerda, A., Veloso, A. and Ziviani, N. (2012). Pareto-efficient hybridization for multi-objective recommender systems, in *Proceedings of the sixth ACM conference on Recommender systems* (ACM), pp. 19–26.

Robertson, S. E. (1977). The probability ranking principle in ir, *Journal of documentation* **33**, 4, pp. 294–304.

Said, A. and Bellogín, A. (2014). Rival: a toolkit to foster reproducibility in recommender system evaluation, in *Proceedings of the 8th ACM Conference on Recommender systems* (ACM), pp. 371–372.

Said, A., Fields, B., Jain, B. J. and Albayrak, S. (2013). User-centric evaluation of a k-furthest neighbor collaborative filtering recommender algorithm, in *Proceedings of the 2013 Conference on Computer Supported Cooperative Work*, CSCW '13 (ACM, New York, NY, USA), ISBN 978-1-4503-1331-5, pp. 1399–1408, doi:10.1145/2441776.2441933, http://doi.acm.org/10.1145/2441776.2441933.

Shani, G. and Gunawardana, A. (2011). Evaluating Recommendation Systems, in F. Ricci, L. Rokach, B. Shapira and P. B. Kantor (eds.), *Recommender Systems Handbook* (Springer), ISBN 978-0-387-85819-7, pp. 257–297.

Shi, L. (2013). Trading-off among accuracy, similarity, diversity, and long-tail: A graph-based recommendation approach, in *Proceedings of the 7th ACM conference on Recommender systems* (ACM), pp. 57–64.

Shi, Y., Zhao, X., Wang, J., Larson, M. and Hanjalic, A. (2012). Adaptive diversification of recommendation results via latent factor portfolio, in *Proceedings of the 35th International ACM SIGIR Conference on Research and Development in Information Retrieval*, SIGIR '12 (ACM, New York, NY, USA), ISBN 978-1-4503-1472-5, pp. 175–184, doi:10.1145/2348283.2348310, http://doi.acm.org/10.1145/2348283.2348310.

Smyth, B. and McClave, P. (2001). Similarity vs. diversity, in *Proceedings of the 4th International Conference on Case-Based Reasoning: Case-Based Reasoning Research and Development*, ICCBR '01 (Springer-Verlag, London, UK), ISBN 3-540-42358-3, pp. 347–361, http://dl.acm.org/citation.cfm?id=646268.758890.

Steck, H. (2011). Item popularity and recommendation accuracy, in B. Mobasher, R. D. Burke, D. Jannach and G. Adomavicius (eds.), *Proceedings of the 2011 ACM Conference on Recommender Systems, RecSys 2011, Chicago, IL, USA, October 23-27, 2011* (ACM), ISBN 978-1-4503-0683-6, pp. 125–132, doi:10.1145/2043932.2043957.

Su, R., Yin, L., Chen, K. and Yu, Y. (2013). Set-oriented personalized ranking for diversified top-n recommendation, in *Proceedings of the 7th ACM Conference on Recommender Systems*, RecSys '13 (ACM, New York, NY, USA), ISBN 978-1-4503-2409-0, pp. 415–418, doi:10.1145/2507157.2507207, http://doi.acm.org/10.1145/2507157.2507207.

Takács, G. and Tikk, D. (2012). Alternating least squares for personalized ranking, in P. Cunningham, N. J. Hurley, I. Guy and S. S. Anand (eds.), *Sixth ACM Conference on Recommender Systems, RecSys '12, Dublin, Ireland, September 9-13, 2012* (ACM), ISBN 978-1-4503-1270-7, pp. 83–90, doi:10.1145/2365952.2365972.

Toms, E. (2000). Serendipitous Information Retrieval, in *Proceedings of DELOS Workshop: Information Seeking, Searching and Querying in Digital Libraries* (ERCIM Workshop Proceedings - No. 01/W001).

Vargas, S., Baltrunas, L., Karatzoglou, A. and Castells, P. (2014). Coverage, redundancy and size-awareness in genre diversity for recommender systems, in *Proceedings of the 8th ACM Conference on Recommender systems* (ACM), pp. 209–216.

Vargas, S. and Castells, P. (2011). Rank and relevance in novelty and diversity metrics for recommender systems, in *Proceedings of the fifth ACM conference on Recommender systems* (ACM), pp. 109–116.

Vargas, S. and Castells, P. (2014). Improving sales diversity by recommending users to items, in A. Kobsa, M. X. Zhou, M. Ester and Y. Koren (eds.), *Eighth ACM Conference on Recommender Systems, RecSys '14, Foster City, Silicon Valley, CA, USA - October 06-10, 2014* (ACM), ISBN 978-1-4503-2668-1, pp. 145–152, doi: 10.1145/2645710.2645744.

Wang, J. and Zhu, J. (2009). Portfolio theory of information retrieval, in *Proceedings of the 32Nd International ACM SIGIR Conference on Research and Development in Information Retrieval*, SIGIR '09 (ACM, New York, NY, USA), ISBN 978-1-60558-483-6, pp. 115–122, doi:10.1145/1571941.1571963, `http://doi.acm.org/10.1145/1571941.1571963`.

Zhang, L. (2013). The definition of novelty in recommendation system, *Journal of Engineering Science & Technology Review* **6**, 3.

Zhang, M. and Hurley, N. (2008). Avoiding monotony: improving the diversity of recommendation lists, in *Proceedings of the 2008 ACM conference on Recommender systems* (ACM), pp. 123–130.

Zhang, Y., Callan, J. and Minka, T. (2002). Novelty and redundancy detection in adaptive filtering, in *Proceedings of the 25th annual international ACM SIGIR conference on Research and development in information retrieval* (ACM), pp. 81–88.

Zhang, Y. C., Séaghdha, D. Ó., Quercia, D. and Jambor, T. (2012). Auralist: introducing serendipity into music recommendation, in E. Adar, J. Teevan, E. Agichtein and Y. Maarek (eds.), *Proceedings of the Fifth International Conference on Web Search and Web Data Mining, WSDM 2012, Seattle, WA, USA, February 8-12, 2012*, ISBN 978-1-4503-0747-5, pp. 13–22, doi:10.1145/2124295.2124300, `http://doi.acm.org/10.1145/2124295.2124300`.

Zheng, Q., Chan, C. and Ip, H. H. S. (2015). An unexpectedness-augmented utility model for making serendipitous recommendation, in *Advances in Data Mining: Applications and Theoretical Aspects - 15th Industrial Conference, ICDM 2015, Hamburg, Germany, July 11-24, 2015, Proceedings, Lecture Notes in Computer Science*, Vol. 9165 (Springer), pp. 216–230, doi:10.1007/978-3-319-20910-4_16, `https://doi.org/10.1007/978-3-319-20910-4_16`.

Zhou, T., Kuscsik, Z., Liu, J.-G., Medo, M., Wakeling, J. R. and Zhang, Y.-C. (2010). Solving the apparent diversity-accuracy dilemma of recommender systems, *Proceedings of the National Academy of Sciences* **107**, 10, pp. 4511–4515.

Ziegler, C.-N., McNee, S. M., Konstan, J. A. and Lausen, G. (2005). Improving Recommendation Lists through Topic Diversification, in A. Ellis and T. Hagino (eds.), *Proceedings of the 14th International Conference on World Wide Web, WWW 2005* (ACM), ISBN 1-59593-046-9, pp. 22–32.

Chapter 11

Scalability and Distribution of Collaborative Recommenders

Evangelia Christakopoulou
Computer Science & Engineering
University of Minnesota, Minneapolis, MN
evangel@cs.umn.edu

Shaden Smith
Computer Science & Engineering
University of Minnesota, Minneapolis, MN
shaden@cs.umn.edu

Mohit Sharma
Computer Science & Engineering
University of Minnesota, Minneapolis, MN
mohit@cs.umn.edu

Alex Richards
Department of Computer Engineering
San José State University
alexander.richards@sjsu.edu

David Anastasiu
Department of Computer Engineering
San José State University
david.anastasiu@sjsu.edu

George Karypis
Computer Science & Engineering
University of Minnesota, Minneapolis, MN
karypis@cs.umn.edu

Recommender systems are ubiquitous; they are foundational to a wide variety of industries ranging from media companies such as Netflix to e-commerce companies such as Amazon. As recommender systems continue to permeate the marketplace, we observe two major shifts which must be addressed. First, the amount of data used to provide quality recommendations grows at an unprecedented rate. Secondly, modern computer architectures display great processing capabilities that significantly outpace memory speeds. These two trend shifts must be taken into account in order to design recommendation systems that can efficiently handle the amount of available data by distributing computations in order to take advantage of modern parallel architectures. In this chapter, we present ways to scale popular collaborative recommendation methods via parallel computing.

11.1. Introduction

Recommender systems are ubiquitous; they are foundational to a wide variety of industries ranging from media companies such as Netflix to e-commerce companies such as Amazon. Their popularity is attributed to their ability to effectively navigate users through a plethora of product options which would otherwise go unexplored. As recommender systems continue to permeate the marketplace, we observe two major shifts which must be addressed.

First, the amount of data used to provide quality recommendations grows at an unprecedented rate. For example, companies such as Netflix stream millions of movies each day. Secondly, modern computer architectures forego great changes. The last two decades have seen available processing capabilities significantly outpace memory speeds. This disparity has shifted the cost of many computations from mathematical operations to data movements. In consequence, algorithm designers must now expose large amounts of parallelism while reducing data movement costs. These two trend shifts must be taken into account in order to design recommendation systems that can efficiently handle the amount of available data by distributing computations in order to take advantage of modern parallel architectures.

Unfortunately, designing successful recommendation systems that can effectively utilize modern parallel architectures is not always straightforward. Most popular recommendation algorithms are inherently unstructured, meaning that data accesses are not known apriori because they are determined by the structure of the input data. The unstructured nature of the underlying computations is further complicated by non-uniform distributions exhibited by real-world ratings datasets. A small number of popular items and prolific users cause the data to follow a power-law distribution, which presents challenges when partitioning work to processing cores in a balanced manner.

In this chapter, we present ways to scale popular collaborative recommendation methods via parallel computing. The methods we present span both of the primary recommendation tasks: the top-N recommendation task, where we are interested in whether a user will likely purchase an item, and the rating prediction task, which focuses on determining how much a user would enjoy a product. The nearest neighbor methods presented in Section 11.2 are used for both tasks. The sparse linear methods presented in Section 11.3 are oriented towards the top-N recommendation task. Finally, the matrix and tensor factorization methods presented in Section 11.4 can be used for both tasks, but for the purposes of this chapter, they are primarily viewed in the context of rating prediction.

The presented methods are very different from each other and exhibit different parallelization opportunities. For each method, an overview is presented, along with a discussion of how it can be scaled and experimental results showing the runtimes and speedup achieved on the MovieLens 10M (ML10M) dataset.

The rest of the chapter has the following structure: Subsections 11.1.1 and 11.1.2 present the notation and evaluation methodology used in the rest of the chapter. Section 11.2 presents methods to efficiently identify neighbors in the nearest-neighbor approaches. Section 11.3 describes sparse linear methods, where both the neighbors to a target item and their similarities are estimated through solving an optimization problem and discusses their scalability. Finally, Section 11.4 gives an overview and discusses the efficiency of matrix and tensor factorization approaches.

11.1.1. *Notation*

In the rest of the chapter, we will use bold capital letters to refer to matrices (e.g., $\mathbf{A}$) and bold lower case letters to refer to vectors (e.g., $\boldsymbol{p}$). The vectors are assumed to be column vectors. Thus, $\boldsymbol{a}_i$ refers to the ith column of matrix $\mathbf{A}$ and we will use the transpose ($\boldsymbol{a}_i^T$) to describe row vectors.

We note the rating matrix, which contains the feedback given by n users to m items as $\mathbf{R}$. The dimensions of $\mathbf{R}$ are $n \times m$. We will use the term u to note a user, and i to note an item. The rating given by a user u to an item i will be noted by r_{ui} and with a slight abuse of notation will be used to show both the explicit rating that user u gave to item i and/or the implicit feedback (purchase/like) from u to i. We will use the notation $\hat{r}_{ui}$ to refer to the predicted rating. Finally, we use the notation $|\cdot|$ to refer to the number of non-zeros in the corresponding matrix or vector.

11.1.2. *Evaluation*

Dataset Throughout this chapter, we will demonstrate the efficiency and effectiveness of different recommender methods using the MovieLens 10 Million (ML10M) [Harper and Konstan (2015)] ratings dataset. ML10M consists of 10 million ratings provided to 10677 movies by 69878 users. Each rating is accompanied by a timestamp. The timestamps span 158 months and are used in tensor factorization approaches in Section 11.4.

Evaluation In order to evaluate the performance of the methods, we employ a leave-one-out validation scheme. For every user, we leave out one rating and we train the model on the rest of the user' ratings. All the left-out ratings comprise the test set. We repeat this procedure three times, thus resulting in three folds (three train sets and three associated test sets). In the end, for every method, we report the average time taken and the average performance of the folds.

As the exact rating is not used in top-N methods, the implicit feedback of the ML10M dataset is used instead in Section 11.3. In this scenario, non-zero rating values in $\mathbf{R}$ are replaced with 1s, signifying the existence of a rating given by a user to an item in the original data. To evaluate top-N recommendation methods, we need to investigate whether the item of the user that belongs to the test set is included in the list of top-N recommendations to the user, and in which position. Therefore, we use two performance metrics: HR and ARHR, where:

$$HR = \frac{\#hits}{\#users}, \tag{11.1}$$

and

$$ARHR = \sum_{i=1}^{\#hits} \frac{1}{p_i}. \tag{11.2}$$

The $\#hits$ denotes the number of times that the test items were included in the top-N recommendation list and p_i is the position of the test item in the recommended list, with $p_i = 1$ being the top position. In this chapter, we have evaluated the top-N recommendation methods by computing HR and ARHR for $N = 10$.

In the rating prediction methods, the explicit ratings are used. Also, when presenting the tensor factorization methods in Section 11.4, the associated timestamps are used beyond the ratings. To evaluate the rating prediction methods, we need to see how similar or different the predicted values of the ratings are with the actual ratings. We employ the Root Mean Squared Error (RMSE) to do that:

$$RMSE = \sqrt{\frac{\sum_u \sum_i (r_{ui} - \hat{r}_{ui})^2}{|\mathbf{R}|}}. \tag{11.3}$$

More details on RMSE, HR, ARHR and other evaluation measures can be found in Chapter 9. A thorough explanation of the difference between explicit and implicit feedback can be found in Chapter 7.

System characteristics For consistency and comparison purposes, all experiments are executed on the same system[1] consisting of identical nodes equipped with 64 GB RAM and two twelve-core 2.5 GHz Intel Xeon E5-2680v3 (Haswell) processors. Each core is equipped with 32 KB L1 cache and 256 KB of L2 cache, and the 12 cores on each socket share 30 MB of L3 cache.

11.2. Scaling up nearest-neighbor approaches

A number of recommender systems rely on finding nearest neighbors among users or items as an integral part of deriving recommendations or training a predictive recommendation model. More details on nearest-neighbor approaches can be found in Chapter 1. A neighbor is defined as a user/item who is similar to the target user/item, based on a similarity notion (e.g. the target user and the neighbor user have co-rated a lot of items). The symbol $\mathcal{N}_i(u)$ represents the set of neighbors of the item i rated by the user u. Similarly, the symbol $\mathcal{N}_u(i)$ represents the set of neighbors of user u, who have rated item i. In this section, we discuss approaches to speed up neighborhood identification, which directly affects recommendation efficiency.

[1] https://www.msi.umn.edu/content/mesabi

11.2.1. *Use of neighborhoods in Recommender Systems*

Collaborative filtering based recommenders, such as the user-based neighborhood method [Konstan *et al.* (1997)] or item-based neighborhood methods [Sarwar *et al.* (2001); Deshpande and Karypis (2004)], first identify a set of neighbors and then use the ratings associated with those neighbors to derive the predicted rating for the user in question. User-based methods identify a set of users similar to target user u and predict the rating of user u on item i as

$$\hat{r}_{ui} = \mu_u + \frac{\sum_{v \in \mathcal{N}_u(i)} \text{sim}(u,v)(r_{vi} - \mu_v)}{\sum_{v \in \mathcal{N}_u(i)} \text{sim}(u,v)},$$

where $\text{sim}(u,v)$ denotes the similarity between the users u and v, and μ_u and μ_v are the means of the ratings provided by users u and v, respectively.

Item-based neighborhood methods, however, derive the rating of user u on target item i by considering other items that have been rated (generally high) by the user u. The predicted rating of u on i is given by

$$\hat{r}_{ui} = \mu_i + \frac{\sum_{j \in \mathcal{N}_i(u)} \text{sim}(i,j)(r_{uj} - \mu_j)}{\sum_{j \in \mathcal{N}_i(u)} \text{sim}(i,j)}.$$

where $\text{sim}(i,j)$ denotes the similarity between the items i and j, and μ_i and μ_j are the means of the ratings received by items i and j, respectively.

A number of different recommenders can be designed given different choices in applying the formulas above with respect to user and item representations, similarity function, or neighborhood construction. The standard approach is to represent user u as the sparse vector of ratings for items rated by the user (row u in the ratings matrix $\mathbf{R}$) and item i as the vector of all ratings given to item i by users (column i in $\mathbf{R}$). Given a vector representation of users and items, the similarity between users or between items is most often computed as the cosine similarity or Pearson correlation coefficient between their respective vectors. Finally, the neighborhoods $\mathcal{N}_u(i)$ and $\mathcal{N}_i(u)$ can be constructed by finding all neighbors with a similarity above some minimum threshold ϵ, or one may consider only the k closest neighbors to the target user or item.

Beyond deriving recommendations by relying on similar users or items, some optimization-based recommenders learn a recommendation model by focusing only on the ratings of similar users or items during the learning process (e.g., the fs-SLIM model [Ning and Karypis (2011)]).

Naïve approaches will compare each user to every other user, thus leading to quadratic complexity in the number of computed similarities. In the remainder of this section, we will discuss efficient methods that identify nearest neighbors given user-item ratings. The methods rely on aggressive pruning of the search space by identifying pairs of users or items that cannot be similar enough based on theoretic upper bounds on their computed similarity. Additional performance gains are achieved via efficient use of the memory hierarchy of modern computing systems and shared-memory parallelism.

We focus our discussion on two nearest neighbor problems useful in the recommender systems context. The ϵ-nearest neighbor graph (ϵNNG) construction problem, also known as *all-pairs similarity search* (APSS) or *similarity join*, identifies, for each user/item in a set, all other users/items with a similarity of at least ϵ. On the other hand, the k-nearest neighbor graph (kNNG) construction constrains each identified neighborhood to the k users/items closest to the target user/item. To simplify the discussion, we will describe the methods in the context of constructing item neighborhoods. The same methods can be applied to find user-based neighbors.

11.2.2. *ϵ-nearest neighbor graph construction*

Recently, several methods have been proposed that efficiently construct the ϵNNG by filtering (or ignoring) pairs of items that cannot be neighbors [Bayardo *et al.* (2007); Anastasiu and Karypis (2014, 2015b); Anastasiu (2017)]. Item rating profile vectors are inherently sparse, as few users may consume and rate each item. The proposed methods take advantage of this sparsity and use data structures and processing strategies designed to eliminate unnecessary memory accesses and multiplication operations. The L_2-norm All Pairs (L2AP) [Anastasiu and Karypis (2014)] and Parallel L_2-norm All Pairs (pL2AP) [Anastasiu and Karypis (2015b)] methods construct an *exact* neighborhood graph, finding the same neighbors as those found by a brute-force method that compares each user/item against all other users/items. Cosine Approximate Nearest Neighbors (CANN) [Anastasiu (2017)], on the other hand, finds most but not necessarily all of the neighbors with a similarity of at least ϵ.

These methods use an *inverted index* data structure to eliminate unnecessary comparisons. The inverted index is represented by the sparse user rating profiles. It consists of a set of lists, one for each user, such that the uth list contains pairs (i, r_{ui}) for all items i that have a non-zero r_{ui} rating. Many unnecessary memory accesses and similarity computations can be avoided by only comparing an item against the set of items found in the inverted index lists for the users that rated it. In this way, two items that have not been rated by any common user will never have their similarity computed.

Additional savings can be achieved by taking advantage of the input similarity threshold ϵ. Note that cosine similarity measures the cosine of the angle between the two vectors and is thus independent of the vector lengths. A standard preprocessing step in computing cosine similarity is to normalize the vectors with respect to their L_2-norm, which reduces computing the cosine similarity of two vectors to finding their dot-product. The methods compute only part of the dot-product of profile vectors for most pairs of items, e.g., using only the tail-end features in the profile vector. Several theoretic upper bounds of vector dot-products are used to estimate the portion of the dot-product for the leading features. If the sum of the estimate and computed portions of the dot-product is below ϵ, the items cannot be similar enough and are pruned.

Many item comparisons are completely avoided through a *partial indexing* strategy. Only a few of the leading features of the profile of an item i are indexed, enough to ensure that any item j with a similarity of at least ϵ will be found by traversing the partial index. This strategy leads to a two phase process for constructing the exact ϵNNG. First, partial similarities (dot-products) are computed using values stored in the inverted index lists for users that rated item i, which are called *candidates*. In the second phase, the un-indexed portion of each of the candidate profile vectors is used to finish computing similarities only for those items with non-zero similarity after the first phase. In both phases, additional similarity upper-bounds are used to eliminate candidates that cannot be similar enough.

Parallelization of pL2AP focuses on a cache-tiling strategy that aims to fit critical data structures used during similarity search in the high-speed yet limited cache memory of the system. The method splits the set of items such that each subset has a partial inverted index that can fit in cache memory. Each core is then assigned small sets of neighborhood searches for 20 consecutive items, which could be independently executed. Additionally, a small-memory footprint hash table data structure is proposed which is uniquely suited to the memory access patterns in pL2AP and provides fast access to profile vector values and meta-data necessary for computing similarity upper bounds. Algorithm 1 provides a sketch of the pL2AP processing pipeline. Additional details for the algorithm and the different similarity upper-bounds used in the filtering process in pL2AP can be found in [Anastasiu and Karypis (2014, 2015b); Anastasiu (2017)].

Algorithm 1 pL2AP

```
Normalize profiles for every item i.
for all items i in parallel do
    Identify prefix to be indexed.
end for
Split items to index into tiles based on cumulative prefix size.
for all index tiles in parallel do
    Create partial inverted index for assigned tile items.
end for
for all index tiles do
    for all items not in already processed index tiles in parallel do
        Use partial inverted index to identify candidate neighbors.
        Filter some candidates based on similarity upper-bound estimates.
        Use un-indexed portion of candidate vectors to finish computing
        their similarity, while continuing to filter those with estimates be-
        low ε.
        Output un-filtered candidates with similarity ≥ ε.
    end for
end for
```

Table 11.1 and Figure 11.1 show the runtime and parallel speedups of pL2AP, when building the ϵNNG for items in the ML10M training datasets, given ϵ ranging between 0.1 and 0.9. The method pL2AP is compared against pij, a baseline that uses similar cache-tiling strategies as pL2AP but does not prune the search space. Instead, it computes similarities for all items co-rated by at at least one user. The left graph of Figure 11.1 shows execution times in seconds, averaged over all three training folds, while the right one shows strong scaling results for the two methods, measuring the speedup of each method against their own serial execution. Strong scaling is when the problem size remains the same but the amount of parallelism increases. By effectively eliminating unnecessary similarity computations, pL2AP is able to achieve 4.24x–29.27x speedup over pij for different similarity thresholds ϵ.

11.2.3. *k-nearest neighbor graph construction*

One potential problem with using the ϵNNG to derive recommendations is that, given a high enough value for ϵ, some neighborhoods may not

Table 11.1. Runtime and speedup of pL2AP over pij on the ML10M dataset, when executed using 24 cores.

Method	$\epsilon = 0.1$	0.2	0.3	0.4	0.5	0.6	0.7	0.8	0.9
					time (s)				
pij	3.22	3.22	3.22	3.22	3.22	3.22	3.22	3.23	3.22
pL2AP	0.76	0.55	0.42	0.34	0.26	0.21	0.15	0.13	0.11
					speedup				
pL2AP	4.24	5.85	7.67	9.38	12.38	15.60	21.47	24.82	29.27

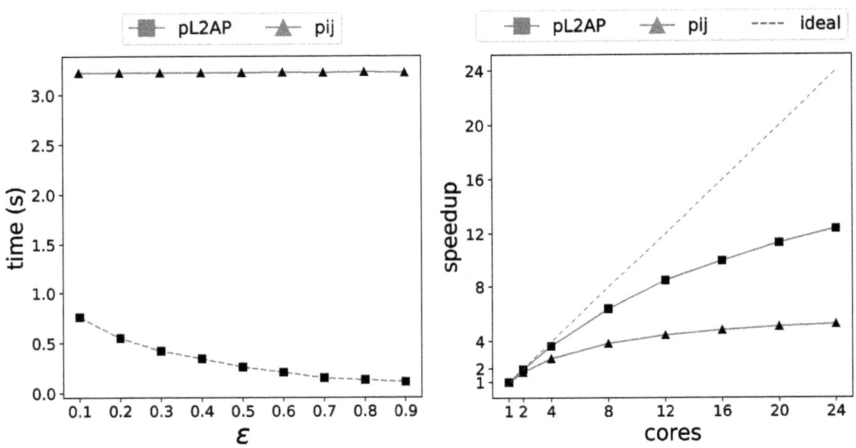

Fig. 11.1. Runtime (left) and strong scaling (right) of pL2AP and the naïve baseline pij when executed on the ML10M dataset.

contain any neighbors. The kNNG provides a guaranteed estimate of local preferences by retrieving the k nearest neighbors for each item in the set. L2Knng [Anastasiu and Karypis (2015a)] and pL2Knng [Anastasiu and Karypis (2016)] have been proposed for the purpose of efficiently constructing the *exact* kNNG. The main idea in L2Knng is to bootstrap the similarity search with a quickly constructed approximate graph. The minimum similarities in the approximate neighborhoods can then be used as filtering criteria in a search framework similar to the one in L2AP.

In the first phase of constructing the kNNG, L2Knng efficiently finds most, but not necessarily all of the k items closest to each target item, heuristically choosing a finite set of comparison items that are likely to be in the exact neighborhood. First, L2Knng identifies items that have high-value ratings in common with the target item, building an initial approximate kNNG. This graph is then iteratively improved by looking for neighbors among the neighbors of current neighbors. Finally, a filtering framework similar to the one described in Section 11.2.2 is employed to construct the exact kNNG. Unlike L2AP, L2Knng does not have an input threshold ϵ that could be used for pruning. Instead, it relies on the idea that any item that has the potential to be in the exact neighborhood of the target must have a similarity greater than the minimum similarity of the target with any item in its current approximate neighborhood. These minimum neighborhood similarities are used to forgo most of the item pair comparisons.

Similar to pL2AP, parallelization in pL2Knng focuses on cache-tiling and strategies for maximizing load balance among the cores. Unlike pL2AP, neighborhood searches are not independent. Given that cosine similarity is commutative, a neighbor j that the method finds for an item i could also benefit from the search if item i is not yet in j's neighborhood and the similarity between i and j is greater than the minimum neighborhood similarity in j's neighborhood. A lock-free update strategy is used for the in-memory shared neighborhood graph to address the potential resource contention encountered when items i and j are being processed by different cores. Algorithm 2 provides a high-level sketch of the pL2Knng method. Additional details regarding the initial approximate graph construction and filtering used to build the exact kNNG solution can be found in [Anastasiu and Karypis (2015a, 2016); Anastasiu (2017)].

Table 11.2 and Figure 11.2 show the efficiency of the parallel method, pL2Knng, when building the kNNG for items in the ML10M training datasets, given k ranging between 5 and 50. Our method, pL2Knng, is compared against pkij, a similar baseline to pij that uses similar cache-tiling strategies but does not prune the search space. The left graph of Figure 11.2 shows execution times in seconds, averaged over all three training folds, while the right one shows strong scaling results for the two methods, measuring the speedup of each method against their own serial execution. By effectively eliminating unnecessary similarity computations, pL2Knng is able to achieve 2.2x–2.97x speedup over pkij for different k values. Given larger datasets, such as one containing 1M pages from the English Wikipedia Web site, containing almost half a billion non-zero values, pL2Knng has been shown to outperform pkij by 7.3x–11.5x for $k \leq 50$ [Anastasiu and Karypis (2015b)]. The results show the value of pruning the search space as a means to improve the efficiency of nearest neighbor identification.

Algorithm 2 pL2Knng

Normalize profiles for every item i.
for all items i **in parallel do**
    Choose $\mu \geq k$ candidates highly co-rated with i by some user.
    Compute candidate similarities and keep top-k candidates.
end for
for all items i **in parallel do**
    Choose μ candidates from the neighborhoods of the current k neighbors.
    Compute candidate similarities and update ith neighborhood as necessary.
end for
Use minimum neighborhood similarities to define partial inverted index tiles.
for all index tiles **do**
    for all items i not in already processed index tiles **in parallel do**
        Use partial inverted index to identify candidate neighbors.
        Filter some candidates based on similarity upper-bound estimates.
        Use un-indexed portion of candidate vectors to finish computing their similarity, while continuing to filter those with estimates below minimum similarities in the candidate or i's neighborhoods.
        Update neighborhoods of i and un-filtered candidates as necessary.
    end for
end for

Table 11.2. Execution times and speedup of pL2Knng over pkij on the ML10M dataset, when executed using 24 cores.

Method	k=5	10	20	30	40	50
	time (s)					
pKij	3.23	3.235	3.237	3.198	3.24	3.247
pL2Knng	1.089	1.136	1.228	1.324	1.408	1.479
	speedup					
pL2Knng	2.97	2.85	2.64	2.42	2.3	2.2

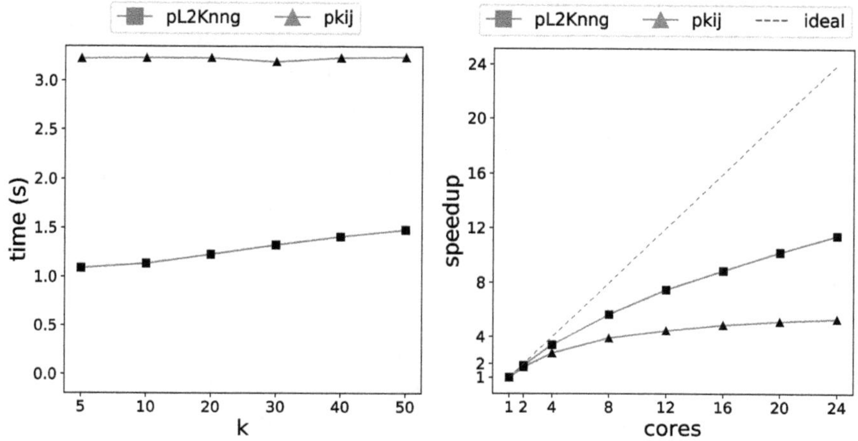

Fig. 11.2. Runtime (left) and strong scaling (right) of pL2Knng and the naïve baseline pkij when executed on the ML10M dataset.

11.3. Efficiently Estimating Item-Item Similarities by solving an Optimization Problem

11.3.1. *Sparse LInear Methods for Top-N Recommendation (SLIM)*

In contrast to the standard item-based methods which use a predefined similarity measure like cosine or Pearson correlation, Sparse LInear Methods (SLIM) [Ning and Karypis (2011)] learn the item-item relationships from the user-item feedback matrix **R** instead. SLIM is a popular method for top-N recommendation, as it has been shown to provide high-quality recommendations [Ning and Karypis (2011)]. In SLIM, the rating for an item is predicted as a sparse aggregation of the existing ratings provided by the user:

$$\hat{r}_{ui} = \boldsymbol{r}_u^T \boldsymbol{s}_i,$$

where $\boldsymbol{r}_u^T$ is the uth row of the rating matrix **R** and $\boldsymbol{s}_i$ is a sparse vector containing non-zero aggregation coefficients over all items. The sparse aggregation coefficient matrix **S** of size $m \times m$, capturing the item-item relationships is estimated by solving the following optimization problem:

$$\begin{aligned} \underset{\mathbf{S}}{\text{minimize}} \quad & \frac{1}{2}||\mathbf{R} - \mathbf{RS}||_F^2 + \frac{\beta}{2}||\mathbf{S}||_F^2 + \lambda||\mathbf{S}||_1 \\ \text{subject to} \quad & \mathbf{S} \geq 0 \\ & diag(\mathbf{S}) = 0. \end{aligned} \tag{11.4}$$

The optimization problem of Equation 11.4 tries to minimize the training error, denoted by $||\mathbf{R} - \mathbf{RS}||_F^2$, while also regularizing the matrix $\mathbf{S}$. The problem uses two regularizers. The first one is the Frobenius norm of the matrix $\mathbf{S}$ (noted by $||\mathbf{S}||_F^2$), which is controlled by the parameter β, in order to prevent overfitting. The other regularizer is the l_1 norm of the matrix $\mathbf{S}$ (noted by $||\mathbf{S}||_1$), which is controlled by the parameter λ, in order to promote sparsity [Tibshirani (1996)]. Larger values of β and λ leads to more severe regularization. The use of both l_F and l_1 regularization makes the optimization problem of Equation 11.4 an elastic net problem [Zou and Hastie (2005)].

The non-negativity constraint on $\mathbf{S}$ imposes the item-item relations to be positive. The constraint diag($\mathbf{S}$) = 0 is added to avoid trivial solutions (e.g., $\mathbf{S}$ corresponding to the identity matrix) and ensure that r_{ui} is not used to compute $\hat{r}_{ui}$.

11.3.1.1. *Parallelizing SLIM*

Equation 11.4 can be accelerated by learning similarities in parallel for every target item i, as every column of $\mathbf{S}$ can be learned independently from the other columns. Then the optimization problem of Equation 11.4 changes to a set of optimization problems of the form:

$$\begin{aligned} \underset{\boldsymbol{s}_i}{\text{minimize}} \quad & \frac{1}{2}||\boldsymbol{r}_i - \mathbf{R}\boldsymbol{s}_i||_2^2 + \frac{\beta}{2}||\boldsymbol{s}_i||_2^2 + \lambda||\boldsymbol{s}_i||_1 \\ \text{subject to} \quad & \boldsymbol{s}_i \geq 0 \\ & s_{ii} = 0, \end{aligned}$$

which allows us to estimate the ith column of $\mathbf{S}$, noted by $\boldsymbol{s}_i$. The term $\boldsymbol{r}_i$ refers to the ith column of the training matrix $\mathbf{R}$. The problem is solved with the use of coordinate descent and soft thresholding [Friedman *et al.* (2010)].

The software implementation of SLIM provided by the author Xia Ning[2] utilizes the property that different columns of the sparse aggregation coefficient matrix can be solved independently and allows the users to specify which columns of the sparse aggregation coefficient matrix they would like to estimate. The software is implemented with the use of the Bound Constrained Least Squares (BCLS) library[3].

In order to fully utilize the benefits from the parallel estimation of different columns of $\mathbf{S}$, we use a multithreaded implementation of SLIM which

[2] http://www-users.cs.umn.edu/~xning/slim/html/
[3] http://www.cs.ubc.ca/~mpf/bcls/index.html

relies on OpenMP. This allows us to have parallelism within a multi-core node. Each thread is assigned a set of columns i and estimates the associated sparse aggregation coefficient vectors $\boldsymbol{s}_i$. After all the threads have estimated the set of $\boldsymbol{s}_i$ vectors, the vectors are combined into the overall sparse aggregation coefficient matrix $\mathbf{S}$. We will refer to the multithreaded implementation of SLIM, as mt-SLIM. Figure 11.3 shows the speedup achieved by mt-SLIM on the ML10M dataset, with respect to the serial runtime (cores = 1). The results shown correspond to the time taken for model estimation and they correspond to the average of three folds.

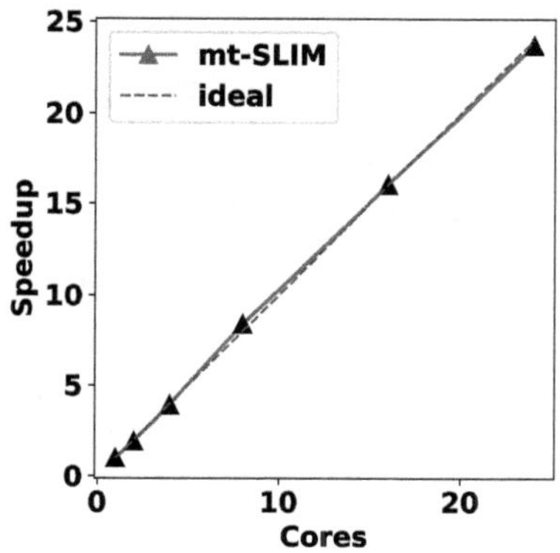

Fig. 11.3. The speedup achieved by mt-SLIM on the ML10M dataset, while increasing the number of cores (strong scaling).

As both the rating matrix and the estimated sparse aggregation coefficient matrix are sparse, they are stored in CSR (Compressed Sparse Row) format, in which three one-dimensional arrays are stored, that contain the non-zero values, with their associated row and column indices.

11.3.1.2. *Accelerating the training time of SLIM during parameter search*

In order to find the pair of regularization parameters β and λ that give the best results, a parameter search needs to be conducted. However, the number of models to estimate increases quadratically with the number of values of the regularization parameters β and λ explored. In order to be able to estimate the models more efficiently, mt-SLIM utilizes 'warm-start'. This means that with the exception of the model estimated with the very first choice of parameters, every subsequent model is initialized with the previous model estimated with a different choice of regularization parameters, instead of being initialized with zero values.

Figure 11.4 compares the time spent by mt-SLIM without initialization and mt-SLIM with warm start, for the same number of cores and and for the same choice of regularization parameters ($\beta = 10$, $\lambda = 1$). We can see that mt-SLIM with warm start is on average 15 times faster than mt-SLIM with no initialization.

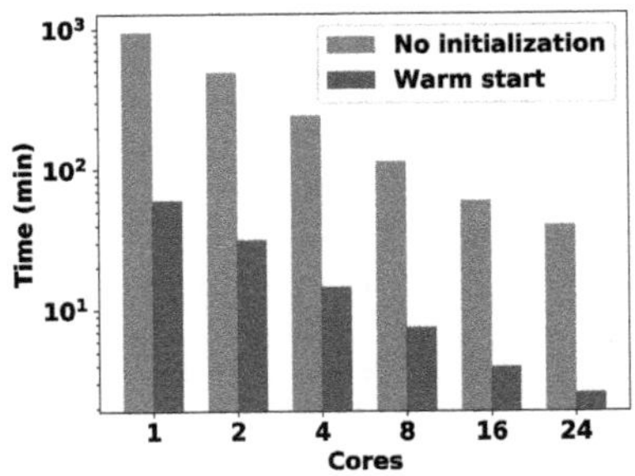

Fig. 11.4. The time in minutes achieved by mt-SLIM with and without warm start on the ML10M dataset for $\beta = 10$ and $\lambda = 1$, while increasing the number of cores (strong scaling).

By evaluating the performance of mt-SLIM with no initialization and with warm start, we get the same performance results, which shows that with warm start, we gain in estimation times, without compromising the quality of the performance.

11.3.2. *Global and Local Sparse LInear Methods for Top-N Recommendation (GLSLIM)*

A limitation of SLIM is that it estimates only a single model for all the users. In many cases, there are differences in users' behavior, who can have diverse preferences. These cannot be captured by a single model. Recently, GLSLIM [Christakopoulou and Karypis (2016)] was proposed, which utilizes both user-subset specific models and a global model, and was shown to improve the top-N recommendation quality. The models, (which are estimated with SLIM) are jointly optimized and combined in a personalized way. Also, GLSLIM automatically identifies the appropriate user subsets. If we note the global model as $\mathbf{S}$ and the local user-subset specific models as $\mathbf{S}^{p_u}$, where $p_u \in \{1..k\}$ denotes the user subset, then the predicted rating of user u, who belongs to subset p_u for item i, will be estimated as:

$$\hat{r}_{ui} = \sum_{l \in R_u} g_u s_{li} + (1 - g_u) s_{li}^{p_u}, \tag{11.5}$$

where s_{li} shows the global item-item similarity between the lth item rated by the user u and the target item i and $s_{li}^{p_u}$ shows the p_u user-subset specific similarity between the lth rated item by u and the target item i. The term g_u is the personalized weight which controls the interplay between the global and the local model and ranges between 0 and 1.

GLSLIM is an iterative algorithm which jointly optimizes the global and local models, the user assignment and the personalized weights. The global and local models are estimated by solving an elastic net optimization problem. Following SLIM, GLSLIM can estimate the columns of its global and local models independent of the other columns. Separate regularization is enforced on the global and on the local models, in order to allow more flexibility in model estimation: we thus have the global l_2 regularization parameter β_g, the global l_1 regularization parameter λ_g, the local l_2 regularization parameter β_l and the local l_1 regularization parameter λ_l. Initially, the users are assigned to clusters. In each iteration, every user is assigned to the subset that resulted in the smallest training error, and his personalized weight is updated accordingly. The models and the user assignment with the personalized weights are updated iteratively, until convergence (the algorithm converges when the users switching subsets are less than one percent). An overview of the algorithm is shown in Algorithm 3.

After having completed the training, the top-N recommendation is performed in the following way: for user u, the ratings of all the unrated items i are estimated with Equation 11.5, and the items with the N highest ratings are recommended to the user.

11.3.2.1. *Parallelizing GLSLIM*

We can see from Algorithm 3 that every iteration has two parts: estimating the global and local models (line 4) and user refinement (lines $5-11$). Both parts allow for parallelization, each in its own way. The model estimation part can be parallelized with respect to the items, since every column of the models can be estimated independently of the others. The user refinement part can be parallelized with respect to the users, as provided the models are fixed, the assignment and personalized weight of each user does not depend on the other users.

[4]http://glaros.dtc.umn.edu/gkhome/cluto/cluto/overview

Algorithm 3 GLSLIM

Assign $g_u = 0.5$, to every user u.
Compute the initial clustering of users with CLUTO[4].
while number of users who switched clusters $> 1\%$ of the total number of users **do**
  Estimate $\mathbf{S}$ and $\mathbf{S}^{p_u}$, $\forall p_u \in \{1, \ldots, k\}$. The estimation is initialized in all iterations but the first one with the corresponding matrices $\mathbf{S}$ and $\mathbf{S}^{p_u}$, $\forall p_u \in \{1, \ldots, k\}$ computed in the previous iteration.
  for all user u **do**
    for all cluster p_u **do**
      Compute g_u for cluster p_u by minimizing the squared error.
      Compute the training error.
    end for
    Assign user u to the cluster p_u that has the smallest training error and update g_u to the corresponding one for cluster p_u.
  end for
end while

Taking advantage of these parallelization opportunities, there is an MPI-based GLSLIM software[5], which we use for our subsequent experiments. GLSLIM relies on MPI, instead of OpenMP which was used for mt-SLIM, as it requires more computations than SLIM. SLIM solves one elastic net problem for the whole training matrix $\mathbf{R}$, while GLSLIM is iterative and in each iteration, a new elastic net problem is solved for the global matrix and for all user subsets. Thus, the distributed framework MPI is employed, which allows model estimation and user refinement to be done in a distributed way, thus taking advantage of multiple nodes (where each node consists of cores).

Figure 11.5 shows the speedup achieved by GLSLIM on different nodes, with respect to the time taken by GLSLIM on one node (which consists of 24 cores in our shown results), for the ML10M dataset.

11.3.2.2. *Accelerating the training time of GLSLIM during parameter search*

GLSLIM has many parameters, for which a parameter search needs to be conducted in order to find the set of them that gives the best performance: the regularization parameters β_g, λ_g, β_l, λ_l, and the number of user subsets

[5] http://www-users.cs.umn.edu/~evangel/code.html

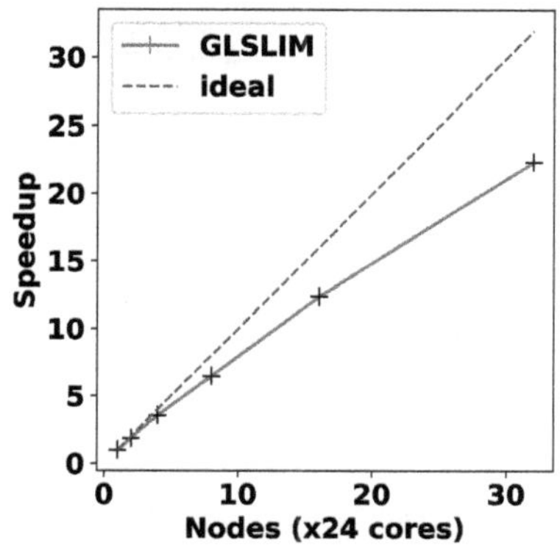

Fig. 11.5. The speedup achieved by GLSLIM on the ML10M dataset, while increasing the number of nodes. The speedup is computed with respect to the time of running GLSLIM on one node.

k. We can see that the cost of finding the best possible set of parameters increases exponentially. It is thus crucial to be able to run GLSLIM as efficiently as possible.

In order to do so, we again employ warm start. This is done in the following way: When estimating a model with a new choice of parameters, we use another model learned with a different choice of parameters as its initialization. Thus, the only time it is needed to estimate a model with no initialization is when estimating the very first model for this dataset (model of the first iteration with the first choice of parameters). After it is estimated, the models of the subsequent iterations get initialized with the models of the previous iterations. Then, when moving on to a new choice of parameters, the model of the first iteration is initialized with the model estimated with the previous choice of parameters and so on.

Figure 11.6 shows the time taken in minutes to run GLSLIM on ML10M with 'warm start' and with 'no initialization'. Figure 11.6 shows the total time for all iterations when run with $k = 5$ user subsets and with l_2 regularizations parameters $\beta_g = \beta_l = 10$ and l_1 regularization parameters $\lambda_g = \lambda_l = 1$. Note that four iterations were needed until convergence. Also note that the greatest part of the time shown corresponds to the model estimations, as the user refinement does not take more than a couple of seconds (in this example, the user refinement part took fourteen seconds when run on one node). A speedup of $4\times$ is achieved by employing warm start.

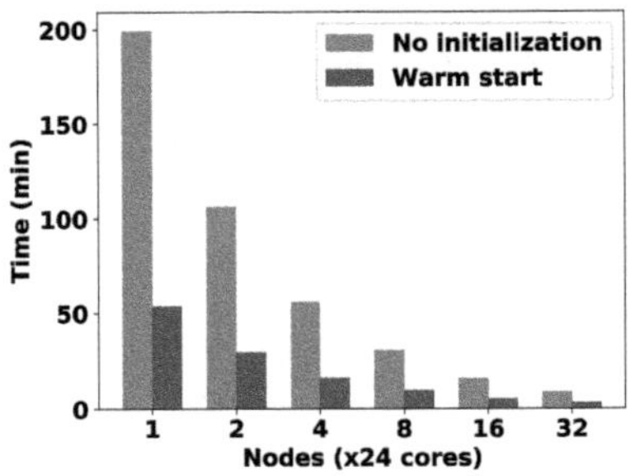

Fig. 11.6. The total time in minutes achieved by GLSLIM with and without warm start on the ML10M dataset, while increasing the number of nodes.

Table 11.3. Comparison of SLIM with GLSLIM in terms of top-N performance and training time.

Method	HR	ARHR	Time (min)
SLIM	0.310	0.152	2.56
GLSLIM	0.336	0.167	51.72

Time corresponds to 'warm start' time in minutes, and corresponds to the time taken on one node (24 cores). GLSLIM time corresponds to total time of all iterations until convergence.

Table 11.3 shows the top-N recommendation performance and training times of SLIM and GLSLIM with warm start, when run with the same parameters $\beta = \beta_g = \beta_l = 10$ and $\lambda = \lambda_g = \lambda_l = 1$. Five user subsets were used for GLSLIM. The top-N performance is measured in terms of HR (Equation 11.1) and ARHR (Equation 11.2). The reported time corresponds to running SLIM and GLSLIM on one node (24 cores). This is done for fairness of comparison between the two methods. The shown times correspond to the warm-start right-most column of Figure 11.4 and the warm-start left-most column of Figure 11.6. We can see that GLSLIM has an average performance gain of 9.5% over SLIM, while requiring more time-consuming training; although higher number of nodes used allows for great decrease in running time.

11.4. Scaling up latent factor approaches

Latent factor approaches are a class of methods that map users and items to vectors in a common low-rank space known as the *latent space*. A detailed overview of latent space approaches can be found in Chapter 2. Latent factor approaches are perhaps the most popular techniques used for rating prediction. The success of these approaches has led to a wealth of research on developing algorithms to facilitate high-quality recommendations from massive training datasets. These algorithms exhibit complex tradeoffs in terms of computational characteristics, convergence rate, and available parallelism.

11.4.1. *Overview of matrix and tensor factorization*

Matrix factorization approaches [Koren (2008)] are state-of-the-art collaborative filtering methods and have gained high popularity since the Netflix Prize [Koren (2009); Takács *et al.* (2009)]. They assume that the user-item rating matrix $\mathbf{R}$ is low rank and can be computed as a product of two matrices known as the user and the item latent factors (denoted $\mathbf{P}$ and $\mathbf{Q}$, respectively). Rows of $\mathbf{P}$ and $\mathbf{Q}$ are F-dimensional vectors which represent the corresponding user or item. The value F is referred to as the *rank* of the factorization.

An item's latent factor, denoted $\mathbf{q}_i$, represents a few characteristics of the item, and a user's latent factor, denoted $\mathbf{p}_u$, signifies how much a user weights these characteristics. The predicted rating for the user u on the item i is given by

$$\hat{r}_{ui} = \mathbf{p}_u^T \mathbf{q}_i.$$

The completed matrix $\hat{\mathbf{R}} = \mathbf{P}\mathbf{Q}^T$ is used to serve the recommendation to the user for the items for which their preferences were unknown in the original matrix $\mathbf{R}$.

The user and the item latent factors are estimated by minimizing a regularized squared loss

$$\underset{\mathbf{P},\mathbf{Q}}{\text{minimize}} \quad \frac{1}{2} \sum_{r_{ui} \in \mathbf{R}} \left(r_{ui} - \mathbf{p}_u^T \mathbf{q}_i \right)^2 + \frac{\beta}{2} \left(||\mathbf{P}||_F^2 + ||\mathbf{Q}||_F^2 \right), \tag{11.6}$$

where the parameter β controls the Frobenius norm regularization to prevent overfitting.

Additionally, instead of optimizing for rating predictions, one can optimize for ranking performance by substituting a ranking loss function instead of the squared error loss function. For example, Bayesian Personalized Ranking (BPR) [Rendle and Schmidt-Thieme (2010)], Collaborative Less-is-More Filtering (CLiMF) [Shi *et al.* (2012)] and CofiRank [Weimer *et al.* (2008)] optimize approximation of different ranking metrics to estimate the user and the item latent factors for better ranking performance.

Ratings are often accompanied by contextual information associated with the ratings. For example, the ML10M dataset provides both *timestamps* and *tags* which can be used to improve recommendation quality.

The traditional ratings matrix can be extended to include contextual information in the form of a *tensor*, which is the generalization of a matrix to higher orders. For example, associating each rating with a timestamp would result in a third-order tensor whose modes represent users, items, and time. Latent factor approaches can be extended to include higher-order data provided by tensors. The canonical polyadic decomposition (CPD) is a popular model for tensor factorization which has be used successfully for rating prediction. The CPD seeks to model a ratings tensor $\mathcal{R}$ as the combination of user factor $\mathbf{P}$, item factor $\mathbf{Q}$, and context factor $\mathbf{C}$. The resulting optimization problem closely resembles that of matrix factorization:

$$\underset{\mathbf{P},\mathbf{Q},\mathbf{C}}{\text{minimize}} \quad \frac{1}{2} \sum_{r_{uik} \in \mathcal{R}} \left(r_{uik} - \sum_{f=1}^{F} p_{uf} q_{if} c_{kf} \right)^2 + \frac{\beta}{2} \left(||\mathbf{P}||_F^2 + ||\mathbf{Q}||_F^2 + ||\mathbf{C}||_F^2 \right).$$

The estimation of user and item latent factors by solving Equation 11.6 is one method of solving a problem referred to as *matrix completion*. It is a non-convex and computationally expensive problem. Several optimization algorithms have been successfully applied for matrix completion on large scale datasets.

Experimental environment. In the remaining discussion, we evaluate three latent factor approaches that solves matrix completion problem. Each algorithm is iterative in nature, though by convention we refer to these iterations as *epochs*. We define an epoch as the work required to update the latent factors one time using all available rating data. Convergence is detected when the RMSE does not improve for twenty epochs. We fix F, the rank of the factorization, to 40. All presented results are collected using SPLATT [Smith and Karypis (2015)], a publicly available[6] toolkit for high-performance sparse tensor factorizations. While optimized for tensors, SPLATT supports matrix factorizations because a matrix is equivalent to a two-mode tensor. SPLATT has also been integrated into the Spark+MPI framework [Anderson *et al.* (2017)], achieving over 10× speedup over pure Spark solutions.

[6] http://cs.umn.edu/~splatt/

Algorithm 4 Matrix factorization via alternating least squares (ALS)

Initialize $\mathbf{P}$ and $\mathbf{Q}$ randomly.
while $\mathbf{P}$ and $\mathbf{Q}$ have not converged **do**
    for all user u in parallel **do**
        $\mathbf{H}_u \leftarrow \mathbf{0}$.
        For each rating r_{ui} in $\mathbf{r}_u$, append row $\mathbf{q}_i$ to $\mathbf{H}_u$.
        $\mathbf{p}_u \leftarrow \left(\mathbf{H}_u^T\mathbf{H}_u + \beta\mathbf{I}\right)^{-1}\mathbf{H}_u^T\mathbf{r}_u$.
    end for
    for all item i in parallel **do**
        $\mathbf{H}_i \leftarrow \mathbf{0}$.
        For each rating r_{ui} in $\mathbf{r}_i$, append row $\mathbf{p}_u$ to $\mathbf{H}_i$.
        $\mathbf{q}_i \leftarrow \left(\mathbf{H}_i^T\mathbf{H}_i + \beta\mathbf{I}\right)^{-1}\mathbf{H}_i^T\mathbf{r}_i$.
    end for
end while

11.4.2. *Alternating Least Squares (ALS)*

ALS was one of the first matrix completion algorithms applied to large scale data [Zhou *et al.* (2008)]. ALS is based on the observation that if we solve Equation 11.6 for one latent factor at a time, the solution has a linear least squares solution. ALS is an iterative algorithm which first minimizes with respect to $\mathbf{P}$ and then $\mathbf{Q}$. The process is repeated until convergence.

Let $\mathbf{r}_u$ be the vector of all ratings supplied by user u. $\mathbf{H}_u$ is an $|\mathbf{r}_u|\times F$ matrix whose rows are the feature vectors $\mathbf{q}_i$, for each item i rated in $\mathbf{r}_u$. Similarly, $\mathbf{r}_i$ is the vector of all ratings supplied for item i, and $\mathbf{H}_i$ is an $|\mathbf{r}_i|\times F$ matrix. ALS proceeds by updating all $\mathbf{p}_u$ followed by all $\mathbf{q}_i$:

$$\begin{aligned}\mathbf{p}_u &\leftarrow \left(\mathbf{H}_u^T\mathbf{H}_u + \beta\mathbf{I}\right)^{-1}\mathbf{H}_u^T\mathbf{r}_u, \quad \forall u \in 1,\ldots,m\\ \mathbf{q}_i &\leftarrow \left(\mathbf{H}_i^T\mathbf{H}_i + \beta\mathbf{I}\right)^{-1}\mathbf{H}_i^T\mathbf{r}_i, \quad \forall i \in 1,\ldots,n.\end{aligned} \tag{11.7}$$

The full procedure is outlined in Algorithm 4. Extending Equation 11.7 to tensors changes the construction of the $\mathbf{H}_u$ matrices [Shao (2012)]. For example, the row of $\mathbf{H}_u$ associated with rating r_{uik} is the elementwise multiplication of the corresponding feature vectors $\mathbf{q}_i$ and $\mathbf{c}_k$. The $\mathbf{H}_i$ and $\mathbf{H}_k$ matrices are constructed similarly.

Each row in Equation 11.7 is independent and thus can be computed in parallel [Zhou *et al.* (2008)]. The simplicity of this approach has led ALS to be optimized for high-performance shared- and distributed-memory systems[Karlsson *et al.* (2015); Smith *et al.* (2017)], GPUs [Gates *et al.* (2015);

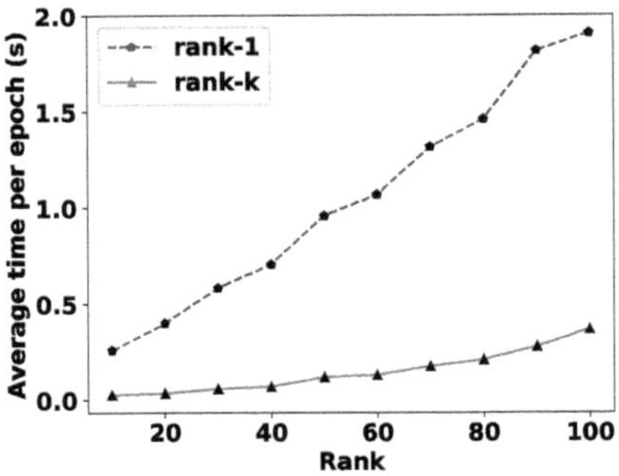

(a) Matrix

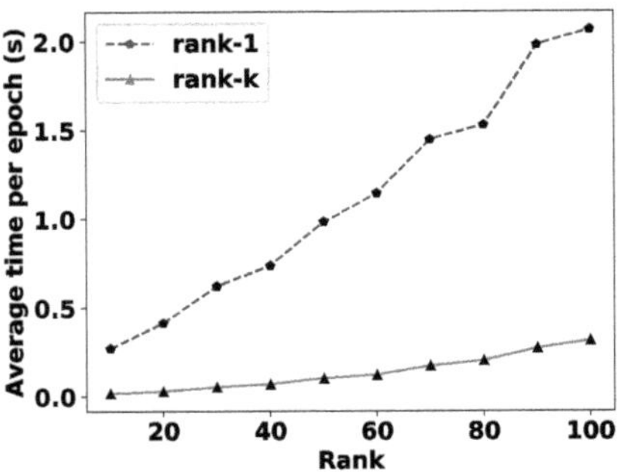

(b) Tensor

Fig. 11.7. Average time per epoch when using rank-1 and rank-k updates during ALS. Execution is on 24 cores and an epoch is counted as updating each factor matrix once.

Tan *et al.* (2016)], Hadoop [Shin and Kang (2014)], and is implemented in Spark's MLlib[7]. Successful approaches distribute the ratings data in a one-dimensional fashion such that all of the ratings required to construct an $\mathbf{H}$ matrix are located on the same node. By distributing the data in this fashion, none of the partially-constructed $\mathbf{H}$ matrices need to be communicated or aggregated. However, this distribution requires that each node stores potentially the entire latent factors. Fortunately, in practice this is not prohibitive on most modern systems.

The computational complexity of ALS is $\mathcal{O}\left(F^2|\mathbf{R}| + F^3(m+n)\right)$ per epoch. In practice, the $\mathcal{O}(F^2)$ computation per rating that comes from constructing the various $\mathbf{H}$ matrices dominates the computation. A common implementation strategy is to process one rating at a time and accumulate directly into $\mathbf{H}_u^T\mathbf{H}_u$ and $\mathbf{H}_u\mathbf{r}_u$ instead of explicitly constructing $\mathbf{H}_u$. However, this strategy ignores the details of modern hardware architectures in which memory movement is more expensive than floating-point operations. Each rating produces an accumulation that is a rank-1 update performing $\mathcal{O}(F^2)$ work on F^2 data. Alternatively, performing a single rank-k update by explicitly forming $\mathbf{H}_u$ instead performs $\mathcal{O}(|\mathbf{r}_u|F^2)$ work on $(|\mathbf{r}_u|F + F^2)$ data [Gates *et al.* (2015); Smith *et al.* (2017)]. While the final amount of work is the same, the rank-k update fetches less data from memory and is thus better suited for modern processors. We explore this phenomenon in Figure 11.7, which illustrates runtime per epoch as F is increased. Using rank-k updates can be over 10× faster than the more common rank-1 updates.

[7]https://spark.apache.org/mllib/

Algorithm 5 Matrix factorization via stochastic gradient descent (SGD)

```
1: Initialize P and Q randomly.
2: while P and Q have not converged do
3:     Shuffle the permutation of ratings.
4:     for all rating r_ui do
5:         e_ui ← r_ui − p_u^T q_i
6:         p_u ← p_u + η (e_ui q_i − β p_u).
7:         q_i ← q_i + η (e_ui p_u − β q_i).
8:     end for
9: end while
```

11.4.3. *Stochastic Gradient Descent (SGD)*

SGD is an optimization algorithm that trades a large number of epochs for a low computational complexity. An epoch consists of processing all ratings one-at-a-time in random order and updating the factorization based on the local gradient. Updates are of the form:

$$\begin{aligned} e_{ui} &\leftarrow r_{ui} - \mathbf{p}_u^T \mathbf{q}_i, \\ \mathbf{p}_u &\leftarrow \mathbf{p}_u + \eta \left(e_{ui} \mathbf{q}_i - \beta \mathbf{p}_u \right), \\ \mathbf{q}_i &\leftarrow \mathbf{q}_i + \eta \left(e_{ui} \mathbf{p}_u - \beta \mathbf{q}_i \right), \end{aligned} \tag{11.8}$$

where η is a hyperparameter representing the learning rate. The complexity of Equation 11.8 is linear in F, resulting in a total complexity of $\mathcal{O}(F|\mathbf{R}|)$ per epoch. The low complexity and simple implementation of SGD has led to it being widely adopted by researchers and industry alike. The details of SGD are outlined in Algorithm 5.

SGD is less straightforward than ALS to parallelize. Since processing a rating updates rows of both **P** and **Q**, special care must be taken to prevent the same rows from being modified at the same time (called a *race condition*). There are two broad approaches for parallelizing SGD.

Stratified methods are based on the observation that if two ratings do not overlap (i.e., they have neither a row nor a column in common) then they can be updated with Equation 11.8 in parallel. This strategy was introduced by DSGD [Gemulla *et al.* (2011)], which imposes a grid on **R** to identify blocks that can be processed in parallel. Stratification is illustrated in Figure 11.8. Stratification has proven to be an effective strategy for parallelizing SGD and has been extended in works on multithreaded environments, distributed systems, and GPUs [Zhuang *et al.* (2013); Yun *et al.* (2014); Xie *et al.* (2017)].

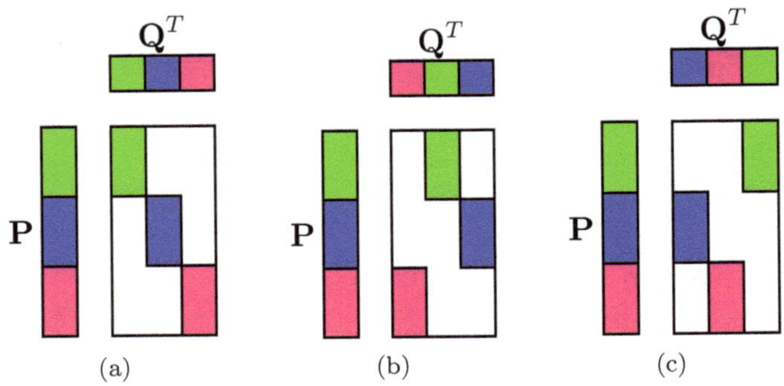

Fig. 11.8. Stratified SGD with three workers. Colored blocks represent independent sets of ratings and the rows of **P** and **Q** which model them. Each colored block of ratings and their corresponding rows can be processed in parallel.

Asynchronous methods rely on the stochastic nature of SGD to allow overlapping updates. This technique was popularized by Hogwild [Recht *et al.* (2011)] on shared-memory systems. The key idea is to simply allow race conditions to occur without attempting to avoid them. Convergence is still achieved due to the iterative nature of SGD and infrequent overlaps from the high level of sparsity in **R**. Teflioudi *et al.* later introduced asynchronous SGD (ASGD) [Teflioudi *et al.* (2012)] for distributed computing environments. During ASGD, nodes maintain locally modified copies of **P** and **Q** and updates are asynchronously communicated several times per epoch. Overlapping updates are averaged with the master copy and sent to workers.

Extending the formulation of SGD to tensors is straightforward and again only requires additional elementwise multiplications [Shao (2012)]. However, parallelization becomes a significant challenge when a contextual mode is added to the data. The number of blocks in a stratified SGD algorithm increases *exponentially* with the number of tensor modes despite the work per rating only increasing linearly, and thus the time of synchronization and communication quickly dominate the factorization time. Asynchronous methods also suffer because the number of unique contexts is typically much smaller than the number of users or items, resulting in more frequent update conflicts. A hybrid of stratification and asynchronous SGD addresses these challenges, but the hybrid is still bested by ALS at large numbers of cores [Smith *et al.* (2017)].

Algorithm 6 Matrix factorization via coordinate descent (CCD++)

Initialize **P** and **Q** randomly.
while P and **Q** have not converged **do**
  for all column f **do**
    Pre-compute all error terms: $(r_{ui} - \mathbf{p}_u^T \mathbf{q}_i)$.
    for all user u in parallel **do**
      Update p_{uf} following (11.9).
    end for
    for all item i in parallel **do**
      Update q_{if} following (11.9).
    end for
  end for
end while

11.4.4. *Coordinate Descent (CCD++)*

Coordinate descent is a class of optimization algorithms which update one parameter of the output at a time. CCD++ is a column-oriented descent algorithms for matrix completion [Yu *et al.* (2012)]. CCD++ updates columns of **P** and **Q** in sequence, with a single parameter update taking the form

$$p_{uf} \leftarrow \frac{\sum_{r_{ui} \in \mathbf{R}} (r_{ui} - \mathbf{p}_u^T \mathbf{q}_i + p_{uf} q_{if}) q_{if}}{\beta + \sum_{r_{ui} \in \mathbf{R}} q_{if}^2}. \tag{11.9}$$

The full procedure is outlined in Algorithm 4. If all $(r_{ui} - \mathbf{p}_u^T \mathbf{q}_i)$ are pre-computed, CCD++ has a complexity of $\mathcal{O}(F|\mathbf{R}|)$ per epoch, matching SGD. The extension of CCD++ to tensors follows that of ALS and SGD, in which additional elementwise multiplications are introduced to the formulation [Karlsson *et al.* (2015)].

Similar to ALS, each column entry is independent and can thus be computed in parallel. CCD++ has accordingly been parallelized on shared- and distributed-memory systems [Yu *et al.* (2012); Karlsson *et al.* (2015); Smith *et al.* (2017)] and GPUs [Nisa *et al.* (2017)]. However, unlike ALS, the communication cost from aggregating partial computations is only of constant size per column as opposed to the larger **H** matrices of ALS. The lower communication volume affords more flexible partitionings of the ratings. Recent work has shown that a Cartesian (i.e., grid) distribution of the data is an effective formulation and has been scaled to over sixteen thousand cores [Smith *et al.* (2017)].

11.4.5. *Evaluation of optimization algorithms*

Parallel scalability. We examine the parallel scalability of ALS, SGD, and CCD++ for matrix and tensor completion in Figure 11.9. ALS scales notably better than the competing methods. The scalability of ALS comes from being rich in dense linear algebra kernels which effectively use the floating-point hardware found in each core, instead of being bound by memory bandwidth which is a shared resource. In contrast, SGD and CCD++ perform a factor of F fewer floating-point operations per rating processed, resulting in heavy reliance on available memory bandwidth. Interestingly, the scalability of CCD++ and SGD is also more dependent on the size and characteristics of the ratings dataset. Figure 11.10 shows parallel scalability on a tensor of 210 million Yahoo! music ratings with timestamps from the 2011 KDD cup [Dror *et al.* (2012)]. CCD++ achieves perfect speedup on this significantly larger and more sparse dataset.

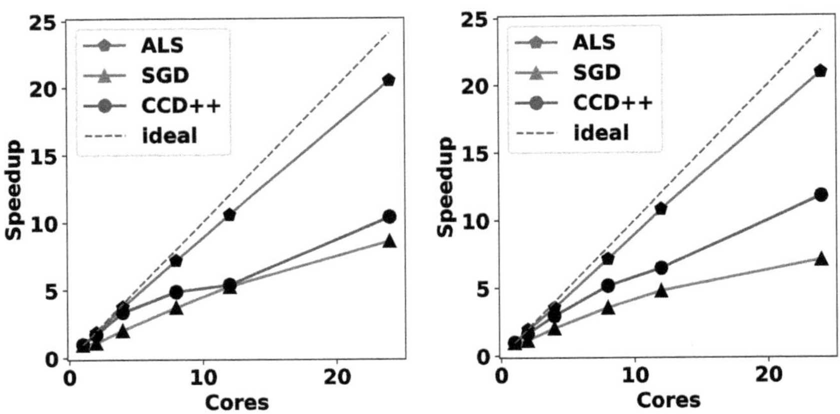

(a) Matrix (b) Tensor

Fig. 11.9. Speedup on ML10M dataset scaling from 1 to 24 cores. SGD is parallelized using Hogwild [Recht *et al.* (2011)].

Time to solution. Finally, we compare solution qualities and convergence times for the latent factor approaches in Table 11.4. In the matrix case, SGD arrives at the lowest RMSE while being competitive in runtime to ALS. CCD++ obtains the lowest RMSE in the tensor case, but at 5× the runtime of the similar-quality ALS. Utilizing the timestamp for tensor completion notably improves the RMSE for ALS and CCD++, but not SGD.

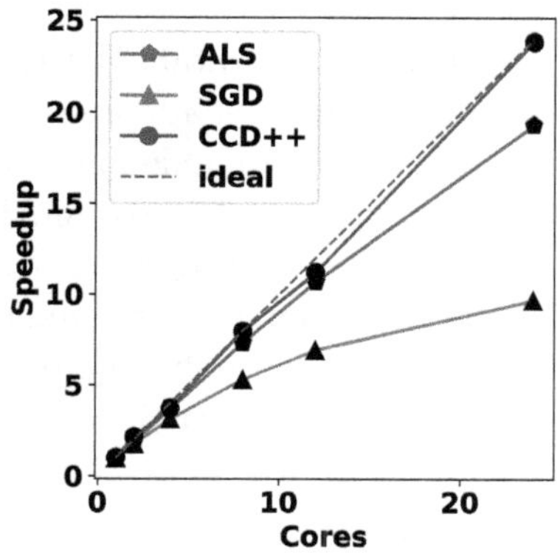

Fig. 11.10. Parallel speedup on a three-mode tensor made from 210 million Yahoo! music ratings.

Table 11.4. Comparison of solution quality and time.

	Matrix		Tensor	
Algorithm	**RMSE**	**Time (s)**	**RMSE**	**Time (s)**
ALS	0.8805	4.82	0.8662	50.26
SGD	0.8747	11.54	0.9107	17.19
CCD++	0.8955	435.61	0.8622	256.25

RMSE is evaluated on the test dataset, averaged over three folds. **Time** measures the time to convergence.

Since timestamps are grouped by month (133 months in total), the number of independent months is significantly more limited than the number of independent users or items. Thus, there are frequent overlapping updates to the **C** latent factor. Lastly, we note that the time-to-solution for CCD++ is longer in the matrix case than the tensor case, despite performing less work and arriving at a higher RMSE. While the tensor completion algorithm performs more work than the matrix equivalent, in practice we find that it converges in fewer epochs.

11.4.6. *Singular Value Decomposition (SVD)*

The key idea of SVD-based models is to factorize the user-item rating matrix to a product of two lower rank matrices, one containing the user factors and the other containing the item factors. Since conventional SVD is undefined in the presence of missing values, *PureSVD* [Cremonesi *et al.* (2010)] treats all the missing values as zeros prior to the application of the standard

SVD method. PureSVD is shown to be suitable for the top-N recommendations task. The better top-N recommendation performance of PureSVD in comparison to the standard matrix completion-based approaches that are optimized for rating predictions, can be attributed to the fact that it considers all the items present in the catalog rather than considering only the items rated by the user. Additionally, for ranking purposes, it does not need to predict the exact ratings but only requires to achieve a correct relative ordering of the predictions for a user.

PureSVD estimates the rating matrix $\mathbf{R}$ as

$$\hat{\mathbf{R}} = \mathbf{U}\mathbf{\Sigma}\mathbf{Q}^T,$$

where $\mathbf{U}$ is a $n \times F$ orthonormal matrix, $\mathbf{Q}$ is an $m \times F$ orthonormal matrix and $\mathbf{\Sigma}$ is an $F \times F$ diagonal matrix, containing the F largest singular values. It can be noted that the matrix $\mathbf{P}$ representing the user factors can be derived by

$$\mathbf{P} = \mathbf{U}\mathbf{\Sigma}.$$

The matrices $\mathbf{U}$, $\mathbf{\Sigma}$ and $\mathbf{Q}$ can be estimated by solving the following optimization problem with orthonormal constraints

$$\begin{aligned} \underset{\mathbf{U},\mathbf{Q},\mathbf{\Sigma}}{\text{minimize}} \quad & \frac{1}{2}||\mathbf{R} - \sum_{i=1}^{F} \sigma_i u_i q_i^T||_F^2 \\ \text{subject to} \quad & \mathbf{U}^T\mathbf{U} = \mathbf{I} \\ & \mathbf{Q}^T\mathbf{Q} = \mathbf{I}, \end{aligned} \tag{11.10}$$

where $\mathbf{I}$ is the identity matrix, σ_i denotes the ith largest singular value, u_i represents the ith column vector of $\mathbf{U}$ and q_i denotes the ith column vector of $\mathbf{Q}$. The application of PureSVD on large scale sparse matrices can be optimized with the *Golub-Kahan-Lanczos Bidiagonalization* [Golub and Kahan (1965); Lanczos (1950)] approach. It computes the SVD of given matrix $\mathbf{R}$ in two steps. First, it bidiagonalizes $\mathbf{R}$ using Lanczos procedure as,

$$\mathbf{R} = \mathbf{P}\mathbf{B}\mathbf{Q}^T, \tag{11.11}$$

where $\mathbf{P}$ and $\mathbf{Q}$ are unitary matrices, and $\mathbf{B}$ is an upper bidiagonal matrix. The Lanczos procedure can take advantage of optimized sparse matrix-vector multiplications and efficient orthogonalization. Then, it uses an efficient method [Demmel and Kahan (1990)] to compute the singular values of $\mathbf{B}$ without computing $\mathbf{B}^T\mathbf{B}$ as,

$$\mathbf{B} = \mathbf{X}\mathbf{\Sigma}\mathbf{Y}^T. \tag{11.12}$$

Now, using Equations 11.11 and 11.12 we can compute left singular vectors, i.e., $\mathbf{U} = \mathbf{PX}$, and right singular vectors, i.e., $\mathbf{V} = \mathbf{QY}$, of matrix $\mathbf{R}$.

Modern randomized matrix approximation techniques [Halko *et al.* (2011)] can be used to compute a faster but approximate SVD of the rating matrix. We will refer to these approximation techniques as *Randomized SVD*. Essentially, Randomized SVD technique is carried out in two steps. First, it tries to find $\mathbf{Q}$ with F orthonormal columns such that,

$$\mathbf{R} \approx \mathbf{Q}\mathbf{Q}^T\mathbf{A}. \tag{11.13}$$

Next, it constructs $\mathbf{B} = \mathbf{Q}^T\mathbf{A}$, and since $\mathbf{B}$ has relatively smaller number of rows, i.e., F, we can employ standard methods to efficiently compute SVD of $\mathbf{B}$ as,

$$\mathbf{B} = \tilde{\mathbf{U}}\mathbf{\Sigma}\mathbf{V}^T. \tag{11.14}$$

Thus, left singular vector of $\mathbf{R}$ can be approximated by, $\mathbf{U} = \mathbf{Q}\tilde{\mathbf{U}}$, and right singular vectors can be approximated by $\mathbf{V}$. Randomized SVD can take advantage of efficient sparse matrix-matrix multiplication to find $\mathbf{Q}$ and to compute $\mathbf{B}$.

For our experiments, we used the optimized and parallel implementation of Golub-Kahan-Lanczos Bidiagonalization approach available in SLEPc[8] [Hernandez *et al.* (2007)] for PureSVD, and utilized the implementation of Randomized SVD available in RedSVD[9]. These implementations rely on well-studied sparse and dense linear algebra operations, that are further optimized for efficient usage on high-performance computers [Anderson *et al.* (1990)]. Figure 11.11 shows the speedup and total time achieved by PureSVD and Randomized SVD on ML10M dataset with increasing number of cores. As can be seen in the figure, the parallel implementation of PureSVD achieves better speedup than Randomized SVD with increase in the number of cores. Also, the time taken by Randomized SVD is lower in comparison to that of PureSVD on a single core. Table 11.5 presents the results for the best ranking performance achieved by both the methods on ML10M dataset. As can be seen in the table, PureSVD and Randomized SVD do not outperform SLIM for top-N recommendation but the time taken by PureSVD and Randomized SVD is lower than that of SLIM. Furthermore, PureSVD outperforms Randomized SVD for top-N recommendation performance. We should note though that the performance of Randomized SVD is comparable to that of PureSVD, and therefore Randomized SVD can serve as an alternative to PureSVD under time and

[8]slepc.upv.es
[9]https://github.com/ntessore/redsvd-h

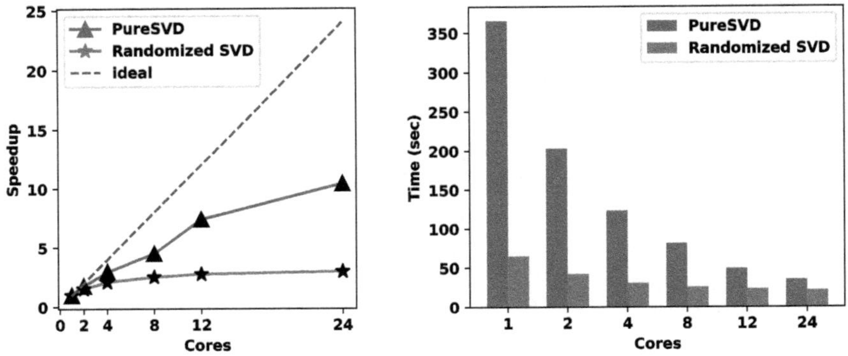

Fig. 11.11. Speedup (left) and total time in seconds (right) achieved by PureSVD and Randomized SVD on the ML10M dataset ($F = 500$).

Table 11.5. Comparison of PureSVD with Randomized SVD in terms of top-N performance and training time.

Method	HR	ARHR	Rank (F)	Time (sec)
PureSVD	0.292	0.139	60	9.24
Randomized SVD	0.247	0.112	400	15.46

Time corresponds to the time taken on one node (24 cores).

compute resource constraints. Also, Randomized SVD needs higher rank to achieve its best performance.

11.5. Conclusion

In this chapter, we presented different methods which speed up popular collaborative recommenders, by taking advantage of modern parallel multi-core architectures. We discussed ways to efficiently identify neighbors in k-nearest neighbor approaches in Section 11.2. We investigated how to parallelize the sparse linear methods well-suited for the top-N recommendation task, presented in Section 11.3 and how to speed up their parameter search. Finally, in Section 11.4, we showed ways to scale up the latent factor approaches, which could extend to tensor factorization approaches. In each section, we also presented experimental results on the popular ML10M dataset, illustrating the runtimes and speedup achieved in comparison to serial core implementations. Overall, the goal of this chapter is to illustrate that modern popular collaborative recommenders, although of very different nature, are able to be parallelized. We believe that research that focuses on ways to distribute and scale popular collaborative recommenders is crucial, as it leads to faster training times without sacrificing recommendation quality.

References

Anastasiu, D. C. (2017). Cosine approximate nearest neighbors, in P. Haber, T. Lampoltshammer and M. Mayr (eds.), *Data Science – Analytics and Applications*, iDSC 2017 (Springer Fachmedien Wiesbaden, Wiesbaden), ISBN 978-3-658-19287-7, pp. 45–50.

Anastasiu, D. C. and Karypis, G. (2014). L2ap: Fast cosine similarity search with prefix l-2 norm bounds, in *30th IEEE International Conference on Data Engineering*, ICDE '14, pp. 784–795, doi:10.1109/ICDE.2014.6816700.

Anastasiu, D. C. and Karypis, G. (2015a). L2knng: Fast exact k-nearest neighbor graph construction with l2-norm pruning, in *24th ACM International Conference on Information and Knowledge Management*, CIKM '15 (ACM, New York, NY, USA), ISBN 978-1-4503-3794-6, pp. 791–800, doi: 10.1145/2806416.2806534, http://doi.acm.org/10.1145/2806416.2806534.

Anastasiu, D. C. and Karypis, G. (2015b). Pl2ap: Fast parallel cosine similarity search, in *Proceedings of the 5th Workshop on Irregular Applications: Architectures and Algorithms*, IA3 '15 (ACM, New York, NY, USA), pp. 8:1–8:8.

Anastasiu, D. C. and Karypis, G. (2016). Fast parallel cosine k-nearest neighbor graph construction, in *2016 6th Workshop on Irregular Applications: Architecture and Algorithms (IA3)*, pp. 50–53, doi:10.1109/IA3.2016.013.

Anderson, E., Bai, Z., Dongarra, J., Greenbaum, A., McKenney, A., Du Croz, J., Hammarling, S., Demmel, J., Bischof, C. and Sorensen, D. (1990). Lapack: A portable linear algebra library for high-performance computers, in *Proceedings of the 1990 ACM/IEEE Conference on Supercomputing*, Supercomputing '90 (IEEE Computer Society Press, Los Alamitos, CA, USA), ISBN 0-89791-412-0, pp. 2–11, http://dl.acm.org/citation.cfm?id=110382.110385.

Anderson, M., Smith, S., Sundaram, N., Capotă, M., Zhao, Z., Dulloor, S., Satish, N. and Willke, T. L. (2017). Bridging the gap between HPC and Big Data frameworks, *Proceedings of the VLDB Endowment (PVLDB '17)*.

Bayardo, R. J., Ma, Y. and Srikant, R. (2007). Scaling up all pairs similarity search, in *Proceedings of the 16th International Conference on World Wide Web*, WWW '07 (ACM, New York, NY, USA), pp. 131–140.

Christakopoulou, E. and Karypis, G. (2016). Local item-item models for top-n recommendation, in *Proceedings of the 10th ACM Conference on Recommender Systems* (ACM), pp. 67–74.

Cremonesi, P., Koren, Y. and Turrin, R. (2010). Performance of recommender algorithms on top-n recommendation tasks, in *Proceedings of the Fourth ACM Conference on Recommender Systems*, RecSys '10 (ACM, New York, NY, USA), ISBN 978-1-60558-906-0, pp. 39–46, doi:10.1145/1864708.1864721, http://doi.acm.org/10.1145/1864708.1864721.

Demmel, J. and Kahan, W. (1990). Accurate singular values of bidiagonal matrices, *SIAM Journal on Scientific and Statistical Computing* **11**, 5, pp. 873–912.

Deshpande, M. and Karypis, G. (2004). Item-based top-n recommendation algorithms, *ACM Transactions on Information Systems (TOIS)* **22**, 1, pp. 143–177.

Dror, G., Koenigstein, N., Koren, Y. and Weimer, M. (2012). The yahoo! music dataset and kdd-cup'11. in *KDD Cup*, pp. 8–18.

Friedman, J., Hastie, T. and Tibshirani, R. (2010). Regularization paths for generalized linear models via coordinate descent, *Journal of statistical software* **33**, 1, p. 1.

Gates, M., Anzt, H., Kurzak, J. and Dongarra, J. (2015). Accelerating collaborative filtering using concepts from high performance computing, in *Big Data (Big Data), 2015 IEEE International Conference on* (IEEE), pp. 667–676.

Gemulla, R., Nijkamp, E., Haas, P. J. and Sismanis, Y. (2011). Large-scale matrix factorization with distributed stochastic gradient descent, in *Proceedings of the 17th ACM SIGKDD international conference on Knowledge discovery and data mining* (ACM), pp. 69–77.

Golub, G. and Kahan, W. (1965). Calculating the singular values and pseudo-inverse of a matrix, *Journal of the Society for Industrial and Applied Mathematics, Series B: Numerical Analysis* **2**, 2, pp. 205–224.

Halko, N., Martinsson, P.-G. and Tropp, J. A. (2011). Finding structure with randomness: Probabilistic algorithms for constructing approximate matrix decompositions, *SIAM review* **53**, 2, pp. 217–288.

Harper, F. M. and Konstan, J. A. (2015). The movielens datasets: History and context, *ACM Trans. Interact. Intell. Syst.* **5**, 4, pp. 19:1–19:19.

Hernandez, V., Roman, J., Tomas, A. and Vidal, V. (2007). Restarted lanczos bidiagonalization for the svd in slepc, *STR-8, Tech. Rep.*

Karlsson, L., Kressner, D. and Uschmajew, A. (2015). Parallel algorithms for tensor completion in the cp format, *Parallel Computing.*

Konstan, J. A., Miller, B. N., Maltz, D., Herlocker, J. L., Gordon, L. R. and Riedl, J. (1997). Grouplens: Applying collaborative filtering to usenet news, *Commun. ACM* **40**, 3, pp. 77–87, doi:10.1145/245108.245126, http://doi.acm.org/10.1145/245108.245126.

Koren, Y. (2008). Factorization meets the neighborhood: a multifaceted collaborative filtering model, in *Proceedings of the 14th ACM SIGKDD international conference on Knowledge discovery and data mining* (ACM), pp. 426–434.

Koren, Y. (2009). The bellkor solution to the netflix grand prize, *Netflix prize documentation* **81**, pp. 1–10.

Lanczos, C. (1950). *An iteration method for the solution of the eigenvalue problem of linear differential and integral operators* (United States Governm. Press Office Los Angeles, CA).

Ning, X. and Karypis, G. (2011). Slim: Sparse linear methods for top-n recommender systems, in *2011 IEEE 11th International Conference on Data Mining* (IEEE), pp. 497–506.

Nisa, I., Sukumaran-Rajam, A., Kunchum, R. and Sadayappan, P. (2017). Parallel CCD++ on GPU for matrix factorization, in *Proceedings of the General Purpose GPUs (GPGPU)* (ACM), pp. 73–83.

Recht, B., Re, C., Wright, S. and Niu, F. (2011). Hogwild: A lock-free approach to parallelizing stochastic gradient descent, in *Advances in Neural Information Processing Systems*, pp. 693–701.

Rendle, S. and Schmidt-Thieme, L. (2010). Pairwise interaction tensor factorization for personalized tag recommendation, in *Proceedings of the third ACM international conference on Web search and data mining* (ACM), pp. 81–90.

Sarwar, B., Karypis, G., Konstan, J. and Riedl, J. (2001). Item-based collaborative filtering recommendation algorithms, in *Proceedings of the 10th international conference on World Wide Web* (ACM), pp. 285–295.

Shao, W. (2012). *Tensor Completion*, Master's thesis, Universität des Saarlandes Saarbrücken.

Shi, Y., Karatzoglou, A., Baltrunas, L., Larson, M., Oliver, N. and Hanjalic, A. (2012). Climf: learning to maximize reciprocal rank with collaborative less-is-more filtering, in *Proceedings of the sixth ACM conference on Recommender systems* (ACM), pp. 139–146.

Shin, K. and Kang, U. (2014). Distributed methods for high-dimensional and large-scale tensor factorization, in *Data Mining (ICDM), 2014 IEEE International Conference on*, pp. 989–994.

Smith, S. and Karypis, G. (2015). SPLATT: the Surprisingly Parallel spArse Tensor Toolkit, http://cs.umn.edu/~splatt/.

Smith, S., Park, J. and Karypis, G. (2017). Hpc formulations of optimization algorithms for tensor completion, *Parallel Computing.*

Takács, G., Pilászy, I., Németh, B. and Tikk, D. (2009). Scalable collaborative filtering approaches for large recommender systems, *Journal of machine learning research* **10**, Mar, pp. 623–656.

Tan, W., Cao, L. and Fong, L. (2016). Faster and cheaper: Parallelizing large-scale matrix factorization on gpus, in *Proceedings of the 25th ACM International Symposium on High-Performance Parallel and Distributed Computing* (ACM), pp. 219–230.

Teflioudi, C., Makari, F. and Gemulla, R. (2012). Distributed matrix completion, in *Data Mining (ICDM), 2012 IEEE 12th International Conference on* (IEEE), pp. 655–664.

Tibshirani, R. (1996). Regression shrinkage and selection via the lasso, *Journal of the Royal Statistical Society. Series B (Methodological)*, pp. 267–288.

Weimer, M., Karatzoglou, A., Le, Q. V. and Smola, A. J. (2008). Cofi rank-maximum margin matrix factorization for collaborative ranking, in *Advances in neural information processing systems*, pp. 1593–1600.

Xie, X., Tan, W., Fong, L. L. and Liang, Y. (2017). CuMF_SGD: Parallelized stochastic gradient descent for matrix factorization on gpus, in *Proceedings of the 26th International Symposium on High-Performance Parallel and Distributed Computing (HPDC)*, pp. 79–92.

Yu, H.-F., Hsieh, C.-J., Dhillon, I. *et al.* (2012). Scalable coordinate descent approaches to parallel matrix factorization for recommender systems, in *Data Mining (ICDM), 2012 IEEE 12th International Conference on* (IEEE), pp. 765–774.

Yun, H., Yu, H.-F., Hsieh, C.-J., Vishwanathan, S. V. N. and Dhillon, I. (2014). Nomad: Non-locking, stochastic multi-machine algorithm for asynchronous and decentralized matrix completion, *Proc. VLDB Endow.* **7**, 11, pp. 975–986, doi:10.14778/2732967.2732973, http://dx.doi.org/10.14778/2732967.2732973.

Zhou, Y., Wilkinson, D., Schreiber, R. and Pan, R. (2008). Large-scale parallel collaborative filtering for the netflix prize, in *Algorithmic Aspects in Information and Management* (Springer), pp. 337–348.

Zhuang, Y., Chin, W.-S., Juan, Y.-C. and Lin, C.-J. (2013). A fast parallel sgd for matrix factorization in shared memory systems, in *Proceedings of the 7th ACM conference on Recommender systems* (ACM), pp. 249–256.

Zou, H. and Hastie, T. (2005). Regularization and variable selection via the elastic net, *Journal of the Royal Statistical Society: Series B (Statistical Methodology)* **67**, 2, pp. 301–320.

Chapter 12

Robustness and Attacks on Recommenders

Neil J. Hurley
Insight Centre for Data Analytics,
School of Computer Science,
University College Dublin
neil.hurley@ucd.ie

How robust is a recommender to the presence of corrupt data in the training dataset? To what extent will the recommendation algorithm be affected by such data? These are the basic questions of a analysis of recommender algorithms. They are particularly relevant in the context of maliciously injected corrupt data, where the data injection is designed to deliberately distort the recommendation output for particular sets of users and items. In the context of online recommenders, where the creation of fake identities may be relatively low-cost, such a robustness attack against a recommender system is of genuine concern. Robustness is also of some concern in the context of privacy preserving recommender algorithms in which user rating profiles are distorted in order to protect private data. In this chapter, we review research carried out since the early 2000s on robustness attacks and practical defences against them. We introduce a new library for robustness analysis incorporated into the `Ranksys` recommender systems framework, in which the main attacks proposed in the state-of-the-art are implemented and can be simply tested against a range of recommender algorithms.

12.1. Introduction

Recommender algorithms are concerned with delivering a personalised experience to their end-users. The implicit contract between end-user and system is that users are willing to share personal information, namely their preferences towards particular products or services, in order for the system to better find new instances that are likely to be relevant to them. As much as the user depends on the quality of the recommender algorithm, the algorithm itself depends on the quality of data shared with it. Recently, the possibility and consequences of obtaining poor recommendations has percolated into the public consciousness. We are highly aware that "fake news" may have had significant impact on issues of great societal impact. Discussions abound about the possibility of "filter bubbles" reinforcing polarised views. While enjoying the benefits of data-driven services, people are more and more aware that such services can perform poorly. Recommender system designers need to be sensitive to people's trust in the quality of their recommendations and poor performance can seriously damage a system's reputation.

In this context, understanding a system's robustness to poor quality data is an important task for system designers. Even more important is to understand its robustness to *maliciously tailored* corrupt data. We can imagine many motivations for malicious users to deliberately attempt to distort the output of a recommendation algorithm. Attackers may simply wish to destroy a system's reputation by degrading the overall quality of its recommendations; or attackers may be motivated to deliberately promote or demote particular recommendations to or from recommendation lists of particular users or groups of users. As a recommendation system relies entirely on data provided through user interaction, the possibility exists to carry out such distortion without explicitly hacking the algorithm itself. Instead, attackers can focus on the system database and inject corrupted data into it. Particularly for online recommendation, where the creation of fake user identities, (sometimes referred to as *sybils*) may be of relative low-cost, attackers can tailor the preferences of such sybils, in such a way that they influence the recommendation output for other genuine users. Such *active* attacks, that focus on the injection of fake user profiles into the dataset, in order to distort the recommendations to genuine users, are the main focus of this chapter. It is worth noting that this sort of *profile inejection* attack may have other motivations—for instance, in order to learn the preferences of genuine users—but in this chapter we focus on rating distortion.

Profile injection attacks to distort recommender outputs were first studied in O'Mahony *et al.* (2002). Afterwards, a general framework for the design of attack profiles was presented in the work of Burke *et al.* (2005). Several attack strategies were proposed within this framework and the leading collaborative filtering algorithms were analysed with respect to their susceptibility to such attacks. In following works, measures to counter such attacks were proposed, in particular attack detection algorithms and modifications to recommender algorithms to enhance robustness. The main findings of this work are summarised in the following sections. However, differently to some of the earlier studies of robustness, we fix our focus on top-*N* recommendation and ranking, rather than rating prediction and study the attacks and defences in the context of metrics relevant to this setting. In particular, we have implemented the main attacks and defences as a package[1] which can be added to the `RankSys`[2] recommender system framework. This `Java 8` framework contains implementations of leading recommender algorithms, including user- and item-based kNN algorithms and matrix factorisation algorithms. Moreover, it provides a set of evaluation metrics relevant to the ranking task. Our new `robustness` package implements the general profile attack framework allowing analysts to explore different attack strategies and assess their impact when applying different recommender algorithms on different datasets.

Notation: Let U represent a set of users, such that $n = |\mathrm{U}|$ and I a set of items, such that $m = |\mathrm{I}|$. We write $\mathscr{D}$ for a database consisting of a set of user-item pairs with associated preference scores, denoted as r_{ui} for $u \in \mathrm{U}$ and $i \in \mathrm{I}$ and write $r_{ui} = \emptyset$ to denote a missing rating. Given a user, u, the goal of a top-N recommender system is to rank a set of $N > 0$ items, for which user u's preferences are unknown, in order of their predicted relevance to user u. The system algorithm is trained using a database $\mathscr{D}$. We write R_u for the recommender list produced by the system, or $R_u(\mathscr{D})$ to emphasise its dependence on the training set. Furthermore, we use k_{ui} to denote the rank of item i in the recommender list, where low ranks correspond to more preferable items. This paper is concerned with algorithms that augment the training set with spam data. We write $\mathscr{A}$ to denote a profile injection attack and $\mathscr{D}_{\mathscr{A}}$ to denote the dataset augmented with profiles produced by attack $\mathscr{A}$. We write P_u to denote a user's profile, i.e. the set of items for which u's preferences are available in the training set $\mathscr{D}$. We write N_u for a neighbourhood of user u and similarly N_i for a neighbourhood of item i, corresponding, respectively, to sets of users or items that are similar to user u or item i. Finally, denote by T_u a hold-out test set of items known to be relevant to user u.

12.2. Robustness Analysis Context

We distinguish and discuss two contexts. The first is robustness or stability of recommender algorithms to unbiased, non-malicious noise. Robustness in this context has received relatively less attention in the literature, but is important nonetheless. As several studies have shown, explicit ratings gathered directly from users are very prone to error. Users tend not to rate consistently. Their evaluations can depend on the context in which the rating is provided, for instance the quality of other recently explored items and their preferences at the time of rating, which may well change over time. For recommender algorithms that rely on implicit ratings, the correlation between implicit interaction data and a user's true preferences is never exact. Hence we should consider the data on which algorithms

[1] `https://github.com/neilhurley/RankSysRobustness.git`
[2] `https://github.com/RankSys/RankSys`

are trained to be noisy observations of users' true ratings, and indeed matrix factorisation models, for instance, are developed from this perspective. Moreover, recent approaches to privacy preservation has explored techniques, e.g. based on differential privacy, which amount to the injection of tailored noise into the system algorithm, in order to afford some protection against privacy attacks. In order for such methods to be effective, the underlying inference algorithm must still have some utility in the presence of such noise. It is worth noting that personalisation in the context of noise is challenging, as the measured impact of personalised algorithms over non-personalised can be small and thus a small drop in utility in absolute terms can still be large relative to a non-personalised baseline comparator. For example, for the Movielens 1M dataset, measuring top 20 precision on a 20% holdout dataset, personalised algorithms record a performance between 20% and 22% precision, but the non-personalised most-popular item algorithm yields a performance of 12% precision. Relative to this baseline, a drop of even a few percentage points in precision is quite significant.

The second context is the injection of tailored, purposeful data, with a specific attack intention in mind. In the state-of-the-art, the intention of deliberately downgrading the recommendations of a specific target item or set of items, by causing the system to lower its predicted ratings for the target, or to rank it lower in its recommendation lists, is referred to as a *nuke* attack. Attacks whose purpose is to increase targeted items' predicted ratings or ranked positions are referred to as *push* attacks [O'Mahony *et al.* (2004a)]. Such attacks amount to designing special user profiles that can influence in the required manner and injecting these into the system database. It is generally assumed in this analysis that malicious users who mount such attacks operate by creating fake user identities and that these fake identities inject preferences in the same way as genuine users, through interaction with the system. Of course, depending on the system, such interactions may be more or less costly. A product recommendation system that requires a purchase before accepting a rating is clearly more difficult to attack in this way, than one that accepts ratings without purchase. Systems that operate a secure sign-up protocol before accepting user interactions, again pose a difficulty for the attacker. Robustness analysis however has tended not to focus on such practical interventions and instead poses the abstract problem of the algorithm: assuming that malicious users can, by whatever means, successfully inject malicious preferences into the system database, how susceptible is the recommendation algorithm to being influenced by such preferences? The attacker's effort is measured purely in terms of the number of preferences that must be injected to successfully mount a particular attack. It is concluded that an algorithm that can be effectively influenced by a small attack, relative to the size of the full training dataset, is weak with respect to robustness.

12.2.1. *Attack Knowledge*

While overcoming the practical barriers to is seen as outside the scope of a robustness analysis, attack analysis does consider the amount of knowledge that an attacker must have in order to mount a particular attack. There are two types of knowledge to consider. Firstly there is knowledge of users' preferences. Generally, the challenge for the attacker is to create profiles that are particularly influential, which requires that the attacker choose which items should be rated by the sybil profiles and what ratings they should be given. Knowledge of how other genuine users have rated can greatly aid the attacker in making these decisions. In the state-of-the-art, when attacks are referred to as "*high-knowledge*" or "*low-knowledge*", it is knowledge of the rating distributions of genuine users is considered. Most malicious injection attacks discussed in the literature require some level of rating knowledge, from, at the lowest level, knowledge of the set of globally most popular items, to knowledge of the global mean and standard deviation of ratings in the dataset, to knowledge of the mean and standard deviation of the ratings for each item in the dataset. Again, depending on the particular system, it can be reasonable to expect that attackers are able to attain or accurately guess such information. Generally, when carrying out a robustness analysis, we assume the worst case that such information is available exactly to the attacker.

The second type of knowledge, is knowledge of the recommender algorithm, and, more particularly, knowledge of the parameters of the algorithm. Attacks mounted against particular recommender algorithms have been referred to in the literature as "*informed*" attacks. Profiles employed in these attacks are specially tailored using knowledge of the exact way in which database preferences are used to form the recommendation. While it may be argued that such detailed knowledge is much harder to attain than knowledge of the rating distribution, such informed attacks follow the philosophy of Kerckhoff's principle from cryptography, that a system should be secure even when everything is known about the system, except for the keys used to secure it. Similarly an informed attack provides a worst-case analysis of the intrinsic robustness of a recommender algorithm, to being misled by overly-influential profiles. We note in stating this that such a robustness analysis gives us some insight into how fairly recommender algorithms treat the rating information available to them. If each preference available to the system is equally influential on the output of the recommender system, then the attacker will need to inject sufficient preferences to swamp those already in the database. On the other hand, if it is possible to design profiles that have greater impact on the recommender output than an average profile, then this provides the weakness in the algorithm that an attacker can exploit.

12.2.2. *Toy Example*

To fix ideas, we present a small toy example of how profile injection can influence recommender output. In Figure 12.1, an attack has been mounted to target item i_1, with the purpose of promoting its recommendation. Consider a simple user-based algorithm over a transaction database where users have indicated items that they like and dislike. For each user, a set of two neighbours is chosen and the user is recommended the most popular item in the neighbourhood that is not already in their profile. Neighbours are chosen based on a similarity function $s(u_j, u_k)$ that computes the difference between the number of items that the pair of users agree on minus the number they disagree on. The user-user similarities are shown in the figure and genuine neighbours are indicated by the green cells representing the two most similar users to each user. The two attack profiles, tailored to be sufficiently similar to users $\{u_2, u_4, u_6\}$, can displace the genuine users from their neighbourhoods, resulting in a promotion of i_1 to these users.

	i_1	i_2	i_3	i_4	i_5	i_6	i_7	Similarity $s(u_i,u_j)$						
u_1	×						✓	-	0	1	0	1	0	Genuine User profiles
u_2		✓	✓	✓	×			0	-	-2	2	0	2	
u_3	×		×	×				1	-2	-	-1	1	-1	
u_4		✓		✓		×		0	2	-1	-	0	2	
u_5		✓	×			✓	✓	1	0	1	0	-	-1	
u_6		✓	✓			×		0	2	-1	2	-1	-	
a_1	✓	✓		✓	×	×		-1	3	-2	3	0	2	Sybil profiles pushing item i_1
a_2	✓	✓	✓	✓		×		-1	3	-3	3	-1	3	

Fig. 12.1. Toy system database showing genuine user profiles with a number of sybil profiles inserted. In this example, two sybils are sufficient to disrupt the recommendation to three targeted users.

12.3. Attack Profile Nomenclature

In the Mobasher *et al.* (2007) framework for attack profile construction (see Figure 12.2), an attack profile is considered to consist of the following item subsets:

(1) *Selected Items* This is a sub-set of items that are chosen to support the specific purpose of the attack. We write this subset as $I^S \subseteq$ I.

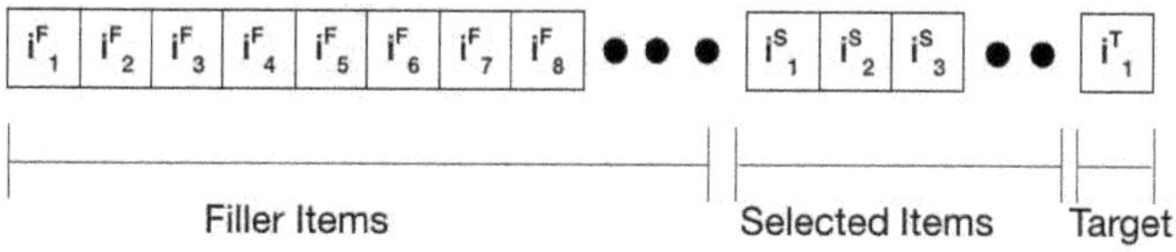

Fig. 12.2. Attack Profile Item Subsets.

(2) *Filler Items* This is a sub-set of items chosen to fill out the profile, in order to give it the appearance of a genuine profile. We write this subset as $I^F \subseteq$ I.
(3) *Targeted Items* This is the sub-set of items that are the specific focus of the attack i.e. the items that the attack aims to push or to nuke. We write this subset as $I^T \subseteq$ I. For the basic attacks discussed below, we usually assume that $I^T = \{i^T\}$, a single attacked item.

In some attacks, the preferences of *both* filler and selected items are set so as to support the purpose of the attack, so it is perhaps more correct to say that these two subsets are distinguished only by the manner in which their ratings are set. Generally, the ratings of selected items are set in order to focus the attack towards a particular set of genuine users on whom the attack is mounted. For instance, the selected items may be a set of items known to be liked by the targeted user group and their preferences set high, so that the fake profile is similar to members of this group. When the role of the filler items is just to fill in the profile to obfuscate its purpose, their values may be set using the rating distribution of genuine preferences, but some attacks are more effective when filler items are set to influence the recommendation. For instance, low ratings given to filler items can have the effect of pushing the recommendations of the targeted items.

In this framework, different attacks are defined by

(1) How the selected, filler and targeted items are chosen.
(2) How the ratings for each of these sub-sets are set.

12.4. Attack Strategies

12.4.1. *Noise Injection*

As discussed earlier, it is interesting to understand the stability of a recommender algorithm's output in the face of noise in the rating database. Hence, we include for evaluation two noise "attacks". The first, which we refer to as "Random Rating Injection" (RRI) is the insertion of additional user-item-rating triples, where the user, item and rating are all chosen uniformly at random. This has the effect of diluting the database with non-informative ratings, which we would expect will destroy the recommender utility, once these ratings swamp the genuine ones. Later, we examine whether different algorithms perform differently with respect to the amount of such dilution they can resist. The second type of noise attack, which we refer to as "Noise Injection" (NI) is the distortion of the existing ratings in the dataset with random centred noise of a given variance. A more refined version of this attack, not considered further in this chapter, would be make the noise variance dependent on the rating value, which could model rating behaviour in which users are more or less certain about what they like, compared to what they do not like.

12.5. Purposeful Attack Strategies

The main non-informed attacks proposed in the literature are summarised in Table 12.1. Item ratings are chosen as the minimum rating $r_{\min}$, maximum rating, $r_{\max}$, or randomly from distributions based on statistics of the rating dataset. In the table, μ_g, σ_g refer to the global mean and standard deviation of all ratings in the dataset and μ_i, Σ_i are the vectors of means and standard deviations of ratings over each item. Items for each of the subsets I^F and I^S are chosen uniformly at random (indicated with "Unif"); randomly from the top $x\%$ of most popular items (indicated with "From Most Pop"); in order of the most popular or least popular items (indicated by "Most pop" and "Least pop") respectively; or amongst the items that tend to be liked ("low-rated") or disliked ("high-rated").

Table 12.1. Non-informed Purposeful attacks.

Attack	**Purpose**	I^T	I^F		I^S		Reference
		Rating	Rating	Selection	Rating	Selection	
Random	Push	$r_{\max}$	$\mathcal{N}(\mu_g,\sigma_g)$	Unif	-	$\emptyset$	Lam and Riedl (2004)
Random	Nuke	$r_{\min}$	$\mathcal{N}(\mu_g,\sigma_g)$	Unif	-	$\emptyset$	Lam and Riedl (2004)
Average	Push	$r_{\max}$	$\mathcal{N}(\mu_i,\Sigma_i)$	Unif	-	$\emptyset$	Lam and Riedl (2004)
Average	Nuke	$r_{\min}$	$\mathcal{N}(\mu_i,\Sigma_i)$	Unif	-	$\emptyset$	Lam and Riedl (2004)
Avg Over Popular (AoP)	Push	$r_{\max}$	$\mathcal{N}(\mu_i,\Sigma_i)$	From Most Pop	-	$\emptyset$	Hurley *et al.* (2009)
Avg Over Popular (AoP)	Nuke	$r_{\min}$	$\mathcal{N}(\mu_i,\Sigma_i)$	From Most Pop	-	$\emptyset$	Hurley *et al.* (2009)
Segment	Push	$r_{\max}$	$r_{\min}$	Unif	$r_{\max}$	By segment	Mobasher *et al.* (2005)
Bandwagon	Push	$r_{\max}$	$\mathcal{N}(\mu_i,\Sigma_i)$	Unif	$r_{\max}$	Most pop	Mobasher *et al.* (2007)
Reverse Bandwagon	Nuke	$r_{\max}$	$\mathcal{N}(\mu_i,\Sigma_i)$	Unif	$r_{\min}$	Least pop	Mobasher *et al.* (2007)
Love/Hate	Nuke	$r_{\min}$	$r_{\max}$	Unif	-	$\emptyset$	Mobasher *et al.* (2007)
Popular	Push	$r_{\max}$	$r_{\min}$	Low-rated	$r_{\min}+1$	High-rated	O'Mahony *et al.* (2003)

The segment attack is focused on a particular user segment identified by the attacker and "By Segment" refers to a set of items that are popular among users in that segment. Such a segment could correspond to users who like particular genres in movie recommendation, for example. By rating popular items within the user segment highly, the attacker ensures that the attack profile is similar to the profiles of users within the segment.

Early work on profile injection compared the Random and Average attacks in terms of their ability to shift the rating predictions made for the target item. This work demonstrated that, due to its higher knowledge requirements, the Average attack was more effective than the Random attack on user-based kNN algorithms. Item-based and model-based algorithms were shown to be more robust in comparison to user-based, though they were still somewhat vulnerable to more focused attacks, such as the segment attack. The work of Hurley *et al.* (2009) showed the importance of choosing filler items according to the item-choice distribution of genuine users, in order to increase undetectability. We replay this analysis in the evaluation section below, but focus only on performance metrics relevant to the top-*N* problem.

One attack which we do not include in Table 12.1 is the probe attack [Mobasher *et al.* (2007)]. In this attack, it is assumed that the recommender system generates predicted ratings, so that an attack profile can be built by iteratively querying the system to provide ratings for items to be added to the attack profile. A top-*N* recommender does not provide such feedback and hence we do not consider this attack in our analysis.

The original Popular attack of O'Mahony *et al.* (2002) is an informed attack directed specifically at the user-based rating prediction algorithm proposed in Resnick *et al.* (1994) that uses Pearson similarity. The attack exploits the fact that this user-based algorithm computes user-user similarities using only those items that have been rated by *both* users and that correlations computed over small itemsets are likely to attain extreme values. Hence it is possible to build small (i.e. low cost) attack profiles, that correlate strongly with and are therefore likely to be in the neighbourhood of many users. In the most extreme case, if a genuine and attack user have only two items in common, a correlation of 1 is attained provided they agree on the ordering of the item ratings, and -1 otherwise. An effective push attack is then constructed by forming filler and selected item sets such that the filler items are likely to have received lower ratings by users than the selected items. The filler items are given a rating value of $r_{\min}$ and the selected items are given a rating of $r_{\min}+1$. These low ratings ensure that the difference between the pushed target item rating and the attack profile's average rating is as large as possible, as this difference is used by Resnick's algorithm to compute the rating for the target. Some other informed attacks have been proposed, targeting other algorithms, such as matrix factorisation, but we do not consider these further in this chapter.

Each of the above attacks is parameterised by the number of sybil profiles added to the dataset and by the sizes of the three subsets of items added to each sybil's profile. Writing U_s as the set of sybil profiles, we have that the total attack size J is given by

$$J = \sum_{u \in U_s} |I^T(u)| + |I^F(u)| + |I^S(u)|.$$

Typically evaluations are carried out in a context in which a single item is attacked, i.e. $|I^T| = 1$. Moreover, the number of selected items is typically a small fixed number (e.g. < 5), so that the size of each sybil profile is determined by the filler size. As discussed in Section 12.9, if detectability is a concern, then the attacker will choose a filler size that is similar to the sparsity of the dataset. Hence, in evaluations, the effect of an attack will be reported as a function of the number of sybil users, for a fixed filler size.

12.6. `robustness` library

The `robustness` library defines an `interface` called `ProfileInjector`, containing a single function `inject` that returns a stream of preferences, consisting of <`user`, `item`, `rating`> tuples.

To allow for evaluation against unbiased noise, two noise injection classes that implement the interface are provided. The `inject` method of the `RandomRatingInjector` class creates a stream of <`user`, `item`, `rating`> selected uniformly at random. The `NoiseInjectorClass` creates a stream in which the user-item pairs in the stream are the same as those in the original dataset, but the rating is modified by adding noise of a given variance, generated from a Gaussian distribution.

Attacks are also provided as `classes` that implement this interface. Two general purpose classes are provided:

- `TargetedFillInjector` is a class for creating attack profiles containing targeted items, I^T and filler items, I^F. The strategies for selecting the filler items and the setting their ratings can be specified through the constructor.
- `TargetedFillSelectInjector` is a specialisation of `TargetedFillInjector` that also allows selected items I^S to be specified.

In the constructor of these classes, it is necessary to specify whether the attack is a `push` or a `nuke`. Each of the attacks discussed below is implemented as a specialisation of these classes, restricted to particular item selection and rating specification strategies.

12.7. Measuring Robustness

A number of measures relevant to robustness are defined in Table 12.2. In this table, I^T represents a set of targeted items and U the set of users. $\mathscr{D}$ represents an unattacked dataset on which an algorithm is trained. $\mathscr{D}_{\mathscr{A}_i}$ represents a dataset modified by a particular injection attack $\mathscr{A}$ focused on a particular targeted item i. When averaging over such attacks, we assume that each attack is carried out independently and separately, so that the *avgRank* and *HitRatio* represent average performance, over a set of separately executed attacks on each item in the targeted set. The *HitRatio* can be used to assess the percentage of recommender lists that the targeted item reaches, while the *avgRank* informs as to the average position attained by the items when they succeed in making it into the top N list.

Table 12.2. Robustness Measures.

Definition	Description
$HitRatio(i, \mathscr{D}_{\mathscr{A}_i}) = \frac{1}{\lvert U \rvert} \sum_{u \in U} \mathbb{1}(i \in R_u(\mathscr{D}_{\mathscr{A}_i}))$	The average number of users for whom attacked item i appears in a top-N recommendation.
$\overline{HitRatio}(\mathscr{D}, \mathscr{A}) = \frac{1}{\lvert I_T \rvert} \sum_{i \in I_T} HitRatio(i, \mathscr{D}_{\mathscr{A}_i})$	The average hit ratio over all attacked items I^T.
$avgRank(\mathscr{D}, \mathscr{A}) = \frac{1}{\lvert U \rvert} \frac{\sum_{i \in I_T} \sum_{u \in U} k_{ui}(\mathscr{D}_{\mathscr{A}_i})}{\sum_{i \in I_T} \sum_{u \in U} \mathbb{1}(i \in R_u(\mathscr{D}_{\mathscr{A}_i}))}$	The average rank of attacked items in the recommender lists, counting only those items that make it into R_u i.e. $k_{ui} = 0$ if $i \notin R_u$.
$Overlap(\mathscr{D}, \mathscr{A}) = \frac{1}{N \lvert U \rvert \lvert I_T \rvert} \sum_{u \in U} \sum_{i \in I_T} \lvert R_u(\mathscr{D}) \cap R_u(\mathscr{D}_{\mathscr{A}_i}) \rvert$	The average overlap between top-N sets before and after an attack.

Depending on whether the purpose of the attack is to push or nuke an item, a successful attack corresponds to one in which the *HitRatio* and *avgRank* is greater than (resp. less than) the corresponding measure applied to the unattacked dataset (when $\mathscr{A}$ is the null attack of injecting no profiles). The *Overlap* measure allows for the stability of a recommendation algorithm to be evaluated against noisy training data.

None of the measures in Table 12.2 measures the utility of the recommender, in terms of its ability to recommend relevant items. Two evaluation measures appropriate for top-N recommendation, precision and NDCG are listed in Table 12.3.

12.8. Evaluation

The `RankSys` framework contains a number of different top-N recommender algorithms. In this chapter, we investigate the following algorithms for producing

Table 12.3. Utility Measures.

Definition	Description
$\text{Prec}(\mathscr{D})@N = \sum_{u \in U} \frac{\lvert R_u(\mathscr{D}) \cap T_u \rvert}{N \lvert U \rvert}$	The precision of the top-N recommendation.
$\text{DCG}(\mathscr{D})@N = \mathbb{1}(i_1 \in T_u) + \sum_{j=2}^{N} \frac{\mathbb{1}(i_j \in T_u)}{\log_2(j+1)}$	The discounted accumulative gain ranking measures
$\text{NDCG}(\mathscr{D})@N = \text{DCG}(\mathscr{D})@N / \text{IDCG}$	Normalised DCG, by dividing by ideal DCG

a score s_{ui} on which to rank items for recommendation:

- `ub`: A user-based kNN recommendation algorithm. Here a recommendation is formed for user u by constructing a neighbourhood N_u of the top $k > 0$ most similar users to u, then computing a score for each item rated by the neighbours, as follows:

$$s_{ui} = \sum_{\{v \in N_u | i \in P_v\}} \text{sim}_{uv} r_{vi}$$

- `ib`: An item-based kNN recommendation algorithm. Here a recommendation is formed for user u by constructing a neighbourhood N_i of the top $k > 0$ most similar items to each item in P_u. A score is then computed for the union of all neighbourhoods as

$$s_{ui} = \sum_{\{j \in P_u | i \in N_j\}} \text{sim}_{ij} r_{uj}$$

- `hkv`: This is the implicit matrix factorisation algorithm of Hu *et al.* (2008) which, given the dimension of a factor space $k \ll n, m$, factorises the interaction matrix into an $n \times k$ user matrix $\text{P} = \{p_{u\ell}\}$ and $m \times k$ item matrix, $\text{Q} = \{q_{i\ell}\}$ such that

$$s_{ui} = \sum_{\ell} p_{u\ell} q_{i\ell} \,. \tag{12.1}$$

  P and Q are obtained by an alternating minimisation of a weighted least squares loss function.
- `plsa`: This is the probabilistic latent semantic analysis (pLSA) algorithm of Hofmann (2004) in which the scores are based on the probability of relevance of an item given the user, which is modelled as a product of user and item probabilities over a set of latent aspects, leading to a scoring function of the same form as Equation (12.1), but learned through an expectation maximisation algorithm that maximises the log likelihood of the data.

Given the scores s_{ui}, those items not in P_u are ranked according to s_{ui} and the top N items are recommended.

The user- and item-based algorithms depend on the similarity measure used to form the neighbourhood. A number of similarity functions are provided in `RankSys`, including cosine and Jaccard similarity. We present results on the cosine similarity below. Moreover, these algorithms depend on some parameters, in particular, k, the size of the neighbourhoods or factor space dimension. For `ub`, we set $k = 100$, for `ib`, we set $k = 20$ and for `hkv` and `plsa`, we set $k = 20$. These parameters are chosen to given high utility on the evaluated dataset. Further exploration of the impact of parameters on the effectiveness of attacks can be easily carried out using the `robustness` library. We apply our evaluation to the Movielens 1M dataset [Harper and Konstan (2015)], consisting of approximately 1 million ratings given by 6040 users to 3952 movies on a 5-point rating scale. The dataset is divided in a 80:20 ratio into training and test set.

12.8.1. *Unbiased Noise Injection*

The impact of adding unbiased noise to the ratings is presented in Figure 12.3a, where NDCG@20 is plotted against the strength of the added noise. Injected noise is measured as the *noise to signal ratio* (NSR) in decibels (dB), where

$$\text{NSR} = 10\log_{10}(\sigma_n^2/\sigma_i^2)\,, \tag{12.2}$$

σ_n is the standard deviation of the noise added to the ratings and σ_i is the variance of the ratings for the item to which noise is added. The figure shows a non-personalised baseline—the `pop` algorithm—that recommends items in order of their global popularity. Note that `pop` has been applied to the unattacked dataset. An NSR of 20dB means that the noise is two orders of magnitude greater than the rating variance. For the kNN algorithms, we see a gradual fall-off in performance, but with still stronger performance than the non-personalised baseline at 20dB of added noise. The matrix factorisation algorithms are more brittle and fail more suddenly at a lower level of noise. In Figure 12.3b, we see the effect of adding random ratings to the dataset. NDCG@20 is shown against the number of added profiles, as a percentage of the original dataset size. We have added up to six times the number of ratings as were in the original dataset. While performance has again degraded, all algorithms maintain more utility than the popularity baseline.

These results have implications for designers of privacy-preserving algorithms, using differential privacy techniques, for example. Such techniques require the injection of noise into the recommendation process, so it is important that the algorithm be somewhat robust to such noise. The popularity baseline is a threshold below which performance should not drop, since, at this point, the algorithm is failing to usefully exploit any personal information in the dataset.

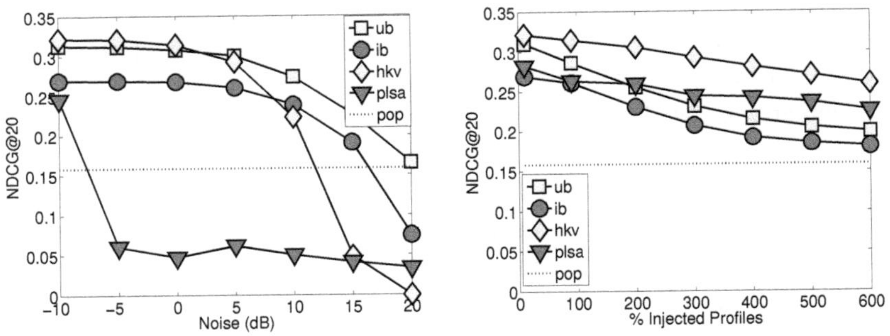

(a) Noise Injection (b) Random Rating Injection

Fig. 12.3. Performance under unbiased noise injection.

See Afsharinejad and Hurley (2018) for more analysis on a differential privacy algorithm and its impact on top-N recommendation performance.

12.8.2. *Targeted Push Analysis*

We compare the performance of the following attack variants:

(1) *Average*: An Average push attack, with filler items chosen uniformly at random (**RU**);
(2) *Average over Popular*: An AoP push attack where filler items are chosen uniformly from the top 20% of most popular items (**AoP@20**);
(3) *Random with Most Popular*: A push attack where the filler data is chosen in same way as the Average attack, but I^F are chosen as the set of top most popular items in the dataset (**RMP**).

This study therefore, emphasises the importance of the choice of I^F on the effectiveness of an attack. We fix $N = 20$ and the filler size to 5% of the total number of items, which is close to the sparsity of the Movielens dataset. It is worth noting that, from the point of view of detectability (see Section 12.9), **RMP** is highly detectable, since all sybil profiles consist of the same set of items, **RU** is also detectable by a detector that analyses its filler placement, while **AoP@20** is the hardest to detect. Twenty items were chosen at random among those items that receive *no* top-20 recommendations from the unattacked `plsa` algorithm applied to the entire user-base. These poorly performing items are ones which could be expected to benefit most from a push attack. Setting each item in turn as the target, i^T, each of the four recommendation algorithms is attacked. The resulting *hitRatio*, *avgRank* and *Overlap* against the attack size, measured as a percentage

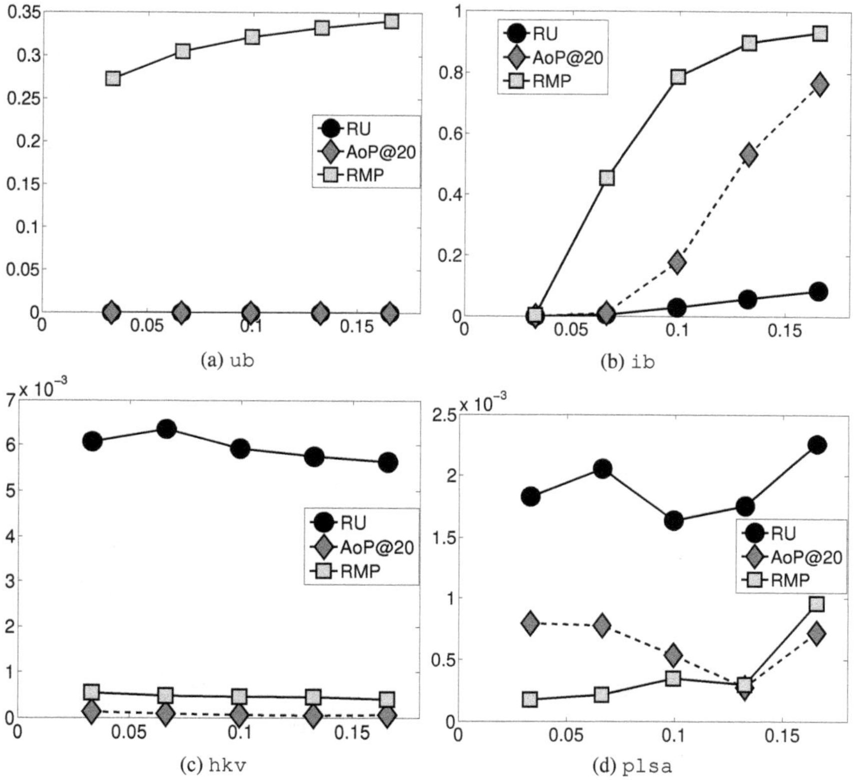

Fig. 12.4. $\overline{HitRatio}$ over 20 items.

of the total number of genuine users in the dataset, are shown in Figures 12.4, 12.5 and 12.6, respectively. Note carefully the scale of the y-axis in each of these plots.

For the kNN algorithms, we note the benefit of building attack profiles from popular items. In fact, `ub` remains largely invariant under the attack, unless it is formed from the most popular items, **RMP**. In this case, even the weakest attack is able to push the target item into the top 3 items of over 25% of the user-base. The **AoP@20** attack, that draws randomly from among the top quintile of popular items is still not strong enough to shift the recommendations significantly. The `ib` algorithm is somewhat more susceptible to attack than the `ub` algorithm. Initially, a somewhat larger attack is required to yield a significant push effect, but eventually the **RMP** attack is able to push the targeted item into 90% of recommender lists, and, on average, it appears in rank position 2. More seriously, a fairly stealthy attack such as `AoP` can push into 80% of the recommendation at an average rank of 6, when the attack size is 16%.

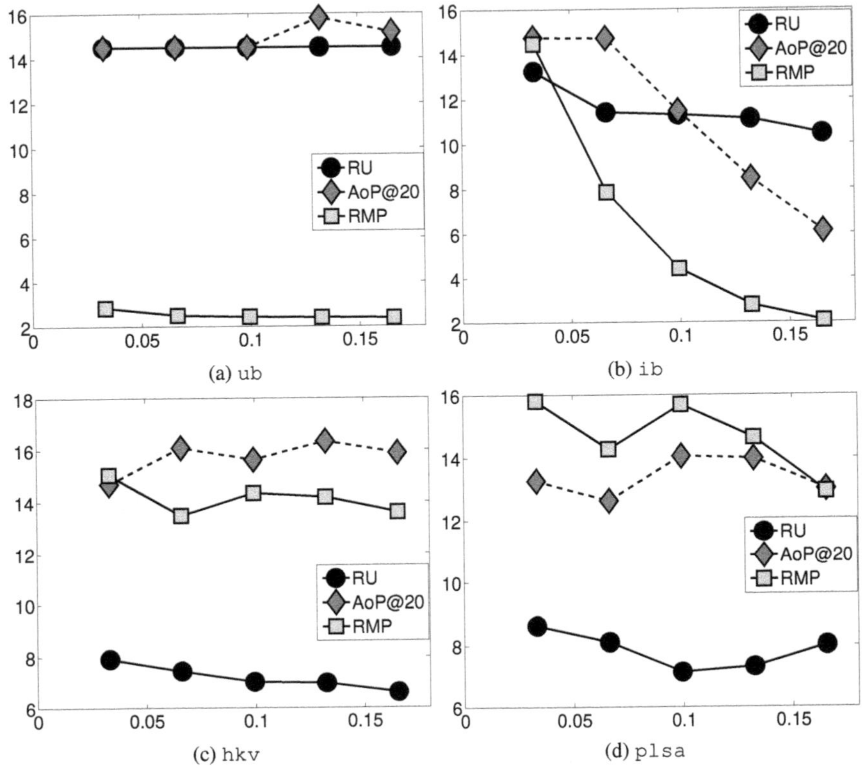

Fig. 12.5. *avgRank* over 20 items.

Note that this is somewhat contrary to previous work on robustness in which it has been stated that item-based algorithms are less susceptible to attack that user-based algorithms. However, it is important to bear in mind that the two kNN algorithms discussed here are top-*N* recommendation algorithms, rather than rating prediction algorithms. The attacked item is likely to be in the neighbourhood of many other items, particularly when it is connected, via sybil profiles to other popular items. It is harder for the item to influence the user-based algorithm, as its weight in the score is dependent on the user-similarity between the sybils and genuine users. In order to make this similarity large, the sybils need to contain extremely popular items.

The matrix factorisation algorithms are significantly more robust, which agrees with observations in the literature. It is interesting to note however, that in the case of these algorithms, the attack is *more* effective when I^F is chosen uniformly from the item-set, **RU**. All three attacks are successful on only a small

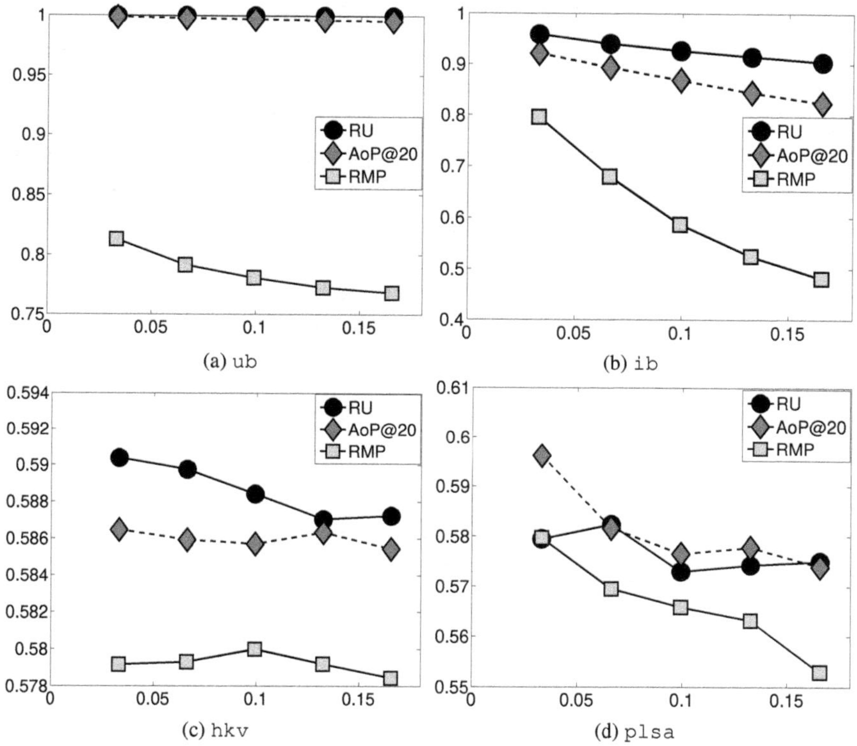

Fig. 12.6. *Overlap* over 20 items.

percentage of the user-base. When **RU** is successful, the item is pushed to rank position 7, on average, while the other attack variants only squeeze into the top 20, at high ranks of 15 or above. It is also noteworthy that increasing the attack size does not greatly improve the effectiveness of the attack.

Investigating further, we plot in Figure 12.7, the *HitRatio* against *fillersize* for the hkz algorithm when attack sizes of 3% and 16% are applied. We note that *smaller* attack profiles yield more effective results and that when the filler size is just 0.1% of the profile length, the strongest attack on hkv now succeeds in pushing into over 10% of all recommendation lists. On Movielens this corresponds to a profile containing just 4 items. We can conclude that small profiles may be overly influential in matrix factorisation, although such profiles could be easily identified by a filtering process.

The impact of the attack algorithms on general recommender performance may be partially assessed by examining the *Overlap* in Figure 12.6. We see that, for the kNN algorithms, the top $N = 20$ list is distorted somewhat, even by the lowest level of attack. *Overlap* is less useful for the matrix factorisation algorithms, since model fitting involves some randomisation and we can expect variation between different runs, regardless of an attack. We therefore show the precision@20 in Figure 12.8, for the **RMP** and **AoP@20** attacks, when a single item is attacked. The **RMP** attack leaves precision largely unchanged. A similar pattern may be observed for NDCG. A strong negative impact on results is only seen for the **RMP** attack on the kNN algorithms.

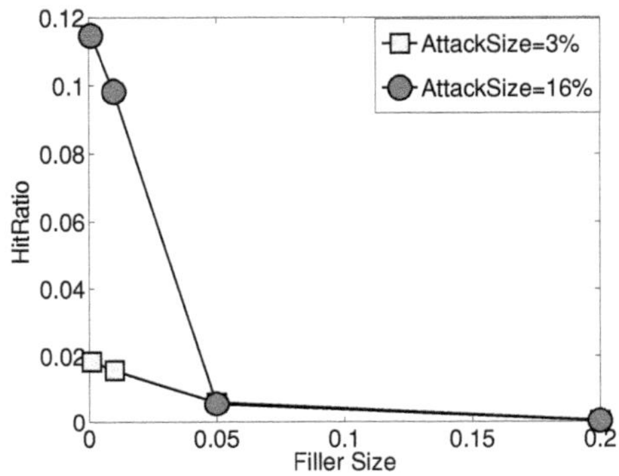

Fig. 12.7. $\overline{HitRatio}$ vs filler size, **RU** attack on `hkv` algorithm.

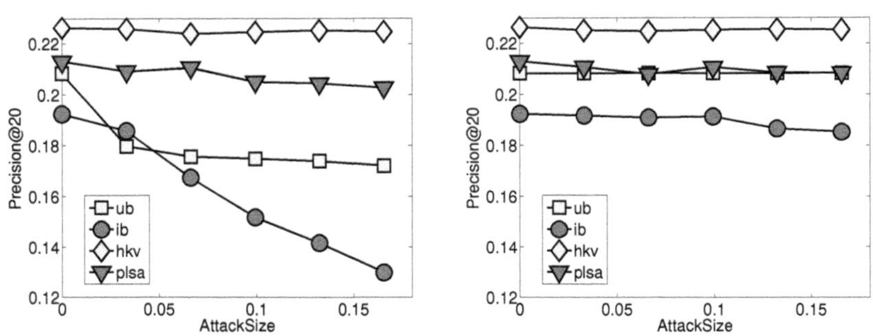

(a) **RMP** (b) **AoP@20**

Fig. 12.8. Precision of algorithms under different attacks.

12.9. Detecting Profile Injection

An obvious strategy to protect against profile injection is to apply filtering to remove such profiles from the database before training the system. An attack is effective if relatively few rating injections have a large impact on the system's output. This suggests that sybil profiles need to be more influential than typical genuine profiles and this difference provides a basis for detection and filtering of such profiles. On the other hand, if a filter removes all "unusual" profiles, leaving only "typcial" behaviour, this goes against the very principal of personalisation on which recommender systems are founded. So a filtering strategy needs to find a right balance, where truly malicious data is removed, without, at the same time, removing genuine but unusual profiles from the dataset and lowering the system's personalisation capabilities.

Classifying a given user profile as a sybil profile is a binary classification problem, with two categories that we can call *Sybil* or *Genuine*. The usual metrics appropriate for binary classification can be used to measure the success of any filtering strategy. In the following, we will concentrate on Receiver Operating Curve (ROC) analysis, as this allows the full capability of the detection algorithm to be analysed, across all possible detection thresholds. The ROC plots the probability of good detection, $\mathrm{p_d}$, i.e. the probability of correctly identify a sybil profile, against the probability of false alarm, p_f, i.e. the probability of incorrectly labelling a genuine profile as a sybil, as the threshold of the detection algorithm is varied. The Area under the ROC Curve (AUC) provides a good overall measure of the success of the detector.

Detection algorithms can be categorised as *supervised* or *unsupervised* depending on whether or not the algorithm is trained using instances of the two classes. While unsupervised approaches hold the promise of a general filtering strategy, independent of any given attack type, supervised algorithms have potential for greater performance if a good model of the classes can be devised. In this regard, it is worth noting that, as detection will typically be carried out by the system owner, there is access to a large source of instances from the genuine class at least, from which good models can be designed. In the following section we discuss one supervised and one unsupervised algorithm which demonstrate the possible approaches.

12.9.1. *Supervised Classification using Neyman-Pearson Statistical Detection Theory*

Supervised detection methods have been considered in [Burke *et al.* (2006)] and [Chirita *et al.* (2005)] in which feature vectors based on statistics of the filler and other sub-sets of the profile are used in classifiers such as kNN, C4.5 and SVM. We elaborate on the approach proposed in [Hurley *et al.* (2009)] where attack detection is treated as a statistical detection problem, using *Neyman-Pearson* (N-P) detection. Let $\mathbf{y} \in \mathscr{Y}$ be an observation i.e. a user profile, over the space $\mathscr{Y}$ of all possible profiles. A detection test is formed based on $f_{\mathbf{Y}|H_1}(\mathbf{y})$ and $f_{\mathbf{Y}|H_0}(\mathbf{y})$, the probability distribution functions (pdf) of $\mathbf{y}$ under the assumption that H_1 or H_0 is true, where H_1, H_0 denote the hypotheses that the profile is a sybil or genuine, respectively. The Neyman-Pearson (N-P) test is a likelihood ratio test, of the form:

$$l(\mathbf{y}) \triangleq \frac{f_{\mathbf{Y}|H_1}(\mathbf{y})}{f_{\mathbf{Y}|H_0}(\mathbf{y})}. \tag{12.3}$$

12.9.1.1. *Modelling Random and Average Attacks*

Define θ_i as an indicator variable such that $\theta_i = 1$ if $y_i = \phi$ and $\theta_i = 0$ otherwise. The pdf, $f_{\mathbf{Y}|H_1}(\mathbf{y})$, of $\mathbf{y}$ under hypothesis H_1 that it is an attack profile is given by

$$\prod_{i=1}^{m} \Pr[Y_i = y_i] = \prod_{i=1}^{m} \Pr[Y_i = \emptyset]^{\theta_i} \left(\Pr[Y_i = y_i | Y_i \neq \emptyset] \Pr[Y_i \neq \emptyset]\right)^{1-\theta_i}. \tag{12.4}$$

Furthermore, assuming that the continuous values generated by the Gaussian distribution are rounded to the nearest integer, so that the probability of obtaining the rating value y_i is the probability that a Gaussian random number lies in the range $[y_i - \frac{1}{2}, y_i + \frac{1}{2}]$, we can write

$$\Pr[Y_i = y_i | y_i \neq \emptyset] = \mathcal{Q}\left(\frac{y_i - \frac{1}{2} - \mu_i}{\sigma_i}\right) - \mathcal{Q}\left(\frac{y_i + \frac{1}{2} - \mu_i}{\sigma_i}\right)$$

where $\mathcal{Q}(.)$ is the Gaussian $\mathcal{Q}$-function[3]. For Random attack profiles, we take $\mu_i = \mu_g$ and $\sigma_i = \sigma_g$.

$\Pr[Y_i \neq \emptyset]$ can be approximated as $|I^F|/m$, since filler items are chosen with equal probability from the item set. Other attacks in which the profile is determined largely by the filler, such as Bandwagon and Love/Hate attacks can be modelled using this model, or a small variant of it.

12.9.1.2. *Modelling Genuine Profiles*

A model that is sufficient to distinguish genuine profiles from Random and Average attack profiles can use Equation (12.4) for the pdf $f_{\mathbf{Y}|H_0}(\mathbf{y})$, of genuine profiles, but with $\Pr[Y_i \neq \emptyset]$ set to model the rating behaviour of genuine users, since it is the placement of the filler items, more so than their value, that distinguishes these attacks. For each item i, we set $\Pr[Y_i \neq \phi] = p_i$, where p_i, the probability that an item is rated, can be estimated from a training set of genuine profiles, as the fraction of profiles that rated the item. We will refer to the detector that forms a Neyman-Pearson test using only filler selection as **N-P Filler**.

---

[3]i.e. $\mathcal{Q}(x) = 1/(2\sqrt{\pi}) \int_x^\infty e^{-\frac{t^2}{2}} dt$.

Note that the AoP attack was designed to deliberately thwart this detection mechanism by placing item fillers with probability according to item popularity. In this case, more sophisticated models of the user profile are required, that take the covariance between item values into account, since in the Average and AoP attacks, all values are selected independently.

12.9.2. *Unsupervised Attack Detection using Profile Clustering*

A number of unsupervised algorithms that try to identify groups of attack profiles have been proposed [O'Mahony *et al.* (2004b); Su *et al.* (2005); Mobasher *et al.* (2006); Bryan *et al.* (2008); Mehta and Nejdl (2009)]. Generally, these algorithms rely on clustering strategies that attempt to distinguish clusters of attack profiles from clusters of genuine profiles. As a prototypical example of this approach, we discuss the method proposed in Mehta and Nejdl (2009). This method exploits the fact that attacks consist of profiles that are highly correlated to each. A profile clustering algorithm is therefore likely to place all attack profiles in one or just a few clusters. Mehta and Nejdl apply this intuition by clustering profiles using either pLSA or principal component analysis (PCA). The PCA method attempts to choose a cluster where the sum of the pairwise covariance between profiles in the cluster is maximised and this is achieved by sorting profiles using a score formed from the sum of the components of a few (typically three) eigenvectors of the covariance matrix.

An alternative approach, discussed in Hurley *et al.* (2009), models the set of Genuine and Sybil profiles as a Gaussian Mixture Model and applies an expectation-maximisation algorithm to learn the mean and covariance of the user profiles in each cluster. These parameters are then used to form a Neyman-Pearson test, with multivariate Gaussian likelihoods, which we refer to as **N-P Gauss**.

12.9.3. *Discussion*

We evaluate on Movielens, and set the attack filler size to 5%, the overall sparsity of the this dataset. For the PCA clustering strategy, the ROC is generated by varying the attack cluster size, from 1 to m and computing p_f and p_d for each size. This detector is labelled as **PCA Z1** in Table 12.4 and clearly performs extremely well. The **NP Filler** detector also performs well on the Random and Average attacks. Note that this detector only exploits the difference in filler selection. In fact, implicitly, the PCA detector is also exploiting the difference in filler selection. The PCA detection strategy calculates the principal components of the covariance matrix of the ratings dataset, converted to z-scores. The z-score

Table 12.4. Area under the ROC curve for attack detection.

Attack	Detector	AUC
Random	PCA Z0	0.979
Random	PCA Z1	**0.999**
Random	N-P Filler	0.975
Average	PCA Z0	0.81
Average	PCA Z1	**0.997**
Average	N-P Filler	0.956
AoP@60	PCA Z1	0.847
AoP@60	N-P Gauss	**0.874**
AoP@20	PCA Z1	0.736
AoP@20	N-P Gauss	**0.951**

for a user profile is obtained by subtracting the profile mean μ_u and dividing by the standard deviation σ_u:

$$z_{ui} = \frac{r_{ui} - \mu_u}{\sigma_u}.$$

Given that a typical profile contains many unrated items, the z-score depends strongly on how missing values are dealt with. One strategy is to ignore them and calculate μ_u and σ_u based on the rated items only. Call the resulting matrix of z-score ratings Z_0. A second strategy is to treat missing values as zeros and compute μ_u and σ_u across the entire profile. Call the resulting matrix Z_1. As observed in Table 12.4, using Z_0 is extremely unsuccessful. In fact, the success of the **PCA Z1** strategy is largely explained by the fact that the covariance is dominated by the missing value distribution.

This analysis shows that an attacker should choose filler items according to their overall popularity among the genuine user base. The AoP attack obfuscates the Average attack by choosing filler items with equal probability from the top $x\%$ of most popular items, rather than from the entire catalogue of items, where x is chosen to ensure that the profiles are undetectable by the PCA detector. In order to thwart such an attack, it becomes necessary to design a more accurate model of the Genuine and Sybil profiles. In Table 12.4, we see that **N-P Gauss** obtains significantly better detection quality than **PCA Z1** on the AoP@20 attack.

The conclusion that may be drawn is that, faced with sophisticated detection techniques, attackers need to devise obfuscation strategies to render their attack profiles undetectable. However, the more like genuine profiles that attack profiles become, the less powerful they will be in terms of influencing the recommender output.

12.10. Recommender Robustness and the Spread of Fake News

Since September 2016, towards the end of the American presidential campaign the term 'fake news' started to trend in popular discourse. There is now heightened awareness that social media platforms can filter stories to users whose veracity is questionable. More ominously, such platforms can be manipulated so that particular stories and opinions appear on the news feeds of targeted users, in order to influence their opinions [Bessi and Ferrara (2016)]. A good survey of the use of false information on the web and ways to detect it is provided in Kumar and Shah (2018). In this survey, the authors point out the distinction between *misinformation*, which is accidentally erroneous and *disinformation*, which is maliciously incorrect [Fallis (2014)] and also distinguish between opinion-based and fact-based false information. They discuss ways that false information is spread, using bots and online identities created for deception, sometimes referred to as *sock-puppets*. Bessi and Ferrara, for instance, found that almost one fifth of Twitter political chatter is from bot accounts. Kumar and Shah point to research such as Ott *et al.* (2011) that shows that humans are poor at judging the veracity of information. Malicious actors focus not just on spoofing directly targeted users, but also apply strategies to promote the trustworthiness and status of the false accounts. They take account of the social network by, for instance, selecting influential users to attack, in order to encourage the disinformation to be spread more widely by genuine users who have been deceived. This social aspect is key to disinformation spreading virally.

The work discussed in this chapter clearly falls into the category of dissemination of opinion-based disinformation but our focus is directed towards the characteristics of the recommendation algorithms, rather than the social aspects of disinformation spread. Our concern has been to determine whether collaborative, personalisation algorithms have inherent weaknesses in the way that they process user data, which can be leveraged by malicious attackers. This provides just one point of attack for a malicious actor, within a complex web of possible vulnerabilities that exists when automated tools are engaged in the filtering of information to end users. In Kumar and Shah, it is pointed out that personalisation plays a key role in the creation of "echo chambers", in which a single point of view echoes in a network of users, where conflicting views are rarely heard. Our study of recommender algorithm robustness gives a sense of how easily an algorithm can suppress certain recommendations through being influenced to promote others. In some sense, algorithms that are weak from a robustness point-of-view are also more susceptible to the creation of echo chambers. Our analysis has focused on collaborative recommender systems that exploit explicit ratings or implicit feedback. However, if the recommended items correspond to news articles or include user reviews, there is a much richer channel of textual opinion through which to influence end users, that has not been the focus of our study. Automated recommendation tools still largely rely on the collaborative signal obtained through explicit or implicit feedback to filter and rank the content presented to end users, so it is important to understand this point of weakness. Equally, it is important to observe that addressing algorithm robustness is only one aspect in the larger question of tracking and preventing the dissemination of disinformation.

12.11. Conclusion

In this paper we have reviewed the state-of-the-art in recommender robustness. Moreover, as a practical contribution to practitioners in this field, we have implemented the main attack strategies discussed in the literature into a robustness analysis library in the `Ranksys` framework. The evaluation presented in this chapter has focused on the top-N recommendation problem, rather than on rating prediction and the `Ranksys` framework provides easy access to a number of top-N algorithms, so that their vulnerability to attack of top-N algorithms can be assessed. We have demonstrated the use of this library through an analysis of two kNN top-N algorithms and two matrix factorisation algorithms.

Acknowledgements

This project has been funded by Science Foundation Ireland under Grant No. SFI/12/RC/2289.

References

Afsharinejad, A. and Hurley, N. (2018). Performance analysis of a privacy constrained knn recommendation using data sketches, in *Proceedings of the Eleventh ACM International Conference on Web Search and Data Mining, WSDM 2018, Los Angeles, USA, February 6-8, 2018* (ACM).

Bessi, A. and Ferrara, E. (2016). Social bots distort the 2016 u.s. presidential election online discussion, *First Monday* `http://firstmonday.org/ojs/index.php/fm/article/view/7090`.

Bryan, K., O'Mahony, M. and Cunningham, P. (2008). Unsupervised retrieval of attack profiles in collaborative recommender systems, in *RecSys '08: Proceedings of the 2008 ACM conference on Recommender systems* (ACM, New York, NY, USA), ISBN 978-1-60558-093-7, pp. 155–162, doi:http://doi.acm.org/10.1145/1454008.1454034.

Burke, R., Mobasher, B. and Williams, C. (2006). Classification features for attack detection in collaborative recommender systems, in *Proceedings of the 12th International Conference on Knowledge Discovery and Data Mining*, pp. 17–20.

Burke, R., Mobasher, B., Zabicki, R. and Bhaumik, R. (2005). Identifying attack models for secure recommendation, in *Beyond Personalization: A Workshop on the Next Generation of Recommender Systems*.

Chirita, P. A., Nejdl, W. and Zamfir, C. (2005). Preventing shilling attacks in online recommender systems, *In Proceedings of the ACM Workshop on Web Information and Data Management (WIDM'2005)*, pp. 67–74.

Fallis, D. (2014). A functional analysis of disinformation, in *iConference 2014* (Berlin, Germany), `http://hdl.handle.net/2142/47258`.

Harper, F. M. and Konstan, J. A. (2015). The movielens datasets: History and context, *ACM Trans. Interact. Intell. Syst.* **5**, 4, pp. 19:1–19:19, doi:10.1145/2827872, `http://doi.acm.org/10.1145/2827872`.

Hofmann, T. (2004). Latent semantic models for collaborative filtering, *ACM Trans. Inf. Syst.* **22**, 1, pp. 89–115, doi:10.1145/963770.963774, `http://doi.acm.org/10.1145/963770.963774`.

Hu, Y., Koren, Y. and Volinsky, C. (2008). Collaborative filtering for implicit feedback datasets, in *Proceedings of the 2008 Eighth IEEE International Conference on Data Mining*, ICDM '08 (IEEE Computer Society, Washington, DC, USA), ISBN 978-0-7695-3502-9, pp. 263–272, doi:10.1109/ICDM.2008.22, `http://dx.doi.org/10.1109/ICDM.2008.22`.

Hurley, N., Cheng, Z. and Zhang, M. (2009). Statistical attack detection, in *Proceedings of the Third ACM Conference on Recommender Systems*, RecSys '09 (ACM, New York, NY, USA), ISBN 978-1-60558-435-5, pp. 149–156, doi:10.1145/1639714.1639740, `http://doi.acm.org/10.1145/1639714.1639740`.

Kumar, S. and Shah, N. (2018). False Information on Web and Social Media: A Survey, *ArXiv e-prints* arXiv:1804.08559.

Lam, S. K. and Riedl, J. (2004). Shilling recommender systems for fun and profit, *In Proceedings of the 13th International World Wide Web Conference*, pp. 393–402.

Mehta, B. and Nejdl, W. (2009). Unsupervised strategies for shilling detection and robust collaborative filtering, *User Modeling and User-Adapted Interaction* **19**, 1-2, pp. 65–97, doi:http://dx.doi.org/10.1007/s11257-008-9050-4.

Mobasher, B., Burke, R., Bhaumik, R. and Williams, C. (2005). Effective attack models for shilling item-based collaborative filtering system, *In Proceedings of the 2005 WebKDD Workshop (KDD'2005)*.

Mobasher, B., Burke, R., Bhaumik, R. and Williams, C. (2007). Toward trustworthy recommender systems: An analysis of attack models and algorithm robustness, *ACM Transactions on Internet Technology* **7**, 4.

Mobasher, B., Burke, R. D. and Sandvig, J. J. (2006). Model-based collaborative filtering as a defense against profile injection attacks, in *AAAI* (AAAI Press).

O'Mahony, M. P., Hurley, N. J., Kushmerick, N. and Silvestre, G. C. M. (2004a). Collaborative recommendation: A robustness analysis, *ACM Transactions on Internet Technology (TOIT), Special Issue on Machine Learning for the Internet* **4**, 4, pp. 344–377.

O'Mahony, M. P., Hurley, N. J. and Silvestre, C. C. M. (2004b). An evaluation of neighbourhood formation on the performance of collaborative filtering, *Artificial Intelligence Review* **21**, 1, pp. 215–228.

O'Mahony, M. P., Hurley, N. J. and Silvestre, G. C. M. (2002). Promoting recommendations: An attack on collaborative filtering, in A. Hameurlain, R. Cicchetti and R. Traunmüller (eds.), *DEXA*, *Lecture Notes in Computer Science*, Vol. 2453 (Springer), ISBN 3-540-44126-3, pp. 494–503.

O'Mahony, M. P., Hurley, N. J. and Silvestre, G. C. M. (2003). An evaluation of the performance of collaborative filtering, *In Proceedings of the 14th Irish International Conference on Artificial Intelligence and Cognitive Science (AICS'03)*, pp. 164–168.

Ott, M., Choi, Y., Cardie, C. and Hancock, J. T. (2011). Finding deceptive opinion spam by any stretch of the imagination, in *Proceedings of the 49th Annual Meeting of the Association for Computational Linguistics: Human Language Technologies - Volume 1*, HLT '11 (Association for Computational Linguistics, Stroudsburg, PA, USA), ISBN 978-1-932432-87-9, pp. 309–319, `http://dl.acm.org/citation.cfm?id=2002472.2002512`.

Resnick, P., Iacovou, N., Suchak, M., Bergstrom, P. and Riedl, J. (1994). Grouplens: An open architecture for collaborative filtering of netnews, *In Proceedings of the ACM Conference on Computer Supported Cooperative Work (CSCW'94)*, pp. 175–186.

Su, X.-F., Zeng, H.-J. and Chen, Z. (2005). Finding group shilling in recommendation system, in *WWW '05: Special interest tracks and posters of the 14th international conference on World Wide Web* (ACM, New York, NY, USA), ISBN 1-59593-051-5, pp. 960–961, doi:http://doi.acm.org/10.1145/1062745.1062818.

Chapter 13

Privacy in Collaborative Recommenders

Qiang Tang
Luxembourg Institute of Science and Technology
5 Avenue des Hauts-Fourneaux
4362 Esch-sur-Alzette, Luxembourg
qiang.tang@list.lu

Today, recommender systems are ubiquitous in our daily life and are becoming an indispensable decision supporting tool. Unfortunately, these systems are extremely greedy for (sensitive) personal data and they incur enormous economic and political incentives for malicious players to misbehave. An increasing number of privacy breaches and system abuses have been disclosed over the time. In this chapter, we aim at a comprehensive investigation into the privacy issues associated with recommender systems, with a focus on the subtle conflicting-yet-complementary relationships between privacy, robustness, and transparency. Taking several existing privacy-preserving solutions, including cryptographic and differential privacy based ones, we comparatively analyse their strengths and weaknesses. At the end, we outline some research directions to design pragmatic privacy-preserving recommender systems which also facilitate other important properties such as robustness and transparency.

13.1. Introduction

With an ever-expanding sea of data, predictive analytics are playing an increasingly important role in both professional and personal spheres. They are indispensable tools to discover knowledge from the historical data and provide predictions for future decision making. In practice, they have offered us with unprecedented efficiency, in the sense that many decisions can be made by one click. Today, predictive analytics have been intensively used in targeted advertisement, precision medicine and disease screening, credit scoring, fraud detection and prevention, traffic analysis, homeland security, and so on. Among all, recommender systems are the key enabler for various personalized services, and they are at every corner of our life.

When recommenders continue to penetrate into our life and extract every detail possible, more and more concerns have emerged. Worries against such systems root in their greedy demand for personal data and the uncontrolled data usage inside and outside recommendation computation. Since the beginning, privacy has been a concern for recommender customers who are afraid of personal information disclosure, and nowadays it becomes more worrisome due to its cascaded effects. For example, using Google as an example, Newman [Newman (2014)] presented in-depth discussions on the consequences of lost privacy by focusing on consumer harm and economic inequality. So far, many privacy-preserving recommender systems have been proposed for the purpose of minimizing information leakages from some perspectives. Unfortunately, privacy is not a stand-alone requirement, and it is entangled with other fundamental requirements, such as robustness and transparency. This implies that most existing privacy-preserving solutions will in-fact result in a lot of *unexpected* problems in practice, even (surprisingly) downgrade the actual privacy guarantees when deployed.

In this chapter, we aim at a deeper understanding about the privacy concerns for both end users and the service provider, with a focus on the tension with the robustness and transparency requirements in recommender systems. As a result, we wish to identify future directions to design privacy-preserving solutions which can align the expectations for different requirements and the incentives of different players.

13.1.1. *Recommender in a Nutshell*

In most recommender systems, there is a service provider (RecSys) which facilitates the interactions among users, shown in Figure 13.1. The RecSys will implement the recommender algorithms and collect users' data implicitly (e.g. tracking users' behaviour) and/or explicitly (e.g. collecting users' rating values on previously consumed items), and then predict what the users will prefer. A less common setting is that the user population is partitioned into two (or more) service providers which are responsible for computing recommendations for their own users, shown in Figure 13.2.

Let's assume that the item set is denoted by $\mathbf{I} = (1, 2, \cdots, b, \cdots, M)$, and a user x's ratings are denoted by a vector $\mathbf{R}_x = (r_{x,1}, \cdots, r_{x,b}, \cdots, r_{x,M})$. The rating value is often an integer from

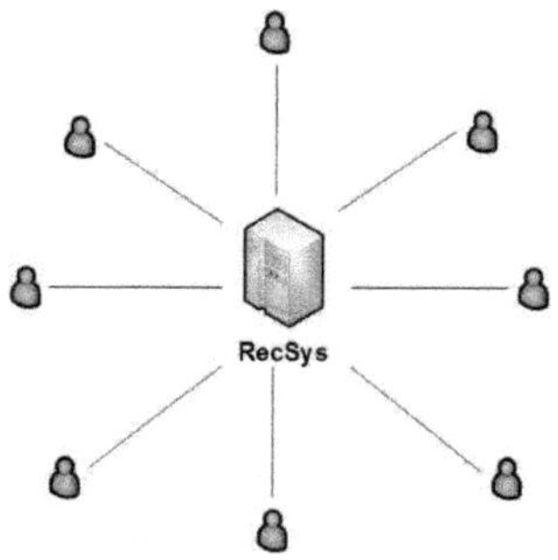

Fig. 13.1. Standard Setting.

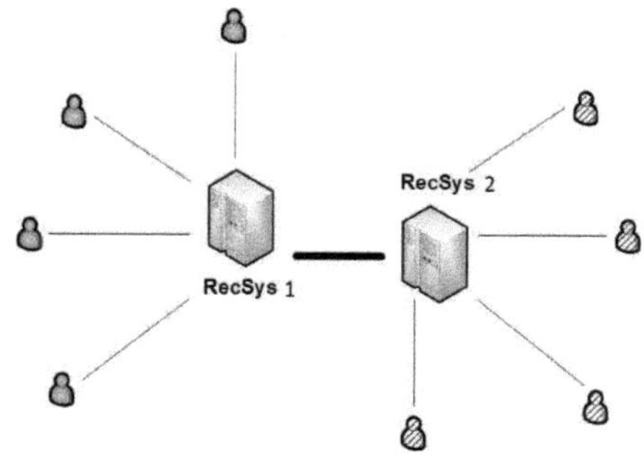

Fig. 13.2. Partitioned Setting.

$\{0, 1, 2, 3, 4, 5\}$. If item i has not been rated, then $r_{x,i}$ is set to be 0. The ratings are often organized in a rating matrix, as shown in Table 13.1. In its simplest form, the functionality of a recommender system is to predict the unrated $r_{x,i}$ values based on rated ones.

Table 13.1. Rating Matrix.

	Item 1	$\cdots$	Item b	$\cdots$	Item M
User 1 ($\mathbf{R}_1$)	$r_{1,1}$	$\cdots$	$r_{1,b}$	$\cdots$	$r_{1,M}$
User 2 ($\mathbf{R}_2$)	$r_{2,1}$	$\cdots$	$r_{2,b}$	$\cdots$	$r_{2,M}$
$\vdots$	$\vdots$	$\vdots$	$\vdots$	$\vdots$	$\vdots$
User N ($\mathbf{R}_N$)	$r_{N,1}$	$\cdots$	$r_{N,b}$	$\cdots$	$r_{N,M}$

In the literature, a lot of generic recommender algorithms have been proposed, as detailed in [Adomavicius and Tuzhilin (2005); Shani and Gunawardana (2011)]. Among all, two categories of well-known algorithms are neighborhood-based and matrix factorization based, which we will briefly exemplify below for the self-containment of this chapter. More detailed description for recommender algorithms appears in **Part I: Algorithms** of this book.

User-based Neighborhood. Neighborhood-based algorithms can be further divided into user-based and item-based. We only introduce the former in this chapter by assuming a Cosine similarity metric. For two rating vectors $\mathbf{R}_x, \mathbf{R}_y$, their Cosine similarity is denoted as $Sim_{x,y}$, where

$$Sim_{x,y} = \frac{\langle \mathbf{R}_x, \mathbf{R}_y \rangle}{\|\mathbf{R}_x\| \times \|\mathbf{R}_y\|} = \langle \frac{\mathbf{R}_x}{\|\mathbf{R}_x\|}, \frac{\mathbf{R}_y}{\|\mathbf{R}_y\|} \rangle$$

Suppose we want to compute predictions for a user x based on the rating data from a user group $\mathcal{G}$, then the formula is as follows.

$$p_{x,i} = \overline{r_x} + \frac{\sum_{y \in \mathcal{G} \wedge r_{y,i} \neq 0} Sim_{x,y}(r_{y,i} - \overline{r_y})}{\sum_{y \in \mathcal{G} \wedge r_{y,i} \neq 0} Sim_{x,y}}$$

Matrix Factorization. Given a user set $\mathcal{U} = \{1, 2, \cdots, N\}$ and their rating vectors $\mathbf{R}_x$ for $x \in \mathcal{U}$, let $\mathcal{R}$ denote the set of (x, j) such that $r_{x,j} \neq 0$. One of the most popular collaborative filtering algorithms is based on low-dimensional factor models, which derive two feature matrices $\mathbf{U}$ and $\mathbf{V}$ from the rating matrix. The feature vector $\mathbf{u}_x$ denotes user x's interest and the feature vector $\mathbf{v}_j$ denotes item j's characteristics. Every feature vector has the dimension k, which is often a much smaller integer than M and N.

$$\mathbf{U} = \begin{bmatrix} \mathbf{u}_1 \\ \mathbf{u}_2 \\ \vdots \\ \mathbf{u}_N \end{bmatrix} \quad \text{and} \quad \mathbf{V} = \begin{bmatrix} \mathbf{v}_1 \\ \mathbf{v}_2 \\ \vdots \\ \mathbf{v}_M \end{bmatrix}$$

In practice, $\mathbf{U}$ and $\mathbf{V}$ are often computed by minimizing the following Regularized least Squares Error (RSE) function:

$$\min_{\mathbf{U},\mathbf{V}} \frac{1}{|\mathcal{R}|} \sum_{(x,j) \in \mathcal{R}} (r_{x,j} - \langle \mathbf{u}_x, \mathbf{v}_j \rangle)^2 + \lambda \sum_{x \in \mathcal{U}} ||\mathbf{u}_x||_2^2 + \mu \sum_{j \in \mathbf{I}} ||\mathbf{v}_j||_2^2 \tag{13.1}$$

for some positive parameters λ, μ. Using the standard gradient descent method, $\mathbf{U}$ and $\mathbf{V}$ can be learned through recursively applying the updating rules.

$$\mathbf{u}_x^{(t)} = \mathbf{u}_x^{(t-1)} - \gamma \nabla_{\mathbf{u}_x} F(\mathbf{U}^{(t-1)}, \mathbf{V}^{(t-1)}) \tag{13.2}$$

$$\mathbf{v}_j^{(t)} = \mathbf{v}_j^{(t-1)} - \gamma \nabla_{\mathbf{v}_j} F(\mathbf{U}^{(t-1)}, \mathbf{V}^{(t-1)}) \tag{13.3}$$

where $\gamma > 0$ is a small gain factor and

$$\nabla_{\mathbf{u}_x} F(\mathbf{U}, \mathbf{V}) = -2 \sum_{j:(x,j) \in \mathcal{R}} \mathbf{v}_j (r_{x,j} - \langle \mathbf{u}_x, \mathbf{v}_j \rangle) + 2\lambda \mathbf{u}_x \tag{13.4}$$

$$\nabla_{\mathbf{v}_j} F(\mathbf{U}, \mathbf{V}) = -2 \sum_{x:(x,j) \in \mathcal{R}} \mathbf{u}_x (r_{x,j} - \langle \mathbf{u}_x, \mathbf{v}_j \rangle) + 2\mu \mathbf{v}_j \tag{13.5}$$

In practice, these generic recommender algorithms are often combined in different ways to overcome certain obstacles, e.g. the cold-start problem. To this end, one of the most famous example is the BellKor's solution in the Netflix competition [Koren (2009)]. Comprehensive reviews about recommender systems can be found in a number of survey papers such as [Adomavicius and Tuzhilin (2005); Su and Khoshgoftaar (2009); Shani and Gunawardana (2011); Shi *et al.* (2014)]. Recently, deep learning has become a very powerful tool and has been used to numerous applications. Zhang, Yao, and Sun [Zhang *et al.* (2017)] gave a very comprehensive survey to the applications of deep learning in recommender systems.

It is worth stressing that, in practical deployment, a wide spectrum of auxiliary information (e.g. those in the first column of Table 13.2 from Wang and Tang [Wang and Tang (2015)]) may be incorporated to improve performances from different perspectives, like mitigating the cold-start problem, increasing accuracy, enhancing stability. Shi, Larson and Hanjalic [Shi *et al.* (2014)] gave more detailed discussions in this direction. Due to the availability of datasets, most academic papers consider only the rating matrix (i.e. explicit feedback) as input and evaluate their solutions with respect to some publicly well-known datasets such as MovieLens[1] and Netflix[2]. It is easy to understand that the more data used in recommenders the more challenges for privacy protection! We elaborate on this in Section 13.5.

Table 13.2. Features used in different Recommenders ([Wang and Tang (2015)]).

	News	Tourism	Movies&Video	Music	Books	Social
Explicit feedback		⋆	⋆	⋆	⋆	⋆
Implicit feedback	⋆	⋆	⋆	⋆	⋆	⋆
Time	⋆	⋆	⋆			
Content	⋆	⋆	⋆	⋆	⋆	⋆
Cost		⋆				
Location	⋆	⋆	⋆	⋆		⋆
Social interaction	⋆	⋆	⋆	⋆		⋆
Demography	⋆	⋆	⋆	⋆	⋆	⋆
Tags		⋆	⋆	⋆		⋆
Emotion			⋆	⋆		

13.1.2. *Contribution and Organization*

Privacy issues and solutions in recommender systems have been incrementally surveyed by Ramakrishnan *et al.* [Ramakrishnan *et al.* (2001)], Lam, Frankowski, and Riedl [Lam *et al.* (2006)], Beye *et al.* [Beye *et al.* (2013)], Friedman *et al.* [Friedman *et al.* (2015)]. In particular, the work [Friedman *et al.* (2015)] has offered high-level discussions on the privacy risks in a very

[1]`https://grouplens.org/datasets/movielens/`

[2]`https://www.kaggle.com/netflix-inc/netflix-prize-data`

broad sense, while it considers a privacy violation when any of the following principles has been violated: collection limitation, data quality, purpose specification, use limitation, security safeguards, openness, individual participation, and accountability. It also provides a a high-level review on a broad set of solutions, ranging from cryptographic ones to policy-based ones.

Different from the above works, in this chapter, we focus on information leakage-oriented privacy breaches and offer an in-depth discussion and comparison with respect to representative privacy-preservation solutions. We first reflect on the privacy requirement in recommender systems and emphasizing its relationship with other requirements in Section 13.2. In Section 13.3 and Section 13.4, we review some privacy-preserving solutions in the cryptographic setting and data-obfuscation setting, respectively. We finally conclude the chapter in Section 13.5. In this chapter, due to the fact that we are unable to get the source codes for most published solutions, we compared their performances by directly using the experimental results in the original papers.

13.2. Privacy Concerns in Recommenders

For users and the RecSys in a recommender system, they have their own privacy concerns. We categorize them as follows.

- *User privacy* is about controlling the information that an honest user discloses to the RecSys and other users in the system. One privacy concern is that some input data is sensitive by itself because such data can tell a lot about personality and behaviour. Even the input may not look sensitive in itself in some scenarios, it can result in invasive inferences to breach privacy. For instance, Weinsberg *et al.* [Weinsberg *et al.* (2012)] demonstrated that what has been rated by a user can potentially help an attacker identify this user.
- *RecSys privacy* is about controlling the information that the RecSys discloses to the users. For the RecSys, the proprietary system design, the actual algorithms in use and the corresponding parameters are very sensitive information and they may imply a lot of business secrets. In practice, the RecSys often collects a lot of behavioural information and feeds it to the system as auxiliary input, for better performances (e.g. accuracy). Naturally, such auxiliary data will be regarded as business asset and kept as secret as well.

Roughly, *user privacy* concerns can be divided into two levels, namely membership privacy and data privacy. As implied by Weinsberg *et al.* [Weinsberg *et al.* (2012)], these two levels of privacy concerns are tightly connected in the sense that breaching one will likely affect the other. Moreover, information leakage can happen both in the process of computation and the recommendation results. We summarize the privacy-related questions in different dimensions in Table 13.3, and correspondingly define four scenarios $\mathcal{S}_1, \cdots, \mathcal{S}_4$. In contrast, we treat *RecSys privacy* at the level of data privacy only since there is no membership concern for the RecSys.

Table 13.3. Dimensional Privacy Concerns.

	Membership Privacy	Data Privacy
Leakage in the process of computing predictions	$\mathcal{S}_1$. Whether a user has been involved in the computation?	$\mathcal{S}_2$. What is revealed about a user's input in the computation?
Leakage in the outputs	$\mathcal{S}_3$. What can be inferred about a user's involvement from the outputs?	$\mathcal{S}_4$. How much information about a user's input can be inferred from the outputs, given that this user is involved in the computation?

The literature has told us that privacy risks are real. Related to scenario $\mathcal{S}_1$, Narayanan and Shmatikov [Narayanan and Shmatikov (2008)] presented de-anonymization attacks and applied them to anonymized Netflix dataset, where the anonymization means removal of customer identifying information. They showed that an attacker, with a small number of rating values (some of them are possibly wrong) and their approximate generation dates, can link it to the original record with very high probability. This implies that simple anonymisation is not adequate to prevent de-identification attacks in publishing rating datasets in recommenders. Related to scenario $\mathcal{S}_4$, Calandrino *et al.* [Calandrino *et al.* (2011)] proposed information reference attacks against item-based neighborhood recommenders. In an attack, the attacker registers a number of fake user accounts to receive recommendation outputs, and then monitors changes in the public outputs of recommender systems (i.e. item similarity lists or cross-item correlations) over a period of time. By combining the dynamic information from monitoring and some auxiliary information about some of the transactions of a particular "target" user, the paper proposed an algorithm to infer the target user's private input.

In most existing privacy-preserving solutions, *RecSys privacy* has not been a concern because the recommender algorithms and parameters are often assumed to be public. One exception is expert-based recommenders, e.g. [Tang and Wang (2017); Ahn and Amatriain (2010)], where the RecSys serves users based on a model learned from experts' dataset. In this case, the RecSys has a strong incentive to protect its model. It is worth mentioning that *RecSys privacy* or *model privacy* has become an emerging topic for cloud-based machine learning platforms which employ a *machine learning model as a service* business model, e.g. [Tramèr *et al.* (2016)].

13.2.1. *Entanglement with Robustness and Transparency*

Recommendation accuracy and efficiency are two ultimate objectives of recommender systems, that incentivize all players to participate and benefit. Following the axiom of data analytics, satisfying these requirements crucially relies on *high-quality* input data from the users. Unfortunately, the privacy requirement seemingly sits in the opposite direction because it aims at hiding users' input as much as possible and even hiding who the users are. This has lead to the well-known dilemma between privacy and these two fundamental objectives. As such, in the literature, the practicality of privacy-preservation solutions has been evaluated based on their degradation to the recommendation accuracy (in case noise is added to achieve privacy) and efficiency of the underlying systems. We argue that the existing practicality validation approach is far from realistic, because two other important requirements, i.e. *system robustness* and *algorithmic transparency*, have not been taken into account. This is particularly the case from the perspective of the potential degradation to recommendation accuracy. The readers can refer to Chapter 12 for more details about the robustness properties.

Informally, *system robustness* is about controlling the effect of manipulated inputs. Clearly, it is a prerequisite for guaranteeing recommendation accuracy. The attackers can be malicious users and even the RecSys. From the economic perspective, there are strong incentives for robustness attacks. Early in 2004, Amazon disclosed that many authors rated their own books to gain popularity[3]. In countries like China where e-commerce are extremely popular nowadays, robustness attacks (or, generally trust

[3]`http://www.nytimes.com/2004/02/14/us/amazon-glitch-unmasks-war-of-reviewers.html?scp=6&sq=amazon+book+reviews&st=nyt`

fraud) are a major threat to the business[4]. It is worth emphasizing that the malicious behaviors in robustness attacks are not captured in any privacy models (where the objective is to hide the input of any user, no matter honest or malicious)!

Informally, *algorithmic transparency* is about the openness of how personalized predictions have been computed. A transparent system allows its users to see why they have received the contents or decisions. For recommenders, it is related to the explainability requirement although it can mean more than that. Arguably, a transparent system opens the door for the users to collaboratively assess a broad range of desirable properties such as fairness and anti-discrimination. It is worth mentioning that algorithmic transparency for data analytics at large has been in the centre of public attention[5]. Notably, the new General Data Protection Regulation (GDPR) of EU creates a "right to explanation" privilege (i.e. some kind of algorithmic transparency) for users in the process of algorithmic decision-making [Goodman and Flaxman (2016)].

We believe that the ultimate difficulty in addressing the privacy concerns in recommender systems comes from the entanglements among different requirements, as shown in Figure 13.3 and elaborated below. The subtle relationships among them indicate that it is unwise to emphasize one while ignoring others. Unfortunately, most of existing solutions fall into this trap. Interestingly, even with a competing nature, these requirements do contribute to each other, and a pragmatic solution should respect all of them simultaneously.

(1) User privacy and system robustness do not get along naturally. In order to achieve user privacy, we may want to protect users' inputs against the RecSys. However, in order to make the system robust, the RecSys needs to clean users' inputs and exclude the suspicious users. In existing solutions, such operations require to access the raw data and possibly know a user's identity. Clearly, there is a tension here. Interestingly, on the other hand, user privacy is crucial to achieve system robustness. When honest users' data is compromised by an attacker, it becomes much easier for this attacker to tailor its inputs to bias the outputs to these honest users. This has become an important research direction in adversarial machine learning, see [Huang *et al.* (2011)]. Lam *et al.* The paper [Lam *et al.* (2006)] already gave a discussion between data privacy and robustness properties ten years ago.

[4]`https://en.wikipedia.org/wiki/Internet_Water_Army`

[5]`https://www.nytimes.com/2015/07/10/upshot/when-algorithms-discriminate.html`

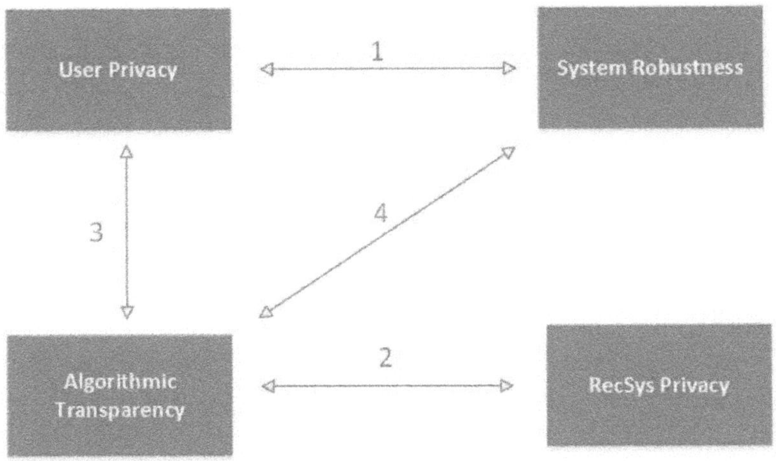

Fig. 13.3. Requirements Correlations.

(2) There is clearly a tension between algorithmic transparency and RecSys privacy. If the RecSys is asked to publish the details of the deployed algorithms and parameters to provide algorithmic transparency, then it loses privacy. Note that the RecSys may also need to publish the auxiliary data it collects.

(3) There is an embedded conflict between user privacy and algorithmic transparency. The link lies in the fact that, in most collaborative filtering recommender systems, the output to an individual not only depends on this user's input but also heavily relies on the inputs from other users. This immediately means that, in order to provide algorithmic transparency to a party X, both the underlying algorithm and relevant users' inputs need to be made available in some manner to this party X. As such, there is a tension between the two requirements.

(4) There is a natural tension between algorithmic transparency and system robustness. It is based on a simple fact: the more the attacker knows about the details of the algorithms, parameters, and some information about other users' data; the easier for it to bias the outputs to honest users.

13.2.2. *High-level Literature Review*

In the literature, most privacy-preserving cryptographic solutions assume an honest-but-curious or semi-honest adversary model, where RecSys, users and other third parties (e.g. crypto service provider — CSP) are assumed to follow the protocol specification all the time. An adversary might look at the communication records and try to infer the private information about others, while different adversaries (e.g. RecSys and CSP) will not collude. As to the users, some of solutions do assume them to be malicious, which means that a group of users can collude to infer the private information about other honest users and other parties without following the protocol specification. On one hand, the semi-honest assumption dramatically simplifies the solution design. Nevertheless, we will show that even with this assumption, it is not a trivial task to have a secure solution. On the other hand, this assumption ignores a lot of potential risks in the perspective of privacy protection. Moreover, it creates barriers to fulfill other requirements such as robustness and transparency (when the RecSys can be malicious). We expand on these comments in the rest of this chapter.

Existing privacy-preserving solutions can be generally divided into two categories, and we give a brief summary in Table 13.4. We analyse some representatives in Section 13.3.

Table 13.4. Literature Summary.

	Neighborhood	Matrix Factorization	Hybrid
Cryptographic	[Polat and Du (2005a)], [Basu *et al.* (2012)], [Tang (2012)], [Veugen *et al.* (2015)]†, [Tang and Wang (2015)], [Tang and Wang (2016)]	[Canny (2002b)], [Canny (2002a)], [Han *et al.* (2009)]*, [Nikolaenko *et al.* (2013)]†, [Kim *et al.* (2016)]†	[Aïmeur *et al.* (2008)], [Jeckmans *et al.* (2012)], [Tang and Wang (2017)]†
Data-obfuscation (ad hoc)	[Polat and Du (2003)], [Polat and Du (2006)], [Shokri *et al.* (2009)]	[Polat and Du (2005b)], [Ioannidis *et al.* (2014)]	
Data-obfuscation (dp)	[McSherry and Mironov (2009)],[Wang and Tang (2017)]	[Wang *et al.* (2015)],[Berlioz *et al.* (2015)], [Liu *et al.* (2015)]	

Cryptographic Solutions. They essentially model recommender system as a multi-party computation protocol, and the main objective is to secure users' inputs rather than the information leakage in the outputs. In another word, these solutions mainly consider scenario $\mathcal{S}_2$ in Table 13.3. Often, attackers are defined to be a semi-honest RecSys or users, and the standard setting (shown in Figure 13.1) is considered. Some solutions have also been proposed for partitioned datasets (shown in Figure 13.2), and they are marked with the symbol *. Note that in the partitioned setting, users are often assumed to fully trust their local RecSys. To improve efficiency, some solutions have introduced additional semi-honest servers, and they are marked by the symbol † in Table 13.4.

Data-obfuscation Solutions. They rely on adding noise to the original data or computation results to protect users' inputs. They mainly consider scenario $\mathcal{S}_3$ in Table 13.3. We term some solutions as *ad hoc* due to the fact that they do not use a rigorous security notion. In contrast, differential privacy based approach has gained a lot of popularity because it provides mathematically sound privacy notions. Often, these solutions assume a trusted curator, e.g. RecSys, which will calibrate noises to aggregated users' data and then publish the noisy results.

To the end of security analysis, several works have been published, related to scenario $\mathcal{S}_2$ defined in Table 13.3. These attacks raise questions on the trust assumptions (i.e. who should be trusted to protect some entity's privacy and to which extent the trust can be assumed) that should be used in both data-obfuscation and cryptographic solutions.

Zhang *et al.* [Zhang *et al.* (2006)] showed how to recover perturbed ratings in the solutions by Polat and Du [Polat and Du (2003, 2005b)]. With the solutions from [Polat and Du (2003, 2005b)], a user first disguises his rating vector by perturbing every rating value using uniform or Gaussian noise and then sends the disguised rating vector (referred to as z-scores) to the RecSys, which will compute recommendations based on the disguised values from all users. We note that this approach is similar to the differential privacy approach by Shen and Jin [Shen and Jin (2014, 2016)], although it assumes the RecSys to be the attacker while the latter solutions assume users to be the attacker. Zhang *et al.* [Zhang *et al.* (2006)] proposed two attacks. In one, the RecSys can recover the original rating values of a user based on the disguised rating vector, using simple k-means clustering method. In the other, the RecSys can recover all z-scores of all users by using SVD-based method. It remains as a challenge to apply data-obfuscation methods to secure individual record against the RecSys (or, the curator in the case of differential privacy), because the attacker can always apply machine learning techniques to infer the original data. To this end, the differential privacy approach does not help much.

Tang and Wang [Tang and Wang (2015)] proposed an attack against the cryptographic solution by Jeckmans *et al.* [Jeckmans *et al.* (2013)]. Through the attack, an attacker can first recover the private key of the RecSys, and then recover a lot of private information about honest peers. Such an attack can succeed in practice due to two facts. One is that it is always possible for a small group of users to collude, or a single malicious user can set up a few fake user accounts. It might make sense to assume the RecSys to be semi-honest, but it does not to assume all users to be semi-honest. The other is that the underlying homomorphic encryption scheme is vulnerable to key recovery attacks, where a malicious attacker can recover the private key of RecSys by asking the RecSys to decrypt some ciphertexts of his choice. In fact, recent research has shown that most homomorphic encryption schemes suffer from this type of attacks (see [Chenal and Tang (2014, 2015)]). The attack from [Tang and Wang (2015)] shows that we should be very careful to make semi-honest assumptions in recommenders where there is often no rigorous security mechanism to enforce honest behaviours. Moreover, we should also be careful to rely on homomorphic encryption for straightforward security. Special attention is required to mitigate the vulnerabilities of potential key recovery attacks.

Generally speaking, both types of solutions affect the detection of robustness attacks. Cryptographic solutions makes it impossible for the RecSys to run any existing detectors because no plaintext input is directly shared by the users. While for data-obfuscation solutions, the RecSys will still be able to run its detectors but the performance will be (seriously) affected due to the perturbations or added noise.

13.3. Examining some Cryptographic Solutions

To construct cryptographic recommenders, many building blocks can be employed. Some elementary building blocks, such as digital signatures and message authentication codes, can be found in the books [Menezes *et al.* (1996)] and [Katz and Lindell (2007)]. The most widely-used ones are *Homomorphic Encryption (HE)* and *Garbled Circuit (GC)*. We briefly review them below.

HE is a concept that dates back to Rivest, Adleman and Dertouzos [Rivest *et al.* (1978)]. An encryption is fully homomorphic if one can evaluate any circuit on the ciphertexts without decryption (i.e. one can perform any number of additions and multiplications), while an encryption scheme is somewhat homomorphic if one can only evaluate circuits with a limited depth (i.e. only allowing limited number of additions and multiplications). Many somewhat homomorphic encryption schemes have been proposed so far (e.g. BGV scheme [Brakerski *et al.* (2012)] and YASHE scheme [Bos *et al.* (2013)]), and they can be made fully homomorphic via the bootstrapping technique by Gentry [Gentry (2009)]. In practice, the most widely-used library is Simple Encrypted Arithmetic Library (SEAL) from Microsoft [Dowlin *et al.* (2017)], which is an optimized implementation of the YASHE scheme [Bos *et al.* (2013)]. Another popular library is HElib [Halevi and Shoup (2014)], which is an optimized implementation of the BGV scheme [Brakerski *et al.* (2012)]. Formally, a somewhat homomorphic encryption scheme in the asymmetric setting can be described by four algorithms $(\mathsf{Keygen}, \mathsf{Enc}, \mathsf{Dec}, \mathsf{Eval})$.

- $\mathsf{Keygen}(\lambda, L)$: With the security level λ, the multiplication depth L, this algorithm outputs a key tuple (PK, SK, EVK).
- $\mathsf{Enc}(PK, m)$: this algorithm outputs a ciphertext c.
- $\mathsf{Dec}(SK, c)$: this algorithm outputs a plaintext m.
- $\mathsf{Eval}(EVK, circuit, ciphertexts)$: this algorithm outputs a ciphertext, which encrypts the evaluation result of *circuit* on the plaintexts in *ciphertexts*.

When using HE, special attention is required for potential key recovery attacks due to the fact that most HE schemes are vulnerable. It becomes trickier when considering a semi-honest model, as we have shown in the previous section about the work in [Tang and Wang (2015)].

GC is a cryptographic primitive introduced by Yao to solve the millionaires' problem [Yao (1986)], namely to determine who is richer between two millionaires. Intuitively, the idea is as follows. Suppose the millionaires are denoted as A and B, and their wealth is denoted by two integers I_A and I_B (modeled as binary strings) respectively.

(1) Millionaire A can construct a circuit for comparing two integers I_A and I_B, where every gate has two input wires/bits and one output wire/bit. This circuit can be a standard one, known to both millionaires. Intuitively, the circuit can be considered as a tree, where the leaf nodes represent I_A and I_B and the root node is the comparison result (denoted as 0 or 1).
(2) To protect the input privacy, millionaire A selects two random secret keys for every wire (either input or output) and makes a lookup table which encrypts all the output wires. The table is referred to as a GC.

(3) Millionaire A sends the random keys, which represents I_A, and the lookup table to millionaire B.
(4) Millionaire B runs an oblivious transfer (OT) protocol to retrieve the secret keys corresponding to I_B. The OT ptotocol guarantees millionaire B learns nothing more, while Millionaire A learns nothing.
(5) Millionaire B can then evaluate the garbled circuit by decrypting the look-up table using the received keys from millionaire A and the retrieved keys by himself. The decrypted final result will indicate who is richer by a bit 0 or 1.
(6) If necessary, millionaire B can notify millionaire A the result.

13.3.1. *Extra Third-party enabled Solutions*

Under the semi-honest assumption, Nikolaenko *et al.* [Nikolaenko *et al.* (2013)] proposed a GC-based protocol for privacy-preserving matrix factorization, as shown in Figure 13.4. This protocol can be used as a main building block for privacy-preserving solutions. Roughly, the protocol works as follows.

(1) The crypto service provider (CSP) generates a key pair (pk, sk) for a somewhat homomorphic encryption scheme.
(2) Each user encrypts his rating vector with pk and send the ciphertexts to the RecSys. Note that the data transmission should be through a secure channel with confidentiality protection, in order to hide the encrypted ratings from the CSP.
(3) Based on the homomorphic property of the underlying encryption scheme, the RecSys re-randomizes the encrypted rating vectors for all users. Then it sends the re-randomized ciphertexts to the CSP.
(4) The CSP decrypts the re-randomized ciphertexts and obtains the re-randomized rating vectors for every user. Based on the decrypted values, the CSP prepares a garbled circuit for matrix factorization, where the input wires stand for the randomness used by the RecSys in Step 3.
(5) Referring to the brief introduction to GC above, the RecSys can retrieve the corresponding secret keys for its randomness and evaluate the garbled circuit afterwards. At the end, the RecSys obtains the items' feature matrix $\mathcal{V}$ (referring to Section 13.1.1).

Intuitively, the encryption in Step 2 prevents information leakage to the RecSys, while the re-randomization by the RecSys in Step 3 prevents information leakage to the CSP. However, note that the above protocol allows the RecSys to learn $\mathcal{V}$ in plaintext, which may leak information about users' rating values. By applying additional randomization procedures on the values in the above protocol, Nikolaenko *et al.* [Nikolaenko *et al.* (2013)] constructed an interactive solution to allow every user to learn his predictions while neither the RecSys nor CSP learns anything. In addition, the authors also gave a sketch on how to defeat malicious adversaries.

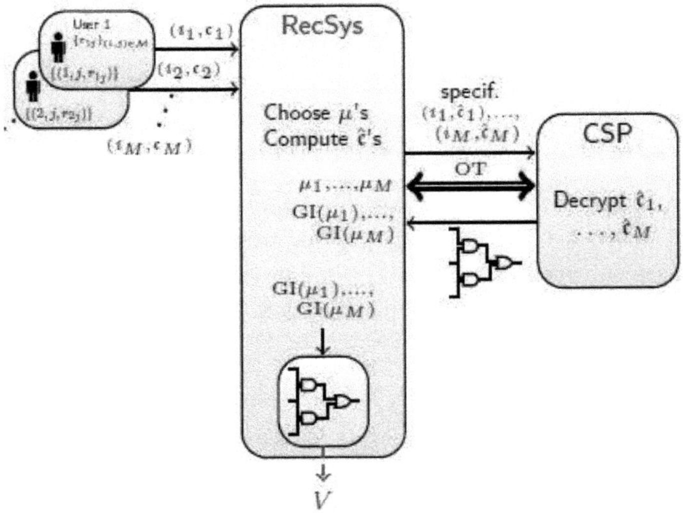

Fig. 13.4. Solution from [Nikolaenko *et al.* (2013)].

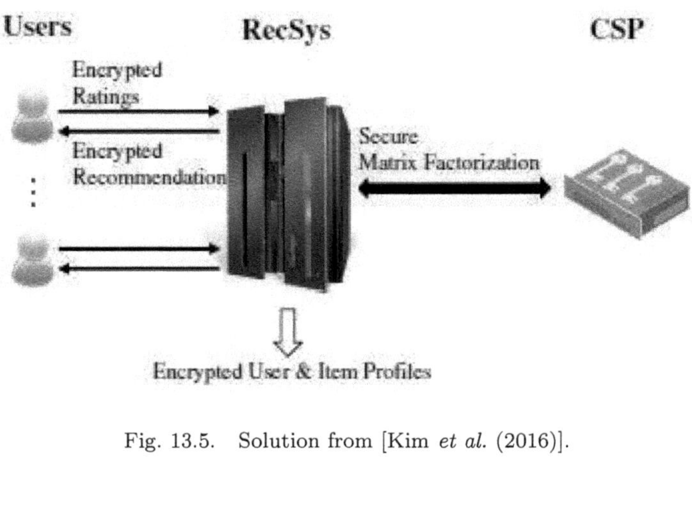

Fig. 13.5. Solution from [Kim *et al.* (2016)].

Kim *et al.* [Kim *et al.* (2016)] proposed a HE-based protocol for privacy-preserving matrix factorization, as shown in Figure 13.5. Roughly, the protocol works as follows.

(1) The crypto service provider (CSP) generates a key pair (pk_1, sk_1) for a somewhat homomorphic encryption scheme and another key pair (pk_2, sk_2) for another somewhat homomorphic encryption scheme (this scheme only needs to support additions).
(2) Each user encrypts his rating vector with pk_2 and send the ciphertexts to the RecSys. Note that the transmission should be through a secure channel with confidentiality protection, in order to hide the encrypted ratings from the CSP.
(3) Based on the homomorphic property of the underlying encryption scheme, the RecSys re-randomizes the encrypted rating vectors for all users. Then it sends the randomized ciphertexts to the CSP.
(4) The CSP uses sk_2 to decrypt the re-randomized ciphertexts and obtains the randomized rating vectors for every user. It encrypts the randomized rating vectors with pk_1 and send them back to RecSys.
(5) The RecSys and the CSP interactively execute gradient descent algorithm (referring to Section 13.1.1) until some stopping conditions are met. At the end, the RecSys obtains the user and item feature matrices.

Similar to the GC-based approach, privacy-preserving solutions can be constructed based on the above protocol to allow every user to learn his predictions while neither RecSys nor CSP learns anything.

It is worth noting that, besides introducing the CSP, both solutions from [Nikolaenko *et al.* (2013)] and [Kim *et al.* (2016)] applied additional innovative techniques to improve the computational efficiency. The authors from [Kim *et al.* (2016)] performed a comparison test based on a single machine with 3.4GHz 6-cores 64GB RAM. We directly duplicate their results in Figure 13.6 and Figure 13.7. It is clear that the HE-based solution is more efficient than the GC-based solution with respect to both computation and communication, regardless the fact that the HE-based solution requires more rounds of communications due to the interactive execution of gradient descent algorithm.

Despite the security claims and the possible enhancements, we want to point out that the above solutions and alike have many drawbacks in reality. First of all, assuming two or more non-colluding third parties is not as realistic as it is claimed to be, even if incentives can be given to these

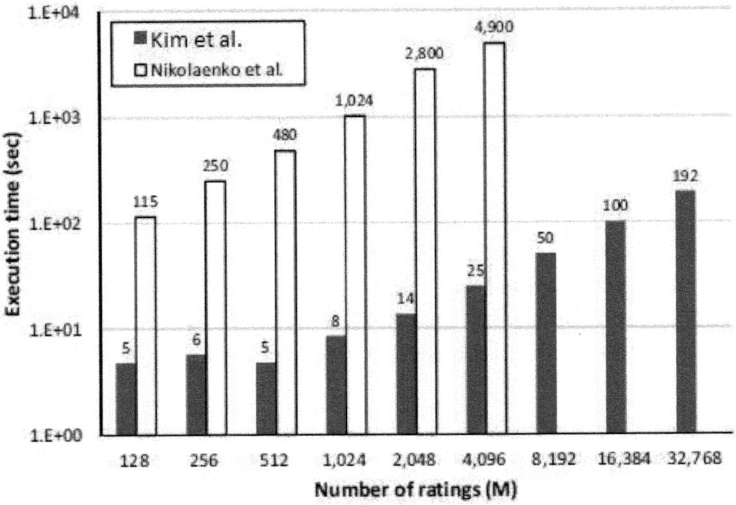

Fig. 13.6. Computational Complexity.

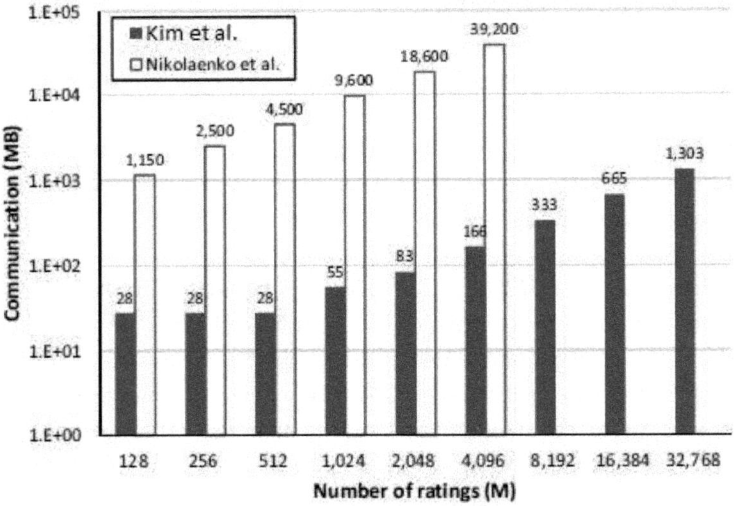

Fig. 13.7. Communication Complexity.

entities for them to follow the protocol specification. One obvious risk is that one server might be compromised, say the CSP. Suppose the attacker cannot compromise the other server, then the attacker can just release the private key to the public (much like the *leaks stories nowadays), then the RecSys will be able to decrypt everything and gain all the secret information about the users. Secondly, semi-honest model is too weak to be justified. One risk along this line is that a "curious" RecSys or CSP may slightly deviate from the protocol specification in order to get some private information from a targeted user. To this end, there are many tricks, e.g. use special randomness or "collude" with some users (note that recommender is often an open system without strong authentication, particularly true for privacy-preserving ones). Such attack can be carried out without affecting the utility of anyone, because recommender algorithms are often non-deterministic so that there is no single right output for a user. Using cryptographic techniques to prevent malicious attackers is notoriously expensive due to the scale of recommenders. Thirdly, these solutions make it impossible for detecting robustness attacks. Normally, the RecSys should clean the input data and exclude the abnormal ones before computing recommendations. With privacy-preserving solutions deployed, in order to perform the tasks, new cryptographic protocols need to be proposed and deployed, which will further downgrade the efficiency and increase information leakage. Finally, some subtle technical considerations might make the situation harder, e.g. the reusability issue for garbled circuit [Goldwasser *et al.* (2013)] and the key recovery attacks against homomorphic encryption schemes [Chenal and Tang (2014, 2015)].

13.3.2. *Auxiliary Information enabled Solution*

Different from assuming extra third parties, Tang and Wang [Tang and Wang (2015, 2016)] proposed privacy-preserving friendship-based recommender systems, which leverage the background social network information among users to reduce the computational costs as shown in Figure 13.8. The intuition comes from the folklore that similar users (e.g. friends) share similar tastes. Following the principle of neighborhood-based approach, only considering the inputs from friends can already produce reasonably good predictions. However, the efficiency gain does not come for free. When the user population is small, the information leakage from the output can be significant in scenario $\mathcal{S}_4$ in Table 13.3. Even friends trust each other more than strangers, they may not want to share everything with each other. To reduce the risk in this direction, Tang and Wang proposed the idea of introducing randomly chosen strangers in the computation.

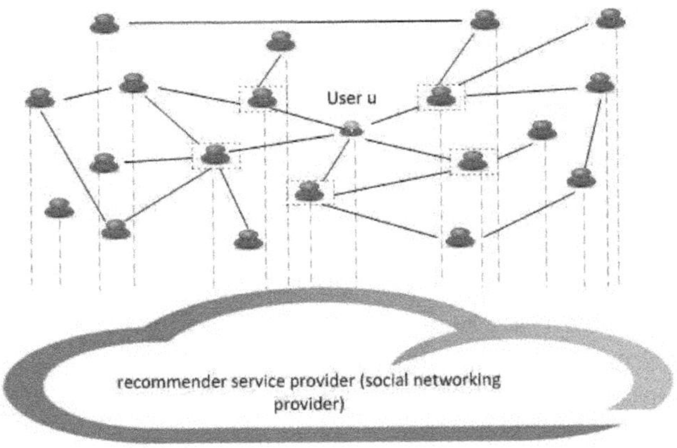

Fig. 13.8. System Structure in the View of User u.

Two types of protocols are proposed in [Tang and Wang (2015, 2016)], and the recommender algorithm is an adapted one for the sake of efficiency from the neighborhood-based algorithm in Section 13.1.1. One is single prediction protocol which allows a user to test whether or not a predicted rating value is larger than a threshold. The other is Top-k recommendation, which allows a user to learn the k unrated items that have the highest predicted rating values. With a semi-honest RecSys, the Top-k protocol works as follows. Suppose the user is Alice.

(1) Alice generates a key pair (pk, sk) for a somewhat homomorphic encryption scheme.
(2) Alice sends encrypted weights to the RecSys, which will forward the values to Alice's friends and some randomly-chosen strangers. Different from those in Section 13.3.1, encryption is done with respect to Alice's public key.
(3) Each friend or stranger computes his contribution (encrypted under Alice's public key pk) to the final predictions. Note that a tailored neighborhood-based recommender algorithm is used, and it allows a friend or stranger to compute his contribution without knowing Alice's rating values. The contributions are sent to the RecSys.
(4) Based on the received values and also some other encrypted values from Alice, the RecSys can compute the predictions for all items in encrypted form. Note that each prediction is in the form of an encrypted nominator and denominator pair.

(5) The RecSys and Alice can then interactively run a ranking protocol to rank the predictions in encrypted form.
(6) Finally, the RecSys can send the encrypted Top-k predictions to Alice.

Tang and Wang [Tang and Wang (2015, 2016)] have done experiments based on simulated and real datasets to evaluate the accuracy of the solutions. It is shown that the proposed solutions only require $\frac{1}{10}$ of the user population in contrast to a standard neighborhood algorithm. Based on an Intel(R) Core(TM) i7-5600U CPU 2.60GHz PC, the computational costs for the two types of protocols are shown in Table 13.5. The experiment assumes a user population of 80, while the MovieLens dataset has a user population 1000. For the Top-k protocol, the main source of the computational costs for the RecSys and the user is from the ranking of the unrated items. The *Top-k (no ranking)* row of the table indicates the costs when the encrypted predictions are sent to Alice directly. In this case, more information is leaked to Alice. We note that the MF-based solutions (e.g. [Nikolaenko *et al.* (2013); Kim *et al.* (2016)] described in the previous subsection) face the same situation, in the sense that similar costs will be incurred if the users are only allowed to receive the Top-k predictions. Such costs are not included in Figure 13.6.

Table 13.5. Execution time of Tang-Wang Solutions (in seconds).

	Friend	Stranger	RecSys	Alice
Single	1.12	1.00	0.72	74.17
Top-k	141.22	140.55	1726446	236424
Top-k (no ranking)	141.22	140.55	1562	141.58

Regarding these solutions and alike, one concern is that they do not generalize to broader settings when appropriate social information is unavailable. In addition, friendship information may be sensitive in the first place, so that these solutions may have their inherent privacy drawbacks. On the positive side, unless *membership privacy* (shown in Table 13.3) is strongly enforced, friendship information can always be collected in practical recommender systems and such information is also important to fulfill other requirements such as robustness. Note also that friendship can be defined based on a variety of metrics, not necessarily on the real-world social network links. We conjecture that many practically deployed recommenders may have already heavily made use of such information. The other concern is about the scalability. MF-based solutions has an advantage, because the learned model can serve all the users while, in neighborhood-based systems, predictions need to be computed for each user separately (i.e. the protocol needs to be executed once for each user). This advantage is more obvious when the dataset or user population is very large. However, we need to keep in mind that the advantage comes with the extra third party CSP. Without such additional party, privacy-preserving matrix factorization will be computationally infeasible (to the best of our knowledge). When the user population is relatively small and friendship information can be easily constructed, the above friendship-based solutions will be a better fit. Even in this case, standard privacy-preserving neighborhood-based solutions can still have a scalability problem because they need to process data from all users. Concerning trust assumptions, the friendship-based solutions have further potentials to be exploited. For example, friendship information can be leveraged for users to validate each other's public keys, similar to the web of trust paradigm of PGP[6]. To some extent, this is superior to the somehow artificial CSP. In addition, by asking each participant to directly compute his contributions instead of sending his rating values will also reduce the risk of privacy breach in case the some user's private key is compromised.

As a remark, the work of Tang and Wang [Tang and Wang (2015, 2016)] is related to the concept of trust-aware collaborative filtering recommender systems proposed by Massa and Avesani [Massa and Avesani (2004, 2007)], where trust metrics between users are proposed and taken into account when computing recommendations. Different from [Tang and Wang (2015, 2016)], no rigorous security analysis has been provided in [Massa and Avesani (2004, 2007)]. O'Donovan and Smyth [O'Donovan and Smyth (2005)] showed that trust-based defence can be defeated if the attacker can access the trust values. This implies that the confidentiality of these values need to be guaranteed. Cheng and Hurley [Cheng and Hurley (2009)] proposed the concept of informed model-based attacks against trust-aware solutions, where the attacker is given the model parameters, and showed that such attacks are more efficient than others in the decentralized privacy-preserving recommender system by Canny [Canny (2002a)]. They demonstrated the trade-off between privacy and robustness. We note that these attacks do not apply to the solutions from [Tang and Wang (2015, 2016)] due to the fact that similarity values (equivalent to the trust metrics in [Massa and Avesani (2004, 2007)]) are kept secret from all the players.

[6]https://en.wikipedia.org/wiki/Pretty_Good_Privacy

13.4. Examining some Differentially Private Solutions

Differential privacy [Dwork (2006); Dwork *et al.* (2006)] is a rigorous privacy notion, proposed to model inference attacks in releasing statistical databases. In a short period of time, it has been applied to a wide range of application scenarios, including recommenders.

Definition 13.1. A random algorithm $\mathcal{M}$ is (ϵ, σ)-differentially private if for all $\mathcal{O} \subset Range(\mathcal{M})$ and for any $(\mathcal{D}_0, \mathcal{D}_1)$ which only differs on one single record, i.e. $||\mathcal{D}_0 - \mathcal{D}_1|| \leq 1$, the following inequality holds.

$$Pr[\mathcal{M}(\mathcal{D}_0) \in \mathcal{O}] \leq exp(\epsilon) Pr[(\mathcal{M}(\mathcal{D}_1) \in \mathcal{O}] + \sigma$$

We say $\mathcal{M}$ guarantees ϵ-differential privacy if $\sigma = 0$.

The parameter ϵ measures the privacy loss, by bounding the difference of algorithm $\mathcal{M}$'s output for any $(\mathcal{D}_0, \mathcal{D}_1)$. Lower ϵ indicates stronger privacy protection, with $\epsilon = 0$ for perfect privacy.

Laplace Mechanism [Dwork *et al.* (2006)] is a common approach to realise differential privacy guarantee by calibrating additive noise sampled from Laplace distribution: $\mathcal{M}(\mathcal{D}) \triangleq f(\mathcal{D}) + Lap(0, \frac{\Delta\mathcal{F}}{\epsilon})$. The $\Delta\mathcal{F}$ stands for the L_1-sensitivity, namely $\Delta\mathcal{F} = \max_{(\mathcal{D}_0, \mathcal{D}_1)} ||f(\mathcal{D}_0) - f(\mathcal{D}_1)||_1$. Recently, *sampling* from the posterior distribution of a Bayesian model with bounded log-likelihood has been proven to be differentially private [Wang *et al.* (2015)]. It is essentially an *exponential mechanism* [McSherry and Talwar (2007)]. Formally, suppose that we have a dataset of $\mathcal{L}$ i.i.d examples $\mathcal{X} = \{x_i\}_{i=1}^{\mathcal{L}}$ which we model using a conditional probability distribution $p(x|\theta)$ where θ is a parameter vector, with a prior distribution $p(\theta)$. If $p(x|\theta)$ satisfies $sup_{x \in \mathcal{X}, \theta \in \Theta} |\log p(x|\theta)| \leq B$, then releasing one sample from the posterior distribution $p(\theta|\mathcal{X})$ with any prior $p(\theta)$ preserves $4B$-differential privacy. Alternatively, ϵ-differential privacy can be achieved by simply re-scaling the log-posterior distribution by a factor of $\frac{\epsilon}{4B}$, under the regularity conditions where asymptotic normality (or, the Bernstein-von Mises theorem) holds.

13.4.1. *Global Differential Privacy Approach*

Following the blueprint of differential privacy paradigm, Berlioz *et al.* [Berlioz *et al.* (2015)] compared three different ways of noise calibration for MF-based recommender systems. As shown in Figure 13.9, noise can

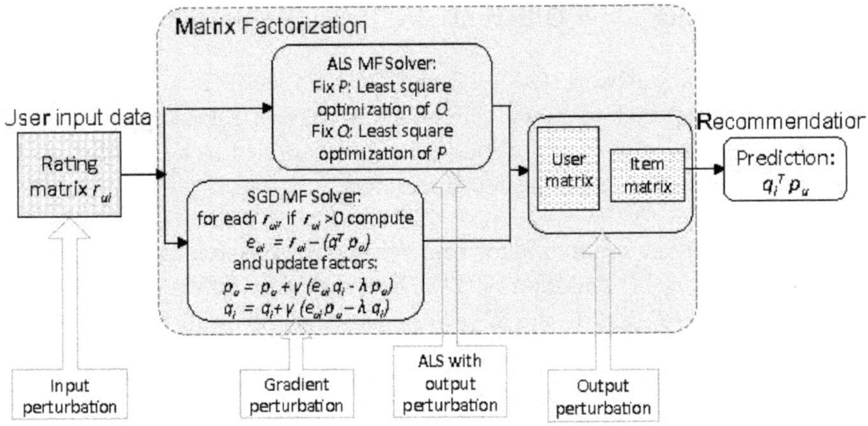

Fig. 13.9. Various Approaches to Achieve Differential Privacy [Berlioz *et al.* (2015)].

be calibrated in the preparation of rating matrix (referred to as input perturbation), in the stochastic gradient descent training process, and in the output. The experimental results from [Berlioz *et al.* (2015)] showed that the input perturbation approach preserves the most accuracy at the same level of privacy guarantee.

Instead of directly calibrating noise, Wang, Fienberg, and Smola [Wang *et al.* (2015)] proposed the concept of differential privacy via posterior sampling for MF-based recommenders, where differential privacy is achieved based on scaling the gradient in the model training. They showed that this approach outperforms the standard differential privacy approach (e.g. direct noise calibration) with respect to Bayesian learning. One concern of this approach is that the distance between the distribution where the samples are from and the true posterior distribution might not be the same or even close, which means the privacy guarantee might not be truly what is claimed to be. Fortunately, there are some theoretical results for bounding the distance (e.g. [Sato and Nakagawa (2014)]), although additional efforts are required to interpret them in individual application scenarios. Wang and Tang [Wang and Tang (2017)] evaluated the two methods (noise calibration and posterior sampling) with respect to a probabilistic neighborhood-based recommender algorithm from [Wang and Tang (2016)]. Their experimental results show that both methods can obtain quite good accuracy when $\epsilon \approx 1$; while, for the same privacy guarantee, posterior sampling method seems to result in higher accuracy. In general, they showed that differentially private MF solutions are more accurate when privacy loss is large while probabilistic neighborhood-based solutions are better when we want to reduce the privacy loss to a moderately small range.

Guerraoui *et al.* [Guerraoui *et al.* (2015)] proposed the concept of distance-based differential privacy for neighborhood-based recommender algorithms. With this approach, the curator (i.e. RecSys) first collects the rating vectors for all users, and then builds an *AlterEgo* profile for every user (with a certain probability p a rating item will be replaced with a random item, otherwise it will be replaced with a similar item), and finally any neighborhood-based algorithm can be applied to compute the final recommendations. Their experimental results are evaluated with respect to classification accuracy metrics, so that it is difficult to compare them with other solutions, e.g. that from [Wang and Tang (2017)].

13.4.2. *Local Differential Privacy Approach*

One obvious drawback of many differentially private recommenders is the assumption that the curator accumulates all users' input and then enforces the privacy. This process forces all users to fully trust the curator, essentially the RecSys. In order to relax this strong trust requirement, researchers have investigated the concept of *local differential privacy*. Notably, Banerjee, Hegde, and Massoulie [Banerjee *et al.* (2015)] investigated the local differential privacy concept in the context of recommender systems from an information theoretic perspective. They established lower bounds for sample complexity (i.e. local community size) in order to cluster items (i.e. learn the item similarities). Shen and Jin [Shen and Jin (2014, 2016)] proposed the concept of instance-based approach for differentially private MF-based recommenders. With this approach, every user randomizes their rating vector before sending it to the RecSys, and recommendations are computed based on the perturbed data from all users. In these works, performances are evaluated against self defined measures only, therefore it is unclear how it is compared to other solutions.

Clearly the local differential privacy concept mitigates the aforementioned drawback (i.e. reliance on a fully trusted curator), but it may bring undesirable effects on recommendation accuracy due to the fact that many privacy-savvy users will add a large amount of noise in their input. The heavily perturbed inputs will not only downgrade privacy-savvy users' accuracy (acceptable due to the privacy-utility tradeoff), but also downgrade the accuracy of those who have added very little noise in order to get accurate recommendations. In addition, local differential privacy adds a formidable barrier for fighting against robustness attacks due to the added noises from all users.

13.4.3. *Comparison to Cryptographic Solutions*

Referring to the scenarios in Table 13.3, cryptographic and data-obfuscation solutions target different scenarios. Most cryptographic solutions only target the scenario $\mathcal{S}_2$, namely data leakage in the computations. They usually assume a semi-honest RecSys, which is not trusted to see users' plaintext inputs. On the other hand, most data-obfuscation solutions only target the scenario $\mathcal{S}_3$, namely membership information leakage in the outputs. They often assume a fully trusted RecSys, which collects inputs from all users and then adds privacy protection afterwards. Naturally, these solutions might also provide some limited privacy protection for the scenario $\mathcal{S}_4$, in the sense that not knowing whether a user is involved or not somehow implies not knowing the input of this user. Taking differentially-private recommender as an example, and assume $\epsilon = 1$ for record-level privacy. For any honest user who is involved in the protocol, an attacker can only confirm this fact with the probability $\rho = \frac{e^{-\epsilon}}{1+e^{-\epsilon}}$. Note that ρ is not negligible in the cryptographic sense. There is no guarantee that with this ρ probability, how much the attacker can learn Alice's input.

In Table 13.6, we briefly summarize the comparison. Note that it captures the general properties of most solutions in these two categories, while exceptions do exist.

Table 13.6. Comparison Summary.

	Cryptographic Solutions	**Data-obfuscation Solutions**
Types of Attacker	RecSys and End Users	End Users
Privacy Coverage	$\mathcal{S}_2$ (leakage from computation)	$\mathcal{S}_3$ and $\mathcal{S}_4$ (leakage and inference from outputs)
Trust Setup	Semi-honest RecSys	Honest RecSys
Security Guarantee	Computational Assumptions	Unconditional

13.5. Future Directions

Based on our previous analysis, we can conclude that privacy is an extreme important yet tricky requirement for recommender systems. Its entanglement with other (equally important) requirements challenges the assumptions made in most existing solutions so far. In the following, we outline our vision for a comprehensive solution which addresses all three requirements, and identify some promising technical directions.

Analog to employing *collaborative* filtering approach for recommendation accuracy, we envision a collaborative and crowdsourced approach for achieving robustness and transparency. For consistency/robustness reasons, users should commit their encrypted inputs to a distributed ledger [Narayanan *et al.* (2016)], and these inputs will be fed to the privacy-preserving recommender protocols which are transparency and robustness friendly by design. In an off-line manner, any group of users can run lightweight secure multi-party computation protocols to detect robustness attacks and test transparency properties, and then log the results on the ledger. The distributed ledger gives control to users and glue protocols together. The overall solution is illustrated in Figure 13.10.

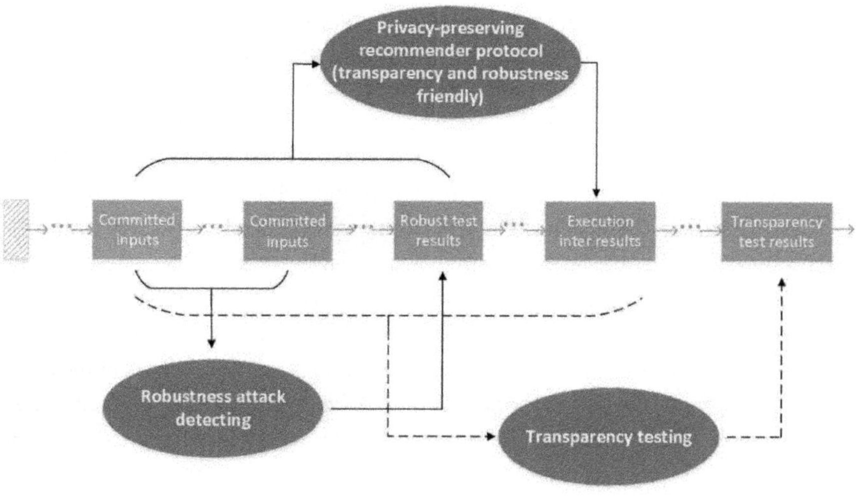

Fig. 13.10. Integrated Solution.

Despite the enormous innovations on the way, efficiency has been the main bottleneck for rigorous cryptographic solutions. Data obfuscation approach is efficient, however it relies on a very strong trust assumption on the RecSys. Obviously, the desire to address three requirements instead of only one creates additional barriers for achieving efficiency. In addition to continue improving existing building blocks, we highlight some promising directions to realise an efficient solution.

As to fulfilling the privacy requirement, we can leverage the inherent uncertainty characteristics inside a recommender and reduce the input size of the recommender algorithm. The spirit of leveraging uncertainty is inspired by the works by Aggarwal [Aggarwal (2008)] and Tang and Wang [Tang and Wang (2015, 2016)]. In the former, Aggarwal proposed an uncertain version of k-anonymity model based on the inherent uncertainty property of underlying data inputs in data mining services; while in the latter, Tang and Wang proposed to reduce the privacy leakage by randomly sampling some users in the computation. The advantage of this method is that it does not need to heavily manipulate the data and has very little effect on the utility. In some sense, this is a relaxation of the worst case scenario of differential privacy approach, where the attacker is assumed to know every user's input. Reducing the input size is a crucial direction to improve efficiency, due to the fact that the complexity is linear to the input size. As demonstrated in [Tang and Wang (2015, 2016)], a careful implementation will not affect the recommendation accuracy. It is an interesting future work to integrate the above methods for matrix factorization based recommenders. For such recommenders, an additional challenge is to satisfy RecSys's privacy requirement (i.e. prevent unnecessary information leakage of model information to individual users). We foresee the differential privacy concept may play an important role in tackling this challenge.

As to fulfilling the robustness requirement, we need to design privacy-aware hierarchical robustness testing protocols. Existing robustness detection solutions assume the users' inputs are in plaintext, and even in this case they can be computation-intensive if the user population is large. With privacy protection for the inputs (e.g. encryption), it will be unrealistic to directly transform these solutions to their privacy-preserving variants. Alternatively, it is a promising direction to investigate statistical detectors with two new properties. One property is that they can run among smaller user groups and their outputs can be aggregated to serve as the global output. This enables efficient cryptographic variant of statistical detectors to be executed in groups in parallel. The other property is privacy-preserving, which means the output from individual group should leak a negligible amount of information at both group and individual level.

As to fulfilling the transparency requirement, we can realise users' expectations via computed consensus. From our previous discussion, it is clear that it is neither desirable nor enough to publish the algorithms and parameters in order to guarantee transparency. We foresee a two-step approach. First, transparency indicators should be defined for the recommender algorithms. To this end, some valuable references are the works by Zliobaite [Zliobaite (2017)] and Datta, Sen, and Zick [Datta *et al.* (2016)]. Then, any group of users can run some cryptographic protocols to evaluate the indicators without disclosing their private data, and vote on the transparency property of the system.

Acknowledgements

When this chapter was prepared, the author was supported by a CORE (junior track) grant from the National Research Fund, Luxembourg. The author would like to thank Prof. Jiuyong Li from University of South Australia for his comments.

References

Adomavicius, G. and Tuzhilin, A. (2005). Toward the next generation of recommender systems: A survey of the state-of-the-art and possible extensions, *IEEE Trans. Knowl. Data Eng.* **17**, 6, pp. 734–749.

Aggarwal, C. C. (2008). On unifying privacy and uncertain data models, in *Proceedings of the 24th International Conference on Data Engineering, ICDE 2008*, pp. 386–395.

Ahn, J. and Amatriain, X. (2010). Towards fully distributed and privacy-preserving recommendations via expert collaborative filtering and restful linked data, in *2010 IEEE/WIC/ACM International Conference on Web Intelligence, WI 2010*, pp. 66–73.

Aïmeur, E., Brassard, G., Fernandez, J. M. and Onana, F. S. M. (2008). Alambic: a privacy-preserving recommender system for electronic commerce, *Int. J. Inf. Secur.* **7**, pp. 307–334.

Banerjee, S., Hegde, N. and Massoulie, L. (2015). The price of privacy in untrusted recommender systems, *IEEE Journal of Selected Topics in Signal Processing* **9**, pp. 1319–1331.

Basu, A., Vaidya, J., Kikuchi, H., Dimitrakos, T. and Nair, S. K. (2012). Privacy preserving collaborative filtering for saas enabling paas clouds, *Journal of Cloud Computing: Advances, Systems and Applications* **1**, 1, pp. 1–14.

Berlioz, A., Friedman, A., Kaafar, M. A., Boreli, R. and Berkovsky, S. (2015). Applying differential privacy to matrix factorization, in *Proceedings of the 9th ACM Conference on Recommender Systems*, pp. 107–114.

Beye, M., Jeckmans, A., Erkin, Z., Tang, Q., Hartel, P. and Lagendijk, I. (2013). *Social Media Retrieval*, chap. Privacy in Recommender systems (Springer), pp. 263–281.

Bos, J. W., Lauter, K. E., Loftus, J. and Naehrig, M. (2013). Improved security for a ring-based fully homomorphic encryption scheme, in *Cryptography and Coding – 14th IMA International Conference*, pp. 45–64.

Brakerski, Z., Gentry, C. and Vaikuntanathan, V. (2012). (leveled) fully homomorphic encryption without bootstrapping, in *Proceedings of the 3rd Innovations in Theoretical Computer Science Conference*, pp. 309–325.

Calandrino, J. A., Kilzer, A., Narayanan, A., Felten, E. W. and Shmatikov, V. (2011). You Might Also Like: Privacy Risks of Collaborative Filtering, in *32nd IEEE Symposium on Security and Privacy, S&P 2011*, pp. 231–246.

Canny, J. (2002a). Collaborative filtering with privacy via factor analysis, in *Proceedings of the 25th Annual International ACM SIGIR Conference on Research and Development in Information Retrieval*, pp. 238–245.

Canny, J. F. (2002b). Collaborative filtering with privacy, in *IEEE Symposium on Security and Privacy*, pp. 45–57.

Chenal, M. and Tang, Q. (2014). On key recovery attacks against existing somewhat homomorphic encryption schemes, in *Progress in Cryptology – LATINCRYPT 2014*, pp. 239–258.

Chenal, M. and Tang, Q. (2015). Key recovery attacks against ntru-based somewhat homomorphic encryption schemes, in *Information Security - 18th International Conference, ISC 2015*, pp. 397–418.

Cheng, Z. and Hurley, N. (2009). Trading robustness for privacy in decentralized recommender systems, in *Proceedings of the Twenty-First Conference on Innovative Applications of Artificial Intelligence*, pp. 3–15.

Datta, A., Sen, S. and Zick, Y. (2016). Algorithmic transparency via quantitative input influence: Theory and experiments with learning systems, in *37th IEEE Symposium on Security and Privacy, S&P 2011*, pp. 598–617.

Dowlin, N., Gilad-Bachrach, R., Laine, K., Lauter, K. E., Naehrig, M. and Wernsing, J. (2017). Manual for using homomorphic encryption for bioinformatics, *Proceedings of the IEEE* **105**, 3, pp. 552–567.

Dwork, C. (2006). Differential privacy, in M. Bugliesi, B. Preneel, V. Sassone and I. Wegener (eds.), *Automata, Languages and Programming, 33rd International Colloquium, ICALP 2006, LNCS*, Vol. 4052 (Springer), pp. 1–12.

Dwork, C., McSherry, F., Nissim, K. and Smith, A. (2006). Calibrating noise to sensitivity in private data analysis, in S. Halevi and T. Rabin (eds.), *Theory of Cryptography, Third Theory of Cryptography Conference, LNCS*, Vol. 3876 (Springer), pp. 265–284.

Friedman, A., Knijnenburg, B. P., Vanhecke, K., Martens, L. and Berkovsky, S. (2015). *Recommender Systems Handbook*, chap. Privacy Aspects of Recommender Systems (Springer), pp. 649–688.

Gentry, C. (2009). Fully homomorphic encryption using ideal lattices, in *Proceedings of the Forty-first Annual ACM Symposium on Theory of Computing*, pp. 169–178.

Goldwasser, S., Kalai, Y. T., Popa, R. A., Vaikuntanathan, V. and Zeldovich, N. (2013). Reusable garbled circuits and succinct functional encryption, in *Symposium on Theory of Computing Conference, STOC'13*, pp. 555–564.

Goodman, B. and Flaxman, S. (2016). European union regulations on algorithmic decision-making and a right to explanation, https://arxiv.org/abs/1606.08813.

Guerraoui, R., Kermarrec, A., Patra, R. and Taziki, M. (2015). D2P:distance-based differential privacy in recommenders, *PVLDB* **8**, 8, pp. 862–873.

Halevi, S. and Shoup, V. (2014). Algorithms in helib, in *Advances in Cryptology – CRYPTO 2014*, pp. 554–571.

Han, S., Ng, W. K. and Yu, P. S. (2009). Privacy-preserving singular value decomposition, in Y. E. Ioannidis, D. L. Lee and R. T. Ng (eds.), *Proceedings of the 25th International Conference on Data Engineering* (IEEE), pp. 1267–1270.

Huang, L., Joseph, A. D., Nelson, B., Rubinstein, B. I. P. and Tygar, J. D. (2011). Adversarial machine learning, in *Proceedings of the 4th ACM Workshop on Security and Artificial Intelligence, AISec 2011, Chicago, IL, USA, October 21, 2011*, pp. 43–58.

Ioannidis, S., Montanari, A., Weinsberg, U., Bhagat, S., Fawaz, N. and Taft, N. (2014). Privacy tradeoffs in predictive analytics, in *The 2014 ACM International Conference on Measurement and Modeling of Computer Systems*, pp. 57–69.

Jeckmans, A., Peter, A. and Hartel, P. H. (2013). Efficient privacy-enhanced familiarity-based recommender system, in J. Crampton, S. Jajodia and K. Mayes (eds.), *Computer Security - ESORICS 2013 - 18th European Symposium on Research in Computer Security, LNCS*, Vol. 8134 (Springer), pp. 400–417.

Jeckmans, A., Tang, Q. and Hartel, P. (2012). Privacy-preserving collaborative filtering based on horizontally partitioned dataset, in *2012 International Symposium on Security in Collaboration Technologies and Systems (CTS 2012)*, pp. 439–446.

Katz, J. and Lindell, Y. (2007). *Introduction to Modern Cryptography (Chapman & Hall/Crc Cryptography and Network Security Series)* (Chapman & Hall/CRC).

Kim, S., Kim, J., Koo, D., Kim, Y., Yoon, H. and Shin, J. (2016). Efficient privacy-preserving matrix factorization via fully homomorphic encryption: Extended abstract, in *Proceedings of the 11th ACM on Asia Conference on Computer and Communications Security*, pp. 617–628.

Koren, Y. (2009). The bellkor solution to the netflix grand prize, http://www.netflixprize.com/assets/.

Lam, S. K., Frankowski, D. and Riedl, J. (2006). Do you trust your recommendations? an exploration of security and privacy issues in recommender systems, in G. Muller (ed.), *Emerging Trends in Information and Communication Security, LNCS*, Vol. 3995 (Springer), pp. 14–29.

Liu, Z., Wang, Y. X. and Smola, A. (2015). Fast differentially private matrix factorization, in *Proceedings of the 9th ACM Conference on Recommender Systems* (ACM), pp. 171–178.

Massa, P. and Avesani, P. (2004). Trust-aware collaborative filtering for recommender systems, in R. Meersman and Z. Tari (eds.), *On the Move to Meaningful Internet Systems 2004: CoopIS, DOA, and ODBASE, OTM Confederated International Conferences, Agia Napa, Cyprus, October 25-29, 2004, Proceedings, Part I, LNCS*, Vol. 3290 (Springer), pp. 492–508.

Massa, P. and Avesani, P. (2007). Trust-aware recommender systems, in J. A. Konstan, J. Riedl and B. Smyth (eds.), *Proceedings of the 2007 ACM Conference on Recommender Systems* (ACM), pp. 17–24.

McSherry, F. and Mironov, I. (2009). Differentially private recommender systems: building privacy into the Netflix prize contenders, in *Proceedings of the 15th ACM SIGKDD international conference on Knowledge discovery and data mining*, pp. 627–636.

McSherry, F. and Talwar, K. (2007). Mechanism design via differential privacy, in *Proceedings of the 48th Annual IEEE Symposium on Foundations of Computer Science*, pp. 94–103.

Menezes, A. J., Vanstone, S. A. and Oorschot, P. C. V. (1996). *Handbook of Applied Cryptography*, 1st edn. (CRC Press, Inc.).

Narayanan, A., Bonneau, J., Felten, E. W., Miller, A. and Goldfeder, S. (2016). *Bitcoin and Cryptocurrency Technologies - A Comprehensive Introduction* (Princeton University Press).

Narayanan, A. and Shmatikov, V. (2008). Robust de-anonymization of large sparse datasets, in *2008 IEEE Symposium on Security and Privacy (S&P 2008)*, pp. 111–125.

Newman, N. (2014). The costs of lost privacy: Consumer harm and rising economic inequality in the age of google, *William Mitchell Law Review* **40**, 2.

Nikolaenko, V., Ioannidis, S., Weinsberg, U., Joye, M., Taft, N. and Boneh, D. (2013). Privacy-preserving matrix factorization, in *Proceedings of the 2013 ACM SIGSAC Conference on Computer & Communications Security*, pp. 801–812.

O'Donovan, J. and Smyth, B. (2005). Trust in recommender systems, in *Proceedings of the 10th International Conference on Intelligent User Interfaces*, pp. 167–174.

Polat, H. and Du, W. (2003). Privacy-preserving collaborative filtering using randomized perturbation techniques, in *Proceedings of the Third IEEE International Conference on Data Mining*, pp. 625–628.

Polat, H. and Du, W. (2005a). Privacy-preserving collaborative filtering on vertically partitioned data, in *Knowledge Discovery in Databases: PKDD 2005, 9th European Conference on Principles and Practice of Knowledge Discovery in Databases*, pp. 651–658.

Polat, H. and Du, W. (2005b). Svd-based collaborative filtering with privacy, in *Proceedings of the 2005 ACM Symposium on Applied Computing*, pp. 791–795.

Polat, H. and Du, W. (2006). Achieving private recommendations using randomized response techniques, in *Proceedings of the 10th Pacific-Asia Conference on Advances in Knowledge Discovery and Data Mining*, pp. 637–646.

Ramakrishnan, N., Keller, B., Mirza, B. and Grama, A. Y. (2001). Privacy risks in recommender systems, *Internet Computing, IEEE* **5**, pp. 54–63.

Rivest, R., Adleman, L. and Dertouzos, M. (1978). On data banks and privacy homomorphisms, *Foundations of Secure Computation*, pp. 169–179.

Sato, I. and Nakagawa, H. (2014). Approximation analysis of stochastic gradient langevin dynamics by using fokker-planck equation and itô process, in *Proceedings of the 31st International Conference on International Conference on Machine Learning*, pp. 982–990.

Shani, G. and Gunawardana, A. (2011). Evaluating recommendation systems, in F. Ricci, L. Rokach, B. Shapira and P. B. Kantor (eds.), *Recommender Systems Handbook* (Springer), pp. 257–297.

Shen, Y. and Jin, H. (2014). Privacy-preserving personalized recommendation: An instance-based approach via differential privacy, in *Proceedings of the 2014 IEEE International Conference on Data Mining*, pp. 540–549.

Shen, Y. and Jin, H. (2016). Epicrec: Towards practical differentially private framework for personalized recommendation, in *Proceedings of the 2016 ACM SIGSAC Conference on Computer and Communications Security*, pp. 180–191.

Shi, Y., Larson, M. and Hanjalic, A. (2014). Collaborative filtering beyond the user-item matrix: A survey of the state of the art and future challenges, *ACM Comput. Surv.* **47**, 1, pp. 3:1–3:45.

Shokri, R., Pedarsani, P., Theodorakopoulos, G. and Hubaux, J. (2009). Preserving privacy in collaborative filtering through distributed aggregation of offline profiles, in *Proceedings of the third ACM conference on Recommender systems (RecSys '09)*, pp. 157–164.

Su, X. and Khoshgoftaar, T. M. (2009). A survey of collaborative filtering techniques, *Adv. in Artif. Intell.* **2009**, pp. 4:2–4:2.

Tang, Q. (2012). Cryptographic framework for analyzing the privacy of recommender algorithms, in *2012 International Symposium on Security in Collaboration Technologies and Systems (CTS 2012)*, pp. 455–462.

Tang, Q. and Wang, H. (2017). Privacy-preserving hybrid recommender system, in *The Fifth International Workshop on Security in Cloud Computing (SCC)* (Springer), pp. 59–66.

Tang, Q. and Wang, J. (2015). Privacy-preserving context-aware recommender systems: Analysis and new solutions, in G. Pernul, P. Y. A. Ryan and E. R. Weippl (eds.), *Computer Security - ESORICS 2015, LNCS*, Vol. 9327 (Springer), pp. 101–119.

Tang, Q. and Wang, J. (2016). Privacy-preserving friendship-based recommender systems, *IEEE Transactions on Dependable and Secure Computing (TDSC)*, to appear.

Tramèr, F., Zhang, F., Juels, A., Reiter, M. K. and Ristenpart, T. (2016). Stealing machine learning models via prediction apis, in *25th USENIX Security Symposium, USENIX Security 16*, pp. 601–618.

Veugen, T., de Haan, R., Cramer, R. and Muller, F. (2015). A framework for secure computations with two non-colluding servers and multiple clients, applied to recommendations, *IEEE Transactions on Information Forensics and Security* **10**, 3, pp. 445–457.

Wang, J. and Tang, Q. (2015). Recommender systems and their security concerns, http://eprint.iacr.org/2015/1108.

Wang, J. and Tang, Q. (2016). A probabilistic view of neighborhood-based recommendation methods, in *IEEE International Conference on Data Mining Workshops, ICDM Workshops 2016*, pp. 14–20.

Wang, J. and Tang, Q. (2017). Differentially private neighborhood-based recommender systems, in *ICT Systems Security and Privacy Protection - 32nd IFIP TC 11 International Conference* (Springer), pp. 459–473.

Wang, Y., Fienberg, S. E. and Smola, A. J. (2015). Privacy for free: Posterior sampling and stochastic gradient monte carlo, in *Proceedings of the 32nd International Conference on Machine Learning*, pp. 2493–2502.

Weinsberg, U., Bhagat, S., Ioannidis, S. and Taft, N. (2012). Blurme: inferring and obfuscating user gender based on ratings, in *Sixth ACM Conference on Recommender Systems*, pp. 195–202.

Yao, C. C. (1986). How to generate and exchange secrets, in *Proceedings of the 27th Annual Symposium on Foundations of Computer Science*, pp. 162–167.

Zhang, S., Ford, J. and Makedon, F. (2006). Deriving private information from randomly perturbed ratings, in *Proceedings of the Sixth SIAM International Conference on Data Mining* (SIAM), pp. 59–69.

Zhang, S., Yao, L. and Sun, A. (2017). Deep learning based recommender system: A survey and new perspectives, https://arxiv.org/abs/1707.07435.

Zliobaite, I. (2017). Measuring discrimination in algorithmic decision making, *Data Min. Knowl. Discov.* **31**, 4, pp. 1060–1089.

Chapter 14

TV and Movie Recommendations: The Comcast Case

Shahin Sefati, Jan Neumann and Hassan Sayyadi
Comcast Applied AI Research
Washington DC, USA
shahin_sefati@comcast.com, jan_neumann@comcast.com,
hassan.sayyadi@comcast.com

With the ever-increasing number of TV shows and movies available to the users, facilitating content discovery is key toward creating a better user experience for an entertainment platform. Comcast is one of the largest cable providers in the United State with millions of customers. In this chapter, we provide an overview of the large scale recommendation system and its building blocks that power Comcast X1 entertainment platform. We discuss some of the challenges in building a TV and movie recommendation system and review a set of product features for content discovery on Comcast X1 platform. We briefly describe the historical evolution of the recommendation system at Comcast and provide some insights and results on evaluating the recommendation system with offline and online metrics. We conclude this chapter with some recent areas of research and development.

14.1. Introduction

In recent years, personalized recommendations have become a key value driver across different industries. Personalized recommendation systems are ubiquitous in most online services, from ecommerce to entertainment, and recently Comcast has brought that experience to the television.

For many households, the television is the central entertainment hub, and the average TV viewer spends about half of her or his leisure time in front of a TV. In addition to the content that is available on linear TV at any moment in time, we may now pick our favorite show from DVR recordings, or from on-demand video services that offer the desired content whenever a customer wants to access it. At any given moment, a customer has hundreds or even thousands of entertainment choices available, which makes some sort of automatic, personalized recommendations mechanism desirable to help them navigate a growing number of choices.

Comcast's X1 platform is a premium video experience enjoyed by millions of Comcast customers throughout the United States. In this chapter, we first review a number of the algorithms we developed, which form the building blocks that power the X1 recommender system. We then review a number of product features and how the challenges associated with each feature are addressed. We further discuss several offline and online metrics that are employed for evaluation. We conclude this chapter by exploring the future directions of research and development for the X1 recommender system.

14.2. Developed Algorithms

In this section we review number of building blocks and algorithms that were developed in-house for TV and movie recommendations.

14.2.1. *Collaborative Filtering*

There are two main techniques for building a recommender system: *Collaborative Filtering* (CF), and *Content-Based Filtering* (CBF). CF-based (CF-based) recommendation systems usually provide superior performance compared to content-based recommendation, but face the so-called "cold start problem" — what to recommend when no customer feedback data is available. Hybrid recommender systems that combine CF-based and content-based recommendations have proven to be the most successful approach. In this section, we review how a CF-based recommender system can be built for TV and movie recommendations. In the following section we will discuss how additional content meta-data (such as genre tags) can be incorporated to build a hybrid recommender system with improved performance.

CF-based recommendation systems leverage users' past interactions with items to generate item recommendations [Chapter 1]. Latent-based approaches such as matrix factorization (and its numerous variations) are among the most effective techniques. For example, [Mnih and Salakhutdinov (2008); Hu *et al.* (2008); Koren *et al.* (2009)] utilize matrix factorization to represent each user and item by a low-dimensional vector. The affinity between users and items can then be computed as the inner product between the user's and item's respective vector representations. An alternative to this approach is to directly model the item-item similarity matrix. In item-based collaborative filtering, a critical intermediate step to personalized recommendations is the definition of an item-similarity metric. The item similarity can be computed simply from the user-to-item observation matrix (e.g. ratings, item consumption statistics, etc.) with similarity measures such as cosine and Jaccard coefficient being popular choices for their simplicity.

Several years ago Comcast developed a novel probabilistic approach for modeling the item-item similarity, [Jojic *et al.* (2011)], which outperformed the state of the art in internal benchmarks.

The algorithm was developed to address a few issues with commonly used item similarity measures:

- Common similarity measures use a priori knowledge about items, but treat all user inputs as equal. For the purpose of computing item similarities, what weight should be assigned to a user who likes ten items, compared to a user who likes a thousand items?
- Item-to-item similarity is generally an asymmetrical relationship. For example, consider the case of a war movie, "W", and a movie about a love story happening during a war, "WL." Many users that enjoy "W" might also enjoy "WL" because "W'L" has a *war* component. In contrast, some of the users that enjoyed "WL" because of its *love* story might not enjoy "W." Overall, "W" is more similar to "WL" than "WL" is similar to "W."

In a recommendation task, the co-occurrences between two items are usually defined as the number of users that have liked/consumed both items. In our probabilistic approach, the similarity between two items is defined as the ratio of the actual number of co-occurrences, to the number of co-occurrences that would be expected if a user's likes/dislikes were assigned randomly to movies [Jojic *et al.* (2011)]. This algorithm has been powering the Comcast X1 personalized recommendations system and is a core component that enables many X1 personalization features, such as user-to-item personalized recommendations ("For You Sort" — see Section 14.3.1 for more details), and single item recommendations ("People Also Watched" — see Section 14.3.2 for more details).

To generate personalized recommendations, two types of relevance feedback are potentially available: 1) *explicit feedback* (for example, ratings of movies provided by users), and 2) *implicit feedback* (for example, the history of movies and TV shows watched, and interactions with the UX interface). While explicit feedback *may* more directly reflect the user's taste, users often do not provide explicit feedback, because it takes additional effort. On the other hand, for TV and movies recommendations, usually a vast amount of implicit feedback is available, and the relevance of a given movie can be indirectly inferred from implicit feedback.

14.2.2. *Hybrid Recommendation System*

When enough implicit relevance feedback is present, collaborative filtering techniques are the most effective approach for building a recommender engine. Nonetheless, CF-based recommendation algorithms suffer from the aforementioned cold-start problem, when a new user subscribes to the system or when a new item is added to a catalog and implicit feedback is not yet available.

On the other hand, content-based filtering leverages meta-data associated with items (e.g. genre and synopsis for movies). Leveraging meta-data in addition to implicit relevance feedback facilitates the content discovery process. This leads to better user engagement by providing more relevant and personalized results for the customer, and also addresses the cold start problem [Chapter 4]. For these reasons, when both implicit relevance feedback and content meta-data is available, hybrid recommender systems that combine CF-based and content-based recommendation algorithms are often preferred. Moreover, content meta-data is key to support an improved browse experience by allowing the creation of genre-specific menu collections, such as "Science-Fiction" movies, or "Comedy-Drama" movies.

Comcast's meta-data platform makes several types of meta-data about TV shows and movies available, such as their title, description, year of first airing, and descriptive tags (Genre, Theme, Tone, Keyword, etc.). In what follows, we describe how tag meta-data is refined and used together with implicit relevance feedback to generate better content recommendations.

14.2.2.1. *Tag weights*

Usually, descriptive meta-data tags are assigned to TV shows and movies by human editors, and a tag is either associated with an item or not. Thus, these tags are binary in nature and relatively limited in terms of the information they contain, because they do not reflect the degree to which a certain tag actually applies to the item.

For example, it is likely that a movie such as "Shrek" is tagged as "Action and Adventure", but "Shrek" is not considered a typical "Action and Adventure" movie such as "The Avengers". In this example, we can either remove this tag for the movie "Shrek", or, alternatively, we can assign it a weight corresponding to the degree that the tag matches the movie (and we would expect that weight to be relatively low in this case). Asking human editors to add a weight to each tag would be very time consuming and likely error prone.

In this section, we describe how with the use of collaborative filtering we can expand the tags for a given program, and how we can transform the binary tag data to more appropriate and continuously valued weights.

The average number of tags for a popular program is a few dozen; across all content, Comcast's meta-data database contains a few thousands unique tags. Less popular TV shows and movies may have only two or three tags, because it is not expected that human editors homogeneously assign tags to all available programs. This can impact the extent to which less popular programs are discoverable.

We designed an automated method to expand the number of tags that are associated with each program, and assigned program-specific weights to each tag. This approach takes as input the binary, human-generated tags, as well as the item-item similarities that we calculated using collaborative filtering (see Section 14.2.1). The idea is that these item-item similarities implicitly reflect the tag similarities between programs. For example, if a program, P, is "more similar" (according to the usage data) to the subset of programs that are tagged with, T, then the pair (P, T) should have a strong weight. In nutshell, two factors play a role: 1) the fraction of the set of programs that are similar to program, P that are tagged with Tag, T, 2) the fraction of the set of all programs that are tagged with Tag, T, that are similar to program P. In Table 14.1, the top 5 weights assigned to existing tags in few tag category are shown for the movie "Shrek". We see that after refinement, the "Action and Adventure" has a only small Genre score, which is what we wanted to accomplish.

14.2.2.2. *Tag importance*

While we have a few thousand unique tags, and a handful of tag categories to describe each program, not all of these descriptors are equally discriminative for a recommendation task. For example, two programs that have the tag "Drama" in common might not be as similar as two programs

Table 14.1. Top 5 weights assigned to existing tags in each tag category for the movie Shrek.

Genre	Keyword	Tone
Animated 0.2693	Friendship 0.0696	Fanciful 0.0872
Family Entertainment 0.2519	Hero 0.0498	Humorous 0.0851
Children's Family Entertainment 0.2045	Danger 0.0461	Rousing 0.0475
Fantasy 0.1142	Partner 0.0266	Stylized 0.0279
Action and Adventure 0.0652	Fantasy-world 0.0231	Witty 0.0081

which have the tag "Basketball" in common. Some techniques, such as TF/IDF (Term Frequency/Inverse Document Frequency), suggest that the popularity of a tag has an inverse correlation with its discriminative power. However, our data shows this is not always the case and pairwise similarities between programs derived from implicit relevance feedback can be used to determine the importance of each tag and its category. Tag importance plays a major role in computing a purely tag-based similarity score which is helpful when a new program becomes available and we need to deal with the cold-start problem.

14.2.3. *Implicit and Explicit Favorites (I know what I want to watch)*

The recommendation task is not limited to recommending only *new* relevant content, such as a new TV show or movie, to users. Weekdays when you return home from work, you may find yourself tuning to the same group of channels on TV because you know where you can find the content you want to watch, such as your local news channel, your favorite sitcom, etc., or you know which channels suit your taste. You may know when a new episode of your favorite show becomes available on demand, or perhaps you are binge-watching a new favorite show. Sports fans may want to know anytime their favorite teams are playing — when, and on what channels.

What is common in all these cases is that *you know what you want to watch* and it is our task to get you to the programming you love with as little effort on your part as possible. Enabling customers to find their favorite channels, programs, and sports teams is of great importance in facilitating the content discovery. At Comcast, there are two ways to identify favorite content for our customers:

(i) explicit favorites: a feature that enables a customer to mark channels, programs, or sport teams explicitly as their favorites;

(ii) implicit favorites: a feature that automatically learns a customer's favorites by identifying patterns in the programs the customer chose to watch that we then use to deliver personalized viewing experiences in accordance with our privacy policies and all applicable laws.

Favorite content is currently being surfaced in several personalization features both on the Comcast X1 platform and the mobile app. In Section 14.3, we review some use cases that benefit from identifying favorite content in more detail.

14.2.4. *Predictive Popularity*

We continuously research additional signals to improve our recommender system and to further facilitate the content discovery experience for our customers. In addition to the previously mentioned signals, the popularity of entities such as channels, programs, actors, etc. and their variance over time is another important signal for our search, browse, and recommendation algorithms. Since popularity varies over time, we developed a predictive popularity forecasting method to capture the seasonalities in our time series data.

An example of a seasonality is when a new season of a TV show airs: The search and browse popularity for that TV show may exhibit a weekly pattern that peaks on the weekday that the show is aired, and for video-on-demand consumption it would peak the day after the initial airing due to catch-up viewing.

Browsing and discovering popular TV shows and movies is an effective way to watch new content. By leveraging signals such as Video-On-Demand data, we developed a predictive model for program popularity. The program popularity signal helps us to surface popular programs in genre menus, or boost the rank of popular items when returning personalized recommendations.

For live TV, while we currently have a service that captures which programs are currently trending in real-time, a predictive model for popularity is still desirable, so that we can incorporate the predicted popularity signal into recommendations for upcoming programs. In addition to program popularity, the predictive model can also take advantage of what is trending on social media to further boost content that is noteworthy and likely to increase viewership in the near future, e.g., exciting sporting or breaking news events.

14.2.5. *Day Parting*

Another application of the time-varying popularity signal is day parting. Watching content on the TV is usually a shared experience, and we do not know who is actually in front of the screen. In multi-person households, family members watch categorically different content and have distinct tastes. Indeed, individual customers' viewing patterns may differ depending on mood and time-of-day. Our day parting pipeline employs time-series models and collaborative filtering to capture the following temporal patterns in usage data [Sayyadi-Harikandehei (2018)]:

- Daily patterns (content watched now is more similar to content consumed at the same time of the day than at other times)
- Weekly patterns (content watched now is more similar to content consumed on the same day of the week than on other days)
- Recency (content watched now is more similar to content consumed recently than content consumed further in the past)

This day-parting feature allows us to personalize the X1 experience without forcing our customers to create different profiles.

14.3. Addressed Challenges and Problems

In this section we will review some of the personalization features of the Comcast X1 platform, and discuss a few of the challenges we faced and addressed while building our TV and Movie content discovery and recommendations platform.

14.3.1. *For You: Personalized Recommendation for TV Shows and Movies*

As consumers are increasingly faced by a proliferation of new video content choices and different ways to watch, video recommendations have become an important tool for helping users discover content that they are likely to enjoy. The "For You" service provides personalized recommendations for TV shows and movies, potentially grouped by Genre, Free / Premium, etc. if so desired. The recommendations are generated by combining collaborative filtering, meta-data similarity via tag weights, day-parting as well as predictive popularity forecasting.

14.3.2. *People Also Watched*

For each TV show or movie in the catalog, the X1 interface provides an entity page that displays details about the selected content item, such as the plot summary, ways to watch (linear TV or VOD), and the cast and crew of the show. Entity pages are popular destinations during the content discovery process and users may land on an entity page via several paths, including browse and search via text or the X1 voice remote. The "People Also Watched" feature provides a convenient way for the user to find the content that is similar to the TV show or the movie that they currently selected. This feature makes direct use of the item-item similarity scores that are computed from both implicit relevance feedback and meta-data (see Section 14.2.2 for more details about our hybrid recommender system).

14.3.3. *Because You Watched*

Recommender systems often behave like black boxes and provide no reasoning to support why a recommended item is being presented to a user. Research has shown that providing explanations for recommendations improves the trust of users in a recommender system and creates a better experience for the users [Herlocker *et al.* (2000)]. The "Because You Watched" feature explicitly presents content that is similar to a movie or TV show that the user has watched in the past and by presenting the source item, the service explains why the recommended items were selected. This feature makes use of the "People Also Watched" service (see Section 14.3.2) as well as a list of favorite TV shows and movies the user has identified.

14.3.4. *What Should I Watch?*

"What Should I Watch" is a voice-enabled feature that provides the user with different types of personalized and trending content. The goal is to generate a personalized *landing page* for our customers that combines content from all the available sources arranged in thematic rows. Some of the row examples include: live TV, movie collections (filtered by Genre, Free / Premium, etc.), Because You Watched, What is Trending on TV and social media at the moment, implicit and explicit favorites, recently recorded programs on digital video recorder (DVR), etc. We combine both batch and real-time data processing to build this personalized experience for Comcast's customers.

In what follows we describe some of the challenges with generating personalized recommendations for live TV and discuss how the service is built. Despite the rise in video-on-demand (VOD) consumption, live TV is still the most popular way people consume video entertainment. At Comcast, we are developing novel ways to make it easy for our customers to access live TV content that is interesting and relevant for them at the current moment. The content available on linear TV channels is constantly changing and the information about it is often only available at the time of airing, which leads to cold start challenges. In addition, we often consume TV in groups of varying and unknown composition (household vs individual), which makes building taste profiles and modeling consumer behavior very challenging. The recommendations in the TV row of the "What Should I Watch" feature are generated by combining the output of several personalization services, including:

- Predicted popularity and real-time trending
- Favorite programs, teams, and stations
- Day parting
- Collaborative filtering

Finally, we built an ensemble machine learning model to blend the scores from the personalization services mentioned above, to compute a relevance score for each channel, based on what is being aired and time of the day.

In Figure 14.1, schematic of this service is depicted.

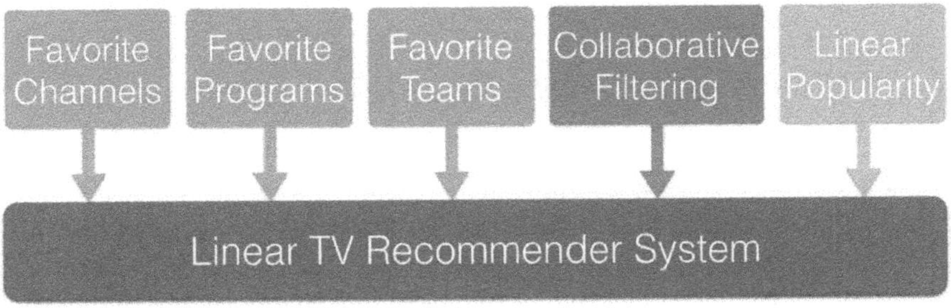

Fig. 14.1. "What should I Watch" for live TV.

14.3.5. *Taste-based menus: Menu Personalization*

Movies are associated with meta-data tags such as "Genre" or "Tone," to name two of many types. These tags exist at different levels of granularity, e.g., "Drama" and a number of its sub-genres, such as "Comedy-Drama", "Political-drama", etc. Based on these tags, thousands of menu collections can be created for movies and TV shows, by filtering lists of recommended items based on the presence or absence of tags and their combinations. The goal of these taste-based menus is to provide a personalized set of menus for each customer, which reflects their taste. The underlying algorithm leverages the history of what a customer has chosen to watch to determine the movies that they have watched in the past, from each possible menu, as well as a personalized score for each menu. Without limiting the types of menus that can be recommended to a customer, a probabilistic model will generate a set of recommended menus for a customer every time they visit the interface.

14.3.6. *Personalized Search*

Search is still an essential component of our content discovery platform, with a goal of helping customers find the content they are looking for — as quickly as possible. Customers perform a search query to find different types of entities such as programs, sport teams, TV channels, actors, etc. Our goal was to build a personalized search experience that minimizes the number of interactions required to find the desired content. The personalized search experience is an ensemble of the output of several algorithms, including a predictive popularity model for every term in the title, and a predictive popularity model for every entity, as computed from the individual and aggregate search histories. For example, even though the official name of "CNN" is "Cable News Network", virtually nobody searches for the word "Cable" when they want to watch "CNN". The predictive model for every term in the title identifies the importance of each word in the title, as well as frequently used other terms that customers have used in their searches before tuning to "CNN". A personalized search feature is also important to distinguish between ambiguous search queries. For example, "Chicago Fire" is the name of both a TV series and a soccer team that is based in Chicago. In this case, the personalized search engine ranks the TV series and the soccer club differently for different customers, depending on their previous searches and resulting tune-in behavior.

14.4. Implementation Resources and Historical Evolution and Versions

The implementation of our recommendations system underwent a number of evolutions over the years. The first version was implemented in Java using the Hadoop big data processing framework, and consisted of a series of batch jobs that ran every day on multi-node clusters. The jobs started with a daily ingest of the tune activity logs, which were then joined with the TV schedule (for live TV watching activity) and program metadata to create viewing sessions for each device. Another Hadoop job then took these viewing sessions and computed the item-to-item similarity matrix, as explained in Section 14.2.2. Both the viewing sessions and sparse item-to-item similarities are stored in a key-value cache, for fast access at the time of the recommendations API call.

At time of prediction, the system combines viewing data and the item-to-item similarity information to compute the candidate list of relevant items for the customer. This list is then re-ranked using the predicted popularity and day parting signals using optimized Java code.

For the next evolution of our pipeline, we re-implemented the batch processing steps using the Spark framework which reduced the latency and overall processing time of our pipelines noticeably.

For the current version of our pipeline, we are focused on converting it from a batch process into a real-time streaming pipeline. We used Kafka for the streaming data transport and storage, and the Spark Streaming framework to process the rich set of implicit feedback signals with minimal latency.

14.5. Evaluation

14.5.1. *Offline and Online metrics*

It is very important to accurately evaluate the performance of a content discovery system during both development and production. We use a number of offline and online metrics to evaluate the performance of our system. These metrics include recall, click-through rate, position of selected items on the page, watch rate, average search clicks, and Spearman's rank correlation coefficient. While the offline metrics will ideally correlate highly with the online metrics, running systematic A/B tests in production is essential to verify the assumptions behind the offline metrics. Measurable online metrics in A/B tests may sometimes reveal counter-intuitive facts about the actual impact of changes to the platform which contradict the hypothesized behavior.

In one experiment, we conducted an A/B test for comparing two sorting algorithms for movie recommendations: 1) popularity-based recommendations, 2) personalized recommendations by leveraging users' viewing choices, and we noticed that the watch-rate for the personalized recommendations increased by 13% compared to the popularity-based recommendations.

14.6. Lessons Learned and Future Directions

With the ever-increasing number of TV shows and movies available to the customers, facilitating content discovery is key toward creating a better user experience for an entertainment platform. In this chapter, we reviewed some of the challenges in building large scale recommendation systems for contents available on linear TV and movies. In what follows, we describe some of our recent approaches to improve the recommendation system.

14.6.1. *Deep Learning-based Recommender System*

In Section 14.2.2, we described how tag metadata can be used, in addition to collaborative filtering, to build a hybrid recommender system. Content metadata is not limited to genre or theme tags, though. Comcast's metadata platform provides other types of metadata, such as title, description, year, and even poster art. Incorporating additional meta-data into collaborative filtering is a challenging problem. Recently, deep learning-based collaborative filtering algorithms produced superior performance, compared to prior state of the art approaches, such as matrix factorization. Moreover, deep learning models provide more flexible frameworks for leveraging multi-modal data in applications, including the personalized recommendation services we described in this chapter. At Comcast, we are continuously adding deep learning models to the X1 recommender system.

14.6.2. *Automatic Content Analysis*

As a customer, we want the services we use to consume entertainment to guide and assist us, in a personalized way, so that we can find something to watch that we like and that fits our current mood. Traditionally, the content we could choose from consisted only of content entities as a whole.

Nowadays, and in part due to increased consumption of content on mobile devices, we often do not want to consume whole shows anymore, but only the "best" parts, i.e., the segments that are (or social networks deem to be) most relevant and interesting to us.

To enable such personalized entertainment experiences, we need to know more about the content shown than just the metadata at the asset level, like titles, credits, keywords. This additional information could be the semantic segments and moments of a program, who is on screen and when, the theme or topic of the segment, and how relevant a segment is based on a customer's taste profile. In addition, if the information is aligned with the timeline of the video, it allows for advanced navigational experiences within and between assets, and can be indexed to enable relevant search and recommendations within videos.

To build the entertainment experiences of the future, at Comcast Applied AI Research, we are exploring automatic content analysis solutions that combine video, audio and text processing with machine learning algorithms to identify relevant moments, segments and their descriptions potentially with very limited human interaction. The generated metadata could then be used by other applications to provide consumers with a more interactive and personalized experience. Examples are more accurate program (segment) recommendations, in and between program navigation, and enhanced search capabilities, such as free-form queries about specific segments, using our voice interfaces.

Acknowledgement

The authors would like to thank Oliver Jojic, Leemay Nassery and many other members of the team who contributed to the algorithms, features, and material covered in this chapter.

References

Herlocker, J. L., Konstan, J. A. and Riedl, J. (2000). Explaining collaborative filtering recommendations, in *Proceedings of the 2000 ACM conference on Computer supported cooperative work* (ACM), pp. 241–250.

Hu, Y., Koren, Y. and Volinsky, C. (2008). Collaborative filtering for implicit feedback datasets, in *Proceedings of the 2008 Eighth IEEE International Conference on Data Mining*, ICDM '08 (IEEE Computer Society, Washington, DC, USA), ISBN 978-0-7695-3502-9, pp. 263–272, doi: 10.1109/ICDM.2008.22, `http://dx.doi.org/10.1109/ICDM.2008.22`.

Jojic, O., Shukla, M. and Bhosarekar, N. (2011). A probabilistic definition of item similarity, in *Proceedings of the fifth ACM conference on Recommender systems* (ACM), pp. 229–236.

Koren, Y., Bell, R. and Volinsky, C. (2009). Matrix factorization techniques for recommender systems, *Computer* **42**, 8.

Mnih, A. and Salakhutdinov, R. R. (2008). Probabilistic matrix factorization, in *Advances in neural information processing systems*, pp. 1257–1264.

Sayyadi-Harikandehei, H. (2018). Personalized content recommendations based on consumption periodicity, US Patent 9,942,609.

Chapter 15

Music Recommendations

Dietmar Jannach[a], Iman Kamehkhosh[b] and Geoffray Bonnin[c]
[a]AAU Klagenfurt, Austria, [b]TU Dortmund, Germany, [c]Loria, Nancy, France
dietmar.jannach@aau.at, iman.kamehkhosh@tu-dortmund.de, bonnin@loria.fr

Today's online music services like Spotify provide their listeners with different types of music recommendations, e.g., in the form of weekly recommendations or personalized radio stations. Such recommendations are often based, at least in parts, on collaborative filtering techniques. In this chapter, we first review the different types of music recommendations that can be found in practice and discuss the specific challenges of the domain. Next, we discuss technical approaches for the problems of music discovery and next-track recommendation in more depth, with a specific focus on their practical application at Spotify. Finally, we further elaborate on open challenges in the field and revisit the specific problems of evaluating music recommendation systems in academic environments.

15.1. Introduction

Music was one of the first application fields of collaborative filtering (CF) recommender systems. The *Ringo* system presented in [Shardanand and Maes (1995)] went online as early as 1994 and was designed to recommend albums and musical artists to users, initially as an email-based service. It was based on explicit rating information provided by users for artists to construct preference profiles. Recommendations in that system were then made based on a user-to-user or item-to-item neighborhood scheme as described in Chapter 1 and sent via email to Ringo's users.

Today, more than twenty years later, the social web has led to new dimensions of social, "word of mouth" information filtering. At the same time, the way we listen to music and how we discover new artists or albums has dramatically changed. Millions of music tracks are nowadays available to us instantaneously through various online music streaming services. As a result, almost all of today's major online music services, including

Spotify or the services by Google or Microsoft, provide some form of music recommendation functionality.

15.1.1. *Music Recommendation Tasks*

The Ringo system supported three main tasks: (i) suggest artists/albums, (ii) list artists/albums that the user will most probably dislike, and (iii) predict the rating of a user for an album. At the time when Ringo was online, users could consider the recommendations the next time they visited a music store or when they placed a mail order at a catalog company.

Of the three tasks supported by the Ringo system, today mainly the artist and album recommendation services are common on online music platforms. However, the recent possibility to provide recommendations and play music instantaneously opened new opportunities for other recommendation scenarios. Spotify, one of today's market leaders in the online music streaming industry, as of 2018, provides recommendations through a number of features of their platform and apps. Specifically, they for example support *discovery* through personalized recommendations [Johnson (2014)]. They furthermore generate user-specific playlists ("mix tapes") for listeners on a weekly basis and provide a "release radar" to point users to newly released tracks that might be interesting to them. In addition to these features, non-personalized recommendations, e.g., of trending tracks, are common on most music platforms.

While some types of music recommendations are nowadays already common on music-related sites and apps, some additional scenarios for music recommendation have been explored in the literature in the past. Table 15.1 provides an overview of common recommendation scenarios. We can categorize the different recommendation tasks into different groups.

- First, recommendations can be *non-personalized* and simply consist of currently trending tracks, artists, albums, concerts, etc. Another common (non-personalized) way of pointing users to something interesting is to provide them with *curated* playlists, which represents a convenient way for users to browse the catalog [Johnson and Newett (2014)].
- Second, recommendations can be *contextualized*, but *non-personalized*. On Last.fm, for example, virtually endless radio stations can be created based on a seed track or seed artist. In the research literature, a number of approaches to support users during

manual playlist construction were proposed, where the system's task is to recommend additional tracks given a playlist beginning or a set of user-specified constraints.

- The third category consists of *personalized*, but *non-contextual* recommendations. Spotify's Discover Weekly feature is an example of such a functionality, where the system provides a set of recommended tracks based on the user's observed behavior.
- The final category comprises approaches that are both *personalized and contextualized*. An example would be a system that recommends items during playlist construction where these items both match the current playlist and the user's general tastes. Approaches in that category can mostly be found in the academic literature.

In this chapter, our main focus will be on approaches that are either contextualized, personalized, or both. We will specifically discuss collaborative filtering techniques and hybrid approaches that combine collaborative information with additional data, e.g., the musical features of the tracks.

15.1.2. *Specific Challenges of Music Recommendation*

From a computational perspective, some of the tasks listed in Table 15.1 seem quite similar to recommendation tasks in other domains, like e-commerce. In particular, the *personalized and non-contextualized* recommendation scenarios can in principle be addressed with collaborative filtering approaches that are designed for relevance prediction and learning-to-rank scenarios, where the final goal is to create a ranked list of objects that are supposed to be *generally relevant* for a user. We will discuss a variant of such a standard collaborative filtering technique, as used by Spotify, later in Section 15.2. Similarly, the problem of providing a virtually endless playlist given the user's recently played tracks can be found in a comparable form as a *session-based recommendation* scenario in e-commerce [Hidasi *et al.* (2016a); Jannach and Ludewig (2017); Quadrana *et al.* (2018)]. We will discuss the specifics of the problem setting for music recommendation in a later section as well.

A number of aspects are however very specific to music recommendation, and some others are at least more relevant for music than for other application areas of recommender systems. These aspects relate both to technical and non-technical issues and include, among others, the following.

Table 15.1. Examples of Music Recommendation Tasks.

	Non-personalized	**Personalized**
Non-contextualized	• Trending List: Provide a list of currently trending (or currently being played) items, e.g., tracks, albums, artists, concerts etc. Often used as a baseline for experiments in the research literature, e.g., [Chen *et al.* (2016); Pálovics *et al.* (2014); Vasile *et al.* (2016)]. • Similar Objects: Find similar tracks or artists, available, e.g., on Spotify. This type of recommendation can often be found in the Music Information Retrieval literature, e.g., [Germain and Chakareski (2013); Hartono and Yoshitake (2013); Moore *et al.* (2012)]. • Curated Playlists: Help users discover things through curated (editorial) playlists, e.g., on 8tracks.com. • Broadcasting Radios: Usually made by professional disc jockeys. Such playlists often contain popular tracks, and are often targeting specific audiences [Ekelund *et al.* (2000)].	• Track, Artist Discovery: Help users to find something new that matches their general preferences, as implemented, e.g., by Spotify in the "Discover Weekly" and "Release Radar" features [Johnson (2014)]. • Album Discovery: Recommend albums to listen to. Such a functionality can be found on general e-commerce sites like Amazon.com as well. • Enjoyment Prediction: Provide an assessment if the user will like a certain track, artist or album; also, create a list of things that the user will presumably *not* like. • Static Playlist Generation: Generate a personalized playlist based on user tastes; like Spotify's "Mix Tape" feature. • Personalized Recommendation of Curated Playlists: Suggest hand-made playlists to users that are likely to generally match their taste, e.g., on Deezer. Rarely studied in the literature, see, e.g., [Loni *et al.* (2016)]. • Recommendation of Radio Streams: Recommend broadcasting radio stations to users based on their profile and feedback. Also rarely studied, see, e.g., [Moling *et al.* (2012)].
Contextualized	• Virtual Radio Station: Create a virtually endless playlist, given a seed track or seed artist. To be found on Spotify, Deezer, Pandora, and other popular services, as well as in the research literature, see, e.g., [Oliver and Flores-Mangas (2006); Cliff (2006); Moens *et al.* (2010); Jylhä *et al.* (2012)]. • Playlist Construction Support: Generate a playlist based on seed tracks or other information regarding the current session, like the user's mood; or provide suggestions of tracks during manual playlist creation. • Contextualized Playlist Recommendation: Recommend a curated playlist based, e.g., on the time of the day, day of the week, or season.	• Personalized Radio (next-track recommendation): Generate a virtually endless radio based on the last played tracks, while possibly taking into account the user immediate feedback (e.g., "like", "skip", and "ban" actions). • Personalized Playlist Construction Support: Generate a playlist based on seed tracks or other information, like the user's mood and past preferences.

Catalog aspects: While larger e-commerce shops can easily have hundred thousands of catalog items, the number of recommendable items on Spotify, as of 2018, is over 35 million tracks. This can make the application in particular of academic approaches challenging. Furthermore, constantly new tracks are released and added to the catalog. And, at least for some musical genres, the freshness or recency of the recommendations might be an important quality factor to consider. Moreover, the meta-data of the tracks in the catalog can be very noisy and incomplete, and significant efforts might be required in order to clean it and infer missing information, in particular as a huge number of new tracks is released every year.

Preference information: Users of the Ringo system were asked to indicate their preferences for 125 artists (popular ones and random ones). While also some of today's systems (such as Microsoft Groove) ask users to provide an initial set of preferences regarding artists and genres, music recommenders often have to rely on mostly implicit preference signals in terms of listening logs, sometimes in combination with explicit like statements or "skip" actions. Besides the problem of correctly interpreting very large amounts of implicit feedback, an additional challenge in that context is that preferences can change over time.

Repeated recommendations: Many recommendation algorithms, and in particular those that are based on the matrix completion problem abstraction, aim to predict the relevance of *unseen* items. In the music domain, repeatedly listening to the same tracks is however common. If such repeated consumptions should be supported, algorithmic approaches have to be able both to decide *which* tracks to recommend repeatedly and *when* to recommend these tracks.

Immediate consumption and feedback: Differently from many e-commerce domains, the recommendations provided on a music streaming service can be immediately "consumed" by the listeners, e.g., using a personalized radio station. A main challenge in that context is that the system should be able to provide the user with a means to "correct" recommendations or give feedback (e.g., in terms of a like or dislike button). Moreover, this feedback should be immediately taken into account in the recommendation process.

Mainstream might not be enough: In some sense, music is "more niche" than movies [Johnson (2014)]. While in movie recommendation there are many blockbusters that are safe to recommend to a major fraction of the users, there are many musical genres which have their specific audiences (like jazz, classical music, or pop), and recommending generally popular items might easily lead to a bad user experience.

Context-dependence and time variance: Which music we want to listen to can be highly context-dependent. The relevant contextual factors can include, for example, the user's mood, the time of the day, or if the user listens to the music alone or as part of a group. Being able to capture and consider these contextual factors can be crucial for the quality perception and acceptance of a recommender.

Purposes of music listening: One related specificity of music is the fact that one often listens to music with a very specific purpose in mind: create a particular ambiance for a party, getting some motivation to wash the dishes, enhance the experience of reading a good book, getting relaxed before going to bed, etc. This means that the recommended items not only have to fit the current context, but also fit the purpose of the user.

Musical taste and stated preferences can be socially influenced: Which music we like and listen to is not only affected by our own mood, it can also be substantially affected by our social environment ("social bonding") and/or trends in the community as a whole. For some scenarios it can therefore be particularly helpful to consider a user's social environment and corresponding behavior in the past in the recommendation process. At the same time, when users share their tastes and preferences on social networks, it is not always clear if people actually listen to what they publicly "like" or if they merely use their public profiles to create a desired image of themselves.

15.1.3. *Chapter Outline*

In the remainder of this chapter, we will first discuss selected algorithmic approaches for music recommendation problems. Specifically, we will first discuss the application of matrix factorization techniques for the Discover Weekly feature of Spotify and will then review recent approaches of adaptive playlist generation (next-track music recommendation). Afterwards, we will outline open challenges both from a practical and academic perspective. We will then report how a music recommendation service is deployed and tested in industrial environments, again based on publicly available information about Spotify's solution. The final sections of the paper will be devoted to questions of how to evaluate music recommenders in practice and which resources are available for researchers in academic environments. The chapter ends with a summary of the lessons learned and open challenges.

15.2. Computational Tasks and Algorithms

Various algorithmic approaches have been proposed over the years for music-related recommendation scenarios. Three types of approaches generally exist: (1) collaborative techniques, i.e., recommending music liked by similar users, (2) content-based techniques, i.e., recommending music whose content (musical features, lyrics, etc.) is similar to the content of the tracks the user liked in the past, and (3) hybrid techniques, i.e., combining both previous approaches. Content-based approaches are mostly studied in the field of Music Information Retrieval (MIR). Correspondingly, the MIR literature often focuses on the problem of deriving such features from the low-level musical signal.[1] This chapter will not cover these aspects and will mostly focus on collaborative and hybrid techniques.

Companies usually do not publicly reveal the details of how they generate their recommendations or on which data they are based. Some discussions about the inner workings of the used machinery are sometimes revealed in public presentations. In many cases, presenters from industry report that they use a variety of algorithmic approaches for the different recommendation tasks. The streaming service *Pandora*, for example, in addition to its unique database of manually annotated musical tracks[2] which is used for *content-based* recommendations, relies also on collaborative techniques [Bieschke (2014)].

15.2.1. *Implicit Matrix Factorization for Discovery and Item Search*

In this section, we will briefly review how collaborative filtering was — among other techniques — applied at Spotify based on public presentations around the year 2014 [Johnson and Newett (2014); Johnson (2014)]. In more recent presentations, such as [Steck *et al.* (2015)], the authors report that Spotify uses an ensemble of different techniques (including NLP models and Recurrent Neural Networks) as well as explicit feedback signals (thumbs up / down), and also audio features for certain recommendation tasks.

---

[1]The various aspects of automated music data analysis are discussed, e.g., in [Weihs *et al.* (2016)].

[2]Music Genome Project `https://en.wikipedia.org/wiki/Music_Genome_Project`

15.2.1.1. *Distributed Matrix Factorization based on Listening Logs*

At its core, the algorithm that was used at Spotify for its discovery feature, is based on a standard user-item rating matrix, where users are rows and the columns represent tracks. There are, however, a number of specific aspects to be addressed in this domain, in particular the following two:

(1) In comparison to the well-researched movie recommendation domain, the number of items is much larger and there are more than 35 million songs that can potentially be recommended. Also, the number of users is substantial, in particular when compared to public datasets that are used in academic research.
(2) The entries in the matrix are often not explicit item ratings, but are based on the users' listening histories. As users often listen to tracks many times, the resulting play counts can be used as an implicit preference signal. Exploiting play counts instead of explicit ratings can actually lead to more accurate recommendations [Jawaheer *et al.* (2010)].

Let us recall a common formulation of the optimization problem for matrix factorization techniques as described, e.g., in [Koren (2008)].

$$\min_{q*,p*} \sum_{(u,i)\in K} (r_{ui} - \mu - b_u - b_i - q_i^T p_u)^2 + \lambda(\|q_i\|^2 + \|p_i\|^2 + b_u^2 + b_i^2) \quad (15.1)$$

Here, the goal is to minimize the squared prediction error for all (u, i) pairs in the training set (K), where the rating prediction r_{ui} for a user u and an item i is based on the rating average μ, some user and item bias factors b_u and b_i, and the user and item latent factors q_i and p_i; see also Chapter 2.

In the problem encoding of what is called "implicit matrix factorization" at Spotify, as proposed in [Hu *et al.* (2008)], the entries in the input matrix are zeros and ones, where a one in a cell indicates that a user has streamed a track at least one time. Multiple streaming events for the same track are therefore not considered in the encoding. However, the assumed higher preference for a track that was played multiple times is captured in the optimization function. Instead of optimizing the root mean squared error (RMSE), a weighted version of the RMSE is optimized, where a certain weight factor c_{ui} is used, which is based on the stream counts for a given user and track. Ignoring the average rating μ, the optimization function is therefore defined as [Hu *et al.* (2008); Johnson (2014)]:

$$\min_{q*,p*} \sum_{(u,i)\in K} c_{ui}(r_{ui} - b_u - b_i - q_i^T p_u)^2 + \lambda(\|q_i\|^2 + \|p_i\|^2 + b_u^2 + b_i^2) \quad (15.2)$$

Details of how the weight function is calculated are not revealed. To find the optimal parameters, the Alternating Least Squares (ALS) method is applied. Due to the huge amount of required computations, a distributed computing architecture based on the map-reduce scheme (using Hadoop[3]) was utilized at that time. The exact details of the distributed computations are not available. Nevertheless, distributed ALS methods as described, e.g., in [Zhou *et al.* (2008)] are nowadays available as off-the-shelf components, e.g., within the Apache Mahout[4] software. The required infrastructure, however, seems to be substantial and Steck *et al.* (2015) report that Spotify, as of 2015, used a Hadoop cluster consisting of 1,000 nodes and collected one terabyte of user data each day.

Once the latent user and item vectors are computed, they can be used for the two following tasks:

(1) *Relevance prediction*: As usual, the relevance of an item can be determined by computing the dot product of the user and the item vectors.
(2) *Item or user similarity assessment*: Two items or users can be compared using their coordinates in the latent factor space by computing the cosine similarity of the item and vectors, respectively.

15.2.1.2. *Finding Similar Objects with Approximate Nearest Neighbors*

For this latter task — finding similar objects, which is a key non-personalized functionality of Spotify — the problem is again that there are many users and items. Finding the exact set of the most similar objects ("neighbors") is therefore computationally challenging. As a consequence, an "Approximate Nearest Neighbor" method called *Annoy* was used [Bernhardsson (2015)].

Technically, the algorithm works by recursively splitting up the set of objects into two classes. In each step, two random points are chosen and the hyperplane that is equidistant to the points is used to split the data. This procedure is recursively repeated until a certain stopping criterion is

[3] `http://hadoop.apache.org/`
[4] `http://mahout.apache.org/`

met, leading to a binary tree where each object is assigned to exactly one leaf nodes. As a stopping criterion, one could for example define that each leaf node has at most K objects assigned.

Figure 15.1 shows an example of such a tree, which has the property that most objects in the same leaf node have a higher probability to be similar to each other than to objects of another leaf node. When looking for the neighbors of a certain user or item, one can walk down the tree from the root node by assessing on which side of the separating hyperplane the input item will be. Such a simple algorithm, however, has its limitations. One can, for example, easily end up with a situation where there are (a) only very few objects in the resulting leaf node and (b) some nodes are actually close but not part of the leaf node.

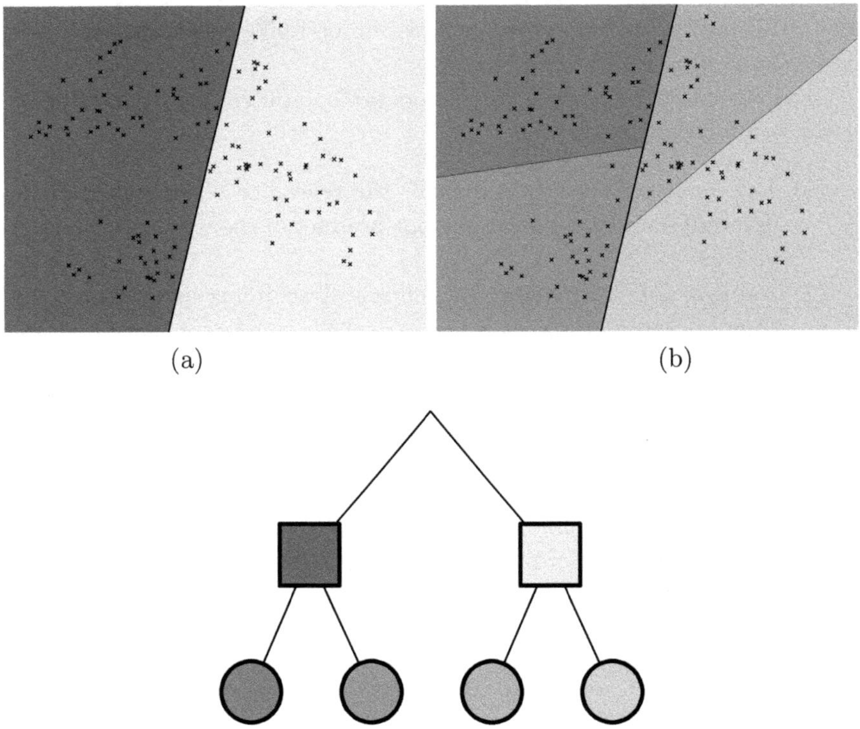

Fig. 15.1. Building the binary tree in *Annoy*, adapted from [Bernhardsson (2015)]. The set of points is first split into two halves according to the first selected hyperplane (a). Both halves are then split again using the same principle (b).

Two techniques are proposed to deal with these potential limitations:

(1) An algorithm is used that explores both child nodes when a certain closeness threshold is met.
(2) Instead of building only one decomposition tree, multiple trees are randomly generated and the set of neighborhood candidates is determined by collecting the elements of all leaf nodes for the given target items.

For the resulting set of candidate neighbors, the similarity values are computed and the N most similar ones are returned. While some of the true nearest neighbors might be missed by the procedure, the neighborhoods can be computed fast. Using the different algorithm parameters (e.g., the number of trees used), one can furthermore balance accuracy and computational costs.

Generally, *Annoy* is not the first method that combines clustering and nearest-neighbor search. Different other approximate techniques were proposed in the literature before, for instance, balltrees [Omohundro (1989)]. In principle, the proposed technique is only one of several possible ways of constructing so-called k-d trees, i.e., trees that partition points in the space using hyperplanes. According to the authors of *Annoy*[5], the method however has the advantage of having a small memory footprint and that it is engineered to be used in a distributed environment as index files can be shared between processes.

15.2.2. *Adaptive Playlist Generation*

The matrix factorization approach presented in the previous section is mainly designed to help users discover items that are assumed to be *generally* relevant or interesting to them. However, as discussed in Chapter 1, users of a recommendation service often have a specific intent or goal when they visit a website or use an app. In the music domain, the goal or intention of a user might be influenced by his or her current mood or contextual situation (e.g., being at a party, or doing sports). One common functionality of music streaming services therefore is to provide users with a means to play a virtual endless list of suitable tracks based only on some limited amounts of initial input.

[5]See the documentation at `https://github.com/spotify/annoy/`

15.2.2.1. *Spotify's Adaptive Radio*

On some platforms this functionality is called "Radio". Given, e.g., an artist or a track as an input, a playlist is automatically created by the service. Often, such playlists are based on tracks by related artists or by tracks that are in some sense similar to the seed track. In some cases, such radio stations are non-personalized, i.e., with the same input, every user receives more or less the same set of recommendations. In principle, however, the selection of tracks could be personalized as well, in case the listening history or some other preferences information about the current user is known.

Since in this scenario the system's recommendations are immediately consumed in the order determined by the system, the users should be enabled to give feedback on the recommendations and this feedback should be immediately processed by the recommender. Spotify's *Radio* feature supports both of these functionalities, personalized playlists and immediate feedback, in some form.

Bernhardsson (2013) summarizes the main principle of how Spotify optimizes their "artist radio" or "song radio" as sketched in Figure 15.2. The general idea is to use a number of different algorithms (which probably also use different optimization measures) and combine the recommendations in an ensemble with Gradient Boosted Decision Trees [Steck *et al.* (2015)], with explicit thumbs data and random negatives as an input for the optimization process. The outcome is a pool of tracks that can in principle be played on the current radio. The final ranking of the tracks is then done by computing scores for each possible next track. The following factors are taken into account. However, how the different aspects are combined, is not revealed in detail.

- The "global rank" of a track in the track pool, as determined by the algorithms;
- the estimated relevance of each track for a user based on the latent factor model;
- the estimated relevance to the user based on the given thumbs;
- the diversity of the tracks in the session in terms of artists and albums.

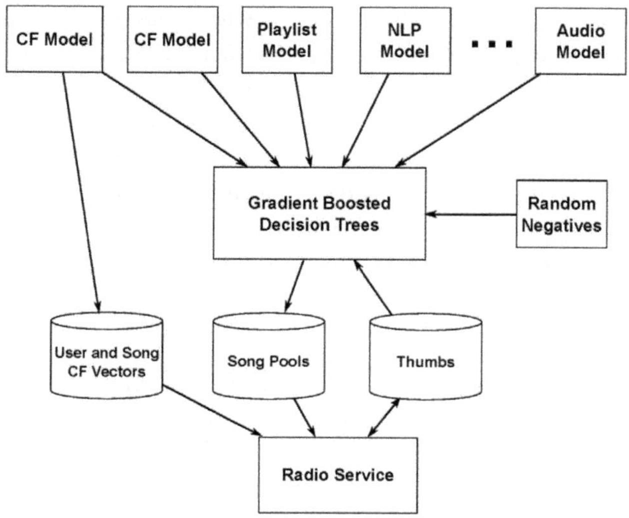

Fig. 15.2. Conceptual Model of the Radio Service, adapted from [Steck *et al.* (2015)].

15.2.2.2. *CF-based Next-Track Music Recommendation in the Literature*

Academic research has been conducted on various forms of music recommendation. The addressed application scenarios for example include non-contextualized item (e.g., track, artist, or album) recommendation for discovery [Bachrach *et al.* (2014)], similar object retrieval [Chen *et al.* (2016)], the automated construction of finite-length playlists or playlists with some length bounds [Pauws *et al.* (2008)], as well as next-item recommendation (playlist continuation) [Vasile *et al.* (2016)].

In this section, we will focus on applying collaborative filtering techniques for next-item recommendation scenarios, i.e., on situations where the collective behavior of a larger user community is considered in the recommendation process. The general computational task in these settings is to compute a ranked list of tracks to be played next given some seed information like an artist or a start track. There are two main applications where such a functionality is needed. First, next-item recommendations can be the basis for a virtually endless radio station like Spotify's Radio feature. Second, this type of recommendations can be used to assist users when they manually create a finite-length playlist, as implemented, e.g., in Apple's iTunes Genius.

In the following, we consider scenarios where the provided seed information is one or more tracks, representing a playlist beginning or a set of recently played tracks. Given this ordered list of tracks and possibly some auxiliary information, e.g., track metadata, musical features of the tracks, or the user's past preferences, the goal is to compute a ranked list of tracks to be recommended as a playlist continuation. Generally, the setting therefore corresponds to a session-aware or session-based recommendation scenario, as discussed in Chapter 1 and in more detail in [Quadrana *et al.* (2018)].

Non-Personalized Approaches. A large part of the published approaches in the research literature are non-personalized, i.e., given a playlist beginning, they return the same set of tracks for every user. Technically, a variety of approaches from different algorithm families has been explored [Bonnin and Jannach (2014)]. Generally, we can differentiate between two types of approaches:

(1) *Sequence-agnostic techniques*: These algorithms only consider the co-occurrence of items in the current session and past listening sessions or playlists.
(2) *Sequence-aware techniques*: Algorithms of this type also consider the order of the tracks, both in the current as well as in the past sessions.

Sequence-agnostic Approaches: A basic approach would be to apply frequent pattern rule mining techniques and look for tracks that often appear together in the listening logs of users in the past. These association rules can then be used to make Amazon-like recommendations of the form "Customers who bought this item also bought these items". A more effective method, however, is to apply a session-based k-nearest-neighbor (kNN) approach as discussed in [Hariri *et al.* (2012)] or [Bonnin and Jannach (2014)].

The kNN method takes the sequence of the last played tracks as an input and then in a first step determines the k most similar past sessions in the logs of all user sessions. Given the current session s, the set of the k nearest neighbors N_s, and a function $sim(s_1, s_2)$ that returns a similarity score for two sessions s_1 and s_2, the score of a recommendable item i is defined as

$$score_{\text{KNN}}(i, s) = \Sigma_{n \in N_s} sim(s, n) \times 1_n(i) \qquad (15.3)$$

where $1_n(i) = 1$ if n contains i and 0 otherwise, see also [Bonnin and Jannach (2014)]. Technically, different distance (similarity) measures can be used to compare two sessions or playlists, e.g., binary cosine similarity. To ensure scalability, neighborhood sampling can be applied, e.g., based on the recency of the listening logs of the community [Jannach and Ludewig (2017)].

Sequence-aware Techniques: A number of sequence-aware techniques were explored in the literature as well. *Sequential patterns* are a counterpart of frequent pattern techniques and were investigated, e.g., in [Bonnin and Jannach (2014)]. In a related approach, Hariri *et al.* (2012) aimed at the identification of topic sequences in playlists instead of only looking at track sequences. Their results showed that considering these topic transitions leads to a performance improvement over a kNN technique. In another work, Park *et al.* (2011) presented a modified collaborative filtering method that uses session information to capture sequence and repetition patterns in listening sessions. They showed that their proposed session-based CF approach can outperform a basic CF method. A number of alternative and comparably simple sequence-based techniques were recently evaluated in the music domain and other domains in [Kamehkhosh *et al.* (2017); Ludewig and Jannach (2018)]. The investigated methods include, for example, sequential patterns of size two or neighborhood-based techniques that use a similarity functions which consider the order of the tracks in a session.

More elaborate techniques are based on *sequence learning* models. Some approaches of that type that are mentioned in the literature for example include Recurrent Neural Networks (RNNs) [Bernhardsson (2014)]. In principle, however, any form of algorithm that can be used for sequence-aware session-based recommendation can be applied to the music domain as well, including, for example, ones that use Markov-models in some form [Rendle *et al.* (2010)] or that rely on special types of RNNs [Hidasi *et al.* (2016a); Hidasi and Karatzoglou (2017)]. A potential drawback of such more sophisticated models is that they can easily become computationally challenging. At the same time, today's more sophisticated models are not necessarily better than the simple techniques mentioned above in terms of the prediction accuracy, as discussed in [Ludewig and Jannach (2018)].

Personalized Approaches. While *item discovery* approaches are almost always personalized, the work presented in [Jannach *et al.* (2017)] represents one of the few attempts in the literature where the process of creating a playlist continuation is personalized. The personalization features in this work are embedded within a multi-faceted track scoring scheme. First, a baseline relevance score for each recommendable track is computed based on a k-nearest-neighbor method as described above. Then, a variety of other relevance signals are considered in a weighted approach.

Technically, the overall relevance score $score_{overall}$ for a possible next track t^*, given the playlist beginning h, is computed as follows [Jannach *et al.* (2017)]:

$$score_{overall}(h, t^*) = w_{base} \cdot score_{base}(h, t^*) + \sum_{pers \in P} w_{pers} \cdot score_{pers}(h, t^*) \quad (15.4)$$

where P is a set of personalization strategies, each with a different weight w_{pers}, and w_{base} is the weight of the baseline. The functions $score_{base}$ and $score_{pers}$ compute the baseline score and the scores of the individual personalization components, respectively.

The following signals were considered for the personalization step:

(1) Favorite Tracks
(2) Favorite Artists
(3) Topic Similarity
(4) Extended Neighborhood
(5) Social Friends

Favorite Tracks: It is quite common in the domain that users listen to their favorite tracks again and again. Therefore, it is reasonable to also recommend tracks that the user has already heard (several times) in the past. Different strategies to select tracks for repeated recommendations are possible. One can, for example, repeatedly recommend tracks that are generally popular; or, one can recommend tracks that the user has listened to at the same time of the day in the past. Yet another approach is to repeatedly recommend tracks of artists that the user is listening to in the current session.

Since the literature suggests that users tend to repeatedly consume things that they have experienced more recently [Anderson *et al.* (2014)], one can furthermore give more weight to tracks that were more recently played by the user. Finally, the effectiveness of repeated recommendations can be further increased when the system can guess if the user is currently in the mood of discovering something new or rather prefers to listen to his or her favorites, as discussed in [Kapoor *et al.* (2015)].

Favorite Artists: Many music lovers have their favorite artists and it therefore might be a comparably safe strategy to recommend tracks of these artists to users, even when this might lead to limited discovery effects.[6] Technically, to compute an artist-based score, one can inspect the current and the past listening session of the user and assign higher scores to tracks that are performed by the artists that also appeared in these past listening sessions.

Topic Similarity: The assumption of this personalization score is that some users are interested in certain types of music, for example, mostly sad ballads or instrumental music. Therefore, recommendable tracks that in some respect are "content-wise" similar to those that the user listened to in the past should receive a higher rank. Generally, one can use all sorts of information to determine the similarity of objects, e.g., based on musical features. Since such information is not always available, one can also look at the publicly available tags that users assign to certain artists or tracks, for example, on Last.fm. Technically, again different similarity measures can be applied, for example, by using the cosine similarity between two TF-IDF encoded track representations.

Extended Neighborhood: The kNN-method described above merely looks for sessions that are similar to the current one. To consider the long-term personal preferences of the user, one can however also consider sessions that are similar to the *past* sessions of the user, maybe with a lower weight.

Social Friends: Finally, the last personalization approach considered in [Jannach *et al.* (2017)] takes the musical preferences of the user's social friends into account. Our musical tastes and preferences, as mentioned in the introduction, can to some extent be determined by the tastes of our social environment. One possible technical approach is therefore to recommend the favorite tracks of the user's social friends, giving more weight to social friends that are generally more popular and, e.g., have more followers.

Overall, experimental evaluations in [Jannach *et al.* (2017)] showed that all these personalization features can have a positive effect on the quality of the resulting recommendations. The best results are usually achieved when multiple signals are considered in parallel, which however requires that the importance weights are fine-tuned. Generally, the applicability of some approaches depends on the availability of the corresponding data, e.g., the information about the user' social connections.

[6]We will discuss questions of recommendation quality later on in more depth.

15.3. Challenges

Building a successful music recommender system can be challenging in a variety of dimensions. A key challenge clearly is to understand what makes a good recommendation in the first place, e.g., because the perception of music is highly subjective and context-dependent, as will be discussed in Section 15.4.

15.3.1. *Data-Related Aspects*

Huge amounts of data can be available for providers of a music streaming service. For larger services, there are millions of recommendable items and even more users.[7] Furthermore, there can be billions of streaming events, which can potentially be leveraged for building better recommendation models. As a result, scalability aspects can be a main concern when choosing or designing a recommendation algorithm. For the case of the implicit matrix factorization approach at Spotify, for example, only the play counts for the tracks were considered when optimizing the models. Obviously, however, one could also consider the temporal sequences of the events or even the musical features of the individual tracks to end up with better recommendations [Cai *et al.* (2007); Su *et al.* (2010); Dias and Fonseca (2013); Johnson (2014)].

Another data-related issue is concerning potential biases in the available data. There is a very long tail of musical tracks available on music platforms that have been barely listened to by anyone and many modern algorithms might mostly focus on the more popular items [Celma and Cano (2008)]. This might in turn lead to a "rich-get-richer" (popularity reinforcement) effect as a small fraction of the catalog receives most of the (mostly positive) feedback. Furthermore, when using listening logs as input, the recorded events might be, to some larger extent, influenced by a recommendation functionality that was provided by the platform.

15.3.2. *User Interaction Aspects*

The recommendation techniques discussed in this chapter are mostly based on implicit feedback, i.e., play events for the tracks. However, music platforms also provide different mechanisms for users to explicitly state their

[7] As of 2018, Spotify reports to have over 70 million paying subscribers and almost 160 million active users of the service (`https://press.spotify.com/es/about/`, accessed 8 March 2018).

preferences. Most platforms, for example, give the users the opportunity to rate individual tracks (recommendations), typically using a binary feedback scale. In addition, users can often skip individual tracks, which represents another form of explicit feedback.[8] At least on some platforms, like Microsoft Groove, users are initially asked to provide their preferences for certain artists or genres.

In the context of explicit preference statements, different challenges can arise and various design decisions have to be made.[9] The initial preference acquisition process, for example, should not represent a burden for the user. And, users should be given the opportunity to revise their statements later on. Regarding the ratings for individual tracks, we commonly see thumbs-up/thumbs-down approaches, probably because many users tend to give only extreme ratings or would find it too tedious to give fine-grained feedback, e.g., on a five-point scale. A specific phenomenon in that context mentioned in [Steck *et al.* (2015)] is that the provided explicit preference statements might not be fully reliable, and users sometimes use their preference statements to create a public image of themselves, but then in fact listen to other types of music more frequently.

Finer-grained forms of giving feedback are provided, however, on other media sites. Figure 15.3, for example, shows how users could give feedback to individual video recommendations at YouTube around the year 2015. Here, the users could not only state that they did not like a certain item, but also state that they either had seen it before or that they are not interested in a certain channel. Clearly, while such an approach gives more control to the users, it also increases the complexity and required user effort. At the same time, such fine-grained feedback forms can be difficult to operate on mobile devices, which are often used to consume streaming music.

In order to acquire short-term listening preferences, music platforms, as described above, for example let the user provide some initial track or artist, and base the subsequent tracks on that seed information. Since (a) the system might misinterpret the user's intentions or (b) users sometimes only has a vague idea of what they want to hear [Steck *et al.* (2015)], at least some form of allowing the user to fine-tune the playlists has to be provided. For the Spotify app, it was decided to provide thumbs-up/thumps-down buttons for the user. A particular challenge in that context, however, is that the

[8]In fact, skipping a track can be considered both as implicit and explicit feedback.

[9]For a recent review on user interaction aspects for recommender systems, see [Jugovac and Jannach (2017)].

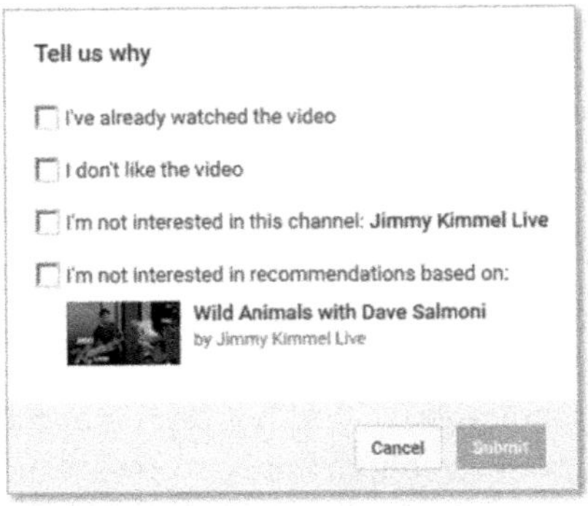

Fig. 15.3. Feedback on the YouTube platform (replaced with a simpler form as of 2018).

user's reactions should be "immediately gratifying", i.e., the playlist should be refreshed instantaneously, at least when there is negative feedback on the track. This might in turn lead to computational challenges and limited research exists in the literature that discusses algorithmic approaches to incorporate such feedback on the fly.

In addition, as mentioned in [Johnson (2014)], most of the provided feedback is positive and obviously only available for tracks that were actually recommended to users. As a result, this can easily lead to some algorithmic bias when optimizing only using feedback that was given to tracks that were presented to users.

Generally, not much research exists on how to design the user interfaces of music apps that feature a recommendation component. One typical aspect that is mentioned as a way to increase the adoption of the recommendations is the provision of explanations, so that the users have a chance to *understand* why a certain item was recommended [Lamere and Celma (2011)]. Other open questions include how to determine the actual users' context, e.g., their mood, activities, surroundings etc., either through explicit elicitation methods or through additional sensor information.

15.3.3. *Incorporating Song Feature Information*

Relying mostly or even solely on the collective behavior of a community when recommending using collaborative filtering techniques can have certain limitations, e.g., low prediction accuracy when there are limited amounts of feedback signals for recently added tracks (*item cold-start*). Therefore, in this domain, considering the musical features of the recommendable tracks can be helpful to avoid unsuitable elements in a playlist, e.g., high-energy tracks within a chill-out playlist. This is in particular important since such unsuitable recommendations can have a measurable negative effect on the user's quality perception of the service [Chau *et al.* (2013)].

Traditional *content-based* recommendation approaches typically suffer less from such problems. The basic idea of such algorithms is to recommend items that are similar to the ones that the user has liked in the past. Technically, this is usually achieved by representing both users and items in terms of the features of the musical tracks. The features can include both musical characteristics as well as track meta-data such as release years. The shared form of representation can then be used to compute the match (similarity) between a given user profile and a recommendable track. A common way of representing the items is to use vectors whose dimensions correspond to content or meta-data features, such as the tempo, the loudness, the different possible genres, etc. Users are also represented based on their preferences, which either can be obtained by asking the user to explicitly provide some preference information (e.g., using a questionnaire), or can be extracted from the music that the user has listened to in the past.

In principle, a large number of musical features can be used, including low-level ones related to the timbre or rhythmic aspects as well as more high-level ones like the instrumentation or the "danceability" of a track. Pandora.com, as mentioned above, relies on manual annotations by experts. An alternative is to automatically extract the information from various sources, including the audio signal as well as other data sources like social annotations or lyrics. Extracting such features is a key topic in the field of music information retrieval and significant progress has been made in the last years, see, e.g., [Casey *et al.* (2008); Müller (2015); Weihs *et al.* (2016)]. The extraction process can, however, lead to inaccurate results in some cases. At the same time, the process can be computationally challenging, given the huge number of available tracks.

Since also content-based methods have limitations, e.g., their tendency to recommend "more of the same", a large variety of *hybrid* algorithms were proposed in the literature on recommender systems. Such hybrids combine different techniques and are usually designed in order to overcome the limitations of the individual techniques. A key challenge in that context is how to merge the available types of information in the best possible way.

An early hybrid method was proposed by Yoshii *et al.* (2006), who used a probabilistic model in the form of a Bayesian network to integrate user ratings collected from Amazon and content data represented as mel-frequency cepstral coefficients. Their results showed that the hybrid method achieves higher accuracy and more diverse recommendations in terms of artists than the two underlying methods. In a more recent work, Wang and Wang (2014) developed an approach to combine a collaborative filtering method based on probabilistic matrix factorization [Salakhutdinov and Mnih (2007)] and content features that were learned automatically via a deep belief network [Hinton *et al.* (2006)].

Generally, a number of ways exist to combine the recommendations of different algorithms, e.g., by weighting the individual prediction scores or by using the recommendations generated by one algorithm as a pre-filtered input to another. Alternatively, machine learning models like Factorization Machines can be applied that combine the different types of input data in an integrated model.

Overall, in many application domains of music recommendation hybrid techniques can be considered the method of choice. When creating a playlist for a virtually endless radio station, for example, the application of collaborative filtering techniques can increase the probability that the recommendations include tracks that are new to the user ("discovery"). Using content-based techniques in parallel can, at the same time, help to ensure that the tracks played in the future do not deviate too much from the seed tracks in terms of their musical features. A general challenge in the context of such hybrid systems, however, is how to combine the different techniques in the best possible way.

15.4. Evaluation

The evaluation and comparison of different music recommendation strategies can be challenging, both for music service providers and for academic researchers. After discussing the determination of the relevant quality criteria in general and how to balance these criteria, this section will focus on the assessment of performance in real world settings and in academic environments.

15.4.1. *Quality Criteria*

Generally, the acceptance of a music recommendation service depends on the quality and utility perception of its users. Regarding the specific recommendation scenario, a number of different factors can have an influence on the users' perceptions. The following list gives examples of such factors, as mentioned also in [Bonnin and Jannach (2014)] in the context of the playlist generation problems. In general, these factors can be *music-related* and *purpose-oriented* [Jannach and Adomavicius (2016)].

Examples of *music-related* factors:

- *Diversity:* The recommendations (e.g., a list of tracks to be played next) should be sufficiently different from each other, e.g., in terms of their artist [Slaney and White (2006); Lee *et al.* (2011); Kamalzadeh *et al.* (2012); Puthiya Parambath *et al.* (2016)].
- *Homogeneity:* At the same time (e.g., in the context of playlist construction support), it is often desirable that the next tracks are not too different from each other, for instance, in terms of their genre, tempo, or mood [Logan (2002); Balkema and van der Heijden (2010); Jannach *et al.* (2015)].
- *Coherence:* Also, it is often important that the recommended tracks do not only fit to each other, but also represent coherent continuations to the previously played tracks [Kamehkhosh and Jannach (2017)].
- *Transitions:* In some situations, for example, when creating playlists for parties, the differences between subsequent tracks, e.g., in terms of their tempo, should in general not be too high [Flexer *et al.* (2008); Sarroff and Casey (2012)].

Examples of *purpose-oriented* factors:

- *Discovery / Novelty*: The recommendations could, for example, be designed to be helpful for users to discover new artists, tracks, or genres [Celma (2010); Zhang *et al.* (2012)].
- *Agreeableness*: In particular, when a set of music track recommendations should be listened to by a group of users (e.g., at a party), it might be important that the music is enjoyed by the majority, which can, for example, be achieved by mainly focusing on generally popular tracks [Popescu and Pu (2012)].
- *Context-fit*: As described in [Bonnin and Jannach (2014)], playlists are often created for a certain "theme" or context (e.g., sports workout). The quality of the recommendations then depends on their fit for the given theme or context.

As can be easily seen, the different quality criteria can be antagonistic, e.g., agreeableness vs. discovery or homogeneity vs. diversity, and trade-offs have to be found. How to balance these aspects can depend on the specific application scenario.

15.4.2. *Balancing Different Quality Factors*

Given these potentially conflicting goals, it is often insufficient to only consider the relevance of a certain item in isolation. Instead, the suitability of an entire list of tracks to be played next has to be taken into account. Usually, however, considering additional quality factors other than prediction accuracy leads to a trade-off situation as multiple, possibly competing goals have to be balanced.

In the research literature on recommender systems, a number of approaches were proposed to deal with such trade-off situations and to improve recommendations by considering additional quality factors, e.g., in [Adomavicius and Kwon (2012); Ziegler *et al.* (2005); Bradley and Smyth (2001); Zhang and Hurley (2008)], or [Vargas and Castells (2011)].

Most of these approaches however (i) consider only two quality factors (e.g., accuracy vs. diversity) and (ii) do not consider the user's individual tendencies (e.g., with respect to diversity in general). Some more recent works that try to overcome these limitations can be found in [Kapoor *et al.* (2015); Shi *et al.* (2012); Oh *et al.* (2011)], and [Ribeiro *et al.* (2014)]. These approaches are however often designed for a very specific quality factor (e.g., diversity). Or, they implement the balancing strategy within their own recommendation algorithm so that existing algorithmic frameworks and approaches cannot be reused.

An alternative, two-phase approach that can be combined with any existing item-ranking algorithm was proposed in the context of next-track music recommendation problems in [Jannach *et al.* (2015)]. Similar to works like [Adomavicius and Kwon (2012)], the general idea of the method is to take a limited set of k (e.g., $k = 30$) most relevant items according to an arbitrary item-ranking or scoring algorithm and then to re-rank these top items in a way that selection of the top-10 items optimizes some quality criterion. In the context of the "radio" problem, this could mean to take the top 30 tracks according to their predicted suitability as a continuation for the last played tracks and then re-rank the items in order to maximize the artist diversity within the 10 next tracks. To compute such a re-ranked list, a greedy approach can be applied, which can compute solutions in real time that are very close to the true optimum.

A general question in such problem settings, however, is to determine the *right amount* of artist diversity. Some approaches use a combined quality measure that considers both accuracy and diversity. This, however, assumes that there is a globally accepted level of, e.g., diversity. In reality, the appropriate level of diversity (or homogeneity etc.) can be dependent on different factors, including the diversity of current playlist or even the general long-term preferences of the particular user. The method proposed in [Jugovac *et al.* (2017)] takes such considerations into account. Specifically, the implemented optimization procedure can be configured to be guided by the characteristics of a "seed" of items and the optimization goal then consists of *minimizing the difference* between the seed items' characteristics and the characteristics of the recommendation list.

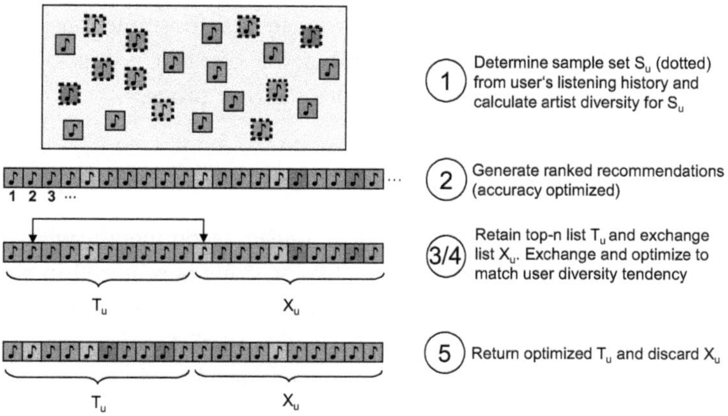

Fig. 15.4. Illustration of the re-ranking scheme, adapted from [Jugovac *et al.* (2017)].

Figure 15.4 illustrates the general idea based on the artist diversity problem. The elements in dotted lines in the upper rectangle (marked with ①) represent the selected seed items and the different colors represent different artists. The seed set can for example be taken from the set of the user's last played tracks or be a subset of the user's favorite tracks. Based on this seed set, the user's individual *tendency* towards artist diversity can be computed. In the next step, the ranked list of recommendations (next tracks for a given playlist history) is computed with any algorithm. Again, different colors in step ② represent different artists. Since the top-10 items of the list have a lower artist diversity than the items in the seed set, the algorithm then starts exchanging elements from the top of the list with elements from the end of the list, which probably have a slightly lower predicted relevance but help to improve the diversity of the top-10 list. Improving in that context means that the algorithm tries to minimize the difference between the diversity level of the top-10 list and the seed tracks. Therefore, if a user generally prefers lists with high diversity, the re-ranking will lead to higher diversity. Vice versa, a user who usually listens to various tracks by the same artist in a session will receive recommendations with a lower artist diversity. As a result, the definition of a globally desired artist diversity level can be avoided.

Generally, there can be multiple and possibly conflicting optimization goals that should be considered in parallel, e.g., high artist diversity, homogeneous tempo, smooth transitions. Such situations can be considered in the approach as long as the relative importance of the difference factors can be specified. The acceptable compromises on recommendation accuracy can also be fine-tuned by determining the size of the top-k list from which items can be picked during the optimization process.

15.4.3. *Performance Assessment in Real-World Settings*

In order to assess the performance of a music recommendation service, an understanding about its expected utility or value is required in the first place. Recommendation services on music platforms are primarily designed to support the user during a certain task like playlist construction or to provide some other functionality like a personalized radio station. While such services often do not directly lead to additional revenue for the provider as in the e-commerce domain, recommendation services on music platforms can represent an additional value for the consumer (e.g., in terms of better a user experience or an easier discovery of new things). This can, in turn, lead to a higher customer retention and an indirect business value.

In order to quantify the effectiveness of a recommendation service, different basic measurements and analyses can be made. First, one can measure the *adoption* of the service, i.e., determine which fraction of the users actually use the service over an extended period of time to a significant extent. For certain services like a personalized radio station, one can also analyze the specific user behavior and their feedback to the recommendations. Do users, for example, skip certain recommended tracks or are they continuing to listen to the recommended tracks most of the time? Such measurements regarding the acceptance of the recommendation can be made and compared in field (A/B) tests, possibly accompanied by user surveys. Finally, since the user interface (UI) design can have a significant impact on the adoption and effectiveness of some recommendation services [Steck *et al.* (2015)], the UI design should be evaluated as well through laboratory studies, A/B tests, or both.

The indirect business value of a music recommendation service can typically only be assessed through field tests and the specific measurement depends on the business model of the provider. For flat-rate subscription-based services like Spotify or others, user engagement and customer retention (subscription renewals) might be a measure that one wants to optimize. Another potential measure is the conversion of users of a free service to premium (paying) users.[10] If the music recommendation service on the other hand is used to recommend items in a pay-per-stream or pay-per-download model, the overall revenue or profit usually represent the target measures.

A main challenge in that context is that for many complex machine learning approaches a number of parameters have to be optimized, and that not all parameter settings can be tested in field studies. Therefore, algorithms are usually optimized regarding some *proxy* measures in an offline process (e.g., precision or recall). The challenge however is that for many of these proxy measures, it is unclear if they truly correlate with the target measure that one wants to optimize in reality (e.g., customer retention). In fact, a number of recent works suggest that success measures often used in academia, in particular the prediction accuracy, are often not good indicators of the true success of a recommendation service, see also the discussion of Netflix's recommendation service in [Gomez-Uribe and Hunt (2015)]. Another challenge when optimizing accuracy measures in an offline process is that the observed data (e.g., thumbs) can be biased in different ways. Often, the large majority of the signals consists of only positive feedback. And, which items are actually listened to (and rated) by consumers can itself be heavily influenced by the existing recommendation service or other functionalities of the platform, e.g., lists of trending items.

15.4.4. *Comparing Algorithms in Academic Environments*

Similar to other application domains of recommenders, the predominant form of evaluating recommendation algorithms in the music domain is based on offline experiments, using either explicit user preference data or recorded user activity logs.

[10]See [Bernhardsson (2017)] for a discussion of potential issues when computing conversion rates.

Datasets. In case the goal is to predict user preferences (i.e., the relevance of individual items) for a user given a common user-item preference matrix, "standard" evaluation schemes for the matrix completion setup from other application domains can be applied.

For many application scenarios of music recommenders — and in particular for personalized radio stations — using recorded listening logs as a basis for offline experiments however seems more appropriate. A number of public datasets are available online. Some of them contain the listening histories of thousands of users over an extended period of time. Often, these logs were obtained through the public APIs of music services like Last.fm. Table 15.2 gives an overview of a number of public datasets.

Finally, some research works on collaborative music recommendation are based on publicly shared "hand-crafted" playlists (mixtapes), i.e., sequences of tracks that were put together and shared by users of music services for a certain purpose, e.g., for a road trip or a sports workout. Different datasets containing such playlists are publicly available today, see also Table 15.2.

Evaluation Protocols and Metrics. Comparing the prediction accuracy of different music recommendation strategies in terms of information retrieval or machine learning measures is common in the academic literature. In case sequential user activity logs or playlists are the basis for the evaluation, a typical evaluation procedure is to use a "session-wise" approach as discussed in Chapter 1. For each listening session or playlist, a defined number of tracks from the beginning of a session are revealed and the task of the recommender is to predict the next (hidden) track(s). Such a research approach was, for example, used by Hariri *et al.* (2012) or by Bonnin and Jannach (2014), where standard information retrieval measures were applied. An alternative evaluation approach based on the *Average Log-Likelihood* was proposed by Mcfee and Lanckriet (2011), a measure which can be used to assess how likely a system is to generate the tracks of a given playlist or listening session. This measure, however, has certain limitations, as discussed in [Bonnin and Jannach (2014)].

---

[11]The Echo Nest has been acquired by Spotify in 2014. A similar Web API has since been made available by Spotify, see `https://developer.spotify.com/web-api`
[12]`http://webscope.sandbox.yahoo.com`
[13]`https://labrosa.ee.columbia.edu/millionsong/tasteprofile`
[14]`http://www.kkbox.com`
[15]`https://www.kaggle.com/c/kkbox-music-recommendation-challenge/leaderboard`
[16]`http://www.artofthemix.org`

Table 15.2. Selection of public datasets.

Type	Name	Description
Music meta-data	Million Song Dataset	Made available in 2011 by researchers from the Music Information Retrieval community [Bertin-Mahieux *et al.* (2011)]. The dataset consists of meta-data for 1 million tracks of 44,745 artists, including tags, date of release and several acoustic features from The Echo Nest.[11]
Rating data	Yahoo! Music	Artist and track ratings from data sets published by Yahoo! Research.[12]
	Taste Profile subset	An addition to the Million Song Dataset that contains real user play counts for a subset of the tracks (384,546 tracks) and was released by The Echo Nest.[13]
	Amazon product co-purchasing network metadata	Made available in [Leskovec *et al.* (2007)]. It contains various reviews and review information such as ratings, number of votes, etc. on 548,552 items of different types (books, music CDs, DVDs and VHS video tapes).
	InCarMusic	Constructed for the work presented in [Baltrunas *et al.* (2011)]. It contains user ratings of genres and car-related context information, such as driving style (relaxed or sport), road type (city, highway or serpentine), etc. for about 140 tracks of 10 different genres.
Listening logs	Last.fm	Celma (2010) published two datasets containing the listening logs of about 1,000 and 360,000 users, respectively, collected from Last.fm API.
	30Music	A collection of listening and playlists data retrieved from Internet radio stations through Last.fm API. The datasets consists of 31 million user play events, 2.7 million user play sessions, and 4.1 million user "love" statements and was published by Turrin *et al.* (2015).
	KKBOX	KKBOX is a music streaming service provider in East Asia.[14] In the context of a machine learning competition[15], listening logs of over 34,000 users were published together with some information about over 400,000 individual tracks.
Listening logs extracted from microblogs	MMTD	The Million Musical Tweet Dataset includes listening histories based on more than 1 million tweets referring to 133,968 unique tracks by 25,060 different artists created by 215,375 users. The dataset was introduced by Hauger *et al.* (2013).
	#nowplaying	Contains about 40 million listening events extracted from music-related tweets of users on Twitter. The dataset is enriched with additional information about the artist, the track title, and metadata about the tweet and was published by Zangerle *et al.* (2014).
Music playlists	Art of the Mix	A dataset of hand-crafted playlists made available by McFee and Lanckriet (2012). It contains all playlists (about 100,000) of the Website Art of the Mix[16] created before June 2011. Each playlist contains artist names, track names, creator name, a categorical label (e.g., "Reggae") and the identifiers of the tracks that could be found in the Million Song Dataset.
	Kollect.fm	Contains about 35,000 playlists with feedback about them from about 16,000 users retrieved from kollect.fm, which is a music discovery website where users can receive recommendation of playlists.

Focusing solely on accuracy measures has its limitations also in the music domain. Predicting mostly very popular tracks in fact proves to be a very competitive strategy. In [Bonnin and Jannach (2013)], the authors proposed a popularity-based method called "Collocated Artists – Greatest Hits". This method simply takes the artists who appear in the given listening session and then recommends to play the most popular tracks of these artists and of similar artists, where the artist similarity is determined based on their co-occurrence in past listening sessions. Experimental evaluations show that this method is particularly effective in guessing the immediate next tracks.

Recommendation lists that include only very popular tracks might however not be satisfying for many users, e.g., because these recommendations lead to limited discovery and low artist diversity. Therefore, researchers sometimes apply multi-metric evaluation schemes which consider both prediction or ranking accuracy and other quality factors like the homogeneity of the recommended tracks, artist diversity, coherence with the previous tracks, or the transitions between the tracks [Jannach *et al.* (2016, 2017)]. As discussed above, different techniques can then be applied to balance the given quality factors, see [Jugovac *et al.* (2017)] for a comparison of recent approaches, which also showed that considering additional quality factors can even lead to an improvement in terms of ranking accuracy. Generally, the problem in these situations is not only to assess the relevance of the individual recommended items, but to consider quality factors that are determined by the characteristics of the recommendation list as a whole.

Research works that are based on recorded user activity logs have additional limitations. In particular, it is not clear for datasets that are obtained, e.g., from Last.fm, to which extent the users' listening activity is "natural" in the sense that the users selected one track after the other when listening. In reality, larger parts of log entries might be the result of the existing "radio" service of the platform. Evaluating algorithms in terms of predicting the next track in the log then amounts to predicting what the existing recommendation service would play. Also, in some datasets, we can observe that users listen to entire albums and the sequence of the tracks then often corresponds to the order of the tracks on the album. Predicting the next track is then comparably trivial and an algorithm that is capable of detecting album listening sessions will achieve high prediction accuracy values for these cases.

Offline evaluation approaches have their limitations, as discussed above, in particular as it is not always clear if the chosen computational metrics are good estimators of (a) the quality perception of users and (b) of the business success of the recommendation service. User studies on music recommendation approaches are comparably rare. Examples of such studies can be found in [Barrington *et al.* (2009)] and, more recently, in [Kamehkhosh and Jannach (2017)] and [Kamehkhosh *et al.* (2018)]. In particular, the work presented in [Kamehkhosh and Jannach (2017)] indicated that using handcrafted playlists can represent a reasonable "gold standard" for evaluating playlist generation techniques. The work, however, also revealed potential limitations of user studies in terms of familiarity effects, i.e., users prefer recommendations when they contain tracks they already know.

15.5. Lessons Learned, Open Challenges, and Outlook

Collaborative filtering based recommendations are nowadays a common functionality on several music platforms. And, as described in the chapter, at least some of these platforms rely on elaborate algorithms, e.g., based on matrix factorization, to create personalized recommendations.

Several challenges of building such recommendation services have been discussed throughout the chapter, including in particular the problems of scalability, the importance of considering short-term trends, or the general problem of finding good proxy measures for offline evaluation scenarios.

Pure collaborative filtering approaches, in general, have a number of limitations and this applies also for the music domain, where we for example have the phenomenon that constantly new items are available on the platform. To recommend such items, relying on "content" information (e.g., musical signals, metadata, lyrics) is a viable approach to assess the relevance of new tracks for individual users. How to combine the multitude of different signals in the best possible way in hybrid recommendation techniques is in our view an area which requires additional research. Recently, Jannach *et al.* (2017) investigated the use of rich user models that combine collaborative filtering techniques with such content information and social information, showing that considering all of these signals can be beneficial. This work was based only on a very limited set of musical features and simple weighting schemes. More elaborate approaches to derive and process a variety of signals, e.g., based on deep learning approaches, already exist today and should be further explored in the future [van den Oord *et al.* (2013); Wang *et al.* (2015); Hidasi *et al.* (2016b)].

Another music recommendation scenario that is not fully investigated in the literature is group-based music recommendation. Consider, for example, playlist creation for a party or music selection in public places. In such scenarios, the recommended items are consumed by a group of listeners rather than by individuals and the recommendations should, therefore, satisfy the entire group as a whole. Generally, group-based recommendation strategies either aggregate individual user models to build a group profile, or aggregate individual predictions for users to generate group-based recommendations [Berkovsky and Freyne (2010)].

More research is also required in terms of understanding which factors influence the quality perception by users. A number of questions are open in that context, for example: How much diversity, e.g., in terms of artists or genres, is appropriate? To which extent is the diversity level depending on the current user's context, e.g., mood? How do we quantify diversity with a computational metric and how do we know that the computational metric corresponds to the user's quality perception? How problematic are "bad" recommendations — can the recommendation of one or a few unsuitable items have a significant negative effect on the acceptance of the service?[17]

Finally, many of the mentioned quality factors can depend on the user's current context. Limited research exists on the contextualization of recommendations, see, for instance, [Hariri *et al.* (2012)] or [Kapoor *et al.* (2015)], which mainly try to derive the user's context and short-term preferences from their behavior. So far, to the best of our knowledge, only the InCarMusic dataset [Baltrunas *et al.* (2011)] contains additional information about the users' context.

References

Adomavicius, G. and Kwon, Y. (2012). Improving Aggregate Recommendation Diversity Using Ranking-Based Techniques, *IEEE TKDE* **24**, 5, pp. 896–911.

Anderson, A., Kumar, R., Tomkins, A. and Vassilvitskii, S. (2014). The Dynamics of Repeat Consumption, in *WWW '14*, pp. 419–430.

Bachrach, Y., Finkelstein, Y., Gilad-Bachrach, R., Katzir, L., Koenigstein, N., Nice, N. and Paquet, U. (2014). Speeding Up the Xbox Recommender System Using a Euclidean Transformation for Inner-Product Spaces, in *RecSys '14*, pp. 257–264.

Balkema, W. and van der Heijden, F. (2010). Music Playlist Generation by Assimilating GMMs into SOMs, *Pattern Recognition Letters* **31**, 11, pp. 1396–1402.

[17]See [Chau *et al.* (2013)] for a user study on the topic of bad recommendations.

Baltrunas, L., Kaminskas, M., Ludwig, B., Moling, O., Ricci, F., Aydin, A., Lüke, K.-H. and Schwaiger, R. (2011). InCarMusic: Context-Aware Music Recommendations in a Car, in *EC-Web '11*, pp. 89–100.

Barrington, L., Oda, R. and Lanckriet, G. R. G. (2009). Smarter than Genius? Human Evaluation of Music Recommender Systems, in *ISMIR '09*, pp. 357–362.

Berkovsky, S. and Freyne, J. (2010). Group-based Recipe Recommendations: Analysis of Data Aggregation Strategies, in *RecSys '10*, pp. 111–118.

Bernhardsson, E. (2013). Music Recommendations at Spotify, Online `https://de.slideshare.net/erikbern/collaborative-filtering-at-spotify-16182818`.

Bernhardsson, E. (2014). Recurrent Neural Networks for Collaborative Filtering, Online `https://erikbern.com/2014/06/28/recurrent-neural-networks-for-collaborative-filtering.html`.

Bernhardsson, E. (2015). Approximate Nearest Neighbor Methods and Vector Models, Online https://de.slideshare.net/erikbern/approximate-nearest-neighbor-methods-and-vector-models-nyc-ml-meetup.

Bernhardsson, E. (2017). Conversion Rates — You Are (Most Likely) Computing Them Wrong, Online `https://erikbern.com/2017/05/23/conversion-rates-you-are-most-likely-computing-them-wrong.html`.

Bertin-Mahieux, T., Ellis, D. P., Whitman, B. and Lamere, P. (2011). The Million Song Dataset, in *ISMIR '11*, pp. 591–596.

Bieschke, E. (2014). Pandora, presentation at MLconf2013, Online `https://de.slideshare.net/SessionsEvents/eric-bieschke-slides`.

Bonnin, G. and Jannach, D. (2013). Evaluating the Quality of Generated Playlists Based on Hand-Crafted Samples, in *ISMIR '13*, pp. 263–268.

Bonnin, G. and Jannach, D. (2014). Automated Generation of Music Playlists: Survey and Experiments, *Computing Surveys* **47**, 2, pp. 26:1–26:35.

Bradley, K. and Smyth, B. (2001). Improving Recommendation Diversity, in *AICS '01*, pp. 75–84.

Cai, R., Zhang, C., Zhang, L. and Ma, W.-Y. (2007). Scalable Music Recommendation by Search, in *MM '07*, pp. 1065–1074.

Casey, M. A., Veltkamp, R., Goto, M., Leman, M., Rhodes, C. and Slaney, M. (2008). Content-Based Music Information Retrieval: Current Directions and Future Challenges, *Proceedings of the IEEE* **96**, 4, pp. 668–696.

Celma, Ò. (2010). *Music Recommendation and Discovery in the Long Tail* (Springer).

Celma, O. and Cano, P. (2008). From hits to niches?: Or How Popular Artists can Bias Music Recommendation and Discovery, in *NETFLIX '08*, pp. 5:1–5:8.

Chau, P. Y. K., Ho, S. Y., Ho, K. K. W. and Yao, Y. (2013). Examining the Effects of Malfunctioning Personalized Services on Online Users' Distrust and Behaviors, *Decision Support Systems* **56**, pp. 180–191.

Chen, C.-M., Tsai, M.-F., Lin, Y.-C. and Yang, Y.-H. (2016). Query-based Music Recommendations via Preference Embedding, in *RecSys '16*, pp. 79–82.

Cliff, D. (2006). hpDJ: An Automated DJ with Floorshow Feedback, in *Consuming Music Together*, pp. 241–264.

Dias, R. and Fonseca, M. J. (2013). Improving Music Recommendation in Session-Based Collaborative Filtering by Using Temporal Context, in *ICTAI '13*, pp. 783–788.

Ekelund, R. B., Ford, G. S. and Koutsky, T. (2000). Market Power in Radio Markets: An Empirical Analysis of Local and National Concentration, *The Journal of Law and Economics* **43**, 1, pp. 157–184.

Flexer, A., Schnitzer, D., Gasser, M. and Widmer, G. (2008). Playlist Generation Using Start and End Songs, in *ISMIR '08*, pp. 173–178.

Germain, A. and Chakareski, J. (2013). Spotify Me: Facebook-Assisted Automatic Playlist Generation, in *MMSP '13*, pp. 25–28.

Gomez-Uribe, C. A. and Hunt, N. (2015). The Netflix Recommender System: Algorithms, Business Value, and Innovation, *Transactions on Management Information Systems* **6**, 4, pp. 13:1–13:19.

Hariri, N., Mobasher, B. and Burke, R. (2012). Context-Aware Music Recommendation Based on Latent Topic Sequential Patterns, in *RecSys '12*, pp. 131–138.

Hartono, P. and Yoshitake, R. (2013). Automatic Playlist Generation from Self-Organizing Music Map, *Journal of Signal Processing* **17**, 1, pp. 11–19.

Hauger, D., Schedl, M., Kosir, A. and Tkalcic, M. (2013). The Million Musical Tweet Dataset - What We Can Learn From Microblogs, in *ISMIR '13*, pp. 189–194.

Hidasi, B. and Karatzoglou, A. (2017). Recurrent Neural Networks with Top-k Gains for Session-based Recommendations, *CoRR* **abs/1706.03847**.

Hidasi, B., Karatzoglou, A., Baltrunas, L. and Tikk, D. (2016a). Session-based Recommendations with Recurrent Neural Networks, in *ICLR '16*.

Hidasi, B., Quadrana, M., Karatzoglou, A. and Tikk, D. (2016b). Parallel Recurrent Neural Network Architectures for Feature-rich Session-based Recommendations, in *RecSys '16*, pp. 241–248.

Hinton, G. E., Osindero, S. and Teh, Y.-W. (2006). A Fast Learning Algorithm for Deep Belief Nets, *Neural Computation* **18**, 7, pp. 1527–1554.

Hu, Y., Koren, Y. and Volinsky, C. (2008). Collaborative Filtering for Implicit Feedback Datasets, in *ICDM '08*, pp. 263–272.

Jannach, D. and Adomavicius, G. (2016). Recommendations with a Purpose, in *RecSys '16*, pp. 7–10.

Jannach, D., Kamehkhosh, I. and Bonnin, G. (2016). Biases in Automated Music Playlist Generation, in *UMAP '16*, pp. 281–285.

Jannach, D., Kamehkhosh, I. and Lerche, L. (2017). Leveraging Multi-Dimensional User Models for Personalized Next-Track Music Recommendation, in *SAC '17*, pp. 1635–1642.

Jannach, D., Lerche, L. and Kamehkhosh, I. (2015). Beyond "Hitting the Hits" – Generating Coherent Music Playlist Continuations with the Right Tracks, in *RecSys '15*, pp. 187–194.

Jannach, D. and Ludewig, M. (2017). When Recurrent Neural Networks meet the Neighborhood for Session-Based Recommendation, in *RecSys '17*, pp. 306–310.

Jawaheer, G., Szomszor, M. and Kostkova, P. (2010). Comparison of Implicit and Explicit Feedback from an Online Music Recommendation Service, in *Workshop on information heterogeneity and fusion in recommender systems*, pp. 47–51.

Johnson, C. (2014). Algorithmic Music Discovery at Spotify, Online `https://de.slideshare.net/MrChrisJohnson/algorithmic-music-recommendations-at-spotify`.

Johnson, C. and Newett, E. (2014). From Idea to Execution: Spotify's Discover Weekly, Online `https://de.slideshare.net/MrChrisJohnson/from-idea-to-execution-spotifys-discover-weekly/12-Insight_users_spending_more_time`.

Jugovac, M. and Jannach, D. (2017). Interacting with Recommenders — Overview and Research Directions, *ACM Transactions on Intelligent Interactive Systems (ACM TiiS)* **7**.

Jugovac, M., Jannach, D. and Lerche, L. (2017). Efficient Optimization of Multiple Recommendation Quality Factors According to Individual User Tendencies, *Expert Systems With Applications* **81**, pp. 321–331.

Jylhä, A., Serafin, S. and Erkut, C. (2012). Rhythmic Walking Interactions with Auditory Feedback: an Exploratory Study, in *AM '12*, pp. 68–75.

Kamalzadeh, M., Baur, D. and Möller, T. (2012). A Survey on Music Listening and Management Behaviours, in *ISMIR '12*, pp. 373–378.

Kamehkhosh, I. and Jannach, D. (2017). User Perception of Next-Track Music Recommendations, in *UMAP '17*, pp. 113–121.

Kamehkhosh, I., Jannach, D. and Bonnin, G. (2018). How Automated Recommendations Affect the Playlist Creation Behavior of Users, in *Proceedings of the Workshop on Intelligent Music Interfaces for Listening and Creation at IUI '18*.

Kamehkhosh, I., Jannach, D. and Ludewig, M. (2017). A Comparison of Frequent Pattern Techniques and a Deep Learning Method for Session-Based Recommendation, in *Workshop on Temporal Reasoning in Recommender Systems at RecSys '17*, pp. 50–56.

Kapoor, K., Kumar, V., Terveen, L., Konstan, J. A. and Schrater, P. (2015). "I Like to Explore Sometimes": Adapting to Dynamic User Novelty Preferences, in *RecSys '15*, pp. 19–26.

Koren, Y. (2008). Factorization Meets the Neighborhood: A Multifaceted Collaborative Filtering Model, in *KDD '08*, pp. 426–434.

Lamere, P. and Celma, O. (2011). Music Recommendation and Discovery Remastered, Tutorial at ACM RecSys 2011, Online `https://musicmachinery.com/2011/10/24/music-recommendation-and-discovery-remastered-a-tutorial/`.

Lee, J. H., Bare, B. and Meek, G. (2011). How Similar Is Too Similar?: Exploring Users' Perceptions of Similarity in Playlist Evaluation, in *ISMIR '11*, pp. 109–114.

Leskovec, J., Adamic, L. A. and Huberman, B. A. (2007). The Dynamics of Viral Marketing, *ACM Trans. Web* **1**, 1.

Logan, B. (2002). Content-Based Playlist Generation: Exploratory Experiments, in *ISMIR '02*, pp. 295–296.

Loni, B., Pagano, R., Larson, M. and Hanjalic, A. (2016). Bayesian Personalized Ranking with Multi-Channel User Feedback, in *RecSys '16* (ACM), pp. 361–364.

Ludewig, M. and Jannach, D. (2018). Evaluation of Session-Based Recommenation Algorithms, arXiv:1803.09587 [cs.IR], https://arxiv.org/abs/1803.09587.

Mcfee, B. and Lanckriet, G. (2011). The Natural Language of Playlists, in *ISMIR '11*, pp. 537–541.

McFee, B. and Lanckriet, G. R. (2012). Hypergraph Models of Playlist Dialects, in *ISMIR '12*, pp. 343–348.

Moens, B., van Noorden, L. and Leman, M. (2010). D-Jogger: Syncing Music With Walking, in *SMC '10.*

Moling, O., Baltrunas, L. and Ricci, F. (2012). Optimal Radio Channel Recommendations with Explicit and Implicit Feedback, in *RecSys '12*, pp. 75–82.

Moore, J. L., Chen, S., Joachims, T. and Turnbull, D. (2012). Learning to Embed Songs and Tags for Playlist Prediction, in *ISMIR '12*, pp. 349–354.

Müller, M. (2015). *Fundamentals of Music Processing: Audio, Analysis, Algorithms, Applications* (Springer International Publishing), doi:10.1007/978-3-319-21945-5.

Oh, J., Park, S., Yu, H., Song, M. and Park, S. (2011). Novel Recommendation Based on Personal Popularity Tendency, in *ICDM '11*, pp. 507–516.

Oliver, N. and Flores-Mangas, F. (2006). MPTrain: A Mobile, Music and Physiology-Based Personal Trainer, in *MobileHCI '06*, pp. 21–28.

Omohundro, S. M. (1989). Five Balltree Construction Algorithms, Tech. rep., International Computer Science Institute, Berkeley, California.

Pálovics, R., Benczúr, A. A., Kocsis, L., Kiss, T. and Frigó, E. (2014). Exploiting Temporal Influence in Online Recommendation, in *ACM '14*, pp. 273–280.

Park, S. E., Lee, S. and Lee, S.-G. (2011). Session-Based Collaborative Filtering for Predicting the Next Song, in *CNSI '11*, pp. 353–358.

Pauws, S., Verhaegh, W. and Vossen, M. (2008). Music Playlist Generation by Adapted Simulated Annealing, *Information Sciences* **178**, 3, pp. 647–662.

Popescu, G. and Pu, P. (2012). What's the Best Music You Have?: Designing Music Recommendation for Group Enjoyment in Groupfun, in *CHI '12*, pp. 1673–1678.

Puthiya Parambath, S. A., Usunier, N. and Grandvalet, Y. (2016). A Coverage-Based Approach to Recommendation Diversity On Similarity Graph, in *RecSys '16*, pp. 15–22.

Quadrana, M., Cremonesi, P. and Jannach, D. (2018). Sequence-Aware Recommender Systems, *ACM Computing Surveys.*

Rendle, S., Freudenthaler, C. and Schmidt-Thieme, L. (2010). Factorizing Personalized Markov Chains for Next-basket Recommendation, in *WWW '10*, pp. 811–820.

Ribeiro, M. T., Ziviani, N., Moura, E. S. D., Hata, I., Lacerda, A. and Veloso, A. (2014). Multiobjective Pareto-Efficient Approaches for Recommender Systems, *ACM Transactions on Intelligent Systems and Technology* **5**, 4, pp. 1–20.

Salakhutdinov, R. and Mnih, A. (2007). Probabilistic Matrix Factorization, in *Advances in Neural Information Processing Systems 20, Proceedings of the Twenty-First Annual Conference on Neural Information Processing Systems, Vancouver, British Columbia, Canada, December 3-6, 2007*, pp. 1257–1264.

Sarroff, A. M. and Casey, M. (2012). Modeling and Predicting Song Adjacencies In Commercial Albums, in *SMC '12*.

Shardanand, U. and Maes, P. (1995). Social Information Filtering: Algorithms for Automating "Word of Mouth", in *CHI '95*, pp. 210–217.

Shi, Y., Zhao, X., Wang, J., Larson, M. and Hanjalic, A. (2012). Adaptive Diversification of Recommendation Results via Latent Factor Portfolio, in *SIGIR'12*, pp. 175–184.

Slaney, M. and White, W. (2006). Measuring Playlist Diversity for Recommendation Systems, in *AMCMM '06*, pp. 77–82.

Steck, H., von Zwol, R. and Johnson, C. (2015). Interactive Recommender Systems, Online `https://de.slideshare.net/MrChrisJohnson/interactive-recommender-systems-with-netflix-and-spotify`.

Su, J.-H., Yeh, H.-H., Yu, P. S. and Tseng, V. S. (2010). Music Recommendation Using Content and Context Information Mining, *IEEE Intelligent Systems* **25**, 1, pp. 16–26.

Turrin, R., Quadrana, M., Condorelli, A., Pagano, R. and Cremonesi, P. (2015). 30Music Listening and Playlists Dataset, in *RecSys '15 Posters*.

van den Oord, A., Dieleman, S. and Schrauwen, B. (2013). Deep Content-Based Music Recommendation, in *NIPS '13*, pp. 2643–2651.

Vargas, S. and Castells, P. (2011). Rank and Relevance in Novelty and Diversity Metrics for Recommender Systems, in *RecSys '11*, pp. 109–116.

Vasile, F., Smirnova, E. and Conneau, A. (2016). Meta-Prod2Vec: Product Embeddings Using Side-Information for Recommendation, in *RecSys '16*, pp. 225–232.

Wang, H., Wang, N. and Yeung, D.-Y. (2015). Collaborative Deep Learning for Recommender Systems, in *KDD '15*, pp. 1235–1244.

Wang, X. and Wang, Y. (2014). Improving Content-based and Hybrid Music Recommendation Using Deep Learning, in *Proceedings of the 22Nd ACM International Conference on Multimedia*, MM '14 (ACM, New York, NY, USA), ISBN 978-1-4503-3063-3, pp. 627–636, doi:10.1145/2647868.2654940, `http://doi.acm.org/10.1145/2647868.2654940`.

Weihs, C., Jannach, D., Vatolkin, I. and Rudolph, G. (eds.) (2016). *Music Data Analysis: Foundations and Applications* (CRC Press).

Yoshii, K., Goto, M., Komatani, K., Ogata, T. and Okuno, H. G. (2006). Hybrid Collaborative and Content-Based Music Recommendation Using Probabilistic Model with Latent User Preferences, in *ISMIR '06*, pp. 296–301.

Zangerle, E., Pichl, M., Gassler, W. and Specht, G. (2014). #Nowplaying Music Dataset: Extracting Listening Behavior from Twitter, in *WISMM Workshop at MM '14*, pp. 21–26.

Zhang, M. and Hurley, N. (2008). Avoiding Monotony: Improving the Diversity of Recommendation Lists, in *RecSys '08*, pp. 123–130.

Zhang, Y. C., Séaghdha, D. O., Quercia, D. and Jambor, T. (2012). Auralist: Introducing Serendipity into Music Recommendation, in *WSDM '12*, pp. 13–22.

Zhou, Y., Wilkinson, D., Schreiber, R. and Pan, R. (2008). Large-Scale Parallel Collaborative Filtering for the Netflix Prize, in *AAIM '08*, pp. 337–348.

Ziegler, C.-N., McNee, S. M., Konstan, J. A. and Lausen, G. (2005). Improving Recommendation Lists Through Topic Diversification, in *WWW '05*, pp. 22–32.

Chapter 16

Contact Recommendations in Social Networks

Javier Sanz-Cruzado and Pablo Castells
Universidad Autónoma de Madrid,
Avda. Francisco Tomás y Valiente 11, 28049 Spain,
javier.sanz-cruzado@uam.es, pablo.castells@uam.es

The increasingly fast development and expansion of recommender systems technology over the last two and a half decades, along with the exponential growth of online social networks in the last few years, has given place to the concurrence of the two areas in several directions. The present chapter focuses on a specific area within this confluence: the recommendation of people to connect with in social networks. We analyze the specifics of contact suggestion as a very particular recommendation task, where both the target users and the target items are people. We give an overview of the most relevant state of the art algorithms in this area, including methods that were originally developed with slightly different problems in mind. We present a global empirical comparison of the reviewed algorithms in order to get a perspective of their comparative performance. We conclude discussing future possible directions for research and development in this area.

16.1. Introduction

The increasingly fast development and expansion of recommender systems technology over the last two and a half decades, along with the massive growth of online social networks in the last few years, has given place to the concurrence of the two areas in several directions. Most commercial social platforms today incorporate automatic friend suggestions functionalities of some sort [25], and link recommendation is now a widely addressed research topic in recommender systems and network science [30,48,52]. Research in different fields has been developed somewhat independently though, and the equivalence or nuances in solutions and problem statements may not have always been obvious.

The problem of finding suitable links to add to a social network (or any complex network for that matter) can be traced back to the early work on network growth models [8], the purpose of which is to understand (model, predict) how a complex network may form and develop globally, to acquire specific shapes and properties. Suitable link in that context would mean to be an as good match of reality as possible, in terms of the resulting network structures and characteristics. Later on link prediction took shape as a more specific problem in its own [48], where guessing the right individual edges mattered, the purpose being to uncover, as accurately as possible, specific links that were present but unobserved in a network, or would eventually form as the network evolves. The problem took on new and more consequential meaning with the emergence of massive online networks, along with the continued blooming of recommender system technologies, where link prediction could turn a more direct attention towards the users involved in the predicted edges, and what the new edge may signify particularly for them from a personalized recommendation perspective [5].

This chapter gives an overview of current advances in the area of contact recommendation in social networks, aiming to provide a comprehensive and integrative perspective of the area. After briefly discussing the nature of the problem in a broader context, we provide a wide overview of the most relevant state of the art algorithms, including methods that were originally developed with slightly different problems in mind, but which can be straightforwardly applied in a recommendation task. In order to get a perspective of their comparative performance, we present a global empirical comparison of the reviewed algorithms.

16.2. The Contact Recommendation Task

Contact recommendation in online social networks aims to help people enhance their social connectivity in a number of ways. It may help speed up the process of transferring already existing offline relationships to an online social platform. It may also point users towards people they do not necessarily know, or had forgotten about, but may wish to establish contact with. In general, the aim of recommending people can be to find relationships that a) already exist, b) would form naturally in the future, or c) may form as a consequence of the recommendation [16]. These distinctions shall not make a particular difference as far as our analysis and description of methods go, at the level of generality we shall address them in the present chapter. It is useful to keep the distinction in mind though as to the potential use and example applications of the discussed techniques.

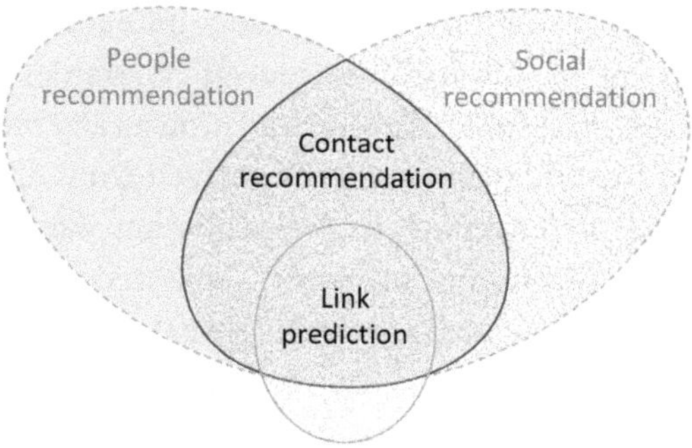

Figure 16.1. Relationship between the relevant areas.

Table 16.1. Specifics of the related tasks.

	Suggests	Assumes a social network	Primary task type
Social recommendation	Items or people	Yes	Ranking
People recommendation	People	No	Ranking
Link prediction	Links in a graph	Yes	Classification
Contact recommendation	People	Yes	Ranking

Contact recommendation has close links to related tasks and notions, which we start by discussing here. We illustrate these relationships in Figure 16.1, in our own proposed perspective, and summarize some main aspects in Table 16.1. As a recommender system task, contact recommendation is a quite peculiar particular case, where the users and the items are exactly the same set — in other words, the graph defined by the rating matrix is not bipartite. Ratings represent explicit or implicit interactions of some form between users. Interactions can consist of actions such as the formation of stable social ties (friendship, following, etc.), the direct exchange of data or information (directed messages), indirect interaction on user-authored content (liking, forwarding, etc.), and so forth. The inter-user ties in the social network can have arbitrarily rich associated information (tie type, timestamps, exchanged content, etc.).

As a general functionality, people recommendation can be conceived without the presence of a social network, as in e.g. expert recommendation [56] or online dating [21,41,66,67]. Reciprocally, it also makes sense to recommend other things than people in a social environment, such as posts, events, groups, or any other entity users interact with in the social platform (see [30] for an extensive review of recommender systems in social media). If we define social recommendation as any recommendation approach that uses specific data structures derived from direct interaction between users [75], then contact recommendation can be seen as the intersection of people recommendation and social recommendation.

A classic and tightly related, almost equivalent problem to contact recommendation is that of link prediction [48]. One clear difference between both problems is that link prediction can be put forward in complex networks that are not necessarily social, such as biological networks [13,74]. Hence some link prediction problems are not contact recommendation problems. Other than this, the difference — whereby we would not consider contact recommendation as just a subset of link prediction — is rather subtle. On the one hand, link prediction, as the name suggests, was originally aimed to find unobserved links that exist or will form in the future [48]. There is nothing however that prevents from applying the exact same prediction techniques (or close variations) to find links that may not exist and never form spontaneously, but can be useful for the involved people. Using this as a distinction from contact recommendation would thus seem artificial. We might also distinguish the tasks by their ultimate purpose and to whom the output is targeted: link prediction aims to uncover hidden parts of a network, or predict how the network will grow, whereas contact recommendation aims to satisfy or help network users by suggesting them people to link to. This difference seems more subjective than substantial, and it does not really change by itself the problem statement or the design of solutions.

A subtle but somewhat more meaningful difference is that link prediction is generally stated as a binary link classification problem, whereas contact recommendation is rather a user ranking problem. A link prediction method would essentially output two score values for each link, representing the probability that the link belongs to the "exists" or the "does not exist" classes (or "useful", or whatever the specific goal is), respectively. In the usual case, the scores have a global scope and define a ranking of links by likelihood of belonging to the class of interest (unobserved existing links). The scores can be processed in different ways (thresholding, ranking, etc.) depending on the decision to be made: advise the most likely case (exists or does not exist) for each link, select the most likely unobserved links that may exist in the network, etc.

In contrast, the output of a contact recommender is essentially a set of recommended user rankings (one for each target user). It is always possible to use a global link prediction ranking to produce a set of recommended user rankings,[a] but the reverse is not necessarily true — the local ranking scores, returned by a contact recommender for a given target user, commonly involve monotonic, rank-preserving transformations (e.g. removal of constants) that depend on the target user, so that merging the recommendations into a global link ranking would generally not make appropriate sense. To this extent we may consider that link prediction methods, when applied to people in social networks, are simply (equivalent to) a subset of contact recommendation methods, but the reverse is not true. The relation between contact recommendation and link prediction is comparable to the relation between information retrieval (IR) and binary classification: a document classifier into the relevant and not relevant classes given a query can be used as an IR system by ranking the documents by decreasing classifier score for the relevant class for each query. But the rankings of an IR system may not provide an immediate basis to make a decision as to which documents are relevant for which queries — it depends on how the IR scores are elaborated, and whether they make global sense over different queries.

---

[a]Theoretically, a classification method might apply monotonic transformations on its scoring function for a fixed link (e.g. removal of the prior link probability in Naïve Bayes) thereby losing a global meaning, but such transformations are generally optional and not strictly required by the method.

16.3. Contact Recommendation Algorithms

From a recommender system point of view, contact recommendation can be seen as a standard problem where the set of items is equivalent to the set of users, and the ratings matrix is the adjacency matrix of the social network. In this perspective, state of the art approaches such as neighborhood-based methods or matrix factorization algorithms could be used straightforwardly to recommend people in social networks. However, since social networks are a quite distinctive application domain, people recommendation has been addressed by quite particular perspectives, leading to a wide variety of specialized algorithms. In this section, we review some of the most notable approaches. Figure 6.2 shows a summary of the algorithms introduced in the next sections and their taxonomy.

Throughout the rest of this chapter we shall use the following notation. We denote by $\mathcal{G} = \langle \mathcal{U}, E \rangle$ the graph structure of a social network, where $\mathcal{U}$ is the set of users in the network, and $E \subset \mathcal{U}_*^2$ represents the relations between the users.[b] For each user $u \in \mathcal{U}$, we denote by $\Gamma(u) \subset \mathcal{U}$ the set of people the user is connected to in the network. Given a target user $u \in \mathcal{U}$, the contact recommendation task consists then in finding the people that u may find most benefit in connecting to. The recommendation generally consists of a ranked subset of $\mathcal{U} \setminus \Gamma(u)$. The ranking can be equivalently represented as a scoring function $f_u: \mathcal{U} \setminus \Gamma(u) \to \mathbb{R}$, the values of which induce an order (by decreasing score value) over the candidate users.

[b]We use the shortcut notation $\mathcal{U}_*^2$ to refer to the pairs of different users, i.e. $\mathcal{U}_*^2 = \mathcal{U}^2 \setminus \{(u,u) \in \mathcal{U}^2 \mid u \in \mathcal{U}\}$.

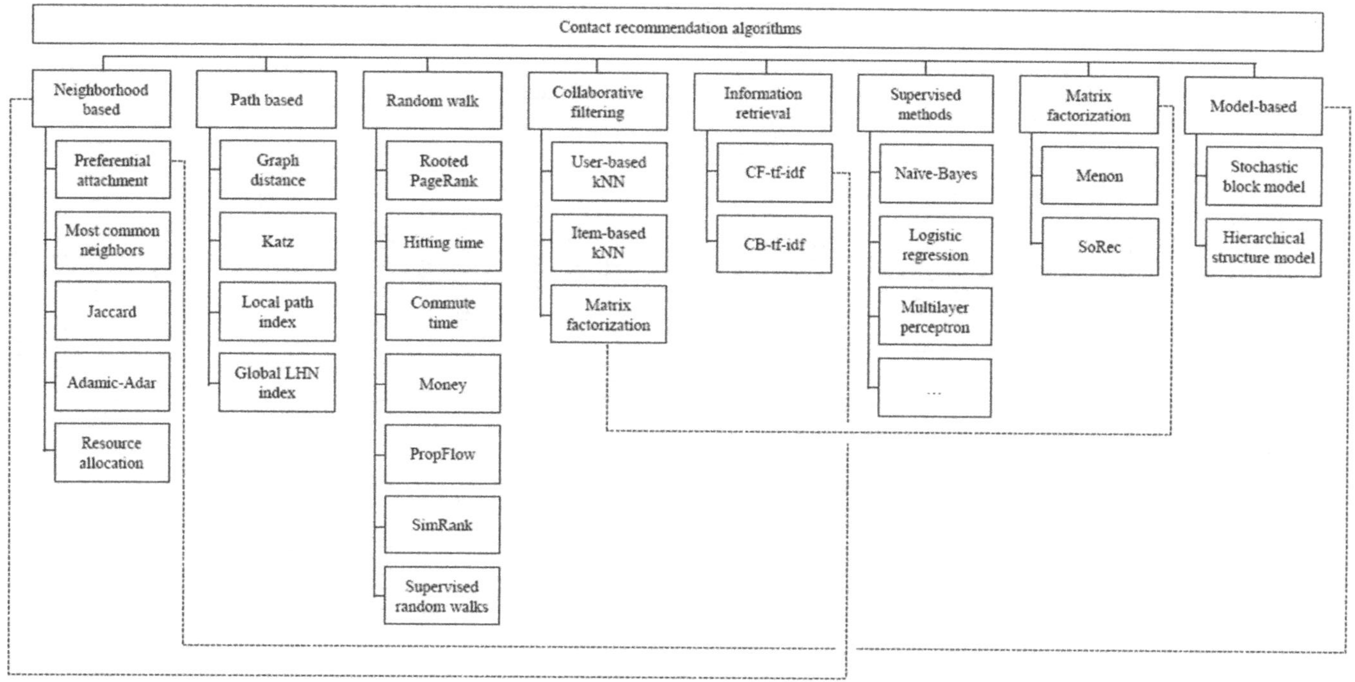

Figure 16.2. Contact recommendation algorithm taxonomy. Dashed lines show alternative categorizations for some of the algorithms.

16.3.1. *Neighborhood-based methods*

Contact recommendation algorithms within this category consider that the most informative elements in the network for predicting new links are the sets of neighbors of the two endpoints of the candidate links. Many algorithms have been developed in the field that can be classed in this category, a representative set of which we describe next.

Preferential attachment. This algorithm takes the preferential attachment phenomenon [8] as a model for prediction. The idea is that if a network evolves following a certain growth model, new links can be accurately predicted based on the model. The preferential attachment model has been observed as a trend in many real networks, where nodes connect to nodes with high degree (more specifically, neighbors are selected by new nodes with proportional probability to their degree). Accordingly, a link prediction algorithm can be defined considering that a link between two nodes is most likely to appear when both have a large degree [48]. That principle motivates the following mathematical formulation for the score:

$$f_u(v) = |\Gamma(u)||\Gamma(v)|$$

When used as a contact recommendation method, $f_u(v) \propto |\Gamma(v)|$ ranks the recommended contacts by decreasing degree, and amounts to a plain non-personalized popularity-based recommendation, which can also be read as ranking recommended contacts by the prior probability that a random person is connected to the candidate contact. The method is rather trivial but still achieves a basic effectiveness level that makes it suitable as a sanity-check baseline, or even a minimum-cost recommendation approach in commercial applications [4].

Most common neighbors (MCN). This simple method recommends the users with the highest number of common neighbors with the target user [61]:

$$f_u(v) = |\Gamma(u) \cap \Gamma(v)|$$

The ranking function here can be read as rank-equivalent to the (joint or, rank-equivalently, conditional) probability that a neighbor of the target user is a common neighbor with the candidate user.

Though the previous formula is straightforward for undirected networks, directed networks admit different particularizations for the directionality of the neighborhoods, which can be taken as: the set of users that follow the user (which we will denote as $\Gamma_{\text{in}}(u)$), the set of people the user follows, $\Gamma_{\text{out}}(u)$, or the union of both, $\Gamma_{\text{und}}(u) = \Gamma_{\text{in}}(u) \cup \Gamma_{\text{out}}(u)$.

Applying the principle of homophily, i.e. considering that new bonds are more likely to be created between similar people [57], Golder *et al.* [26] considered two different variations of this algorithm: recommending users who share interests with the target user i.e. people with common followees: $f_u(v) = |\Gamma_{\text{out}}(u) \cap \Gamma_{\text{out}}(v)|$; and recommending users who share audiences with the target user, i.e. people with common followers: $f_u(v) = |\Gamma_{\text{in}}(u) \cap \Gamma_{\text{in}}(v)|$. In addition to the previous options, they explore as an additional alternative recommending the followees' followees of the target user, i.e. people at distance 2 from the target user: $f_u(v) = |\Gamma_{\text{out}}(u) \cap \Gamma_{\text{in}}(v)|$.

Jaccard. In networks where the degree distribution is highly skewed, the MCN algorithm tends to promote the most highly connected people (since the probability that the target user shares common friends increases with the number of connections of the candidate user), and hence the algorithm becomes highly similar to preferential attachment recommendation. Several approaches have been considered to mitigate this effect. The best-known is the Jaccard coefficient [36,48,70], which measures the probability that a random neighbor of the target or the candidate user is common to both of them:

$$f_u(v) = \frac{|\Gamma(u) \cap \Gamma(v)|}{|\Gamma(u) \cup \Gamma(v)|}$$

Many other approaches have been proposed for normalizing the common neighbor score, such as the Sørensen index [72], the local Leicht-Holme-Newman index [44], Hub Promoted and Depressed indexes [68,81] or the Salton index [70]. We omit them here, as they are very similar in terms of formulation and performance.

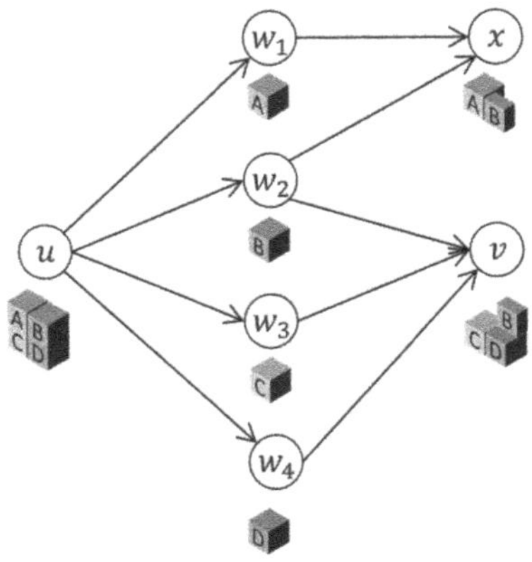

Figure 16.3. Resource allocation.

Adamic-Adar. This algorithm promotes users with a highly overlapping friendship neighborhood to the target user [2,48], similarly to the common neighbor methods. But differently from the latter, this approach gives more importance to common friends with a low degree, as being in a way more "unique" to both friendship circles than more popular people who have a lower discriminative power in comparing overlapping neighborhoods. The formula for this method is the following:

$$f_u(v) = \sum_{w \in \Gamma(u) \cap \Gamma(v)} \frac{1}{\log|\Gamma(w)|}$$

which amounts to counting common neighbors, weighted by the inverse of the neighbor popularity, with the logarithm as a damping function to modulate the penalization of highly connected common neighbors.

Resource allocation. This method [81] takes inspiration in physical processes involving the distribution of finite resources through a social network. Consider that a person u has a definite amount of a certain resource, and he wants to spread it through the network. This resource is impossible to copy, so it is divided into equal parts, each of which is distributed to a single neighbor, who repeats the process. Then one may want to know the fraction of the initial resource that arrives to the candidate user in two steps. Figure 16.3 illustrates this process in an example, showing the fractions of the initial resource from person u that reach each user in the network after two steps. Based on this metaphor, a contact recommendation is defined where candidate users v are ranked by decreasing order of the amount of resource that each candidate will receive from u after two iterations:

$$f_u(v) = \frac{1}{|\Gamma(u)|} \sum_{w \in \Gamma(u) \cap \Gamma(v)} \frac{1}{|\Gamma(w)|}$$

Since $1/|\Gamma(u)|$ is a constant that does not affect the ranking for a given target user u, we can see that this approach is equivalent to Adamic-Adar without the logarithmic damping, that is, highly popular common neighbors will have a heavy penalization and hence have little effect on the resulting recommendations.

16.3.2. *Path-based methods*

Expanding the vision of the recommendation algorithms to the full network allows taking advantage of several graph properties which are invisible at the local level, such as the existing paths between pairs of nodes in the network. In this section, we introduce the most relevant path-based methods.

Graph distance. Following the notion of small world network, in which individuals are related by short chains [79], this algorithm ranks recommended contacts by the inverse length of the shortest path between the target user and the candidate contacts:

$$f_u(v) = -\delta(u, v)$$

where $\delta(u, v)$ denotes the shortest-path distance between u and v.

Katz. The approach was originally proposed as a method for computing the status of a node in a social graph [39], and later adopted as a method for link prediction [48]. This method seeks to recommend the most densely connected and proximate people to the target user by counting the number of paths between the target and the candidates, exponentially damped by the length of the path to give more importance to short paths. The most general formulation is the following:

$$f_u(v) = \sum_{l=1}^{\infty} \beta^l \text{paths}_l(u, v)$$

where $\beta \in (0,1)$ penalizes for the path lengths (and ensures convergence of the infinite sum), and $\text{paths}_l(u, v)$ represents the number of paths of length l between u and v. An exact expression can be found by using the properties of the adjacency matrix A of the network. The n-th power of the adjacency matrix A^n has the number of paths of length n between every two nodes. Using this we can computing the sum as a geometric series, and we get:

$$f_u(v) = \sum_{l=1}^{\infty} \beta^l A_{uv}^l = ((I - \beta A)^{-1} - I)_{uv}$$

The matrix inversion can be computed by numerical methods. It is also possible to estimate a fair approximation of the infinite sum by summing a few of the first summands, since the tail of the geometric series tends to zero very fast.

A derivate index has also been proposed, called **local path index** [51] that ignores paths of length 1. The expression for this index is the following:

$$f_u(v) = \sum_{l=0}^{n} \beta^l \text{paths}_{l+2}(u, v)$$

where $n + 2$ is the maximum length of the path.

Global Leicht-Holme-Newman index. This index was created as a similarity measure between vertices in networks [44]. The measure is based on the concept that two vertices are similar if their immediate neighbors in the network are themselves similar.

The starting point of the algorithm is a generalized version of the Katz formula, where each term A_{uv}^l has its own weight C_{uv}^l. This weight is the inverse of the expected number of paths of length l between u and v in a configuration model [63]:

$$\frac{1}{C_{uv}^{l}} = \mathbb{E}[A_{uv}^{l}] = \frac{|\Gamma(u)||\Gamma(v)|}{2m}\lambda_1^{l-1}$$

where $m = |E|$ is the total number of links in the network, and λ_1 is the principal eigenvalue of the adjacency matrix. This leads to the following expression:

$$f_u(v) = \frac{2m\lambda_1}{|\Gamma(u)||\Gamma(v)|}\left[\left(I - \frac{\phi}{\lambda_1}A\right)^{-1}\right]_{uv}$$

where $\phi \in (0,1)$ is a free parameter that essentially controls how fast similarity decays with distance in network paths (the lower ϕ the fastest similarity decays).

16.3.3. *Random walk methods*

Social interactions can also be modeled by random walks. These methods use transition probabilities between connected nodes to determine the probability that a random walker is at a certain node of the network at a random point in time.

Rooted PageRank. PageRank [12] exploits the network link structure to generate scores for each network node. The score for a node u represents the stationary probability of being in that node for a user who "walks" randomly across the network edges. Originally proposed for ranking web pages based on the hyperlink topology, it is one of the best-known random walk algorithms.

The importance of a node relies on three factors: the number of nodes that point to it, their importance, and the number of outgoing edges from those nodes. If a node is linked to by a high number of nodes, the importance of that node increases. The importance transmitted by an in-link is recursively proportional to the importance of the link source. Moreover, the importance transmitted by the link is equally divided and distributed to the outgoing edges from the source. This leads to the following recursive equation:

$$f_u(v) = \frac{r}{|\mathcal{U}|} + (1-r) \sum_{w \in \Gamma_{\text{in}}(v)} \frac{f_u(w)}{|\Gamma_{\text{out}}(w)|} + \frac{(1-r)}{|\mathcal{U}|} \sum_{w:|\Gamma_{\text{out}}(w)|=0} f_u(w)$$

where r is a parameter that represents the probability that the random walker teleports to a random node (ignoring links). The rightmost term in the equation helps handle the sinks in the network, ensuring that that $\sum_w f_u(w) = 1$. The values for each node are iteratively computed, using the expression above.

We can easily realize that the resulting recommendation method is non-personalized, since no information about the target user is taken into account. It is however possible to personalize the algorithm in different ways. Among many possibilities, the most common is probably the approach proposed by White and Smith [80].

The personalized refinement is based on the teleport probability. In the basic random walk version, every user can teleport to any node in the network with uniform probability. In the personalized version, users can only teleport to the target user. The effect derived from this variation is that the importance of a node does not only depend on the link structure of the network, but also on how near the node is to the target user (as the walk is frequently reinitialized to the target user as a restarting point). The resulting formula is the following:

$$\begin{aligned} f_u(v) = r\delta_{uv} + (1-r) \sum_{w \in \Gamma_{\text{in}}(v)} \frac{f_u(w)}{|\Gamma_{\text{out}}(w)|} \\ + (1-r)\delta_{uv} \sum_{w:|\Gamma_{\text{out}}(w)|=0} f_u(w) \end{aligned}$$

where δ_{xy} is the Kronecker delta function ($\delta_{xy} = 1$ if $x = y$ and $\delta_{xy} = 0$ if $x \neq y$). This algorithm has been used in link prediction under the name of **rooted PageRank** [48], since the target user acts as a "root" of the random walk.

Hitting time. Also known as *mean first passage time*, the hitting time from a node u to a node v is the expected number of steps (or equivalently, the expected time, if each step takes the same time) required for a random walk starting at u to reach node v:

$$H(u,v) = \sum_{t=0}^{\infty} t\big(p_{u\to v}(t) - p_{u\to v}(t-1)\big)$$

where $p_{u\to v}(t)$ represents the probability of reaching v starting from u in time smaller or equal to t. The hitting time can be used to recommend people by ease of reach, that is, by inverse order of $H(u,v)$:

$$f_u(v) = -H(u,v)$$

Another related measure for predicting links is called **commute time**, and is defined as the expected time for the random walker to travel from u to v and returning from v to u [48].

$$f_u(v) = -H(u,v) - H(v,u)$$

The easiest way to compute the matrix H is the following [59]. First, the PageRank transition matrix T is computed,[c] which represents $p_{u\to v}(t)$ for all user pairs. Then, H is defined as:

$$H = \big[I - B^{\#} + JB^{\#}_{dg}\big]\Pi^{-1}$$

where $B = I - T$, and $B^{\#}$ is the pseudoinverse matrix of B [23]. B_{dg} represents the matrix B with 0 out of the main diagonal, J is a matrix with all its terms equal to 1 and Π is a diagonal matrix whose entries represent the stationary probability of each node in the random walk. The reader can find more details in [59].

Money. The Money algorithm was reported as one of the main components in the Twitter contact recommendation service, "Who-to-Follow" [25,29]. This algorithm exploits the link structure of the network using the SALSA (Stochastic Approach for Link Structure Analysis) [45] random-walk algorithm. To cope with the massive scale of social networks and combat spam users, the Money approach starts by building a much smaller bipartite graph, known as the consumer-producer graph, for each user in the network. The set of link sources of this bipartite graph, known

---

[c]As is well known, the PageRank transition matrix [43] is $T = r\bar{A} + (1-r)J$, where J is a $|U| \times |U|$ matrix with $J_{uv} = 1$ for all user pairs, and $\bar{A}$ is the adjacency matrix normalized to sum 1 by rows and has uniform priors for sink rows.

as the set of consumers or hubs, contain a so-called circle of trust of the target user. To find this circle of trust, an "egocentric" random walk algorithm, very similar to personalized PageRank, is run over the network [7,25,29]. The subset of users who achieve the top k values for this algorithm are selected as the target user's circle of trust. The set of link destinations of the bipartite graph, known as the set of producer or authorities, includes every user followed by the users in the circle of trust.

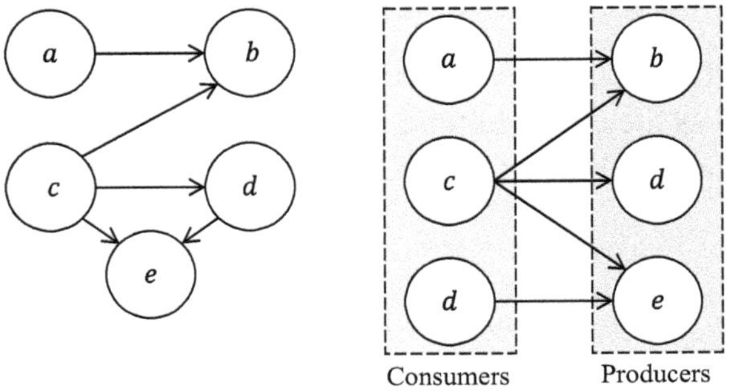

Figure 16.4. Consumer-producer graph.

An example of the consumer consumer-producer bipartite graph is illustrated in Figure 16.4. It should be observed that, if the selected size k for the circle of trust is greater than the number of nodes in the network, this bipartite graph is the same for each user in the network: the hub set would be comprised of all nodes in the network, and the set of authorities would contain all nodes with at least one incoming link.

Once the bipartite graph is built, a personalized version of the SALSA algorithm is applied over it. Since SALSA is only applied over the reduced graph, only nodes in the circle of trust and their followees can be recommended. SALSA defines two scores for each node in the network, which are formulated as two random walks: one on the authorities, and one on the hubs. The walks are made of double steps, where state transitions traverse two links in a row: one link forward and one link backward, or vice versa. At each step, if the random walker has d neighboring nodes to which he can transition, a uniform fraction of the score, $1/d$, is transferred to each of those nodes. Put formally, the scores are computed by:

$$a(v) = \sum_{w \in \Gamma_{\text{in}}(v)} \frac{h(w)}{|\Gamma_{\text{out}}(w)|} = \sum_{w \in \Gamma_{\text{in}}(v)} \sum_{x \in \Gamma_{\text{out}}(w)} \frac{a(x)}{|\Gamma_{\text{out}}(w)||\Gamma_{\text{in}}(x)|}$$

$$h(v) = \sum_{w \in \Gamma_{\text{out}}(v)} \frac{a(w)}{|\Gamma_{\text{in}}(w)|} = \sum_{w \in \Gamma_{\text{out}}(v)} \sum_{x \in \Gamma_{\text{in}}(w)} \frac{h(x)}{|\Gamma_{\text{in}}(w)||\Gamma_{\text{out}}(x)|}$$

This formulation of the algorithm has two main limitations: first, it is only personalized in that the circle of trust is different for each target user, and, second, the authority score has been proved to be proportional to the in-degree of the user, and the hubs score proportional to its out-degree, in such a way that the resulting recommendations can be quite similar to a preferential attachment approach as described in Section 16.3.1 [45]. A more personalized approach can be defined just by adding a user-centered teleport vector to the hubs [7]:

$$f_u^h(v) = h_u(v) = \alpha \delta_{uv} + (1 - \alpha) \sum_{w \in \Gamma_{\text{out}}(v)} \frac{a_u(w)}{|\Gamma_{\text{in}}(w)|}$$

$$f_u^a(v) = a_u(v) = \sum_{w \in \Gamma_{\text{in}}(v)} \frac{h_u(w)}{|\Gamma_{\text{out}}(w)|}$$

In this personalized variant, hub scores are taken as a representation of the similarity between the different users in the circle of trust and the target user, and authority values as a measure of the relevance of a producer for the target user. Hence, authority scores $f_u^a(v)$ are suggested as the most sensible option for selecting the recommended candidates [25,29]. However, according to the homophily principle [57], it may also make sense to apply hub scores $f_u^h(v)$ for recommendation [29]. In the experiments we report in section 16.7, we shall see that none of the two options is necessarily always best, as it may depend on the dataset.

PropFlow. The PropFlow algorithm [50] computes the probability $f_u(v)$ that a restricted random walk starting at u ends at v in l steps or fewer using link weights as transition probabilities. For unweighted networks, all links can be considered to just have a constant weight. The walk terminates upon reaching v or upon revisiting any node (including u). The probabilities $f_u(v)$ can be computed by a simple BFS traversal of the network starting at u. Starting with $f_u(u) = 1$, the procedure updates the probabilities of all the neighbor nodes v of the node x currently visited in the BFS, as follows:

$$f_u(v) \leftarrow f_u(v) + f_u(x) \frac{\text{weight}(x, v)}{\sum_{w \in \Gamma(x)} \text{weight}(x, w)}$$

PropFlow is thus similar to rooted PageRank, but it is cheaper to compute, since only a breadth first search with maximum depth l is needed. The depth restriction is intended to bound the potential cumulative noise that can result from far regions of the network.

SimRank. SimRank [38] seeks to recommend similar users following the recursive intuition that two users are similar if they are followed by similar users. The similarity between users is thus defined as:

$$f_u(v) = \begin{cases} 1 & \text{if } v = u \\ \gamma \dfrac{\sum_{a \in \Gamma(u)} \sum_{b \in \Gamma(v)} f_a(b)}{|\Gamma(u)||\Gamma(v)|} & \text{if } v \neq u \text{ and } |\Gamma(u)||\Gamma(v)| \neq 0 \\ 0 & \text{otherwise} \end{cases}$$

This algorithm can be interpreted in terms of random walks as the expected number of steps at which two random walkers are expected to meet at the same node if they started at nodes u and v, and randomly walked the network backwards.

Supervised random walks: This algorithm combines random walk algorithms with supervised link prediction approaches [5] (which will be described later in Section 16.3.6). This algorithm applies a weighted version of the rooted PageRank algorithm previously described. In the original version, when a node is visited, the probability of travelling from that node to another one without teleport was uniform among the different outgoing edges. In this weighted version, it is proportional to the weight of the edge, as follows:

$$p_u(v) = r\delta_{uv} + (1 - r) \sum_{w \in \Gamma_{\text{in}}(v)} \frac{\text{weight}(w, v)}{\sum_{w \in \Gamma_{\text{out}}(v)} \text{weight}(v, w)} p_u(w)$$

Using profile information (age, place of birth of the users, etc.) and structural information of the network (degree, number of friends, etc.), as well as link formation data, the weights for the random walk are learned by a supervised algorithm so that a random walk is more likely to visit the nodes the target user will create new links to.

16.3.4. *Collaborative filtering approaches*

Contact recommendation can also be stated as a particular case of classical recommendation. In that perspective, users play the role of both users and items, and the network adjacency matrix A is the rating matrix. Collaborative filtering can be hence applied straightforwardly by this mapping for recommending users in social networks. As an example, we show the reformulation of neighborhood-based collaborative filtering algorithms, also known as nearest neighbors (kNN), as one of the most widely known collaborative filtering approach examples. kNN approaches are based on the principle that similar users prefer similar items, and similar items are preferred by similar users [64]. The algorithm selects the top k most similar users (or items) to the target user (or candidate items), and computes the recommendation scores as a linear combination of the neighbors' ratings. We show next the adaptation for the user-based and item-based variants.

User-based kNN. The user-based variant generates scores using the ratings for the candidate items provided by other users similar to the target user u [64]. This set of users is known as the neighborhood of the target user, $\mathcal{N}(u)$ — where "neighbor" here means a similar user, as opposed to network neighbors $\Gamma(u)$ meaning explicitly connected users in the social network. A simple and effective version computes the scores as a weighted linear combination of the neighbors' ratings for the candidate items.

In contact recommendation, the network adjacency matrix A is interpreted as a rating matrix, hence the "rating" of a user v for an "item" (again a user) w is A_{vw}, which is equal to 1 when $(v, w) \in E$ and 0 otherwise. Substituting that in the traditional recommendation scheme, user-based kNN for contact recommendation gets defined by:

$$f_u(v) = \sum_{w \in \mathcal{N}(u)} \mathrm{sim}(u,w) A_{wv} = \sum_{\substack{w \in \mathcal{N}(u) \\ (w,v) \in E}} \mathrm{sim}(u,w)$$

The similarity between users can be assessed in multiple ways. In collaborative filtering, the similarity between users is defined in terms of their interactions with items, which in our context translates to outgoing social connections of the target user. For instance, using the cosine similarity:

$$\mathrm{sim}(u,v) = \frac{|\Gamma_{\mathrm{out}}(u) \cap \Gamma_{\mathrm{out}}(v)|}{\sqrt{|\Gamma_{\mathrm{out}}(u)||\Gamma_{\mathrm{out}}(v)|}}$$

Item-based kNN: The item-based variant exploits the similarity between the items in the system to generate recommendations. Given a candidate item, the algorithm takes a neighborhood, of items with which common users have interacted with the neighbors and the target user. The scores are defined as the weighted linear combination of the ratings the target user has provided for the neighbor items. By the same mapping as in the user-based variant, we get:

$$f_u(v) = \sum_{w \in \mathcal{N}(v)} \mathrm{sim}(v,w) A_{uw} = \sum_{\substack{w \in \mathcal{N}(v) \\ (u,w) \in E}} \mathrm{sim}(v,w)$$

The similarity function can be the same as applied in the user-based approach, except for the neighborhood direction: to be coherent with the classical formulation, item-based kNN should use the incoming neighborhoods $\Gamma_{\mathrm{in}}(u), \Gamma_{\mathrm{in}}(v)$ in the equation for cosine similarity.

16.3.5. *Information retrieval approaches*

Classical text information retrieval (IR) methods [6] have also been adapted for contact recommendation. Hannon *et al.* [31] explored adaptations of the vector space model (VSM) [70] to generate recommended contact rankings, by having users play the dual role of documents (to be retrieved), and the query (a description of what we are looking for). Depending on what elements play the role of terms in text search, two different approaches are proposed: content-based and collaborative-filtering.

In the collaborative filtering approach, the user's neighbors play the role of terms. The "bag-of-terms" representation of a given (target or recommended) user assigns a binary "frequency" of 1 for the user's neighbors, and 0 for the rest of users in the network. The user neighborhood is thus taken as the equivalent of a term representation of the user. Similarly to the MCN algorithm, in directed graphs it is possible to define the neighborhoods in different ways according to the direction of the edges: incoming neighborhood $\Gamma_{\text{in}}(u)$, outgoing neighborhood $\Gamma_{\text{out}}(u)$, or the union of both $\Gamma_{\text{und}}(u) = \Gamma_{\text{in}}(u) \cup \Gamma_{\text{out}}(u)$ [31].

In the content-based approach, terms are actually extracted from text, specifically from user-generated content. The target user vector is built by concatenating user-generated text (e.g. tweets on Twitter) into a single document. Different content selection policies have been explored to build the user term vector, such as all the text produced by the user, or all the text produced by contacts of the user (incoming, outgoing, or both).

Once the terms (or "terms") for the user vectors have been determined by either of the two above described approaches, the terms can be weighted by any usual approach in the VSM. For instance using the classic tf-idf weighting scheme:

$$w(t,u) = \text{tf-idf}(t,u) = \text{tf}(t,u) \cdot \text{idf}(t)$$

where $\text{tf}(t,u)$ is a monotonically increasing function of the frequency of term t in the bag-of-terms representation of user u, and $\text{idf}(t)$ measures the specificity of the term. For instance, using one of the common instantiations of the tf-idf scheme:

$$\text{tf}(t,u) = \begin{cases} 1 + \log_2 \text{freq}(t,u) & \text{if freq}(t,u) > 0 \\ 0 & \text{otherwise} \end{cases}$$

$$\text{idf}(t) = \log_2 \frac{|\mathcal{U}|}{|\{u \in \mathcal{U} | \text{freq}(t,u) > 0\}|}$$

Finally, using the weighted user vector representations, recommendation scores can be computed by the cosine similarity — or any variation thereof within the VSM — between the target user and the candidate users:

$$f_u(v) = \frac{\sum_{t \in u \cap v} \text{tf-idf}(t, u) \cdot \text{tf-idf}(t, v)}{\sqrt{\sum_{t \in v} \text{tf-idf}^2(t, u)}}$$

Throughout the rest of this chapter, we shall denote the collaborative filtering variant of tf-idf as CF-tf-idf, whereas the content-based approach shall be named CB-tf-idf.

16.3.6. *Supervised methods*

Link prediction (and, as a particular case, contact recommendation) can also be addressed from a supervised machine learning (ML) perspective, i.e. applying classifiers such as Naïve-Bayes [3,50], logistic regression [78], support vector machines [3] , multilayer perceptrons [3] or random forests [18,50] to solve the problem. This perspective is commonly referred to as supervised link prediction. Instead of ranking the links according to their probability of existence or formation, the algorithms address link prediction as a classification problem over the set of all pairs of nodes with two classes: the presence or absence of the link. Several issues, specific to the task, have to be addressed:

- **Pattern ranking:** A classifier seeks to determine the most likely class for a pattern, which in link prediction would mean finding out whether each possible link (a pattern) is more likely to be present or not (binary class determination). We can hence think of a classifier as a class ranker given a pattern. However the contact recommendation task goes the other way around, seeking to rank links (the patterns) by their membership likelihood to the "existing link" class. Adapting a classifier for this task is usually straightforward as long as the classifier outputs class scores. It is simply a matter of using the scores to organize rankings differently, and revising any rank-equivalent operations — given a pattern — that may have been applied to the development of the pattern-class scoring function.
- **Class imbalance:** Users typically establish connections with a small fraction of the people in a social network [22]. This fact, along with the upper limit established by some social network applications to the

number of relations allowed for a single user (e.g. to prevent spam bots [60]), causes real social networks to be very sparse. This can pose an extreme unbalance problem for supervised approaches, since most links in the network do not exist or will never form. It is possible to overcome this problem using common approaches in machine learning for dealing with class imbalance [15], such as undersampling the set of examples [50], creating new artificial patterns of the minority class [50], or using specific machine learning algorithms adapted for imbalanced datasets [3].

- **Feature selection:** Selecting a reduced and informative set of features for the classification is key for the good performance of the classifier. Many possibilities have been explored for generating those patterns, such as using network analysis metrics [3], features related to the content published in the social networks (e.g. the number of papers published by two authors in a citation network, the keyword count of each author, or the number of common keywords [3]), or the output of unsupervised link prediction methods [18,50].

16.3.7. *Matrix factorization*

The effectiveness and popularity of matrix factorization techniques as collaborative filtering algorithms [42] has led to several link prediction and contact recommendation algorithms that rely on their principles. One of the most notable approaches is proposed by Menon *et al.* [58], consisting of factorizing the network adjacency matrix as a function of the product of two different matrices, as follows:

$$A \approx L(W^T \Lambda W)$$

In this decomposition, $W \in \mathbb{R}^{k \times n}$ is a matrix that contains the latent vectors for each user in the network (each column w_u represents the different k latent features for a user u), $\Lambda \in \mathbb{R}^{k \times k}$ is a square matrix that is needed to handle directed networks (the adjacency matrix of undirected networks is symmetric and can be factorized into a function of just two matrices WW^T), and $L(\cdot)$ is a link function. Once this decomposition is obtained, the recommendation ranking score is computed as:

$$f_u(v) = L(w_u^T \Lambda w_v)$$

where w_u is the latent vector associated to user u. To find the corresponding factorization, Menon *et al.* propose a method that seeks to minimize the area under the ROC curve [24].

Other matrix factorization techniques have been proposed that exploit social network data to recommend items rather than users [37,53], but can also be adapted to recommend contacts as well. The SoRec algorithm [53] is one such example. This algorithm simultaneously factorizes the network adjacency matrix and a user-item rating matrix R as the products of two matrices:

$$A \approx W^T Z$$

$$R \approx W^T V$$

where W represents the user latent feature space, V represents the item latent feature space, and Z is a factor matrix for the network. Though primarily conceived for recommending items using the approximated rating matrix R, the approximation to the adjacency matrix A can also be applied to recommend contacts, using the following score function:

$$f_u(v) = w_u^T z_v$$

where w_u and z_v are, respectively, the rows of the matrices W, Z associated to users u and v.

In addition to the matrix factorization approaches specifically devised for link prediction, any collaborative filtering algorithm — and hence in particular any matrix factorization algorithm for recommendation [35,42] — can be potentially adapted for recommending links, as stated in Section 16.3.4.

16.3.8. *Model-based approaches*

This family of algorithms supposes that the structure of the social network follows an underlying probabilistic model [28]. Under this assumption, the algorithms fit as closely as possible the probabilistic model to the structure and properties of the real network. Links are then predicted using the probability of being added to the network according to the fitted model. The most notable examples of these algorithms are the hierarchical structure model [17] and the stochastic block model [28]. The former assumes that social networks have an underlying hierarchical structure, while the latter considers that nodes in a network are divided into several partitions or blocks, and the probability that a link exists only depends on the groups that its endpoints belong to. The preferential attachment method, can also be classed in the model-based category, though it is also a neighbor-based approach and hence we described it as such in Section 16.3.1.

16.4. Contact recommendation in online social network platforms

By the late 00's, mainstream online social networks such as Facebook, LinkedIn and Twitter started providing contact recommendation functionalities on their platforms. Public knowledge about the algorithmic internals and specifics is naturally limited, we may observe the outside functionality or even some internal details that have been disclosed to some extent.

Facebook: The "People You May Know" algorithm[d] relies on many sources. Facebook has stated to recommend users mostly according to the existence of common friends between the target and candidate users, but the system also recommends people belonging to common groups or affiliations with the target user (using information about the school, university or work, entered in the user profiles), or people tagged in the same photos or contents.

LinkedIn: Contacts are recommended according to commonalities between users[e], based on information contained in user profiles, common connections between the target and candidate users, the industries or companies both users have worked in, etc.

---

[d]https://www.facebook.com/help/336320879782850

[e]https://www.linkedin.com/help/linkedin/answer/29/people-you-may-know-feature-overview

Tumblr:[f] Tumblr, a social blogging platform, recommends new blogs (a rough equivalent of "user") by considering a mixture of the topics the user prefers and the network structure [1]. At sign-in, users are asked about their preferences over a set of predetermined topics organized in a taxonomy. Based on this, a topical profile is created for each user, which is used to cope with the initial cold-start. As the number of contacts grows, new blogs are recommended by the triad closure principle (i.e. friend-of-a-friend algorithms).

Twitter: In contrast to other social sites, Twitter did disclose some details about the inner workings of the "Who to Follow" service [25,29], which we summarized in Section 16.3.3. In addition to this algorithm, Twitter has also mentioned experiments with several other algorithms [25,29], such as rooted PageRank and MCN, and the closure algorithm, which recommends target user followers at distance two, and a personalized version of the HITS algorithm, labeled "Love" [40].

16.5. Available resources

The massive growth of online social network platforms, their widespread use, and their interest as an area of study has given rise to the availability of a wide array of resources of both broad and specific interest to the area addressed in this chapter. We briefly describe a selection of those resources, comprising datasets, online social network data access APIs, and software libraries for network analysis and manipulation, and link prediction.

16.5.1. *Datasets*

A good number of network datasets of different types, domains, and sizes are publicly available and can be used for the needs of evaluating link recommendation algorithms. Particularly worthy of mention are the following:

- **Stanford Large Network Dataset Collection:**[g] This collection includes a wide variety of large datasets, including samples from social

[f]https://www.tumblr.com
[g]https://snap.stanford.edu/data/index.html

networks such as Twitter, Google+ or Facebook, Web graphs, collaboration networks such as ArXiv, signed networks, and more.

- **Ben-Gurion University Social Networks Security Research Group Datasets:**[h] A collection of datasets that includes anonymized versions of directed, undirected and multi-label social networks datasets. It includes data from Facebook, Google+, Academia.edu, and other sources.
- **CASOS:**[i] Dataset collection from the Center for Computational Analysis of Social and Organizational Systems (CASOS) of the Carnegie-Mellon University. It includes data from social, semantic, communication, and other types of networks.

16.5.2. *APIs*

Some social network sites provide different APIs supporting the development of applications on top of their platform, and access to their data: users, relations between users, user-generated content, etc. Such APIs are a valuable source of data for experimentation with contact recommendation systems, by obtaining data to perform offline tests, or by creating applications that directly operate over the platform [31].

Twitter: Twitter provides several APIs[j]; the most important ones are the REST API and the Streaming API, which allow users to crawl different elements in the network.

- The REST API provides access to most of the functionalities of Twitter: retrieving information from user timelines, obtaining interactions between users, posting new tweets, following users, etc. However, free access is notably limited by the number of calls that can be made to the API.
- The Streaming API is much less restricted as to the number of calls, though it can only be used for obtaining tweets in real time (up to 1% of the published tweets).

---

[h]http://proj.ise.bgu.ac.il/sns/datasets.html

[i]http://www.casos.cs.cmu.edu/tools/data.php

[j]https://developer.twitter.com/en/docs

The Twitter APIs can be accessed directly via URLs, but libraries are also available in different programming languages including C++, Java, Python and .NET, which simplify the development work around the APIs.

Tumblr: This blogging platform offers a API[k] granting access to blogs, posts and relations between them. Tumblr furthermore provides official API clients for JavaScript, Ruby, PHP, Java, Python, Objective-C and Go.

Facebook: Facebook provides an API[l] to work with their data and develop applications upon its social network. This API is however much more restrictive than Twitter or Tumblr in terms of the data access conditions: it requires the explicit permission of users to retrieve their information (the API only authorizes retrieving data from people who use the application).

16.5.3. *Software libraries*

In addition to the previous resources, there are many useful libraries for developing contact recommendation algorithms, or related applications. We may consider two different types of libraries:

- **General network libraries** such as Jung[m] (Java), SNAP[n] [47] (C++, Python) or iGraph[o] (R, Python, C/C++) provide implementations of generic graph manipulation and network analysis, which can simplify the development and testing of contact recommendation applications.
- **Link prediction libraries:** Specific libraries are also available providing explicit link prediction method implementations. In this scope we may highlight LPMade[p] [49], a framework for link prediction that implements some of the most important algorithms, such as most common neighbors, Adamic-Adar, Jaccard and PropFlow, and includes software support for evaluation. Other graph libraries such as

[k]https://www.tumblr.com/docs/en/api/v2
[l]https://developers.facebook.com/docs/graph-api/overview
[m]http://jung.sourceforge.net
[n]http://snap.stanford.edu/index.html
[o]http://igraph.org
[p]https://github.com/rlichtenwalter/LPmade

NetworKit[q] and NetworkX[r] also provide some link prediction algorithm implementations.

16.6. Practical aspects

When building a contact recommendation system, there are some aspects to consider. In this section, we provide an overlook of some of them, like the evaluation of this systems, the directionality of the edges or the scalability of the different algorithms.

16.6.1. *Edge direction*

The development of contact recommendation methods was originally largely oriented to undirected networks. However, the success of many asymmetrical online social networks such as Tumblr, Twitter or Instagram motivated further attention to directed networks. In many cases, contact recommendation algorithms can be generalized to directed versions by a quite straightforward adaptation, as is the case for the methods described in Section 10. A trivial adaptation is to process all directed edges as if they were undirected, but finer alternatives may achieve better accuracy [31]. Some algorithms, such as Katz, can in fact be applied without modification to directed networks, just operating on an asymmetric adjacency matrix. However, for some other algorithms, such as most neighborhood-based approaches described in Section 16.3.1, we need to be specific in what we mean by a node u being a neighbor of another node v. As illustrated in Figure 16.5, there are three different possible definitions of neighborhood: incoming $\Gamma_{\text{in}}(u) = \{v \in \mathcal{U} | (v,u) \in E\}$, outgoing $\Gamma_{\text{out}}(u) = \{v \in \mathcal{U} | (u,v) \in E\}$ or undirected neighborhood $\Gamma_{\text{und}}(u) = \{v \in \mathcal{U} | (u,v) \in E \wedge (v,u) \in E\}$.

The neighborhood choice applies to both target and candidate users [26,31], thus giving rise to six different possible alternative combinations for each algorithm. For instance, Hannon *et al.* [31] chose one of the three possible directions (either incoming, outgoing or undirected) for both

---

[q]https://networkit.iti.kit.edu

[r]https://networkx.github.io

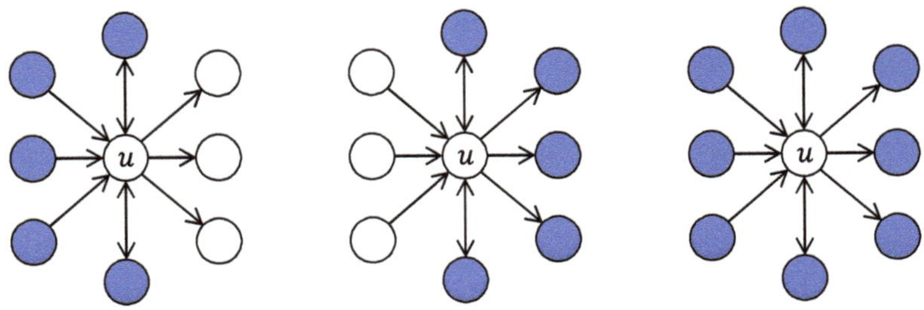

a) Incoming neighborhood b) Outgoing neighborhood c) Undirected neighborhood

Figure 16.5. Possible neighborhoods for a user in directed graphs. Nodes in blue belong to the neighborhood, while nodes in white do not.

nodes in their tf-idf algorithm, while Golder *et al.* [26], as we explained in Section 16.3.1, considered three different configurations for the directed MCN model, where scores were defined as the intersection of the incoming neighbors of the target and candidate users, the intersection of their outgoing neighbors, and the intersection of the outgoing neighbor of the target user and the incoming of the candidate one.

16.6.2. *Scalability*

One critical aspect in the implementation and design of general recommender systems and, as a particular case, social recommendation, is dealing with data at the common massive scale of current online platforms, and social media in particular, with active monthly users counted by the billions [60]. Running content recommendation at such a scale poses challenges in terms of both time and hardware resources. The time and memory requirements of contact recommendation algorithms mainly depend on the number of users in the network. If we were to consider every missing link in a directed network as a candidate recommendation, we would need to compute a total of $|\mathcal{U}|(|\mathcal{U}| - 1) - |E|$ values, essentially a quadratic cost with respect to the network size, for the typical sparse network [33,60]. Even for a single user, this may involve computing billions of scores.

Some methods avoid this bottleneck simply by their own nature. For instance, neighborhood-based methods such as MCN, Jaccard or Adamic-Adar, only consider candidate contacts having common friends with the target user, whereby the algorithm complexity (in an efficient implementation) is in fact linear in the number of edges. Other algorithms, like PropFlow or local path index limit the distance from the target user via their parameters. It is also possible to apply this idea in an artificial way to many other algorithms. For instance, Lichtenwalter *et al.* [50] only generated scores at distance smaller or equal to 5 from the target user. Other algorithms take a reduced set of candidate recommendations for each target user by more sophisticated methods. The Money algorithm [29] described in Section 16.3.3, for instance, first computes a personalized circle of trust around each target user, and then applies a more costly algorithm as a second step within this subnetwork of candidates and the supervised random walk approach in Section 16.3.3 only considers neighborhoods at distance 2.

16.6.3. *Evaluation*

Recommender system evaluation is known to be a non-trivial problem, and contact recommendation, as a particular case, shares similar general issues as well as specific aspects of its own. A primary question to be elucidated in evaluation is whether the recommendation task should be viewed as a classification or a ranking task. The classification perspective evaluates contact recommendation as a binary classification problem, where a link is correctly classified as "interesting" (according to the goal in the particular application domain) when the target user accepts the recommended link, or in an observational perspective, when a link between the target and the recommended user is eventually formed (or, in an offline experiment, is present in some held out test set). In this view, the algorithms can be evaluated by common metrics such as the area under the curve (AUC), confusion matrices, etc. [24], taking classification thresholds if needed. This has been the main evaluation approach in link prediction [5,17,18]. An example is illustrated in Figure 16.6. The ranking-based perspective is more representative of real scenarios though, where a recommendation is a ranking of contacts delivered to the target user. From this point of view, the quality of recommendations is assessed by ranking-oriented IR metrics such as precision, recall, nDCG, etc. [5,31], built around the concept of relevance [6]. A recommended connection is

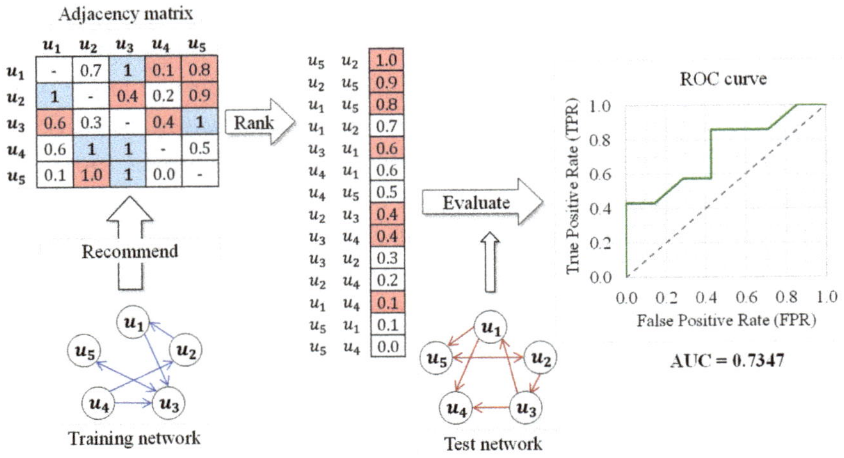

Figure 16.6. Toy example of classification-oriented evaluation (ROC curve and AUC). Cell color in the adjacency matrix (left) represents a network split into training links (light blue) and test links (red), corresponding to the training and test networks, while white cells correspond to non-existing links. The values in the red and white cells represent fictitious recommendation algorithm scores, for exemplification. The table in the middle illustrates the resulting ranking of link predictions based on the algorithm scores, where red cells indicate correctly guessed links (test links), and white cells correspond to incorrect predictions (non-existing links). In the right-hand-side graphic, the green line shows the resulting ROC curve for the algorithm, and the grey line corresponds to random classification.

considered relevant if the target user actually creates a link to the recommended person. Figure 16.7 illustrates this perspective.

An additional issue in offline evaluation is the effect of the data sample. Data collection procedures are known to be a sensitive point in experimental practice in virtually any experimental discipline [34]. Social network data sampling introduces additional complexities when the nodes and edges of network can only be sampled by traversing the network itself, as is commonly the case [1]. Moreover, the massive size of current online social networks causes a major scale gap between the total network and the samples thereof that are commonly feasible to obtain and cope with in iterative experiments. This may further compromise the representativeness of the sample with respect to whole network, and hence how translatable to the latter are the results obtained on the former. In addition to the scale

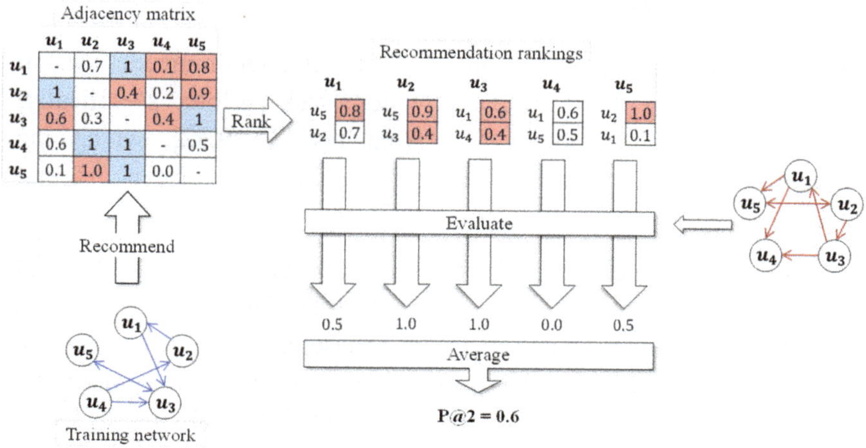

Figure 16.7. Toy example of ranking-based evaluation (precision at $k = 2$). The network data and algorithm scores in the adjacency matrix (left) have the same meaning as in Fig. 6. In the recommendation task, the link ranking is broken down into a ranking for each target user (middle). The test links are used as relevance judgments (in information retrieval evaluation terminology) for metric computation over the rankings. The evaluation metric (P@2 in this toy example) is computed separately for each ranking, and then averaged into a single final value.

issues, some network sampling methods are known to introduce biases, failing to match the properties of the real network [46,54]. For example, breadth-first sampling or degree sampling show a bias towards retrieving high-degree and high-PageRank nodes [54]. Experimental practice in the contact recommendation practice would benefit from further awareness of the effects such biases can have in the outcome of evaluation.

16.7. Empirical observations

In order to analyze and compare the performance of the contact recommendation algorithms described in the previous sections, we present here some empirical observations in offline experiments over network data.

16.7.1. *Experimental setup*

The experiments we report here are conducted in a recommender system perspective, rather than a classification task. More specifically, recommendation is tested as a ranking task [9,19,32], which the community is coming to find is more representative of our experimental setting.

Datasets. We run the experiments over data obtained from real social networking sites, namely Facebook and Twitter. The Facebook data is taken from the Stanford Large Network Dataset Collection (see Section 16.5.1). In particular, we use the ego-Facebook dataset, which contains the friend lists of ten Facebook users, and all the friendship links among them, amounting to a total of 4,039 users and 88,234 edges [55].

For Twitter, we downloaded data by network exploration in a snowball sampling approach [27] using the public REST API. The snowball sampling starts from a seed user and explores, breadth-first, a certain subset of the outgoing links of each visited user. Specifically, we explore all the interaction links in the last 200 tweets posted by the user before August 2nd 2015. Interaction links are defined by the content of tweets, i.e. $(u, v) \in E$ if u mentioned, retweeted or replied v in some of its posted tweets collected in our exploration. The network traversal stops when a desired total number of users are discovered (10,000 in our experiments). At this point, we complete the dataset by obtaining all the remaining interaction edges between the discovered users that may not have been directly traversed in the exploration (this essentially involves the outgoing links from users at the exploration frontier). The resulting network contains 10,000 users and 164,653 edges.

In addition to the interaction network thus built, we obtain an additional network with the same set of users, but where the edges correspond to stable follows links, i.e. $(u, v) \in E$ if u follows v on Twitter. We do so by simply obtaining all the follows relations between the users in the interaction network, via the REST API. The follows network has 582,172 edges and, naturally, as many users as the interaction network (just a few users with zero follows links are dropped — see details in Table 16.2).

Table 16.2. Network dataset details.

		Twitter		
		Interaction	Follows	**Facebook**
	Directed	Yes	Yes	No
Total	Nr. users	10,000	10,000	4,039
Total	Nr. edges	164,653	582,172	88,234
Total	Clustering coef.	0.0967	0.1829	0.5192
Total	Diameter	13	10	8
Training	Nr. users	9,796	9,964	4,020
Training	Validation edges	111,392	427,568	56,333
Training	Training edges	137,850	475,530	70,566
Test	Nr. users	5,652	8,180	3,652
Test	Nr. edges	21,598	98,519	17,644

Experimental procedure. To evaluate an algorithm, we split the network data into a training and a test subgraph. The training subgraph is supplied as input to the recommendation algorithms under evaluation. The test subgraph is used as ground truth to compute accuracy metrics over the output (candidate user rankings for each target user) returned by the algorithms. Specifically, the test network edges are taken as positive relevance judgments in metrics such as precision, recall or nDCG, considering the edges in the test network as "relevant links" for each user, and any other link as "non-relevant".

We apply a random data split for the Facebook network, and a temporal split in the Twitter networks (as the edges have explicit or implicit timestamps). In the random split for Facebook, each network edge is (independently) assigned to the training or test subgraph with a probability equal to the desired split ratio (0.8 training data in our experiments).

The Twitter interaction network is split by dividing the set of tweets from which the network was built. Tweets are split according to their timestamp into two subsets dated, respectively, before and after a certain split time point. The time point chosen in such a way that a desired split ratio (again, 0.8 training data here) is obtained. Based on this split, the training and test networks include all the interaction links induced by the tweets before and after the split point respectively. Finally, the edges that are present at both sides of the split are removed from the test set, in order to get a disjoint split as required for a fair evaluation.

For supervised algorithms, we made a split of the training graphs for both Facebook and Twitter interaction network, using the same procedures explained above. 75% of the training data is used to create training patterns for the supervised algorithms, using the remaining 25% of links as validation data.

For the follows network, we used three different link downloads: the first download was used to generate the training patterns for supervised algorithms; the new links in the second snapshot of the network, taken four months after the first one, is used as validation data. Finally, the new edges in the third snapshot, downloaded two years later, are used as test data for evaluation.

Finally, in both Twitter datasets, we exclude the recommendation of reciprocating links (i.e. the recommendation of people who link back to the target user). This means such links are removed from (or more simply, not requested to be placed in) the rankings returned by the algorithms, and are also removed from the test network (as they are not expected to be recommended). This is for two reasons. First, Twitter already notifies users every time someone interacts with them or starts following them — hence an additional recommendation to pay attention to such new followers or interactors would be redundant. Second, the reciprocating ratio on Twitter is quite high, and this would bias the evaluation in rewarding algorithms that are prone to trivially recommend reciprocal links. This issue is absent from Facebook as the concept of link reciprocation is only defined in directed networks. Table 16.2 shows the size of the training and test sets of each network dataset, as well as the subset for supervised approaches.

Algorithms. We evaluate and compare the different algorithms detailed in Section 16.3. As collaborative filtering algorithms (Section 16.3.4), we include the implementation of kNN (user-based and item-based) and matrix factorization (algorithm for implicit feedback proposed by Hu *et al.*

[35]) provided in the RankSys[s] public framework. We do not have results for the content-based version of the tf-idf algorithms [31] on the Facebook dataset, since no side information of any kind is available there. On the Twitter data, we have used the original tweets (not retweets) posted by each user before the time of the split as the content from which the user representation is built.

We also test two supervised algorithms: Gaussian Naïve Bayes and Logistic Regression [10], using the implementation provided by the Scikit-learn[t] Python package. In order to run those algorithms, we need to generate patterns for every missing link in the network. Supervised methods in the literature typically use other predictors, such as Jaccard or Adamic-Adar as features, hence essentially building ensembles of link prediction algorithms [50], or use data features of very specific domains (such as the number of co-authored articles in collaboration networks [3]). In order to test an approach as generic as possible, we shall use network-related properties associated to the edge we aim to predict. We consider two types of features:

- **Individual features**, related to individual users in the network. We take such features for both endpoints of the edges (target and candidate user). For directed graphs, we include: the in-degree, out-degree, the local clustering coefficient, closeness, betweenness and (not personalized) PageRank. In undirected graphs, we simply take the degree in place of in-degree and out-degree.
- **Paired values,** computed for pairs of users in the network. We use the number of common neighbors between the edge endpoints, and the embeddedness. Furthermore, we include binary features that indicate whether or not the link crosses network communities. In our experiments, we use three such features, corresponding to three different community detection algorithms: Louvain [11], Infomap [69] and Leading Vector [62].

Finally, we add random recommendation as a sanity-check baseline to the comparison. Parameter settings and configurations for the algorithms are shown in Table 16.3. The algorithm parameters are tuned by a simple

---

[s]http://ranksys.org
[t]http://scikit-learn.org/stable

Table 16.3. Optimal parameters for the different algorithms and networks.

	Algorithm	Twitter: Interaction	Twitter: Follows	Facebook
Neighborhood-based	Pref. attachment	$\Gamma_{\text{in}}(v)$	$\Gamma_{\text{in}}(v)$	-
	MCN	$\Gamma_{\text{und}}(u), \Gamma_{\text{in}}(v)$	$\Gamma_{\text{und}}(u), \Gamma_{\text{in}}(v)$	-
	Jaccard	$\Gamma_{\text{und}}(u), \Gamma_{\text{in}}(v)$	$\Gamma_{\text{und}}(u), \Gamma_{\text{in}}(v)$	-
	Adamic-Adar	$\Gamma_{\text{und}}(u), \Gamma_{\text{in}}(v), \Gamma_{\text{und}}(w)$	$\Gamma_{\text{und}}(u), \Gamma_{\text{in}}(v), \Gamma_{\text{out}}(w)$	-
	Res. allocation	$\Gamma_{\text{und}}(u), \Gamma_{\text{in}}(v), \Gamma_{\text{und}}(w)$	$\Gamma_{\text{und}}(u), \Gamma_{\text{in}}(v), \Gamma_{\text{und}}(w)$	-
Path-based	Katz	$\beta = 0.1$	$\beta = 0.4$	$\beta = 0.1$
	Local path index	$\beta = 0.1, l = 3$	$\beta = 0.1, l = 3$	$\beta = 0.1, l = 3$
	Global LHN	$\lambda = 0.1$	$\lambda = 0.1$	$\lambda = 0.1$
Random walk	Rooted PageRank	$r = 0.4$	$r = 0.9$	$r = 0.99$
	PropFlow	$l = 4$	$l = 2$	$l = 2$
	Money	Auth., $\alpha = 0.99$	Auth., $\alpha = 0.99$	Hubs, $\alpha = 0.99$
	Hitting Time	$r = 0.9$	$r = 0.9$	$r = 0.1$
	Commute Time	$r = 0.9$	$r = 0.9$	$r = 0.1$
IR	CF-tf-idf	Γ_{und}	Γ_{in}	-
	CB-tf-idf	Own tweets	Γ_{in} tweets	-
CF	User-based kNN	$k = 100$	$k = 50$	$k = 30$
	Item-based kNN	$k = 290$	$k = 280$	$k = 300$
	Implicit MF	$\alpha = 40$, $\lambda = 150$, $k = 300$	$\alpha = 40$, $\lambda = 150$, $k = 270$	$\alpha = 40$, $\lambda = 150$, $k = 280$

grid search optimizing the P@10 metric. Moreover, in the directed Twitter network, the edge direction was one more configuration setting, which we likewise tune for the best result. All path-based algorithms are sensitive to link direction, except Global LHN which needs a symmetric adjacency matrix. In the tf-idf algorithms, similarly to the original algorithmic proposal [31], we use the same neighborhood direction for both target and candidate nodes.

We do not include in our experiments the algorithms described in Sections 16.3.7 and 16.3.8, as well as the SimRank and supervised random walks algorithms in Section 16.3.3, as they involve complexity and scalability challenges that are out of the scope of the present study; moreover, some of these methods (e.g. SoRec) are very specific of networks with certain properties and or/side information that our datasets do not include.

16.7.2. *Results*

Table 16.4 shows the experimental results obtained for the best version of each algorithm on the different datasets. Along with accuracy, we adapt and report novelty and diversity metrics from the recommender systems field [14,76,77], such as the popularity complement (PC), profile distance (PD), intra-list dissimilarity (ILD) and the complement of the Gini index (1 – Gini) for the distribution of recommended user.

Accuracy. We shall focus on of ranking-based metrics, rather than classification-oriented, as the natural perspective for a recommendation task (even when solved by the adaptation of a classification algorithm). We shall compute the metrics over users who have test data (corresponding to the "TI" evaluation setting option as described in [9], i.e. people who have followed at least one new user in the test set — which is quite a standard option in recommender systems evaluation). We report P@10 as a simple and representative ranking-oriented accuracy metric — other metrics and deeper cutoffs show similar trends as we shall report here.

We can see in Table 16.4 some differences between datasets, which can be attributed to the different nature and properties of the corresponding networks. In terms of P@10, we observe that algorithms exhibit a similar behavior over the two Twitter datasets, with classic collaborative filtering standing out among the rest, followed by neighborhood-based link prediction approaches. However, a different behavior is observed in the Facebook dataset, where personalized random walks and neighborhood-based link prediction become the most accurate recommenders.

The reason for that difference might be the differences between networks in terms of clustering coefficient [63], as shown in Table 16.2. A large value of this metric means that most pairs of triplets are transitive (i.e. if there is a link from u to v and a link from v to w, it is likely that a

Table 16.4. Empirical observation. The algorithms in each group (CF, Neighborhood, etc.) are ordered by decreasing P@10. Each column is colored from white (worst) to blue (best values), and the best value of each metric for each dataset is highlighted in bold. All metrics are computed at a @10 cutoff.

		Twitter Interaction					Twitter Follows					Facebook				
		P	PC	PD	ILD	Gini	P	PC	PD	ILD	Gini	P	PC	PD	ILD	Gini
CF	Implicit MF	**0.0542**	0.9840	0.9005	0.9643	0.0587	**0.1011**	0.9646	0.9557	0.9648	0.1187	0.1611	0.9828	0.1343	0.1090	0.2747
CF	User-based kNN	0.0515	0.9722	**0.9681**	0.9608	0.0508	0.0982	0.9525	0.9694	0.9586	0.0768	0.1731	0.9829	0.1342	0.1232	0.3879
CF	Item-based kNN	0.0388	0.9888	0.9511	0.9435	0.1809	0.0933	0.9544	0.9677	0.9543	0.0870	0.1536	0.9851	0.1181	0.0643	0.2845
Neighborhood	Adamic-Adar	0.0533	0.9815	0.8999	0.9639	0.1403	0.0866	0.9448	0.9577	0.9645	0.0656	0.1803	0.9817	0.1397	0.1333	0.3548
Neighborhood	Res. Allocation	0.0481	0.9839	0.9011	0.9667	0.1895	0.0848	0.9557	0.9571	0.9651	0.1367	0.1862	0.9825	0.1446	0.1451	0.3966
Neighborhood	MCN	0.0507	0.9828	0.8982	0.9613	0.1269	0.0844	0.9439	0.9580	0.9641	0.0641	0.1752	0.9815	0.1401	0.1290	0.3318
Neighborhood	Jaccard	0.0304	0.9976	0.8866	0.9500	0.3810	0.0647	0.9900	0.9536	0.9614	0.3838	0.1649	0.9894	0.1243	0.1064	0.6030
Neighborhood	Pref. Attachment	0.0225	0.9388	0.9117	0.9660	0.0010	0.0343	0.8695	0.9693	0.9698	0.0012	0.0136	0.9128	0.9693	0.8932	0.0025
Random Walk	Money	0.0477	0.9688	0.9006	0.9653	0.0526	0.0840	0.9220	0.9576	0.9642	0.0281	0.1861	0.9821	0.1418	0.1412	0.3886
Random Walk	Rooted PageRank	0.0336	0.9788	0.9046	0.9720	0.0844	0.0677	0.9639	0.9598	0.9690	0.0805	**0.1862**	0.9816	0.1439	0.1467	0.3870
Random Walk	PropFlow	0.0320	0.9821	0.9046	0.9720	0.0764	0.0659	0.9669	0.9597	0.9693	0.0898	0.1857	0.9825	0.1445	0.1450	0.4001
Random Walk	Hitting Time	0.0204	0.9433	0.9134	0.9747	0.0010	0.0287	0.8901	0.9709	0.9778	0.0011	0.0128	0.9232	**0.9726**	0.9478	0.0024
Random Walk	Commute Time	0.0204	0.9433	0.9134	0.9747	0.0010	0.0287	0.8901	0.9709	0.9778	0.0011	0.0139	0.9232	0.9657	**0.9479**	0.0024
Path-based	Local Path Index	0.0424	0.9787	0.8996	0.9627	0.0631	0.0706	0.9235	0.9588	0.9609	0.0190	0.1495	0.9761	0.1602	0.1058	0.1477
Path-based	Distance	0.0136	0.9937	0.9027	**0.9750**	0.2943	0.0089	0.9846	0.9641	**0.9805**	0.3944	0.0144	0.9902	0.5180	0.5253	0.7314
Path-based	Katz	0.0075	0.9848	0.9145	0.9653	0.0083	0.0090	0.9775	**0.9738**	0.9643	0.0034	0.0188	0.9818	0.5565	0.2864	0.0772
Path-based	Global LHN	0.0004	**0.9995**	0.9216	0.9507	0.0009	0.0064	**0.9996**	0.9291	0.9452	0.0905	0.0316	**0.9983**	0.4381	0.0809	0.1282
IR	CF-tf-idf	0.0268	0.9985	0.8588	0.9194	0.4660	0.0478	0.9934	0.9520	0.9609	0.4726	0.1717	0.9888	0.1245	0.1131	0.6053
IR	CB-tf-idf	0.0251	0.9966	0.9001	0.9459	0.3311	0.0513	0.9919	0.9528	0.9600	0.4500	-	-	-	-	-
ML	Logistic Regression	0.0325	0.9685	0.8592	0.9200	0.0490	0.0480	0.8830	0.9667	0.9689	0.0042	0.1426	0.9482	0.2658	0.3989	0.1731
ML	Naïve-Bayes	0.0209	0.9549	0.9094	0.9683	0.0064	0.0359	0.9139	0.9659	0.9691	0.0119	0.0808	0.9529	0.4177	0.6121	0.1745
	Random	0.0004	0.9984	0.9016	0.9713	**0.7347**	0.0015	0.9948	0.9643	0.9792	**0.7961**	0.0014	0.9913	0.9389	0.9309	**0.8130**

link from u to w exists in the network). Therefore, recommending people at distance 2 from the target user, with many common neighbors, is more likely to work on networks with high clustering coefficient than in networks with low clustering. The most accurate algorithms in Facebook are the ones that recommend this type of users, thus performing better than on the Twitter networks.

Several common patterns can be identified otherwise across all three datasets. One notable observation is the good performance of neighborhood-based link prediction algorithms, which consistently achieve competitive results every time. Among them, the Adamic-Adar algorithm stands out, ranking among the top 5 best algorithms in all networks. To a lesser extent, other alternatives such as resource allocation or MCN also achieve good accuracy. Also in this family, preferential attachment is quite suboptimal in all datasets, and the Jaccard coefficient, though far better than popularity, fails to improve over its unnormalized version (MCN). The CF-tf-idf approach, which can also be classed in this group, shows a similar behavior to Jaccard in all datasets. This method has very similar accuracy to the content-based IR approach.

On the other hand, path-based approaches are far from the best results: Katz, global LHN and distance-based recommendation are always among the worst four algorithms in the comparison (random recommendation aside) in terms of precision, along with the hitting time and commute time algorithms. The only exception to this observation is the local path index, which stands as a mid-packer in all three datasets. The three other random walk methods — Money, PageRank and PropFlow — seem to work quite well on the Facebook undirected network — they are in fact the best three algorithms — while they are far from being any good on Twitter. Money, the algorithm devised by researchers at Twitter, seems to be the best of all three.

The poor accuracy of CF-tf-idf, Jaccard and global LHN index can be related to a common factor in all three algorithms: a heavy popularity penalization. It would appear that some degree of popularity is needed to achieve a basic accuracy level.

Finally, the direct adaptation of classical collaborative filtering algorithms is among the best-performing approaches. On Twitter, matrix factorization is the top algorithm, and user-based kNN ranks in the head pack as well. They both achieve decent — though not top — accuracy also on Facebook. We further observe that item-based kNN is slightly behind the two other approaches, particularly on the interaction network.

Novelty and diversity. In addition to the accuracy metrics, we have evaluated the algorithms using complementary perspectives, namely novelty and diversity. The novelty perspective measures how different the recommended users are with respect to the current or previous social experience of the candidate user, whereas diversity refers to the recommendation of people who are different from each other [14]. For novelty, we measure the popularity complement (PC) [14,76], and the profile distance (PD) [14,76] of recommendations. For diversity we use the intra-list distance [14,76] and the Gini coefficient [14,77].

The popularity complement (PC) [14,76] rewards the recommendation of long-tail, low-degree users. Since it is the opposite of the objective function of preferential attachment, it is easy to see why popularity yields the lowest value for this metric. Random-walk algorithms hitting time and commute time follow close by, which indicates that both methods have a strong bias towards recommending popular users. Algorithms like Jaccard, CF-tf-idf and the global LHN index yield high values in this metric. This is mainly due to the strong penalization applied to popular candidate items. This is particularly noticeable in LHN, which even outperforms random recommendation in this metric. On the Twitter datasets, CB-tf-idf also achieves good results in PC.

We assess recommendation unexpectedness with the profile distance metric (PD) [14,76], which measures how different the recommended contacts are from people the target user is already connected to. The metric requires a distance measure, based on user features. Depending on the dataset we evaluate, we consider two different types of features In the Twitter datasets, we take as features the unique terms in the text of tweets posted by users. We weight the terms by tf-idf [70] on the concatenated tweets (as a single document), and we measure the distance between users as the complement of the cosine similarity between the tf-idf vectors. We find many differences between datasets in the results, but also some commonalities. User-based and item-based kNN as well as Katz stand out, reaching high PD values in both datasets. Popularity, hitting time and commute time also achieve fair results. On the other extreme, tf-idf variants are among the worst approaches, along with neighborhood-based algorithms.

Since the Facebook dataset does not include any side information to extract user features from, we take an alternative distance notion. Specifically, we partition the network into communities, which we take as user features. We use three different community detection algorithms for this purpose: leading vector [62], Louvain [11] and Infomap [69], which provide a 3-dimensional feature vector for each user, formed by the communities the user belongs to in each of the three respective network partitions. Using these features, we take the complement of the Jaccard similarity of the users' respective communities as the distance measure. We observe that popularity, hitting time and commute time stand out among the rest, showing the most novel recommendations. Far from them, distance-based and machine learning approaches obtain moderate results, but still much better than the rest of the algorithms.

We use the same distance measures (tf-idf on Twitter and community Jaccard on Facebook) to assess the intra-list dissimilarity (ILD), i.e. how different are recommended users from each other [14,76]. We observe that, in the Twitter networks, distance-based recommendation, hitting time and commute time achieve the best ILD results, followed by popularity, PageRank and Money. On the opposite extreme, item-based kNN, CF-tf-idf and LHN yield the worst values. Despite the differences, we observe some similar results in the Facebook dataset: hitting time and commute time provide the most diverse recommendations, followed by popularity. Machine learning approaches, along with Katz and distance-based recommendation, though far from the best methods, obtain good ILD results.

We nonetheless find a considerable variation from one dataset to the other: on the Twitter interaction network, hitting time and commute time provide the most diverse results, along with supervised Naïve-Bayes; on the Twitter follows network, path-based approaches stand out; preferential attachment, Naïve-Bayes and distance-based recommendation achieve the most diverse recommendations on the Facebook network. In all cases, kNN produces the worst results, while random recommendation trivially produces highly diverse recommendations, as is well-known [14].

Finally, the Gini coefficient measures how evenly distributed are recommendations over the network population [14,77]. Preferential attachment naturally produces, by definition, the most concentrated recommendations, since it is not personalized and predicts essentially (except for excluding existing contacts and reciprocating links for each target user) the same list of contacts to everyone. Hitting time and commute time also show remarkably low scores for this metric, showing that they do not provide very personalized predictions. Excluding random recommendation, which trivially achieves the highest value for this metric in all datasets, Jaccard, distance-based recommendation and CF-tf-idf obtain the best values in all three datasets, because they either penalize or ignore popularity in their ranking function. Interestingly, penalizing popularity alone is not a guarantee for achieving a high Gini diversity: the LHN algorithm seems to concentrate recommendations over a reduced set of users who are not popular — the reason being that the target user has little effect on the ranking score of the algorithm, hence the reduced degree of personalization of the resulting recommendations.

Conclusions. Overall, from the point of view of recommendation accuracy, which has been the focus of the vast majority of research efforts in the field, we can say that state of the art collaborative approaches from the recommender systems field, such as kNN [64] and matrix factorization [35], work very effectively when adapted to this very particular recommendation problem. We also confirm the effectiveness (most consistently across datasets) of one of the most classical and long-standing approaches, Adamic-Adar [2,48,57]. When considering a wider perspective of recommendation utility, we can point at Jaccard [36,48,70] as a particularly balanced option, which procures better novelty and diversity than the most accurate algorithms (again, consistently over metrics and datasets), while still retaining a fair accuracy level.

Let the reader keep in mind that the empirical observations reported here are a particular example experiment from which we cannot extract general claims, but we hope it serves as illustration of the different behavior displayed by some variety of algorithms drawn from work in different contexts, which we test and compare here under an integrative experimental setup.

16.8. Future directions

The recommendation of contacts in a social network is a relatively new problem and many new algorithms and techniques will certainly see the light in the coming years. Algorithmic design, to begin with, can be expected to be an active area of innovation and research for many years to come. More effective and efficient algorithms will foreseeably be developed, and specific questions and aspects therein (such as link directionality in algorithm instantiations) will likely remain open for research improvement in the mid or long term.

Moreover, if the evaluation of recommender systems remains an area with open research problems today, this is more so when recommending people, as the human complexity of the target user gets multiplied by the complexity of a human target item. Further research is needed to understand the motivations, goals and assumptions of contact recommendation, and the different angles that should be considered. Notions such as engagement, reciprocity [66,67], dimensions such as novelty and diversity [14,16] may bear rich meanings when referred to people interacting with people.

Another major and barely addressed prospect is considering what link recommendation brings not just to each individual user, but to the network itself. As personalized contacts suggestions gain presence in social platforms, they may play an increasingly important active role in the growth of the network, and hence in shaping its evolution. This effect should be better understood to avoid undesired drifts (e.g. hub hypertrophy, community bubbles, starving nodes, etc. [65]), and to take the opportunity to draw further benefit from the action of recommenders, with a broader outlook on the network properties, performance and value. This perspective may lead to new and rich connections between recommender system technology and long-standing research on social network analysis, theory and metrics [1,20,71,73].

References

1. Aiello, L., Barbieri, N. (2017). Evolution of Ego-networks in Social Media with Link Recommendations, *Proceedings of the 10th ACM International Conference on Web Search and Data Mining* (WSDM 2017), ACM, pp. 111–120.
2. Adamic, L.A. and Adar, E. (2003). Friends and neighbors on the Web, *Social Networks* 25(3), pp. 211–230.
3. Al Hasan, M., Chaoji, V., Salem, S., and Zaki, M. (2006). Link Prediction using Supervised Learning, *Proceedings of SDM 06 Workshop on Link Analysis, Counterterrorism and Security at the 6th SIAM International Conference of Data Mining* (SDM 2006), IEEE, pp. 1828–1832.
4. Amatriain, X. (2012). Mining Large Streams of User Data for Personalized Recommendations, *ACM SIGKDD Explorations Newsletter* 14(2), pp. 37–48.
5. Backstrom, L. and Leskovec, J. (2011). Supervised Random Walks, *Proceedings of the 4th ACM International Conference on Web Search and Data Mining* (CIKM 2011), ACM, pp. 635–644.
6. Baeza-Yates, R. and Ribeiro-Neto, B. (2011) *Modern Information Retrieval: the concepts and technology behind search*, 2nd Ed. (Addison-Wesley Publishing Company, USA).
7. Bahmani, B., Chowdury, A. and Goel A. (2010). Fast incremental and personalized PageRank, *Proceedings of the VLDB Endowment* 4(3), pp. 173–184.
8. Barabàsi, A-L. and Albert, R. (1999). Emergence of scaling in random networks, *Science* 286(5439), pp. 509–512.
9. Bellogín, A., Castells, P. and Cantador I. (2017). Statistical Biases in Information Retrieval Metrics for Recommender Systems, *Information Retrieval Journal* 20(6), pp. 606–634.
10. Bishop, C. (2006). *Pattern Recognition and Machine Learning* (Springer, New York, USA).
11. Blondel, V., Guillaume, J., Lambiotte, R. and Lefebvre, E. (2008). Fast Unfolding of Communities in Large Networks. *Journal of Statistical Mechanics: Theory and Experiment* 2008(10).
12. Brin, S. and Page, L. (1998). The Anatomy of a Large-Scale Hypertextual Web Search Engine, *Proceedings of the 7th International Conference on World Wide Web* (WWW 1998), Elsevier Science Publishers B.V., pp. 107–117.
13. Cannistraci, C.V., Alanis-Lobato, G., and Ravasi, T. (2013). From link-prediction in brain connectomes and protein interactomes to the local-community-paradigm in complex networks, *Scientific Reports* 3.
14. Castells, P., Hurley, N. and Vargas, S. (2015) *Recommender Systems Handbook*, 2nd Ed., eds. Ricci, F., Rokach, L., Shapira, B., Chapter 26 "Novelty and Diversity in Recommender Systems" (Springer, New York, USA), pp. 881–918.

15. Chawla. N., Bowyer, K., Hall, L. and Kegelmeyer, W. (2002). SMOTE: Synthetic Minority Over-sampling Technique, *Journal of Artificial Intelligence Research* 16, pp. 321–357.
16. Chen, J., Geyer, W., Dugan, C., Muller, M. and Guy, I. (2009). Make new friends, but keep the old, *Proceedings of the 27th International Conference on Human Factors in Computing Systems* (CHI 2009), ACM, pp. 201–210.
17. Clauset, A., Moore, C. and Newman, M.E.J. (2008). Hierarchical structure and the prediction of missing links in networks, *Nature* 453, pp. 98–101.
18. Cukierski, W., Hamner, B. and Yang, B. (2011). Graph-based features for Supervised Link Prediction, *Proceedings of the 2011 International Joint Conference on Neural Networks* (IJCNN 2011), IEEE, pp. 1237–1244.
19. Cremonesi, P., Koren, Y. and Turrin, R. (2010). Performance of Recommender Algorithms on Top-N Recommendation Tasks, *Proceedings of the 4th ACM Conference on Recommender Systems* (RecSys 2010), ACM, pp. 39–46.
20. Daly, E. M., Geyer, W. and Millen, D.R. (2010). The network effects of recommending social connections, *Proceedings of the 4th ACM Conference on Recommender Systems* (RecSys 2010), ACM, pp. 301–304.
21. Diaz, F., Metzler, D. and Amer-Yahia, S. (2010). Relevance and Ranking in Online Dating Systems, *Proceedings of the 33rd Annual International ACM SIGIR Conference on Research and Development in Information Retrieval* (SIGIR 2010), ACM, pp. 66–73.
22. Dunbar, R.I.M. (1993). Coevolution of neocortical size, group size and language in humans. *Behavioral and Brain Sciences* 16, pp. 681–735.
23. Erdelyi, I. (1967). On the Matrix Equation Ax = λBx, *Journal of Mathematical Analysis and Applications* 17(1), pp. 119–132.
24. Fawcett, T. (2006). An introduction to ROC analysis, *Pattern Recognition Letters* 27(8), pp. 861–874.
25. Goel, A., Gupta, P. Sirois, J., Wang, D., Sharma, A. and Gurumurthy, S. (2015). The Who-to-follow system at Twitter: Strategy, algorithms and revenue impact, *Interfaces* 45(1), pp. 98–107.
26. Golder, S.A., Yardi, S., Marwick, A. and boyd, d. (2009). A Structural Approach to Contact Recommendation in Online Social Networks, *Workshop on Search in Social Media (SSM 2009) at the 32n Annual International ACM SIGIR Conference on Research and Development in Information Retrieval (SIGIR 2009).*
27. Goodman, L. (1961). Snowball Sampling, *Annals of Mathematical Statistics* 32 (1), pp. 148–170.
28. Guimera, R. and Sales-Pardo, M. (2009). Missing and spurious interactions and the reconstruction of complex networks, *Proceedings of the National Academy of Sciences* 106, pp. 22073–22078.
29. Gupta, P, Goel, A., Lin, J., Sharma, A., Wang, D. and Zadeh, R. (2013). WTF: The Who To Follow service at Twitter, *Proceedings of the 22nd International Conference on World Wide Web* (WWW 2013), ACM, pp. 505–514.

30. Guy, I. (2015) *Recommender Systems Handbook*, 2nd Ed., eds. Ricci, F., Rokach, L., Shapira, B., Chapter 15 "Social Recommender Systems" (Springer, New York, USA), pp. 881–918.
31. Hannon, J., Bennet, M. and Smyth, B. (2010). Recommending Twitter users to follow using content and collaborative filtering approaches, *Proceedings of the 4thACM Conference on Recommender Systems* (RecSys 2010), ACM, pp. 199–206.
32. Herlocker, J., Konstan, J., Terveen, L. and Riedl, J. (2004). Evaluating Collaborative Filtering Recommender Systems, *ACM Transactions on Information Systems* 22(1), pp. 5–53.
33. Hill, R. and Dunbar, R. (2003). Social Network Size in Humans, *Human Nature* 14(1), 53–72.
34. Hox, J.J. and Boeije, H. (2005). *Encyclopedia of Social Measurement, Volume 1*, eds. Kempf-Leonard, K. "Data Collection, Primary vs. Secondary" (Elsevier, Amsterdam, The Netherlands), pp. 593–599.
35. Hu, Y., Koren, Y. and Volinsky, C. (2008). Collaborative Filtering for Implicit Feedback Datasets, *Proceedings of the 8th IEEE International Conference on Data Mining* (ICDM 2008), IEEE, pp. 263–272.
36. Jaccard, P. (1901). Étude comparative de la distribution florale dans une portion des Alpes et du Jura, *Bulletin de la Société Vaudoise des Sciences Naturelles* 37 (142), pp. 547–579.
37. Jamali, M. and Ester, M. (2010). A Matrix Factorization Technique with Trust Propagation for Recommendation in Social Networks, *Proceedings of the 4th ACM Conference on Recommender Systems* (RecSys 2010), ACM, pp. 135–142.
38. Jeh, G. and Widom, J. (2002). SimRank: A Measure of Structural-Context Similarity, *Proceedings of the 8th ACM SIGKDD International Conference on Knowledge Discovery and Data Mining* (KDD 2002), ACM, pp. 538–549.
39. Katz, L. (1953). A new status index derived from sociometric analysis, *Psychometrika* 18(1), pp. 39–43.
40. Kleinberg, J.M. (1999). Authoritative Sources in a Hyperlinked Environment, *Journal of the ACM* 46(5), pp. 604–632.
41. Koprinska, I. and Yacef, K. (2015) *Recommender Systems Handbook*, 2nd Ed, eds. Ricci, F., Rokach, L., Shapira, B., Chapter 16, "People-to-People Reciprocal Recommenders" (Springer, New York, USA), pp. 545–568.
42. Koren, Y., Bell, R., and Volinsky, C. (2009). Matrix Factorization Techniques for Recommender Systems. *Computer* 42(8), pp. 30–37.
43. Langville, A.N. and Meyer, C.D. (2006) *Google's PageRank and Beyond: The Science of Search Engine Rankings* (Princeton University Press, New Jersey, USA).
44. Leicht, E.A., Holme, P. and Newman, M.E.J. (2006). Vertex similarity in Networks, *Physical Review E* 73.
45. Lempel, R. and Moran, S. (2001). SALSA: The Stochastic Approach for Link-Structure Analysis. *ACM Transactions on Information Systems* 19(2), pp. 131–160.

46. Leskovec, J. and Faloutsos, C. (2006). Sampling from Large Graphs, *Proceedings of the 12th ACM SIGKDD International Conference on Knowledge Discovery and Data Mining* (KDD 2006), ACM, pp. 631–636.
47. Leskovec, J and Sosič, R. (2016). SNAP: A General-Purpose Network Analysis and Graph-Mining Library, *ACM Transactions on Intelligent Systems and Technology* 8(1).
48. Liben-Nowell, D. and Kleinberg, J. (2003). The Link Prediction Problem for Social Networks, *Proceedings of the 12th ACM International Conference on Information Knowledge and Management* (CIKM 2003), ACM, pp. 556–559.
49. Lichtenwalter, R.N and Chawla, N.V. (2011). LPMade: Link Prediction Made Easy, *Journal of Machine Learning Research* 12, pp. 2489–2492.
50. Lichtenwalter, R.N, Lussier, J.T. and Chawla, N.V. (2010). New Perspectives and Methods in Link Prediction, *Proceedings of the 16th ACM SIGKDD International Conference on Knowledge Discovery and Data Mining* (KDD 2010), ACM, pp. 243–252.
51. Lü, L., Jin, C. and Zhou, T. (2009). Similarity Index for Link Prediction of Complex Networks, *Physical Review E* 80(4).
52. Lü, L and Zhou, T. (2010). Link Prediction in Social Networks: A Survey, *Physica A* 390(6), pp. 1150–1170.
53. Ma, H., Yang, H., Lyu, R. and King, I. (2008). SoRec: Social Recommendation Using Probabilistic Matrix Factorization, *Proceedings of the 17th ACM International Conference on Information and Knowledge Management* (CIKM 2008), ACM, pp. 931–940.
54. Maiya, A. and Berger-Wolf, T. (2011). Benefits of Bias: Toward Better Characterization of Network Sampling, *Proceedings of the 17th ACM SIGKDD International Conference on Knowledge and Data Mining* (KDD 2011), ACM, pp. 105–113.
55. McAuley, J. and Leskovec, J. (2012). Learning to Discover Social Circles in Ego Networks, *Advances in Neural Information Processing Systems* 25 (NIPS 2012). Curran Associates, Inc., pp. 539–547.
56. McDonald, D. and Ackerman, M. (2000). Expertise Recommender: A Flexible Recommendation System and Architecture, *Proceedings of the 2000 ACM Conference on Computer Supported Cooperative Work* (CSCW 2000), ACM, pp. 231–240.
57. McPherson, M., Smith-Lovin, L. and Cook, J-M. (2001). Birds of a Feather: Homophily in Social Networks, *Annual Review of Sociology* 27(1), pp. 415–444.
58. Menon, A. and Elkan, C. (2011). Link Prediction via Matrix Factorization, *Proceedings of the European Conference on Machine Learning and Knowledge Discovery in Databases* (ECML PKDD 2011), Springer-Verlag, pp. 437–452.
59. Meyer, C.D. (1975). The Role of Group Generalized Inverse in the Theory of Finite Markov Chains. *SIAM Review* 17(3), pp. 443–464.

60. Myers, S.A, Sharma, A., Gupta, P. and Lin, J. (2014). Information Network or Social Network? The Structure of the Twitter Follow Graph, *Proceedings of the 23rd International Conference on World Wide Web* (WWW Companion 2014), ACM, pp. 493–498.
61. Newman, M.E.J. (2001). Clustering and preferential attachment in growing networks, *Physical Review E* 64.
62. Newman, M.E.J. (2006). Finding community structure in networks using the eigenvalues of matrices, *Physical Review E* 74.
63. Newman, M.E.J. (2010) *Networks: An Introduction* (Oxford University Press, Oxford, United Kingdom).
64. Ning, X., Desrosiers, C. and Karypis, G. (2015) *Recommender Systems Handbook*, 2nd Ed, eds. Ricci, F., Rokach, L., Shapira, B. Chapter 1 "A Comprehensive Survey of Neighborhood-based Recommendation Methods" (Springer, New York, USA), pp. 37–76.
65. Pariser, E. (2011) *The Filter Bubble: How the personalized web is changing what we read and how we think* (Penguin Press, London, United Kingdom).
66. Pizzato, L., Rej, T., Chung, T., Koprinska, I. and Kay, J. (2010). RECON: A Reciprocal Recommender for Online Dating, *Proceedings of the 4th ACM Conference on Recommender Systems* (RecSys 2010), ACM, pp. 207–214.
67. Pizzato, L., Rej, T., Akejurst, J., Koprinska, I., Yacef, K. and Kay, J. (2013). Recommending people to people: the nature of reciprocal recommenders with a case study in online dating, *User Modeling and User-Adapted Interaction* 23(5), pp. 447–488.
68. Ravasz, E., Somera, A.L., Mongru, D.A., Oltvai, Z.N. and Barabàsi, A-L. (2002). Hierarchical Organization of Modularity in Metabolic Networks, *Science* 297, pp. 1551–1555.
69. Rosvall, M. and Bergstrom, C. (2008). Maps of random walks on complex networks reveal community structure, *Proceedings of the National Academy of Sciences* 105(4), pp. 1118–1123.
70. Salton, G. and McGill, M.J. (1983) *Introduction to Modern Information Retrieval.* (McGraw-Hill, Inc., New York, USA).
71. Sanz-Cruzado, J., Pepa, S.M. and Castells, P. (2018). Structural Novelty and Diversity in Link Prediction, *Companion Proceedings of the The Web Conference 2018* (WWW 2018), IW3C2, pp. 1347–1351.
72. Sørensen, T. (1948). A method of establishing groups of equal amplitude in plant sociology based on similarity of species content and its application to analyses of the vegetation on Danish commons, *Biologiske Skrifter* 5(4), pp. 1–34.
73. Su, J., Sharma, A. and Goel, S. (2016). The Effect of Recommendations on Network Structure, *Proceedings of the 25th International Conference on World Wide Web* (WWW 2016), IW3C2, pp. 1157–1167.
74. Sulaimany, S., Khansari, M., Zarrineh, P., Daianu, M., Jahanshad, N., Thompson, P.M. and Masoudi-Nejad, A. (2017). Predicting brain network changes in

Alzheimer's disease with link prediction algorithms, *Molecular BioSystems* 13(4), pp. 725–735.

75. Tang, J., Hu, X., Liu, H. (2013). Social recommendation: a review, *Social Network Analysis and Mining* 3(4), pp. 1113–1133.
76. Vargas, S. and Castells, P. (2011). Rank and Relevance in Novelty and Diversity Metrics for Recommender Systems, *Proceedings of the 5th ACM Conference on Recommender Systems* (RecSys 2011), ACM, pp. 109–116.
77. Vargas, S. and Castells, P. (2014). Improving Sales Diversity by Recommending Users to Items, *Proceedings of the 8th ACM Conference on Recommender Systems* (RecSys 2014), ACM, pp. 145–152.
78. Wang, C., Satuluri, V. and Parthasarthy, S. Local Probabilistic Models for Link Prediction, *Proceedings of the 7th IEEE International Conference on Data Mining* (ICDM 2007), IEEE, pp. 322–331.
79. Watts, D. and Strogatz, S. (1998). Collective dynamics of 'small-world' networks, *Nature* 393, pp. 440–442.
80. White, S. and Smyth, P. (2003). Algorithms for estimating relative importance in networks, *Proceedings of the 9th ACM SIGKDD International Conference on Knowledge Discovery and Data Mining* (KDD 2003), ACM, pp. 266–275.
81. Zhou, T., Lü, L. and Zhang, Y. (2009). Predicting missing links via local information, *The European Physical Journal B* 71(4), pp. 623–630.

Chapter 17

Job Recommendations: The XING Case

Katja Niemann, Daniel Kohlsdorf and Fabian Abel
XING SE, Hamburg, Germany
{katja.niemann, daniel.kohlsdorf, fabian.abel}@xing.com

There are several reasons why people are interested in getting job recommendations. They might be actively looking for a new job or want to stay informed about their own value in the job market, etc. Additionally, recruiters aim for the right people to see their job postings to find suitable candidates for open positions. Real world job recommender systems face several challenges, though. Job postings are only interesting as long as they are up-to-date and, thus, have to be recommended to the right users as soon as they are published. Furthermore, a job recommender system should aim to support the users in their professional growth and needs to find jobs that do not necessarily match the users' current profile but have the potential to form the next step in the users' career. In this chapter, we first introduce the professional social network XING and discuss the challenges it has to tackle when generating job recommendations. Thereafter, we present the recommender's functionality and discuss how we evaluate new features or extensions. Finally, we conclude with an insight in the recommender system's architecture that has to serve thousands of requests per minute.

17.1. Introduction

17.1.1. *XING*

XING[1] is a social network for professionals with more than 20 Million—primarily German-speaking—users. XING offers various products and services supporting social networking, tools enabling companies and recruiters to attract, find, and recruit talent as well as a jobs marketplace amongst others. The social network helps users to approach potential business partners, network with colleagues or find coaches. In order to do so, users create

[1] `https://xing.com`

and maintain a profile page holding information about their curriculum vitae (CV) including details about their current employment, previous career steps, education as well as tags describing their skills and interests. Figure 17.1 shows an excerpt of one of the author's profile page. An individual CV entry typically provides information about the job role (including name of the position, discipline, seniority level), company (including the company name, industry, company size) and the period of time that a user spent in the given position. Additionally, users can add information only visible to recruiters like their aspired salary, companies they are interested in or their willingness to travel or move to another city. Recruiters can use XING's recruiting tools to create job postings or define detailed search profiles and directly contact potential candidates.

In the subsequent sections, we will focus on XING's job marketplace[2] and will discuss the problem of recommending jobs to users [Abel (2015)]. The job marketplace typically features between 650,000 and 1 Million job postings. Some of these job postings are automatically crawled from the Web while others are manually created by recruiters on XING. Companies and recruiters can also buy packages to promote their job ads and therewith increase their visibility.

Users can either search for specific postings using XING's search service or browse and receive job recommendations through multiple channels, e.g. on XING's start page, mobile apps, emails, etc. In the job marketplace itself, recommendations are presented as cards in a matrix with three columns and two rows which can be expanded to see more recommendations (see Figure 17.2(a). These cards list the job title, the company, the city and the age of the postings. Furthermore, a score is shown that indicates to what extent a recommendation matches the profile of the user. The job recommendation engine also allows for actively notifying users about new incoming job postings. Those types of *push recommendations* are highlighted with an additional *NEW* label that is shown at the top right of the recommendations. The number of unread push recommendations is moreover shown in the menu bar on the left.

The top two postings are also presented on the start page of XING as depicted in Figure 17.2(b). Emails (see Figure 17.2(c) and mobile apps for Android, Windows or iOS (see Figure 17.2(d) feature channels for distributing job recommendations as well.

[2]https://www.xing.com/jobs

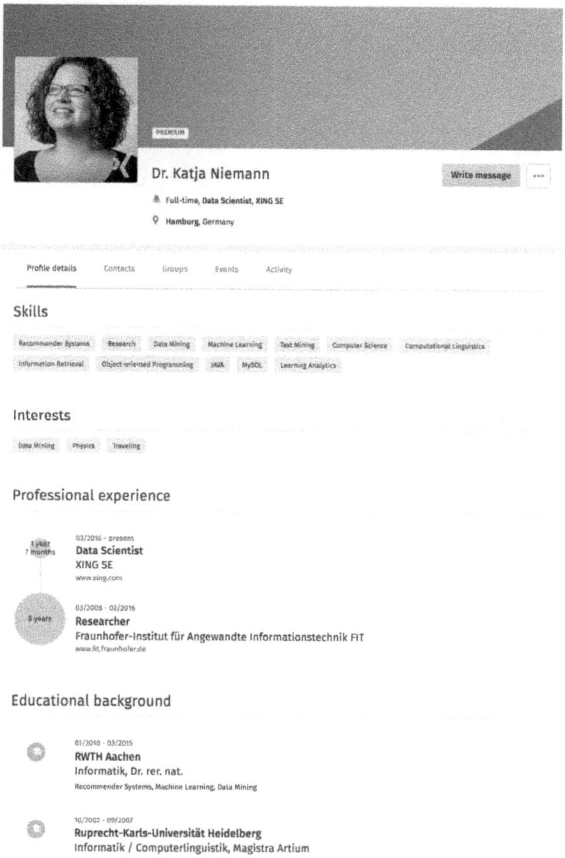

Fig. 17.1. Example of a XING profile with skills, interests, details about current/previous employments and the educational background of the user.

Users have several ways of interacting with a job posting. Clicking a posting's card opens the detailed view of the job posting with a full text description of the advertised opening. It is also possible to bookmark the posting in order to return to it easily or to apply to the opening directly on the page. On the other hand, users can delete job recommendations from their recommendation list.

In order to explain our recommendations and give users the option to give feedback, we provide another view on the job recommendation called *Is this job for me?* that is shown in Figure 17.3. This view lists the changes from the user's current job to the possible future job as advertised by the

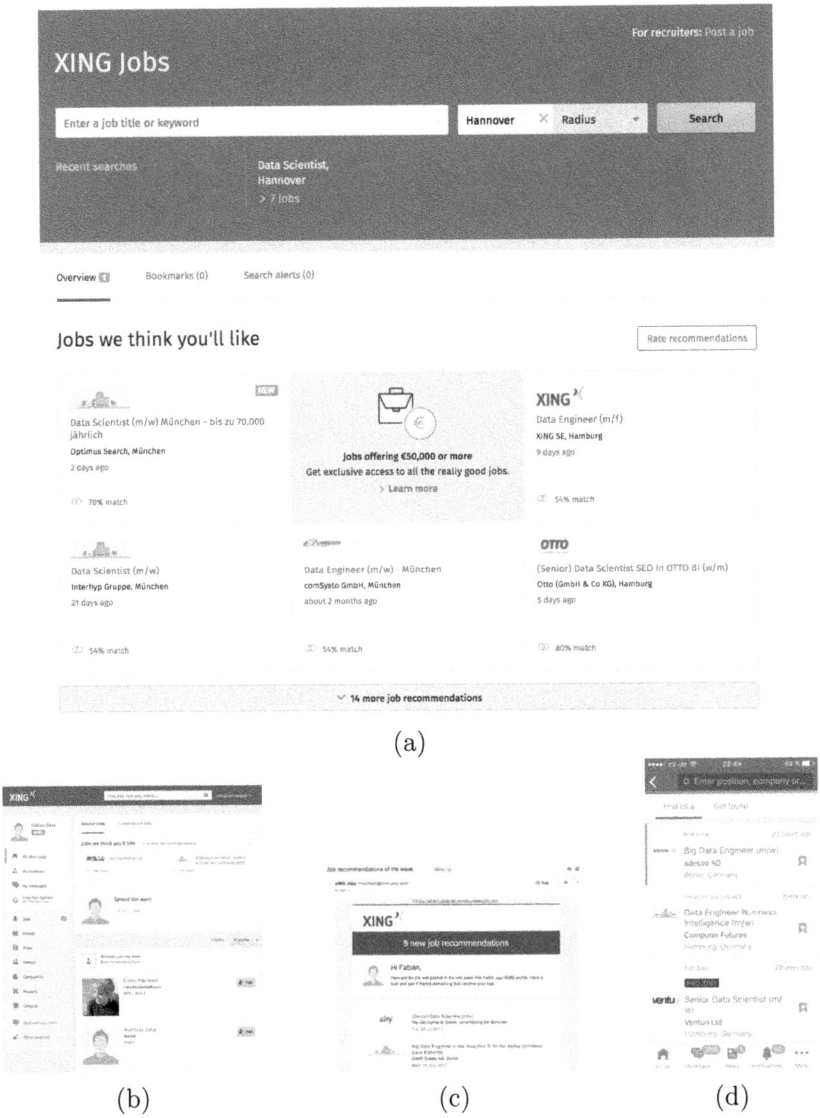

Fig. 17.2. Example channels which are used for distributing job recommendations. (a) Jobs market (b) Startpage (c) Email (d) Mobile app.

posting and highlights how well the profile of the user matches with the open position. Users can provide their personal feedback by rating the recommendation on a five-star rating scale.

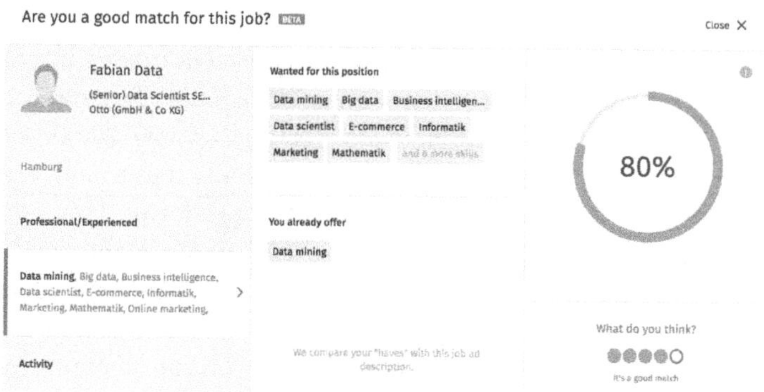

Fig. 17.3. Is this job for me? Explaining recommendations and giving users means to rate the relevance of recommendations on a 5-star rating scale.

Building recommender systems for our jobs market often requires to build and customize algorithms that are able to handle some unique challenges arising in our domain. In the next section, we will describe the most important challenges our algorithms need to overcome.

17.1.2. *Challenges*

There are various challenges XING's recommender system has to tackle. Some of these challenges arise from XING's business interests and use cases while some of the challenges are related to the data extracted from users and items.

17.1.2.1. *Data sparsity and new user problem*

Another challenge for our recommender system are new and incomplete user profiles [Su and Khoshgoftaar (2009)]. For example, profiles without a seniority level or job title will make it difficult to find appropriate jobs. As soon as users start interacting on XING, their implicitly mined interests may compensate for incomplete profiles. However, for newly registered users there is often little information available which makes the task of recommending job advertisements difficult and calls for actively starting a dialog for obtaining a better understanding about the users' demands and interests.

17.1.2.2. *Data sparsity and new item problem*

Since job postings are very short lived, there are often not many interactions for each item. Furthermore, the item collection is quite large and diverse since XING is regularly crawling new job postings. 5,000–20,000 new job ads are typically added to the inventory on a daily basis and a similar amount of items is also removed again from the platform. Users who are actively seeking for a new job would like to be informed about such new incoming jobs as fast as possible. In fact, the vast majority of the job recommendations that are actively pushed to users so that users receive a notification (either per email, in the mobile app or on the Web platform, cf. *NEW* label in Figure 17.2) are pushed immediately after the corresponding job ads were added to the inventory of the jobs marketplace and thus did not receive a single interaction at the time they are pushed. These sparsity issues render many of the collaborative filtering approaches difficult.

17.1.2.3. *Information extraction*

Since collaborative filtering is not applicable in all situations we also apply content-based recommendation strategies. However, another challenge arising from the crawled postings is that usually no meta data is available describing them.

The postings from the crawler are raw HTML documents. Therefore, our system needs to extract important information for our content-based recommendation algorithms including the actual full-text of the job ad, the job title, the location, the company, the seniority label and many more. During the extraction of these fields errors can occur that might subsequently lead to bad recommendations and in turn a loss in trust.

17.1.2.4. *Trust*

One of the hardest challenges is to keep the users' trust and make the recommendations transparent [Konstan and Riedl (2012)]. Since a person's career is a fundamental part of life, a bad job recommendation can be viewed as insulting or anger the user to a point where the user's trust in the recommender system is completely lost. One example of a mistake that often leads to disappointed users is a mismatch regarding the seniority of the user. For example, a recommendation with the job role *Internship Assistant to the CEO* for a user with the job role *CEO* will most probably be interpreted in a negative way.

17.1.2.5. *Ambiguity and false information*

People may sometimes use the same words but in fact refer to different concepts. For example, a job role such as *Sales Manager* leaves quite some room for interpretation. It may refer to a call agent, traveling salesman, shop assistant and many more. Moreover, user profiles and job advertisements may both contain false information. For example, job ads are sometimes written by recruiters who are lacking detailed background information about the actual job role and may thus not correctly describe skills that are required for a given position. On the contrary, users may list skills in their profiles that do not properly reflect their experience.

17.1.2.6. *Evolving careers*

Most users are not looking for a job that is equivalent to their current job. They want to evolve in their career or find a job in which they can gather new skills or competencies. Thus, the profile of a user might describe the present and past of the users but not necessarily their future for which we aim to recommend jobs [Li *et al.* (2017); Xu *et al.* (2016)].

17.1.2.7. *Paid content vs. non-paid content*

One of XING's businesses is to sell job advertisements to recruiters. By paying for job advertisements, recruiters benefit from an increased number of impressions for their job ads. In other words, these postings require a boosted score when ordering recommendations. We want to recommend the most suitable job postings to our users, though. Sometimes, these goals can be conflicting and there might be a trade-off between the quality and the boosting. This is especially true for postings where only very few potential candidates exist.

17.1.2.8. *User interests vs. recruiter interests*

Ideally, job recommendations take two aspects into account: (a) the recommended job ad should be of interest to the user and (b) the user should be an appropriate candidate for the given job. Balancing these two types of interests is not easy. A user may be strongly interested in a given job, but may in fact—from the recruiter's perspective—not be a good candidate for the job opening.

Furthermore, the two parties recruiters and candidates can have different goals. For example, companies might want to reach as many qualified candidates as possible in order to increase their postings' visibility. However, for job roles with a high demand, the candidates might be then flooded with messages from recruiters or notifications about new job recommendations. Particularly when it comes to actively notifying users about new matching job recommendations, one requires smart mechanisms that decide about sending or not sending such notifications.

In the rest of the paper, we will present how we tackle these challenges at XING in a production environment. Additionally, some of these challenges, especially the data sparsity as well as the user vs. recruiter interests, were addressed as part of the ACM Recommender Systems Challenges 2016 and 2017 [Abel *et al.* (2016, 2017); Said (2016)] which were co-organized by XING. In both editions of the challenge, the winning solutions were ensembles of recommender algorithms that exploited hundreds or even thousands of features. Additionally, both solutions made use of gradient-boosted decision trees to learn models combining those features [Volkovs *et al.* (2017); Xiao *et al.* (2016)]. To get a more detailed overview of the various solutions submitted to the Challenges, we refer the reader to the proceedings of the corresponding ACM Recommender Systems Challenges [rec (2016, 2017)].

17.1.3. *Structure of XING's Job Recommender System*

In order to account for the different requirements from our business and our data sources, we implement XING's job recommender system as an ensemble of various recommendation strategies, filters, and rankers. The recommender ensemble consists of multiple components $C = c_1, c_2, ..., c_N$ such as collaborative filtering, more-like-this, or content-based recommender systems. Each of these components c_i returns a list of k recommendations with scores $r_{i1}, r_{i2}, ..., r_{ik}$. To compute the final score of an item j its scores are combined linearly over all components and then squished through a logistic function.

$$s(j) = \beta_0 + \sum_{i \in N} \mathbb{I}(item_j \in C_i) * \beta_j * r_{ij} \qquad (17.1)$$

$$p(item = j) = \frac{1}{1 + e^{-s(j)}} \qquad (17.2)$$

The function $\mathbb{I}(item_j \in C_i)$ is an indicator function that returns 1 if the item was returned from component c_i and 0 otherwise. The weights β are learned using a logistic regression model. During learning, we try to predict a user's click using the components' scores as features. After the final recommendation list is computed using the output of all components, we apply several filtering and ranking algorithms that account for our business rules.

17.2. Data Processing Pipelines

17.2.1. *Users and Items*

In order to better understand the meaning of the information shared by the users in their profiles as well as to understand the job postings and match them with the users, we pre-process the user profiles and job postings and map them to the same knowledge space as described in the following sections.

17.2.1.1. *User profiles*

Each user has a profile page which enables her to provide a detailed overview about her professional career as well as about her competencies, education, and interests. The profile page is thus divided in several sections according to the different information presented with the most important and prominently placed sections being *haves*, *interests*, *professional experience*, and *educational background* (cf. Figure 17.1).

The *haves* and *interests* sections hold the users' skills and competencies and additionally their main purpose for using the platform in a tag-like form. The most common *haves* are hard and soft skills like *'online marketing'*, *'quality management'*, *'C++'* as well as *'reliability'*, *'flexibility'*, and *'creativity'* while the *interests* section expresses what the users are looking for, e.g. *'new contacts'*, *'new challenges'*, or *'exchanging ideas'*.

The *professional experience* section comprises the work experiences of the users, i.e. their previous and current positions. Each work experience holds mandatory information, i.e. the job title and the company name which are both free-text as well as the employment type (e.g. *'full-time employee'*, *'part-time employee'*, *'intern'*, *'freelancer'*, or *'volunteer'*) and the industry (e.g. *'Banking and Financial Services'* or *'IT and Internet'*) which can both be selected using a drop-down menu. Additionally, a work experience can hold optional information about the discipline, the career level, or the number of employees in the company as well as its start and end date. Finally, a user can mark a work experience as primary which forces the job title and company name to be shown next to the profile picture of the user at the top of the web page.

The *educational background* section gives an overview of the university-level education of a user with the only mandatory information being the university which is given by the user as free text. Optional specifications are the field of study, the (future) degree as well as the duration. Similarly to the work experiences, an education entry can be selected as primary to be shown next to the user's profile picture.

No inferred or processed information is shown on a user's profile page but only data explicitly given by the user. However, the users are encouraged to fill in their profiles as detailed as possible and also to keep them up-to-date by utilizing tools like the profile wizard which asks the users about profile entries still being valid and suggests skills that match the current profile or the job recommendation optimizer which enables users to specify what they are looking for in a new job.

17.2.1.2. *Job postings*

Each posting holds free text and optionally images that describe the offered job and the company. Above the posting, additional information about the job are summarized in a structured way, i.e. the job's title, the company, the location, the date when it was posted, the job type (which is similar to the employment type of the user), the career level, and the industry.

XING offers companies a graphical user interface as well as an API to submit their paid job postings. This way, the companies can directly choose several features like the career level or the industry from XING provided lists. Additionally, XING uses a crawler to gather (non-paid) job postings from different web sites and extracts or infers these features automatically.

17.2.1.3. *Interactions of users and job postings*

When a user gets a recommendation for a job posting, she can choose to ignore it, to *bookmark* it in order to create an overview of jobs she is interested in, to *delete* it in order to make the recommendation disappear and show a new one, or to *click* on it to study the job posting in detail. After clicking on a job posting, she can still *bookmark* the job posting or create a *search alert* based on the features automatically extracted from the job posting which can be further refined. In the "Is this job for me" section of the posting (see Figure 17.3), the user has the option to *rate* the posting using stars on a scale from 1–5.

Finally, the user has several opportunities to *reply* to a job posting that vary based on the chosen posting, e.g. paid postings usually hold a

'I'm interested' button that enables the users to notify the job poster with just one click if they are interested in the job. Another *reply* option is the *'Apply'* button that opens a user's mail client with the contact e-mail address of the posting or forwards the user to the application web site of the respective company.

Users can also choose to give general feedback about their job posting recommendations. They then gets a list of up to 20 recommendations that they can rate on a scale from 1–5. The rating should represent the users' preferences with regard to job title match, location match and other fields. Additionally, the users can add free text comments about their job recommendations in general.

17.2.2. *Entity Recognition*

A lot of the data considered by XING's recommender engines is in natural language, e.g. the job titles, the posting descriptions as well as the skills and interests of the users. These fields hold very rich information, however, natural language is ambiguous. This is to say, different users and companies might use different terms to refer to the same job role or skill. Additionally, even though XING solely focuses on the German-speaking market, it has to deal with different languages as many user profiles and job postings hold information in English and sometimes also in other languages like French or Spanish. Thus, XING developed and maintains a highly specified ontology for understanding the job roles, skills and fields of study as stated by the users and job postings. Currently, the ontology holds more than 2,000 job roles and over 20,000 skills with alternative labels in several languages (mostly English and German). An example for an often detected job role is *Java Developer* which comprises 246 different alternative labels, e.g. *Application Architect J2EE* and *Java Entwickler* (which is German). The skills are more diverse than the job roles and hold hard skills as well as personal characteristics as presented in the user profile.

The basic approach that tries to detect the so-called ontology entities by finding complete matches between the text and the ontology is based on the AhoCorasick algorithm [Aho and Corasick (1975)]. However, depending on the field that is currently processed, e.g. job title, discipline, or job description, more specialized algorithms are applied that allow partial matches as well.

17.2.3. *User Modeling*

The users already provide a lot of information in their profiles which are further extended using the XING ontology. However, some user profiles are rather sparse or out-dated. Additionally, most of the information given by the users refers to the past (e.g. education and previous positions) as well as to the present time (e.g. current position and skills). In order to better understand the interests of the users and in which direction they want to evolve in the future, we infer further attributes based on their behavior, e.g. interaction-based interest profiles, their career levels as well as their interest in finding a new job which we describe in more detail in the following paragraphs.

The interest profile of a user is solely based on her interactions with job postings and includes her preferred job roles, skills, industries, and cities whereas each entry is weighted according to the number of times it was present in the clicked postings. Additionally, a time-based component makes sure that the weight of a clicked posting slowly declines. This way we might for example find skills a user does not hold yet but wants to acquire.

Additionally, we infer suitable career levels for the users by applying a rule-based approach. This is done because the career levels stated by the users are not always comparable. For example, if a student is the CEO of her own one-person company, we do not want to provide her with CEO positions of large companies but rather with entry-level positions. Here, the rules take into account the career level explicitly stated by a user, the current work experiences and education entries as well as the size of the company the user is working for and many more.

As XING wants to present only relevant content to its users, it is crucial to understand if a user is at least slightly open to changing her job. Thus, we follow a similar approach as Wang *et al.* [Wang *et al.* (2013)] and estimate the users' *willingness to change jobs* (wtcj) score that is used to pick the number of push job recommendations a user receives amongst others. The wtcj score is calculated using gradient-boosted decision trees that consider over 100 features such as the number of log-ins and clicks on job postings, the last time the profile picture was updated and the duration of the current position in comparison to the average duration in the given job role. Furthermore, the features are weighted according to the common behavior of the respective user to make behavioral changes explicit. For example, the number of login-ins in the last three month is normalized by dividing it by the number of log-ins in the last year.

17.3. Job Recommender System

In the following sections, we will describe our recommendation strategies as well as some of our filtering and ranking algorithms in detail. As discussed in Section 17.1.3 the job recommender system is an ensemble of multiple strategies whose results are combined into a single recommendation list using a logistic regressor. The final list is then filtered and ranked.

17.3.1. *Recommender Engines*

All applied recommender strategies have in common that they build a query based on different features which is then run against an Elasticsearch[3] search engine indexing the job postings. This way, potential recommendations are returned that can be further refined and we do not need to consider all user-posting pairs when applying the different approaches but focus on a small subset which can be done in real-time.

17.3.1.1. *Association Rule Mining for Collaborative Filtering*

Collaborative filtering approaches are very popular for the creation of recommendations as they can utilize the *word of mouth principle* and can, thus, take information into account that are not given by any textual descriptions. Furthermore, they do not require the users to explicitly share any information. However, XING's job recommender system has to be able to recommend job postings as soon as they are published. Additionally, job postings that were posted a few days ago might already feel outdated for actively-looking users. Thus, when only considering user interactions with these short-lived items, the system has to deal with data that is too sparse to apply classical collaborative filtering approaches.

The users' interactions provide very valuable insights about our users' needs, though. In order to utilize these insights, this recommender engines applies interaction-based association rules. Here, the current job title, discipline, industry, location (city), and career level as well as the tags (found in the *haves* section of the user's profile) are taken into account and are combined with the information extracted from the postings the user positively interacted with to build a combined profile. Based on these profiles and the interaction data we calculate the lift for each tag, source, and posting. It is important to mention that when collecting the data for calculating the lift, the posting a user interacted with is not yet part of the user's profile.

[3]https://www.elastic.co/

This way, we might find for a given job posting that users with the jobtitle "Marketing Manager" have a 56th higher probability to click on it compared to all users while another job posting might be especially interesting for users with the skill *Big Data* (lift 45) or the discipline *Analytics* (lift 33). The new item problem remains, though, which is the reason why XING additionally applies content-based filtering approaches as well as item-to-item recommendations.

For creating the recommendations, the combined user profile is used to create an Elasticsearch query and the returned postings are weighted according to the user profile and the pre-calculated association rules.

17.3.1.2. *Content-based filtering*

We apply two content-based strategies, the first one is based on the explicit user profiles (i.e. the information the users shared on their profile pages) while the other one is based on the interest profiles (i.e. the information extracted from the job postings the users interacted with). Both profiles are partially enriched with other inferred information like the most suitable career level and matching entries from XING's ontology, see 17.2.3

Based on these user profiles we create two independent search queries for each user and run them against the Elasticsearch search engine. The result lists are both ranked by their TFxIDF-based similarity to the query, i.e. the user profile.

17.3.1.3. *Item-to-item recommendations: More-Like-This*

The more-like-this recommender is a content-based filtering between items consisting of two components. The first component creates an Elasticsearch query using a user's bookmarks and reply intentions to search for similar postings across all active postings. Here, the similarity is calculated by solely considering the job roles, the skills and the locations in which the job roles and skills are represented by the respective ontology entries.

The second component utilizes word2vec [Mikolov *et al.* (2013)] for ranking the candidate recommendations generated in the first step. Word2vec is a dense, low dimensional (100 dimensions in our case) representation of words. Vectors of words that are semantically similar will also be close in the latent space. For example, the words "developer" and "programmer" are close in this space, while the words "developer" and "logistics" are not. We use Continuous Bag of Words (CBOW) to train these vectors and the latent space of the postings is spanned by the average of all word vectors in each posting. All postings from the candidate list of the first step as well as the postings from the user's previous positive interactions are transformed into a joint latent space using word2vec to then calculate their cosine similarity and rank the recommendations accordingly.

17.3.2. *Filtering and Ranking Components*

Currently, we apply 13 filter and 11 ranking components. The purpose of these strategies is to either support XING's business model, e.g. by boosting paid job postings or to correct common pitfalls of the recommender engines. We describe selected components in this section.

17.3.2.1. *Less-like-this*

For the generation of the job recommendations we apply a more-like-this component that tries to learn from the job postings a user liked in the past what she will like in the future. Similarly, we have a less-like-this component as filter that learns from past interactions like the deletion of recommended job postings what kind of job recommendations a user does not like and removes them from the list of recommendation candidates.

17.3.2.2. *Career Level*

It is important to only recommend job postings with suitable career levels to keep the users' trust as described in Section 17.1.2.4. The career level of a job posting thus has to hold a similar or slightly higher career level as the one of the target user. Here, we consider the users' career levels that were inferred as described in Section 17.2.3, i.e. we do not only use the information about the career levels that the users explicitly shared with us but apply a rule-based approach to infer comparable career levels.

17.3.2.3. *Location*

The location of a job is important for most of the users. Thus, we filter the recommendations in a way that at least a third of the shown job recommendations for a user are close to her current location. We only make an exception if we do not find enough suitable jobs in the proximity of a user. Furthermore, jobs postings from locations where a user has many contacts receive a boost.

17.3.2.4. *Diversity*

We do not try to create recommendation lists holding job postings that are as diverse as possible, as most users are looking for a specific job role. We try to diversify the companies of the job postings, though, to prevent that one company dominates the recommendations of a user, even if this company is actively looking to fill many similar positions.

17.3.2.5. *Rating prediction*

This component predicts how a user would rate a job posting on a scale from 1 to 5 based on a gradient-boosted decision tree model (XGBoost [Chen and Guestrin (2016)]). It then discards all recommendations with a predicted rating that is below a given threshold and boosts recommendations with a high predicted rating. The idea for this model resulted from the ACM Recommender Systems Challenge 2016 [Abel *et al.* (2016)] and particularly from the solution developed by Xiao *et al.* [Xiao *et al.* (2016)].

For training the model we utilize the users' explicit ratings of job postings. Attributes that are taken into account are the job roles, skills, study subjects, disciplines, industries, and locations amongst others. Based on these attributes we hand-engineered features that each aim to capture the overlap between a specific part of the user's profile and the job posting. For example, one feature captures the overlap of skills detected in the *haves* section of a user profile and in the posting, one feature captures the probability of the transition from the user's current job role to the job role offered in the posting and one feature captures if the user's current discipline matches with the job's discipline, etc. This is to say, we calculate matches between the different user and job posting attributes while taking into account the different sections in which they were identified, e.g. current and previous job positions for users or title and body for job postings. Additionally, we consider transitions between jobs as well as common jobs for users who studied a given subject or hold a specific skill.

However, we assume that not all matches hold the same informative value. For example, a match of a skill like *'key account management'* or *'Scala'* seems to be more expressive than a match of skills like *'flexibility'* or *'openness'*, i.e. there are matches in which we can put a higher trust than in other. Accordingly, we weight all matches based on the *trustworthiness* of the matched entities.

In order to calculate the *trustworthiness* of the different entities we first calculate the number of all rated recommendations with a match of the given entity between the user's profile and the job posting. We then take the number of matching recommendations that hold a rating higher than 1 and divide it by the number of all matching recommendations.

17.4. Evaluation

In this section we first describe the evaluation strategies we implement at XING to then present two case studies. In the first case study, we compare the impact of the different recommendation strategies while in the second case study we show how we evaluate a new filtering component.

17.4.1. *Evaluation Setup*

When creating a new model, we usually have a ground truth that can be used for training and offline testing. Here, the kind of ground truth as well as the applied evaluation metrics vary largely depending on the task. Based on the results of the offline evaluation we then pick the most promising model and perform an online evaluation in form of A/B-Testing.

17.4.1.1. *Offline Evaluation*

Common offline evaluation metrics that we apply when working with implicit rating data like clicks and bookmarks are precision@k and recall@k [Herlocker *et al.* (2004)]. When training the rating prediction component using explicit rating data, we utilize the root mean squared error (RMSE) as evaluation metric. However, in practice it does not matter if we are able to correctly predict if a user would rate a job posting with one or two stars but to predict that the user will not like that job posting and discard it [McNee *et al.* (2006)]. Additionally, it is important to not discard job postings a user will like, though, as these can be quite rare. Thus, we came up with an additional way to perform the final evaluation by calculating for each rating (i.e. the five classes 1–5) the percentage of postings in the test set that would be discarded for varying thresholds, see Section 17.4.3. Additionally, a common test before deploying a new recommendation strategy, filter, or ranker is to measure its impact by calculating the difference between the old and new recommendation lists.

17.4.1.2. *Online Evaluation*

The online evaluation is the most important part as a component is only useful when it performs well in production. The online evaluation is performed in the form of A/B-Testing. Therefore, we split all users into two equal subsets named A and B and play out the original recommendations to group A and the new recommendations to group B. The online evaluation metrics differ greatly from the offline evaluation metrics, additionally, they are usually the same independent of the task, i.e. the Click-Through-Rate (CTR) and the User Success. The CTR measures the percentage of clicked job recommendations in relation to all job recommendations while the user success can be divided in the absolute user success, i.e. the total number of users who clicked on a job recommendation in a given time frame as well as the relative user success, i.e. the percentage of users who clicked on at least one job recommendation in relation to all users who received at least one job recommendation.

17.4.2. *Case: Recommender Engines*

As discussed in Section 17.3.1, we apply four different recommender engines that produce the recommendation candidates that are then further filtered and ranked. These recommender engines are collaborative filtering, content-based filtering utilizing the explicitly given user information, content-based filtering utilizing the automatically generated interest profiles of the users, and item-to-item recommendations. Here, we want to discuss how strongly the different recommender engines contribute to the success of the job recommender system. We therefore took a random sample of around 1.5 Million XING users and analyzed the more than 40 Million tracking events that were collected for these users in September 2017. Tracking events include the display of postings as recommendation to users, i.e. *impressions*, positive interactions of users with recommended job postings, i.e. *clicks*, *bookmarks*, or *replies*, and negative interactions, i.e. *deletions*.

17.4.2.1. *User success and CTR*

Figure 17.4(a) gives an overview over the normalized success rates of the four recommender engines. The (relative) user success is the percentage of users who interacted with at least one job recommendation from the specified recommender engine in relation to all users who received a job recommendation from that recommender engine. The CTR measures the percentage of clicked job recommendations from one recommender engine in relation to all job recommendations from that recommender engine. Here, each user-job combination is only counted once, this is to say if a job is recommended several times to the same user from the same recommender engine it is still counted as just one recommendation for that engine. Additionally, a job recommendation can be created by several recommender engines and is then counted separately for each engine. Finally, these success rates are normalized by the least successful strategy.

The item-to-item recommender is the most successful approach in terms of CTR with a score about twice as high as the score of the collaborative filtering which is the least successful approach and thus receives a normalized score of 1. The implicit content-based filtering is the second-best approach with a normalized score of about 1.5, followed by the explicit content-based filtering which performs similarly to the collaborative filtering.

When considering the user success, the item-to-item recommender again is the most successful approach with a score that is almost four times as high as the user success of the explicit content-based filtering which is the least successful approach. The implicit content-based filtering is the second-best approach with a normalized score of about 1.8, followed by the collaborative filtering with a normalized score of about 1.3.

Summing up, when only considering user success and CTR, the item-to-item recommender is the clear winner followed with some distance by the implicit content-based approach while the explicit content-based filtering and the collaborative filtering share the last place.

17.4.2.2. *Reach*

When analyzing the reach of the different recommender engines, see Figure 17.4(b), one gets a more clear picture, though. The reach compares the absolute number of positive interactions that could be initiated by a recommender engine and the absolute number of users that clicked on a recommendation from the specified recommender engine, respectively. As for the success rates, the reach is normalized by the least successful strategy.

The collaborative filtering is the most successful approach in terms of interaction-based reach and holds a score that is 1.8 times higher than the score of the explicit content-based approach which is the least successful approach. The collaborative filtering is followed by the implicit content-based filtering which performs similarly with a normalized score of about 1.7 and the item-to-item recommender with a normalized score of 1.4.

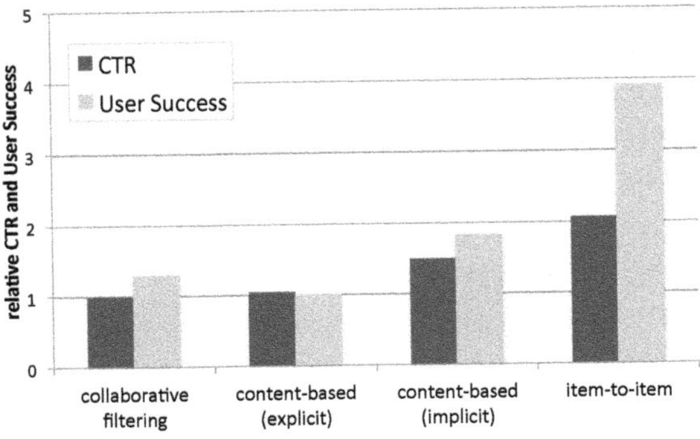

(a)

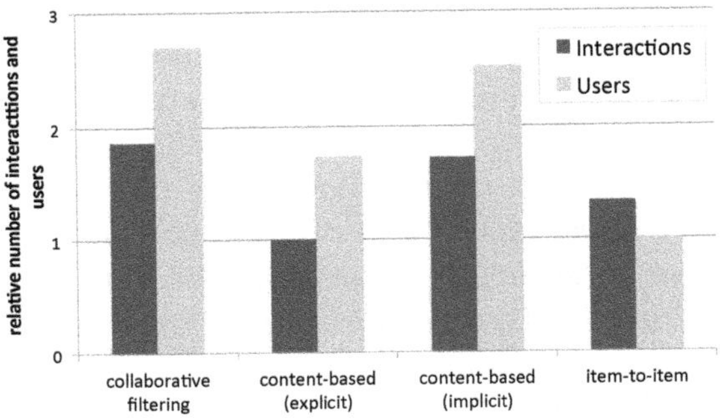

(b)

Fig. 17.4. (a) Success rates: Click-through rate (CTR) and fraction of users who interacted with recommendations (User Success) per recommender engine. Success rates are normalized by the least successful strategy. (b) Reach: Relative number of positively interacted recommendations (Interactions) and relative number of users who interacted with recommendations (Users).

Similarly, the collaborative filtering is also the best performing approach in terms of user-based reach with a score that is about 2.7 times higher than the score of the item-to-item recommender which performs worst. Again, the implicit content-based filtering is the second best approach and performs similarly to the best approach with a normalized score of about 2.5. The explicit content-based approach still performs clearly better than the item-to-item filtering with a score of about 1.8.

Summing up, as for the success rates, there is one approach that performs best for both the user- and the item-based evaluation metric, i.e. the collaborative filtering which is closely followed by the implicit content-based filtering. The explicit content-based approach as well as the item-to-item recommender both lag behind.

17.4.2.3. *Discussion*

The success of the recommender engines varies greatly depending on the evaluation metric. This is also known as precision/recall trade-off. While the item-to-item recommender achieves the highest success rates which means its recommendations are very precise, its reach is very low in comparison to the other approaches, i.e. it has a low recall. In contrast, the reach is very high for the collaborative filtering while it also offers more recommendations that seem to be of no interest for the users.

The approach that appears to be the best trade-off between success rates and reach is the implicit content-based filtering which combines all user interactions to build a meaningful profile about the users' career goals. The explicit content-based filtering lags behind for both evaluations metrics. An explanation for this might be the domain. As already mentioned, many or even most users want to evolve in their career, thus, they are usually not looking for a position that is already mentioned in their profile which might soon represent their past.

Overall, we can state that the approaches complement each other. Even though the explicit content-based approach seems to be the least successful one, it is important if users do not hold any interaction data.

17.4.3. *Case: Rating Prediction*

In order to create rating predictions to filter non-suitable recommendations, i.e. recommendations holding a rating prediction lower than a given threshold, we apply a gradient-boosted decision tree model. For evaluation purposes, we implement a baseline model using linear regression. Both models are evaluated independently in an offline environment to find the best combinations of features and hyper-parameter settings. Thereafter, we test the final models online in an A/B-Test.

17.4.3.1. *Offline Evaluation*

We perform a 5-fold cross validation and calculate for each of the two rating predictors and each rating (i.e. the five classes 1–5) the percentage of postings in the test set that would be discarded for varying thresholds, see Figure 17.5. The tree-based rating predictor performs way better than the linear regression as it is able to better distinguish between *good* and *bad* recommendations. For example, using a threshold of 2.5, 86% of the recommendations rated with only one star are discarded while 85% of the

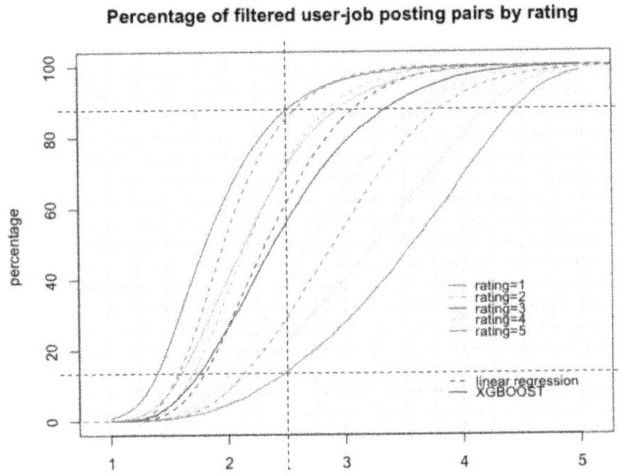

Fig. 17.5. Offline Evaluation for rating prediction-based filtering. With a threshold of 2.5, we can filter 86% of the recommendations holding a 1-star rating while falsely deleting 15% of the recommendations holding a 5-star rating using XGBoost.

recommendations rated with 5 stars are retained using the tree-based model. The linear regression discards just slightly less of the 1-star recommendations but also only keeps 70% of the 5-star ratings. Thus, we decide to use the tree-based model for the application in production.

We use this diagram not only to choose a model but also to choose a default threshold for removing the not suitable recommendations. The threshold is then further refined for each user, e.g. the threshold will be higher for users holding a lot of recommendations with high rating predictions but can be lower for users for which we generate only a few recommendations. As can be seen in the diagram, when choosing the right threshold one has to consider the trade-off of removing bad and keeping good recommendations. For example, by varying the threshold to e.g. 1.5 we are able to keep almost all good recommendations while only removing about 30% of the bad recommendations, by varying the threshold to 3.5 we are able to discard almost all bad recommendation but keep less than 50% of the good recommendations.

17.4.3.2. *Online Evaluation*

We run an A/B-Test with this new filtering component on the platform and analyze the tracking data to compare the results based on our evaluation metrics for the two groups. The test ran for two weeks and involved over two Million users. Overall, we observe for the success rates that the CTR increased by 23% and relative user success increased by 16%. In terms of reach we figured out that the amount of users with recommendations decreased by 26% which seems like a lot, however, the user-based reach increased by 7% and the interaction-based reach increased by 6%. This is to say, even though we target less users, more users click on our recommendations which means that we correctly filter out the bad recommendations and push the good recommendations to the top of the recommendation lists. Furthermore, the rating average increased by 20%.

17.5. A complex recommender system in production

Whenever a user visits XING's start page or the job marketplace, a request for receiving the user's job recommendations is send to the job recommender's REST API. Overall, the job recommender receives a few thousand requests per minute that need to be answered as fast as possible. Here, we aim for the recommender to send a response in less than 200ms for 99% of the requests. In this chapter, we will briefly present the architecture of XING's recommender system as well as the interplay of batch and online processing and finally discuss what it takes to include a new component into the system.

17.5.1. *Architecture*

When the job recommender system's REST API is called, it issues several queries to other REST APIs, database systems and the Elasticsearch search engine to collect information about the user and the potential job recommendations. Doing this in sequential order would take up to several seconds, thus, we do as much parallel processing as possible. Furthermore it is important to allow different components to share the gathered information. Figure 17.6 shows the architecture for calculating online information which also enables the easy integration of new components.

First, the identifier of a user is send to the job recommender system's REST API which then collects all needed information about the user in parallel in *stage 1*. Thereafter, the different recommender engines use this information in *stage 2* to independently create job recommendations which are then combined (see Section 17.1.3). In *stage 3*, the information about these potential job recommendations are gathered from various data sources and used by the filters in *stage 4* and the final rankers in *stage 5*. In contrast to the filters, the rankers have to run in sequential order.

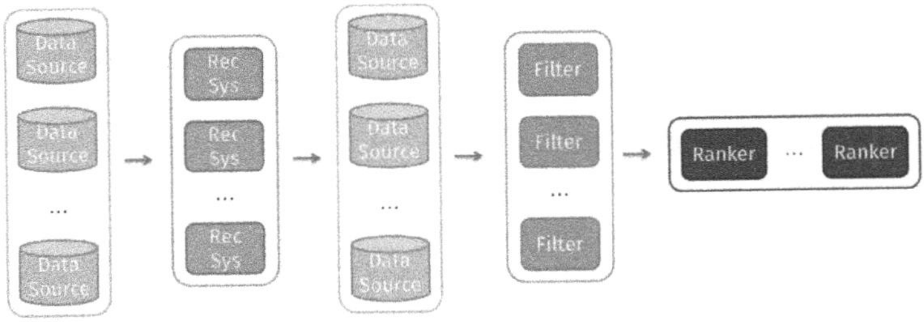

Fig. 17.6. Job Recommender System Architecture.

17.5.2. *Online and Batch Processing*

When calculating the recommendations, some of the data taken into account has to be up-to-date. For example, if a user added a new skill or rated a job recommendation negatively, the job recommender has to adapt its recommendations immediately. Thus, the recommendations cannot be (completely) pre-computed but the recommender system has to get the most current information about the user (in *stage 1*) and use it for finding the most suitable job postings. These look-ups and calculations have to be performed in milliseconds.

However, there is also data that takes minutes or hours to compute. Examples are the complete interaction profiles of the users, the association rules for collaborative filtering or some inferred information about the users like the comparable career level. Here, we apply batch-processing and pre-calculate the information several times a day and export it to a database where it can be easily looked-up.

17.5.3. *Re-training and Deployment of Components*

We use several components that were automatically trained like the rating prediction-based filter or the word2vec-based topic modeling. These models have to be re-trained periodically to reflect the current user behavior or XING's job posting base. Additionally, some external changes, like an update of XING's ontology, might require a re-training of specific components.

However, each re-training has to be triggered and then the new model has to be carefully evaluated offline. Additionally, if the new model produces different recommendations for many users, we have to think about A/B-Testing the new model before pushing it to production. Especially with a growing number of automatically trained models, this is a very expensive process.

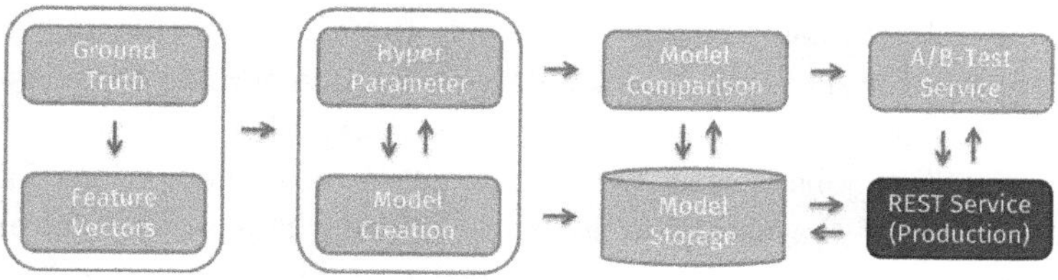

Fig. 17.7. Automatically re-train and deploy models.

We thus created workflows that when triggered do not only learn a new model but also test it against the current model and finally deploy it to production or start an A/B test when the differences to the current model reach a given threshold, as shown in Figure 17.7

First, the ground truth, i.e. training and test set are calculated to then create the respective feature vectors which are used the train the model. The model is trained several times using different hyper-parameters that are determined using stratified random search. The best model is selected using a pre-defined metric, e.g. the f1-measure@20 for the rating predictions, and stored in our model storage. Thereafter, the new model is compared to the current model. Here, we take a random sample of e.g. 10,000 users and calculate two rating lists for each of them using the two models. The recommendation lists are then compared based on the number of recommendations per user or the average distance of the jobs to the location of the respective user. If the lists' differences are below a given threshold, the model is directly pushed into production. Otherwise, an A/B-Test is automatically started to compare the two models. Currently, the analysis of this A/B test and the final decision of pushing it to production if an A/B test was started is done manually.

17.6. Conclusion

In this chapter, we presented XING's job recommender system that aims to support its users in finding new jobs and evolve in their career. We showed that the system has to tackle some common challenges like the new user or the new item problem and discussed some challenges that are specific to the job domain. Especially the transience of the items makes it difficult to apply pure collaborative-based approaches. Thus, we introduced association rules that are automatically re-calculated several times a days and additionally focus on content-based approaches which are combined into an ensemble to complement each other. In order to handle ambiguous job descriptions or skill names, we deployed a domain-specific ontology that is utilized by almost all recommender engines. Another challenge specific to recommending jobs is that we do not create recommendations for the current profiles of the users but for their future. This is why the users' interactions are especially important in order to understand the direction in which a user wants to evolve which is also shown by the high success rates of the more-like-this recommender. Additionally, we consider common career paths or job transitions when filtering the recommendation lists. In order to keep and increase the users' trust in the system, we apply several filters like the rating prediction-based filter to only keep recommendations in which the recommender system has a high confidence. As shown in the second case study in which we presented the evaluation of the rating prediction-based filter, we can increase the reach of the system dramatically by recommending less items and only keeping the most promising ones.

After discussing the challenges and providing an insight in the functionality of the recommender system we presented how we evaluate new components before they go live in production. It is important to remember that we first evaluate several models offline to pick the most promising one which is then evaluated online in an A/B-Test to make the decision if the component should be pushed to production or not. It is not unusual that a promising approach does not perform as expected in a live environment and needs to be revised. Thus, it is crucial to choose suitable offline evaluation metrics to pick the right model in the first place.

Finally, we shortly presented the extensible architecture that allows XING's recommender to serve millions of requests per day in just a few hundred milliseconds each. This is possible by combining batch processing that is conducted offline and parallel processing of the information that has to be gathered and combined in real-time.

References

(2016). *RecSys Challenge '16: Proceedings of the Recommender Systems Challenge* (ACM, New York, NY, USA), ISBN 978-1-4503-4801-0.

(2017). *RecSys Challenge '17: Proceedings of the Recommender Systems Challenge 2017* (ACM, New York, NY, USA), ISBN 978-1-4503-5391-5.

Abel, F. (2015). We know where you should work next summer: job recommendations, in *Proceedings of the 9th ACM Conference on Recommender Systems* (ACM), pp. 230–230.

Abel, F., Benczúr, A., Kohlsdorf, D., Larson, M. and Pálovics, R. (2016). Recsys challenge 2016: Job recommendations, in *Proceedings of the 10th ACM Conference on Recommender Systems*, RecSys '16 (ACM, New York, NY, USA), ISBN 978-1-4503-4035-9, pp. 425–426, doi:10.1145/2959100.2959207, http://doi.acm.org/10.1145/2959100.2959207.

Abel, F., Deldjoo, Y., Elahi, M. and Kohlsdorf, D. (2017). Recsys challenge 2017: Offline and online evaluation, in *Proceedings of the Eleventh ACM Conference on Recommender Systems*, RecSys '17 (ACM, New York, NY, USA), ISBN 978-1-4503-4652-8, pp. 372–373, doi:10.1145/3109859.3109954, http://doi.acm.org/10.1145/3109859.3109954.

Aho, A. V. and Corasick, M. J. (1975). Efficient string matching: An aid to bibliographic search, *Commun. ACM* **18**, 6, pp. 333–340, doi:10.1145/360825.360855, http://doi.acm.org/10.1145/360825.360855.

Chen, T. and Guestrin, C. (2016). Xgboost: A scalable tree boosting system, in *Proceedings of the 22Nd ACM SIGKDD International Conference on Knowledge Discovery and Data Mining*, KDD '16 (ACM, New York, NY, USA), ISBN 978-1-4503-4232-2, pp. 785–794, doi:10.1145/2939672.2939785, http://doi.acm.org/10.1145/2939672.2939785.

Herlocker, J. L., Konstan, J. A., Terveen, L. G. and Riedl, J. T. (2004). Evaluating collaborative filtering recommender systems, *ACM Trans. Inf. Syst.* **22**, 1, pp. 5–53, doi:10.1145/963770.963772, http://doi.acm.org/10.1145/963770.963772.

Konstan, J. A. and Riedl, J. (2012). Recommender systems: From algorithms to user experience, *User Modeling and User-Adapted Interaction* **22**, 1–2, pp. 101–123, doi:10.1007/s11257-011-9112-x.

Li, L., Jing, H., Tong, H., Yang, J., He, Q. and Chen, B.-C. (2017). Nemo: Next career move prediction with contextual embedding, in *Proceedings of the 26th International Conference on World Wide Web Companion*, WWW '17 Companion (International World Wide Web Conferences Steering Committee, Republic and Canton of Geneva, Switzerland), ISBN 978-1-4503-4914-7, pp. 505–513, doi:10.1145/3041021.3054200, https://doi.org/10.1145/3041021.3054200.

McNee, S. M., Riedl, J. and Konstan, J. A. (2006). Being accurate is not enough: How accuracy metrics have hurt recommender systems, in *CHI '06 Extended Abstracts on Human Factors in Computing Systems*, CHI EA '06 (ACM, New York, NY, USA), ISBN 1-59593-298-4, pp. 1097–1101, doi:10.1145/1125451.1125659, http://doi.acm.org/10.1145/1125451.1125659.

Mikolov, T., Sutskever, I., Chen, K., Corrado, G. S. and Dean, J. (2013). Distributed representations of words and phrases and their compositionality, in C. J. C. Burges, L. Bottou, M. Welling, Z. Ghahramani and K. Q. Weinberger (eds.), *Advances in Neural Information Processing Systems 26* (Curran Associates, Inc.), pp. 3111–3119.

Said, A. (2016). A short history of the recsys challenge, *AI Magazine* **37**, 4.

Su, X. and Khoshgoftaar, T. M. (2009). A survey of collaborative filtering techniques, *Adv. in Artif. Intell.* **2009**, pp. 4:2–4:2, doi:10.1155/2009/421425, http://dx.doi.org/10.1155/2009/421425.

Volkovs, M., Yu, G. W. and Poutanen, T. (2017). Content-based neighbor models for cold start in recommender systems, in *Proceedings of the Recommender Systems Challenge 2017*, RecSys Challenge '17 (ACM, New York, NY, USA), ISBN 978-1-4503-5391-5, pp. 7:1–7:6, doi:10.1145/3124791.3124792, http://doi.acm.org/10.1145/3124791.3124792.

Wang, J., Zhang, Y., Posse, C. and Bhasin, A. (2013). Is it time for a career switch? in *Proceedings of the 22nd International Conference on World Wide Web*, WWW '13 (ACM, New York, NY, USA), ISBN 978-1-4503-2035-1, pp. 1377–1388, doi:10.1145/2488388.2488509, http://doi.acm.org/10.1145/2488388.2488509.

Xiao, W., Xu, X., Liang, K., Mao, J. and Wang, J. (2016). Job recommendation with hawkes process: An effective solution for recsys challenge 2016, in *Proceedings of the Recommender Systems Challenge*, RecSys Challenge '16 (ACM, New York, NY, USA), ISBN 978-1-4503-4801-0, pp. 11:1–11:4, doi:10.1145/2987538.2987543, http://doi.acm.org/10.1145/2987538.2987543.

Xu, H., Yu, Z., Yang, J., Xiong, H. and Zhu, H. (2016). Talent circle detection in job transition networks, in *Proceedings of the 22nd ACM SIGKDD International Conference on Knowledge Discovery and Data Mining*, KDD '16 (ACM, New York, NY, USA), ISBN 978-1-4503-4232-2, pp. 655–664, doi:10.1145/2939672.2939732, http://doi.acm.org/10.1145/2939672.2939732.

Chapter 18

Academic Recommendations: The Mendeley Case

Maya Hristakeva, Daniel Kershaw, Benjamin Pettit,
Saúl Vargas and Kris Jack

Elsevier, AlphaBeta Building, London, EC2A 1BR, United Kingdom

Mendeley Suggest is a recommender system that helps researchers keep up to date with articles in their field and fill gaps in their knowledge. It leverages usage data from the Mendeley reference management system, in which millions of users have compiled personal libraries of relevant articles. In this chapter, we explain how recommendations are generated using user-based collaborative filtering. Much of the development of the live system has focused on keeping users engaged as they repeatedly interact with the system and adapting to their research interests as they add new articles to their libraries. The product has evolved over time through taking into account user feedback, experimenting with new approaches, and making data-driven decisions. As Mendeley Suggest continues to develop, we expect to incorporate additional domain-specific features to supplement collaborative filtering and provide even more useful recommendations.

18.1. Introduction

The rate at which scientific research is performed and published poses a challenge for researchers and students who need to keep up to date with an ever increasing amount of literature [Larsen and von Ins (2010)]. Mendeley [Henning and Reichelt (2008); Vargas *et al.* (2016)] is a platform that helps researchers organise and discover relevant scientific literature, showcase their work, and connect with their peers through the Mendeley[1] reference manager and a personalised article recommender system Mendeley Suggest[2]. As with other reference management tools, Mendeley helps researchers and students to manage their reading lists and libraries of relevant

[1] `https://www.mendeley.com`

[2] `https://www.mendeley.com/suggest/`

research articles. Each user's library is an implicit expression of interest in a set of articles, and the combined libraries of millions of users enables Mendeley Suggest to deliver high quality recommendations through collaborative filtering (CF).

A number of information retrieval and discovery tools have been developed to aid the exploration of increasingly large online research catalogues. These include CiteULike[3] for managing libaries as well as Scopus[4] and Google Scholar[5] for searching and mapping research. For a comprehensive survey of research article recommender systems, we refer the readers to [Beel *et al.* (2016)]. The methods for creating recommendations in the academic literature domain include generic approaches such as content-based filtering [Beel *et al.* (2013)], folksonomies [Jomsri *et al.* (2010)], collaborative filtering [McNee *et al.* (2002)], and hybrid techniques [Wang and Blei (2011)], as well as domain-specific methods such as the use of citation networks [Küçüktunç *et al.* (2012)].

All of these methods have their pros and cons in generating research article recommendations. For example, content-based recommendations [Beel *et al.* (2013)] utilise features from the metadata and text of the articles, which results in wide coverage of items, but the recommendations might not be of high quality compared to methods that take usage into account [Jannach *et al.* (2015)]. Folksonomies [Jomsri *et al.* (2010)] utilise the power of the crowd through collaborative tagging, which can be powerful if the catalogue coverage is high, but in practice this is rarely the case. Alternatively, collaborative filtering [McNee *et al.* (2002)] builds on the power of implicit feedback from the users, though this also presumes that every interaction a user has is positive (see Chapter 7). Hybrid techniques [Wang and Blei (2011)] combine the best of a number of methods, for example collaborative filtering and content-based, but can be more complex to implement in practice. Citation networks are good for identifying reputable and highly cited papers, however they tend to favour older articles, which is not always best for keeping researchers up to date [Walker *et al.* (2007)]. Additionally, the majority of these methods for recommending research articles have only been implemented on small datasets compared to Mendeley's catalogue.

Ultimately, for Mendeley Suggest we want a system that produces engaging recommendations and scales to millions of users and billions of documents. After evaluating several of the approaches listed above at

---

[3] `http://www.citeulike.org`
[4] `https://www.scopus.com`
[5] `https://scholar.google.com/`

Mendeley scale, we found that the best performance came from CF, specifically user-based collaborative filtering (UBCF), which sources recommendations from the libraries of similar users [Bhowmick *et al.* (2014)]. Not only does it scale well, but it takes advantage of the implicit feedback from user libraries, which is data unique to Mendeley. We improved on this foundation with modifications to the UBCF algorithm, post-processing steps, and by supplementing it with with content-based and discipline-based recommendations for new Mendeley users ("cold users", Section 18.4.5). Therefore the full system is a hybrid recommender, within which UBCF is the primary method. Adding recommendations from the citation network did not result in big gains to either quality or coverage. However, future iterations are likely to combine a broader range of article features, including citations, in a model-based hybrid approach (Section 18.8).

This chapter details how the Mendeley Suggest recommender system (Figure 18.1) has been designed and developed. We explain recommendations are generated through collaborative filtering, based on user activity from the reference manager. We also present the current architecture, pointing out the design choices and technologies we use and some of the remaining challenges. The remainder of this chapter is structured as follows. We start with an overview of the recommender system architecture, including its data sources, batch and real-time components, and event logging (Section 18.2). Next we focus on the collaborative filtering component, including the algorithms used (Section 18.3), the domain-specific challenges (Section 18.4), and the architecture for generating and serving recommendations at scale (Section 18.5). While this describes the latest iteration of Mendeley Suggest, we also give an overview of its historical evolution (Section 18.6). Ongoing changes to the recommender system are steered by offline and online experimentation, which we explain in Section 18.7.

18.2. Mendeley Suggest Overview

The Mendeley Suggest recommender system is more than just the algorithms that it implements. In fact, it is a collection of core components that are designed to interact with one another in order to meet a set of user needs. Figure 18.1 shows how these core architectural components interact with one another in Mendeley Suggest.

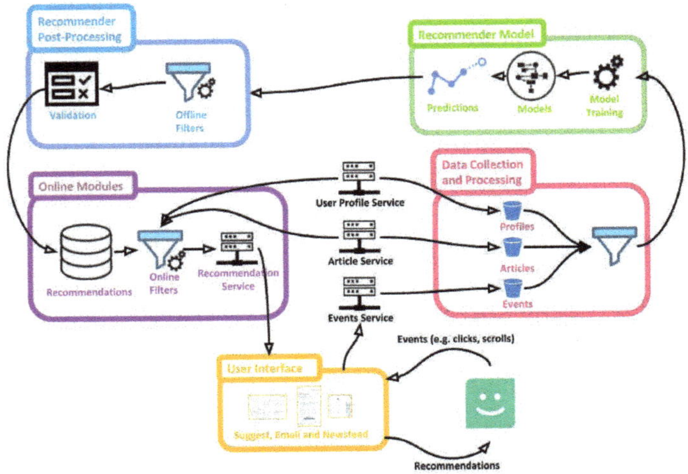

Fig. 18.1. Mendeley Suggest Architecture.

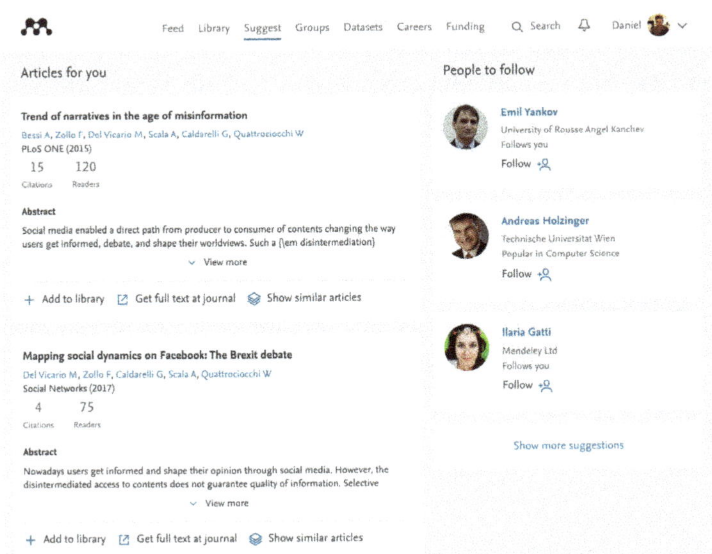

Fig. 18.2. Mendeley Suggest Portal.

18.2.1. *User Interface*

We currently have two main clients that are live: Suggest web portal (Figure 18.2) and email sender (Figure 18.3). Users receive weekly recommendation emails containing three of their top recommendations. If users want

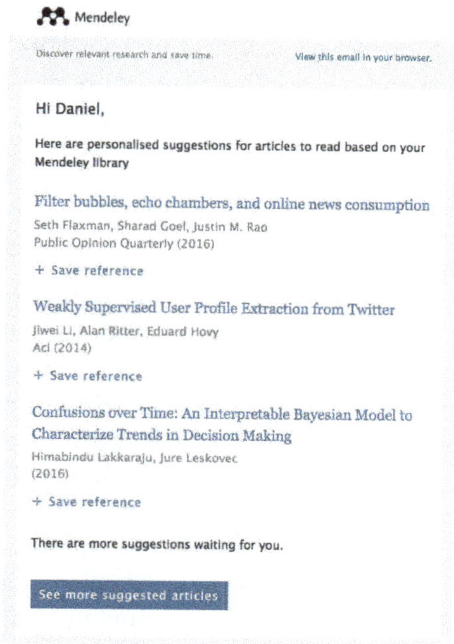

Fig. 18.3. Mendeley Suggest Email.

to view more recommendations they can go to the Suggest portal where we display a list of recommendations that are updated daily. Each client displays a subset of the top recommendations generated by the system, which are ranked and filtered specifically for the client's context. For example, in the emails we don't want to send the same recommendations two weeks in a row, whereas in the portal it might be acceptable to show the same recommendation more than once. Each client records user actions (e.g. recommendation is viewed or added to library) and sends them back to the recommender system so that it can learn from them and improve the recommendations that it serves.

18.2.2. *Data Sources*

Mendeley Suggest's recommendations are generated from a variety of data sources including user profile data, research article data, usage event logs, and user libraries from the Mendeley reference manager. For each user library, we record the ID of each article and the time that the user added it, which forms the interaction data for collaborative filtering and evaluation. Each interaction is an implicit expression of interest by the user in article.

Apart from the usage data, we also have metadata about articles and users. For the articles, we have titles, abstracts and author-defined keywords. Mendeley users can choose a discipline from a list of subject areas, and they can also add research interests to their profile pages in free text form.

18.2.3. *Recommender Approaches*

Mendeley Suggest uses a custom implementation of UBCF, which we explain in Section 18.3. We supplement collaborative filtering with a number of other recommendation approaches, which are particularly useful for user cold start (see Section 18.4.5). We determine which articles are popular in each of Mendeley's subject areas, and also identify articles with a significant and recent uplift in activity and signal them as trending. Another approach is content-based similarity, which allows us to recommend related articles using metadata fields such as title, abstract, keywords, and publication venue. These additional approaches are explained in detail in [Hristakeva *et al.* (2017)]. Mendeley Suggest is therefore a *switching hybrid* recommender system, according to the taxonomy in Chapter 4.

18.2.4. *Recommendation Post-processing*

Once the raw recommendations have been generated, we apply some last minute sanity checking of the recommendations and some business logic aimed at avoiding obvious mistakes. For example, we filter out recommendations that users have already added to their libraries as well as ones that don't meet a metadata quality threshold (Section 18.4.2). We also have some filtering and re-ranking methods that are applied at this point. For example, we typically don't want to recommend the same list of articles to users every day. If a user has already seen a recommendation and not added it to his or her library then we reduce the probability of it being shown again, as we explain in Section 18.3.4.

18.3. Developed Algorithms

Mendeley Suggest consists of multiple algorithmic approaches, filtering and re-ranking methods, and business logic rules, but at its core is a user-based collaborative filtering algorithm that generates candidate recommendations. We decided upon user-based rather than item-based collaborative filtering for two main reasons. First, typically online businesses that use

Table 18.1. Common Notation.

Notation	
D	The set of all papers within the Mendeley catalogue
U	The set of users of Mendeley
d	A document within the catalogue such that $d \in D$
u	A Mendeley users such that $u \in U$
D_u	Returns the set documents (d) in a users library
R_u	Set of recommendations for user u such that $R_u \cap D_u \equiv \emptyset$
$sim(u, U)$	returns the set of 100 most similar users to u
$sim(U, u)$	returns the set of 100 users that user u is most similar to (inverted neighborhood)
r_d^u	The score for recommendation of document d to user u

recommender systems have many more users than items that can be recommended. In our case, the data is an unusual shape in that we have many hundreds of millions of items that can be recommended (i.e. all research articles ever published) and only tens of millions of users (i.e. the research community). User-based collaborative filtering scales better than item-based collaborative filtering in this case. Also, through empirical testing we found that user-based collaborative filtering generated better quality recommendations than the item-based ones (see Section 18.4.1).

We have adapted the vanilla UBCF approach to suit Mendeley Suggest's aim of delivering recommendations that adapt to a user's changing interests. In Section 18.3.3 we incorporate temporal information, forcing it to serve recommendations based on users' recent activity and the recent activity of their neighbors. Additionally, if we repeatedly serve the same recommendations to a user, then over time they will get stale, resulting in the user loosing interest in Suggest. In Section 18.3.4 and 18.3.5 we introduce impression discounting and dithering in order to provide a better user experience.

The set of documents (e.g. papers, journal article and notes) in Mendeley's database is D, such that $d \in D$, and the set of users in Mendeley are U, such that $u \in U$. The set of documents that a user (u) has within his or her library is represented as D_u, which is a subset of all available documents, $D_u \subset D$.

18.3.1. *User-Based Collaborative Filtering*

The idea behind UBCF is that users who have liked the similar items in the past will like the same items in the future. For Mendeley, this would mean if users have added the same articles to their libraries, they would like similar articles in the future. To generate recommendations for users is a two stage process; first we identify similar users to them and then recommend articles which the set of similar users are reading.

To identify similar users we compute the cosine similarity between the sparse vectors that represent users' libraries (D_u):

$$cosine(u, u') = \frac{|D_u \cap D_{u'}|}{\sqrt{|D_u| \times |D_{u'}|}} \tag{18.1}$$

The 100 users with the highest similarity to user u constitute its neighborhood, $sim(u, U)$. An alternative set of users from whom to source recommendations is the inverted neighborhood — any user u' for whom $u \in sim(u', U)$ [Vargas and Castells (2014)]. We denote the inverted neighborhood $sim(U, u)$. In practice, we used the inverted neighborhood because it produced better quality recommendations (see Section 18.7.1).

The set of recommendations for a user (R_u) is then the union of all the neighboring user libraries minus what the user already has in his/her library (D_u) [Sarwar *et al.* (2001)];

$$R_u = \bigcup_{u' \in sim(U,u)} D_{u'} \setminus D_u \tag{18.2}$$

Each document in R_u is then scored (r_d^u) using the sum of user similarities across all neighbors who have the document in their libraries.

$$r_d^u = \sum_{u' \in sim(U,u)} \begin{cases} cosine(u, u'), & \text{if } d \in D_{u'} \setminus D_u \\ 0, & \text{otherwise} \end{cases} \tag{18.3}$$

Using the cosine similarity between users as a measure of trust in the recommendation means that users who have more similar libraries to user u influence the final set of recommendations to a greater extent. This results in a set of recommended documents for each user, with a higher score indicating that the user is more likely to be interested in the document.

18.3.2. *Significance Weighting*

One weakness of using cosine similarity to score neighbors is that it ignores the statistical confidence in the correlation between libraries. For example, a large number of Mendeley users have small libraries, so a neighbor u' may have only one or two articles in common with the focal user u, but nonetheless rank highly in terms of $cosine(u, u')$ because $|D_{u'}|$ is also small. To alleviate this problem, we scale the CF score with a significance weighting computed from the number of articles in common [Ricci *et al.* (2015)]:

$$score(u, u') = \min(1, \frac{|D_u \cap D_{u'}|}{K}) \times cosine(u, u') \tag{18.4}$$

If the users have fewer than K documents in common, then the contribution of neighbor u' to the CF score is scaled down. This means that preference is given to recommendations that are generated from high co-occurrence neighbors, who are more likely to have shared interests with the user. While there are more elegant ways of incorporating uncertainty into the score, this linear weighting with an empirically derived threshold of 5 was a sensible first implementation that improved recommendations in offline evaluation (Section 18.7.1).

18.3.3. *Time Decay*

Researchers' interests are constantly evolving. The aim of Mendeley Suggest is to serve time-appropriate recommendations, allowing users to keep up to date with their current topics of interest. This poses a challenge for traditional CF methods that treat all user-item interactions the same, irrespective of when they took place. We adapted the UBCF approach described above by applying time decay to the user-item interactions.

The input to the CF model can be represented in a sparse format as triples `<user, item, rating>`. In the normal implicit-feedback CF model, any item in a user's library gets a rating of 1 — a binary scale, as discussed in Chapter 7. To bias recommendations toward more recent activity, we modified the ratings using a decay function

$$rating = \beta^{-\Delta t} \tag{18.5}$$

where Δt is the time, in years, since the user added the article to his or her library. This means that articles that were added further back in time will influence recommendations to a lesser extent than more recent additions. We tuned the parameter β through offline and online evaluation (Section 18.7).

This decayed rating can be applied in two ways. The "one-sided" version applies time decay only to the focal user's library D_u, and not to the candidate neighbors $D_{u'}$. The "two-sided" version applies decay to both the focal user and the candidate neighbors, in other words all ratings can be decayed before computing user similarities. We chose the two-sided version as this performed better in offline evaluation (Section 18.7.1).

18.3.4. *Impression Discounting*

If a user is repeatedly shown a recommendation and doesn't interact with it (e.g. doesn't add it to his or her library), Mendeley Suggest treats this as an implicit signal that the user is not interested in the article. Such articles are penalised by reducing their rank in the list of recommended articles or by removing them entirely using *impression discounting*, a technique described in [Lee *et al.* (2014)].

Impression discounting relies on learning a model for users based on their historical exposure to recommendations (i.e. impressions). The model, when applied to a user's list of recommendations will then penalise (i.e. discount) articles that have not been clicked and change the ordering of the recommendations in the list. More formally, on application of impression discounting to a ranked list of recommendations, new scores are generated for the articles based on the product of their original scores and a discounting factor:

$$new\_score = orig\_score * f(g(X)) \tag{18.6}$$

where X is a set of features relevant to the user's impressions of the item (e.g. number of times item was viewed, how long since the item was last viewed), g is a discounting function for individual features, and f is a function that combines the discounted features to get an aggregated discounting factor. Similar to the authors of [Lee *et al.* (2014)], we found that using an exponential decay function for the discounting factor worked best in Mendeley Suggest.

18.3.5. *Dithering*

Since the collaborative filtering recommendations are computed daily, a user could be served the same recommendations on successive visits to Mendeley Suggest and may lose interest in the recommendations. To give the impression of fresh content, *dithering* [Dunning *et al.* (2014)] is applied to re-order the list of recommendations by adding normally distributed random noise to the initial rank of the recommendations (based on the CF scores)

$$new\_score = \log(rank) + N(0, \log \varepsilon) \tag{18.7}$$

where $\varepsilon = \frac{\Delta rank}{rank}$ and typically $\varepsilon \in [1.5, 3]$. The degree of re-shuffling of the original recommendations list depends on the amount of random noise added (i.e. values of ε).

In practice, this is a simple form of an explore-exploit strategy allowing us to learn more about articles that have low ranking positions (according to the model predictions) as they may randomly be pushed up the list being displayed to the user. It can also have the positive side-effect of keeping the list of recommendations looking fresh by surfacing recommendations that were previously hidden lower down the list.

18.4. Addressed challenges and problems

When developing recommender systems a number of challenges and problems need to be addressed, from being able to generate the recommendations at scale to ensuring user privacy is preserved. The following sections outline some of the challenges we faced when developing and implementing Mendeley Suggest and the solutions we put in place.

18.4.1. *Generating recommendations at scale*

Mendeley generates article recommendations for each user on a daily basis. When generating recommendations, we are primarily interested in two factors: recommendation quality and cost (i.e. runtime), ideally looking for approaches which have high quality and relatively low cost. Typically reducing the cost will also reduce the quality. However, as shown in Chapter 11, it is also possible to implement high quality scalable collaborative systems using parallelisation and distributed technology without having to compromise on quality.

As discussed in Section 18.3, we implemented user-based collaborative filtering over item-based collaborative filtering or matrix factorization as it not only produced higher quality recommendations but also scaled better for our use case where there are many more items to be recommended than users. We spent considerable effort attempting to scale and tune these algorithms through different implementations [Hristakeva (2015)]. By going with a tuned implementation of UBCF (see Figure 18.4), we were able to reduce runtime by close to 50% and improve recommendation quality by 100% when compared to item-based collaborative filtering and alternating least squares matrix factorization with 10000 latent factors. As discussed in Chapter 2, matrix factorization is widely used for collaborative filtering but usually requires denser matrices of user-item interactions to achieve high quality results.

Another optimisation involves handling user libraries that contain a very large number of documents. From a technical perspective, large libraries slow down the distributed nearest neighbor computation and make it unstable. Therefore we downsample large libraries to the N most recently added documents. Although this threshold was chosen for the smooth

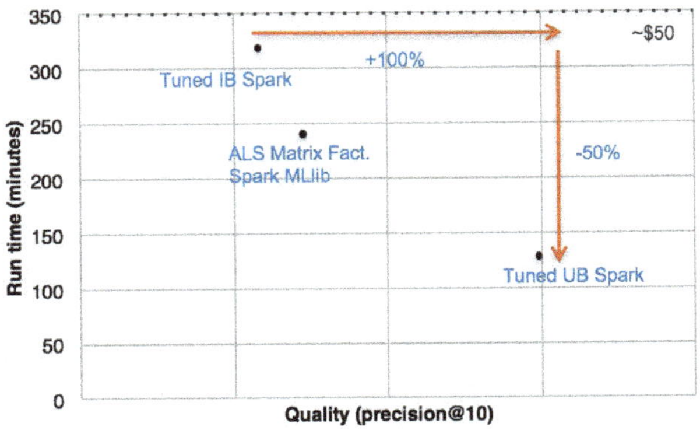

Fig. 18.4. Quality vs runtime trade-off — user-based CF outperformed both item-based CF and matrix factorization in terms of quality of the recommendations and cost.

running of the collaborative filtering job, we reason that it may also benefit recommendation quality. Libraries with thousands of documents are not as representative of a focussed research topic and probably contain more co-occurrences between unrelated documents. The downsampling step reduces the influence of these large libraries, while ensuring that the users who created them can still receive recommendations.

18.4.2. *Recommending high quality articles*

Mendeley's catalogue consists of over one hundred million unique research articles. It is a crowd-sourced collection derived from billions of research articles that individual Mendeley users throughout the world have added to their libraries. As such, it's also a noisy data set and deduplicating these records at scale is challenging [Subasic *et al.* (2016)]. Despite the noise, it is still valuable input for Mendeley Suggest's collaborative filtering algorithms, provided some article quality filters are put in place.

As there is no restriction to what users can add to their Mendeley libraries (phone bills, CVs, presentations, etc.), these items might end up in the user-item interactions on which the recommendations are based. To prevent such items being recommended, we have implemented quality thresholds to identify recommendable articles within the Mendeley catalogue. We perform three checks, first, the articles have to occur in the libraries of at least M users. This quorum rule removes articles that have extremely narrow interest or limited availability. The second check is that articles must be indexed in Scopus[6], a cross-publisher database of research article metadata. The third check is for the completeness of the article metadata, removing articles that have missing titles, abstracts, or authors.

18.4.3. *Respecting users' privacy*

Researchers are inherently protective of their research, thus care must be taken in the generation and explanations of the recommendations to a user, as researchers trust Mendeley to protect their information. As shown in [Mehta and Nejdl (2008)] and [Canny (2002)], collaborative filtering systems can be forced to expose user information through a number of simple algorithm attacks and where the interface provides explanations for the recommendations e.g. "a user you follow has recently read X".

Privacy is maintained in two ways, first by only giving explanations to users in relation to their own activity. Secondly, we only base recommendations on documents that are indexed in Scopus and are therefore widely available, and we check that they occur in multiple user libraries. This means that recommendations will not be based on a single user's library.

18.4.4. *Serving fresh recommendations*

We want users to interact with Mendeley Suggest as part of their research gathering activities, be this daily or weekly. If the recommendations have changed little on subsequent visits then this may deteriorate users' experience and reduce the usefulness of the recommendations. To overcome this we introduced two mechanisms, impression discounting (Section 18.3.4) and dithering (Section 18.3.5) to increase turnover within the recommendations lists.

18.4.5. *Recommending articles to all users*

Mendeley has a constantly growing user base, with new undergraduates, postgraduate researchers and established academics joining. We aim to serve recommendations to users as soon as they join Mendeley, therefore the system must be designed to deal with the *new user cold start* problem discussed in Chapter 8. To address this we implemented a number of recommendation strategies that we can fall back to in the absence of collaborative filtering results:

[6] `https://www.scopus.com`

- **Discipline** — Each user chooses a broad discipline, so even the coldest users can receive a diverse set of documents that are popular among users in the same discipline.
- **Research Interests** — For users who specify research interests, we can offer focused recommendations based directly on the user's self-defined keywords.
- **Recent Activity** — As soon as a user adds the first documents to his or her library, we can immediately recommend articles with related content, even before the daily CF results are generated. These content-based recommendations use the title, abstract, and keywords of the article that the user most recently added or read. This method extends to all articles, even new articles with no previous usage.

Although some article information, such as citations, is not used in this hybrid, the recommendation strategies have been chosen to complement each other. Discipline-based recommendations have the highest user coverage, whereas the third approach, using article content, produces better recommendations than profile information alone. None of these alternatives perform as well as collaborative filtering (see Section 18.7.1). Therefore, to offer high quality recommendations to new users we need them to start building their personal libraries as soon as possible. For this, we rely on a smooth onboarding interface, simple bulk uploads, and prompts to upload articles throughout the product.

18.5. Implementation resources

As we discussed in Section 18.2, Menedeley Suggest's recommender system is made up of a number of core components of varying importance and with different responsibilities. Mendeley's architecture cleanly separates these components out from one another, thus making their relative responsibilities clear. The implementation is robust, allowing it to scale to massive quantities of data and to serve large volumes of traffic.

18.5.1. *Recommendation Generation*

At Mendeley, we have a crowd sourced catalogue of over one hundred million unique research articles, originating from user libraries all of different sizes. Since the recommender system needs to generate recommendations for millions of users from a pool of millions of articles, we need a scalable technology stack. Using scalable technologies means that as the user base grows, our system can scale with it by adding more computational infrastructure. It also means that jobs can be scheduled with relative ease, allowing recommendations to be generated on a daily basis, from fresh Mendeley activity, serving up-to-date recommendations to the end user.

Mendeley Suggest is implemented using Apache Hadoop[7] and Apache Spark[8] as processing engines for the data collection, batch generation and post-processing of the recommendations. For the UBCF algorithm we have two custom implementations, one with Apache Mahout[9] and another with Spark. Although other parallel implementations of neighborhood-based CF are now available (see Chapter 11), we chose these technologies in order to distribute the job across a cluster while integrating with the existing Java codebase. We made the transition from Mahout to Spark to align with the rest of the technology stack as well as for performance benefits. All of these batch processes are orchestrated via Amazon Web Services (AWS)[10] data pipelines and Amazon EMR.

18.5.2. *Recommendation Service*

Mendeley Suggest implements a number of different services all exposed via a single recommender service API and hosted in AWS. Most recommendations are uploaded to HBase or Redis, popular `<key, value>` stores, where the key is the user's profile and the value is the JSON object with his or her recommendations. We also use ElasticSearch[11] (which horizontally scales more gracefully than other systems such as Lucene[12] or Solr[13]) to index research article metadata and return recommendations based on article content and users' research interests.

The API returns a JSON object with three elements: resource being recommended; explicit rank; and trace token. The resource being recommended in this case is the id and metadata of the article that may be interesting for the user to read. The explicit rank is an integer that tells the client the order in which the resources are expected to be of interest to

[7] https://hadoop.apache.org
[8] https://spark.apache.org
[9] http://mahout.apache.org
[10] https://aws.amazon.com
[11] https://www.elastic.co
[12] http://lucene.apache.org/
[13] http://lucene.apache.org/solr/

the user. The trace token is a 64-bit encoded string and is unique for the request that was made. It serves an important purpose, allowing clients that consume recommendations to provide feedback, letting the recommender system know, for example, that a recommended article was received, displayed to the user and within view on the scrollable page.

For a selection of user actions, the client records events and sends them back to the recommender system, so that it can learn from them and improve the recommendations that it serves (e.g. impression discounting). When the client sends these events to the recommender system, they are relayed on to Mendeley's main event handling service, which is built on an inherently scalable framework.

18.6. Historical evolution

Mendeley Suggest has gone through a number of incremental iterations over the years aimed at better serving the users' needs and improving user engagement and retention. These changes can be seen in the various user interfaces as well as the algorithms and data used to generate the recommendations.

Mendeley Suggest was initially released as a feature within the Mendeley desktop client, with recommendations shown in their own tab (Figure 18.5). In order to access their recommendations, users had to use the desktop client installed on their computer. This had some limitations, as

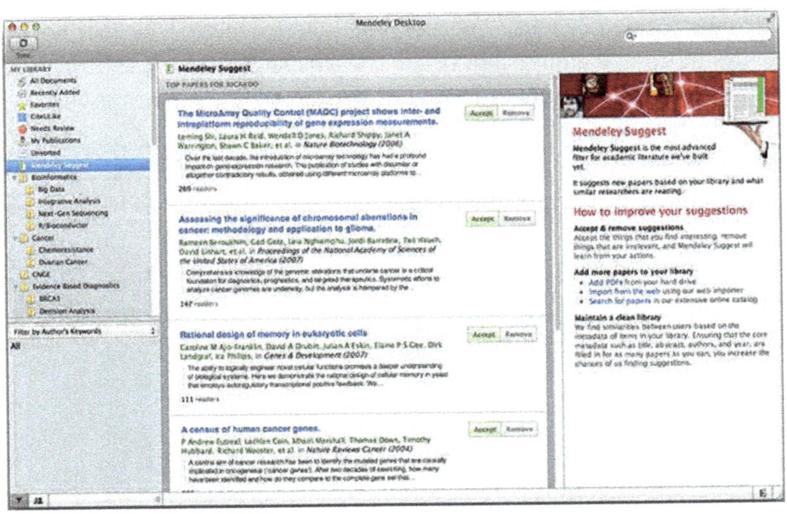

Fig. 18.5. Mendeley Suggest Desktop.

researchers normally engage with online databases and portals when looking for new articles to read. To better serve the use case of keeping up to date, the Mendeley Suggest web portal was built (see Section 18.6). For the first iteration of the portal, a number of recommender approaches were developed in addition to collaborative filtering, such as articles that are popular and trending in the users' disciplines (see Section 18.4.5). This addressed the cold start problem, and also allowed us to compare the recommendation approaches in a live setting. The recommendations from the different approaches were displayed in their own carousels, with the collaborative filtering recommendations being the most prominent at the top and with the subsequent carousels going down the page becoming less personalised. The order of carousels was decided based on the results of offline evaluation (Section 18.7.1).

Through user testing, we received feedback that the carousel layout (Figure 18.6) displayed a lot of recommendations on the screen with not enough detail (e.g. abstract, citations) for the user to decide the usefulness of a recommendation. In order to incorporate more detail for each recommendation, the interface was simplified by removing the carousels and going back to a single list of recommendations (Figure 18.2). The different recommender approaches are still displayed, but rather than showing them in separate carousels we add the approach used as an explanation. We also added more details for each article such as the abstract, the number of Mendeley users who have it in their libraries, and the number of citations the paper has, in order to make it easier for the user to judge the relevance of the recommendation without having to leave Suggest (Figure 18.2).

In addition to user interface changes, the UBCF algorithm at the core of Mendeley Suggest has also gone through a number of iterations and improvements. Initially, the Suggest recommender was launched using a vanilla UBCF implementation. However, we quickly realised from user feedback that even though the recommendations were good, they did not track users' current research interests very well. In order to make the recommendations more reflective of recent activity, we introduced time decay as explained in Section 18.3.3. Once Mendeley Suggest was live, we also noticed that although the recommendations were regenerated daily, they were becoming stale and user engagement was dropping. So we introduced dithering (see Section 18.3.5) and impression discounting (Section 18.3.4) which utilised logged events from user interactions with the Suggest portal. These changes increased the rate at which recommendations refresh over time.

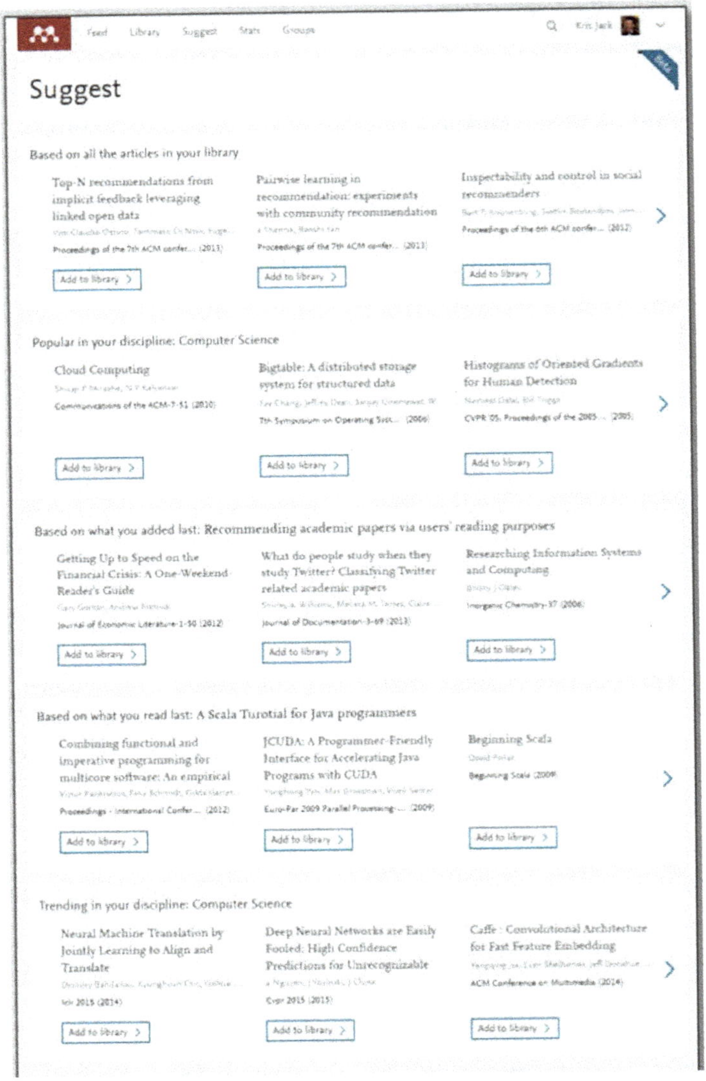

Fig. 18.6. First iteration of Mendeley Suggest web portal.

18.7. Evaluation

With a plethora of possible recommendation approaches, it is essential to be able to compare recommender systems and choose the approaches that deliver the greatest value. Right from the beginning, we placed focus on the evaluation as well as the implementation. This allowed us to make data-driven decisions when designing the recommender system and improving it after release. Agreeing on evaluation metrics up front allowed the recommender team to focus on finding ways to improve quality according to the metrics.

As advocated in Chapter 9, we used a combination of offline and online evaluation. Offline evaluation assesses the recommender system's accuracy in emulating the research work flow of our users; in other words predicting which articles a researcher would add next in the absence of a recommender system. This is useful because we can use historical logs from the reference management system to evaluate many model variations within a few hours. To determine which recommendations users find more relevant, we use online evaluation (A/B testing) to compare click rates by randomly selected groups of users, where each group's recommendations come from a different variant of the recommender system.

Offline and online evaluation complement each other. In general, we use offline evaluation to steer our choice of algorithms and parameters, and then online evaluation to determine which candidates to promote to production. For example, when choosing how strongly to bias recommendations towards users' more recent activity, we ran offline evaluation of a wide range of parameter values, and then tested the winning value online against a control with no recency bias. When first building the system, online evaluation was impossible, so we relied on a combination of offline evaluation, user testing, and qualitative spot checks. On the other hand, offline testing focuses on accuracy rather than novelty and diversity (see Chapter 10), which makes it all the more important to conduct online evaluation and user testing.

18.7.1. *Offline Evaluation*

To compare recommender algorithms, we adopted a time-based evaluation protocol [Gunawardana and Shani (2009)]. We collated all additions to Mendeley user libraries and split them into *test* and *training* sets across a distinct time boundary. Methods were assessed by testing how well recommendations generated from the training set were able to predict what users added to their libraries in the test set. We compared algorithms using precision, recall, F1-score, and mean average precision (MAP). Each of these can be evaluated on the top k recommendations. Recall is more appropriate than precision in this evaluation context, because we know that false negatives are relevant to the user, whereas false positives include relevant articles that the user failed to find. We used MAP as a ranking metric because our test set consisted of binary relevance data rather than graded relevance.

This evaluation procedure is very strict, as it tries to predict exactly what the users would interact with in the next time period, from a catalogue consisting of millions of items. This protocol, however, is close to the real task that a recommender system in production has to tackle. For this reason, precision and recall values are quite low compared to evaluation procedures that do not take into account time, such as repeated holdout or cross validation. In this evaluation setting, low values for precision, recall, etc. do not necessarily indicate low quality. The relative values from different approaches are our main concern and it is important not to be discouraged by the absolute values.

18.7.1.1. *Recommendation approaches*

We compared UBCF to the other recommendation approaches explained in Section 18.2.3. UBCF performed best, followed by the content-based methods that used recently read and recently added articles, respectively (Figure 18.7). User profile data were poorer predictors of which articles users browsed. Of the approaches to using profile data, the best method was to find popular articles that matched key terms entered as research interests. In the absence of research interests, a user's discipline (a broad category chosen from a list) was an even poorer basis for recommendations.

The overall implication is that we should prioritise recommendations based on user libraries rather than profile data. For 'warm' users who have library data, we should display collaborative filtering results with greater prominence than the content-based recommendations.

18.7.1.2. *Variants of user-based collaborative filtering*

When tuning our implementation of user-based nearest-neighbor collaborative filtering, we tested the impact of inverted neighborhoods and significance weighting, as explained in Sections 18.3.2 and 18.3.1. The highest metrics were obtained by using the inverted neighborhood with significance weighting to reduce the contribution of neighbors who had fewer than 5 articles in common (Figure 18.8).

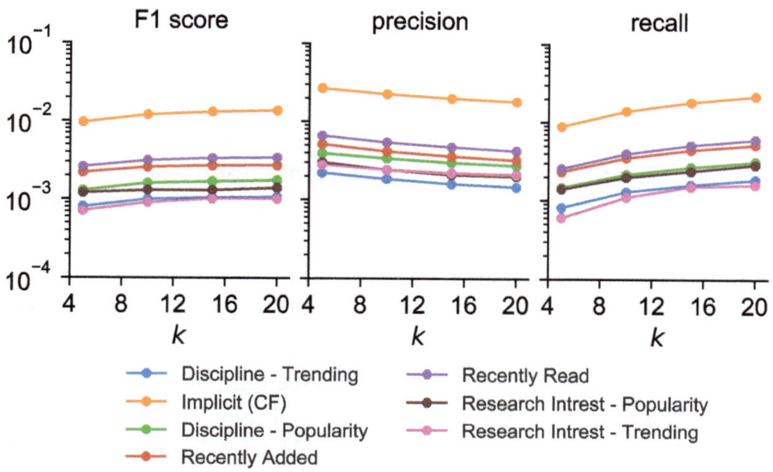

Fig. 18.7. Offline evaluation to compare article recommendation approaches. We calculated F1-score, precision, and recall for the top k recommendations, where $k = 5$, 10, 15, 20. The values of the metrics are shown on a log scale.

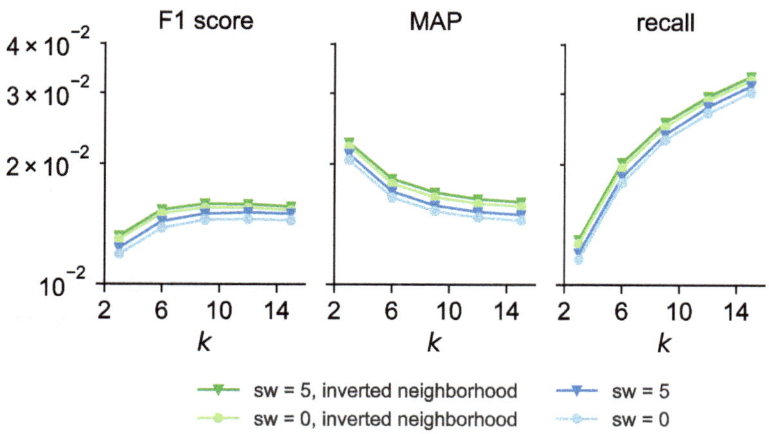

Fig. 18.8. Offline evaluation of collaborative filtering variants. A standard user-based nearest-neighbor method (blue) is compared to inverted neighborhoods (green). Each neighborhood method was tested with significance weighting (sw = 5, triangles) and without (sw = 0, circles). F1-score, MAP, and recall were calculated for the top k recommendations, where $k = 3$, 6, 9, 12, 15.

18.7.1.3. *Time decay*

By applying time decay to collaborative filtering input, we can bias recommendations towards users' more recent activity (Section 18.3.3). This is implemented by applying exponential decay to all user-article interactions (two-sided), or to the interactions of the focal user only and not the candidate neighbors (one-sided). We compared the two variants, in both cases using inverted neighborhoods and significance weighting, as were found to perform best above (Section 18.7.1.2).

The highest MAP@15 was obtained using the two-sided variant with a moderate rate of decay (Figure 18.9). With the optimum value of $\beta = 2$, articles added to the library one year ago have half the weight of articles added now. More extreme time decay with $\beta > 3$ reduced MAP@15 relative to the control. The one-sided variant reduced MAP@15 for all tested values of β. Based on these results, we carried out an A/B test with β in the range $[2, 3]$, and found that it improved click rates on recommended articles sent via email, compared to no time decay.

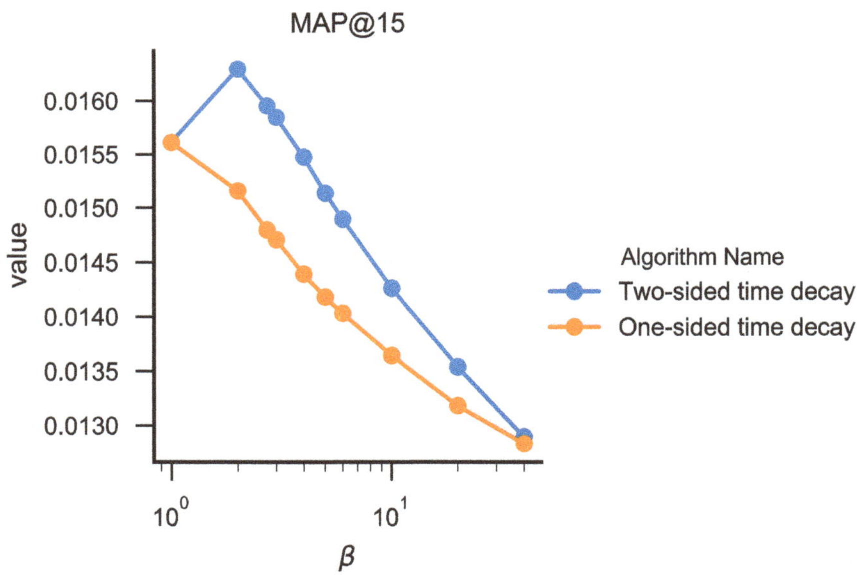

Fig. 18.9. Offline evaluation of collaborative filtering with time decay. For the one-sided and two-sided versions of time decay, we tested time decay constants of $\beta = 1, 2, e, 3, 4, 5, 6, 10, 20, 40$ (shown on a log scale). $\beta = 1$ is the control, with no time decay applied. MAP was calculated for the top $k = 15$ recommendations.

18.7.2. *Online Evaluation*

Our best candidate algorithms from offline evaluation are ultimately compared against the current production model via A/B testing. The experiments are conducted in two separate scenarios: recommendations delivered via email and web-based recommendations. This strategy has resulted in optimizing the recommender system for both of these use cases. For both scenarios we use the same metric as a measure of success, namely click through rates. If a user clicks on a recommendation then we deem this as a positive interaction with the recommender system and if they do not click then it is deemed as negative.

For email-based experiments, statistical inference is based on a classical, fixed-sample hypothesis testing approach. For web-based experiments we use instead a variable-size hypothesis testing approach for statistical inference (sequential testing). A type-I error of 0.05 is typically used in both settings, with Bonferroni correction for multiple comparisons if the test compares more than 2 variants.

The online tests can be classified into user interface changes and content/algorithm changes. An example change to the user interface was to personalise email subject lines, which resulted in 16.7% increase in email open rate. An example change to the algorithm was to apply time decay (see Sections 18.3.3 and 18.7.1.3), which resulted in a 26% increase in the proportion of users who clicked on at least one recommendation sent via email.

Online testing is an essential tool for making data-driven improvements to the recommender system, but it is impossible to test the entire parameter space of possible improvements to the user interface, algorithms, and business logic. Therefore, user testing and qualitative evaluation are necessary to guide our experimentation. Feedback from internal and external users often prompts common-sense changes to the user interface and business logic. Keeping an eye on the actual recommendations, in addition to click rates, helps us get around the fact that A/B testing only compares average engagement. There may be outlying cases of irrelevant recommendations, where a simple business logic change can help us to avoid alienating users.

18.8. Lessons learned and future directions

In this chapter we have introduced Mendeley Suggest, a recommender system for academic literature. We emphasised that the system is not only composed of smart algorithms but also other core components aimed at creating the best user experience possible. For the Mendeley use case, we found that user-based collaborative filtering not only scales better than item-based collaborative filtering and matrix factorization, but it also generates better quality recommendations. We were also able to get significant improvements in the system by introducing two-sided time decay (Section 18.3.3), impression discounting (Section 18.3.4), and dithering (Section 18.3.5).

To take Suggest from a prototype to production serving millions of users, we went through a number of iterations of the algorithms, the technology stack, and the user interface. We quickly learnt that running experiments using the same technologies as the production stack, whenever possible, significantly shortened the time needed to both evaluate and productionize new ideas. From the start of the project, we also focused on setting up an evaluation framework and agreeing on metrics, which allowed us to quantify the impact of different ideas and prioritise accordingly.

A key challenge we still need to address is to identify users' current research interests and adapt the recommendations as their interests change over time, be this a slow drift or a sudden change in direction. In Section 18.3.3, we explained how applying time decay to the input of the collaborative filtering algorithm helped us bias the recommendations towards the recent activity of the user. However, this approach is simplistic and could be improved in several ways. First of all, it applies the same time decay rate to all users, regardless of the user's subject area and usage pattern. Furthermore, the continuous decay function is not a good model for discrete jumps between subject areas, for example when a researcher changes jobs or begins a new collaboration.

So far, the article usage data has proved the most useful for generating recommendations. However, there are many other features available to us that could also factor into an article's relevance to a user, including the article's text, authors, journal, publication date, and its citation links to other articles the user has read. To combine these features with collaborative filtering into a single relevance score, we experimented with adding a learning to rank (LtR) module to Suggest, which takes candidate recommendations from collaborative filtering, enriches them with additional features, and then re-ranks them using a model trained on previous user interactions. Preliminary results have shown that citation-based features had the greatest effect in offline evaluation, particularity if the article was cited within a user's library or if it cited an article from his or her library. An initial A/B test was also promising, so we plan to add LtR to the production system.

Another way to generate hybrid recommendations is with neural networks (see Chapter 3 on Deep learning). As with matrix factorization, neural networks can be used to learn latent dense representations (embeddings) of users and items, based on user-item interactions [Sedhain *et al.* (2015); He *et al.* (2017)]. In addition to usage data, the neural network can handle multimodal inputs, such as user profile features [Covington *et al.* (2016)] and article text [Wang and Blei (2011)]. We plan to experiment with this approach because of its flexibility for leveraging all available information when generating recommendations.

Although a more complex hybrid approach may eventually replace user-based collaborative filtering within our system, the main message of this chapter is that one can use well established methods to implement scalable personalised recommendations for a large and growing user and item base. Regardless of the collaborative filtering approach, additional tweaks are needed to ensure data quality and recommendation freshness. There is much more data we can draw on to further personalise research article recommendations, but using standard collaborative filtering approaches remain powerful and not trivial to improve upon.

References

Beel, J., Gipp, B., Langer, S. and Breitinger, C. (2016). Research-paper recommender systems: a literature survey, *International Journal on Digital Libraries* **17**, 4, pp. 305–338, doi:10.1007/s00799-015-0156-0, `https://doi.org/10.1007/s00799-015-0156-0`.

Beel, J., Langer, S., Genzmehr, M. and Nürnberger, A. (2013). Introducing docear's research paper recommender system, in *Proceedings of the 13th ACM/IEEE-CS Joint Conference on Digital Libraries*, JCDL '13 (ACM, New York, NY, USA), ISBN 978-1-4503-2077-1, pp. 459–460, doi:10.1145/2467696.2467786, `http://doi.acm.org/10.1145/2467696.2467786`.

Bhowmick, A., Prasad, U. and Kottur, S. (2014). Movie Recommendation based on Collaborative Topic Modeling, *satwikkottur.github.io* `https://satwikkottur.github.io/reports/F14-ML-Report.pdf`.

Canny, J. (2002). Collaborative filtering with privacy, in *Proceedings of the 2002 IEEE Symposium on Security and Privacy*, SP '02 (IEEE Computer Society, Washington, DC, USA), ISBN 0-7695-1543-6, pp. 45–, `http://dl.acm.org/citation.cfm?id=829514.830525`.

Covington, P., Adams, J. and Sargin, E. (2016). Deep neural networks for youtube recommendations, in *Proceedings of the 10th ACM Conference on Recommender Systems*, RecSys '16 (ACM, New York, NY, USA), ISBN 978-1-4503-4035-9, pp. 191–198, doi:10.1145/2959100.2959190, `http://doi.acm.org/10.1145/2959100.2959190`.

Dunning, T., Friedman, E. and Ellen Friedman, M. D. (2014). *Practical Machine Learning: Innovations in Recommendation* (O'Reilly Media, Inc.), ISBN 1491915722.

Gunawardana, A. and Shani, G. (2009). A survey of accuracy evaluation metrics of recommendation tasks, *J. Mach. Learn. Res.* **10**, pp. 2935–2962, `http://dl.acm.org/citation.cfm?id=1577069.1755883`.

He, X., Liao, L., Zhang, H., Nie, L., Hu, X. and Chua, T.-S. (2017). Neural collaborative filtering, in *Proceedings of the 26th International Conference on World Wide Web*, WWW '17 (International World Wide Web Conferences Steering Committee, Republic and Canton of Geneva, Switzerland), ISBN 978-1-4503-4913-0, pp. 173–182, doi:10.1145/3038912.3052569, `https://doi.org/10.1145/3038912.3052569`.

Henning, V. and Reichelt, J. (2008). Mendeley - a last.fm for research? in *2008 IEEE Fourth International Conference on eScience*, pp. 327–328, doi:10.1109/eScience.2008.128.

Hristakeva, M. (2015). Sparking Science up with Research Recommendations, in *Spark Summit*, `https://spark-summit.org/eu-2015/events/sparking-science-up-with-research-recommendations/`.

Hristakeva, M., Kershaw, D., Rossetti, M., Knoth, P., Pettit, B., Vargas, S. and Jack, K. (2017). Building recommender systems for scholarly information, in *Proceedings of the 1st Workshop on Scholarly Web Mining*, SWM '17 (ACM, New York, NY, USA), ISBN 978-1-4503-5240-6, pp. 25–32, doi:10.1145/3057148.3057152, `http://doi.acm.org/10.1145/3057148.3057152`.

Jannach, D., Lerche, L., Kamehkhosh, I. and Jugovac, M. (2015). What recommenders recommend: an analysis of recommendation biases and possible countermeasures, *User Modeling and User-Adapted Interaction* **25**, 5, pp. 427–491, doi:10.1007/s11257-015-9165-3, `https://doi.org/10.1007/s11257-015-9165-3`.

Jomsri, P., Sanguansintukul, S. and Choochaiwattana, W. (2010). A framework for tag-based research paper recommender system: An ir approach, in *2010 IEEE 24th International Conference on Advanced Information Networking and Applications Workshops*, pp. 103–108, doi:10.1109/WAINA.2010.35.

Küçüktunç, O., Saule, E., Kaya, K. and Çatalyürek, Ü. V. (2012). Recommendation on academic networks using direction aware citation analysis, *arXiv preprint arXiv:1205.1143*.

Larsen, P. O. and von Ins, M. (2010). The rate of growth in scientific publication and the decline in coverage provided by science citation index, *Scientometrics* **84**, 3, pp. 575–603, doi:10.1007/s11192-010-0202-z, `https://doi.org/10.1007/s11192-010-0202-z`.

Lee, P., Lakshmanan, L. V., Tiwari, M. and Shah, S. (2014). Modeling impression discounting in large-scale recommender systems, in *Proceedings of the 20th ACM SIGKDD International Conference on Knowledge Discovery and Data Mining*, KDD '14 (ACM, New York, NY, USA), ISBN 978-1-4503-2956-9, pp. 1837–1846, doi:10.1145/2623330.2623356, `http://doi.acm.org/10.1145/2623330.2623356`.

McNee, S. M., Albert, I., Cosley, D., Gopalkrishnan, P., Lam, S. K., Rashid, A. M., Konstan, J. A. and Riedl, J. (2002). On the recommending of citations for research papers, in *Proceedings of the 2002 ACM Conference on Computer Supported Cooperative Work*, CSCW '02 (ACM, New York, NY, USA), ISBN 1-58113-560-2, pp. 116–125, doi:10.1145/587078.587096, `http://doi.acm.org/10.1145/587078.587096`.

Mehta, B. and Nejdl, W. (2008). Attack resistant collaborative filtering, in *Proceedings of the 31st Annual International ACM SIGIR Conference on Research and Development in Information Retrieval*, SIGIR '08 (ACM, New York, NY, USA), ISBN 978-1-60558-164-4, pp. 75–82, doi:10.1145/1390334.1390350, `http://doi.acm.org/10.1145/1390334.1390350`.

Ricci, F., Rokach, L. and Shapira, B. (2015). *Recommender Systems Handbook*, 2nd edn. (Springer Publishing Company, Incorporated), ISBN 1489976361, 9781489976369.

Sarwar, B., Karypis, G., Konstan, J. and Riedl, J. (2001). Item-based collaborative filtering recommendation algorithms, in *Proceedings of the 10th International Conference on World Wide Web*, WWW '01 (ACM, New York, NY, USA), ISBN 1-58113-348-0, pp. 285–295, doi:10.1145/371920.372071, `http://doi.acm.org/10.1145/371920.372071`.

Sedhain, S., Menon, A. K., Sanner, S. and Xie, L. (2015). Autorec: Autoencoders meet collaborative filtering, in *Proceedings of the 24th International Conference on World Wide Web*, WWW '15 Companion (ACM, New York, NY, USA), ISBN 978-1-4503-3473-0, pp. 111–112, doi:10.1145/2740908.2742726, `http://doi.acm.org/10.1145/2740908.2742726`.

Subasic, I., Gvozdenovic, N. and Jack, K. (2016). De-duplicating a large crowd-sourced catalogue of bibliographic records, *Program* **50**, 2, pp. 138–156, doi:10.1108/PROG-02-2015-0021.

Vargas, S. and Castells, P. (2014). Improving sales diversity by recommending users to items, in *Proceedings of the 8th ACM Conference on Recommender Systems*, RecSys '14 (ACM, New York, NY, USA), ISBN 978-1-4503-2668-1, pp. 145–152, doi:10.1145/2645710.2645744, `http://doi.acm.org/10.1145/2645710.2645744`.

Vargas, S., Hristakeva, M. and Jack, K. (2016). Mendeley: Recommendations for researchers, in *Proceedings of the 10th ACM Conference on Recommender Systems*, RecSys '16 (ACM, New York, NY, USA), ISBN 978-1-4503-4035-9, pp. 365–365, doi:10.1145/2959100.2959116, `http://doi.acm.org/10.1145/2959100.2959116`.

Walker, D., Xie, H., Yan, K.-K. and Maslov, S. (2007). Ranking scientific publications using a model of network traffic, *Journal of Statistical Mechanics: Theory and Experiment* **2007**, 6, pp. P06010–P06010, doi:10.1088/1742-5468/2007/06/P06010, `http://stacks.iop.org/1742-5468/2007/i=06/a=P06010?key=crossref.fcef8050577d38451490d5f4f0d7df10`.

Wang, C. and Blei, D. M. (2011). Collaborative topic modeling for recommending scientific articles, in *Proceedings of the 17th ACM SIGKDD International Conference on Knowledge Discovery and Data Mining*, KDD '11 (ACM, New York, NY, USA), ISBN 978-1-4503-0813-7, pp. 448–456, doi:10.1145/2020408.2020480, `http://doi.acm.org/10.1145/2020408.2020480`.

Chapter 19

MoocRec.com: Massive Open Online Courses Recommender System

Panagiotis Symeonidis[1] and Dimitrios Malakoudis[2]

[1]*Faculty of Computer Science, Free University of Bolzano, Italy*
email: psymeonidis@unibz.it
[2]*Faculty of Electrical Engineering,*
Aristotle University of Thessaloniki, Greece
email: dmalakoudis@gmail.com

Massive open online courses (MOOCs) have recently gained a huge users' attention on the Web. They are considered as a highly promising form of teaching from leading universities such as Stanford and Berkeley. MoocRec.com is a web site that recommends courses to users so that, they can acquire those skills, that are expected from their ideal job posting. MoocRec's recommendation engine is based on a Matrix Factorization (MF) model, which exploits information from external resources (i.e., job-skill, user-skill, course-skill, etc.), which are extractred from EdX, Coursera and Linkedin. Based on the aforementioned additional matrices we are able to predict course trends and to make rating predictions for users over courses. Based on these rating predictions, we find the similar neighbors of a target user and we provide to her top-N course recommendations.

19.1. Introduction

Massive Open Online Courses (MOOCs) platforms offer thousands of different courses and each course's registration/enrolment can be in the hundreds of thousand students. It would be very useful, if someone could be recommended a course to acquire those skills, that are expected from his ideal job description.

MoocRec.com is a web site that provides to users recommendations of MOOCs. Firstly, users provide some information about their studies and their dream job. Then, MoocRec.com recommends to them related courses, to acquire the required skills for getting their dream job. The heart of the recommendation engine of MoocRec.com is matrix decomposition over a user-course rating matrix R to reduce its dimensions and remove noise from data. To do this, we preserve a small number of k latent features (i.e., dimensions) with the objective to reveal the mainstream users' preferences. For example, in Figure 19.1, we plot users and courses, assuming that k has been tuned to 2.

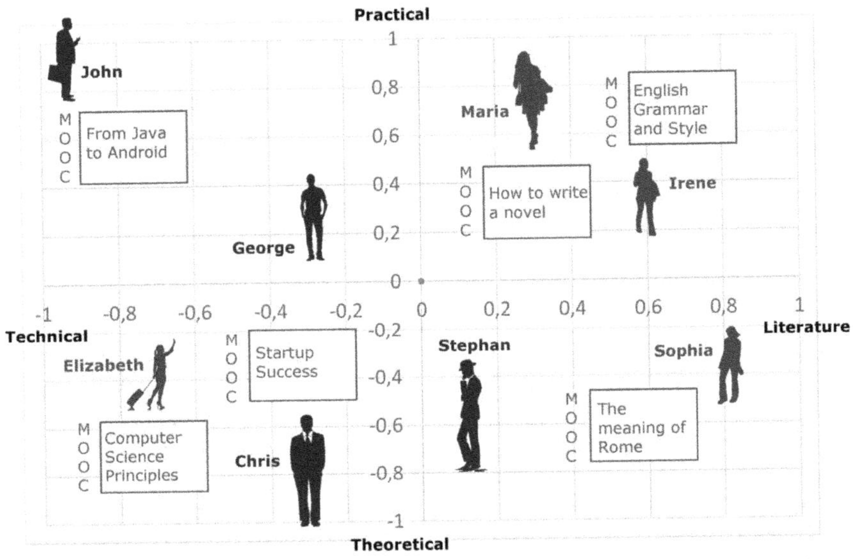

Fig. 19.1. Users and Courses in the 2-D space.

As shown in Figure 19.1, courses/users that are placed in close distance, are the most suitable/similar to each other. As shown, women prefer literature courses, whereas men choose the technical ones. Specifically, the course "English Grammar and Style" can be recommended to Maria and Irene, whereas "From Java to Android" course is more suitable to John. Please notice that matrix decomposition has also revealed a second separation, which takes place among people's preference, towards practical and theoretical types of courses. In MoocRec.com, we predict users' ratings over courses based on matrix factorization (MF) technique, which exploits information from several external resources/matrices. These matrices (i.e., job-skill, user-skill, course-skill, etc.) are constructed after retrieving data about MOOCs (extracted from EdX and Coursera) and the skills that are related to each job description (extracted from Linkedin).

The rest of this chapter is organized as follows. Section 19.2 summarizes the related work. Section 19.3 describes the system's architecture, whereas Section 19.4 describes in details the web site services. Section 19.5 describes the database, whereas Section 19.6 describes three different web crawlers (EdX, Coursera, Linkedin). Section 19.7 describes our recommendation engine and developed algorithms. Finally, Section 19.8 concludes this chapter.

19.2. Related work

Furnas *et al.* (Furnas *et al.*, 1988) proposed Singular Value Decomposition (SVD) in Information Retrieval research field. More specifically, SVD captures latent associations between the terms and the documents. SVD is a well-known factorization technique that factors a matrix into three matrices. An instance of SVD, known as UV-decomposition, searches for two matrices (U and V), whose their multiplication gives an approximation of the original matrix R. A significant improvement on the prediction accuracy of classic MF algorithm may be obtained through the incorporation of implicit feedback into the MF model (Koren, 2008; Koren and Bell, 2011; Paterek, 2007).

Extensions of classic MF algorithm and other methods [Bendakir and Aïmeur (2006); Parameswaran *et al.* (2011)] have been applied for course recommendations. Elbadrawy and Karypis (Elbadrawy and Karypis, 2016) provided top-N course recommendation based on MF, by incorporating into their models additional student and course academic features (e.g., student major, course topic, etc.) and by building multi-granularity student and course groups accordingly. In the MOOC domain, to reduce the high students' drop-out rates, Yang *et al.* (Yang *et al.*, 2014b) provided recommendations of useful forum threads to students based on their blog history inside a MOOC discussion forum. Their model matches forum threads with users based on their previous blog history. To capture the implicit feedback of students, they have incorporated into their model, the consumed content, the social interaction of students and other forum activities. They have shown that learners' preferences over forum threads are almost equal to their preferences for the contents of those threads. Moreover, Yang *et al.* (Yang *et al.*, 2014a) proposed a matrix factorization method that considers also constraints (e.g., students should not be over-burdened with too many questions, etc.) for the task of providing question recommendations in discussion forums that concern a MOOC. To reduce the high students' drop-out rates, Yang *et al.* (Yang *et al.*, 2014b) provided recommendations of useful forum threads to students based on their blog history inside a MOOC discussion forum. Recently, Almutairi *et al.* (Almutairi *et al.*, 2017) exploited temporal and other kinds of contextual information to predict students' grades on courses and recommend courses to reduce student retention. They proposed two methods based on coupled matrix and tensor factorization. The latent factors obtained can be used to predict grades, which can be later used for course recommendation.

Except the MF techniques for course recommendation, there are also other methods. For example, Aher *et al.* (Aher and Lobo, 2013) proposed a combination of a clustering with an association rule algorithm to recommend courses to students based on choice of other students for particular set of courses collected from Moodle. Their approach uses combination of clustering technique (K-means) and association rule algorithm (Apriori) and finds the recommendation results. Finally, Koutrika *et al.* (Garcia-Molina, 2008; Koutrika *et al.*, 2009) proposed FlexRecs that can provide course recommendations to students, who rank them, add comments, and rank the accuracy of each others' comments. FlexRecs is a framework for defining recommendations combining traditional relational operators with other special recommendation operators. They experimented with different recommendation types, such as recommending majors or courses, where users could get flexible (e.g. extensible, novel, etc.) course recommendations.

19.3. System's Architecture

The MoocRec[1] system is a web site that provides recommendations of MOOCs, so that its registered users can acquire new skills that they lack, in order to get their dream job. The process is very simple: They describe us their studies and their dream job and we recommend them related courses/skills. User registration is very simple, through the usage of Wordpress, which is a Content Management System (CMS). Figure 19.2 introduces the architecture of the MoocRec system. As shown, the system consists of the web site, the recommendation engine, the database and the web crawler.

The web crawler scans edX and Coursera web sites and inserts all the MOOCs found in a MySQL database. Moreover, the web crawler gets information from Linkedin about the skills that each job description is

[1] www.moocrec.com

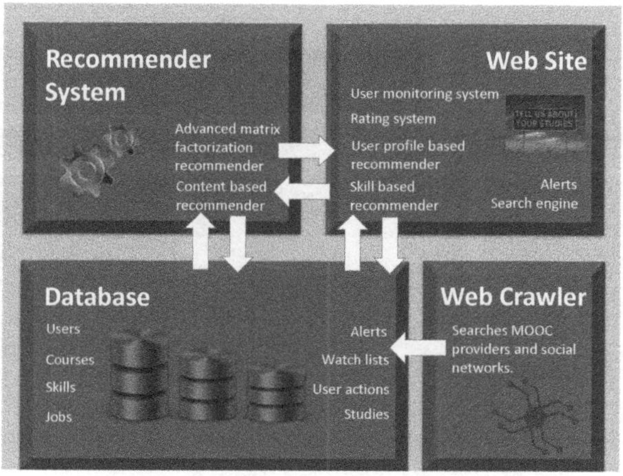

Fig. 19.2. MoocRec System Architecture.

related to. The database holds information that correlates jobs with skills and skills with courses, making the job-driven recommendation possible.

Our website has a search engine for finding MOOCs. It supports two different types of searching (i.e., the open-type search, where you can search for MOOCs similar to the way that google search engine works and the closed-type search, where you have to fill in a web form with several searching criteria).

Our website also incorporates a MOOC recommender system based on the skills of each user and his ideal job characteristics. Specifically, the website provides the user with personalised content according to the skills he wishes to acquire for getting his ideal job. Therefore, we have to connect a course description with the desired skills.

Moreover, using alerts the registered user may be notified of MOOCs when they become available. He may also rate courses and add them to his watch list. A user monitoring system records the user's actions and helps in constructing the user profiles. Please notice that MoocRec members are always able to watch and update the log files and the information which concerns them.

19.4. The Website

19.4.1. *Homepage*

The home page of MoocRec consists of several parts, some of which are shared with other pages. As shown in Figure 19.3, on the upper right we

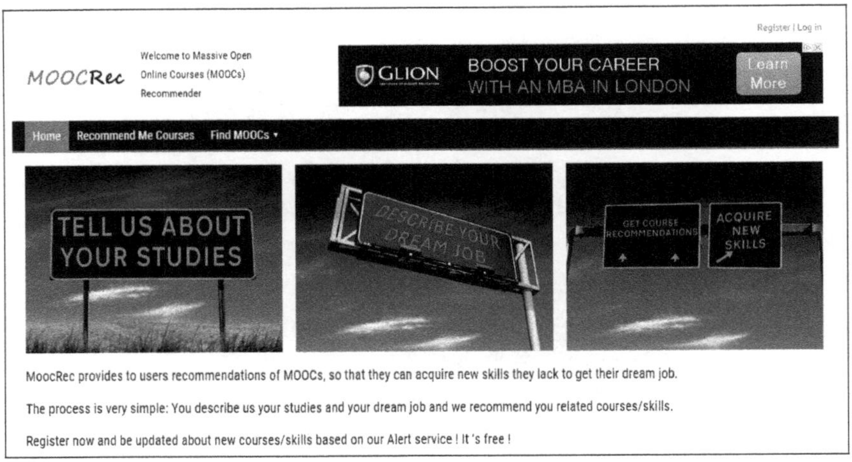

Fig. 19.3. MoocRec homepage.

placed the register and login links. Underneath appears the MoocRec logo, at the left side of an AdSense advertisement. Thereafter, comes the top-menu. For the unregistered visitor, we temporarily allow the usage of the job-driven recommender by clicking on the *Recommend Me Courses* tab. All visitors may also use the search engine by selecting and clicking on *Find MOOCs* choice.

Immediately below the menu, we provide an introduction to the operation of the job-driven recommender. For this reason, we display three images that demonstrate the procedure followed.

19.4.2. *A Skill-aware Recommender System*

The most innovative feature of MoocRec is the job-driven recommender system. When a user visits the *Recommend Me Courses* page, the system asks him to select his studies from a drop-down list. This is the first optional step and it results in a prediction for the skills he possesses based on his education, as pointed out in Figure 19.4. Subsequently, in the second step the candidate, has the options to add skills which he already masters or to discard the missing ones. The reason for the two aforementioned optional steps is that we want to exclude already possessed skills from the recommendation procedure.

Immediately after, as shown in Figure 19.5, the user selects his ideal job from a drop-down list consisting of 78 different top-rated jobs. Then, his predicted missing skills appear on step 4/4 and the user can correct

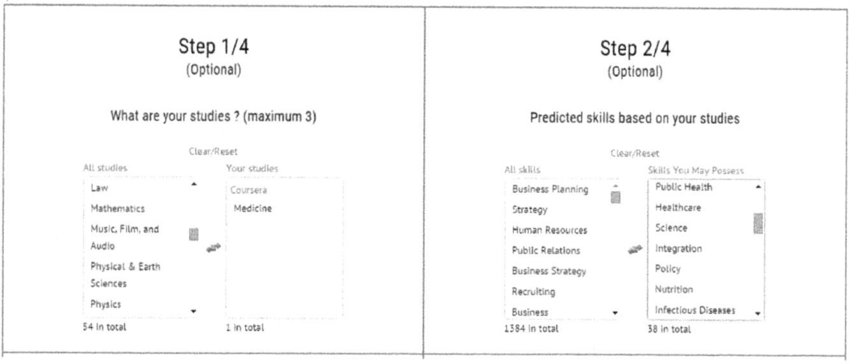

Fig. 19.4. First two optional steps of *Recommend Me Courses* page.

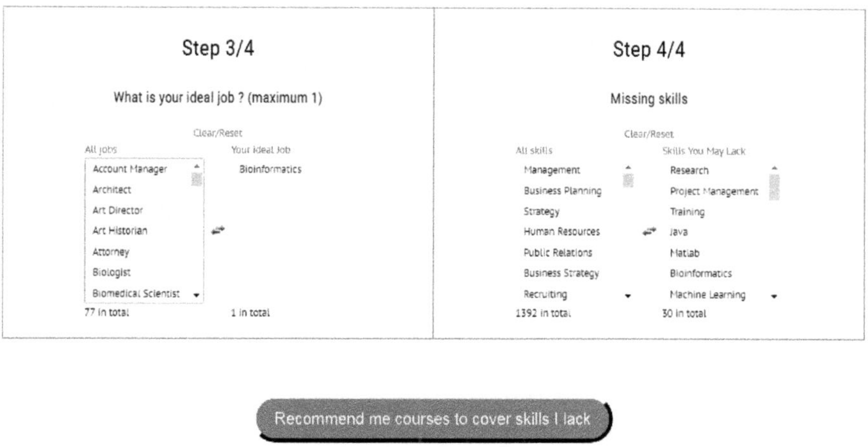

Fig. 19.5. Final two steps and *Recommend Me Courses* button.

them. At this point, as presented in Figure 19.5, MoocRec's visitor is ready to push *Recommend Me Courses* button and get recommendations about courses to acquire the skills he may lack.

19.4.3. *The search engine*

MoocRec system offers two types of search forms, the simple and the advanced. Using the advanced search (Figure 19.6), visitors can search in a specific provider and select to view either active and upcoming courses or extend their results by checking the checkboxes which refer to self-paced and all active courses.

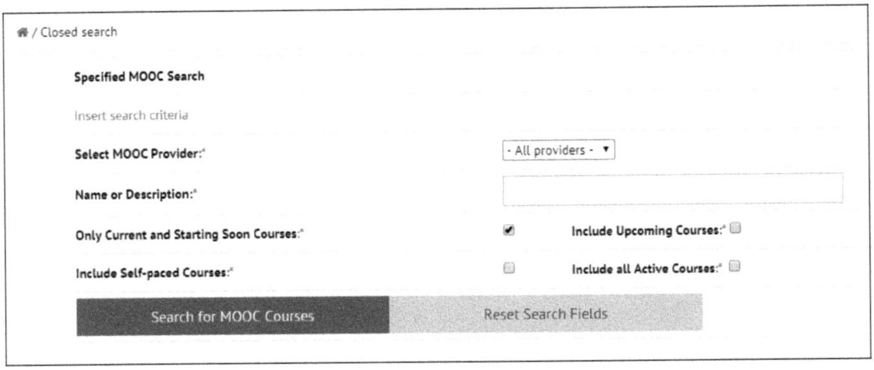

Fig. 19.6. Advanced search form.

As shown in Figure 19.6, we set as default that most of the time our users would like to view results for courses that have already started recently or are going to start in a few days. Of course, a user can also search for upcoming courses or can access past but still active ones in order to download their educational content.

As shown in Figure 19.7, the search results are retrieved from the database with the current and upcoming courses to appear first.

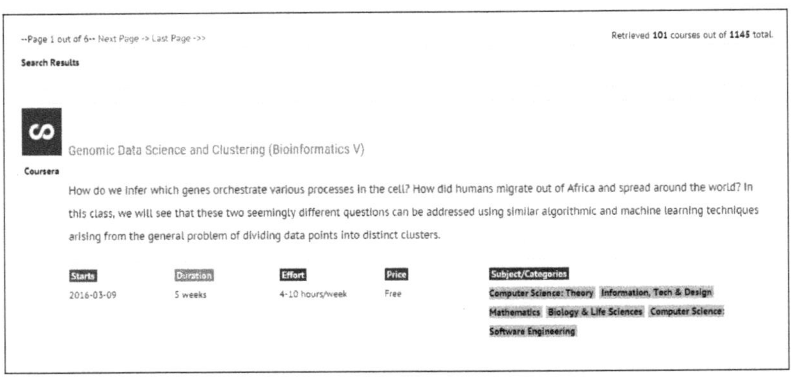

Fig. 19.7. Search results.

19.4.4. *The Alert Creation page*

When a registered user logs in, two more choices are added to the main menu. For example, the *Alert Creation* page is only available for the registered users. The purpose of this page is to allow users to insert their alerts and get informed every time a new MOOC is inserted in the database and matches their alert preferences.

As shown in Figure 19.8, users can import their course preferences and get informed by email as soon as MOOCs like the ones they are interested in, become available.

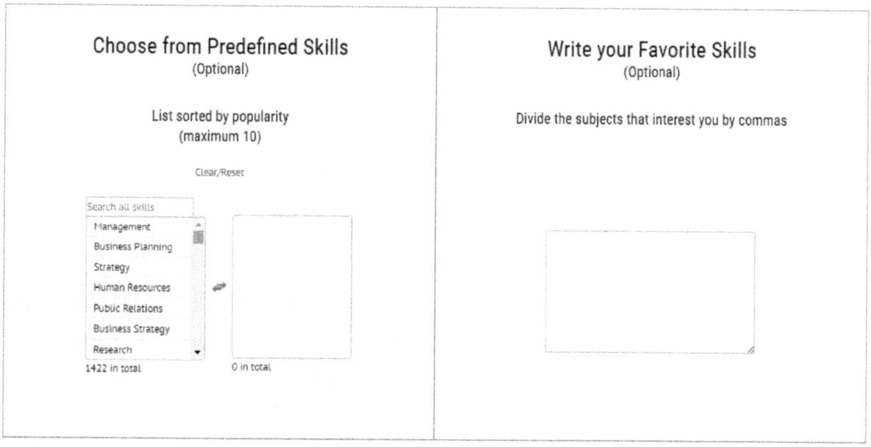

Fig. 19.8. *Alert Creation* page.

For inserting their preferences, they may choose one or both of the following two options:

- **Choose from predefined skills.** This option allows the user to select his desired skills from a similar drop-down list as the one used in the *Recommend Me Courses* page described on Subsection 19.4.2. As it is illustrated in Figure 19.8, skills are disposed in order of importance, meaning that most popular skills are listed first. Of course this helps users discover the most popular skills they may lack, but on the other hand users may want to search for a specific (not popular) skill. Therefore, we have added also a search function.
- **Write their Favorite Skills.** Users can simply write down the skills that they would like to be alerted. Similar to the open-type search, users should divide each skill keyword by comma in order to insert more than one of them.

Finally, a *Create Alert* button has been placed on the bottom of the aforementioned two options. By clicking on it, user views his updated alerts. From that moment and on, each time a new MOOC provides at least one of the desired skills which were inserted in user alerts, we send him an informative email message. A characteristic email message is shown on Figure 19.9.

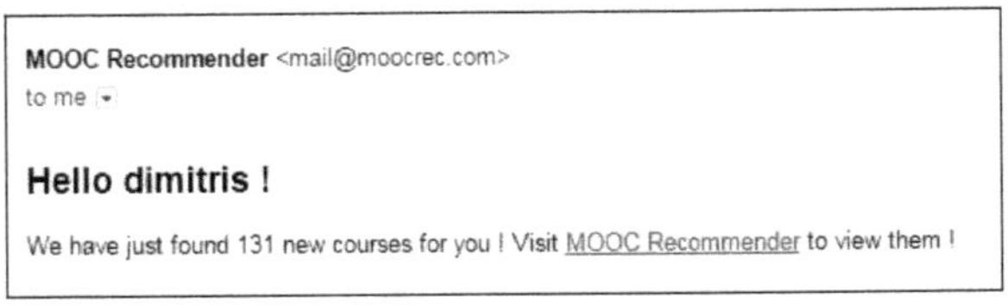

Fig. 19.9. Alert email message.

We have created the email account mail@moocrec.com to send the alert email messages. Thus, the user is prompted to visit the MoocRec system in order to view the MOOCs we have found and proposed. When users click on the MOOC recommender link, he visits his special alerts page.

19.4.5. *Administrator's privileges*

MoocRec web site was implemented with CMS Wordpress. Administrators have access to MoocRec's control panel, which is shown in Figure 19.10. Therefore, they may add pages or menu items, change web site's appearance, modify users data and view statistics. In addition, we have created a Google Analytics web page to capture in more detail users' behavior.

19.5. Database

The database contains all the information necessary for the entities of the MoocRec system and the correlations between them. The database system is MySQL and includes 23 tables. According to the entity — relationship model (ER model) we distinguish the following main participating entities in our system (Figure 19.11):

- **Users.** It is the set of registered users. Each user has his own preferences and by the time he logs in, we collect information concerning his behavior. Our goal is to create a personal user profile and recommend him the most appropriate courses. As it is illustrated in Figure 19.11, a user may (i) get alerted of a course he is

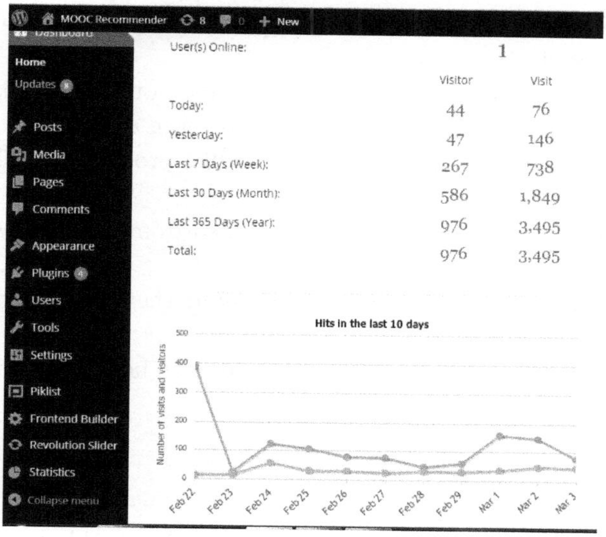

Fig. 19.10. Wordpress control panel.

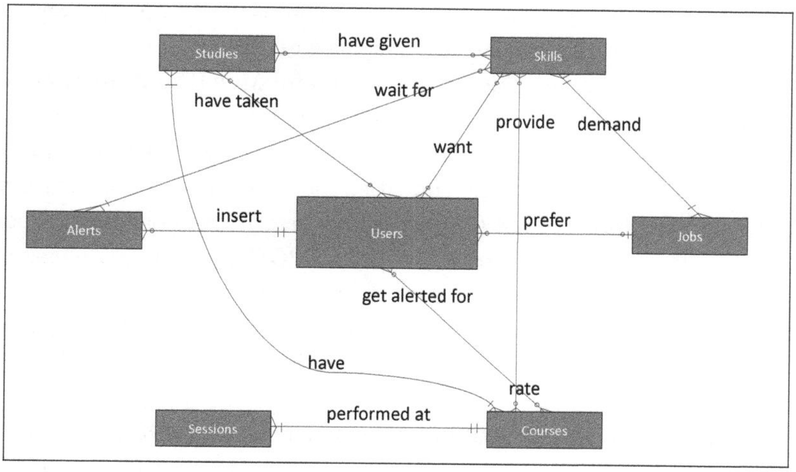

Fig. 19.11. MoocRec database entities and relationships.

interested in, (ii) view a course description, and (iii) rate a course. He can also add a MOOC in his watch list. In addition, he may insert his studies, and find the adequate skills to get his dream job.

- **Courses.** These are the MOOCs. Normally a MOOC is performed at several time periods/session.

- **Sessions.** This means that the same MOOC may be repeated time by time during a year.
- **Studies.** These are actually the subjects or the categories of the available MOOCs. We have connected each course with particular skills. Therefore, we assume that studies provide the skills of their courses.
- **Jobs.** The top-ranked jobs, web-mined from social networks such as Linkedin.
- **Skills.** Users have to acquire the skills they lack in order to get the job they dream of.
- **Alerts.** When issuing an alert, a user is in fact asking and waiting for courses.

19.5.1. *Database Tables*

In the following, we will briefly describe the database tables:

(1) **Users.** The table that saves user information. Normally, this is the Wordpress table wp_users. Column ID is used as primary key and serves as a user index for other database tables.
(2) **Courses.** The main table containing useful MOOC information. We hold the MOOC's id in the provider's database, the provider, the short name of the course which sometimes is used to form the link of the official MOOC page, the title, the short description which we reveal in *Recommend Me Courses* and *Search Results* pages, the about description which we present in the *MOOC Description* page, the studies in which the course is part of, the estimated class workload, the duration, the start and end dates, the price, the link and information about whether MOOC is in the English language and is a self-paced course.
(3) **Sessions.** The current and future sessions of each MOOC. This table is used as an intermediate step to discover the current session of each Coursera MOOC.
(4) **Saved alerts.** The courses for which users have been alerted. User views them in his personal *My Alerts* page.
(5) **Pending alerts.** Intermediate information table which holds new MOOCs, we have found but not yet informed users about.
(6) **User seen MOOCs.** We save the event of a user viewing a MOOC's description page. We also hold the useful information of the time stamp.

(7) **User favorite MOOCs.** We capture the action of a user viewing a MOOC's official page outside the MoocRec System.
(8) **User watch list.** Every user's MOOC watch list and time stamp of the event.
(9) **User rated MOOCs.** The ratings users have given to the MOOCs and the time they have rated them.
(10) **Jobs.** The jobs which the MoocRec system offers as selection options in the drop-down lists.
(11) **User jobs.** The dreamed job of each user.
(12) **Skills.** All the skills we have web mined.
(13) **Skills popularity.** Resulting of the web mining we have conducted, we classify skills according to their popularity. In our drop down lists of *Recommend Me Courses* and *Alert Creation* pages, we present the most popular skills first.
(14) **Skills popularity per job.** We save the skills according to their different popularity for every particular job. Thus, we discover the top desired skills for each one and we present them to users after they select their desired job.
(15) **User skills.** The skills each user lacks and desires to acquire.
(16) **Studies.** The studies which also serve as categories-subjects of the MOOCs.
(17) **User Studies.** The studies users have taken.
(18) **User Studies acquired skills.** The skills users have acquired from their studies.
(19) **Alerts.** The alerts table. We save the names of the skills and the keywords users have inserted while creating their alerts.
(20) **New alert definition.** By default if 6 months have passed since the last alert of a MOOC to a user, we re-alert him. Administrator may change this time period from the selection list of *Manage Alerts* page. The new alert definition will be saved in this table.
(21) **My new alert definition.** Similar to the previous table. The difference is that we save a different new alert definition for each user.
(22) **Course skills.** The skill keywords that are contained in a MOOC's title and short description.
(23) **Study skills.** The skill keywords that are contained in the MOOCs which belong to a particular study.

19.6. Web Crawler

The web crawler is implemented using PHP and SeleniumHQ's[2] PHP web driver. We periodically mine edX and Coursera MOOC providers to update our database with new or updated courses. We have also mined the LinkedIn social network to discover the top jobs and the top skills per each job. Unfortunately, only Coursera provides an API that offers easy access to MOOCs. Instead, edX and LinkedIn do not offer any API's which would have helped us. Moreover, edX and LinkedIn are using JavaScript to load pages, so the usage of SeleniumHQ is obligatory to succeed in getting the data we need. The web crawler was developed as an external module.

In Figure 19.12 we illustrate the graphical user interface of the web crawler. The administrator has the options to insert MOOCs from Coursera and edX, mine skills from LinkedIn, correlate skills to MOOCs and Studies and transfer the whole database to the MoocRec system. Thereafter, we will explain the operation of each selection.

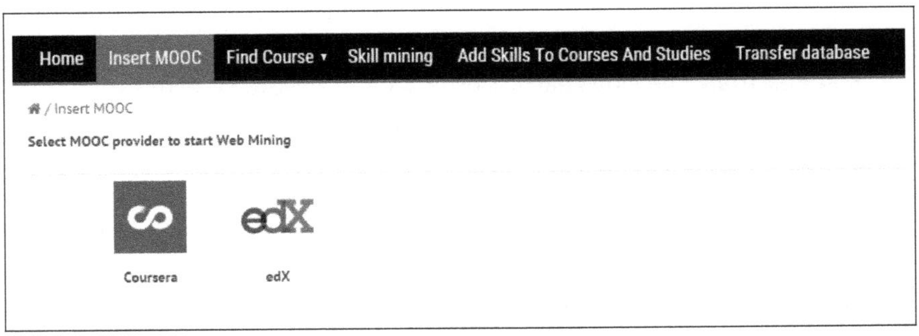

Fig. 19.12. MoocRec's web crawler.

19.6.1. *Coursera mining*

Coursera's catalog API[3] is a RESTful API that uses JSON for the data exchange format. Coursera's web mining process is executed by the following steps:

(1) We start by retrieving the HTML code of the courses URL[4]. We request the fields id, shortName, name, language, shortDescription,

[2] www.seleniumhq.org
[3] tech.coursera.org/app-platform/catalog/
[4] api.coursera.org/api/catalog.v1/courses

aboutTheCourse, targetAudience, estimatedClassWorkload and we ask the inclusion of the categories (Studies) of each course.

(2) Using the well-known from the Linux operating system function *preg-match-all()*, we are able to do pattern matching. Essentially, we manage to export content between HTML tags to be able to keep data such as Coursera's course id, MOOC's short name, title, language, short description, about description etc.
(3) If a MOOC has not yet been inserted in the MoocRec's database, we enter it together with the web mined information.
(4) If a MOOC has been removed from Coursera, we declare it as not existing in our database. This is very important, as MOOCs are related to users and we do not want to miss past user preferences by deleting the MOOC.
(5) We retrieve the HTML code of the sessions URL[5]. We ask for the fields id, courseId, homeLink, status, active, durationString, startDay, startMonth and startYear.
(6) We execute pattern matching and keep data such session's start date.
(7) We update database with new sessions.
(8) We retrieve the HTML code of the categories (Studies) URL[6].
(9) We update the database with new categories (Studies).
(10) We search sessions database table for current, upcoming or just active courses.
(11) We update courses database table and insert MOOC's start date and status (current, upcoming and just active).

In Figure 19.13 we may view the result of a Coursera web mining (March 4, 2016).

19.6.2. *EdX mining*

EdX mining process is accomplished using SeleniumHQ PHP web driver. We present a sample of edX's web mining code in Appendix A.3. EdX mining is managed as follows:

(1) SeleniumHQ connects to firefox and gets edX courses URL[7].
(2) We select English MOOCs automatically with PHP coding from edX's sidebar, so firefox loads the first English MOOCs.

[5] api.coursera.org/api/catalog.v1/sessions
[6] api.coursera.org/api/catalog.v1/categories
[7] www.edx.org/course

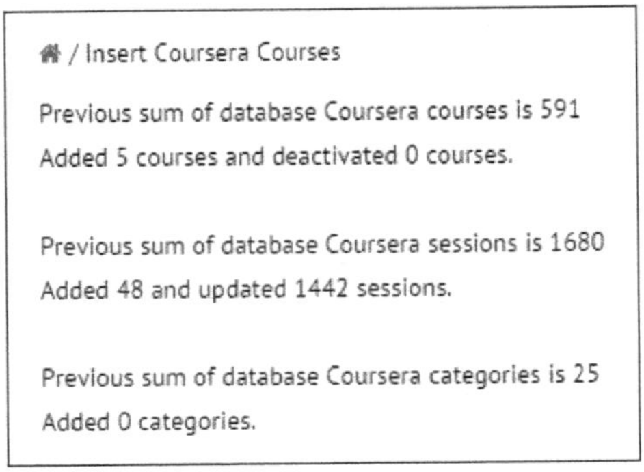

Fig. 19.13. Coursera's web mining result.

(3) We automatically move to the end of the page to load more edX courses. We repeat this step until we load all MOOCs. While we load MOOCs, firefox looks like Figure 19.14.

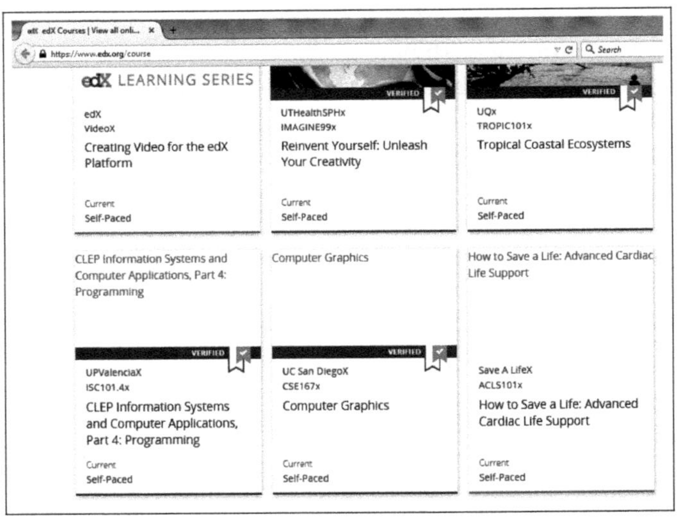

Fig. 19.14. EdX's mining process.

(4) We save the full loaded page in a file and execute pattern matching. We retrieve MOOC's link, code, title, date and all other relevant information.
(5) We test if each MOOC already exists in our database and if not we visit MOOC's link and get the page. Then, we do pattern matching and retrieve course's short description, about description, length, effort,

price and subjects (Studies). We save all the retrieved information in the database.

(6) If a MOOC has been removed from edX, we declare it as not existing in our database. This is very important, as MOOCs are connected with users and we do not want to miss past user preferences by deleting the MOOC.

(7) If a MOOC's link or date has been changed, we update database with the new values.

In Figure 19.15 we may view an excerpt of the result of an edX web mining (March 4, 2016). We present the update of a date and a link of an edX MOOC. As it is depicted, 37 MOOCs were updated. Many times archived courses become active again, therefore we need to always check and update when needed.

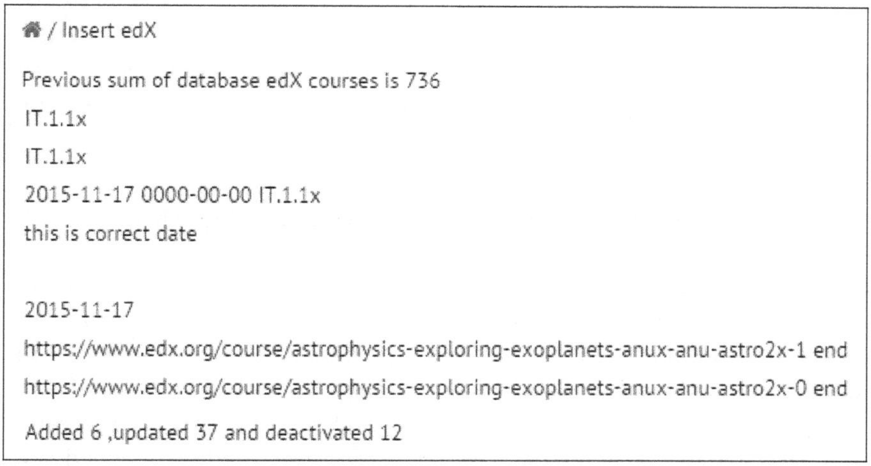

Fig. 19.15. EdX's web mining result.

19.6.3. *LinkedIn mining*

LinkedIn web mining was executed with two different goals. Firstly, we have web mined LinkedIn profiles and retrieved jobs. Thus, we have discovered the most common jobs inside LinkedIn members. The second and most important target was to discover the appropriate skills for each job. We have again used SeleniumHQ because many LinkedIn pages load with JavaScript. The skill mining process is described below:

(1) SeleniumHQ connects to firefox and gets LinkedIn's webpage[8].
(2) We log in LinkedIn automatically using PHP and SeleniumHQ.
(3) We search for those users, who have worked or currently work in the job, we are looking for its required skills.
(4) For each user we find, we visit his personal page and get his skills using pattern matching.
(5) Therefore, for each job we save the relevant skills in the database.
(6) We classify skills according to their total and job-related popularity.

The next step is to correlate skills to courses and studies. By clicking on *Add Skills to Courses and Studies* tab of web crawler's menu (Figure 19.12), we sweep all MOOCs and search if a skill keyword exists inside a MOOC's title or description. We calculate also, how many times a skill exists in and save the information to our database. Afterwards, we download the studies of each MOOC and save the aforementioned information to the database table which connects studies with skills. Finally, we may connect MoocRec's database, using the *Transfer database* option of crawler's menu and transfer the mined data. The connection is secured because we are using Static IP as additional login ID.

19.7. Recommendation Engine and Developed Algorithms

The recommendation engine is a key part of the system, which is interconnected with the web site and the database. It is the part that performs the recommendation algorithms, takes input data from the database and finally displays the results to the users through the web site.

In MoocRec system we use two recommendation algorithms for recommending MOOCs to the users. The algorithm we have already added is a content-based filtering one, while the upcoming one is the enhanced matrix factorization which will be also analyzed. In particular, we describe the two implementations in the next two subsections.

19.7.1. *Content-based filtering*

The purpose of this algorithm is to display the most relevant MOOCs, taking account of the skills each user is interested in. This is accomplished by comparing the skill keywords with the title and the description of the MOOCs. We have preinserted the aforementioned information while

[8]www.linkedin.com

executing the *Correlate skills to courses and studies* function of the web crawler. Thereafter, we access the *Course skills* database table and make an SQL query, ordering the MOOC's according to the number of skills they provide. The execution is really fast because of the preprocessing we have done, after the web mining phase. Consequently, we propose the top 10 MOOCs to the user in *Recommend Me Courses* page which was described in Subsection 19.4.2.

In Figure 19.16, we illustrate the recommendations we provide to a user, after he has clicked on *Recommend Me Courses to cover skills I lack* button, which was presented in Figure 19.5. As it is depicted, the recommendation includes a short description of the course and useful information such as when the course starts, whether the course is self-paced, the total duration, the effort required, the price and the related categories. The user may click on the title and be directed to a page describing the course in more detail.

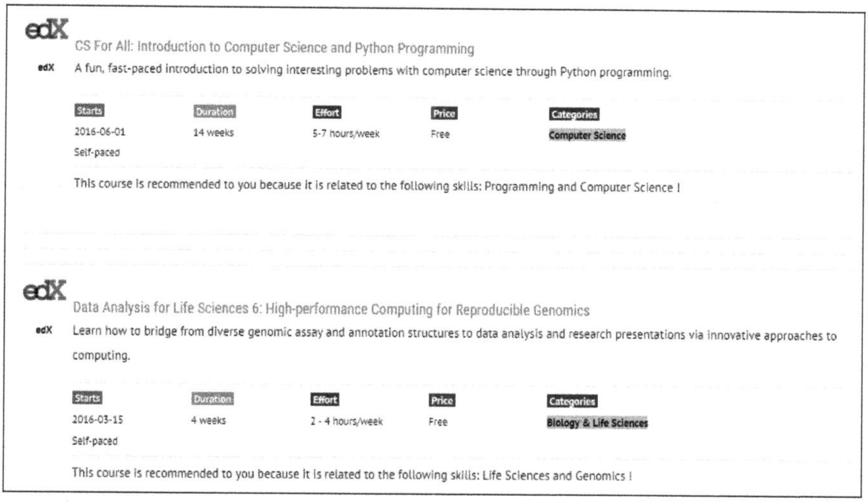

Fig. 19.16. Recommendation appearance and reasoning.

One very important characteristic of the recommendation is that user is informed about the reason he was recommended that course. For example, course Data Analysis for Life Sciences: High-Performance Computing for Reproducible Genomics is recommended because it is related to Life Sciences and Genomics. The content-based recommendation algorithm, places first the courses which provide the greatest number of skills. Moreover, courses that start soon or are self-paced have priority towards others.

19.7.2. *Enhanced Matrix Factorization*

In this Section, we will describe and present with pseudocode the implementation of enhanced matrix factorization algorithm.

First of all, we have to set the requirements of the algorithm. So, the input data which will be used to implement the method are the two-dimensional sparse user-course rating matrix R ($R \in \mathbb{R}^{m \times n}$), the two-dimensional sparse user-skill matrix $US \in \mathbb{R}^{m \times s}$, the two-dimensional sparse course-skill matrix $CS \in \mathbb{R}^{n \times s}$, the objective function G, the cosine similarity function SIM, the learning rates η_1 and η_2, the regularization parameters λ_1, λ_2 and λ_3, the number of total latent-features K and the *steps* of algorithm's predictions. The objective function is shown in the following:

$$G = min \sum_{i,j} \left(r_{ij} - \mu - bu_i - bc_j - \sum_{k=1}^{K} \left(u_{ik} + \frac{\sum_{s=1}^{S} us_{is} us^*_{sk}}{\sum_{s=1}^{S} us_{is}} \right) \left(v_{kj} + \frac{\sum_{s=1}^{S} cs_{js} cs^*_{ks}}{\sum_{s=1}^{S} cs_{js}} \right) \right)^2$$
$$+ \lambda_1 (bu_i^2 + bc_j^2) + \lambda_2 \sum_{k=1}^{K} \left(u_{ik}^2 + v_{kj}^2 \right) + \lambda_3 \sum_{k=1}^{K} \sum_{s=1}^{S} us_{is} (us^*_{sk})^2 + cs_{js} (cs^*_{ks})^2 \tag{19.1}$$

The next step in the implementation is to initialize with random values the two vectors bu and bc and the four matrices U, V, US^* and CS^*, as shown in the first line of Algorithm 1. Due to the fact that our enhanced UV-Decomposition method uses four thin matrices to produce the complete prediction rating matrix $\hat{R}$, one of the dimensions of U, V, US^* and CS^* matrices is K ($k = K$). At line 2 we calculate the mean value of matrix R. Subsequently, as shown at lines 3 to 12 we create an adjacency list *uSkill* to hold user's possessed skills and insert relative information in arrays *uSkillsIndex* and *uSkillsSum*. In the same way, at lines 13 to 22, we construct the same data structures referring to courses.

After all these initializations and data preparations the algorithm is ready to start the repetitive process of prediction. We firstly (lines 26 to 32) compute $\boxed{\frac{\sum_{s=1}^{S} us_{is} us^*_{sk}}{\sum_{s=1}^{S} us_{is}}}$ for each user and save the result in $SCUP$ vector. Also, using the same procedure (lines 33 to 39) we calculate $\boxed{\frac{\sum_{s=1}^{S} cs_{js} cs^*_{ks}}{\sum_{s=1}^{S} cs_{js}}}$ and insert the result in $SCCP$ vector. After that, at the next important step (lines 40 to 43) we find the prediction error e which is the difference between real and predicted rating. The prediction error e is then used to update the two vectors and the four matrices: bu and bc are updated at lines

Algorithm 1 Implementation of our enhanced UV-Decomposition Method in pseudo-code

Require: 2D sparse user-course rating matrix $R \in \mathbb{R}^{m \times n}$, 2D sparse user-skill matrix $US \in \mathbb{R}^{m \times s}$, 2D sparse course-skill matrix $CS \in \mathbb{R}^{n \times s}$, G objective function, similarity function SIM, η_1 and η_2 learning rates, λ_1, λ_2 and λ_3 regularization parameters, K number of total latent-features and *steps* maximum number of algorithm's predictions.

Ensure: Complete prediction matrix $\widehat{R} \in \mathbb{R}^{m \times n}$, Top-N recommendation matrix $TOPN$

```
Initialize bu,bc, U,V,US*, CS*;
Calculate μ = mean(R);
for i = 1 : m do
  uSkillsSum[i] = uSkillsIndex[i] = 0;
  for j = 1 : s do
    if US[i][j] > 0 then
      uSkill[i][uSkillsIndex[i]] = j;
      uSkillsIndex[i]+ = 1;
      uSkillsSum[i]+ = US[i][j];
    end if
  end for
end for
for i = 1 : n do
  cSkillsSum[i] = cSkillsIndex[i] = 0;
  for j = 1 : s do
    if CS[i][j] > 0 then
      cSkill[i][cSkillsIndex[i]] = j;
      cSkillsIndex[i]+ = 1;
      cSkillsSum[i]+ = CS[i][j];
    end if
  end for
end for
repeat
  for i = 1 : m do
    for j = 1 : n do
      for k = 1 : K do
        SCUP[k] = 0;
        for m = 1 : uSkillsIndex[i] do
          SCUP[k]+ = US[i][uSkill[i][m]]US*[uSkill[i][m]][k] ;
        end for
        SCUP[k]/ = uSkillsSum[i];
      end for
      for k = 1 : K do
        SCCP[k] = 0;
        for m = 1 : cSkillsIndex[j] do
          SCCP[k]+ = CS[j][cSkill[j][m]]CS*[k][cSkill[j][m]] ;
        end for
        SCCP[k]/ = cSkillsSum[j];
      end for
      e = R[i][j] − μ − bu[i] − bc[j];
      for k = 1 : K do
        e− = (U[i][k] + SCUP[k])(V[k][j] + SCCP[k]);
      end for
      bu[i]+ = η1(e − λ1bu[i]);
      bc[j]+ = η1(e − λ1bc[j]);
      for k = 1 : K do
        tempU[k] = U[i][k];
        tempV[k] = V[k][j];
        temp = η2 [e(V[k][j] + SCCP[k]) − λ2U[i][k]];
        V[k][j]+ = η2 [e(U[i][k] + SCUP[k]) − λ2V[k][j]];
        U[i][k]+ = temp;
      end for
```

```
        for m = 1 : uSkillsIndex[i] do
          for k = 1 : K do
            skill = uSkill[i][m];
            US*[skill][k]+ =
  η2 US[i][skill] [ (e / uSkillsSum[i]) (tempV[k] + SCCP[k]) − λ3 US*[skill][k] ];
          end for
        end for
        for m = 1 : cSkillsIndex[j] do
          for k = 1 : K do
            skill = cSkill[j][m];
            CS*[k][skill]+ =
  η2 CS[j][skill] [ (e / cSkillsSum[j]) (tempU[k] + SCUP[k]) − λ3 CS*[k][skill] ];
          end for
        end for
      end for
    end for
  until G ceases to improve OR maximum algorithms predictions steps reached
  for i = 1 : m do
    for j = 1 : n do
      r̂[i][j] = μ + bu[i] + bc[j];
      for k = 1 : K do
        Calculate SCUP[k], SCCP[k];
        r̂[i][j]+ = (U[i][k] + SCUP[k])(V[k][j] + SCCP[k]);
      end for
    end for
  end for
  Create vector users ∈ Z^{n'} with all users of dataset and vector courses ∈ Z^{m'} with all
    courses, respectively;
  Create list userRatedCourses to hold the courses each user rated and list
    userCoursesRatings to hold the corresponding ratings;
  for i = 1 : n' do
    for j = 1 : n' do
      Calculate sim[i][j];
    end for
    Quick sort sim[i] in descending order and create neighborhood list N[i];
  end for
  for i = 1 : n' do
    Initialize freq[i] elements at zero;
    for j = 1 : M do
      for m = 1 : userRatedCoursesSize do
        if userCoursesRatings[N[i][j]][m] > P_τ then
          course = userRatedCourses[N[i][j]][m];
          freq[i][course]+ = 1;
        end if
      end for
    end for
    Quick sort freq[i] list in descending order;
    while n < N   AND   freq[i][courseElement] > 0 do
      TOPN[i][n] = courseElement;
      n = n + 1;
    end while
  end for
  return TOPN;
```

44 and 45, while U and V at the following lines 46 to 52, respectively. One can observe that we use temporary vectors in order not to lose previous values of U and V. The reason for this becomes evident at lines 53 to 63, when we use the previous saved values to update US^* and CS^*. The execution of the update rules is really fast, as we have already calculated useful variables, such as the user-skill and course-skill sum (lines 9 and 19) and the skill contribution to user and course profile (lines 26 to 39). The process that has just described above will be repeatedly executed, until objective function G ceases to improve or until the maximum predictions number that we have set is reached (line 67). So far, we have created bu, bc, U, V, US^* and CS^* with the minimum G. The next step is to use them to calculate the prediction rating matrix $\hat{R}$. The $\hat{R}$ matrix is complete without any sparsity. We compute this matrix in lines 68 to 76 and we are ready to start finding the right courses to recommend, using any method we decide.

For the purpose of our work, we have chosen to present the pseudocode of the Most-Frequent item recommendation technique. Please notice that the Most-Frequent item recommendation algorithm counts the majority vote of an item inside the neighborhood of the target user. The neighborhood of the target user consists of his most similar users. In order to save processing time, we construct two vectors holding all the unique users and courses (line 77) and two lists to save the specific courses each user has rated and the related ratings (line 78). The first stage of this technique is to find out each user's neighbors. So, we compute cosine similarity function for every different pair of users (line 81). Subsequently, we quick sort the similarity results and pick up the top ones to create target user's neighborhood (line 83). The neighborhood's size is an external input, illustrated by variable M. Moreover, at line 85, we start the loop to discover the TOP-N courses, which we propose to every user. For every neighbor user (line 87), we scan all of his rated courses (line 88) and if a rating is greater than P_τ, we increase course's frequency inside user's neighborhood. Over and above, we quick sort courses' frequencies (line 95) and finally we select the top N to deliver $TOP-N$ course recommendations to every target user (lines 96 to 101).

19.8. Conclusions

In this chapter, we have described MoocRec.com, which exploits information from external resources (i.e., users' skills, courses' characteristics, job descriptions details, etc.) to provide course recommendations. In future, when there will be an adequate number of registered users, we want to test experimentally our system to check its effectiveness in terms of accurate recommendations. Moreover, we want to explore different ways of explaining our recommendations, so that we make our system more transparent and reliable. Finally, we want to crawl more MOOC platforms and websites that contain information about job descriptions and required skills.

References

Aher, S. B. and Lobo (2013). Combination of machine learning algorithms for recommendation of courses in e-learning system based on historical data, *Knowledge-Based Systems* **51**, pp. 1–14.

Almutairi, F. M., Sidiropoulos, N. D. and Karypis, G. (2017). Context-aware recommendation-based learning analytics using tensor and coupled matrix factorization, *IEEE Journal of Selected Topics in Signal Processing* **11**, 5, p. 729.

Bendakir, N. and Aïmeur, E. (2006). Using association rules for course recommendation, in *Proceedings of the AAAI Workshop on Educational Data Mining*, Vol. 3.

Elbadrawy, A. and Karypis, G. (2016). Domain-aware grade prediction and top-n course recommendation, in *Proceedings of the 10th ACM conference on Recommender systems*.

Furnas, G. W., Deerwester, S., Dumais, S. T., Landauer, T. K., Harshman, R. A., Streeter, L. A. and Lochbaum, K. E. (1988). Information retrieval using a singular value decomposition model of latent semantic structure, in *Proceedings of the 11th annual international ACM SIGIR conference on Research and development in information retrieval*, SIGIR '88 (ACM, New York, NY, USA), ISBN 2-7061-0309-4, pp. 465–480, doi:http://doi.acm.org/10.1145/62437.62487, http://doi.acm.org/10.1145/62437.62487.

Garcia-Molina, H. (2008). Flexible recommendations in courserank, in *OTM Confederated International Conferences "On the Move to Meaningful Internet Systems"* (Springer), pp. 7–7.

Koren, Y. (2008). Factorization meets the neighborhood: a multifaceted collaborative filtering model, in *Proceeding of the 14th ACM SIGKDD international conference on Knowledge discovery and data mining* (ACM), pp. 426–434, http://scholar.google.de/scholar.bib?q=info:oARx59_h04MJ:scholar.google.com/&output=citation&hl=de&as_sdt=0,5&ct=citation&cd=0.

Koren, Y. and Bell, R. M. (2011). Advances in Collaborative Filtering in F. Ricci, L. Rokach, B. Shapira and P. B. Kantor (eds.), *Recommender Systems Handbook* (Springer), ISBN 978-0-387-85819-7, pp. 145–186.

Koutrika, G., Bercovitz, B. and Garcia-Molina, H. (2009). Flexrecs: expressing and combining flexible recommendations, in *Proceedings of the 2009 ACM SIGMOD International Conference on Management of data* (ACM), pp. 745–758.

Parameswaran, A., Venetis, P. and Garcia-Molina, H. (2011). Recommendation systems with complex constraints: A course recommendation perspective, *ACM Transactions on Information Systems (TOIS)* **29**, 4, p. 20.

Paterek, A. (2007). Improving regularized singular value decomposition for collaborative filtering, in *Proceedings of KDD cup and workshop*, Vol. 2007, pp. 5–8.

Yang, D., Adamson, D. and Rosé, C. P. (2014a). Question recommendation with constraints for massive open online courses, in *Proceedings of the 8th ACM Conference on Recommender systems* (ACM), pp. 49–56.

Yang, D., Piergallini, M., Howley, I. and Rose, C. (2014b). Forum thread recommendation for massive open online courses, in *Proceedings of 7th International Conference on Educational Data Mining.*

Chapter 20

Food Recommendations

Christoph Trattner[†] and David Elsweiler[‡]
[†] *University of Bergen, Norway*
[‡] *University of Regensburg, Germany*

The recommendation of food items is important for many reasons. Attaining cooking inspiration via digital sources is becoming evermore popular; as are systems, which recommend other types of food, such as meals in restaurants or products in supermarkets. Researchers have been studying these kinds of systems for many years, suggesting not only that can they be a means to help people find food they might want to eat, but also help them nourish themselves more healthily. This paper provides a summary of the state-of-the-art of so-called food recommender systems, highlighting both seminal and most recent approaches to the problem, as well as important specializations, such as food recommendation systems for groups of users or systems which promote healthy eating. We moreover discuss the diverse challenges involved in designing recsys for food, summarise the lessons learned from past research and outline what we believe to be important future directions and open questions for the field. In providing these contributions we hope to provide a useful resource for researchers and practitioners alike.

20.1. Introduction

Online recommendation systems have proved to be useful in diverse situations by empowering the user to overcome the information overload problem, assisting with the decision making process and serving as a means to change user behavior [Ricci *et al.* (2011)]. One domain, which has historically received comparatively little attention, however, especially when compared to areas relating to leisure and entertainment, is the recommendation of food items. This is surprising given the importance of food for human sustenance, the range of options available, the fact that making food choices is particularly challenging [Scheibehenne *et al.* (2010)], and the high personal and societal costs of poor choices. Worldwide, lifestyle- and diet-related illnesses, such as obesity and diabetes, account for 60% of total deaths [Beaglehole (2016)]. Both are conditions which can be prevented and sometimes even reversed by appropriate dietary choices [Ornish *et al.* (1990)].

As such, health-aware food recommendation systems are often mooted as an important part of the solution to encourage healthier nutritional choices [Freyne and Berkovsky (2010); Freyne *et al.* (2011b); Harvey *et al.* (2012, 2013)].

There are many reasons, however, which make food recommendation challenging, not only in terms of encouraging healthy behaviour, but also in predicting what people would like to eat because this is complex, multi-faceted, culturally determined, not to mention context-dependent. Moreover, when developing food recommendation systems, there are additional issues for practitioners and researchers to consider, which do not arise in other recommendation domains. These include that users may have complex, constrained needs, such as allergies or life-style preferences, such as the desire to eat only vegan or vegetarian food. In such cases, standard approaches work poorly and adequate data sources to filter recipes are not freely available. Other challenges include food items may have multiple names, ingredients can be prepared in different ways and, unlike domains where products or media are recommended, it is not always clear if a recommended item can be prepared or consumed due to the potential for poor availability of ingredients, cooking knowledge or equipment.

This chapter makes two primary contributions. Firstly, we provide a summary of the state-of-the-art in food recommendation systems, highlighting both seminal and most recent approaches to the problem, as well as important specializations, such as food recommendation systems for groups of users or systems which promote healthy eating. We examine which algorithms have been used in the food domain, how systems are typically evaluated, and the resources available to those interested in building or studying recommendation systems in practice. In a second contribution we discuss the diverse challenges involved, as well as a summary of the lessons learned from past research, and an outline of important future directions and open questions. In providing these contributions we hope to provide a useful resource for researchers and practitioners alike.

20.2. Developed Approaches

Despite food recommendation being a comparatively understudied problem in the research community, a decent body of literature exists. Table 20.1 provides a list of important research contributions relating to food recommendation. We list 25 popular, highly cited, recent and relevant papers in chronological order, selected using our experience in the domain in combination with bibliographic tools, such as Google scholar[1], to identify the most relevant for the targeted readership. Special care was taken to identify work relating to different types of food item. As such, the papers cited relate to the recommendation of recipes, meal plans, groceries and menus. Although the problem of recommending restaurants to people, e.g. [Park *et al.* (2008)], is related, especially when the meals served there are taken into consideration, we focus here on research relating to systems directly recommending the food items themselves.

The columns in Table 20.1 relate to dimensions that we believe characterize the nature of different contributions in the area. *Algorithm* defines the various algorithmic approaches that have been tested in the food domain ranging from content based approaches, to collaborative filtering, to machine learning classifiers, some of which involve personalization (*Personalized*). *Recommended Items* describes the food item involved; *Feedback* describes the means by which the system is informed on user preferences and the suitability of any recommendation provided; *Context* provides the context dimension(s) utilised if applicable; *dietary constraint* informs on whether nutrition was considered; *Target* details who the end user(s) of the system was(were); and finally *Dataset* details the proprietary or open dataset utilized. The remainder of the chapter uses Table 20.1 as a structural basis.

In this section, we explain the approaches that have been taken in the literature to implement food item recommenders. In the literature the most prominent form of food recommendation system provides single item recommendations mostly in the form of recipes.

We structure the section around the approaches employed, summarizing content-based, collaborative filtering and hybrid approaches. We continue to show how context information is important and how this has been utilized in practice. Next, we broaden our focus to particular scenarios, which have been addressed, firstly looking at group-recommendations before reviewing research on food recommenders for healthy nutrition.

[1] `http://scholar.google.com`

Table 20.1. Overview of different types of recommendation systems strategies developed for recommending food (recipes, meal plans, groceries and menus) to people sorted in chronological order by publication date.

Author(s)	Algorithm(s)	Person-alized	Rec.Sys Type(s)	Feedback	Context/ Content Feature(s)	Dietary Constrains	Target	Dataset
[Elsweiler *et al.* (2017b)]	Logistic Random Forrest Naive Bayes	no	Recipes	Ratings Binary	Title Image Ingredients Nutrition Pop. & Appr	no	Single User	Allrecipes
[Trattner and Elsweiler (2017)]	LDA WRMF AR SLIM BPR MostPop User- ItemKNN	yes/no	Recipes Meal Plans	Bookmarks Ratings Comments	WHO-FSA health score	no	Single User	Allrecipes
[Cheng *et al.* (2017)]	BPR MostPop	yes/no	Recipes	Ratings	City Size	no	Single User	Kochbar
[Yang *et al.* (2017)]	Learning to Rank	yes	Recipes	Binary	Image Embeddings	yes	Single User	Yummly
[Rokicki *et al.* (2016)]	UserKNN MostPop	yes/no	Recipes	Ratings	Gender	no	Single User	Kochbar
[Ge *et al.* (2015a)]	MF CB	yes	Recipes	Ratings Tags	Tags	no	Single User	Wellbeing Diet Book
[Elsweiler and Harvey (2015)]	SVD-Hybrid	yes	Meal Plans (Set of recipes)	Ratings	Ingredients	yes	Single User	Quizine
[Sano *et al.* (2015)]	UserKNN SVD Hybrid NL-PCA	yes	Groceries	Purchases	Food Categories	no	Single User	Grocery store data
[Trevisiol *et al.* (2014)]	UserKNN CB	yes	Menus (Set of dishes)	Binary	Text Sentiment	no	Single User	Yelp
[Elahi *et al.* (2014)]	MF	yes	Recipes	Ratings Tags	tags	no	Group of Users	Wellbeing Diet Book
[Harvey *et al.* (2013)]	CB, CF Logistic Reg. SVD-Hybrid	yes	Recipes	Ratings	Ingredients etc.	no	Single User	Quizine

Table 20.1. (*Continued*)

Author(s)	Algorithm(s)	Person-alized	Rec.Sys Type(s)	Feedback	Context/ Content Feature(s)	Dietary Constrains	Target	Dataset
[Teng *et al.* (2012)]	SVM	no	Recipes	Ratings	Ingredients Nutrition Cook effort Cook methods	no	Single User	Allrecipes
[Kuo *et al.* (2012)]	Graph-based CB	yes	Menus (Set of recipes)	Tags	Ingredients	no	Single User	Food
[El-Dosuky *et al.* (2012)]	CB KB	yes	Food items	Query	tags	no	Single User	USDA
[Freyne *et al.* (2011a)]	CF	yes	Meal plans (Set of recipes)	Ratings	-	no	Single User	Wellbeing Diet Book
[Ueta *et al.* (2011)]	KB	yes	Recipes	Query	tags	no	Single User	Cookpad
[van Pinxteren *et al.* (2011)]	CB	yes	Recipes	Cooked recipes	Recipe content features	no	Single User	Smulweb
[Freyne and Berkovsky (2010)]	UserKNN CB Hybrid	yes	Recipes	Ratings	Ingredients	no	Single User	Wellbeing Diet Book
[Berkovsky and Freyne (2010)]	UserKNN GroupKNN Hybrid	yes	Recipes	Ratings	-	no	Group of Users	Wellbeing Diet Book
[Aberg (2006)]	CF	yes	Meal Plans (Set of recipes)	Ratings	-	yes	Single User	Unknown
[Khan and Hoffmann (2003)]	CBR	yes	Meal Plans	Query	Nutrition Content	yes	Single User	Unknown
[Mankoff *et al.* (2002)]	CB	yes	Groceries	Purchases	Food groups	no	Single User	Grocery store data
[Lawrence *et al.* (2001)]	AR CF CB	yes	Groceries	Purchases	Product class	no	Single User	Grocery store data
[Hinrichs and Kolodner (1991)]	CBR	yes	Meal Plans	Query	Content	yes	Single User Group of Users	Unknown
[Hammond (1986)]	CBR	yes	Single New Recipe	Query	-	yes	Single User	Unknown

Reflecting the literature as a whole, the majority of section details work recommending recipes to end users. That being said, there are other kinds of food items, which have been studied, albeit to a lesser extent. This is reflected in Table 20.1. As datasets are becoming more readily available (see Section 20.4), we expect interest in other food items to increase. The work published to date has largely employed the same standard approaches that have been applied when studying recipe recommendations. For example, content-based, collaborative filtering and hybrid approaches have been applied to restaurant review data to recommend menus [Trevisiol *et al.* (2014)] and online shopping data to recommend groceries [Lawrence *et al.* (2001); Mankoff *et al.* (2002); Sano *et al.* (2015)].

20.2.1. *Content-Based Methods (CB)*

Content-based approaches have been used as a means to tailor recommendations to the user's individual tastes. Freyne and Berkovsky, for example, made recommendations by breaking recipes down into individual ingredients and scoring based on the ingredients contained within recipes, which users had rated positively [Freyne and Berkovsky (2010); Freyne *et al.* (2011b)]. That is, if tomatoes had been present in recipes a user had reported liking, further recipes containing tomatoes would be predicted to also be liked by the user. Later work progressed this approach by not only accounting for positive ingredient biases, but also negatively weighting recipes based on contained ingredients featuring in recipes the user reported disliking [Harvey *et al.* (2013)].

[Teng *et al.* (2012)] proposed the use of complement and substitution networks as a means to generate accurate predictions. Complement networks of ingredients are constructed via co-occurrence of the same ingredients in the same recipes, while substitute networks are derived from user-generated suggestions for modifications. Experiments show that the use of these networks can predict the user preferences significantly better than approaches that rely on for example ingredient lists as features, cooking method, style, etc.

Other content-based approaches are more applicable to food recommendation systems than other domains. For example, as food decisions are often visually driven [Mormann *et al.* (2012); Schur *et al.* (2009)], the images associated with recipes can be exploited. Yang and colleagues have shown baseline approaches can be outperformed by algorithms designed to extrapolate important visual aspects of food images [Yang *et al.* (2015, 2017)].

In their work, Convolutional Neural Networks (CNN) provide a powerful framework for automatic feature learning. [Elsweiler *et al.* (2017b)] also show that automatically extracted low-level image features, such as brightness, colorfulness and sharpness can be useful for predicting user food preference.

20.2.2. *Collaborative Filtering-Based Methods (CF)*

Collaborative filtering-based recommendation methods (see Chapter 1) for food reccommender systems have also been proposed and evaluated. Freyne and Berkovsky tested a nearest neighbour approach using Pearson correlation on the ratings matrix, which offered poorer performance than the content approach described above [Freyne and Berkovsky (2010)]. [Harvey *et al.* (2013)] showed that SVD outperformed both the content and collaborative filtering approaches suggested in [Freyne and Berkovsky (2010)]. [Ge *et al.* (2015a)] propose a matrix factorization (MF) approach for food recommendation systems that fuses ratings information and user supplied tags to achieve significantly better prediction accuracy than content-based and standard matrix factorization baselines. They also present a mobile interface for the approach as shown in Figure 20.1. These screenshots show how a finer granularity of feedback can be assigned via tags, complementing the standard binary and scaled ratings typically used.

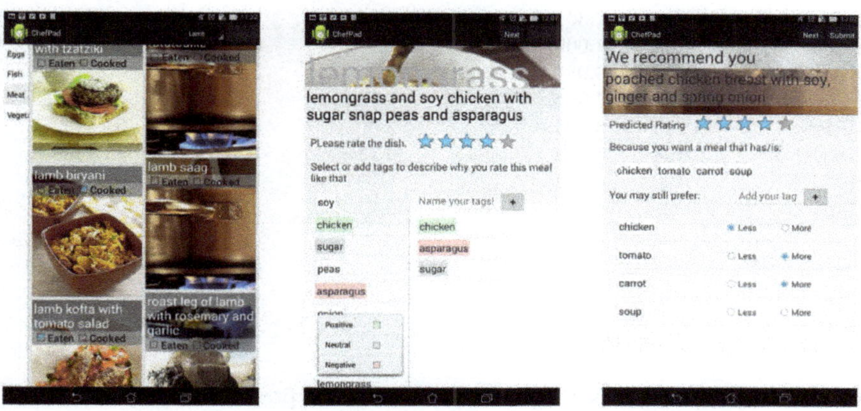

Fig. 20.1. Example of a mobile food recommender interface as proposed by [Elahi *et al.* (2014)] using not only ratings for preference elicitation but also tags at the same time. Taken with permission from the authors.

More recently, [Trattner and Elsweiler (2017)] tested a diverse range of collaborative filtering approaches implemented in the LibRec[2] framework using a large dataset crawled from the online recipe portal allrecipes.com. The highest performing CF approaches were Latent Dirichlet Allocation (LDA) [Griffiths (2002)] and Weighted matrix factorization (WRMF) [Hu *et al.* (2008)]. The results of their experiments are shown in Table 20.2 below.

Table 20.2. Recommender ranking accuracy sorted by nDCG and recommender accuracy post-filtered by FSA scores. The mean FSA scores of the top-5 recommended recipes are also reported along with the different average nutriens of the lists and the according FSA health labels (taken from [Trattner and Elsweiler (2017)]).

Algorithm	nDCG@5	FSA score	Fat (g)	Sat. Fat (g)	Sugar (g)	Sodium (g)
LDA	.0395	9.110	8.70	3.73	8.73	0.32
WRMF	.0365	9.114	9.50	3.89	8.84	0.34
AR	.0343	9.206	9.27	4.12	10.50	0.25
SLIM	.0326	8.907	9.27	3.82	7.91	0.33
BPR	.0325	9.252	8.69	3.82	7.83	0.29
MostPop	.0294	9.004	9.02	3.94	10.01	0.23
UserKNN	.024	8.985	8.96	3.73	7.98	0.31
ItemKNN	.0178	8.652	8.59	3.51	6.03	0.31
Random	.0029	8.486	8.74	3.49	5.71	0.30
		FSA score post-filtered ($score_{u,i,fsa}$)				
LDA	.0321	7.323	6.51	2.42	4.03	0.29
WRMF	.0303	7.361	6.48	2.30	4.75	0.31
SLIM	.0248	7.008	6.20	2.56	2.59	0.24
AR	.0238	6.984	5.64	1.94	3.95	0.28
MostPop	.0228	7.334	5.37	2.02	2.46	0.24
BPR	.0205	6.722	6.42	2.30	4.95	0.26
UserKNN	.0168	6.722	6.88	2.73	3.33	0.33
ItemKNN	.0109	6.124	5.15	1.79	3.51	0.25
Random	.0022	4.305	1.59	0.43	1.45	0.09

20.2.3. *Hybrid Methods (Hybrid)*

Hybrid recommenders (see Chapter 4) have been proposed by other scholars for the recipe recommendation task. For example, [Freyne and Berkovsky (2010)] combined a user-based collaborative filtering method with a content-based method. Moreover, in their follow-up work, which targeted groups of users (described in more detail in Sub-section 27.2.5),

[2]`http://www.librec.net/`

they employed a hybrid approach to combine three different recommendation strategies in a single model, using a switching strategy. The switching was based on the ratio between the number of items rated by a user and overall number of items. This approach was motivated by [Burke (2002)]. Another example of a hybrid approach can be found in the work of [Harvey *et al.* (2013)] who tested various collaborative filtering and content-based approaches. They achieved the best performance in their experiments, however, by combining an SVD approach with user and item biases in a weighted linear model.

20.2.4. *Context-Aware Approaches*

Numerous exploratory data analyses have demonstrated that context (see Chapter 5) is important in food recommendation, with gender [Rokicki *et al.* (2016)], time [Kusmierczyk *et al.* (2015)], hobbies [Trattner *et al.* (2017c)], location [Cheng *et al.* (2017); De Choudhury *et al.* (2016); Zhu *et al.* (2013)] and food availability [De Choudhury *et al.* (2016)] being identified as important variables. All of these studies employed relatively simple filtering techniques to split naturalistic datasets in order to explore how recipes were rated [Freyne *et al.* (2011a)], bookmarked [Trattner and Elsweiler (2017)], shared [Abbar *et al.* (2015)] or related to health statistics [Trattner *et al.* (2017b)].

[Harvey *et al.* (2012)] collected detailed context data encapsulating the ratings participants provided for their dataset, where participants could identify a broad range of factors to justify the rating assigned to a recipe as a meal to cook for dinner that day. Analyzing these with regression modeling showed that factors, such as how well the preparation steps are described, as well as the nutritional properties of the dish, the availability of ingredients and temporal factors such as day of the week have a bearing on the user's opinion of the recommendation.

What is lacking with respect to context is an understanding of which variables are the most important and how best to account for these algorithmically. Despite studying numerous factors, [Harvey *et al.* (2013)], for example, limited their algorithmic efforts to approaches with only nutritional, user and item biases.

In summary, although the problem of improving the precision of recommendations has been attended to by numerous researchers with diverse approaches, the results achieved for the recommendation of recipes to individual users, measured on standard metrics are typically poorer than in other domains. To demonstrate typical results achieved with standard approaches on recipe data, we present the results of our own experiments with standard techniques on a well-known dataset in Table 20.2. These results underline the challenge of predicting which dishes people will like and emphasize that further effort is required.

20.2.5. *Group-Based Methods*

Of course people do not always eat or make food choices alone. Often these are activities done together with friends, families or colleagues. It is well known in social psychology that the social situation within which one will eat (who is present, why they are present, and what their preferences are) influences the food choices taken [Wansink (2006)]. In food recommendation systems, such social contexts are addressed by group recommendation systems (see Chapter 6). In this setting, a list of items is produced for a group of people rather than for an individual user. Despite the pervasiveness of shared food consumption experiences, group-based food recommendation systems research has been limited, even though the earliest efforts can be traced to the early 1990's [Hinrichs and Kolodner (1991)]. [Berkovsky and Freyne (2010)] not only studied different strategies for recommending recipes to a group of people but evaluate these methods with real users in a family scenario. In particular their work introduces four different strategies: a general strategy (which employs a most popular approach to recommend items), an aggregated model (which first combines individual user models into a single model before applying the collaborative filter), aggregated predictions strategies (which first computes CF on the individual user profiles and then combines the predicted rating) and finally a personalized strategy (which exploits a standard CF algorithm). The results show that the personlized version works the best, but it was not possible to create personalized recommendations for all of the users. More recently [Elahi *et al.* (2014)] proposed a mobile interface and algorithm for food recommendation system in a group-context. In addition to improving the prediction algorithm with tags, the authors use group-based preference elicitation, in which users play different roles in the food choice process. One user is designated as the group leader or cook to whom the system delivers meal recommendations based on the group utility score, which aggregates predictions using the tags and ratings of all the group members.

20.2.6. *Health-Aware Methods*

When motivating research on food recommendation systems, health problems and improving nutritional habits are usually mentioned e.g. [Freyne and Berkovsky (2010); Freyne *et al.* (2011b); Harvey *et al.* (2012, 2013)]. Incorporating health into the recommendation, however, has largely been a recent focus [Schäfer *et al.* (2017); Elsweiler *et al.* (2017a, 2016)]. One means of achieving this is to incorporate nutritional aspects into the recommendation approach directly. [Ge *et al.* (2015b)] took this approach by accounting for calorie counts in the recommendation algorithm. They did this based on a so-called "calorie balance function" that accounts for the differences between the calories the user needs and the calories in a recipe.

[Elsweiler *et al.* (2015)] refer to the trade-off for most users between recommending the user what she wants and what is nutritionally appropriate. This is a trade-off applicable for a large proportion of users [Harvey *et al.* (2013)] and should be optimized [Elsweiler *et al.* (2015)]. The authors proposed combining to two aspects linearly as a framework for evaluating different algorithmic approaches to incorporate health in the recommendation process.

The formula (see Equation 20.1) illustrates the simple concept. Here, i is a given recipe, $r\hat{(i)}$ is the estimated rating for recipe i, Max$(r\hat{(i)})$ is the maximum estimated rating over all recipes. $n(i)$ is the nutritional "error" incurred when recommending this recipe (relative to some ideal set of nutritional values). λ is a free parameter that can be set to suit the researcher/practitioner's priorities, although λ=.5 is probably preferable initially as it gives equal weighting to rating and nutrition. Note that all of these estimates are implicitly conditioned on a specific user u.

$$score(i) = \lambda \frac{r(i)}{max(r(i))} + (1 - \lambda) - 1 \times \frac{n(i)}{max(n(i))} \tag{20.1}$$

[Trattner and Elsweiler (2017)] employed a post-filtering (see Equations 20.2 and 20.3) approach to incorporate further nutritional aspects. To post-filter items a straightforward scoring function is applied which reweights the scores of a recipe for a particular user based on the WHO or inverse FSA score, employing a simple multiplication. The $score_{u,i}$ in the equation stands for the score of the item i for user u and who_i, fsa_i denote the health scores for that item. The two nutrition metrics are based on widely accepted nutritional standards from The World Health Organisation (WHO) [WHO (2003)] and the United Kingdom Food Standards Agency (FSA) [FSA (2016)] (see Section 20.4). Their previous work had used these measures to establish the (un)healthiness of recipes from a popular Internet food portal [Trattner *et al.* (2017a)].

$$score_{u,i,who} = score_{u,i} \cdot (who_i + 1) \quad (20.2)$$

$$score_{u,i,fsa} = score_{u,i} \cdot (16 - fsa_i - 4 + 1) \quad (20.3)$$

Table 20.2 describes the performance of 9 prominent recommendation algorithms as implemented in the LibRec framework in Trattner and Elsweiler's experiments. The top and bottom halves of table shows the performance without and with post-filtering respectively. Full details of the experimental setup can be found in [Trattner and Elsweiler (2017)].

These experiments with post-filtering on nutritional properties show that 1) it is possible to balance and potentially optimize the trade-off between recommendation accuracy and the healthiness of recommendations, 2) some recommendation algorithms may be more (e.g., LDA and WRMF) or less suitable (e.g., MostPop and BPR) to this process.

Nevertheless, the results also show that 3) while the approach shows potential benefit and future work should try to optimize the trade-off, the method by itself will not lead to healthy nutrition — at least not with the collection evaluated in this work. Despite offering a significant improvement on the standard approaches, the post-filtered results show that the best FSA and WHO scores achieved were not particularly high and are associated with extremely poor recommendation accuracy. These represent the best health values which can be achieved using an individual item recommendation approach, indicating that complementary ideas are necessary.

One such complementary approach is to combine individual recommended items for a user, such that they meet the recommended intake for that user over a longer period of time (e.g. day, week etc.). [Freyne *et al.* (2011a)] presented an interface, which allowed users to generate their own meal plans from individually recommended dishes. The recommendations were generated using the authors' hybrid approach as described above [Freyne and Berkovsky (2010)]. The interface for such plans evaluated on 5000 people in Australia. To encourage variation in meal plans a decay function was applied to meals appearing regularly in plans. Users manually created plans from lists of recommendations but the lists were filtered such that only meals that could be added and still ensure plans met guidelines featured in the list of recommended items.

[Harvey and Elsweiler (2015)] presented a similar interface, which automated the creation of plans consisting of a combination of breakfast, lunch and dinner plus an allowance for snacks and drinks. The same authors evaluated their planning approach systematically by deriving plans from taste profiles (i.e. from users featuring in naturalistic datasets) combined with diverse personas (simulated user properties, such as height, weight, gender, age, nutritional goal (lose/gain/maintain weight) and activity level (from sedentary to highly active) [Elsweiler and Harvey (2015)]. In a first step, the authors predicted the ratings users with particular profiles might assign to recipes (using approaches like those described above). In a second step, following approaches from nutritional science, the recommended nutritional intake was calculated for the user persona, including the required calories, but also where these should be sourced (proteins, carbohydrates etc.). Lastly, plans were generated for a given user (persona-profile combination) by taking the top-n recommendations from the recommendation system for the taste profile, splitting these into two separate sets, one for breakfasts and one for main meals and performing a full search finds *every combination* of these recipes in the sequence [breakfast, main meal, main meal] meeting the target nutritional requirements as defined above.

Using this method the authors were able to generate plans for 4025/6400 cases (63%) and at least 1 plan was generated for 58 out of the 64 (91%) user profiles and for each of the 100 personas. The authors moreover analyzed the factors, which made the development of plans challenging. When personas required a relatively high calorie intake, e.g. if the persona was tall or wanted to gain weight, the simple approach using 3 meals of fixed portions was often unable to address this properly. Similarly, profiles with little diversity in preferred ingredients were also hard to satisfy.

Substituting meals has been mooted as a further approach to influencing food choices. [Elsweiler *et al.* (2017b)] developed predictive models with the aim of forecasting the choices people will make. After evaluating the models for prediction accuracy using cross-validation, these were used to select recipe replacements such that users were be "nudged" towards making healthier choices. Aligning with the findings reported above, visual and nutritional features were important. A user study found that using the predictive models as the basis for recommendations, participants were significantly more likely to choose a recipe with much less fat content — the opposite of the trend that one typically sees.

Substituting ingredients within recipes has also been proposed to improve the health credentials of individual recipes healthier e.g. [Teng *et al.* (2012); Achananuparp and Weber (2016)]. This approach has, however, yet to be evaluated properly in a nutritional context. Initial steps in this direction were taken by [Kusmierczyk *et al.* (2016)], whose findings illustrate to what extent it may be possible to recommend a user substitute ingredients based on the user's previous recipe uploads and accounting for social-, temporal and geographic-context.

In our experience the standard approaches applied to date in the literature do not work well when dealing with specialist diets, e.g. vegetarian, vegan or allergies[3]. Constraint-based approaches are found surprisingly rarely in the literature.

One exception is [Yang *et al.* (2017)] who had access to the data basis to apply filters based on vegetarian, vegan and gluten-free food. A further example can be found in the nutritional science literature whereby linear programming is used to ensure Malawian children achieve the required nutritional intake recommended by experts [Ferguson *et al.* (2004)]. As comparable datasets exist (see Section 20.4 below), there is no reason why a similar approach cannot be taken to promote healthy eating patterns in other demographics.

20.3. Addressed Challenges and Problems

As should now be clear the food recommendation task brings additional challenges to those in other recommender systems domains. There are also standard challenges, applicable to all domains, which have been addressed, at least to some extent, in food recommendation research. In this section we first relate the generic challenges and how these have been addressed or not in the food domain, before switching focus to the challenges unique to food recommendation.

User preference sources (see Chapter 7). Food recommendation research has mainly exploited explicit sources of user feedback in the form of ratings [Freyne *et al.* (2011a); Freyne and Berkovsky (2010); Harvey *et al.* (2013)], bookmarks [Trattner and Elsweiler (2017)] or shares [Abbar *et al.* (2015)]. Methods of implicit feedback have been used less often, but examples include recipe views [West *et al.* (2013); Wagner *et al.* (2014)] and the sentiment of reviews submitted about recipes [Trattner and Elsweiler (2017)].

User preference scarcity (see Chapter 8). To our knowledge the problems of scarcity of user feedback, illustrated by the cold-start problem and sparse matrices, has not been directly addressed in the food

[3]We have not published our findings, but we have run several test runs with vegetarian, vegan, and gluten allergy user profiles.

recommendation systems literature. Rather standard solutions, which cope well, such as SVD have been applied [Harvey *et al.* (2013); Trattner and Elsweiler (2017)].

Offline and online evaluation of recommendations (see Chapter 9). To our knowledge, evaluation in the food recommendation domain has been almost offline. Typically, as is explained in more detail below, datasets have been created naturalistically e.g. [Yang *et al.* (2017); Trevisiol *et al.* (2014); Trattner and Elsweiler (2017)] or via user studies [Freyne and Berkovsky (2010); Harvey *et al.* (2013)]. These datasets form the basis of offline evaluations in the form of prediction tasks. Other evaluations have taken the form of user studies, where users test interfaces in a semi-controlled [Ge *et al.* (2015a)] or naturalistic environment [Freyne *et al.* (2011a)]. However, full-online evaluations have to our knowledge not yet been published.

Beyond accuracy (see Chapter 10). Accuracy has been the overwhelming focus of research efforts to date but nevertheless, as described above, it remains a challenge, which in the food domain, has yet to be adequately solved. Accuracy, however, is not the only important aspect to consider when recommending food. Novelty and serendipity are both properties of food recommendations, which users appreciate [Harvey *et al.* (2013)], but to our knowledge, these are yet to be studied. [Elsweiler and Harvey (2015)] did acknowledge the importance of dietary diversity in their meal plan work. Moreover, the preference-healthfulness trade-off bears many similarities to traditional work on novelty and serendipity in that it involves recommending non-preferred items while minimizing the loss in precision. While preliminary research in this direction exists [Elsweiler and Harvey (2015); Trattner and Elsweiler (2017)], there is much work to do in order to understand how to optimize this trade-off appropriately.

Recommendation visualizations and explanations. Methods of visualization and the explanation of recommendations have been, at best, implemented in a superficial way within food recommendation research. Examples include the traffic light system employed by [Trattner and Elsweiler (2017)] and the plan meta-data provided in the demo system presented by [Harvey and Elsweiler (2015)]. [Elahi *et al.* (2014)] provide the best example of explanations for the recommendations offered by their system as can be seen in Figure 20.1. Nevertheless, only superficial evaluations of any of these systems have been published.

Other common challenges. Despite their importance generally to recommendation systems, there is nothing to report from the food domain in terms of significant contributions on the issues of privacy and collaborative recommenders, scalability and distribution of collaborative recommenders or issues of robustness or attacks on food recommenders.

Challenges unique to food recommendation systems. We can see from the numerous challenges yet to be addressed in the food domain, that research in this area is still preliminary. That being said, we wish to acknowledge some domain specific challenges, which have been addressed to some extent. Firstly, as Section 20.2 shows, the challenge of tailoring standard approaches to the problem has been tackled.

There have been efforts to better process and understand the content of items to be recommended. These include normalizing ingredients and ingredient quantities [Müller *et al.* (2012); Kusmierczyk *et al.* (2015)]; understanding the role of context in user decision processes (see Section 20.2.4), and understanding which visual features are helpful in guiding these choices [Yang *et al.* (2017); Elsweiler *et al.* (2017b)].

With respect to health, there have been preliminary efforts to model nutritional aspects of the process [Schäfer *et al.* (2017)], which include user requirements [Gibney *et al.* (2002)], user intake [Straßburg (2010)] and the estimation of portion sizes [Zhang *et al.* (2011)]. Other work has pre-processed recipes to establish the nutritional content either by ingredient matching [Müller *et al.* (2012)] or by visually analyzing food images [Chokr and Elbassuoni (2017)]. Finally, as we described in detail above, progress has been made in incorporating health in the recommendation process either by considering nutrition in item recommendation e.g. [Ge *et al.* (2015a)], generating meal plans [Elsweiler and Harvey (2015)] or via algorithmic nudging [Elsweiler *et al.* (2017b)]. It is unclear, however, which method works most effectively.

20.4. Implementation Resources

In this section we summarize resources that can help in the development of food recommendation systems. We summarize (i) datasets typically used to study food consumption patterns and to evaluate algorithmic approaches, (ii) nutrition and health resources, available to implement health-aware recommendation systems. Finally, frameworks typically employed to build these are described.

20.4.1. *Recipe, Meal plan, Menu and Grocery Store Datasets*

To date research in the food recommendation systems domain typically relies on proprietary and none standardized datasets. This contrasts with domains such as movie recommendation domain, where the well-known MovieLens and Netflix datasets have set a standard. The following list highlights datasets usually employed when it comes to the implementation of recipe, meal plan, grocery and menu recommendation systems.

Recipes. Most of the research for recommending recipes relies on Web resources, e.g., Allrecipes[4] or Food.com[5] which comprise rich item and user profiles. Although these offer an extensive basis for conducting research in that direction, most of the datasets cannot be shared as the terms of services of the sites explicitly forbid it. As such, few publicly accessible datasets comprising recipe and user profiles are available. Researchers must typically develop their own crawlers or seek a license agreement with the platform providers. The Australian government agency CSIRO'S Wellbeing Diet Book[6] has been used by Australian researchers [Freyne and Berkovsky (2010)] and connected researchers in Italy [Elahi *et al.* (2014)], but is not readily available to other researchers. Cookpad[7] and Yummly[8] have both supported academic research by providing licensed access to recipe and profile data, and Yummly also supports broad access to restricted data via a no-cost API. One dataset has recently been made available by the Massachusetts Institute of Technology (MIT)[9] comprising of over 1 million recipes including food images and some meta-data. The dataset is limited, however, in that no user profiles or interactions are available, and as such the dataset may not be suitable for evaluating a recipe recommendation system in an offline scenario. The lack of standard collections restricts the reliability and generalizability of research published to date.

Meal plans and restaurant menus. Meal plan recommender research has typically relied on the same recipe datasets as above. To our knowledge no freely available datasets containing meal plans exist. Yelp[10] has been used as a resources to build and evaluate menu recommendation systems.

---

[4] `http://www.allrecipes.com`
[5] `http://www.food.com`
[6] `https://www.csiro.au/en/Research/Health/CSIRO-diets/CSIRO-Total-Wellbeing-Diet`
[7] `http://www.cookpad.com`
[8] `http://www.yummly.com`
[9] `http://im2recipe.csail.mit.edu`
[10] `http://www.yelp.com`

As with recipe datasets, in order to obtain the data one might need to implement a crawling framework as the terms of services of the site to date omit data sharing

Groceries. In the grocery recommendation scenario, to our knowledge, only one dataset is freely available. This dataset was published via Kaggle[11] and contains 3 millions purchases of users on instacart and comprises limited meta-data (such as grocery name) in respect to the groceries bought out in a basket. Table 20.1 refers to some other datasets but these are not available publicly.

20.4.2. *Nutrition & Health Resources*

When it comes to the implementation of food recommendation systems it is not only beneficial to have open-data datasets comprising of user and item profiles (as discussed earlier), but also other external resources that help in building such a system. For instance, to build a health-aware recipe recommendation system, it is essential to know the nutritional values of food items and to what extent these may be healthy or unhealthy. To estimate nutrition, the typical approach is to map ingredients to standard databases, such as those provided by the USDA[12] (US) or the BLS[13] (Germany). As an example, Table 20.3 provides a partial entry for the ingredient 'apple'[14]. The example is far from being complete, as also 'Minerals', 'Vitamins', 'Lipids' and other macro nutrients can be obtained such as 'Caffeine' are also accessible in the database. One of the challenges typically involved in the matching process is the normalization of the ingredients in a recipe, as different names are often used to express the same entity, such as '100g Parmesan cheese' vs '100g of shredded Parmesan cheese'. The method of processing or cooking may additionally influence the nutritional value. Moreover, units are often not expressed using normalized units of quantity. One recipe may refer to 'one cup of water' whereas another may refer to the same item as '235ml water'. Detailed descriptions of the challenges involved can be found in [Müller *et al.* (2012)]. Standard NLP techniques such as stop-word removal, conjunction splitting, string matching, etc. can be applied to address some of these (see for example [Kusmierczyk *et al.* (2015)]). A more practical means to extract this kind of information is though to

[11]`https://www.kaggle.com/c/instacart-market-basket-analysis`
[12]`https://ndb.nal.usda.gov/ndb`
[13]`https://www.blsdb.de`
[14]`https://ndb.nal.usda.gov/ndb/foods/show/2122`

Table 20.3. Example of an nutrition entry for the query 'apple' in the USDA database.

Nutrition	Unit	Value per 100g
Water	g	85.56
Energy	kcal	52
Protein	g	0.26
Total lipid (fat)	g	0.17
Carbohydrate, by difference	g	13.81
Fiber, total dietary	g	2.4
Sugars, total	g	10.39

for instance employ a Web services such as provided by SPOONACULAR[15], whose API is able to extract ingredient names and amounts in a unified way, which can in some cases be accessed for free for purposes of academic research.

Other resources to identify the nutritional properties of a meal (recipe) are provided by [Müller *et al.* (2012)]. These output the nutritional properties for a given German recipe by utilizing the BLS database. Müller employ a multi-step process, first utilising a rule-based infrastructure before a learning to rank approach to identify the most appropriate database entry for a given ingredient. The framework can be obtained from the authors without cost but a license for the BLS is required to use the software. The EDAMAM[16] Web service offers similar functionality for English and Spanish recipes. This service is a commercial product, but as with Spoonacular, can in some cases used without cost for academic purposes.

To estimate the healthiness of a meal [Trattner *et al.* (2017a)], one may rely on standards as set by nutrition scientists. There are many of such standards for different countries and other geographical regions. The ones which have been successfully applied to the food recommendation problem (see [Trattner and Elsweiler (2017); Elsweiler *et al.* (2017b)]) are provided by the Food Standard Agency (FSA) [FSA (2016)] and the World Health Organization (WHO) [WHO (2003)]. Both provide tables based on a 2000kcal diet that contain ranges of nutrients, such as for example Fat, Saturated Fat, Sugar and Sodium (see Table 20.4 and Table 20.5). The WHO guidelines account for macronutrients, such Fiber content, and so on. The FSA guidelines are typically used to derive front of package labels for meals and other food products sold in UK. In addition to the nutrients per portion or per 100g, a traffic light system (red, amber, green) is used

[15]`https://market.mashape.com/spoonacular`
[16]`https://www.edamam.com`

Table 20.4. FSA front of package guidelines as proposed in [FSA (2016)] and as, for example, used in [Trattner and Elsweiler (2017)].

Text	**LOW**	**MEDIUM**	**HIGH**
Color code	**Green**	**Amber**	**Red**
Fat	$\leq$ 3.0g/100g	$>$ 3.0g to $\leq$ 17.5g/100g	$>$ 17.5g/100g or $>$ 21g/portion
Saturates	$\leq$ 1.5g/100g	$>$ 1.5g to $\leq$ 5.0g/100g	$>$ 5.0g/100g or $>$ 6.0g/portion
Sugars	$\leq$ 5.0g/100g	$>$ 5.0g to $\leq$ 22.5g/100g	$>$ 22.5g/100g or $>$ 27g/portion
Salt	$\leq$ 0.3g/100g	$>$ 0.3g to $\leq$ 1.5g/100g	$>$ 1.5g/100g or $>$ 1.8g/portion

Table 20.5. WHO guidelines as originally proposed in [WHO (2003)] and adopted to recipes by [Howard *et al.* (2012)] and as, for example, used in [Trattner and Elsweiler (2017)].

Dietary Factor	**Range (percentage of kcal per meal/recipe)**
Protein	10-15
Carbohydrates	55-75
Sugar	$<$ 10
Fat	15-30
Saturated Fat	$<$ 10
Fiber density (g/MJ)	$> 3.0^{\dagger}$
Sodium density (g/MJ)	$< 0.2^{\ddagger}$

†Based on 8.4 MJ/day (2,000 kcal/day) diet and recommended daily fiber intake of >25g.
‡Based on 8.4 MJ/day (2,000 kcal/day) diet and recommended daily sodium intake of <2g.

to inform the consumer, whether the meal is healthy (green) or unhealthy (red) with respect to a given property. We employed these guidelines in Table 20.2. As the FSA scoring system is rather unpractical to use in a recommender scenario, one might want to use a single metric by following the procedure proposed by [Sacks *et al.* (2009)] who first assign an integer value to each color (green=1, amber=2 and red=3) then sum the scores for each macro nutrient, resulting in a final range from 4 (very healthy) to 12 (very unhealthy). A further health index, which may offer utility is the 'Healthy Eating Index' [HEI (2016)] proposed by the USDA. The index was developed to target the US population. To date it has not been applied in any food recommendation systems project.

Other useful resources for building food recommendation systems are provided by FOODSUBS[17], a food thesaurus service which can suggest food substitutes. This might be helpful to implement food recommendation systems promoting healthier eating (see [Achananuparp and Weber (2016)]) by replacing unhealthy ingredients in a meal with more healthy variants, but also assist people with allergies or intolerances.

[17] `http://www.foodsubs.com`

Food word lists, such as provided by ENCHANTEDLEARNING[18] and WIKIPEDIA[19] provides a rich knowledge base relating to food and cooking and may be used to assist with the normalization process of ingredients.

Finally, one may also employ health data as provided by the Centers for Disease Control and Prevention (CDC) in the US. The reports contain state and county data of diabetes and obesity. As different regions have different impact on what and how people eat (see [Trattner *et al.* (2017b)]), this might be a useful source of information when implementing food recommendation systems for different regions and areas in the US [Said and Bellogín (2014)].

20.4.3. *Food recommendation System Frameworks*

To date, research in the food recommendation systems domain relies mostly on software custom built by researchers themselves explicitly for the purpose of their research. To the best of our knowledge, there is no food recommendation systems framework available that has been shared by the research community or on open-access platforms, such as GITHUB[20]. This makes it challenging not only to progress the research in that area, but also to reproduce or validate findings published already. To counter this trend, in our own research, we have recently started to use publicly available frameworks, such as the well-known LibRec library. The framework is implemented in the Java programming language and comprises a relatively complete set of standard recommendation systems algorithms, such as UserKNN, ItemKNN, BPR, SVD++, and so on, to tackle the rating prediction and item ranking problem. In [Trattner and Elsweiler (2017)] we adopted the framework with pre- and post-filtering functions (as described in the previous section) to re-rank items (in our case) recipes in terms of their healthiness. We are happy to share this code upon request. The framework can also be easily extended to the problem of recommending, e.g., recipes to a group of people as well as generating personalized meal plans. Other examples of frameworks in other programming languages may be found on Graham Jenson's Github page[21] as well as on the Rec.Sys Wiki[22].

[18] `http://www.enchantedlearning.com/wordlist/food.shtml`
[19] `https://en.wikipedia.org/wiki/Lists_of_foods`
[20] `http://www.github.com`
[21] `https://github.com/grahamjenson/list_of_recommender_systems`
[22] `http://www.recsyswiki.com/wiki/Recommendation_Software`

20.5. Historical Evolution and Versions of the System

The earliest examples of food recommendation systems were proposed by the case-based reasoning (CBR) community [Hammond (1986); Hinrichs (1989)]. In contrast to current state-of-the-art food recommendation approaches both employed planning algorithms taking a set of queries e.g. groceries as input to generate meal plans or a single new recipe. Technically speaking these systems bear little relation to modern systems. Later, systems emerged employing simple variants of today's well-known content-based and collaborative filtering recommendation algorithms. Examples include, for instance the works of [Lawrence *et al.* (2001); Mankoff *et al.* (2002); Aberg (2006)].

The first food recommenders built which are directly comparable to modern systems, i.e. which employ standard algorithms such as UserKNN was presented in [Freyne and Berkovsky (2010); Berkovsky and Freyne (2010)]. These were the first examples where recipe datasets were used as a basis and the system was reliably evaluated. Subsequently other works emerged employing more advanced techniques to recommend food to people. Examples include the work of [van Pinxteren *et al.* (2011)], which was the first to derive a similarity metric for recipes to be used for recommending healthful meals; [Ueta *et al.* (2011)] and [El-Dosuky *et al.* (2012)], which employ knowledge-based food recommendation approaches; and [Kuo *et al.* (2012)] which employs tags to derive a knowledge graph to connect recipes and exploit this graph for recommending menus.

Other break through work was performed by [Teng *et al.* (2012)], who proposed the use of ingredient networks to produce recommendations or the work of [Harvey *et al.* (2013)], who proposed a model accounting for food selection biases.

A significant break-through was recently made by [Yang *et al.* (2017)] who were able to develop a constraint-based (with different types of diets) mobile food recommendation system exploring food images to learn about user food preferences. All previous approaches had relied on ratings or to some extent on tags [Ge *et al.* (2015a)].

Behavior-based investigations, which go beyond the classic food recommendation systems papers can also be considered to have progressed the field. We include our own work showing that people typically prefer the unhealthy recipes in this bracket [Trattner and Elsweiler (2017)]. This was the first study in the context that deals with the health-aware recipe recommendation systems problem. Other work in this direction include [Trattner *et al.* (2017c)] (not shown in Table 20.1) and [Rokicki *et al.* (2016)] which illustrate differences in online food consumption with respect to hobbies and gender.

Finally, we would like to highlight our most recent work [Elsweiler *et al.* (2017b)] which investigated to which extent food recommendation systems can nudge people towards healthier food choices.

20.6. Evaluation: Metrics and Methodologies

The methods of evaluation applied to food recommendation systems have evolved over time. The early concept papers found in the literature do not employ any kind of evaluation [Hammond (1986); Hinrichs and Kolodner (1991)]. With the work of [Freyne and Berkovsky (2010)] researchers started to employ evaluation techniques recognized by the community today as standard practices [Herlocker *et al.* (2004); Ricci *et al.* (2011)].

The most commonly taken approach (as can be seen in the summarized literature in Table 20.1) is to perform simulations using historical data (see Section 20.4). The experimental design specifics vary, but typically datasets are split into training and testing subsets to mimic user-profiles and feedback given for recommendations. Similar to other recommender domains, historical datasets are typically split such that 80% of the data is used for training with the remaining 20% held-out for testing. Alternatives are to use k-fold validation [Harvey *et al.* (2013); Trattner and Elsweiler (2017)] or leave-one out protocol [Freyne and Berkovsky (2010)]. The exact means by which collections are sourced varies from using naturalistic collections crawled from the web [Trattner and Elsweiler (2017)] or from donated sources [Trevisiol *et al.* (2014)] to running user studies to collect small sets of data [Harvey *et al.* (2013)].

Different metrics have been applied to measure the performance of algorithms in such systems. These typically reflect the error in the predicted ratings [Freyne and Berkovsky (2010); Harvey *et al.* (2013)] e.g. Mean Absolute Error (MAE) or Root Spare Mean Error (RSME) or the quality of the top-n ranked list of items e.g. Recall, Precision, Mean Average Precision (MAP) and Normalized Discounted Cumulative Gain (NDCG).

Mirroring the developments in the recommendation systems community generally, earlier contributions focused on the rating prediction task whereas more recent and current work treats recommendation as a ranking problem (e.g., [Trattner and Elsweiler (2017); Yang *et al.* (2017); Cheng *et al.* (2017)]).

Assessing the accuracy of recommendation is typically not enough for recommendation systems and in food recommenders is no exception. Diversity of ingredients used in profiles was measured using Simpson and simple diversity metrics [Elsweiler and Harvey (2015)].

Incorporating health-aspects in the process requires additional metrics to be defined. As our own work shows, see [Trattner and Elsweiler (2017)], metrics derived from the guidelines published by governmental bodies or health organizations are appropriate (see again Section 20.4.2).

In addition to calculating a mean over all food items recommended on per user basis (see Table 20.2), we additionally introduced two further measures referred to as ΔFSA and ΔWHO, which capture the difference in healthfulness between test set items of the users and actual predicted items, as shown in the formulae below

$$\Delta WHO = \frac{1}{|U|}\sum_{u=1}^{|U|}\left(\frac{1}{|Train_u|}\sum_{i=1}^{|Train_u|} who_i - \frac{1}{|Pred_u|}\sum_{j=1}^{|Pred_u|} who_j\right) \tag{20.4}$$

$$\Delta FSA = \frac{1}{|U|}\sum_{u=1}^{|U|}\left(\frac{1}{|Train_u|}\sum_{i=1}^{|Train_u|} fsa_i - \frac{1}{|Pred_u|}\sum_{j=1}^{|Pred_u|} fsa_j\right), \tag{20.5}$$

where $|U|$ denotes the total number of users in the dataset, $|Train_u|$ the size of the train set for user u respectively, $|Pred_u|$ the size of the set for the predicted items and who_i, who_j and fsa_i, fsa_j represent the WHO, FSA health scores for items (i and j) in these sets.

These delta measures are useful as they capture whether the recommended items are more or less healthy than those already rated positively by the user. The same procedure can also be applied to calculate a delta between the test and prediction sets to observe whether the recommended items are actually more or less healthy to what the user would actually eat in the future.

Similar to other recommendation domains, studies employing online evaluation protocols, such as A/B testing or laboratory studies for the purpose of testing the performance of food recommendation systems are rare. Among the studies to employ online testing is for instance the work of [Freyne *et al.* (2011a)] who ran two types of meal planners in a live system. The two methods tested were a personalized and a non-personalized algorithm. Over the course of 12 weeks over 5000 users participated in the study. According to the authors an A/B like setup was chosen to refer half of the users to the personalized condition and half of the user to the non-personalized one. Earlier work from the same authors employed also some variant of online experiment to gather ratings from users on recipes by e.g. using Amazon's Mechanical Turk platform [Freyne *et al.* (2011b); Freyne and Berkovsky (2010)]. However, rather than generating the recommendations on the fly to test their validity, in the end, an offline protocol was utilized. A recent study by [Yang *et al.* (2017)] employed not only offline testing but also an online study protocol to evaluate a mobile food recommendation system. In particular they recruited 60 participants through the university mailing list, Facebook, and Twitter. The study, conducted as a online Web service, consisted of three phases. First, each participant was questioned on any dietary restrictions that may apply, such as the need to avoid gluten. Second, each user was asked to express their preferences by highlighting images of food they find appealing. Lastly, 20 meal recommendations were generated of which 10 were shown in a random order and 10 as proposed by the authors' "Yum-me" algorithm. The participants had the task of classifying the 20 recipes as to whether it is appealing or not.

A final work worthy of mention is an online study that has been recently conducted by the authors with the goal of investigating the potential to nudge people towards healthier food choices via recommendations [Elsweiler *et al.* (2017b)]. The work employed three online studies. Similar to the previously mentioned work we implemented a Web service and recruited between 107 and 138 participants per study. By varying the amount of information shown about two algorithmically determined similar recipes, we were able to learn about the choices people make, the users' perception of these recipes and what influenced these. By applying machine learning approaches we were able to predict with relative certainty, which recipe of the two participants would prefer and demonstrate that the models developed can be used to influence the choices made.

None of evaluation strategies applied to date in food recommendation accounts for one of the primary differences of recommending items in this domain. Presented food items are not the actual item that should be evaluated because varying the cook, the ingredients and any number of other contextual features will certainly influence the experience. It is not yet clear how to best account for this challenge.

In summary, no specialized offline protocols exist for the evaluation food recommendation systems. Typically standard metrics are used to determine prediction accuracy and diversity. Furthermore, no standardized or specialized online evaluation protocols exist for food recommendation systems. Current approaches rely on methods that have been previously developed in other recommendation domains such as movies or music (see also Chapter 9). Exceptions are the metrics specifically designed to incorporate healthy nutrition into the process, such as the WHO and FSA scores in [Trattner and Elsweiler (2017)].

20.7. Lessons Learned and Future Directions

Food recommendation is an important domain both for individuals and society. What the work described in this chapter shows is that despite its importance, food item recommendation, in comparison to other domains is relatively under-researched. The work that has been performed to date shows that although user taste predictions for food can be achieved with existing methods, the performance achieved is poorer than in other domains.

This means that preference learning should remain a focus for the food domain because experiments described in the literature have shown that even regardless of the source of user feedback applied (i.e. ratings, tags or comments) standard methods are only capable of producing relatively unsatisfactory performance. It is clear that new methods are required for the food domain and some work has shown promise. Yang and colleague's (2017) work uses images and embeddings (DNNs) to learn user preferences and the results are very promising.

Other key findings in the literature relating to preference prediction are those illustrating the importance of context variables. One promising research direction would be to capture important context variables via different sensors and incorporate these into recommendation models. For example, one could imagine using activity sensors, such as Fitbits, which have become popular to influence recommendations. Do people choose calorie richer food when they have done more exercise?

Relating to context, social situations and recommendation for groups needs to be considered more concretely. The pervasiveness of social culinary experiences and how these influence food choices need to be considered by technological systems.

One particular task in food recommendation systems which, for societal and socio-economic reasons, has become a hot research focus is food recommenders for nutritional health. Researchers have proposed diverse methods of incorporating nutrition (nutritional components in algorithm, meal plans, and nudging), but to date all of these proposals remain preliminary and it is not yet clear, which is the best approach to take.

Specialist diets or users with strong constraints (e.g. vegetarians, vegans or those with food intolerance), have largely been omitted from studies to date. This is challenging both in the case of recommending single food items to individual users, for which we expect standard approaches to offer poor performance and in the case of group recommendations when one member of the group has strong constraints. Should systems attempt to maximize the satisfaction of certain members or minimize the misery of the group on average?

One further aspect which needs to develop in the community is the evaluation of food recommenders and the methods employed to do so. In the literature evaluation has mostly been offline with proprietary collections. As a community we need to work together to achieve standard data collections, standard base-line approaches and importantly, more online studies to understand how our approaches work as live systems used in naturalistic scenarios. We reported the problem in food recommendation that presented items will vary from experience due to factors such as the cooking skills of the user and available ingredients etc. Perhaps using a living-lab setting e.g. [Balog *et al.* (2014)] is one approach that could be tested to address this problem.

As a final note, we emphasized in the the reviewed literature that most of the food recommendation literature has focused on recipe recommendation. Now that the new datasets, such as the Instacart dataset, are being made available to researchers, we hope to see this situation change.

References

Abbar, S., Mejova, Y. and Weber, I. (2015). You tweet what you eat: Studying food consumption through twitter, in *Proceedings of the 33rd Annual ACM Conference on Human Factors in Computing Systems*, pp. 3197–3206.

Aberg, J. (2006). Dealing with malnutrition: A meal planning system for elderly, in *AAAI Spring Symposium: Argumentation for Consumers of Healthcare*, pp. 1–7.

Achananuparp, P. and Weber, I. (2016). Extracting food substitutes from food diary via distributional similarity, *CoRR* **abs/1607.08807**, `http://arxiv.org/abs/1607.08807`.

Balog, K., Elsweiler, D., Kanoulas, E., Kelly, L. and Smucker, M. D. (2014). Report on the cikm workshop on living labs for information retrieval evaluation, in *ACM SIGIR Forum*, Vol. 48 (ACM), pp. 21–28.

Beaglehole, R. (2016). Misunderstaning vs reality, `http://www.who.int/chp/advocacy/MediaFeatures_EN_web.pdf`.

Berkovsky, S. and Freyne, J. (2010). Group-based recipe recommendations: analysis of data aggregation strategies, in *Proceedings of the fourth ACM conference on Recommender systems* (ACM), pp. 111–118.

Burke, R. (2002). Hybrid recommender systems: Survey and experiments, *User modeling and user-adapted interaction* **12**, 4, pp. 331–370.

Cheng, H., Rokicki, M. and Herder, E. (2017). The influence of city size on dietary choices and food recommendation, in *Proceedings of the 25th Conference on User Modeling, Adaptation and Personalization*, UMAP '17 (ACM, New York, NY, USA), ISBN 978-1-4503-4635-1, pp. 359–360, doi:10.1145/3079628.3079641, `http://doi.acm.org/10.1145/3079628.3079641`.

Chokr, M. and Elbassuoni, S. (2017). Calories prediction from food images, in *AAAI*, pp. 4664–4669.

De Choudhury, M., Sharma, S. and Kiciman, E. (2016). Characterizing dietary choices, nutrition, and language in food deserts via social media, in *Proceedings of the 19th ACM Conference on Computer-Supported Cooperative Work & Social Computing*, CSCW '16 (ACM, New York, NY, USA), ISBN 978-1-4503-3592-8, pp. 1157–1170, doi:10.1145/2818048.2819956, `http://doi.acm.org/10.1145/2818048.2819956`.

El-Dosuky, M., Rashad, M., Hamza, T. and El-Bassiouny, A. (2012). Food recommendation using ontology and heuristics, in *International Conference on Advanced Machine Learning Technologies and Applications* (Springer), pp. 423–429.

Elahi, M., Ge, M., Ricci, F., Massimo, D. and Berkovsky, S. (2014). Interactive food recommendation for groups, in *RecSys Posters*.

Elsweiler, D. and Harvey, M. (2015). Towards automatic meal plan recommendations for balanced nutrition, in *Proceedings of the 9th ACM Conference on Recommender Systems*, RecSys '15 (ACM, New York, NY, USA), ISBN 978-1-4503-3692-5, pp. 313–316, doi:10.1145/2792838.2799665, `http://doi.acm.org/10.1145/2792838.2799665`.

Elsweiler, D., Harvey, M., Ludwig, B. and Said, A. (2015). Bringing the "healthy" into food recommenders, in *DMRS*, pp. 33–36.

Elsweiler, D., Hors-Fraile, S., Ludwig, B., Said, A., Schäfer, H., Trattner, C., Torkamaan, H. and Valdez, A. C. (2017a). Second workshop on health recommender systems: (healthrecsys 2017), in *Proceedings of the Eleventh ACM Conference on Recommender Systems, RecSys 2017, Como, Italy, August 27-31, 2017*, pp. 374–375, doi:10.1145/3109859.3109955, `http://doi.acm.org/10.1145/3109859.3109955`.

Elsweiler, D., Ludwig, B., Said, A., Schäfer, H. and Trattner, C. (2016). Engendering health with recommender systems, in *Proceedings of the 10th ACM Conference on Recommender Systems, Boston, MA, USA, September 15-19, 2016*, pp. 409–410, doi:10.1145/2959100.2959203, `http://doi.acm.org/10.1145/2959100.2959203`.

Elsweiler, D., Trattner, C. and Harvey, M. (2017b). Exploiting food choice biases for healthier recipe recommendation, in *Proceedings of the 40th International ACM SIGIR Conference on Research and Development in Information Retrieval*, SIGIR '17 (ACM, New York, NY, USA), ISBN 978-1-4503-

5022-8, pp. 575–584, doi:10.1145/3077136.3080826, http://doi.acm.org/10.1145/3077136.3080826.

Ferguson, E. L., Darmon, N., Briend, A. and Premachandra, I. M. (2004). Food-based dietary guidelines can be developed and tested using linear programming analysis, *The Journal of nutrition* **134**, 4, pp. 951–957.

Freyne, J. and Berkovsky, S. (2010). Intelligent food planning: Personalized recipe recommendation, in *Proceedings of the 15th International Conference on Intelligent User Interfaces*, IUI '10 (ACM, New York, NY, USA), ISBN 978-1-60558-515-4, pp. 321–324, doi:10.1145/1719970.1720021, http://doi.acm.org/10.1145/1719970.1720021.

Freyne, J., Berkovsky, S., Baghaei, N., Kimani, S. and Smith, G. (2011a). Personalized techniques for lifestyle change, *Artificial Intelligence in Medicine*, pp. 139–148.

Freyne, J., Berkovsky, S. and Smith, G. (2011b). Recipe recommendation: Accuracy and reasoning, in *International Conference on User Modeling, Adaptation, and Personalization*, pp. 99–110.

FSA (2016). Guide to creating a front of pack (fop) nutrition label for pre-packed products sold through retail outlets. Available at https://www.food.gov.uk/sites/default/files/multimedia/pdfs/pdf-ni/fop-guidance.pdf last accessed on 20.6.2016.

Ge, M., Elahi, M., Fernaández-Tobías, I., Ricci, F. and Massimo, D. (2015a). Using tags and latent factors in a food recommender system, in *Proceedings of the 5th International Conference on Digital Health 2015*, DH '15 (ACM, New York, NY, USA), ISBN 978-1-4503-3492-1, pp. 105–112, doi:10.1145/2750511.2750528, http://doi.acm.org/10.1145/2750511.2750528.

Ge, M., Ricci, F. and Massimo, D. (2015b). Health-aware food recommender system, in *Proceedings of the 9th ACM Conference on Recommender Systems*, RecSys '15 (ACM, New York, NY, USA), ISBN 978-1-4503-3692-5, pp. 333–334, doi:10.1145/2792838.2796554, http://doi.acm.org/10.1145/2792838.2796554.

Gibney, M., Vorster, H. and Kok, F. (2002). *Introduction to human nutrition*.

Griffiths, T. (2002). Gibbs sampling in the generative model of latent dirichlet allocation, pp. x–y.

Hammond, K. J. (1986). Chef: A model of case-based planning, in *Proceedings of AAAI*, pp. 267–271.

Harvey, M. and Elsweiler, D. (2015). Automated recommendation of healthy, personalised meal plans, in *Proceedings of the 9th ACM Conference on Recommender Systems* (ACM), pp. 327–328.

Harvey, M., Ludwig, B. and Elsweiler, D. (2012). Learning user tastes: A first step to generating healthy meal plans, in *First international workshop on recommendation technologies for lifestyle change (lifestyle 2012)*, p. 18.

Harvey, M., Ludwig, B. and Elsweiler, D. (2013). You are what you eat: Learning user tastes for rating prediction, in *Proceedings of the 20th International Symposium on String Processing and Information Retrieval - Volume 8214*, SPIRE 2013 (Springer-Verlag New York, Inc., New York, NY, USA), ISBN 978-3-319-02431-8, pp. 153–164, doi:10.1007/978-3-319-02432-5_19, http://dx.doi.org/10.1007/978-3-319-02432-5_19.

HEI (2016). Healthy eating index, https://www.cnpp.usda.gov/healthyeatingindex.

Herlocker, J. L., Konstan, J. A., Terveen, L. G. and Riedl, J. T. (2004). Evaluating collaborative filtering recommender systems, *ACM Transactions on Information Systems (TOIS)* **22**, 1, pp. 5–53.

Hinrichs, T. R. (1989). Strategies for adaptation and recovery in a design problem solver, in *Proceedings of the 2nd Workshop on Case-Based Reasoning*, pp. 115–118.

Hinrichs, T. R. and Kolodner, J. L. (1991). The roles of adaptation in case-based design, in *AAAI*, Vol. 91, pp. 28–33.

Howard, S., Adams, J. and White, M. (2012). Nutritional content of supermarket ready meals and recipes by television chefs in the United Kingdom: cross sectional study, *BMJ* **345**, p. e7607.

Hu, Y., Koren, Y. and Volinsky, C. (2008). Collaborative filtering for implicit feedback datasets, in *Proc. of ICDM'08* (IEEE), pp. 263–272.

Khan, A. S. and Hoffmann, A. (2003). Building a case-based diet recommendation system without a knowledge engineer, *Artificial Intelligence in Medicine* **27**, 2, pp. 155–179.

Kuo, F.-F., Li, C.-T., Shan, M.-K. and Lee, S.-Y. (2012). Intelligent menu planning: Recommending set of recipes by ingredients, in *Proceedings of the ACM multimedia 2012 workshop on Multimedia for cooking and eating activities* (ACM), pp. 1–6.

Kusmierczyk, T., Trattner, C. and Nørvåg, K. (2015). Temporal patterns in online food innovation, in *Proceedings of the 24th International Conference on World Wide Web* (ACM), pp. 1345–1350.

Kusmierczyk, T., Trattner, C. and Nørvåg, K. (2016). Understanding and predicting online food recipe production patterns, in *Proceedings of the 27th ACM Conference on Hypertext and Social Media*, HT '16, ISBN 978-1-4503-4247-6, pp. 243–248.

Lawrence, R. D., Almasi, G. S., Kotlyar, V., Viveros, M. and Duri, S. S. (2001). Personalization of supermarket product recommendations, in *Applications of Data Mining to Electronic Commerce* (Springer), pp. 11–32.

Mankoff, J., Hsieh, G., Hung, H. C., Lee, S. and Nitao, E. (2002). Using low-cost sensing to support nutritional awareness, in *International Conference on Ubiquitous Computing* (Springer), pp. 371–378.

Mormann, M. M., Navalpakkam, V., Koch, C. and Rangel, A. (2012). Relative visual saliency differences induce sizable bias in consumer choice.

Müller, M., Mika, S., Harvey, M. and Elsweiler, D. (2012). Estimating nutrition values for internet recipes, in *Pervasive Computing Technologies for Healthcare (PervasiveHealth), 2012 6th International Conference on* (IEEE), pp. 191–192.

Ornish, D., Brown, S., Billings, J., Scherwitz, L., Armstrong, W., Ports, T., McLanahan, S., Kirkeeide, R., Gould, K. and Brand, R. (1990). Can lifestyle changes reverse coronary heart disease?: The lifestyle heart trial, *The Lancet* **336**, 8708, pp. 129–133, http://www.sciencedirect.com/science/article/pii/014067369091656U.

Park, M.-H., Park, H.-S. and Cho, S.-B. (2008). Restaurant recommendation for group of people in mobile environments using probabilistic multi-criteria decision making, in *Computer-Human Interaction* (Springer), pp. 114–122.

Ricci, F., Rokach, L. and Shapira, B. (2011). *Introduction to recommender systems handbook* (Springer).

Rokicki, M., Herder, E., Kuśmierczyk, T. and Trattner, C. (2016). Plate and prejudice: Gender differences in online cooking, in *Proceedings of the 2016 Conference on User Modeling Adaptation and Personalization*, UMAP '16 (ACM, New York, NY, USA), ISBN 978-1-4503-4368-8, pp. 207–215, doi: 10.1145/2930238.2930248.

Sacks, G., Rayner, M. and Swinburn, B. (2009). Impact of front-of-pack traffic-lightnutrition labelling on consumer food purchases in the UK, *Health promotion international* **24**, 4, pp. 344–352.

Said, A. and Bellogín, A. (2014). You are what you eat! tracking health through recipe interactions, in *RSWeb@ RecSys*.

Sano, N., Machino, N., Yada, K. and Suzuki, T. (2015). Recommendation system for grocery store considering data sparsity, *Procedia Computer Science* **60**, pp. 1406–1413.

Schäfer, H., Elahi, M., Elsweiler, D., Groh, G., Harvey, M., Ludwig, B., Ricci, F. and Said, A. (2017). User nutrition modelling and recommendation: Balancing simplicity and complexity, in *Adjunct Publication of the 25th Conference on User Modeling, Adaptation and Personalization* (ACM), pp. 93–96.

Schäfer, H., Hors-Fraile, S., Karumur, R. P., Calero Valdez, A., Said, A., Torkamaan, H., Ulmer, T. and Trattner, C. (2017). Towards health (aware) recommender systems, in *Proceedings of the 2017 International Conference on Digital Health*, DH '17 (ACM, New York, NY, USA), ISBN 978-1-4503-5249-9, pp. 157–161, doi:10.1145/3079452.3079499, `http://doi.acm.org/10.1145/3079452.3079499`.

Scheibehenne, B., Greifeneder, R. and Todd, P. M. (2010). Can there ever be too many options? a meta-analytic review of choice overload, *Journal of Consumer Research* **37**, 3, pp. 409–425.

Schur, E., Kleinhans, N., Goldberg, J., Buchwald, D., Schwartz, M. and Maravilla, K. (2009). Activation in brain energy regulation and reward centers by food cues varies with choice of visual stimulus, *International journal of obesity (2005)* **33**, 6, p. 653.

Straßburg, A. (2010). Ernährungserhebungen - methoden und instrumente, *Ernährungs Umschau*.

Teng, C.-Y., Lin, Y.-R. and Adamic, L. A. (2012). Recipe recommendation using ingredient networks, in *Proceedings of the 4th Annual ACM Web Science Conference* (ACM), pp. 298–307.

Trattner, C. and Elsweiler, D. (2017). Investigating the healthiness of internet-sourced recipes: implications for meal planning and recommender systems, in *Proceedings of the 26th International Conference on World Wide Web* (International World Wide Web Conferences Steering Committee), pp. 489–498.

Trattner, C., Elsweiler, D. and Howard, S. (2017a). Estimating the healthiness of internet recipes: A cross sectional study, *Frontiers in Public Health* **5**, p. 16, doi:10.3389/fpubh.2017.00016, http://journal.frontiersin.org/article/10.3389/fpubh.2017.00016.

Trattner, C., Parra, D. and Elsweiler, D. (2017b). Monitoring obesity prevalence in the united states through bookmarking activities in online food portals, *PloS one* **12**, 6, p. e0179144.

Trattner, C., Rokicki, M. and Herder, E. (2017c). On the relations between cooking interests, hobbies and nutritional values of online recipes: Implications for health-aware recipe recommender systems, in *Adjunct Publication of the 25th Conference on User Modeling, Adaptation and Personalization*, UMAP '17 (ACM, New York, NY, USA), ISBN 978-1-4503-5067-9, pp. 59–64, doi:10.1145/3099023.3099072, http://doi.acm.org/10.1145/3099023.3099072.

Trevisiol, M., Chiarandini, L. and Baeza-Yates, R. (2014). Buon appetito: Recommending personalized menus, in *Proceedings of the 25th ACM Conference on Hypertext and Social Media*, HT '14 (ACM, New York, NY, USA), ISBN 978-1-4503-2954-5, pp. 327–329, doi:10.1145/2631775.2631784, http://doi.acm.org/10.1145/2631775.2631784.

Ueta, T., Iwakami, M. and Ito, T. (2011). A recipe recommendation system based on automatic nutrition information extraction, in *Proceedings of the 5th International Conference on Knowledge Science, Engineering and Management*, KSEM'11 (Springer-Verlag, Berlin, Heidelberg), ISBN 978-3-642-25974-6, pp. 79–90, doi:10.1007/978-3-642-25975-3_8, http://dx.doi.org/10.1007/978-3-642-25975-3_8.

van Pinxteren, Y., Geleijnse, G. and Kamsteeg, P. (2011). Deriving a recipe similarity measure for recommending healthful meals, in *Proceedings of the 16th International Conference on Intelligent User Interfaces*, IUI '11 (ACM, New York, NY, USA), ISBN 978-1-4503-0419-1, pp. 105–114, doi:10.1145/1943403.1943422, http://doi.acm.org/10.1145/1943403.1943422.

Wagner, C., Singer, P. and Strohmaier, M. (2014). The nature and evolution of online food preferences, *EPJ Data Science* **3**, 1, p. 1.

Wansink, B. (2006). *Mindless eating* (Bantam Books).

West, R., White, R. W. and Horvitz, E. (2013). From cookies to cooks: Insights on dietary patterns via analysis of web usage logs, in *Proceedings of the 22nd international conference on World Wide Web*, pp. 1399–1410.

WHO (2003). Diet, nutrition and the prevention of chronic diseases, *World Health Organ Tech Rep Ser* **916**, i-viii.

Yang, L., Cui, Y., Zhang, F., Pollak, J. P., Belongie, S. and Estrin, D. (2015). Plateclick: Bootstrapping food preferences through an adaptive visual interface, in *Proceedings of the 24th ACM International on Conference on Information and Knowledge Management* (ACM), pp. 183–192.

Yang, L., Hsieh, C.-K., Yang, H., Pollak, J. P., Dell, N., Belongie, S., Cole, C. and Estrin, D. (2017). Yum-me: A personalized nutrient-based meal recommender system, *ACM Transactions on Information Systems (TOIS)* **36**, 1, p. 7.

Zhang, Z., Yang, Y., Yue, Y., Fernstrom, J., Jia, W. and Sun, M. (2011). Food volume estimation from a single image using virtual reality technology, in *Bioengineering Conference (NEBEC), 2011 IEEE 37th Annual Northeast.*

Zhu, Y.-X., Huang, J., Zhang, Z.-K., Zhang, Q.-M., Zhou, T. and Ahn, Y.-Y. (2013). Geography and similarity of regional cuisines in China, *PloS one* **8**, 11, p. e79161.

Chapter 21

Clothing Recommendations: The Zalando Case

Antonino Freno
Zalando SE
Mühlenstr. 25, Berlin 10243, Germany
antonino.freno@zalando.de

21.1. Introduction

Developing a *real-world* recommender system, i.e. for use in large-scale online retail, poses a number of different challenges. Interestingly, only a small part of these challenges are of algorithmic nature, such as how to select the most accurate model for a given use case. Instead, most technical problems usually arise from *operational constraints*, such as: adaptation to novel use cases; cost and complexity of system maintenance; capability of reusing pre-existing signal and integrating heterogeneous data sources. In this chapter, we describe the system we developed in order to address those constraints at Zalando, which is one of the most popular online fashion retailers in Europe. In particular, we explain how moving from a collaborative filtering approach to a learning-to-rank model helped us to effectively tackle the challenges mentioned above, while improving at the same time the quality of our recommendations. A fairly detailed description of our software architecture is also provided, along with an overview of the algorithmic approach. On the other hand, we illustrate the typical workflow through which different versions of our recommender system are experimentally evaluated, by means of both offline and online assessment protocols.

Zalando (`http://www.zalando.com/`) provides a unique hub for digital fashion content in Europe. In the online shop, machine-learned product recommendations are disseminated over many different contexts, ranging from general navigation aid on the homepage to personalized sorting in the catalog and item-to-item recommendations on product detail pages. In order to keep up with an ever-changing market and the constantly growing customer needs, algorithmic progress has always been a crucial requirement for our recommendation services. Yet, when viewed from the trenches of online retail industry, recommender systems development assumes a significantly different shape than we are used to think from an academic perspective.

From a scientific point of view, research on machine-learned recommender systems tries to identify *optimal* recommendation algorithms. Here, optimality is typically defined in terms of *maximum achievable accuracy* (possibly weighted by computational efficiency) with respect to specific recommendation goals, settings, and constraints. Accuracy is measured by a number of different metrics, which formally measure the quality of recommendations in terms of the ranking they induce on a set of candidate items. However, real-world recommender systems, such as those powering all major e-commerce platforms, have to face a sensibly different question, i.e. how to *maximize business value* in the long term. In this context, the notion of business value has at least two different sides. On the one hand, we want to generate value for our customers, e.g. in terms of user engagement, retention, conversion rate, revenue, and related performance indicators. On the other hand, we want to generate internal value, in terms of the long-term profitability of our technology assets. Here, profitability means *operational excellence*, on at least three different dimensions: (*i*) possibility of adapting the recommender system to novel use cases; (*ii*) cost and complexity of the involved maintenance; (*iii*) capability of capitalizing on pre-existing signal and effectively integrating newly available data sources. In this sense, an optimal recommender system is one that achieves operational excellence along at least these three dimensions, while also generating value for customers.

Back in the past, most of the recommendation contexts on Zalando were powered by machine learning models based on collaborative filtering (CF). As we tried to improve on these models and replace them by more sophisticated approaches, we carefully redesigned our system by keeping in mind the three guidelines of operational excellence mentioned above. To this purpose, we moved to a recommendation framework based on learning to rank (L2R). In this chapter, we provide an in-depth overview of our L2R architecture. In particular, we discuss how the chosen framework enabled our recommendation stack to address operational concerns. Our goal is to stress the importance of treating the operational quality of recommender systems as a major criterion for the scientific analysis of those systems. We believe that by achieving a tighter integration of operational and algorithmic concerns, faster and more effective technology transfer can be achieved for the outcome of recommender systems research. In Section 21.2, we describe the main contexts where we serve recommendations in the fashion store. Section 21.3 reviews the theoretical framework underlying our machine learning system, whereas the system architecture is illustrated in Section 21.4. Section 21.6 presents some results from offline and online experiments. Finally, in Section 21.7 we draw some conclusions and sketch directions for future research.

21.2. Use Cases and General Approach

In Sections 21.2.1–21.2.3 we review three different recommendation contexts from the Zalando web store. At the same time, we sketch the backbone of our modeling approach.

21.2.1. *Item-based Recommendations*

The first type of recommendations we serve at Zalando are the ones available on product detail pages, namely the ones usually presented in a "Similar items" carousel. These recommendations are not necessarily personalized, in that they simply model the relevance of additional candidate products with respect to the item that the user is currently checking out (i.e. the main subject of the currently open page). An example is provided in Figure 21.1.

The way we model the relevance of candidate products for a given reference item is via the scoring function s:

$$s(\boldsymbol{i}_i, \boldsymbol{i}_j) = \varphi(\psi^I(\boldsymbol{i}_i, \boldsymbol{i}_j)) \tag{21.1}$$

where $\boldsymbol{i}_i$ is the reference item, and $\boldsymbol{i}_j$ is a candidate product for which we want to compute the relevance score. Once we score all relevant candidates $\boldsymbol{i}_j$, we rank them by score, and then recommend the top k.

The key ingredients to define are on the one hand the scoring function φ, and on the other hand ψ^I, which combines the base article and the candidate recommendation into the feature vector to be fed into φ. We will address the definition of φ in Section 21.3.2. While φ will be estimated by a general L2R routine, ψ^I encodes instead some of our domain knowledge

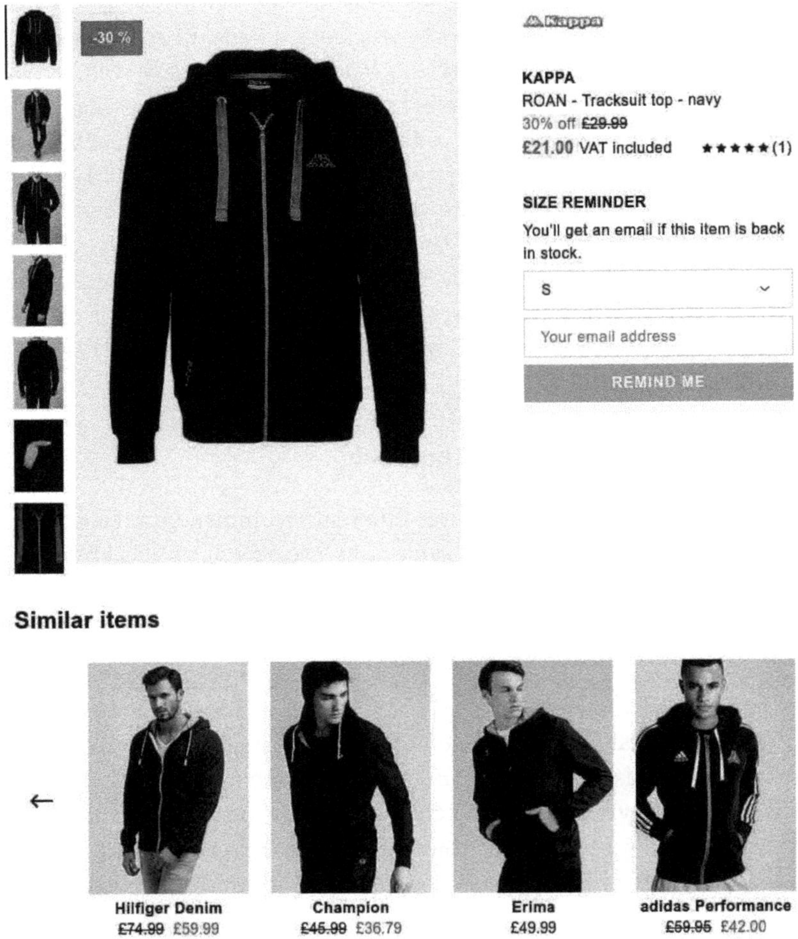

Fig. 21.1. Item-based recommendations on a product detail page.

by specifying how the attributes of $\boldsymbol{i}_i$ and $\boldsymbol{i}_j$ interact to one another. For example, if articles have a color attribute, their interaction $\psi^I(\boldsymbol{i}_i, \boldsymbol{i}_j)$ might contain an attribute specifying whether the two articles have the same color or not. Another interaction attribute might be given by the price difference of the two products, and so on. Therefore, we refer to ψ^I as an *interaction function*. Virtually, there is no limit to the type and quantity of interaction attributes we can encode into our model, as long as the information required to calculate them is available from the attributes of the input items. And of course, whenever relevant to the task at hand, non-relational attributes from the interacting items (e.g. the popularity of candidate articles) can also be incorporated into ψ^I.

21.2.2. *Personalized Recommendations*

General personalized recommendations are served to customers whenever no specific context or intention is assumed for the current context. For example, we serve personalized recommendations on the Zalando home page, usually presenting them in a "Recommended for you" box.

Fig. 21.2. Personalized recommendations on the Zalando home.

The scoring function for personalized recommendations can be modeled as follows:

$$s(\boldsymbol{u}_i, \boldsymbol{i}_j) = \varphi(\psi^U(\boldsymbol{u}_i, \boldsymbol{i}_j)) \tag{21.2}$$

where $\boldsymbol{u}_i$ is a user, and $\boldsymbol{i}_j$ is the candidate product we want to score with respect to $\boldsymbol{u}_i$. The key difference between Eq. 21.1 and Eq. 21.2 is that now, the function ψ^U has to model the interaction between user and article attributes (rather than an article-to-article interaction). For example, if we know the user's favorite brand, we can encode a boolean attribute specifying whether the brand of $\boldsymbol{i}_j$ is the same as $\boldsymbol{u}_i$'s favorite brand. In other words, ψ^U encapsulates the feature-engineering side of our user-to-item recommendation model.

21.2.3. *Personalized Item-based Recommendations*

Another recommendation context arises when we attempt to personalize the recommendations we show in product detail pages, such as the ones that appear in the carousel already displayed in Figure 21.1. In this case, there are three different entities to be accounted for in our scoring function, namely the user $\boldsymbol{u}_i$, the base article $\boldsymbol{i}_j$, and the candidate product $\boldsymbol{i}_k$:

$$s(\boldsymbol{u}_i, \boldsymbol{i}_j, \boldsymbol{i}_k) = \varphi(\psi^{U,I}(\boldsymbol{u}_i, \boldsymbol{i}_j, \boldsymbol{i}_k)) \tag{21.3}$$

To address this context, we have to leverage both item-to-item and user-to-item interactions. For example, we might simply define $\psi^{U,I}$ as a concatenation of the values returned by the previously defined functions ψ^I and ψ^U, i.e. $\psi^{U,I}(\boldsymbol{u}_i, \boldsymbol{i}_j, \boldsymbol{i}_k) = (\psi^U(\boldsymbol{u}_i, \boldsymbol{i}_k), \psi^I(\boldsymbol{i}_j, \boldsymbol{i}_k))$. Also, we might want to include ternary interaction attributes. For example, we can have a boolean feature specifying whether the user's most frequently purchased brand, the base item's brand, and the candidate article's brand are all the same or not, or a real-valued feature calculating the fraction of pairs (x, y) from the set $\{(\boldsymbol{u}_i, \boldsymbol{i}_k), (\boldsymbol{i}_j, \boldsymbol{i}_k)\}$ such that x and y can be assigned to the same brand.

21.3. Developed Algorithm: Learning to Rank

We now review the L2R framework underlying our recommender system. In Section 21.3.1, we summarize some of the vast literature available on the topic, whereas Section 21.3.2 outlines the approach we use to estimate our models from data.

21.3.1. *Related Work*

L2R approaches are usually classified into three broad families: pointwise, pairwise, and listwise methods [Liu (2009)]. Pointwise models attempt to estimate the relevance score of each item as an independent data point. The supervision for learning the relevance score is typically extracted from previously (e.g. manually) labeled lists of candidates, by using either the relevance ratings attached to candidates or their positions within ranked lists as the respective labels. At prediction time, the items returned for a query are sorted according to their scores. Linear or logistic regression are examples of scoring functions used in pointwise methods. Among the earliest pointwise models available from the literature, we can mention [Fuhr (1989)] and [Cooper *et al.* (1992)], which focus on estimating the probability distribution of item relevance given a user/query. Pairwise approaches score ordered pairs of items instead of individual items. Here, the goal is to learn the correct order of these pairs. In other words, the goal is to score the more relevant items higher than less relevant ones. The advantage of this approach over pointwise methods is that it does not require to learn absolute relevance scores. RankSVM [Herbrich *et al.* (2000); Joachims (2002)] is one of the most popular pairwise models. It formalizes ranking as a binary classification problem for item pairs and uses support vector machines as the underlying classifier. Another one is Rank Logistic Regression [Sculley (2009)]. RankBoost [Freund *et al.* (2003)] is also a pairwise model, where the ranking is learned through boosting. The idea is to construct a sequence of "weak" rankers in an iterative fashion, and then to predict ranks using a linear combination of the weak learners. Finally, listwise approaches rely on ranked lists as training examples. In particular, they try to minimize a loss function defined over full lists instead of ordered pairs sampled from those lists. ListNet [Cao *et al.* (2007)] is a listwise ranking algorithm, which performs gradient descent over a loss function based on cross-entropy. AdaRank [Xu and Li (2007)] is also a listwise approach, based instead on boosting. Gradient-boosted trees (GBTs) have recently become quite popular in learning to rank [Burges *et al.* (2011); Mohan *et al.* (2011)]. Although GBTs have been shown to outperform simpler approaches (such as linear models), training them over web-scale datasets can be very expensive. Moreover, scoring latency is a serious issue for GBTs whenever we are not able to precompute and cache predictions. An alternative model, which is very suitable for the large-scale setting, is the WSABIE algorithm [Weston *et al.* (2011)], which relies on low-dimensional embeddings and data sub-sampling. Yet another model is given by ElasticRank [Freno *et al.* (2015)], inspired by a former approach known as LambdaRank [Burges *et al.* (2006)]. ElasticRank inherits from LambdaRank the idea of weighting the loss function by a listwise penalization scheme. The loss is minimized by stochastic gradient descent, where the training algorithm is allowed to perform only one pass through the training data. Moreover, the special focus of this model is given by sparsity-inducing regularization schemes.

21.3.2. *Ranking Loss Minimization*

Let our training sample be a set $\mathcal{X}$ containing the pairs $(\boldsymbol{x}_1^+, \boldsymbol{x}_1^-), \ldots,$ $(\boldsymbol{x}_n^+, \boldsymbol{x}_n^-)$, where each pair $(\boldsymbol{x}_i^+, \boldsymbol{x}_i^-)$ is such that $\boldsymbol{x}_i^+$ should be ranked higher than $\boldsymbol{x}_i^-$. If the setting is given by personalized recommendation, as described in Section 21.2.2, each training pair will be defined as follows:

$$(\boldsymbol{x}_i^+, \boldsymbol{x}_i^-) = (\psi^U(\boldsymbol{u}_j, \boldsymbol{i}^+), \psi^U(\boldsymbol{u}_j, \boldsymbol{i}^-)) \tag{21.4}$$

where item $\boldsymbol{i}^+$ is more relevant to user $\boldsymbol{u}_j$ than item $\boldsymbol{i}^-$. Relevance labeling might be obtained in several different ways, via implicit or explicit feedback. For example, $\boldsymbol{u}_j$ might have clicked (or purchased) $\boldsymbol{i}^+$ but not $\boldsymbol{i}^-$, assuming that both items where available in the same candidate set. Or, $\boldsymbol{u}_j$ might have explicitly labeled some articles, by rating them or placing them in a wishlist. The way we sample user feedback clearly depends on the available data (and especially on the quality and limitations of user feedback tracking), the specific use case, and possibly on computational requirements, and it is tuned for each application both by offline and online experimentation. The definition given in Eq. 21.4 can easily be cast to suitable forms for the item-based and personalized item-based recommendation setting, respectively.

For each training pair $(\boldsymbol{x}_i^+, \boldsymbol{x}_i^-)$, our ranking loss is defined as follows:

$$\ell(\boldsymbol{x}_i^+, \boldsymbol{x}_i^-; \varphi) = \max\{0, \varphi(\boldsymbol{x}_i^-) - \varphi(\boldsymbol{x}_i^+) + \epsilon\} \tag{21.5}$$

for some tunable parameter $\epsilon \geqslant 0$. Different choices are possible other than the hinge loss specified in Eq. 21.5, e.g. the logistic loss $\log\{1+\exp(\varphi(\boldsymbol{x}_i^+) - \varphi(\boldsymbol{x}_i^-))\}$. For our applications, we found the hinge loss completely satisfying.

Given a training pair $(\boldsymbol{x}_i^+, \boldsymbol{x}_i^-)$, we can optimize the parameters in φ by minimizing the ranking loss over $(\boldsymbol{x}_i^+, \boldsymbol{x}_i^-)$. This task can be accomplished via standard (sub-)gradient descent methods. Any choice for the parametric form of φ, ranging from simple linear regression to a multilayer perceptron, will be compatible with the ranking metric we are adopting as long as we are able to compute the corresponding (sub-)gradients. Throughout our applications, we assume a linear form for the scoring function φ, i.e. $\varphi(\boldsymbol{x}) = \boldsymbol{w}^\intercal \boldsymbol{x}$. This is not only practical from the computational point of view, but it also preserves the convexity of the ranking loss defined in (21.5), which is a convenient property for the purpose of parameter optimization.

In order to induce sparsity in the learned scoring function, i.e. to completely drop some components in the weight vector $\boldsymbol{w}$, we can add ℓ_1-based regularization to the loss defined in Eq. 21.5. The resulting objective function will be convex, with both theoretical guarantees and efficient optimization schemes [Negahban *et al.* (2009)]. As previously advocated in the large-scale setting [McMahan *et al.* (2013)], we additionally consider a squared ℓ_2-term, hence leading to the following ranking loss:

$$\ell^*(\boldsymbol{x}_i^+, \boldsymbol{x}_i^-; \varphi_{\boldsymbol{w}}) = \ell(\boldsymbol{x}_i^+, \boldsymbol{x}_i^-; \varphi_{\boldsymbol{w}}) + \lambda_1 \|\boldsymbol{w}\|_1 + \frac{1}{2}\lambda_2 \|\boldsymbol{w}\|_2^2 \tag{21.6}$$

where λ_1 and λ_2 are non-negative hyperparameters determining, respectively, the weight of the ℓ_1 and ℓ_2 penalties within the overall loss. The resulting regularization model is usually referred to as elastic-net [Zou and Hastie (2005)]. It is worth stressing that, for our web-scale application, sparsity is an extremely important requirement for the learned model. The reason is that, to extract useful signal from the available training data, we often need to exploit very high-dimensional feature representations, e.g. by considering all possible pairwise interactions of categorical attributes through a one-hot encoding scheme. This easily leads to feature vectors having billions of components. Therefore, dropping as many irrelevant features as possible is especially useful both to improve generalization and to reduce the memory footprint (and possibly the latency) of the learned model.

In order to learn our parameter vector from a training set $\mathcal{X} = \{(\boldsymbol{x}_1^+, \boldsymbol{x}_1^-), \ldots, (\boldsymbol{x}_n^+, \boldsymbol{x}_n^-)\}$, we use gradient descent to address the following, averaged regularized problem:

$$\min_{\boldsymbol{w}} \frac{1}{n} \sum_{i=1}^{n} \ell(\boldsymbol{x}_i^+, \boldsymbol{x}_i^-; \varphi_{\boldsymbol{w}}) + \lambda_1 \|\boldsymbol{w}\|_1 + \frac{1}{2}\lambda_2 \|\boldsymbol{w}\|_2^2 \tag{21.7}$$

In our applications, the number of training data points is typically in the order of several millions. To cope with the dataset size, we distribute the training load over several workers using the in-memory cluster computing capabilities offered by Apache Spark (`http://spark.apache.org/`). This way, we can easily calculate and sum up the gradient of ℓ^* over the full dataset in no more than a couple of seconds, and hence perform a few hundreds of gradient descent iterations in just a few minutes.

As proposed in [Langford *et al.* (2009)], we can simply enforce sparsity by adding a pruning operation to gradient descent. The pruning step, scheduled every k gradient iterations, simply consists of setting to 0 all the weights w_i such that, for a chosen threshold θ, $|w_i| < \theta$. Clearly, the higher the value we choose for θ, the sparser the model will become. The experiments reported in [Freno *et al.* (2015)] show that this simple technique performs surprisingly well as compared to way more sophisticated approaches available from the literature [Duchi and Singer (2009); Xiao (2010)]. Therefore, we adopt such regularization scheme as our default choice.

21.4. System Architecture

This section provides an overview of our system architecture. Our stack is fully deployed on Amazon AWS (`https://aws.amazon.com/`), and it is organized into a collection of offline jobs on the one hand, and a group of online services on the other hand. Section 21.4.1 describes the former, whereas the web services are presented in Section 21.4.2. A critical outlook is then proposed in Section 21.5.

21.4.1. *Offline Jobs*

Let us look at how we organize all of the necessary batch calculations into a graph of interdependent offline jobs. Here, the goal is to produce and persist all data that will be necessary in order for the live recommendation engine to serve fresh results to our visitors. The jobs architecture is depicted in Figure 21.3.

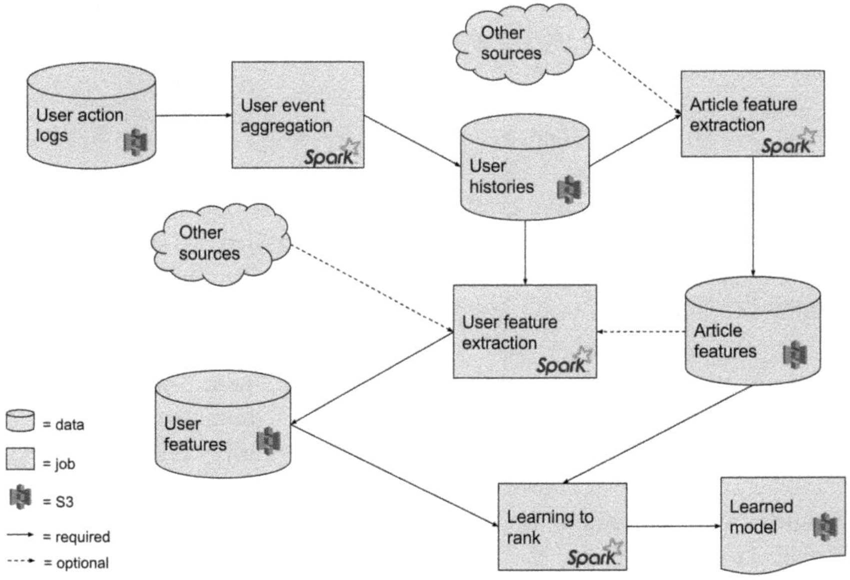

Fig. 21.3. A diagram of our offline jobs: event aggregation, feature extraction, and learning to rank.

Everything starts from user action logs, which record every action performed by our customers on the Zalando web store. The first job aggregates all actions on a per-user basis. The difficulty here is that the volume of server logs is massive and distributed over multiple locations. Therefore, we cannot afford to crunch and re-aggregate all of this data every time we need to retrieve actions for a given customer. Pre-aggregation of customer actions allows subsequent jobs to selectively retrieve the relevant information in a much more convenient way.

Once the user histories have been aggregated, they are used for two purposes. On the one hand, we can extract a number of dynamic article features (i.e. quantities that change over time), such as number of clicks received by a product within a given time window, number of purchases, and so on. Here, we might be tempted to extract such features directly from the (non-aggregated) event logs. The drawback of such an idea is that it would prevent us from filtering out undesirable data points, such as events generated by robots or crawlers, which pollute our data by introducing noise and distorting their distribution. These problematic events can typically be identified only once we analyze the behavior of the respective cookies/user identifiers, and this behavior only materializes once the relevant event sets are aggregated and some basic statistics are derived from them.

On the other hand, we extract a wide range of user attributes, such as time elapsed since the user was last active on the shop, how the user choices are distributed over different product brands, or the price distribution of purchased articles. User feature extraction can also exploit already computed article features. For example, if we want to measure how user purchases are affected by product popularity, then for each purchase, we need to check how popular the corresponding product was at the time it was purchased (e.g. in terms of click-through rate).

Finally, the L2R job estimates the scoring function from the extracted historical data, relying on both user and article features. A dedicated job has to be run for each different recommendation setup (i.e. for each one of the different models specified in Eqs. 21.1–21.3), since they lean on different data representations and features. The learned scoring functions are then deployed to our live ranking services.

The full offline job pipeline is run at short, regular intervals (e.g. daily). This is important to ensure that new events ingested in our browsing logs can contribute to the calculation of user and article features, as well as to the adaptation of the machine-learned scoring function to drifting data distributions.

21.4.2. *Web Services*

The purpose of our recommendation services is to provide a RESTful API [Fielding (2000)], returning sorted sets of results for queries that conform to one of the following use cases: (*i*) get top-k recommendations for item $\boldsymbol{i}_i$; (*ii*) get top-k recommendations for user $\boldsymbol{u}_i$; (*ii*) get top-k recommendations for user $\boldsymbol{u}_i$ and item $\boldsymbol{i}_j$. These patterns correspond to the three use cases described in Section 21.2. A diagram of our online architecture is given in Figure 21.4.

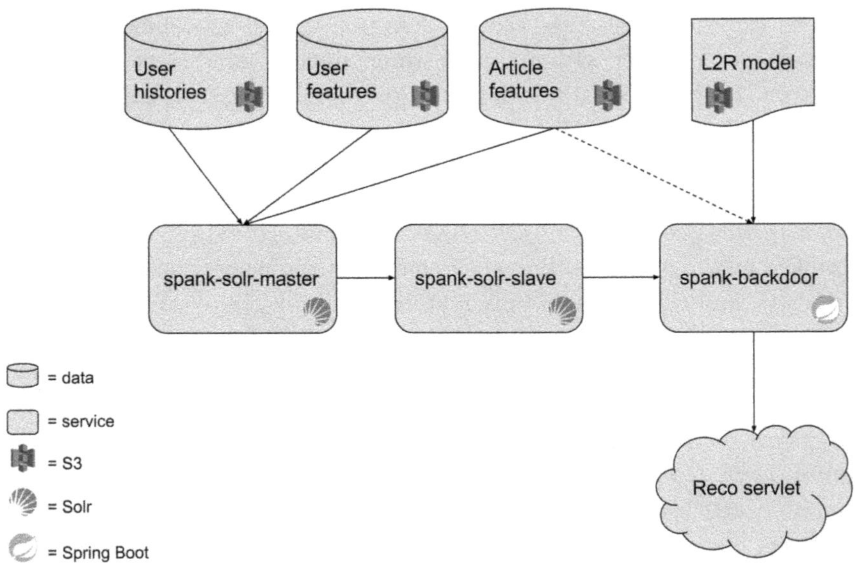

Fig. 21.4. A diagram of our web services: data indexing, real-time ranking, and post-processing. 'Spank' is the name we use to refer to our content-based recommendation stack.

The first step in setting up our live recommendation stack is to make the user and article data available for real-time processing, as they are originally stored in large, distributed batches. To this aim, we index our data into Apache Solr (`http://lucene.apache.org/solr/`), which allows us to retrieve single feature vectors by (user/article) id within an average latency of less than 5 milliseconds. Our Solr service is set up with a master/slave architecture. That is, the master server only takes care of indexing data coming from the offline jobs whenever new batches become available, which is a computationally expensive operation. On the other hand, the slaves mirror the content available from the master by mere data replication, and they expose it to the ranking engine for real-time querying. This way, we make sure that the computational load incurred while indexing new data batches will not affect the latency of Solr queries, which would have a tremendous impact on the responsiveness of the recommendation API.

A second component of the live system is given by the ranking servlet, exposing the recommendation API mentioned above. We refer to this component as the *backdoor engine*, for the following reason. As illustrated in Figure 21.4, the backdoor service does not serve directly customer requests. Such requests go first to the stack labeled as 'Reco servlet', which then forwards the relevant requests to the backdoor engine. The goal behind this choice is to separate the problem of calculating and ranking recommendations from the problem of post-processing (e.g. filtering or rendering) them for the final customers. Such post-processing tasks are both complex to perform and extrinsic to the genuine recommendation problem, as they typically come from *ad hoc* business requirements or from interoperating with external services, such as sales, campaigns, hand-coded rules, advertising, de-duplication, front-end restrictions, and so on. Therefore, the backdoor-based architecture frees up the recommendation engine from the complexity of managing extrinsic logic, hence streamlining both software development and system operation work.

As shown in the diagram, the L2R model artifact can be loaded directly from the location where it is saved by the offline jobs, i.e. it does not need any sort of indexing, since this artifact is nothing but a self-contained weight-vector (possibly enriched with some additional configuration). This happens as soon as the backdoor servlet is started, while the model is also immediately refreshed whenever more recent artifacts become available. On the other hand, article features can be usefully cached in the same servlet, to a more or less significant extent. Caching will be crucial to minimize serving latency. Differently from user data, which have to be queried according to more unpredictable patterns, article queries tend to be generated according to a power-law distribution, because of the popularity patterns which are intrinsic to fashion shopping. Because of this, we can not only achieve relatively high cache-hit rates, but also initialize the cache with the most popular articles whenever new instances are added to the stack, which will be immensely beneficial to the initial responsiveness of these instances.

21.5. Addressed Challenges and Problems

We now proceed to discuss, in Sections 21.5.1–21.5.3, how the system presented so far copes, respectively, with the three operational concerns motivating our general approach.

21.5.1. *Adaptation to Diverse Use cases*

As described in Section 21.2, product recommendation takes different forms as soon as we move through different contexts, e.g. depending on whether personalization is involved, whether a reference article is currently the focus, and so on. Therefore, being able to seamlessly adapt the available infrastructure to the different use cases is crucial to maximize the value of the adopted recommendation technology. This is the main consideration we had in mind when choosing a model such that only one specific component (i.e. the interaction function ψ) has to be modified whenever we tackle a new context. ψ^I, ψ^U, and $\psi^{U,I}$ are self-contained, plug and play modules, which are completely decoupled from anything but the offline L2R job and the online backdoor servlet. Of course, these last two components need to be aware of which interaction function has to be loaded for each specific use case. But the logic of the encompassing system remains completely untouched whenever we move from one context to another.

21.5.2. *System Operation*

Maintaining a recommendation architecture can require significant efforts, for a variety of reasons. The amount of data to be processed, both for offline modeling and for online serving, is typically huge, which requires distributed processing infrastructure on both sides. Also, because of the system complexity and the need to continuously improve over time (in order to tackle new challenges and meet continuously evolving customer needs), the system components need to be easily replaceable and modifiable, while avoiding the risk of breaking the whole system whenever a local change is operated. While the system described in Sections 21.4.1–21.4.2 takes a non-trivial *cognitive* effort to be operated and mastered, it does not require *expensive* operations whenever a component has to be modified in some way. For example, extracting some more user features from the historical data only requires a change in the corresponding feature extraction job, and the outcome will be automatically available for all subsequent stages, independent on when and how the other components will start using the new input. Similarly, adding one more L2R job, e.g. in order to move from non-personalized to personalized item-based recommendations, only requires to implement and add one job to the offline pipeline, without affecting any other component, and independent of whether and when the new model will go live. In other words, while designing the presented architecture, we strived to decouple as much as possible all different system components, so that the cost of operating the whole system could be minimized by isolating the scope of the most typically required interventions. Furthermore, the sharp separation between offline and online stack encapsulates the live system from unpredictable, possibly breaking changes happening in any of the ingested data sources, allowing us to detect the problem from its impact on the offline jobs, and to fix it before it starts affecting our customers through its downstream effects.

21.5.3. *Exploitation of Pre-existing Signal*

Whenever new data sources become available or whenever new signal can be exploited from a (possibly pre-existing) predictive model, the novel information can be homogeneously fed into the L2R engine in the form of additional features. Such an incremental modeling approach can lead to continuous improvement in recommendation quality, due to the seamless exploitation of the novel information. As we will illustrate in Section 21.6, we found this strategy to generate a lot of momentum in the process of replacing the engines currently used in production by increasingly accurate ones.

An extremely simple example of how we incorporated a pre-existing recommender into L2R is given by how we integrated the output of our previous CF engine for item-to-item scoring. Here, we just started to ingest the predicted item-to-item relevance into the article feature-extraction job, so that the similarities were made available to the interaction function ψ^I as pre-calculated features. A more sophisticated example of how we added a new type of signal is offered by the integration of a Word2Vec-based article embedding model [Mikolov *et al.* (2013)]. For the purpose of article recommendation, we were interested in extracting a latent-vector representation of our products, by capturing the contextual relation of article views with the articles explored within the same browsing sessions. To this purpose, we simply model sessions as sequences of article views, and have a Word2Vec learning routine run on these sequences. The job flow is described in Figure 21.5. As shown in the diagram, a simple entry point for

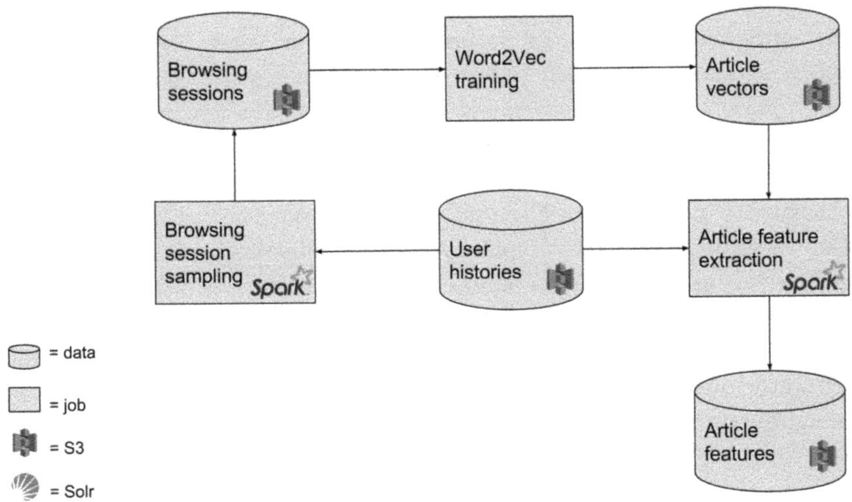

Fig. 21.5. Integration of Word2Vec-based article embeddings into our offline pipeline: the entry point is provided by the article feature extraction job.

the new embeddings into our offline job graph is provided by the feature extraction job for articles. The new data source is thereby made available to the L2R stage.

21.6. Experimental Evaluation

As Zalando is mainly focused on online shopping, the most reliable way for us to choose between competing recommendation algorithms and models is by comparing their live (i.e. online) performance in terms of a number of metrics, such as click-through rate, customer conversion rate, number of sales, generated revenue, effective catalog size [Gomez-Uribe and Hunt (2016)], and additional indicators as well. However, when the candidates for online testing are too many, the only way to prioritize the different options is given by offline testing, i.e. by measuring the performance of the alternative models on historical data. The advantage of offline experiments is not only that the same data can be used to benchmark an arbitrary number of models, but also that poor models can be prevented from hurting our customers by exposing inaccurate recommendations. Moreover, whenever a new machine learning model is developed, extensive hyperparameter tuning and hypothesis testing needs to be done on validation data. Performing this task on online data is simply prohibitive both in terms of time and in terms of business risk. On the other hand, the drawback of offline testing is both that it suffers from bias (as the historical data are generated by the very same models we are trying to supersede), and that it does not capture all subtleties involved in a live system (such as the interaction with other components, the impact on user experience of higher/lower latencies, and so on). Therefore, we usually select between competing models by first deciding, through offline experimentation, which ones we want to benchmark in a live setting, and then by running dedicated A/B tests to determine whether the current production model should be replaced by a new one.

The purpose of this section is to illustrate how we learned one of the lessons discussed in the previous sections, namely the one considered in Section 21.5.3. One point that became clear to us both via offline and online experimentation is that focusing on the feature-based model described so far allows us to significantly speed-up algorithmic innovation, throughout our recommendation use cases, via inherently incremental improvements. The reason is that, without forcing us to go for the operational risk and expense of substituting new systems for the current production model, the adopted L2R framework makes room for a seamless integration of heterogeneous models and data sources, by merging all of them into the final scoring model in the form of additional features. Sections 21.6.1–21.6.2 give an example of how this incremental benefit was detected and brought live in one of our personalized recommendation contexts, namely on the Zalando Home.

21.6.1. *Offline Evaluation*

One way to test the ranking accuracy of a given recommendation engine from historical data is the following. We build our training set by sampling browsing sessions from user action logs, covering whatever time interval is suitable for training our algorithms. Typically, we use data from at least 7 consecutive days, going up at most to 4 consecutive weeks. Then, if our training sample extends to day d at latest, we use day $d+1$ for testing. For hyperparameter tuning, day d is actually held out of the training data, so that it can be used to build a validation set, and the algorithms are trained using data generated no later than on day $d-1$. Such a training sample is enough to provide several millions of user sessions, whereas for the test set we usually subsample about half a million sessions. To construct test queries from a user session, we proceed as follows. All articles that the user clicks on (or purchases) within the session are labeled as relevant items, i.e. as items that the ranking/recommendation algorithm should rank/recommend on top of anything else. Then, we check the position of the relevant items in the top-k recommendation lists produced by the competing algorithms, and we measure the corresponding ranking accuracy.

Our chosen metric for this measurement is normalized discounted cumulative gain (NDCG) at k [Croft *et al.* (2009)]. A formal comparison of different evaluation metrics goes beyond the scope of this chapter. However, it is worth mentioning that, on our datasets, alternative metrics (such as precision or recall at k) usually give consistent indications about which models are most accurate. NDCG is a normalized version of the following metric:

$$DCG@k(\boldsymbol{r}, \boldsymbol{l}) = \sum_{i=1}^{\min\{k,n\}} \frac{2^{l_i} - 1}{\log_2(r_i) + 1} \tag{21.8}$$

where $\boldsymbol{r} = (r_1, \ldots, r_n)$ is the ranking induced by a given algorithm on n candidate products, and l_i is the relevance label of the i-th product in the list. We set the relevance label to 1 if the product was clicked/purchased, whereas $l_i = 0$ otherwise. DCG has non-negative values, and larger DCG values correspond to better rankings. NDCG normalizes the right-hand side of Eq. 21.8, dividing it by the DCG of a perfect ranking, i.e. the ranking induced by the relevance labels themselves.

As we started experimenting with L2R, the very first feature to use was simply the score calculated by the pre-existing CF engine. Basically, the CF engine was an adaptation of the approach presented in [Aiolli (2013)]. The key idea behind that CF approach is to estimate relevance scores from (implicit) binary ratings, which in our case are given by the preferences (implicitly) revealed by user clicks. As a number of static article features were available from our product catalog, such as brand, price, category, colors, and a variety of high-level tags, we simply added all such attributes to the L2R model. To this purpose, we engineered several types of attribute interactions and encoded them into our ψ^U function, using the strategy described in Section 21.2.2. We then compared this first version of our L2R model (L2R #1) to the CF baseline, leading to the NDCG measurements plotted in Figure 21.6.

As the figure shows, L2R #1 achieves some noticeable improvement over CF, although the ranking deteriorates for large values of k. We notice that

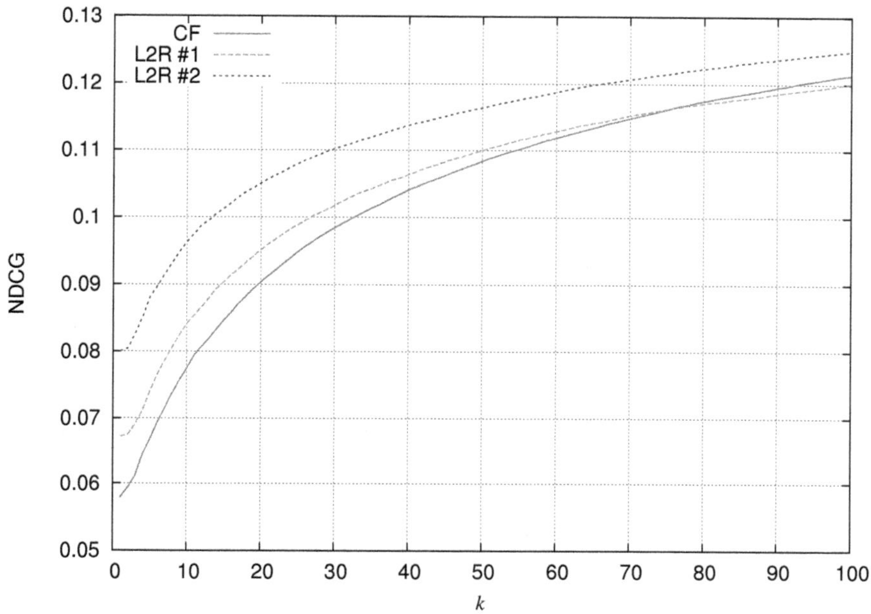

Fig. 21.6. *NDCG@k* measurements on a one-day traffic sample for: (*i*) CF; (*ii*) L2R using CF-based features plus static product attributes (L2R #1); (*iii*) L2R model using all the features from L2R #1 plus Word2Vec-based similarity (L2R #2).

the measurement was seen to be statistically significant, as the experiment was repeated on several occasions, for different data samples. The evidence was enough for us to set up and run an A/B test, which we describe in Section 21.6.2. It is important to note that, in order to see the mentioned improvement in NDCG, it was not enough to add a bunch of features to the CF score, as described above, and then just hope for the learned model to regularize away all noise while learning the optimal scoring. Instead, extensive feature selection was necessary, which we performed by repeated cross-validation on held-out data. Unfortunately, when the features are poorly engineered, or just too noisy, the model accuracy can be hurt to a more or less significant extent, and regularization does not seem to be enough to isolate the good signal from noise.

Once the Word2Vec-based article embeddings were available from our modeling pipeline, we also started to experiment with features extracted from the embeddings. The simplest way to engineer such a feature into ψ^U is, for example, to calculate the cosine similarity between the latent vector representation of the candidate product and a suitable vector representation of the user. As a heuristic, the user vector might be calculated as the average of the article vectors from the relevant shopping history. We then add the new feature to the ones already used in L2R #1, and refer to the resulting model as L2R #2. As plotted in Figure 21.6, L2R #2 significantly improves over L2R #1. Presumably, this happens because the newly engineered feature conveys signal which is not captured by the pre-existing features. This makes the L2R #2 model a second, even more promising candidate for further A/B testing (which is currently work in progress). As the Word2Vec model requires a dedicated training routine to be run, relying on specific hyperparameter tuning, the chosen hyperparameters turn out to have a non-trivial effect on the accuracy of the final L2R model. However, a convenient property of the used L2R framework is that, while we observed a poor hyperparameter setup to make the Word2Vec-based similarity completely inaccurate if used as a stand-alone article-relevance predictor, using it instead as one of several L2R input features makes the final predictions way more robust to inaccuracies in the learned embeddings. In particular, if we refer to the plot in Figure 21.6, what we observed is that, for less accurate settings of the Word2Vec hyperparameters, the NDCG curve of L2R #2 would get closer and closer to L2R #1, without however falling below that level. On the other hand, we did not see a stand-alone prediction model based only on the Word2Vec similarity coming anywhere close to the CF baseline. As a side remark, it is fair to say that the L2R solutions we adopt are still some kind of collaborative approach at their essence. This is because we make crucial use of collaborative features in order to learn the final scoring function, such as the scores calculated by our legacy CF engine or the Word2Vec-based similarity measurements.

21.6.2. *Online Evaluation*

Offline NDCG measurements do not tell us with certainty how the L2R model will impact our business performance indicators in production. Therefore, we ran some A/B tests in order to compare L2R to the CF baseline. The tested L2R model is L2R #1 from the previous section (i.e. not including Word2Vec-based features, which were not yet available when the online tests were run). The tests concern the Zalando Home, where general personalized recommendations are shown to visitors (along with other content) as hints for navigation. The metrics we used for the evaluation were mainly click-through rate (CTR) and conversion rate (CR). The former is defined as the fraction of displayed recommendations that receive at least one-click, while the latter is the fraction of visitors who end up making at least one purchase during the relevant time window. The results are summarized in Table 21.1 for five different countries. To avoid disclosing any sensitive information with respect to our business performance, for CTR we report the relative difference between control (CF) and treatment group (L2R), whereas for CR, instead of mentioning explicit values, we report whether a statistically significant improvement ($p < 0.1$) was achieved or not. For an overview of the A/B testing methodology on the web and the involved significance calculations, we refer the reader to [Kohavi *et al.* (2009)].

As shown in the table, the treatment group consistently outperforms control in CTR. All CTR measurements were statistically significant, where the p values very quickly converged to 0. An interesting point to remark is that, although the gap between L2R and CF is significantly positive everywhere, different countries display a noticeably different behavior in this respect, as the relative difference between the two algorithms exhibits a relatively large variance over different domains. This country-specific behavior is the reason why it is extremely important to run a dedicated test for each involved domain, in order to avoid making false (and financially risky) generalizations. On the other hand, our previous experience with the subtleties of fashion shopping shows that CR, as compared to CTR, is much more difficult to increase by mere algorithmic change. Nevertheless, CR increases significantly in two out of the five reported experiments, i.e. in countries #2 and #5, while it does not deviate significantly from the control in all other experiments. Besides the metrics discussed here, we also monitored additional revenue- and engagement-based measurements,

Table 21.1. Click-through rate and conversion rate results from A/B tests run over five different countries. Country names are masked in order to not disclose possibly sensitive information. L2R is used for treatment, while CF is used for the control group.

Country	**ΔCTR**	**ΔCR**> 0 & $p < 0.1$
#1	+15.22%	×
#2	+20.44%	✓
#3	+33.13%	×
#4	+23.93%	×
#5	+37.88%	✓

which also showed, overall, either an improvement or no statistically significant difference. Therefore, the L2R engine was rolled out to production in all involved domains. In our interpretation, the key strategy by which L2R superseded CF in our production systems is by enriching the signal provided by CF through the integration of additional data sources into the final predictions, rather than discarding the old system and building on entirely different foundations.

21.7. Conclusions and Future Directions

In this chapter, we discussed how the operational challenges involved in applying recommender systems to large-scale online retail provide some fundamental dimensions for a formal analysis of such systems. Our analysis was based on an in-depth overview of the software architecture used at Zalando for machine-learned ranking of clothing recommendations. In order to explain the rationale behind many of our design decisions, we formulated three criteria by which such an analysis can be fruitfully conducted, namely: (*i*) flexibility of the recommender system with respect to different recommendation use cases; (*ii*) cost of the operations involved in maintaining the system; (*iii*) ease of integration of heterogeneous data sources and signal types into a unified machine-learning workflow. According to the experience we made in the clothing recommendation domain, these guiding principles have turned out to be extremely effective in helping us to ensure steady progress and reliable improvement patterns throughout our algorithmic innovation efforts. Therefore, we believe that a strong and explicit focus on operational excellence should never be missing from a scientific investigation of recommender systems.

Although several directions are open for further development, at least one of them deserves to be explicitly mentioned here. Deep neural networks were recently shown to have significant innovation potential also in the recommendation domain [Covington *et al.* (2016)]. In particular, one of their key selling points is that they promise to alleviate the burden of the feature engineering process, by automating and improving signal extraction and transformation to a significant extent. In this respect, our current setup crucially relies on the quality of the feature engineering pipeline, which still involves a non-trivial amount of domain insight and quite a few iterations of manual work in order to deliver satisfying results. Therefore, we are also working in order to be able to extract and ingest within our system additional signal based on deep-learning models. The extent to which a deep learning component can ultimately relieve us from manual work is probably something that only a joint effort from the scientific and industrial community can uncover.

Acknowledgments

The author is extremely grateful to everyone in the Recommendation Team at Zalando for the outstanding support they provided throughout all phases of this project.

References

Aiolli, F. (2013). Efficient top-n recommendation for very large scale binary rated datasets, in Q. Yang, I. King, Q. Li, P. Pu and G. Karypis (eds.), *Seventh ACM Conference on Recommender Systems, RecSys '13, Hong Kong, China, October 12-16, 2013* (ACM), pp. 273–280.

Burges, C. J. C., Ragno, R. and Le, Q. V. (2006). Learning to rank with nonsmooth cost functions, in *Advances in Neural Information Processing Systems (NIPS)*, pp. 193–200.

Burges, C. J. C., Svore, K. M., Bennett, P. N., Pastusiak, A. and Wu, Q. (2011). Learning to Rank Using an Ensemble of Lambda-Gradient Models, in *Proceedings of the Yahoo! Learning to Rank Challenge, held at ICML 2010, Haifa, Israel, June 25, 2010*, pp. 25–35.

Cao, Z., Qin, T., Liu, T.-Y., Tsai, M.-F. and Li, H. (2007). Learning to rank: from pairwise approach to listwise approach, in *Proceedings of the 24th International Conference on Machine learning (ICML 2007)* (ACM, New York, NY, USA), pp. 129–136.

Cooper, W. S., Gey, F. C. and Dabney, D. P. (1992). Probabilistic retrieval based on staged logistic regression, in *Proceedings of the 15th Annual International ACM SIGIR Conference on Research and Development in Information Retrieval*, SIGIR '92 (ACM, New York, NY, USA), pp. 198–210.

Covington, P., Adams, J. and Sargin, E. (2016). Deep neural networks for youtube recommendations, in *Proceedings of the 10th ACM Conference on Recommender Systems*, RecSys '16 (ACM, New York, NY, USA), pp. 191–198.

Croft, B., Metzler, D. and Strohman, T. (2009). *Search Engines: Information Retrieval in Practice* (Addison-Wesley, Boston (MA)).

Duchi, J. C. and Singer, Y. (2009). Efficient online and batch learning using forward backward splitting, *Journal of Machine Learning Research* **10**, pp. 2899–2934.

Fielding, R. T. (2000). *Architectural Styles and the Design of Network-based Software Architectures*, Ph.D. thesis, University of California, Irvine.

Freno, A., Saveski, M., Jenatton, R. and Archambeau, C. (2015). One-Pass Ranking Models for Low-Latency Product Recommendations, in *Proceedings of the 21th ACM SIGKDD International Conference on Knowledge Discovery and Data Mining, Sydney, NSW, Australia, August 10-13, 2015* (ACM), pp. 1789–1798.

Freund, Y., Iyer, R., Schapire, R. E. and Singer, Y. (2003). An Efficient Boosting Algorithm for Combining Preferences, *Journal of Maching Learning Research* **4**, pp. 933–969.

Fuhr, N. (1989). Optimum polynomial retrieval functions based on the probability ranking principle, *ACM Trans. Inf. Syst.* **7**, 3, pp. 183–204.

Gomez-Uribe, C. A. and Hunt, N. (2016). The netflix recommender system: Algorithms, business value, and innovation, *ACM Trans. Management Inf. Syst.* **6**, 4, pp. 13:1–13:19.

Herbrich, R., Graepel, T. and Obermayer, K. (2000). Large Margin Rank Boundaries for Ordinal Regression, in Smola, Bartlett, Schölkopf and Schuurmans (eds.), *Advances in Large Margin Classifiers*, chap. 7 (MIT Press), pp. 115–132.

Joachims, T. (2002). Optimizing search engines using clickthrough data, in *Proceedings of the Eighth ACM SIGKDD International Conference on Knowledge Discovery and Data Mining* (ACM, New York, NY, USA), pp. 133–142.

Kohavi, R., Longbotham, R., Sommerfield, D. and Henne, R. M. (2009). Controlled experiments on the web: survey and practical guide, *Data Mining and Knowledge Discovery* **18**, 1, pp. 140–181.

Langford, J., Li, L. and Zhang, T. (2009). Sparse Online Learning via Truncated Gradient, *Journal of Machine Learning Research* **10**, pp. 777–801.

Liu, T.-Y. (2009). Learning to rank for information retrieval, *Foundations and Trends in Information Retrieval* **3**, 3, pp. 225–331.

McMahan, H. B., Holt, G., Sculley, D., Young, M., Ebner, D., Grady, J., Nie, L., Phillips, T., Davydov, E., Golovin, D. *et al.* (2013). Ad click prediction: a view from the trenches, in *Proceedings of the 19th ACM SIGKDD international conference on Knowledge discovery and data mining (KDD 2013)* (ACM), pp. 1222–1230.

Mikolov, T., Sutskever, I., Chen, K., Corrado, G. S. and Dean, J. (2013). Distributed representations of words and phrases and their compositionality, in C. J. C. Burges, L. Bottou, Z. Ghahramani and K. Q. Weinberger (eds.), *Advances in Neural Information Processing Systems (NIPS 2013)*, pp. 3111–3119.

Mohan, A., Chen, Z. and Weinberger, K. Q. (2011). Web-Search Ranking with Initialized Gradient Boosted Regression Trees, in *Yahoo! Learning to Rank Challenge*, pp. 77–89.

Negahban, S., Ravikumar, P., Wainwright, M. J. and Yu, B. (2009). A unified framework for high-dimensional analysis of M-estimators with decomposable regularizers, in Y. Bengio, D. Schuurmans, J. D. Lafferty, C. K. I. Williams and A. Culotta (eds.), *Advances in Neural Information Processing Systems 22 (NIPS 2009)*, pp. 1348–1356.

Sculley, D. (2009). Large scale learning to rank, in *NIPS Workshop on Advances in Ranking.*

Weston, J., Bengio, S. and Usunier, N. (2011). WSABIE: Scaling Up to Large Vocabulary Image Annotation, in *IJCAI 2011, Proceedings of the 22nd*

International Joint Conference on Artificial Intelligence, Barcelona, Catalonia, Spain, July 16-22, 2011, pp. 2764–2770.

Xiao, L. (2010). Dual Averaging Methods for Regularized Stochastic Learning and Online Optimization, *Journal of Machine Learning Research* **11**, pp. 2543–2596.

Xu, J. and Li, H. (2007). Adarank: A boosting algorithm for information retrieval, in *SIGIR '07: Proceedings of the 30th Annual International ACM SIGIR Conference on Research and Development in Information Retrieval* (ACM, New York, NY, USA), pp. 391–398.

Zou, H. and Hastie, T. (2005). Regularization and variable selection via the elastic net, *Journal of the Royal Statistical Society. Series B* **67**, 2, pp. 301–320.

Index

ϵ nearest neighbor graph (ϵNNG), 375
k nearest neighbor graph (kNNG), 377

A/B testing, 305, 507, 587, 588, 594
A/B tests, 476
accuracy, 557
activation function, 82
active learning, 257, 258, 260, 261, 268, 274, 279, 280, 282–284
Adamic-Adar, 528
adaptive learning rate, 85
algorithmic transparency, 438–440
algorithms, 203, 204, 219, 220, 222
all-pairs similarity search, 375
alternating least squares (ALS), 391
Amazon AWS, 696
Apache Solr, 698
Apache Spark, 695
artificial neuron, 80
association rules, 583, 584, 594, 595
autoencoder, 93
average reciprocal hit rank (ARHR), 372, 388, 400

backpropagation, 81
backpropagation through time, 100
batch gradient descent, 84
beyond-accuracy measures, 20
beyond-accuracy metrics, 508
binary tree, 490
black sheep, 138
bpr-max loss, 108
business success, 510

canonical polyadic decomposition, 390
CARS, 174
CARSKit, 188
cascade hybrid, 131
CBOW, 89
circle of trust, 534
cold start, 233, 253, 255–257, 265, 266, 269, 270, 272, 274, 276, 279, 280, 282–284, 604, 611, 615
cold-start problem, 133
collaborate filtering, 466
collaborative filtering, 1, 127, 220, 225, 537
Comcast, 465–470, 472, 477, 478
commute time, 533
concentration bias, 332, 343–345
contact recommendation algorithms, 524
contact recommendations, 519
content-based recommendation, 127, 501
content-based RS, 12
context, 175
context condition, 177
context dimension, 177
context selection, 178
context situation, 177
context suggestion, 197
context-aware matrix factorization, 187
context-aware recommendation, 483, 486, 512
context-Aware Recommender

Systems, 174
contextual modeling, 184
contextual post-filtering, 183
contextual pre-filtering, 180
convolutional neural network, 96
coordinate descent (CCD++), 395
course recommendation, 627, 629, 630
coverage, 303

data privacy, 437, 439
dataset partition, 298
datasets, 25, 311, 508, 544
day parting, 472, 474, 476
deep collaborative filtering, 92
deep learning, 79
Deep Learning for Recommender Systems workshop, 87
deep learning frameworks, 117
denoising autoencoder, 94
descision trees, 492
differential context modeling, 180
differential context relaxation, 180
differential context weighting, 180
differential privacy, 442
dimensionality reduction, 38
discovery, 487
distributed ledger technology, 457
distributed matrix factorization, 488
distributed memory, 386, 390, 391, 393–395
distributed processing, 700
dithering, 608
diversity, 20, 303, 332–335, 339, 347–352
dropout, 83

edge direction, 547
embedding, 89
ensemble design, 132
entity detection, 581
evaluation, 17, 502, 549
  candidate items generation, 299
  library, 310
  reproducibility, 319
evaluation competition
  CAMRa, 311
  Netflix Prize, 311
  RecSys Challenge, 311
evaluation metrics
  error-based, 295, 301
  non-accuracy, 302
  precision-oriented, 296, 301
evaluation protocol, 17, 20, 508
experiments, 212, 223
explicit feedback, 233, 236, 237

Facebook, 543, 546
feature augmentation hybrid, 131
feature combination, 131
feature extraction, 696, 697, 700–702
feature vector, 89
feedfoward neural network, 80
forward pass, 81

garbled circuit, 443, 445, 449
gated recurrent unit, 103
generative adversarial network, 113
generative models, 112
Gini coefficient, 562
graph distance, 529
gray sheep, 138
group building, 204, 210, 211, 222
group modeling, 204, 205, 209, 211, 212, 218, 223
group recommender systems, 203–205, 219, 225, 226
group-based recommendation, 511
GroupLens, 2
gru4rec, 105

hit rate, 302
hit rate (HR), 372, 388, 400
HITS algorithm, 544
hitting time, 532
homomorphic encryption, 443, 445, 447, 449, 450
hybrid recommendation system, 468
hybrid recommendation technique, 130
hybrid recommender, 128, 601, 604, 612, 623
hybridization, 95, 496, 502, 511

hyperparameter optimization, 119

implicit feedback, 37, 233, 235, 236, 244, 245, 247, 600, 601, 603, 607, 608
impression discounting, 607, 608, 614
incremental updates, 15
Independent Context Similarity, 188
information retrieval, 538
interactional context, 177
Intra-List Diversity (ILD), 303
IR metrics, 549
item embedding, 89
item ranking, 8, 19
item splitting, 181
item-based CF, 6
item-based kNN, 538

Jaccard coefficient, 527
job recommendations, 571, 572, 574, 576–578, 581, 582, 585, 588, 589, 593

Katz, 529
keras, 118
knowledge-based RS, 12

Latent Context Similarity, 188
latent feature, 541
latent feature space, 35, 89
latent features, 35, 39
latent space approaches, 388
learning to rank, 622, 687, 688, 692, 693, 696
Leicht-Holme-Newman index, 530
libraries, 16
limited coverage problem, 36
linear regression, 694
linear TV, 465, 473, 474, 477
link prediction, 520
link recommendation, 519
LinkedIn, 543
listening logs, 488, 510
local path index, 530
long short term memory, 102
Love, 544

massive open online courses, 627
matrix completion, 2, 38
matrix factorization, 38, 388, 487, 541, 627–629, 646
  hybrid approach, 60, 65
  learning to rank, 63
  low rank approximation, 39
  weighted, 48
Mean Absolute Error (MAE), 301
Mean Average Precision (MAP), 302
mean first passage time, 532
Mean Reciprocal Rank (MRR), 302
memory-based CF, 4
meta-level hybrid, 131
metadata, 476–478
mini-batch, 84
missing not at random, 40
mixed hybrid, 130
mixed systems, 132
model-based CF, 4
Money algorithm, 533
monolithic design, 132
MOOC, 627–630, 636, 637, 640, 644
most common neighbors, 526
movie recommendations, 465, 466, 468
MovieLens, 25
multi-objective optimization, 505
Multidimensional Context Similarity, 188
music consumption, 485
music preferences, 485, 496

nearest neighbors, 3, 4, 10, 489, 494, 497, 537
nearest-neighbor approaches, 373
neighborhood-based collaborative filtering, 606, 618
Netflix prize, 4
neural network, 80
new context, 255, 279, 280
new item, 253, 255, 262, 266, 267, 269, 279–281, 284
new user, 253, 255, 257, 259, 262, 266, 267, 279–281, 284

next-track music recommendation, 493
normalized discounted cumulative gain (nDCG), 302
novelty, 303, 336–339, 352–354
novelty and diversity, 560

OCCF, 37
offline evaluation, 18, 20, 617, 619, 620
offline experiments, 154
offline jobs, 696, 698, 699, 701
offline metrics, 465, 466, 476
online evaluation, 617, 621
online experiment, 154
online metrics, 465, 466, 476
operational excellence, 688, 708
over-specialization, 136

pairwise preference elicitation, 247
paragraph2vec, 90
parallel architectures, 370
parallel computing, 370
path-based methods, 529
People You May Know, 543
personalization, 467, 471, 472, 474
pL2AP, 376
pL2Knng, 379
playlist continuation, 496
popularity bias, 331, 332, 343–347
popularity-based recommendation, 526
precision, 301
predictive popularity, 471, 472, 475
preference aggregation, 211, 221
preference elicitation, 233, 234, 236, 246, 253, 282
preferential attachment, 526
prod2vec, 91
profile injection, 407–409
PropFlow algorithm, 535
pros and cons, 12
PureSVD, 397
pytorch, 118

quality factors, 503, 504
quality perception, 510

radio station, 492
random walk, 531
ranking task, 549
rating elicitation, 258, 260, 261, 275, 282
rating matrix, 2
rating prediction, 4, 5, 18, 37, 371, 388
rating scale, 236, 237, 239–242, 247
recall, 301
recency, 15
recommendation purpose, 9
recommendation scenarios, 482
recommender system, 466, 468, 471, 473, 477
recommender systems libraries, 163
recommender systems paradigms, 12
recurrent neural network, 96, 100
regularization, 693–695, 705
repetition, 485
representational context, 177
reproducibility, 120
resource allocation, 528
Ringo, 481, 482
robustness, 405
Root Mean Squared Error (RMSE), 301, 373, 397
Rooted PageRank, 531

SALSA, 533
scalability, 14, 119, 485, 498, 548
semantic pre-filtering, 180
sequence learning, 495
serendipity, 303, 338, 354–356
session-based recommendation, 7, 97
session-based recommender system, 245, 247
shared memory, 379, 382, 391, 393–395
short-term intents, 9
similarity join, 375
SimRank, 536
singular value decomposition, 41
  folding-in, 45

truncated SVD, 43
user and item bias, 54
singular value decomposition (SVD), 397
randomized, 399
skip-gram, 90
social network, 571
social networks, 519
SoRec algorithm, 542
sparse linear methods (SLIM), 381
global and local (GLSLIM), 384
sparsity, 137, 253, 255, 258, 264, 270, 272, 274, 281, 284
stochastic gradient descent, 84
stochastic gradient descent (SGD), 393
strong scaling, 377, 379, 383, 384
supervised methods, 540
supervised random walks, 536
switching hybrid, 130
system design, 233
system robustness, 438–440

tapestry, 2
tensor factorization, 184, 388
tensorflow, 118
tf-idf, 539
theano, 117
time-aware recommender systems, 174
top-N recommendation, 371, 381, 384, 397
top-N recommendation, 87
top-n recommendation, 37
Tumblr, 544, 546
TV recommendations, 465, 466, 468
Twitter, 544, 545

UISplitting, 181
user intention, 491
user interaction, 498
user interface, 15
user model, 235
user modeling, 582, 597
user personality, 257, 268
user preferences, 233, 234, 236, 238, 245
user studies, 23, 306, 510
user-based CF, 4
user-based collaborative filtering, 599, 601, 604, 605, 607, 609, 613, 615, 618
user-based kNN, 537
utility function, 36

value imputation, 41
vanishing gradient problem, 83
vanishing/exploding gradient problem, 100
variational autoencoder, 114
vector space model, 538
video-on-demand, 471, 474

warm start, 383, 387
web services, 696, 698
weighted hybrid, 130
weighted matrix factorization, 48
folding-in, 51, 53
regularization, 49
user and item bias, 43
Who to Follow, 533, 544
word2vec, 89

Lightning Source UK Ltd.
Milton Keynes UK
UKHW021157171218
334145UK00001B/46/P

9 789813 275348